D0845818
UNIVERSITY
OCT 21 1992
DAVIS CENTRE LIBRARY
GOVERNMENT PUBLICATIONS

Withdrawn
University of Waterloo

Topological Aspects of the Dynamics of Fluids and Plasmas

NATO ASI Series

Advanced Science Institutes Series

A Series presenting the results of activities sponsored by the NATO Science Committee, which aims at the dissemination of advanced scientific and technological knowledge, with a view to strengthening links between scientific communities.

The Series is published by an international board of publishers in conjunction with the NATO Scientific Affairs Division

A Life Sciences **B Physics**	Plenum Publishing Corporation London and New York
C Mathematical and Physical Sciences **D Behavioural and Social Sciences** **E Applied Sciences**	Kluwer Academic Publishers Dordrecht, Boston and London
F Computer and Systems Sciences **G Ecological Sciences** **H Cell Biology** **I Global Environmental Change**	Springer-Verlag Berlin, Heidelberg, New York, London, Paris and Tokyo

NATO-PCO-DATA BASE

The electronic index to the NATO ASI Series provides full bibliographical references (with keywords and/or abstracts) to more than 30000 contributions from international scientists published in all sections of the NATO ASI Series.
Access to the NATO-PCO-DATA BASE is possible in two ways:

– via online FILE 128 (NATO-PCO-DATA BASE) hosted by ESRIN, Via Galileo Galilei, I-00044 Frascati, Italy.

– via CD-ROM "NATO-PCO-DATA BASE" with user-friendly retrieval software in English, French and German (© WTV GmbH and DATAWARE Technologies Inc. 1989).

The CD-ROM can be ordered through any member of the Board of Publishers or through NATO-PCO, Overijse, Belgium.

Series E: Applied Sciences - Vol. 218

Topological Aspects of the Dynamics of Fluids and Plasmas

edited by

H. K. Moffatt

Department of Applied Mathematics
and Theoretical Physics,
University of Cambridge, Cambridge, U.K.

G. M. Zaslavsky

Courant Institute of Mathematical Sciences,
New York University, New York, NY, U.S.A.

P. Comte

Institut de Mécanique de Grenoble,
Institut National Polytechnique de Grenoble,
Grenoble, France

and

M. Tabor

Department of Mathematics,
University of Arizona, Tucson, AZ, U.S.A.

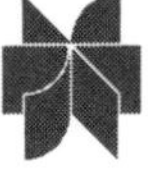

Kluwer Academic Publishers

Dordrecht / Boston / London

Published in cooperation with NATO Scientific Affairs Division

Proceedings of the Program of the Institute for Theoretical Physics, University of California at Santa Barbara, U.S.A. and of the NATO Advanced Research Workshop on Topological Aspects of the Dynamics of Fluids and Plasmas
University of California at Santa Barbara, U.S.A.
November 1–5, 1991

Library of Congress Cataloging-in-Publication Data

Topological aspects of the dynamics of fluids and plasmas / edited by H.K. Moffatt ... [et al.]
p. cm. -- (NATO ASI series. Series E., Applied sciences ; 218)
"Published in cooperation with NATO Scientic Affairs Division."
Proceedings of the program of the Institute for Theoretical Physics, University of California at Santa Barbara, August-December, 1991 and of the NATO Advanced Research Workshop, 1-5 November 1991.
ISBN 0-7923-1900-1 (hb : acid free paper)
1. Fluid dynamics--Mathematics--Congresses. 2. Plasma dynamics--Mathematics--Congresses. 3. Topology--Congresses. 4. Mathematical physics--Congresses. I. Moffatt, H. K. (Henry Keith), 1935- . II. North Atlantic Treaty Organization. Scientific Affairs Division. III. University of California, Santa Barbara. Institute for Theoretical Physics. IV. North Atlantic Treaty Organization. V. Series.
QC138.T67 1992
532'.00151--dc20 92-1668

ISBN 0-7923-1900-1

Published by Kluwer Academic Publishers,
P.O. Box 17, 3300 AA Dordrecht, The Netherlands.

Kluwer Academic Publishers incorporates the publishing programmes of D. Reidel, Martinus Nijhoff, Dr W. Junk and MTP Press.

Sold and distributed in the U.S.A. and Canada
by Kluwer Academic Publishers,
101 Philip Drive, Norwell, MA 02061, U.S.A.

In all other countries, sold and distributed
by Kluwer Academic Publishers Group,
P.O. Box 322, 3300 AH Dordrecht, The Netherlands.

Printed on acid-free paper

Printed in the Netherlands

My soul's an amphicheiral knot
Upon a liquid vortex wrought
By Intellect, in the Unseen residing.
And thine doth like a convict sit,
With marlinspike untwisting it,
Only to find its knottiness abiding...

From a Paradoxical Ode to Hermann Stoffkraft
by James Clerk Maxwell
c. 1873

Preface

This volume contains papers arising out of the program of the Institute for Theoretical Physics (ITP) of the University of California at Santa Barbara, August–December 1991, on the subject "Topological Fluid Dynamics". The first group of papers cover the lectures on Knot Theory, Relaxation under Topological Constraints, Kinematics of Stretching, and Fast Dynamo Theory presented at the initial Pedagogical Workshop of the program. The remaining papers were presented at the subsequent NATO Advanced Research Workshop or were written during the course of the program. We wish to acknowledge the support of the NATO Science Committee in making this workshop possible.

The scope of "Topological Fluid Dynamics" was defined by an earlier Symposium of the International Union of Theoretical and Applied Mechanics (IUTAM) held in Cambridge, England in August, 1989, the Proceedings of which were published (Eds. H.K. Moffatt and A. Tsinober) by Cambridge University Press in 1990. The proposal to hold an ITP program on this subject emerged from that Symposium, and we are grateful to John Greene and Charlie Kennel at whose encouragement the original proposal was formulated.

Topological fluid dynamics covers a range of problems, particularly those involving vortex tubes and/or magnetic flux tubes in nearly ideal fluids, for which topological structures can be identified and to some extent quantified. Just as vortex tubes and flux tubes can reconnect as a result of weak diffusion, so it happens that separate strands of scientific inquiry can "reconnect" in a most fruitful and stimulating way as a result of the sort of inter-diffusion that ITP promotes. The separate disciplines of topology, plasma magnetohydrodynamics (MHD) and high Reynolds number fluid dynamics are linked by certain potent analogies that aid this diffusion; and certain results and techniques from knot theory and from MHD have already stimulated fruitful lines of enquiry in the more classical areas of fluid dynamics, particularly Euler flows, instability, Lagrangian chaos and turbulence.

These areas are all represented in the papers collected in this volume, which we hope will provide the reader with a good picture of much current research in this broad field.

We are extremely grateful to Jim Langer and the staff of ITP, which is supported by NSF Grant PHY89-04035, for hosting this program and for providing an ideal research environment. Our thanks also to Lorraine Shallenberger and Anne Braddock who took good care of all practical arrangements in relation to the workshops, and to Darla Sharp-Fitzpatrick and Donna Freet for their skillful preparation of the camera-ready copy for this volume.

The Editors,
Santa Barbara
December 1991

CONTENTS

PART I

INTRODUCTORY LECTURES

RELAXATION UNDER TOPOLOGICAL CONSTRAINTS

H.K. MOFFATT *
Institute for Theoretical Physics
University of California
Santa Barbara, California 93106-4030, U.S.A.

ABSTRACT. This contribution provides an informal introduction to the technique of magnetic relaxation, whereby an extremely wide family of solutions of the equations of magnetostatics, and of analogous steady solutions of the Euler equations, may be obtained, and their stability investigated. We approach this problem through the simpler, and physically more transparent, problem of gravitational relaxation of an incompressible medium of non-uniform density. We then describe the magnetic relaxation technique which yields solutions of nontrivial field topology, and we discuss the contrasting stability criteria for these magnetostatic states and for the analogous Euler flows. Applications to the theory of vortons (*i.e.*, blobs of propagating vorticity) and to the problem of determining 'energy' invariants of knots and links are then discussed. The chapter concludes with a discussion of alternative relaxation procedures involving artificial modification of the Euler equations in a manner that conserves vorticity topology.

1. Gravitational Relaxation

1.1. SOLID ELASTIC MEDIUM

In order to grasp some elementary concepts, consider first the simple problem depicted in figure 1: a block $\mathcal{D}$ of unstrained elastic material of non-uniform density $\rho_0(\mathbf{x})$ is placed in a uniform gravitational field $\mathbf{g}$, and its boundary $\partial\mathcal{D}$ is fixed. What happens? Evidently the denser parts of the medium will sag slightly under the influence of gravity and the block will reach equilibrium in a strained state. In order to reach the equilibrium, we must suppose that, when the medium is in motion, energy is dissipated (otherwise elastic oscillations would persist forever). We may also assume (for simplicity) that the medium is incompressible, so that volume elements can be distorted but not dilated or compressed during the motion.

Let $\mathbf{v}(\mathbf{x}, t)$ be the velocity field during the relaxation process, and $\rho(\mathbf{x}, t)$ the density field. Then under the assumption of incompressibility, $\nabla \cdot \mathbf{v} = 0$

* Permanent address: DAMTP, Silver Street, Cambridge CB3 9EW, UK.

H. K. Moffatt et al. (eds.), Topological Aspects of the Dynamics of Fluids and Plasmas, 3–28.

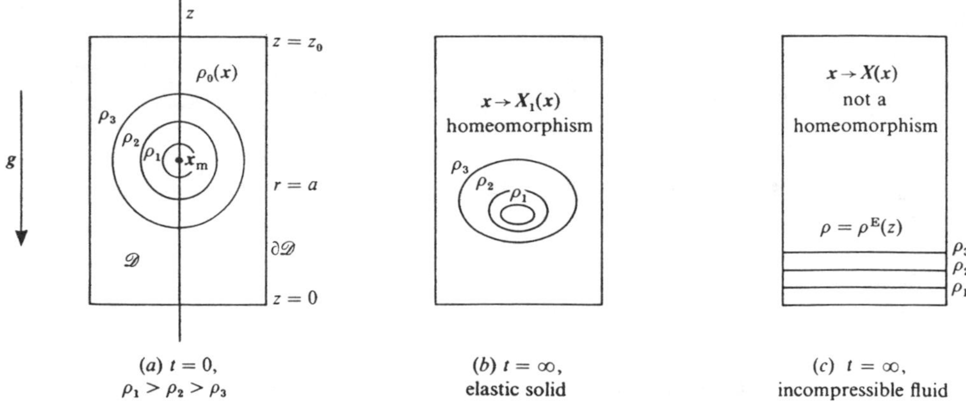

Fig. 1. Gravitational relaxation of an incompressible medium with non-uniform density distribution. When the medium is an elastic solid, the equilibrium state is related by homeomorphism to the initial state. When the medium is fluid, the limit map (as $t \to \infty$) is not a homeomorphism. [From Moffatt, 1985.]

and

$$\frac{D\rho}{Dt} \equiv \frac{\partial\rho}{\partial t} + \mathbf{v}\cdot\nabla\rho = 0. \tag{1.1}$$

This equation simply expresses the fact that the density of each material element is conserved; equivalently the isodensity surfaces $\rho = cst.$ are 'frozen in the medium', so that their topology is conserved, *i.e.*, the picture looks qualitatively the same at time t as it does at time zero, although there is obviously a distortion of each surface. Equation (1.1) however embodies the essential topological constraint that characterizes this problem. Note that the volume $V(\rho)$ inside any closed surface $\rho = cst.$ is conserved during relaxation. Thus the function $V(\rho)$ is a topological invariant of the density field; this function may be described as the *signature* of the field $\rho(\mathbf{x}, t)$ for this problem.

Under the action of the velocity field $\mathbf{v}(\mathbf{x}, t)$, the particle initially at position $\mathbf{x}$ moves to position $\mathbf{X}(\mathbf{x}, t)$ at time t where

$$\frac{d\mathbf{X}}{dt} = \mathbf{v}(\mathbf{X}, t) \quad , \quad \mathbf{X}(\mathbf{x}, 0) = \mathbf{x}. \tag{1.2}$$

If $\mathbf{v}(\mathbf{x}, t)$ is known (which of course it isn't until the dynamical equations governing the relaxation process are prescribed and solved) then (1.2) is a third order nonlinear dynamical system which determines the particle paths $\mathbf{X}(\mathbf{x}, t)$. The particle displacement $\mathbf{x} \to \mathbf{X}(\mathbf{x}, t)$ constitutes a mapping which depends continuously on the parameter t. If $\rho_0(\mathbf{x})$ is a nice smooth function, and we have no reason at this stage to suppose otherwise, and if the elastic medium has uniform and physically sensible elastic properties, then obviously for any t, $\mathbf{X}(\mathbf{x},t)$ is a continuous function of $\mathbf{x}$ and its inverse $\mathbf{x}(\mathbf{X},t)$ is also continuous (neighboring particles at time t came from neighboring sites at time $t=0$). The mapping $\mathbf{x}\to\mathbf{X}(\mathbf{x},t)$ is thus a *homeomorphism* for each t, and the family of mappings $\mathbf{x}\to\mathbf{X}(\mathbf{x},t)(0\le t\le T)$ constitutes an *isotopy* (or flow) for each T. Moreover, for the solution considered, the relaxation process is clearly well under control, and convergent, as $t\to\infty$; we can thus define the limit map

$$\mathbf{x}\to\mathbf{X}^E(\mathbf{x})=\lim_{t\to\infty}\mathbf{X}(\mathbf{x},t), \tag{1.3}$$

which is also a homeomorphism representing the net displacement of particles under the complete relaxation process.

The solution of (1.1) is

$$\rho(\mathbf{X},t)=\rho(\mathbf{x},0)=\rho_0(\mathbf{x}), \tag{1.4}$$

an equation which establishes a *topological equivalence* between the density fields at time t and at time zero. Note that the Jacobian of the volume-preserving mapping $\mathbf{x}\to\mathbf{X}(\mathbf{x},t)$ has value $+1$:

$$J\equiv\left|\frac{\partial(X_1,X_2,X_3)}{\partial(x_1,x_2,x_3)}\right|=+1. \tag{1.5}$$

The relaxed state is evidently one in which the total energy E of the system (gravitational plus elastic strain energy) is a minimum. What does this mean? Simply that, if we consider a virtual volume-preserving displacement $\mathbf{x}\to\mathbf{x}+\xi(\mathbf{x})$ of the medium from its equilibrium position, with $\xi\cdot\mathbf{n}=0$ on ∂D, and with

$$\rho(\mathbf{x}+\xi(\mathbf{x}))=\rho^E(\mathbf{x}), \tag{1.6}$$

where $\rho^E(\mathbf{x})$ is the density distribution in the relaxed state, and if we expand E in powers of ξ in the form

$$E=E_0+\delta^1E+\delta^2E+O(\xi^3), \tag{1.7}$$

then $\delta^1E=0$ and $\delta^2E>0$ for all admissible ξ (*i.e.*, all $\xi(\mathbf{x})$ satisfying the above conditions).

Finally, we note that there is no guarantee that the minimum energy state is unique. We could vary the problem by first imposing a (volume preserving) displacement $\mathbf{x}\to\mathbf{X}_1(\mathbf{x})$ (thus straining the medium), and then release from rest allowing relaxation to proceed along a different path. By this means, different minimum energy configurations may conceivably be attained, characteristic by a spectrum of minimum energies

$$E_0\le E_1\le E_2\le ... \tag{1.8}$$

1.2. INCOMPRESSIBLE FLUID MEDIUM

Consider now the same relaxation problem but in an incompressible (Boussinesq) fluid medium, with the same initial non-uniform density $\rho_0(\mathbf{x})$. Again, for $t > 0$, the denser fluid moves downwards, displacing lighter fluid upwards. For any finite time t, all of the previous considerations apply without change; in particular the density surfaces $\rho = cst.$ move with the flow, and the signature function $V(\rho)$ is invariant. For finite t, $\rho(\mathbf{x}, t)$ is topologically equivalent to $\rho_0(\mathbf{x})$. However this nice state of affairs does not persist in the limit $t \to \infty$. It is physically obvious that, no matter what the initial topology of $\rho_0(\mathbf{x})$ may be, the ultimate state is one in which $\rho = \rho^E(z)$ where z is the vertical coordinate where, for minimum energy (and therefore stability)

$$d\rho^E/dz \leq 0 \quad (\text{all } z). \tag{1.9}$$

The topology of the field $\rho^E(z)$ is very simple: the surfaces $\rho^E = cst.$ are horizontal planes intersecting the boundary $\partial\mathcal{D}$ (figure 1c).

The apparent change in topology as $t \to \infty$ arises because the mapping $\mathbf{x} \to \mathbf{X}(\mathbf{x}, t)$ develops discontinuities for various values of $\mathbf{x}$, *i.e.*, the limit map $\mathbf{x} \to \mathbf{X}^E(\mathbf{x})$ (which exists in the sense that every fluid particle goes somewhere!) is not a homeomorphism. For the simple topology illustrated in figure 1, the limit mapping is obviously discontinuous for $\mathbf{x} = \mathbf{x}_m$ (where $\rho_0(\mathbf{x})$ is maximal) since the fluid particles in an arbitrarily small sphere surrounding $\mathbf{x}_m$ are ultimately spread over the base of the container. A moment's consideration will convince you that the limit mapping is likewise discontinuous for every point $\mathbf{x}$ vertically below $\mathbf{x}_m$. This lack of continuity is present also in the inverse limit map: surfaces $\rho = cst.$ are squeezed together on the fluid boundary as $t \to \infty$ and ultimately coincide in part with the boundary, so that particles ultimately contiguous have inverse images that were widely separated at $t = 0$.

The fluid problem differs from the elastic problem in that there is no counterpart of the elastic strain energy; this lack of constraint is what permits the formation of discontinuities in the limit mapping and the associated simplification in topology of the density field.

The magnetic relaxation problem, to which we now turn, is in some respects intermediate between these two situations. On the one hand, we shall be concerned with minimisation of a single form of energy, namely magnetic energy. On the other hand, this magnetic energy, reflecting as it does the Maxwell tension in the magnetic lines of force in the medium, resembles elastic strain energy, and minimisation of magnetic energy is subject to constraints that are not present for the fluid gravitational relaxation problem. Thus, we shall find that, although discontinuities can appear as $t \to \infty$, nevertheless a nontrivial end-state having topology closely related to that of the critical field, is in general attained.

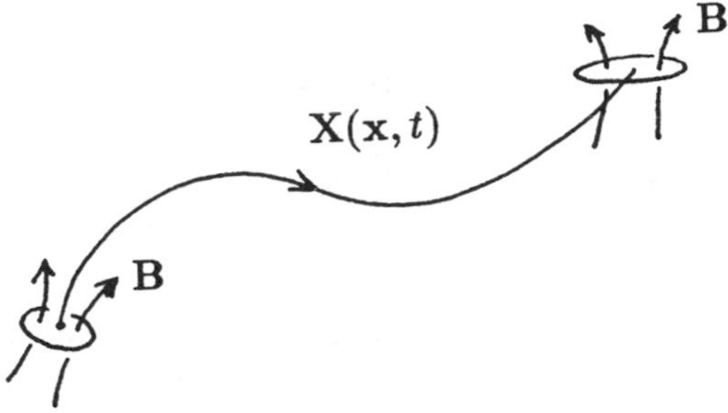

Fig. 2. Particle path and associated deformation of elemental circuit, through which the flux of **B** is conserved.

2. Magnetic Relaxation

2.1. THE MASTER EQUATION

Consider a magnetic field $\mathbf{B}(\mathbf{x}, t)$ in a perfectly conducting incompressible fluid satisfying the solenoidal condition $\nabla \cdot \mathbf{B} = 0$ and the 'frozen field' equation (derivable from Faraday's law of induction)

$$\frac{\partial \mathbf{B}}{\partial t} = \text{curl } (\mathbf{v} \times \mathbf{B}), \tag{2.1}$$

where, as before, $\mathbf{v}(\mathbf{x}, t)$ is the velocity field in the medium. If $\mathbf{X}(\mathbf{x}, t)$ is the mapping given by (1.2), then the (Cauchy) solution of (2.1) is given by

$$B_i(\mathbf{X}, t) = B_j(\mathbf{x}, 0) \frac{\partial X_i}{\partial x_i}, \tag{2.2}$$

an equation which encapsulates the 'frozen-in' character of the field (figure 2). Under this evolution, the flux of **B** through any elemental surface element moving with the fluid is conserved.

Since the **B**-lines move with the fluid, we note immediately that the topology of the **B**-field is conserved by equation (2.1). In particular, any knots or links in the **B**-lines are conserved. Any topological invariant of a knot or link must likewise be an invariant of equation (2.1), if we simply adopt an initial condition that conforms to the knot or link under consideration (see §4 below). In this sense, (2.1) may be described as a 'MASTER equation': all topological invariants of knots and links and indeed of more complex chaotic structures are somehow contained within this equation. Unfortunately this statement does not make it any easier to find such invariants!

2.2. THE HELICITY INVARIANT

There is however one topological invariant which is easily obtained, and which plays a central role in the magnetic relaxation problem, namely the magnetic helicity invariant, defined as follows. Let S be any closed surface moving with the fluid on which the condition $\mathbf{n}\cdot\mathbf{B} = 0$ is satisfied; note that by the frozen field properly, if this condition is satisfied at time zero, then it is satisfied for all t. Let $\mathbf{A}(\mathbf{x}, t)$ be a vector potential for $\mathbf{B}$, *i.e.*, $\mathbf{B} = \text{curl } \mathbf{A}$, and let us for definiteness adopt the Coulomb gauge $\nabla \cdot \mathbf{A} = 0$. We define the magnetic helicity in the volume V inside S by the integral

$$\mathcal{H}_M = \int_V \mathbf{A}\cdot\mathbf{B}dV. \tag{2.3}$$

Now, by 'uncurling' (2.1), we have

$$\frac{\partial \mathbf{A}}{\partial t} = \mathbf{v}\times(\nabla\times\mathbf{A}) - \nabla\varphi \tag{2.4}$$

for some scalar field $\varphi(\mathbf{x}, t)$. Equations (2.1) and (2.4) may be written in the equivalent Lagrangian forms

$$\frac{DB_i}{Dt} = B_j\frac{\partial v_i}{\partial x_j}, \tag{2.5}$$

$$\frac{DA_i}{Dt} = A_j\frac{\partial v_j}{\partial x_i} - \frac{\partial \chi}{\partial x_i}, \tag{2.6}$$

where $\chi = \varphi - \mathbf{v}\cdot\mathbf{A}$. Hence

$$\frac{D}{Dt}(\mathbf{A}\cdot\mathbf{B}) = -(\mathbf{B}\cdot\nabla)\chi = -\nabla\cdot(\mathbf{B}\chi) \tag{2.7}$$

and so

$$\frac{d\mathcal{H}_M}{dt} = \int_V \frac{D}{Dt}(\mathbf{A}\cdot\mathbf{B})dV = -\int_S(\mathbf{n}\cdot\mathbf{B})\chi dV = 0 \tag{2.8}$$

since $\mathbf{n}\cdot\mathbf{B} = 0$ on S. Hence $\mathcal{H}_M$ is indeed invariant.

Note three points: (i) $\mathcal{H}_M$ is in fact gauge-invariant, since replacement of $\mathbf{A}$ by $\mathbf{A} + \nabla\psi$ does not change its value; (ii) the result $\mathcal{H}_M = cst.$ holds also for compressible fluid motion (for which (2.6) hold unchanged while (2.5) holds with $\mathbf{B}$ replaced by $\mathbf{B}/\rho$; (iii) $\mathcal{H}_M$ is a pseudo-scalar (changing sign under a parity transformation from right-handed to left-handed frame of reference).

Let us evaluate $\mathcal{H}_M$ for the very simple 'prototype' linked field shown in figure 3. Here $\mathbf{B} \equiv 0$ except in two linked flux tubes $\mathcal{T}_1$, $\mathcal{T}_2$ centred on unknotted curves C_1 and C_2. We suppose that within each tube, $\mathbf{B}$ has trivial topology, *i.e.*, the $\mathbf{B}$-lines are unlinked closed curves 'parallel' to C_1 and C_2 respectively. If the cross-section of each tube is small, then in the volume integral (2.3) we may make the replacements

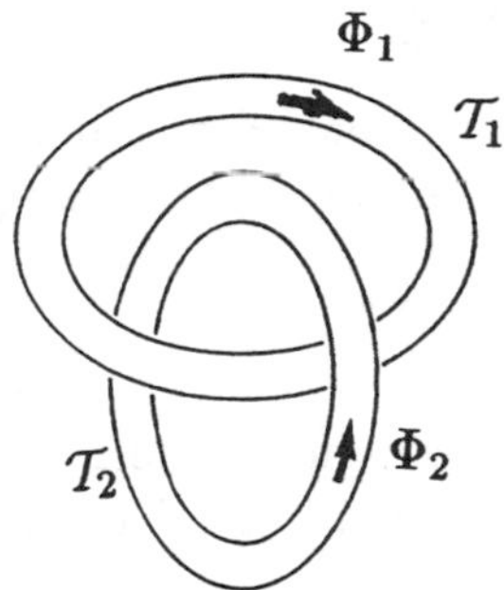

Fig. 3. Prototype linkage of flux tubes $\mathcal{T}_1, \mathcal{T}_2$.

$$\mathbf{B}dV \to \left.\begin{matrix} \Phi_1 d\mathbf{x}_1 \text{ in } \mathcal{T}_1 \\ \Phi_2 d\mathbf{x}_2 \text{ in } \mathcal{T}_2 \end{matrix}\right\} \tag{2.9}$$

and the integral becomes

$$\mathcal{H}_M \sim \Phi_1 \oint_{C_1} \mathbf{A} \cdot d\mathbf{x}_1 + \Phi_2 \oint_{C_2} \mathbf{A} \cdot d\mathbf{x}_2. \tag{2.10}$$

Now

$$\oint_{C_1} \mathbf{A} \cdot d\mathbf{x}_1 = \int_{\Sigma_1} \mathbf{B} \cdot \mathbf{n} dS = \Phi_2,$$

where Σ_1 is an open surface spanning C_1; similarly

$$\oint_{C_2} \mathbf{A} \cdot d\mathbf{x}_2 = \Phi_1.$$

Hence, for the oriented linkage of figure 3,

$$\mathcal{H}_M = 2\Phi_1\Phi_2. \tag{2.11}$$

More generally

$$\mathcal{H}_M = \pm 2n\Phi_1\Phi_2, \tag{2.12}$$

where n is the (Gauss) linking number of C_1 and C_2, and the + or − is chosen according as the linkage is right-handed or left-handed.

Equation (2.12) establishes a bridge between an invariant of the partial differential equation (2.1) and the primitive topological invariant n of the two linked curves. This bridge has been reinforced by Arnold (1974, 1986) who has extended the topological interpretation of $\mathcal{H}_M$ to situations in which the **B** lines do not form conveniently closed curves, but which may wander chaotically in the domain $\mathcal{D}$.

2.3. MINIMUM ENERGY STATES

Now suppose that the fluid is confined to a domain $\mathcal{D}$ and that $\mathbf{B}$ satisfies the condition $\mathbf{B}\cdot\mathbf{n} = 0$ on $\partial\mathcal{D}$. We define the energy of the field $\mathbf{B}$ as

$$M(t) = \frac{1}{2}\int \mathbf{B}^2 dV, \tag{2.13}$$

and we investigate the nature of minimum energy states that can be arrived at under evolution controlled by equation (2.1). That there is in general a positive minimum energy is ensured by the Schwarz inequality

$$\int_D \mathbf{A}^2 dV \int_D \mathbf{B}^2 dV \geq \mathcal{H}_M^2, \tag{2.14}$$

and the Poincaré inequality

$$\frac{\int \mathbf{B}^2 dV}{\int \mathbf{A}^2 dV} \geq q_0^2 > 0. \tag{2.15}$$

[This latter inequality is more familiar in the analogous context in which a current $\mathbf{j}(\mathbf{x})$ confined to a domain $\mathcal{D}$ gives rise to a field $\mathbf{B}$ satisfying $\nabla\times\mathbf{B} = \mathbf{j}$ inside $\mathcal{D}$ and $\nabla\times\mathbf{B} = 0$ outside $\mathcal{D}$ with $\mathbf{B} = O(r^{-3})$ at infinity. The inequality then reads

$$\int \mathbf{j}^2 dV \geq q_0^2 \int \mathbf{B}^2 dV.] \tag{2.16}$$

Multiplying (2.14) and (2.15), we have

$$\int \mathbf{B}^2 dV \geq q_0 |\mathcal{H}_M|, \tag{2.17}$$

and since $\mathcal{H}_M$ is invariant, this clearly provides a lower bound for $M(t)$. As will be clear below, it is the topological barrier implied by nonzero helicity that guarantees this lower bound. As shown by Freedman (1988), even if $\mathcal{H}_M = 0$, a positive lower bound on $M(t)$ exists provided merely that the topology is nontrivial. A simple example of a nontrivial topology having zero helicity is provided by the Whitehead link (figure 4).

2.4. CONSTRUCTION OF A DYNAMICAL PROCESS BY WHICH THE LOWER BOUND MAY BE ATTAINED

Note first that from (2.1), the rate of change of magnetic energy density is given by

$$\frac{\partial}{\partial t}\frac{1}{2}\mathbf{B}^2 = \mathbf{B}\cdot\frac{\partial \mathbf{B}}{\partial t} = \mathbf{B}\cdot\nabla\times(\mathbf{v}\times\mathbf{B}) = (\nabla\times\mathbf{B})\cdot(\mathbf{v}\times\mathbf{B}) + \nabla\cdot[\mathbf{B}\times(\mathbf{v}\times\mathbf{B})] \tag{2.18}$$

so that, integrating over the domain,

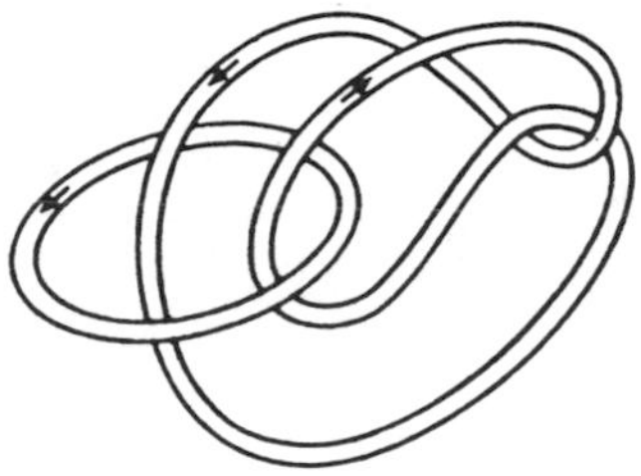

Fig. 4. A simple example of a non-trivial linkage, for which the linking number is zero and the associated helicity is also zero.

$$\frac{d}{dt}M(t) = -\int \mathbf{v}\cdot(\mathbf{j}\times\mathbf{B})dV, \tag{2.19}$$

where $\mathbf{j} = \nabla\times\mathbf{B}$, since $\mathbf{v}\cdot\mathbf{n} = \mathbf{B}\cdot\mathbf{n} = 0$ on $\partial\mathcal{D}$. The right-hand side represents the rate of work of the field on the medium through the action of the Lorentz force $\mathbf{j}\times\mathbf{B}$. We can ensure that $M(t)$ decreases by adopting any dynamical model in which $\mathbf{v}$ 'yields' to this force so that magnetic energy is converted to kinetic energy which can then be dissipated by internal friction. The Navier-Stokes equations for a viscous fluid have this property. A simpler model however, on which we shall focus here, is provided by the instantaneous prescription

$$k\mathbf{v} = \mathbf{j}\times\mathbf{B} - \nabla p, \tag{2.20}$$

where p is a pressure field chosen to ensure that $\nabla\cdot\mathbf{v} = 0$ and $\mathbf{v}\cdot\mathbf{n} = 0$ on $\partial\mathcal{D}$. Thus p satisfies the Neumann problem

$$\left.\begin{array}{r}\nabla^2 p = \nabla\cdot(\mathbf{j}\times\mathbf{B}) \text{ in } \mathcal{D}\\ \mathbf{n}\cdot\nabla p = \mathbf{n}\cdot(\mathbf{j}\times\mathbf{B}) \text{ on } \partial\mathcal{D}\end{array}\right\} \tag{2.21}$$

Substituting the solution back in (2.20), we may write the result in the form

$$k\mathbf{v} = (\mathbf{j}\times\mathbf{B})_s \tag{2.22}$$

where the suffix s means 'solenoidal projection'.

Substituting in (2.19) we now have

$$\frac{d}{dt}M(t) = -k\int \mathbf{v}^2 dV \tag{2.23}$$

(since $\int \mathbf{v}\cdot\nabla p dV = 0$). Thus $M(t)$ decreases monotonically for so long as $v \neq 0$, and, being bounded below, must tend to a limit. Note the importance of the lower bound for the force of this argument.

Since $M(t) \to M^E(cst)$ as $t\to\infty$, it follows from (2.23) that

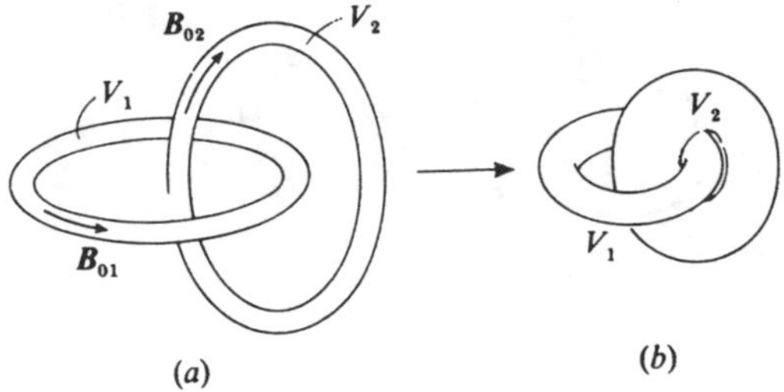

Fig. 5. Magnetic relaxation for the simplest non-trivial topology. (a) the initial state; (b) the stage at which the topological constraint begins to impede the relaxation process. [From Moffatt, 1985.]

$$\int_{\mathcal{D}} \mathbf{v}^2 dV \to 0 \text{ as } t \to \infty. \tag{2.24}$$

Hence, setting aside the extremely unlikely possibility that singularities of $\mathbf{v}$ develop on a set of measure zero in $\mathcal{D}$ (a possibility that is however as difficult to disprove as it is unlikely!) we may conclude that $\mathbf{v} \to 0$ everywhere in $\mathcal{D}$ and that in the asymptotic state the field $\mathbf{B}^E(\mathbf{x})$ satisfies the magnetostatic equation

$$\mathbf{j}^E \times \mathbf{B}^E = \nabla p^E. \tag{2.25}$$

This relaxation process may be best understood with regard to the prototype linked flux tube configuration shown again in figure 5. Just as stretching of B-lines is associated with field intensification (the essence of the dynamo process), so reduction of field energy can occur only through contraction of B-lines. Thus, both flux tubes in figure 5 tend to contract as they release magnetic energy; since the process is volume preserving, the cross sections of the tubes must simultaneously expand, and it is obvious that the process cannot continue indefinitely, since eventually the tubes must make contact with each other. It is the linkage (reflected in this case by a nonzero value of $\mathcal{H}_M$) that provides this topological barrier that guarantees that the energy does indeed have a lower bound.

2.5. THE IDEA OF TOPOLOGICAL ACCESSIBILITY

It is evident from the above example that tangential discontinuities of $\mathbf{B}$ may form in regions where two flux tubes are pressed together by the relaxation process. Hence the limit map $\mathbf{x} \to \mathbf{X}^E(\mathbf{x})$ will not in general be a homeomorphism, and the limit field $\mathbf{B}^E(\mathbf{x})$ will not therefore be topologically equivalent in a strict sense to the original fields. Nevertheless all links and knots are conserved throughout the relaxation process, and the formation of discontinuities simply brings different flux tubes into contact over

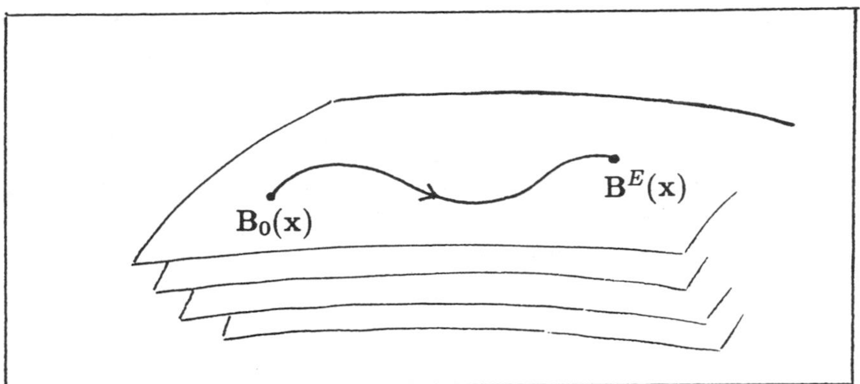

Fig. 6. Isomagnetic foliation of the function-space of solenoidal fields of finite energy. Magnetic relaxation is represented by a trajectory on one sheet of this foliation.

a finite area, without permitting any cutting and reconnecting of **B**-lines. Since the field $\mathbf{B}^E(\mathbf{x})$ is obtained through the convective action of a smooth velocity field $\mathbf{v}(\mathbf{x}, t)$ on the initial field $\mathbf{B}_0(\mathbf{x})$, and since moreover the field $\mathbf{v}$ dissipates a total energy on the time interval $0 > t > \infty$ which is finite, *i.e.*,

$$\int_0^\infty dt \int_{\mathcal{D}} \mathbf{v}^2 dV < \infty, \tag{2.26}$$

we may say that $\mathbf{B}^E(\mathbf{x})$ is *topologically accessible* from $\mathbf{B}_0(\mathbf{x})$.

It is helpful to picture this process in the function space of all bounded square-integrable solenoidal fields (figure 6). This space is foliated through the frozen field property, *i.e.*, folia (or sheets) are distinguished by the property that fields on distinct sheets are topologically distinct. We describe these sheets as 'isomagnetic'. All fields on a given isomagnetic sheet are topologically accessible one from another through frozen field distortion. The magnetic relaxation process may thus be pictured as a trajectory on the isomagnetic sheet determined by the initial field $\mathbf{B}_0(\mathbf{x})$.

2.6. THE STRUCTURE OF EQUILIBRIUM FIELDS $\mathbf{B}^E(\mathbf{X})$

From equation (2.25), it is evident that

$$\mathbf{j}^E \cdot \nabla p^E = 0 \quad \text{and} \quad \mathbf{B}^E \cdot \nabla p^E = 0, \tag{2.27}$$

i.e., both $\mathbf{B}^E$-lines and $\mathbf{j}^E$-lines lie on the family of surfaces $p^E = cst$. It may of course happen that $\nabla p^E \equiv 0$ in some subdomain $\mathcal{D}^p$ of $\mathcal{D}$; then $\mathbf{j}^E \times \mathbf{B}^E = 0$ in $\mathcal{D}^p$, so that

$$\mathbf{j}^E = \alpha(\mathbf{x})\mathbf{B}^E \quad , \quad \mathbf{B}^E \cdot \nabla\alpha = 0 \quad \text{in} \quad \mathcal{D}^p. \tag{2.28}$$

Again, the $\mathbf{B}^E$-lines (which now coincide with the $\mathbf{j}^E$-lines) lie on the family of surfaces $\alpha =$ cst. If, however $\nabla\alpha \equiv 0$ in some further nested subdomain $\mathcal{D}^\alpha \subset \mathcal{D}^p$, then

$$\mathbf{j}^E = \nabla \times \mathbf{B}^E = \alpha \mathbf{B}^E \text{ in } \mathcal{D}^\alpha, \tag{2.29}$$

where α is constant. Fields satisfying the condition (2.28) are described as Beltrami fields. We shall describe the stronger condition (2.29) as the 'strong Beltrami condition', and corresponding fields as strong Beltrami fields. Only if the field is strong Beltrami is it released from the topological constraint of lying on surfaces. The prototype example of a strong Beltrami field is the 'ABC-field'

$$\mathbf{B}^E = (C\sin\alpha z + B\cos\alpha y, A\sin\alpha x + C\cos\alpha z, B\sin\alpha y + A\cos\alpha x) \tag{2.30}$$

for which the condition (2.29) may be easily verified. It is known from the work of Dombre *et al.* (1985) that, when $ABC \neq 0$, at least some of the $\mathbf{B}^E$-lines of this field exhibit chaotic wandering. There are also however islands of regularity within the chaotic sea, within which $\mathbf{B}^E$-lines do lie on surfaces. More general space-periodic fields of the form

$$\mathbf{B}^E = \sum_n \mathbf{B}_n e^{i\boldsymbol{\alpha}_n \cdot \mathbf{x}} \quad , \quad |\boldsymbol{\alpha}_n| = \alpha, \tag{2.31}$$

where the set of vector $\boldsymbol{\alpha}_n$ are symmetrically distributed on the sphere of radius α, have been extensively studied by Zaslavsky *et al.* (1991), and these flows generally exhibit a 'chaotic web' of $\mathbf{B}^E$ lines again containing islands of regularity. It seems likely that the most general space-periodic strong Beltrami field can be expressed in the form (2.31).

Suppose now that the initial field $\mathbf{B}_0(\mathbf{x})$ for our relaxation problem is chaotic in some subdomain $\mathcal{D}^C$ of the fluid domain $\mathcal{D}$. The volume V_C of $\mathcal{D}^C$ (more strictly its measure if it has fractal properties) is conserved under (volume-preserving) relaxation; hence the $\mathbf{B}^E$-lines of the relaxed field are also chaotic within the image domain $\bar{\mathcal{D}}^C$ (under the mapping $\mathbf{x} \to \mathbf{X}^E(\mathbf{x})$); hence $\mathbf{B}^E$ must be a strong Beltrami field in $\bar{\mathcal{D}}^C$.

This result appears quite surprising, because it implies that a strong Beltrami field is always topologically accessible from a chaotic field $\mathbf{B}_0(\mathbf{x})$, no matter what the 'degree of chaos' may be. However it must be recognized that the boundary of $\bar{\mathcal{D}}^C$ may be very highly convoluted, and that complexity of structure of the field $\mathbf{B}_0(\mathbf{x})$ may carry over during relaxation to corresponding complexity of structure of $\bar{\mathcal{D}}^C$, and hence of the strong Beltrami field in $\bar{\mathcal{D}}^C$.

2.7. STABILITY OF MAGNETOSTATIC EQUILIBRIUM

Minimum energy magnetostatic states are, by the nature of the relaxation process, clearly stable: if an equilibrium field $\mathbf{B}^E(\mathbf{x})$ that has been arrived at by magnetic relaxation on an isomagnetic sheet S_0 is perturbed on this sheet, then such perturbation must increase its energy, and within the framework of *any* dynamical model that dissipates kinetic energy, the field will tend to return to its minimum energy state, *i.e.*, to $\mathbf{B}^E(\mathbf{x})$.

Not all magnetostatic equilibria however are the result of a magnetic relaxation process, and it is useful to have an explicit criterion for minimality of magnetic energy with respect to frozen field distortions. Let $\boldsymbol{\xi}(\mathbf{x})$ be a virtual volume-preserving displacement of the medium, which we regard as the mapping associated with a steady velocity field $\mathbf{v}(\mathbf{x})$ acting for a short time interval $0 < t < \tau$, with $\nabla \cdot \mathbf{v} = 0$. It is easily shown that

$$\boldsymbol{\xi}(\mathbf{x}) = \boldsymbol{\eta}(\mathbf{x}) + \frac{1}{2}(\boldsymbol{\eta} \cdot \nabla)\boldsymbol{\eta} + O(\eta^3), \tag{2.32}$$

where $\boldsymbol{\eta}(\mathbf{x}) = \tau \mathbf{v}(\mathbf{x})$, and that the first and second order variation of $\mathbf{B}^E(\mathbf{x})$ under frozen field distortion (*i.e.*, on the isomagnetic sheet S_0) are

$$\delta^1 \mathbf{B} = \nabla \times (\boldsymbol{\eta} \times \mathbf{B}^E) \quad , \quad \delta^2 \mathbf{B} = \frac{1}{2} \nabla \times (\boldsymbol{\eta} \times \delta^1 \mathbf{B}) \tag{2.33}$$

(and generally, $\delta^n \mathbf{B} = \frac{1}{n} \nabla \times (\boldsymbol{\eta} \times \delta^{(n-1)} \mathbf{B})$).

The magnetic energy in the disturbed state is then given by

$$M = M^E + \delta^1 M + \delta^2 M + O(\eta^3), \tag{2.34}$$

where

$$\delta^1 M = \int \mathbf{B}^E \cdot \delta^1 \mathbf{B} dV, \tag{2.35}$$

$$\delta^2 M = \frac{1}{2} \int [(\delta^1 \mathbf{B})^2 + 2\mathbf{B}^E \cdot \delta^2 \mathbf{B}] dV. \tag{2.36}$$

It is not difficult to prove that, under the equilibrium condition (2.25) and the boundary conditions

$$\mathbf{B}^E \cdot \mathbf{n} = 0 \quad , \quad \boldsymbol{\eta} \cdot \mathbf{n} = 0 \text{ on } \partial\mathcal{D}, \tag{2.37}$$

$\delta^1 M = 0$, as is to be expected.

A word of caution is needed here concerning a certain class of displacements $\boldsymbol{\eta}(\mathbf{x})$ for which $\delta^1 \mathbf{B} = 0$ (and therefore $\delta^n \mathbf{B} = 0$ for all $n \geq 2$ also); an example of such displacements are those everywhere parallel to $\mathbf{B}^E$ so that $\boldsymbol{\eta} \times \mathbf{B}^E = 0$. The field is clearly uninfluenced by such displacements, and it seems appropriate to exclude them in a consideration of stability. We therefore define an 'admissible' displacement function $\boldsymbol{\eta}(\mathbf{x})$ by the three conditions

$$\nabla \cdot \boldsymbol{\eta} = 0, \quad \mathbf{n} \cdot \boldsymbol{\eta} = 0 \text{ on } \partial\mathcal{D}, \quad \nabla \times (\boldsymbol{\eta} \times \mathbf{B}^E) \not\equiv 0, \tag{2.38}$$

and we assert that the field $\mathbf{B}^E$ is stable if

$$\delta^2 M > 0 \text{ for all admissible } \boldsymbol{\eta}. \tag{2.39}$$

The *ABC* field provides an interesting and nontrivial illustration of the application of this procedure. With

$$\boldsymbol{\eta}(\mathbf{x}) = \sum_m \boldsymbol{\eta}_m e^{i\mathbf{k}_m\cdot\mathbf{x}} \quad , \boldsymbol{\eta}_m \cdot \mathbf{k}_m = 0, \tag{2.40}$$

and

$$\mathbf{B}^E = \sum_n \mathbf{B}_n e^{i\boldsymbol{\alpha}_n\cdot\mathbf{x}}, \quad |\boldsymbol{\alpha}_n| = |\alpha|, \tag{2.41}$$

direct calculation (Moffatt, 1986) gives

$$\delta^2 M = \frac{1}{2}\sum_{n,m} |\boldsymbol{\eta}_m|^2 |\mathbf{k}_m \cdot \mathbf{B}_n|^2, \tag{2.42}$$

and since this is positive for all admissible $\boldsymbol{\eta}$, the equilibrium is stable. This conclusion still holds when $\boldsymbol{\eta}(\mathbf{x})$ is unrestricted by the incompressibility condition (see Sero-Guillaume and Moffatt, 1992, this volume).

3. Analogous Euler Flows

3.1. TWO USEFUL ANALOGIES

Let us now consider the corresponding situation in relation to solutions of the Euler equations of classical hydrodynamics of an incompressible inviscid fluid. In terms of the velocity field $\mathbf{u}(\mathbf{x}, t)$ and the pressure field $p(\mathbf{x}, t)$ these equation are simply

$$\frac{D\mathbf{u}}{Dt} \equiv \frac{\partial \mathbf{u}}{\partial t} + \mathbf{u}\cdot\nabla\mathbf{u} = -\frac{1}{\rho}\nabla p \quad , \quad \nabla\cdot\mathbf{u} = 0, \tag{3.1}$$

where ρ is the fluid density, assumed uniform. Using the vector identity $\mathbf{u}\times(\nabla\times\mathbf{u}) = \frac{1}{2}\nabla\mathbf{u}^2 - \mathbf{u}\cdot\nabla\mathbf{u}$, the evolution equation may be written in the alternative form

$$\frac{\partial \mathbf{u}}{\partial t} = \mathbf{u}\times\boldsymbol{\omega} - \nabla h, \tag{3.2}$$

where $\boldsymbol{\omega} = \nabla\times\mathbf{u}$ is the vorticity field and $h = p/\rho + \frac{1}{2}\mathbf{u}^2$. The curl of this equation gives the vorticity equation

$$\frac{\partial \boldsymbol{\omega}}{\partial t} = \nabla\times(\mathbf{u}\times\boldsymbol{\omega}). \tag{3.3}$$

We now note two analogies which are quite different in nature and which must not be confused.

Analogy A

First note the analogy between equation (2.1) and (3.3) with the identifications

$$\mathbf{v} \longleftrightarrow \mathbf{u} \quad , \quad \mathbf{B} \longleftrightarrow \boldsymbol{\omega}, \tag{3.4}$$

all four fields being solenoidal. The analogy is not complete, because whereas $\mathbf{u}$ in (3.3) is constrained to be the velocity field associated with $\boldsymbol{\omega}$ through $\boldsymbol{\omega} = \text{curl}\ \mathbf{u}$, there is no corresponding constraint on $\mathbf{v}$ in (2.1). This means that a much wider range of initial conditions can be considered in relation to (2.1); it is this freedom that allows the growth of magnetic modes (*i.e.* dynamo action) that have no obvious counterpart in the Euler flow context.

Despite the imperfection of the analogy however, certain properties of (2.1) do carry over to (3.3). Most important among these is the fact that (3.3) is a 'frozen field' equation so that all topological properties of the vorticity field $\boldsymbol{\omega}$ are conserved under Euler evolution. In particular, there is a helicity invariant analogous to (2.3) namely (Moffatt, 1969)

$$\mathcal{H} = \int_V \mathbf{u} \cdot \boldsymbol{\omega} dV \quad , \quad \boldsymbol{\omega} \cdot \mathbf{n} = 0 \text{ on } \partial V. \tag{3.5}$$

This (kinetic) helicity is conserved for so long as the velocity and vorticity fields remain smooth (C^1) functions of position $\mathbf{x}$. [Other papers in this volume will address the possibility that singularities may develop under Euler evolution within a finite time; if this occurs, then viscous effects must presumably cause reconnection of vortex lines in the neighborhood of such singularities, and helicity (and likewise energy) will then no longer be conserved.]

Similarly, any theorem that is dependent only on the frozen-in character of the field $\mathbf{B}$, will apply equally to the field $\boldsymbol{\omega}$ determined by (3.3). Some such theorems (*e.g.* Kelvin's circulation theorem) were discovered first in the Euler equation context!

However, some statements do *not* carry over: *e.g.* stable magnetostatic equilibria are characterised by minimum magnetic energy, but stable solutions of the Euler equations are *not* characterised by minimisation of enstrophy $\int \boldsymbol{\omega}^2 dV$.

Analogy B

Secondly, note the analogy between the equations of magnetostatic equilibrium (see (2.25))

$$\mathbf{j} \times \mathbf{B} = \nabla p \quad , \quad \mathbf{j} = \nabla \times \mathbf{B}, \tag{3.6}$$

and the *steady* Euler flow equation (see (3.2))

$$\mathbf{u} \times \boldsymbol{\omega} = \nabla h \quad , \quad \boldsymbol{\omega} = \nabla \times \mathbf{u}, \tag{3.7}$$

Here the analogy is evidently between the variables

$$\mathbf{B} \longleftrightarrow \mathbf{u} \quad , \quad \mathbf{j} \longleftrightarrow \boldsymbol{\omega} \quad , \quad p \longleftrightarrow h_0 - h, \tag{3.8}$$

where h_0 is some positive constant. To every solution of (3.6), obtained by whatever means, there corresponds a solution of (3.7) *via* the correspondences (3.8); thus, to every solution of (3.6) obtained by the magnetic relaxation procedure described in section 2, there corresponds a solution of the Euler equations satisfying corresponding boundary conditions (*e.g.* if $\mathbf{n} \cdot \mathbf{B} = 0$ on $\partial \mathcal{D}$, then $\mathbf{n} \cdot \mathbf{u} = 0$ on $\partial \mathcal{D}$ for the corresponding solution). Tangential discontinuities of $\mathbf{B}$ (*i.e.* current sheets) in the magnetostatic solutions become vortex sheets in the Euler flows.

Magnetic relaxation thus provides an important, albeit indirect, technique for obtaining a very wide family of fully three-dimensional solutions of the steady Euler equations. Indeed, since the topology of the initial field $\mathbf{B}_0(\mathbf{x})$ in the relaxation problem is arbitrary, we have here a technique of finding Euler flows of arbitrary streamline topology. But beware! The analogy does *not* extend to considerations of stability: although magnetostatic equilibria obtained by magnetic relaxation are stable, there is no guarantee that the analogous Euler flows are stable, because *unsteady* Euler dynamics conserves the topology not of the $\mathbf{u}$-field but rather of the $\boldsymbol{\omega}$-field, a vital distinction.

The situation is represented in figure 7 which shows again the function space of all bounded square-integrable solenoidal fields, with its isomagnetic foliation that constrains the magnetic relaxation process. To the equilibrium field $\mathbf{B}^E(\mathbf{x})$ there corresponds an Euler flow $\mathbf{u}^E(\mathbf{x})$ with vorticity $\boldsymbol{\omega}^E(\mathbf{x})$. If this flow is perturbed by a weak impulsive pressure field, so that the vortex lines are distorted from the equilibrium configuration, then the vorticity field $\boldsymbol{\omega}(\mathbf{x}, t)$ evolves on the submanifold of fields for which this vorticity topology is conserved. This condition defines a different foliation of the function space, described as the *isovortical* foliation. The stability of the magnetostatic field $\mathbf{B}^E(\mathbf{x})$ is determined by behaviour on the isomagnetic folium; the stability of the analogous Euler flow $\mathbf{u}^E(\mathbf{x})$ by the behaviour on the isovortical folium.

3.2. ARNOLD'S STABILITY CRITERION

The unsteady Euler equations (3.1) conserve kinetic energy

$$K = \frac{1}{2}\rho \int \mathbf{u}^2 dV, \tag{3.9}$$

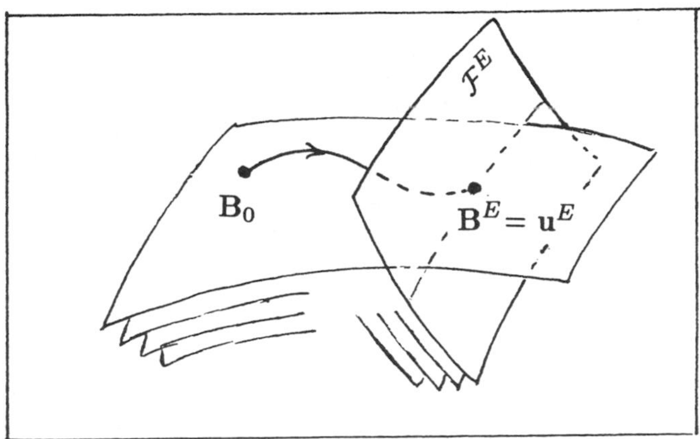

Fig. 7. Isovortical and isomagnetic foliations. Stability of the analogous Euler flow $\mathbf{u}^E(\mathbf{x})$ is determined by the variation of energy on the isovortical folium $\mathcal{F}^E$.

and this further constrains the otherwise free evolution on an isovortical folium. If K has an extremum (maximum or minimum) at the 'point' $\mathbf{u} = \mathbf{u}^E$ in function space, with respect to perturbation on the isovortical folium $\mathcal{F}^E$ through $\mathbf{u}^E$, then the 'curves' of constant K are closed on $\mathcal{F}^E$, so that in this sense (*i.e.* with respect to the energy norm) the flow remains permanently in a neighborhood of $\mathbf{u}^E$, and is therefore stable. This consideration yields a criterion for stability (Arnold, 1966). First note that, with the notation of section (2.7) above, the first and second order variations of vorticity under isovortical perturbation are

$$\delta^1\omega = \nabla \times (\boldsymbol{\eta} \times \omega^E) \quad , \quad \delta^2\omega = \frac{1}{2}\nabla \times (\boldsymbol{\eta} \times \delta^1\omega) \tag{3.10}$$

Hence, using the solenoidal projection as in (2.22), the corresponding velocity perturbations are

$$\delta^1\mathbf{u} = (\boldsymbol{\eta} \times \omega^E)_s \quad , \quad \delta^2\mathbf{u} = \frac{1}{2}(\boldsymbol{\eta} \times \delta^1\omega)_s, \tag{3.11}$$

The first order variation in kinetic energy is

$$\delta^1 K = \rho \int_{\mathcal{D}} \mathbf{u}^E \cdot \delta^1\mathbf{u} dV, \tag{3.12}$$

and it is easily shown, under the conditions $\mathbf{u}^E \cdot \mathbf{n} = \boldsymbol{\eta} \cdot \mathbf{n} = 0$ on $\partial\mathcal{D}$, that $\delta^1 K = 0$, a result due to Kelvin. The second order variation is

$$\begin{aligned} \delta^2 K &= \frac{1}{2}\rho \int \left[(\delta^1\mathbf{u})^2 + 2\mathbf{u}^E \cdot \delta^2\mathbf{u}\right] dV \\ &= \frac{1}{2}\rho \int \left[(\boldsymbol{\eta} \times \omega^E)_S^2 + \mathbf{u}^E \cdot (\boldsymbol{\eta} \times \delta^1\omega)_S\right] dV, \end{aligned} \tag{3.13}$$

and the equilibrium is stable if

$$\left.\begin{array}{lll}\text{either} & \delta^2 K > 0 & \text{for all admissible } \boldsymbol{\eta} \\ \text{or} & \delta^2 K < 0 & \text{for all admissible } \boldsymbol{\eta}\end{array}\right\}. \tag{3.14}$$

Two comments may be made in relation to this criterion. First, although there appears to be complete parity between maximum and minimum energy states in this criterion, there is, as pointed out by Arnold, a distinction if weak dissipative (viscous) effects are taken into account. These act in such a way as to destabilize the states of maximum energy, just as for the simpler prototype problem of free rotation of a rigid body about its centre of mass. In this case, steady rotation about the axis of minimum inertia has (for given angular momentum) maximum kinetic energy; if the body has a cavity containing viscous fluid, then this equilibrium state becomes unstable, viscous dissipation in the cavity always tending to drive the system to its state of minimum kinetic energy (*i.e.* rotation about the axis of maximum inertia).

Secondly, the qualification concerning stability 'with respect to the energy norm' is important. There are some situations in which a flow may be stable in this sense, but which nevertheless exhibit unbounded growth of perturbations. An example considered by Moffatt and Moore (1978) is the spherical vortex of Hill discovered nearly one hundred years ago (Hill, 1894). Axisymmetric perturbations of this vortex decay everywhere except in the immediate neighborhood of the rear stagnation point, where they grow in an unbounded manner (see figure 8, from Pozrikidis, 1986, who followed the disturbance numerically well into the nonlinear regime). This flow is stable to axisymmetric disturbances with respect to the energy norm but unstable with respect to a norm that focuses on this singular region.

The Arnold criterion (3.14) may be used (Moffatt, 1986) to examine the stability of the ABC-flow

$$\mathbf{u}^E = (C\cos\alpha z + B\sin\alpha y, A\cos\alpha x + C\sin\alpha z, B\cos\alpha y + A\sin\alpha x). \tag{3.15}$$

With a virtual displacement

$$\boldsymbol{\eta} = \eta_0(\cos kz, \sin kz, 0), \tag{3.16}$$

we find from (3.13), after some calculation,

$$\delta^2 K = -\frac{1}{4}\eta_0^2\frac{(A^2+B^2)\alpha k^3}{(\alpha^2+k^2)}, \tag{3.17}$$

which is positive or negative according as $\alpha k <$ or > 0. Hence K is neither maximal nor minimal, and we may therefore draw no conclusions as regards stability of the flow (3.15). There is however a rather strong inference that the flow is unstable, and further arguments to this effect have been adduced by Moffatt (1986).

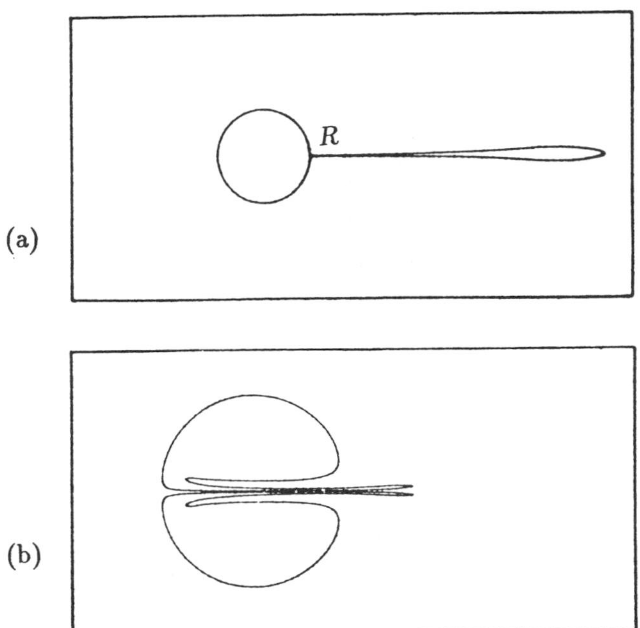

Fig. 8. Nonlinear instability of Hill's vortex. (a) when the initial perturbation of the spherical surface of the vortex is prolate, it sheds a spike from the rear stagnation point R; (b) when the perturbation is oblate, the spike is intrusive, and an irrotational hole is 'drilled' through the vortex. [From Pozrikidis, 1986.]

3.3. VORTONS

We use the word 'vorton' in the sense of Moffatt (1988) to mean a nonlinear rotational disturbance which propagates with constant velocity $\mathbf{U}$ and without change of structure in a fluid of infinite extent. Thus the vorticity field of a vorton has the form

$$\omega(\mathbf{x}, t) = \mathbf{f}(\mathbf{x} - \mathbf{U}t), \tag{3.18}$$

where $\mathbf{f}(\boldsymbol{\xi})$ is a solenoidal field of compact support in $\mathbf{R}^3$. Hill's vortex mentioned above is a classical example of a vorton; further examples are provided by the wide family of spherical vortices with swirl discovered by Hicks (1899).

In a frame of reference moving with this vorton (*i.e.* with velocity $\mathbf{U}$ relative to the fluid at infinity), the flow is steady. We may therefore seek to determine the complete family of vortons by the technique of magnetic relaxation, starting with an initial field of the form

$$\mathbf{B}_0(\mathbf{x}) = \mathbf{B}_{00} + \mathbf{b}_0(\mathbf{x}), \tag{3.19}$$

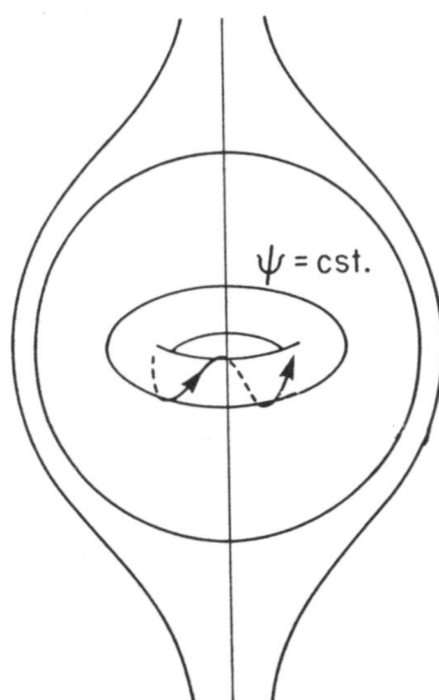

Fig. 9. Axisymmetric vorton topology. The stream-surfaces ψ = cst. are a family of nested tori, and the signature of the flow consists of the pair $\{V(\psi), W(\psi)\}$ where $V(\psi)$ is the volume of the torus with label ψ and $W(\psi)$ is the azimuthal flux within this torus. [From Moffatt, 1988.]

where $\mathbf{B}_{00}$ is uniform (and subsequently to be identified with the velocity $-\mathbf{U}$ in the analogous Euler flow) and $\mathbf{b}(\mathbf{x})$ is a field of compact support of sufficient strength to guarantee the existence of a region $\mathcal{D}$ such that $\mathbf{n}\cdot\mathbf{B}_0$ on $\partial\mathcal{D}$. Inside $\mathcal{D}$, the $\mathbf{B}_0$-lines are either closed curves, or they lie on surfaces (*e.g.* nested tori) or they exhibit chaotic wandering. These (topological) properties are invariant during the relaxation process.

The simplest situation, and probably the most important, is that in which $\mathbf{B}_0(\mathbf{x})$ is axisymmetric. Let $\mathbf{B}_0$ have the poloidal-toroidal decomposition

$$\mathbf{B}_0 = \mathbf{B}_P + \mathbf{B}_T, \tag{3.20}$$

where (in cylindrical polar coordinates (r, ϕ, z))

$$\mathbf{B}_T = (0, B(r,z), 0), \quad \mathbf{B}_P = \left(\frac{1}{r}\frac{\partial\psi}{\partial z}, 0, -\frac{1}{r}\frac{\partial\psi}{\partial r}\right). \tag{3.21}$$

The surfaces ψ = cst. are a family of nested tori within $\mathcal{D}$. Under magnetic relaxation, the volume $V(\psi)$ inside each such torus and the azimuthal flux of field $W(\psi)$ within each such torus are conserved (figure 9). The field relaxes to a magnetostatic equilibrium characterised by the signature $\{V(\psi), W(\psi)\}$, and the analogous Euler flow (or vorton) is likewise characterised by this signature.

A very special situation arises if $\mathbf{B}_{00} = 0$, in which case the analogous Euler flows consists of vortons which do not propagate (*i.e.* $\mathbf{U} = 0$), but persist as stationary rotational 'excitations' in the fluid. Examples of such fields are given in a separate paper in this volume (Chui and Moffatt, 1992).

4. Knots and Links

4.1. THE ENERGY OF A KNOT

The magnetic relaxation procedure lends itself in an interesting way to the theory of knots and links in $\mathbf{R}^3$, and yields the intriguing result that every knot (or link) is characterised by an intrinsic minimum energy which is determined solely by the knot topology (Moffatt, 1990a; Freedman and He, 1991). With each knot K, we associate a magnetic flux tube of volume V carrying flux Φ and such that the total helicity of the field is $\mathcal{H}_M = h\Phi^2$ (see Ricca and Moffatt, 1992, this volume). If this field is allowed to relax to a minimum energy state, then this minimum energy is necessarily given by

$$M_{\min} = m(h)\Phi^2 V^{-1/3}, \tag{4.1}$$

where $m(h)$ is a dimensionless function of the dimensionless variable h (related to the twist of the field within the flux tube). When h is of order unity, the relaxation proceeds essentially through contraction of the axis of the flux tube and expansion of the mean cross-section (thus keeping V constant). The process comes to a halt when the knot 'tightens' on itself (as illustrated for the trefoil knot in figure 10). At this stage, the axial length L is of order $V^{1/3}$ and the mean tube cross-section A is of order $V^{2/3}$.

When $|h| \gg 1$, the opposite occurs! The tube cross-section decreases due to the strong twist component of field around the tube axis; in the case of the unknot (see Chui and Moffatt, 1992), axisymmetric equilibrium is attained when

$$L \sim |h|^{2/3} V^{1/3} \quad , \quad A \sim |h|^{-2/3} V^{2/3}, \tag{4.2}$$

and it seems likely that the same result holds for an arbitrary knot. In this case, the energy function $m(h)$ satisfies

$$m(h) = O(|h|^{4/3}) \quad \text{for} \quad |h| \gg 1. \tag{4.3}$$

For an arbitrary knot, we may then anticipate that the energy function has the form sketched in figure 11, with symmetry about the axis $h = 0$ only for achiral knots. The curve must have a minimum value, m_c say, attained for $h = h_c$, and the real number m_c is determined solely by the knot type. Similar considerations of course apply to links.

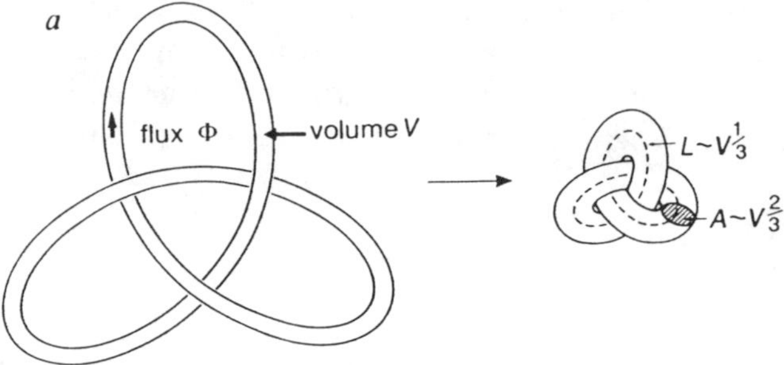

Fig. 10. Relaxation of a flux tube in the form of the trefoil knot. The minimum energy has the form $m(h)\Phi^2 V^{-1/3}$, where Φ and V are the (conserved) flux and volume, and $h = \mathcal{H}_M/\Phi^2$ where $\mathcal{H}_M$ is the conserved helicity. The parameter h can take any value depending on the twist of the field within the tube. The dimensionless function $m(h)$ is determined (in principle) by the topology of the knot. [From Moffatt, 1990].

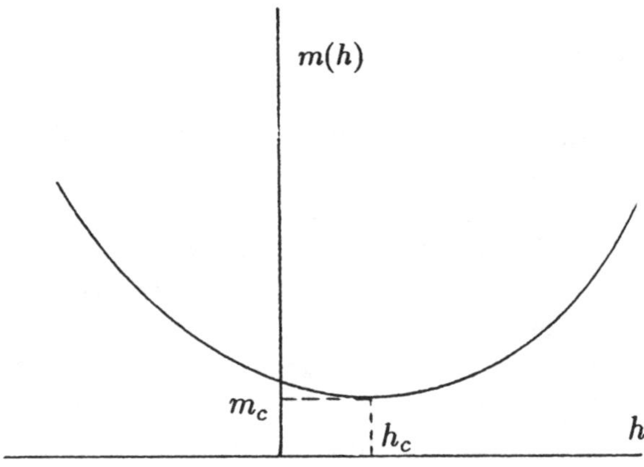

Fig. 11. Expected qualitative form of the energy function $m(h)$ for an arbitrary knot.

The fact that a knot is characterised by a minimum energy, or more correctly a minimum energy function $m(h)$ dependent on the twist (or 'framing') parameter h, is of potential importance in such contexts as polymer physics and molecular biology, where knotted structures may be expected to seek configurations of minimum energy, in the absence of external perturbations. The energy functional $M = \frac{1}{2}\int \mathbf{B}^2 dV$ that we have chosen may however be too specialized for application in these contexts; we adopt a more general approach in the following sub-section.

4.2. ALTERNATIVE ENERGY FUNCTIONALS

The mathematics of the problem is here served by choosing a physical model which is unrealistic but – and this is what matters – self-consistent. Suppose that our conducting fluid is also endowed with magnetic permeability *via* a relationship of the form

$$\mathbf{H} = \mathbf{H}(\mathbf{B}) \quad \text{where} \quad \mathbf{j} = \text{ curl } \mathbf{H}. \tag{4.4}$$

The field $\mathbf{B}$ still satisfies the induction equation (2.1) and the Lorentz force is still given by $\mathbf{j} \times \mathbf{B}$. The only change in obtaining the energy equation (2.23) which lies at the heart of the relaxation process is that the magnetic energy is now replaced by

$$M(t) = \int dV \int \mathbf{H} \cdot d\mathbf{B} \tag{4.5}$$

(which yields (2.13) when $\mathbf{B} = \mathbf{H}$). If, for example, $\mathbf{H} = |\mathbf{B}|^\lambda \mathbf{B}$ for $\lambda > -1$, then

$$M(t) = \frac{1}{\lambda + 2} \int |\mathbf{B}|^{\lambda+2} dV. \tag{4.6}$$

In this case, relaxation of a knotted flux tube leads to a minimum energy which, on dimensional grounds, has the form

$$M_{\text{min}} = m(h)\Phi^{\lambda+2}V^{-(1+2\lambda)/3}. \tag{4.7}$$

The choice $\lambda = -\frac{1}{2}$ is special in that M_{min} is then independent of V, a fact exploited by Freedman and He (1991), who have proved the striking result that $\int |\mathbf{B}|^{3/2} dV$ is bounded below by the least value of the crossing number of the knot (*i.e.* the average over all projections of the number of (unsigned) crossings).

5. Relaxation Using Artificial Dynamics

A technique has been suggested by Vallis, Carnevale and Young (1989) whereby relaxation to steady Euler flows by a process that conserves *vorticity* topology may be forced by the introduction of 'artificial' dynamics. The dynamics *has* to be artificial if energy is to be dissipated while the vorticity field remains frozen in the fluid.

Let us here attempt to construct such a process by appropriate adaptation of the magnetic relaxation procedure. Let $\mathbf{j} =$ curl $\mathbf{B}$, and suppose that $\mathbf{j}$ evolves according to the frozen field equation

$$\frac{\partial \mathbf{j}}{\partial t} = \nabla \times (\mathbf{v} \times \mathbf{j}), \tag{5.1}$$

so that

$$\frac{\partial \mathbf{B}}{\partial t} = \mathbf{v} \times \mathbf{j} - \nabla \phi = (\mathbf{v} \times \mathbf{j})_s \tag{5.2}$$

for some scalar field ϕ. This is an artificial evolution equation, which we may however couple with the same choice of velocity as before,

$$k\mathbf{v} = \mathbf{j} \times \mathbf{B} - \nabla p = (\mathbf{j} \times \mathbf{B})_s, \tag{5.3}$$

which yields an energy equation for $M = \frac{1}{2}\int \mathbf{B}^2 dV$,

$$\frac{dM}{dt} = \int \mathbf{B} \cdot (\mathbf{v} \times \mathbf{j}) dV = -k \int \mathbf{v}^2 dV. \tag{5.4}$$

If either a lower bound or an upper bound can be placed on M, then (choosing $k > 0$ or $k < 0$ respectively) we are driven to a steady state in which $\mathbf{v} = 0$ and $\mathbf{j} \times \mathbf{B} = \nabla p$.

Now make the replacements $\mathbf{B} \to \mathbf{u}, \mathbf{j} \to \boldsymbol{\omega}$, so that (5.2) and (5.3) become

$$\frac{\partial \mathbf{u}}{\partial t} = \frac{1}{k}\left[(\boldsymbol{\omega} \times \mathbf{u})_s \times \boldsymbol{\omega}\right]_s, \tag{5.5}$$

where the suffix s again represents solenoidal projection. With this equation, we associate the energy equation (the exact counterpart of (5.4))

$$\frac{d}{dt}\frac{1}{2}\int \mathbf{u}^2 dV = -k^{-1}\int (\boldsymbol{\omega} \times \mathbf{u})_s^2 dV. \tag{5.6}$$

Thus, provided a bound on the energy exists, the system is driven to a steady Euler flow for which $(\mathbf{u} \times \boldsymbol{\omega})_s = 0$. Moreover, the topology of $\boldsymbol{\omega}$ (the analogue of $\mathbf{j}$) is conserved during this relaxation, *i.e.* the relaxation occurs on an isovortical folium, so that the final state, being an energy extremum on the isovortical folium, may be expected to be stable, by Arnold's criterion.

Unfortunately, everything hinges on the existence of either a lower or an upper bound on the energy. Recall that previously a lower bound was constructed through a product of Schwarz and Poincaré inequalities. Now however these inequalities are

$$\int \mathbf{B}^2 \int \mathbf{j}^2 \geq (\int \mathbf{B} \cdot \mathbf{j})^2 = \text{cst.} \quad , \quad \int \mathbf{j}^2 \Big/ \int \mathbf{B}^2 \geq q_0^2, \tag{5.7}$$

and these cannot be combined to give a lower bound on $\int \mathbf{B}^2 dV$. There is therefore no guarantee that evolution determined by (5.5) will lead to a steady state. If $k > 0$, the energy may decrease to zero, the vorticity field becoming more and more complex although its topology is conserved, while if $k < 0$, the energy may increase to infinity. This difficulty was encountered equally by Vallis *et al.* (1989) within the framework of the particular artificial dynamics that they adopted.

An upper bound can be placed on the energy however when the flow is either two-dimensional (Vallis *et al.*) or axisymmetric without swirl (Moffatt, 1990b). In the two-dimensional case, enstrophy is conserved by (5.5) and the energy has an upper bound (*cf* 5.7b) proportional to this constant enstrophy. The situation for axisymmetric flow with swirl is similar. In these cases, the artificial evolution equation (5.5), with $k < 0$, must drive the system to a *stable* Euler flow for which the topology of the vorticity field is prescribed in advance. What does this mean for the two-dimensional case in which the vortex lines are all parallel to the z-axis? There is still a topology associated with the family of plane curves $\omega(x, y) = \text{cst}$. These isovorticity curves move with the fluid during the evolution process (5.5), and the area $A(\omega)$ inside any such curve is conserved. This is now the appropriate signature which fully describes the topology of the vorticity field, and which characterises the Euler flow that the system approaches as $t \to \infty$.

Since the energy increases during this type of evolution, it is perhaps a misuse of words to describe it as a *relaxation* process; it has more the character of a *spin-up* process which reaches stable equilibrium when the energy is maximal within the family of flows of prescribed signature function $A(\omega)$.

Acknowledgements

I am grateful to Andrew Gilbert who helped to develop the formulation of the 'artificial' relaxation process described in section 5, and who first recognized its applicability to the two-dimensional situation.

References

ARNOL'D, V.I. 1966 Sur un principe variationel pour les ecoulements stationnaires des liquides parfaits et ses applications aux problems de stabilité non-lineaires. *J. Mec.* **5**, 29–43.

ARNOL'D, V.I. 1974 The asymptotic Hopf invariant and its applications. In *Proc. Summer School in Differential Equation.* Erevan: Armenian SSR Academy of Science (English trans. *Sel. Math. Sov.*5, 327–345 (1986).)

CHUI, A.Y.K. AND MOFFATT, H.K. 1992 Minimum energy magnetic fields with toroidal topology. [This volume, pp. 000–000.]

DOMBRE, T., FRISCH, U., GREENE, J.M., HÉNON, M., MEHR, A., AND SOWARD, A.M. 1986 Chaotic streamlines in the ABC flows. *J. Fluid Mech.* **167**, 353–391.

FREEDMAN, M.H. 1988 A note on topology and magnetic energy in incompressible perfectly conducting fluids. *J. Fluid Mech.* **194**, 549–551.

FREEDMAN, M.H. AND HE, Z.-H. 1991 Divergence-free fields: energy and asymptotic crossing number. *Annals of Math.* **134**, 189–229.

HICKS, W.M. 1899 Researches in vortex motion. III. On spiral or gyrostatic vortex aggregates. *Phil. Trans. R. Soc. Lond.* A**192**, 33–101.

HILL, M.J.M. 1894 On a spherical vortex. *Phil Trans. R. Soc. Lond.* A**185**, 213–245.

MOFFATT, H.K. 1969 The degree of knottedness of tangled vortex lines. *J. Fluid Mech.* **36**, 117–129.

MOFFATT, H.K. 1985 Magnetostatic equilibria and analogous Euler flows of arbitrarily complex topology Part I, Fundamentals. *J. Fluid Mech.* **159**, 359–378.

MOFFATT, H.K. 1986 Magnetostatic equilibria and analogous Euler flows of arbitrarily complex topology Part II, Stability considerations. *J. Fluid Mech.* **166**, 359–378.

MOFFATT, H.K. 1988 Generalised vortex rings with and without swirl. *Fluid Dyn. Res.* **3**, 22–30.

MOFFATT, H.K. 1990a The energy spectrum of knots and links. *Nature, Lond.* **347**, 367–369.

MOFFATT, H.K. 1990b Structure and stability of solutions of Euler equations: a Lagrangian approach. *Phil. Trans. R. Soc. Lond.* A**333**, 321–342.

MOFFATT, H.K. AND MOORE, D.W. 1978 The response of Hill's spherical vortex to a small axisymmetric disturbance. *J. Fluid Mech.* **87**, 749–760.

POZRIKIDIS, D. 1986 Nonlinear instability of Hill's vortex. *J. Fluid Mech.* **168**, 337–367.

RICCA, R.L. AND MOFFATT, H.K. 1992 The helicity of a knotted vortex filament. [This volume, pp. 000–000.]

SERO-GUILLAUME, O. AND MOFFATT, H.K. 1992 Stability of space-periodic Beltrami fields. [In preparation.]

VALLIS, G.K., CARNEVALE, G.F. AND YOUNG, W.R. 1989 Extremal energy properties and construction of stable solutions of the Euler equations. *J. Fluid Mech.* **207**, 133–152.

ZASLAVSKY, G.M., SAGDEEV, R.Z., USIKOV, D.A. AND CHERNIKOV, A.A. 1991 *Weak chaos and quasi-regular patterns*, p. 192 *et. seq.*, Cambridge University Press.

KNOT THEORY, JONES' POLYNOMIALS, INVARIANTS OF 3-MANIFOLDS, AND THE TOPOLOGICAL THEORY OF FLUID DYNAMICS

KENNETH C. MILLETT
Department of Mathematics
University of California
Santa Barbara, CA 93106-4030, USA

ABSTRACT. The purpose of these lectures is to present an introduction to some of the fundamental elements of "classical knot theory" surrounding a description of the new families of spatial invariants derived from the 1985 discovery by V.F.R. Jones, [Jon1,2]. These finite integral Laurent polynomial invariants have provided insights into classical questions of knot theory. Their subsequent generalizations have given new invariants of 3-manifolds. Taken together these ideas form the basis of a new combinatorial theory of knots and links and invariants of 3-manifolds. The topics have been selected with an eye towards potential applications in the topological theory of fluid mechanics. In particular, I believe that the construction of satellite knots or links, especially cables, might be of particular interest in the study of closed curves arising in fluid dynamics and have, therefore, made a special effort to highlight those occasions in which this construction arises.

1. Knots and Links in 3-Space

A link in S^3 (One might just as well take the case of ordinary 3-space instead of the sphere!) is smooth embedding of a disjoint family of circles, *i.e.* a collection of disjoint smooth simple closed curves. A knot is the case of a single circle. Equivalently, one may consider submanifolds homeomorphic to this family or piecewise linear embeddings of the family. Although each of these definitions has its special role in the history and development of knot theory, they are all equivalent and we shall move between them depending on the situation. Two links, denoted L and L', are said to be **equivalent** if there is homeomorphism of S^3 taking one configuration to the other. A knot or link is said to be **trivial** if it is equivalent to a family of disjoint concentric planar circles. The efficient determination as to whether an arbitrary configuration of circles in space is equivalent to the trivial configuration is one of the fundamental problems of knot theory and figures in a number of ongoing research projects in topology. For the sake of ease of terminology,

H. K. Moffatt et al. (eds.), Topological Aspects of the Dynamics of Fluids and Plasmas, 29–64.

the spatial equivalence class of the knot or link is also called the knot or link, *i.e.* one does not distinguish between the representative and the equivalence class except by the context of the discussion. Choosing orientations for S^3 and/or the circles constituting the link leads to stronger definitions of equivalence which are also often important. If the homeomorphism giving the equivalence between L and L' preserves the orientation of S^3 there is a continuous family of homeomorphisms of S^3, called an ambient **isotopy**, beginning from the identity and ending with a homeomorphism taking L to L'. A homeomorphism taking one knot to another knot also preserves the complement of the knot. One of the more striking recent results in classical knot theory is the proof by Cameron Gordon and John Luecke that the knot is actually determined by the complement, [GorLu]. This means that if the complements of two knots are equivalent, then the two knots are equivalent.

In this section we shall use a historical description of the effort to classify or enumerate knots as the vehicle for the introduction of the most important concepts, vocabulary and, results in knot theory. The book, "Knots and Links", by Dale Rolfsen, [Rol] is probably the most popular reference for most of these topics. Other alternatives are [BZ, Cr, Fox, Gor, Kau2, Kau3]. Recent articles concerned with the problems of enumeration and identification include Thistlethwaite, [Th1], Sumners, [Su]. Other articles concerned with the computational aspects of the problem include Ewing-Millett, [EM1&2], Millett, [Mi2], and Jaeger-Vertigan-Welsh, [Ja].

Knot theory has its origins in the research of Gauss, [Ga], on inductance in systems of linked circular wires through which an electrical current is flowing. This physical problem was the origin of the linking number determined by an integral formula of fundamental importance. In addition, he developed the notion of a **regular projection** of a knot or link which forms the basis of the most common way in which a knot or link is presented, *via* a knot **diagram**. In the piecewise linear category where knots and links are polygons in 3-space, a regular projection of a knot or link is an orthogonal projection onto a plane so that the images of the vertices are disjoint from the images of the interiors of the edges and the other vertices, the intersection of the images of two edges is a most a point, and the intersection of the images of three or more edges is empty. The forbidden cases are shown in Figure 1.1.

The presentation of the knot or link is usually drawn by breaking the edges of the image of a regular projection so as to indicate under and over crossings of the edges with respect to the given direction of the projection. It is more common to give knot and link presentations with regular projections utilizing smooth curves such as are shown in Figure 1.2 since they are easier to draw. Gauss also described a combinatorial method of representing the diagram. On the circle along which one proceeds, label and list the crossings in the order encountered. For example, the projection associated to the

Fig. 1.1 Non-generic intersections

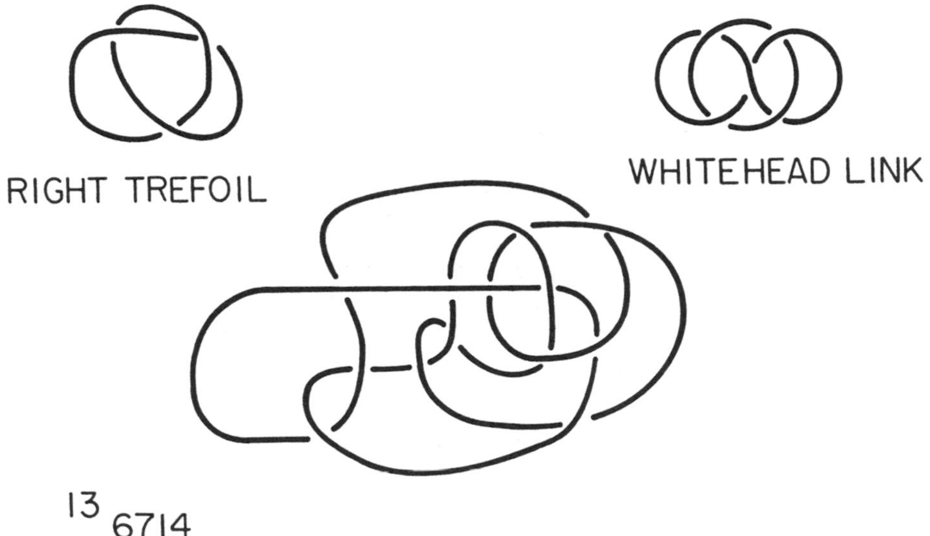

Fig. 1.2 Typical knot and link presentations

right trefoil in Figure 1.2 would be given the sequence ABCABC. Gauss also identified the problem of determining conditions on such sequences of letters in order that they determine a knot projection. This issue has been studied by topologists and graph theorists and algorithms to achieve this determination have figured in the efforts to tabulate knots, [Th1].

The most elementary quantity assigned to a knot or link is its **crossing number**. This is defined to be the minimum number of the crossings which occur in all the possible presentations of the given knot or link. By its very definition, it is an "invariant of the knot or link". Given one presentation of a knot or link, however, there is no systematic method to determine the crossing number except under special restrictions. The crossing number is an excellent example of the "folk theorem" that it is easy to define an invariant of a knot or link but that it is extremely difficult to define one that is both interesting and calculable, cf [R]. The crossing number will play a central role in our discussion.

Fig. 1.3 Figure 8 knot

J.B. Listing, a student of Gauss, devoted some effort to the study of knots and links. He recognized the achirality of the figure 8 knot shown in Figure 1.3. The concept of chirality concerns the relationship between a link, denoted L, and its mirror reflection, denoted $\overline{L}$. Taking the mirror to be the plane of the projection, a presentation of the mirror reflection is obtained from the given presentation by reversing all crossings, as shown in Figure 1.3. If L and $\overline{L}$ are not equivalent, L is said to be a **chiral** link and, if they are equivalent, L is said to be **achiral**. The right trefoil is chiral as are most knots and links. This, however, requires proof and is one of the first easy consequences of the study of the combinatorial invariants. The notion of chirality, a fundamental aspect of the study of the symmetries of knots and links, plays an important role in the application of knot theory to the natural sciences. Listing also worked on development of an effective notation for knot diagrams to be used in the tabulation of knots. His efforts had a significant impact on the work of Tait, [Ta], and Little [Lit1-4], who, with Kirkman, [Kirk] dominated the great 19th century effort at knot tabulation.

A knot or link diagram is said to be **alternating** if it has at least one crossing, all the circles represented in the diagram pass successively under and over at the crossing points as one proceeds to travel around each component, and the diagram contains no **nugatory crossing**, *i.e.* does not contain a crossing of the type shown in Figure 1.4. An **alternating knot or link** is one which admits an alternating diagram. Although the knots and links admitting the simplest diagrams are mostly alternating this is not true for more complicated knots, *i.e.* "the random knot is not alternating". Nevertheless, alternating knots play a fundamental role in the development of knot theory since every non-trivial projection in which there are no nugatory crossings is the projection of a non-trivial alternating knot (one simply needs to change the crossings). Furthermore, it takes a fair bit of work to show that non-alternating knots exist! The spatial movement which elim-

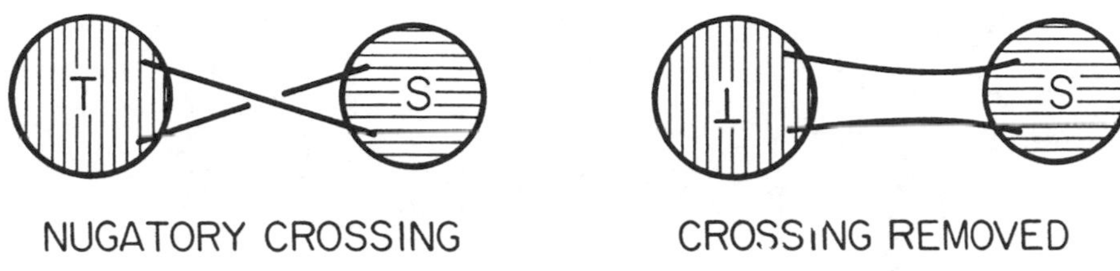

Fig. 1.4 A nugatory crossing

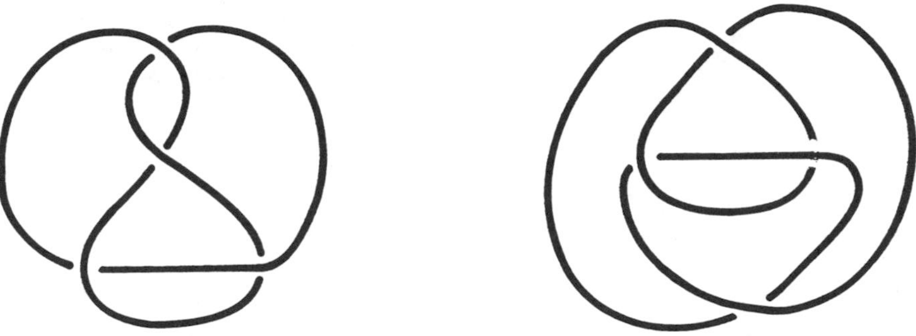

Fig. 1.5 S^2 equivalence of knot diagrams

inates the nugatory crossing gives one example of the type of equivalence that one would like to have for knot or link diagrams. Although we have been discussing projections of a knot or link in 3-space onto a plane, in fact we want to consider projections onto an S^2 and the equivalence of knot diagrams as subsets of S^2. The two diagrams shown in Figure 1.5, presenting the figure 8 knot are considered to be equivalent by means of passing a strand through ∞, *i.e. via* an isotopy of the extended plane. Thus, we will say that two diagrams are equivalent if there is an isotopy of S^2 taking one diagram to the other.

Although equivalence of diagrams is not sufficient to generate all the equivalences between diagrams presenting the same knot or link type, only three additional elementary equivalences, the Reidemeister moves shown in Figure 1.6, are required to generate all such equivalences. Thus,

PROPOSITION 1: Two diagrams represent the same knot or link type if, and only if, there is a finite sequence of transformations, each of which is either an S^2 equivalence or one of the three Reidemeister moves shown in Figure 1.6, taking the one diagram to the other.

From these Reidemeister moves one concludes that any knot having a presentation with only one or two crossings is equivalent to the trivial knot by application of type I moves. Thus the first non-trivial knot can have

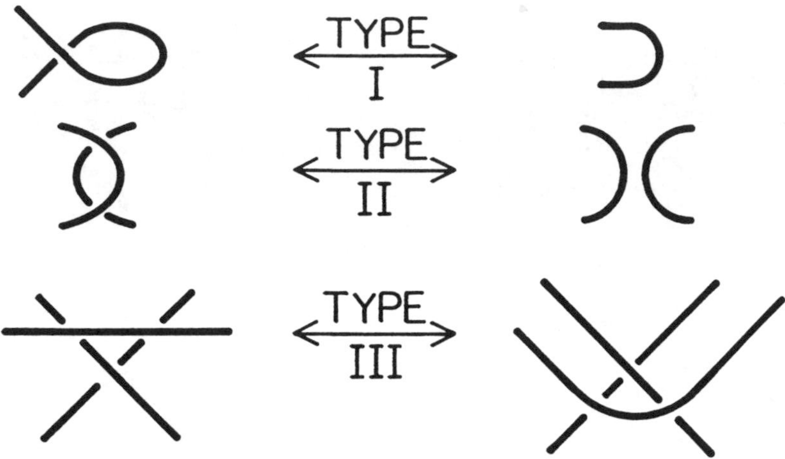

Fig. 1.6 The Reidemeister moves

no fewer than three crossings, *e.g.* the trefoil knot shown in Figure 1.2. By enumeration of all possible cases one can show that the trefoil and its mirror image are the only possibilities for 3 crossings and that the figure 8, shown in Figure 1.3, and its mirror reflection are the only possibilities for 4 crossings. What remains to be shown is that the figure 8 is achiral, this is done directly, and that the right and left trefoils and the figure 8 are actually distinct knots. For this one needs something which can be calculated and which distinguishes between the three cases. This is the purpose of the algebraic spatial invariants that we shall discuss later.

In using diagrams to represent knots and links and to compute spatial invariants of these knots and links from the diagrams it is useful to define another relation between the diagrams. Two link diagrams are **regularly isotopic** if there is a finite sequence of transformations, each of which is either an S^2 isotopy or a second or third Reidemeister move, taking the one diagram to the other. Thus regular isotopy of diagrams forbids only the type I Reidemeister move.

Returning to the historical discussion, the Reverend Thomas P. Kirkman, F.R.S., [Kirk], devoted considerable effort to the classification of alternating knots by means of knot projections. In addition, quoting M.B. Thistlethwaite, [Th1], "He was also the acknowledged master of obscure terminology. His individual style manifests itself admirably in his definition of "knot", at the beginning of his article (Kirk)":

"By a knot of n crossings, I understand a reticulation of any number of meshes of two or more edges, whose summits, all tessaraces ($\alpha k\eta$), are each a single crossing, as when you cross your forefingers straight or slightly curved, so as not link them, and such meshes that every thread is either seen, when the projection of the knot with its n crossings and no more is drawn in double lines, or conceived by the reader of its course when drawn in a single line, to pass alternately under and over the threads to which it comes at successive crossings."

Kirkman developed a systematic reduction process for a knot diagram which Tait and Little subsequently employed in their enumeration efforts and which greatly influenced the efforts of Conway, [Con], and others (*e.g.* Dowker and Thistlethwaite, [Th1]). Kirkman's basic idea is to tabulate irreducible knot diagrams and then to reconstitute all knots from them.

A major consideration in the tabulation program was the decision to consider only irreducible or prime knots or links. A knot or link is said to be reducible if it can be expressed as the **connected sum** of two non-trivial knots or links as shown in figure 1.7. We say that a diagram for a knot or link represents a connected sum if there is a circle in the plane intersecting the diagram only in the interior of two edges and whose complementary regions contain non-trivial diagrams, *i.e.* contain crossings of the diagram. It was not until 1949 that Schubert showed that the decomposition of knots into prime summands was unique, up to the order of the summands in much the same way that one has unique prime factorization of natural numbers. In general, it is very difficult to show that a knot is prime. There are two exceptions to this: alternating knots and unknotting number one knots. In 1984, W. Menasco, [Men], showed that alternating knots are prime if and only if their presentation is prime, *i.e.* one can not find decomposing circle in the alternating presentation. In the same year, M.G. Scharlemann, [Sch], proved that knots having unknotting number (to be defined later in this section) equal to one are also prime. Kirkman's main reduction occurs by the removal of "bigons" appearing in the diagram, see Figure 1.8. Recall that there are no monogons in the presentation, since they can either be removed via a type I Reidemeister move to be considered to be in violation of the requirement irreducibility under connected sum of diagrams. Note that if the edges of the bigon belong to the same curve but have opposite directions, then the result of removing the bigon as indicated in Figure 1.8 results in a diagram having one additional curve. As a consequence, Kirkman tabulated diagrams having more than one curve in order to recursively enumerate knot diagrams associated to alternating knots, the only ones he seems to have considered, through 10 crossings. Despite some errors and omissions, his efforts have had a profound influence upon the tabulation efforts of his successors.

The next knot tabulator was Peter Guthrie Tait, a Scottish physicist.

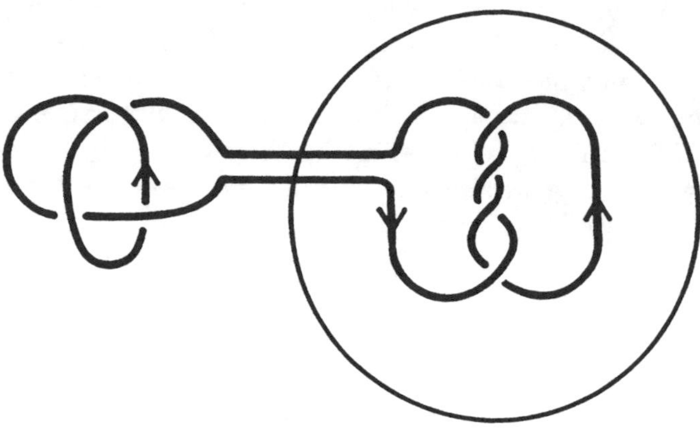

Fig. 1.7 Connected sum

He was originally drawn to the subject by Kelvin's theory of vortex atoms, [Tho], in which different knots were proposed as corresponding to different elements. Unlike Kirkman, Tait had a distinctly 3-dimensional vision of knots which provided the real birthplace of knot theory as it is presently understood. Unfortunately, he was significantly disadvantaged by the lack of "topological invariants", modern topology not yet having been invented. As a consequence he could not be certain that his enumerations did not contain duplications. Tait introduced "2 colorings" of the complementary regions of the diagram in which opposite sides of an edge have distinct colors, defined a graph with vertices in the regions of one color with edges connecting vertices through the crossings of the diagram and, obtained the dual graph by performing the same construction on the regions of the other color. This process is reversed by the construction of the medial graph in which edges are replaced by crossings (or 4-valent vertices) which are then connected to those which are adjacent. Tait also defined the **unknotting number.** This is the minimum number of times that one strand must be passed through another before an unknotted configuration is achieved. To this day the systematic calculation of the unknotting number had withstood the efforts of knot theorists, is known only for the very most simple knots and, questions and conjectures related to its properties form the basis of many research problems and conjectures. Both the trefoil and the figure 8 knots have unknotting number one. Tait also raised several conjectures concerning the presentations of knots, some of which have only recently been solved. The first conjecture is that a prime knot having an alternating presentation without nugatory crossings cannot be presented with fewer crossings. The second conjecture is that any presentation with that number of crossings must be

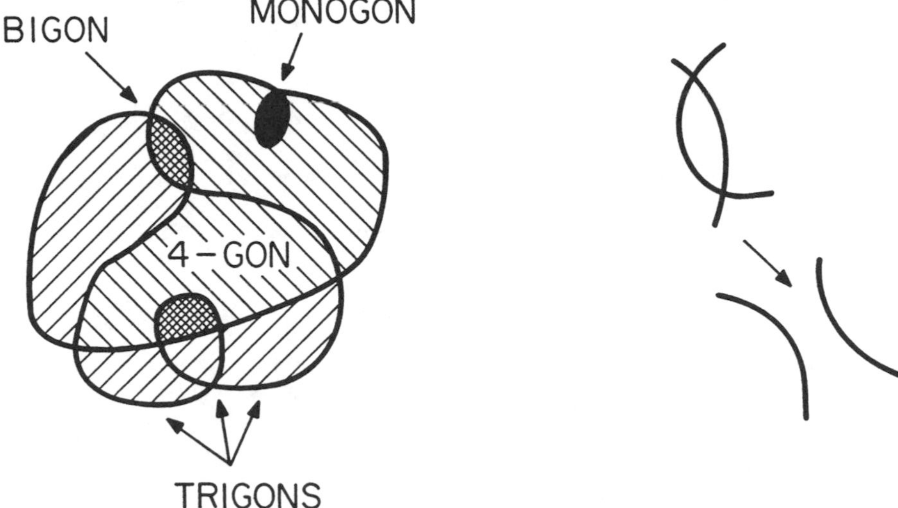

Fig. 1.8 Monogons, bigons, trigons, *etc.*

Fig. 1.9 A Tait flype transformation

alternating. The third Tait conjecture requires some additional vocabulary in order to give its statement. Tait introduced the concept of a **tangle**; the portion of a knot or link presentation lying within a circle intersecting the knot or link diagram in the interior of exactly four edges as shown in Figure 1.9. This is a concept which was to profoundly influence the efforts of later researchers such as Conway, [Con]. A special case of this situation occurs when one can find a circle which intersects the diagram in exactly two edges and a crossing vertex. In this case, Tait defines a spatial transformation, the flype, which moves the crossing from one location to another as shown in Figure 1.9. Recently Menasco and Thistlethwaite, [MTh1], have proved the Tait flype conjecture that flypes generate all the equivalences between min-

Fig. 1.10 Algebraic crossing number convention

imal alternating knot diagrams, *i.e.* if two alternating diagrams of the same knot are given, there exists a sequence of flypes transforming one diagram to the other. Finally, Tait listed all achiral knots up through 10 crossings.

Max Dehn, in 1914, [De], provided the first proof that the trefoil knot is chiral. In her 1917 doctoral dissertation, Mary G. Haseman listed the achiral knots of 12 crossings (with some errors). The historical discussion of tabulation efforts continues with the work of C.N. Little, [Litl-4], who undertook the first serious enumeration of the non-alternating knots. They first appear at 8 crossings, with 3 distinct examples up to mirror reflection, and are far more difficult to classify. Flyping no longer suffices to discover all equivalences and Little, therefore, adds a "2-pass" in which a strand passing over two crossings at the edge of a tangle is passed over the tangle to the analogous position on the other side. This is a generalization of the type III Reidemeister move.

Little also defined the "twist" as the absolute value of the algebraic crossing number. He erroneously thought to be an invariant of the knot since it was unchanged by flypes and passes. The **algebraic crossing number or writhe** of an oriented knot or link in S^3, $\omega(L)$, is defined to be the sum of the ± 1's associated to the crossings in a presentation according to the convention shown in Figure 1.10. The algebraic crossing number associated to an oriented knot or link is unchanged by the second two of the three Reidemeister moves and, therefore, defines an invariant associated to the regular isotopy class, of the presentation of the oriented knot or link. Since changing all the arrows in Figure 1.10 leaves the ± 1 convention fixed, one notes that the writhe of a knot presentation is independent of the orientation of the strand. Note that it reverses sign under mirror reflection. This misconception on the part of Little led to a duplication in the list of 10 crossing non-alternating knots that remained undiscovered until it was noted by Perko in 1974, cf. Figure 1.11. These two presentations are minimal crossing number presentations of the same knot. The left presentation has algebraic crossing number (or writhe) 10 while the right presentation has 8. This demonstrates that the writhe of minimal crossing number presentation of a knot is <u>not</u> an invariant of the knot. It is true that the algebraic crossing number of an alternating knot presentation without nugatory crossings is an

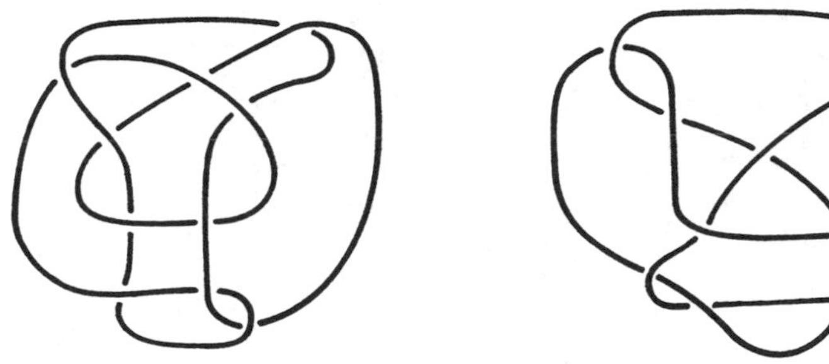

Fig. 1.11 The Perko pair [writhe 10 and 8]

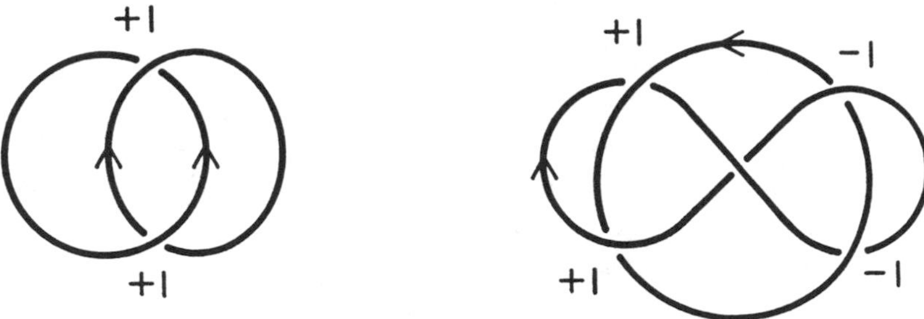

Fig. 1.12 Algebraic linking numbers +1 and 0

invariant, [Mur5]. The knots appear as 10_{161} and 10_{162} in Rolfsen, [Rol].

An important application of the algebraic crossing numbers of an oriented link is the combinatorial calculation of the **Gauss or algebraic linking number**, $lk(K_1, K_2)$, of two components of the link $\{K_1, K_2\}$. It is half the sum of the algebraic crossing numbers of all crossings involving these two components. The linking number reverses sign under mirror reflection. Two examples are shown in Figure 1.12.

In 1927 Alexander and Briggs, [AB], began the first exploitation of algebraic topology to develop spatial invariants associated to a knot, the homology of the cyclic branched covers of S^3 branched along the knot. This was sufficient to accomplish their purposes except in three cases. These were completed by Reidemeister who employed linking numbers in irregular branched covers associated with homomorphisms of the fundamental group of the complement onto dihedral groups. In 1928 Alexander, [A2], described a method by which a finite integral Laurent polynomial in the variable "t" could be calculated from the infinite cyclic branched cover of the knot by means of a simple combinatorial processes which was applied to any pre-

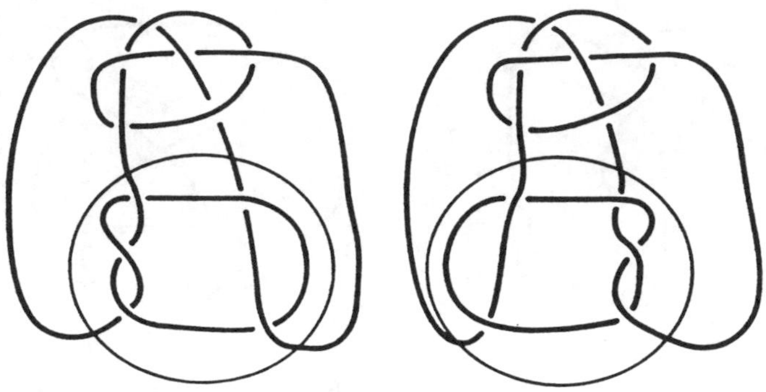

Fig. 1.13 A non-trivial mutation

sentation of the knot. This polynomial is well defined up to multiplication $\pm t^n$ and has served to distinguish between many knots. In the collection of 12965 knots listed by Dowker and Thistlethwaite, [Th1], having fewer than 14 crossings, there are groups numbering as many as fifty and having the same Alexander polynomial. We shall return to the discussion of polynomial invariants in the next section on the various algebraic spatial invariants which can be associated to knots and links.

The next major event was the exploitation of a new notation by J.H. Conway, [Con], to enumerate all knots to 11 crossings and all links to 10 crossings in the 1960's. This was the rebirth of the theory of tangles proposed earlier by Tait and which has now come to play an important role in a variety of aspects of knot theory. One example of this is the concept of **mutation** in which a tangle in the presentation is replaced by a "symmetric tangle", as illustrated in Figure 1.13. Here is shown the first occurrence of a non-trivial mutation, *i.e.* one in which the mutant knots are distinct. Conway also studied other types of possible symmetries, in addition to chirality, *e.g.* the invertibility of a knot which requires that an oriented knot can be isotoped to itself reversing the orientation of the strand. It is easy to see that the trefoil is invertible but it was not until 1964 that Trotter, [Tr], showed the existence of non-invertible knots. Conway also noticed, in the work of Alexander, the existence of a symmetric recursion formula for the calculation of this Alexander polynomial and proposed a reformulation of it, now called the Conway potential function, cf [Kau1]. We will return to this question of the recursion formula for the Alexander polynomial later as it plays an important role in the study of the spatial algebraic invariants. In 1981, Caudron gave a list of the 11 crossing knots, in his Orsay thesis, which corrected the errors in Conway's original calculations. The most recent chapter of the history of knot tabulation is the collaboration of Hugh Dowker

3	4	5	6	7	8	9	10	11	12	13
1	1	2	3	7	21	49	165	552	2176	9988

Table 1.14 The number of prime knots through 13 crossings

and Morwen Thistlethwaite, [Th1], to complete the tabulation of 12 crossing knots, in 1981, and 13 crossing knots, in 1982. All the 12965 knots listed by Thistlethwaite are known to be distinct. There has, however, been no independent verification of the completeness of this list and there remains the possibility that some of the listed knots are not prime.

One of the more useful constructions associated to an oriented knot or link is that of an oriented surface bounding the knot or link. A classical construction, due to Seifert, cf [Rol], demonstrates the existence of such surfaces from any presentation of the knot or link. The construction begins by removing each crossing in the presentation and reconnecting the resulting segments by following the direction given by the orientation. An example is given in Figure 1.15. Here the circles are indicated. The circles define regions, each of which determine discs which are assigned distinct levels according to their inclusions, with the outermost circles at height "0", the next outermost assigned height "1", *etc.* The crossings determine connections between the circles given by twisted bands following the crossings of the knot or link. The bands give the connections between the various levels, shown with different fillings. The resulting surface, F, is oriented. Every oriented surface is equal to a sphere with a number of handles and with holes for each component of the link. The number of handles is called the genus of the surface. It can be determined *via* a calculation of the Euler number as follows:

$1 - 2 \cdot$ Genus $(F) = \chi(F) =$ (# of Seifert circles) - (# of crossings).

In the case illustrated in Figure 1.15, the surface has genus 3 since $1-2\cdot 3 = 4-9$. The **genus of a knot or link** is defined to be the minimum genus among those of all oriented surfaces bounding the knot or link. Except in the case of prime alternating knots or links, it can be rather difficult to compute the genus of a knot or link from a given presentation. The genus of an alternating knot or link, without nugatory crossings, is the genus determined by the application of the Seifert algorithm to the alternating presentation, [Mur2]. The method of Gabai, [Gab] is quite useful in the determination of the genus in many cases, *e.g.* one of the examples of mutants, given in Figure 1.13, has genus 2 while the other has genus 3 demonstrating that they represent distinct knots.

The Seifert surfaces associated to an oriented knot provide an alternative means to calculate linking numbers of the knot with any other oriented curve. If F is a Seifert surface for an oriented knot, K, and L is an oriented

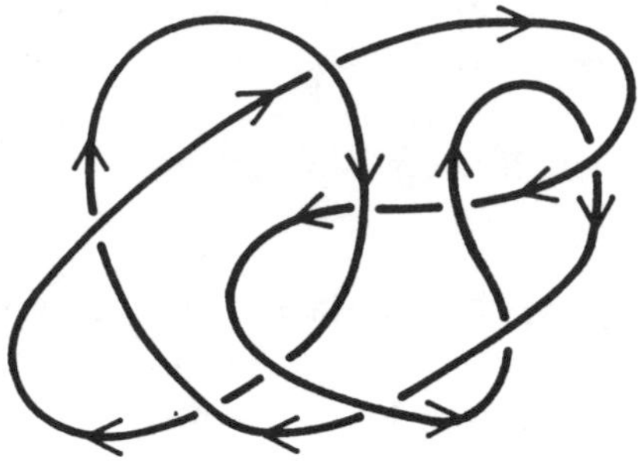

Fig. 1.15

knot disjoint from K, the algebraic linking number, $lk(K, L)$ can be calculated as follows: By a small perturbation of the position of L with respect to F, we may assume that intersections of L with F are perpendicular. Since F is oriented, consistently with its boundary K, it has a preferred normal direction according to the 'right hand rule'. Assign to each intersection point of L with F, either $+1$ or -1 according to whether the direction of L at the intersection agrees with or differs from the direction of the normal to the surface. The algebraic linking number is the sum of the ± 1's.

The Seifert surfaces associated to an oriented knot provide the means to define the **longitude of the knot.** The longitude is a disjoint spatially equivalent oriented knot having zero algebraic linking with the knot. A longitude of a knot, K, can be constructed by first constructing a Seifert surface for the knot. The knot is the boundary of the surface. Taking a parallel copy of the boundary a small distance inside the surface determines a longitude. This longitude can be thought of as the intersection, with the Seifert surface, of the boundary of a small solid torus neighborhood with the knot at its center. A **meridian** of the knot is the boundary of a disk in the solid torus meeting K in a single point. Since K does not intersect the portion of the Seifert surface lying inside the longitude, the linking number of the longitude with K is zero. A meridian will intersect the Seifert surface in exactly one point giving a linking number of ± 1, depending upon its orientation. One traditionally orients the meridian so that the linking number is $+1$.

There is another method of representing knots and links that would appear to be especially relevant to the topological theory of fluid dynamics. It has also proved important in efforts to classify knots and links. This is the presentation of a knot or link as a **braid**. Birman's book, [Bir1], is the standard reference for classical results in the theory of braids, *i.e.* those prior to the 1970's. More recent results can be found in Braids, Proceedings of a Summer Research Conference held July 13-26, 1986, Contemporary Math-

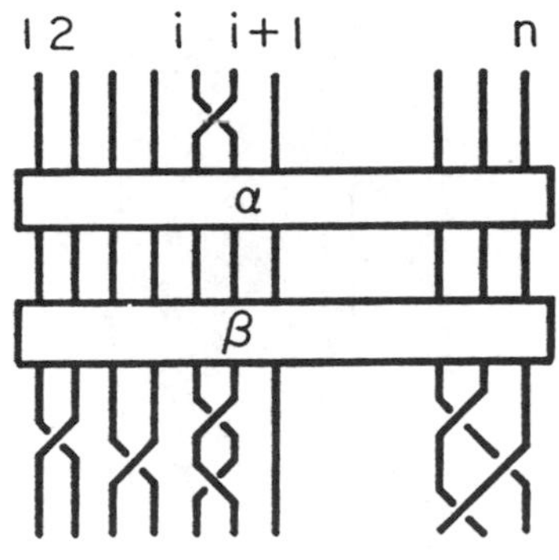

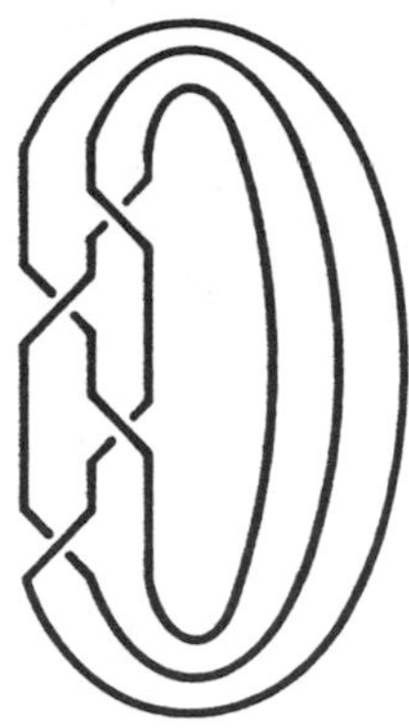

Fig. 1.16 Braid group and braid representation

ematics, vol. 78, published by the American Mathematical Society. The attraction of this method, for the topologist, is the hope to be able to exploit the natural algebraic structure of the braid to provide proofs of topological results. As a result, a great deal of effort has gone into the study of the algebraic structure of the braid group and its representation theory. These efforts have been enriched by the development of the Jones polynomial as well as a new approach to the creation of invariants due to Casson. A **braid of n strings** is a configuration which descends from one level to another monotonically, *i.e.* without local minima or maxima, such as shown in Figure 1.16. The projection to the plane is required to be regular and the crossings are to be at distinct levels. In order to facilitate the algebraic presentation of the braid the various strands are numbered 1 through n and the crossing of the ith strand under the $(i+1)^{st}$ strand is denoted by σ_i. The inverse of σ_i denotes the crossing of the ith strand over the $(i+1)^{st}$ strand. Starting from top to bottom, the braid can be expressed as a word in the σ_i and their inverses. Braids having the same number of strands are multiplied by placing one above the other; thus $\alpha\beta$ denotes the braid defined by taking α then β as the strands descend. Two braids, normalized to have the same length, are equivalent is there is a level preserving isotopy taking one braid to another fixing the top and bottom end points. After normalization for length, one sees that this multiplication is associative and that the product of σ_i and its inverse is equivalent to the configuration of vertical strands, the identity braid, and therefore the set of equivalence classes of n stranded braids is a group. This group, denoted B_n, is called the **n string braid group** and has the following algebraic presentation:

$$B_n = \langle \sigma_1, \sigma_2, \ldots, \sigma_n | \sigma_i\sigma_j = \sigma_j\sigma_i \text{ when } |i-j| > 1,\ \sigma_i\sigma_{i+1}\sigma_i = \sigma_{i+1}\sigma_i\sigma_{i+1} \rangle$$

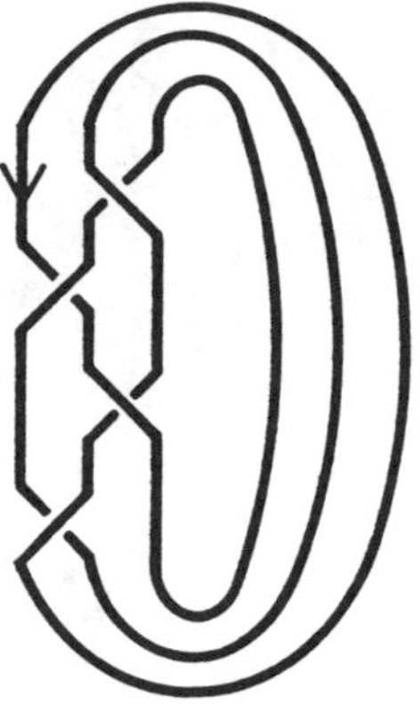 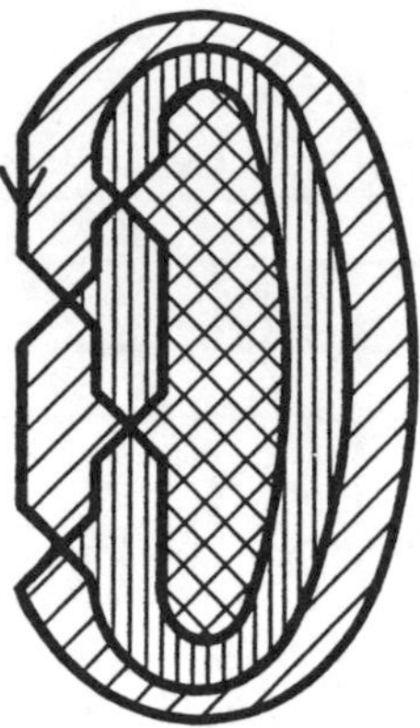

Figure 1.17

To each element of the braid group one may associate a knot or link by means of the closure of the braid shown in Figure 1.17. The knot would be represented by the braid, $\sigma_2\sigma_1^{-1}\sigma_2\sigma_1^{-1} = (\sigma_2\sigma_1^{-1})^2$. Alexander, [A1], proved that any oriented knot or link can be represented by means of just such a braid closure. Note that a braid has a natural orientation, *i.e.* all the strands can be oriented in the descending direction, which is preserved by the closure process. The Seifert algorithm applied to a braid closure gives exactly n circles, *i.e.* the same number as the braid index, cf Figure 1.17. The following theorem of Yamada, [Ya1], extending the Alexander theorem amplifies on this.

THEOREM (Alexander-Yamada) 2: If an oriented knot or link has a presentation with n Seifert circles, then it admits a braid presentation having n strands.

Outline of Vogel's proof [V]: From the Seifert construction applied to the braid presentation one sees that a condition equivalent to being given in a braid presentation is the requirement that the Seifert circles be totally nested. This, however, can be achieved by application of Reidemeister moves to the knot or link presentation. These moves, shown in Figure 1.18, preserve the number of Seifert circles while moving the presentation closer to a braid presentation. The last move is an S^2 equivalence employed to provide an clearer presentation showing that the braid configuration has been achieved.

There are, however, many ways of presenting the same oriented knot or link as a braid. The Markov theorem, [Bir1], states that two braid representations of the same oriented knot or link are related by a sequence

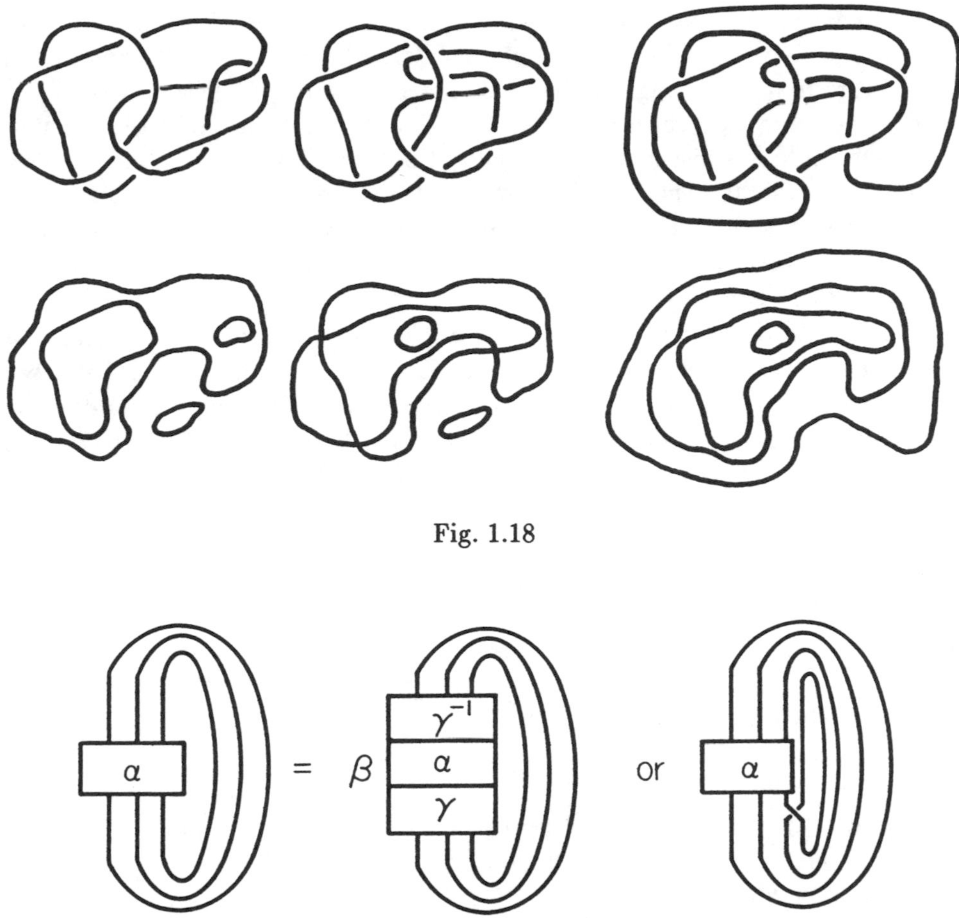

Fig. 1.18

Fig. 1.19 Markov relations

of elementary equivalences of the following two types related to the group structure: $\alpha \approx \beta$ if α and β are conjugate in B_m; $\alpha \approx \beta$ if $\alpha \varepsilon B_m$ and $\beta = \alpha(\sigma_m)^{\pm 1} \varepsilon B_{m+1}$.

The braid structure captures the parallelism associated to nearby stream lines present in a dynamical system such as a fluid flow. Such structures, on a global level, are organized and studied under the name of foliations. In three dimensions there are several major results which have a global bearing on these structures although not directly seen as a part of knot theory. The historical reference for the birth of these topics is Reeb. In the case of periodic flows, *i.e.* every orbit is a closed circle, in three manifolds, the papers of Edwards-Millett-Sullivan, [EMS], Epstein, [E1&2], Epstein-Millett-Tischler, [EMT], Epstein-Vogt, [EV], Millett, [Mil], Sullivan, [Sul], and Vogt, [V1&2], are the fundamental ones. If the setting is compact, the global structure is quite regular and is known as a Seifert fibering. The fact that any periodic flow in a compact three dimensional manifold gives such a structure is a surprisingly difficult and deep result, [E1, EMS, V1]. If the space is not compact, the structure is significantly more complicated, even in R^3. Vogt, [V2], has discovered examples of periodic flows in R^3. The general theory tells us that there is no bound on the periods of the orbits. The great surprise is that such an example can exist at all. Research is currently ongoing to determine the variety of braided structures that can occur in the collection of stream lines.

The **companion/satellite construction** is an important method of constructing new knots and links from old knots and links and introduces a useful hierarchy of complexity among knots and links. Let L be any **geometrically essential** link in the standard solid torus, T, in 3-space (*i.e.* we require that any link in the solid torus equivalent to L interests any meridian disc at least once). Let K be any knot and let T_K be the solid torus neighborhood of K. Any homeomorphism, h, taking T to T_K replaces K by its **satellite** $h(L)$. L is called the **companion** of $h(L)$. Since $h(L)$ depends upon the specific choice of homeomorphism, one must be rather careful in this construction to specify the homeomorphism as well as both L and K. If we wish to require that the longitude and meridian of T go to those of T_K we will say that h is **faithful**. There are several important examples of this construction.

Example: **Connected Sum.** Suppose K_1 and K_2 are two knots and K_1 is placed in the solid torus T as shown in Figure 1.20. Then, for any $h : T \to T_{K_1}$, $h(K_1) = K_1 \# K_2$.

Example: **Doubled Knots.** Suppose L is the knot shown in Figure 1.21. A satellite of a knot, K, is called the untwisted double of K if h is faithful and the q-twisted double if h(longitude) is longitude plus q meridians of K. One needs to make a coherent choice of orientation of longitude and meridian in order to remove sign ambiguities in this definition. Note that a doubled knot will always have unknotting number one but its crossing number can be very large, depending on the choice of K.

Example: **Cabled Knots.** The (p, q)-torus knot, p and q relatively prime integers, is the knot which wraps around just below the boundary of the standard solid torus in the longitudinal direction p times and in the meridional direction q times as shown in Figure 1.22 for the case (2,3), *i.e.* the right trefoil. A faithful satellite of a knot, K is called its (p, q)-cable. Note that the case $p = 0$ gives the unknot (so we require $p \neq 0$) and $p = \pm 1$ gives K (another not very interesting case).

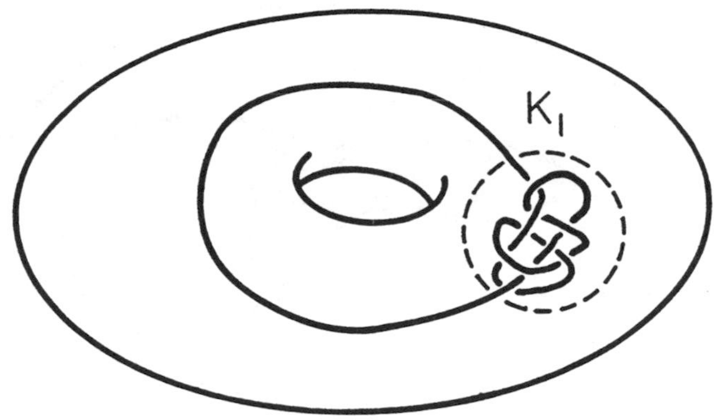

Fig. 1.20

Fig. 1.21

Example: **Parallels of Knots or Links.** The **untwisted k-parallel** of a component of a link is the faithful satellite of the component determined by the k-parallel circle companion shown in Figure 1.23. In the study of invariants we shall wish to use this construction and allow different numbers of parallels on the various components of a link. Since a faithful homeomorphism preserves the algebraic linking between the parallels, the parallel components of the untwisted k-parallel will be algebraically unlinked. The **planar k-parallel** of the right trefoil ($k = 2$) shown in Figure 1.24 is not the untwisted k-parallel since the components may be algebraically linked. If the components are oriented in the 'same' direction, their algebraic linking number is

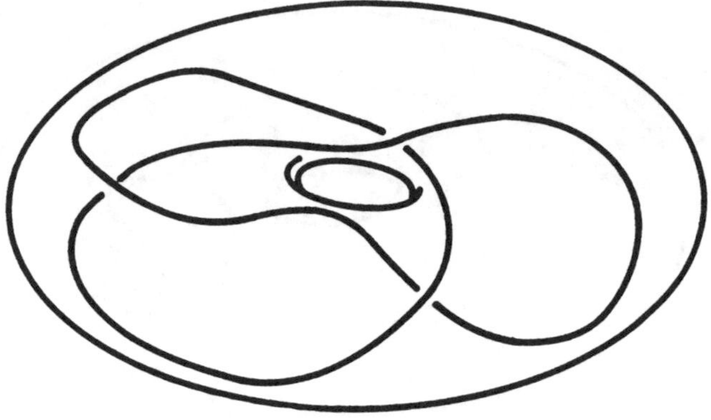

Fig. 1.22

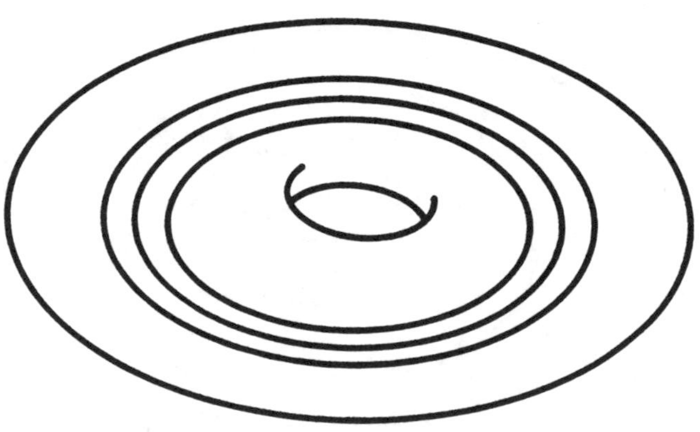

Fig. 1.23

−3. The existence of this linking, a direct result of the writhe of the component, will be exploited in the description of the invariants of 3-manifolds and suggests the possibility of applications in the topological fluid dynamics.

Example: **Braided knots and links.** In the representation of a knot or link as a braid, the braid axis determines a standard solid torus containing the knot or link such as shown in Figure 1.25. If one constructs a satellite based on this model, the result is a braided knot or link. Here also one can exploit the braided knot or link structure to analyze the topological structure of stream lines constrained to lie within a possibly knotted solid torus.

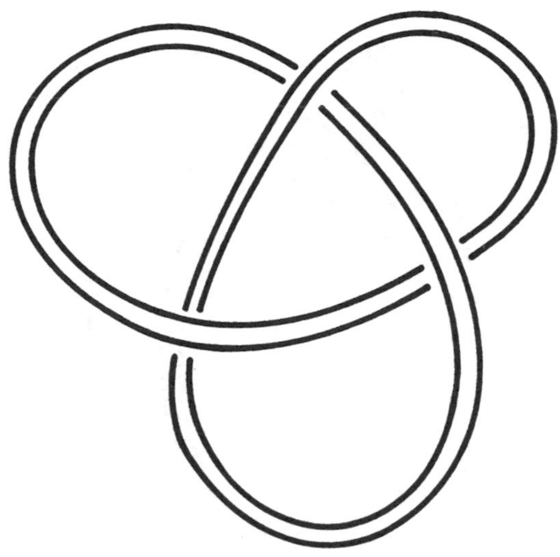

Fig. 1.24

2. The Bracket Polynomial

Although the proof that other polynomial invariants of knots, such as the classical Alexander, [A2], or the recently discovered HOMFLY, [FYHLMO, Hos, LM1, PT], and Kauffman polynomials, [BLM, Ho, Kau4], exist, *i.e.* they depend only upon the knot or link type, is rather technical there is an almost immediate proof of the existence of the original Jones polynomial, [Jon1&2], which was found by L. H. Kauffman, [Kau5]. This approach has also provided a key to some of the more interesting implications of Jones polynomial which we shall describe later. The Jones polynomial of an oriented link, denoted by $V(L)$, is a finite Laurent polynomial in the variable $t^{\frac{1}{2}}$ with integer coefficients. The variable $t^{\frac{1}{2}}$ can be understood as a symbol whose square is t or can, if desired, be understood to be a complex number, *e.g.* a root of unity. The Jones polynomial satisfies the following:

$$V(\text{unknot})=1$$
$$t^{-1}V(L_+) + tV(L_-) + (t^{-\frac{1}{2}} - t^{\frac{1}{2}})V(L_0) = 0$$

where L_+, L_- and, L_0 are oriented link diagrams which are identical except near a single point where they appear as in Figure 2.1, respectively, Kauffman begins the construction by developing a formal calculus of pla-

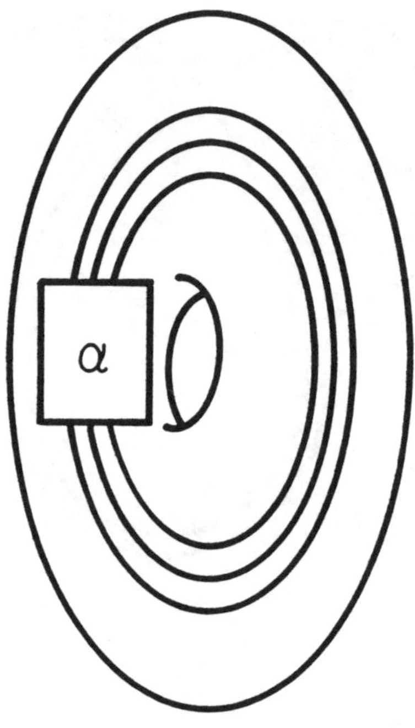

Fig. 1.25

Fig. 2.1

nar diagrams of unoriented links. In this calculus, one associates a Laurent polynomial, $\langle L \rangle$, in one formal variable, A, according to the following three rules:

(i) $\langle \times \rangle = A^{-1}\langle \asymp \rangle + A\langle\,)(\, \rangle$

(ii) $\langle L_{\cup}\mathbf{O} \rangle = (A^{-2} + A^{2})\langle L \rangle$

(iii) $\langle \mathbf{O} \rangle = 1$

The first rule asserts that the bracket of a diagram with a crossing can be replaced by the indicated linear combination of the brackets of the diagrams with the crossing removed as indicated. This equation is well defined because one can choose the orientation of the figure so that the right most strand passes under the left strand. The other such choice gives the same value. The second rule asserts that the bracket of the distant union of a diagram and a trivial circle is the product of the factor $-(A^{-1}+A^2)$ and the bracket of the diagram. The final rule establishes the bracket of a single planar circle as the value 1.

Using these rules, one reduces a knot or link presentation to a finite Laurent polynomial in the abstract variable "A" with integer coefficients. A simple argument shows that this expression is a regular homotopy invariant of the presentation, *i.e.* is unchanged under the Reidemeister moves of types II and III.

Type II Invariance

$$\langle \quad \rangle = \langle \; | \;\; | \; \rangle$$

Proof:

$$\langle \quad \rangle = A\langle \quad \rangle + A^{-1}\langle \quad \rangle$$

$$= A^2\langle \quad \rangle + \langle \quad \rangle + \langle \,)\;(\, \rangle + A^{-2}\langle \quad \rangle$$

$$= \langle \; | \;\; | \; \rangle + \left(A^2 - (A^{-2} + A^2) + A^{-2}\right)\langle \quad \rangle$$

$$= \langle \; | \;\; | \; \rangle$$

Type III Invariance

If $\langle \quad \rangle = \langle \; | \;\; | \; \rangle$, then $\langle \quad \rangle = \langle \quad \rangle$

Proof:

$$\langle \quad \rangle = A\langle \quad \rangle + A^{-1}\langle \quad \rangle$$

$$= A\langle \quad \rangle + A^{-1}\langle \quad \rangle$$

$$= \langle \quad \rangle$$

Type I Reidmaster Moves

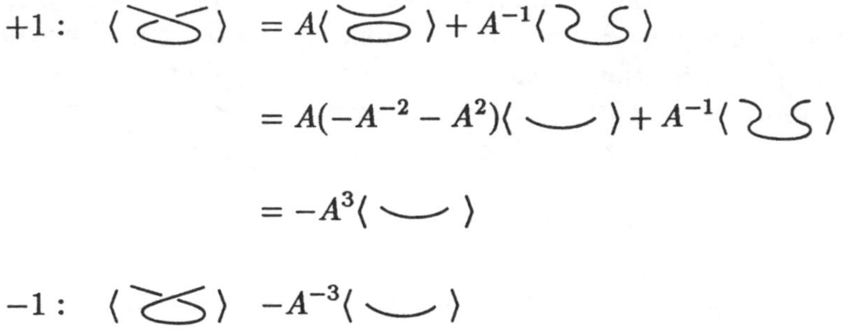

$$+1: \quad \langle \quad \rangle = A\langle \quad \rangle + A^{-1}\langle \quad \rangle$$

$$= A(-A^{-2} - A^{2})\langle \quad \rangle + A^{-1}\langle \quad \rangle$$

$$= -A^{3}\langle \quad \rangle$$

$$-1: \quad \langle \quad \rangle \quad -A^{-3}\langle \quad \rangle$$

Thus, in order to insure invariance under the type I Reidemeister moves we must compensate for this type of twisting. One way in which to accomplish this is to use the writhe of the presentation, *i.e.* the sum of the algebraic crossing numbers in the presentation representing the oriented knot or link, to express an compensating factor so as to give type I invariance. Using such a factor

$$F_L(A) = (-A^3)^{\omega(L)}\langle L\rangle$$

defines a spatial invariant of the oriented knot or link. This simple fact gives a complete proof that $F_L(A)$ is a well defined invariant of the oriented link L. Next we shall see that it is, indeed, the Jones polynomial.

PROPOSITION: $F_L(t^{-1/4}) = V_L(t)$

Proof:

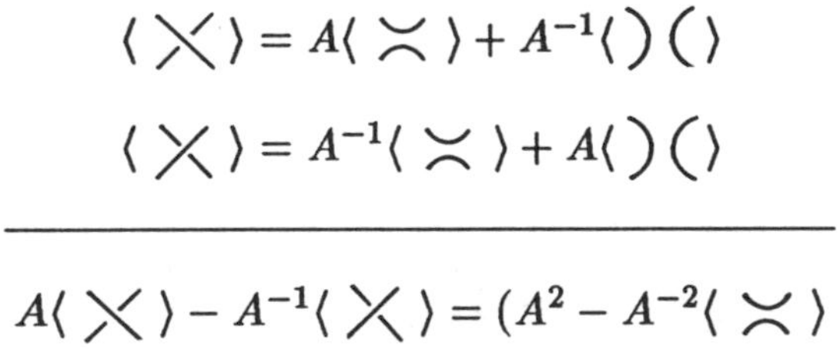

$$\langle \quad \rangle = A\langle \quad \rangle + A^{-1}\langle \;)\,(\; \rangle$$

$$\langle \quad \rangle = A^{-1}\langle \quad \rangle + A\langle \;)\,(\; \rangle$$

$$A\langle \quad \rangle - A^{-1}\langle \quad \rangle = (A^2 - A^{-2}\langle \quad \rangle$$

The writhe of L_0 determines those of $L_\pm$ as follows: $\omega(L_\pm) = \omega(L_0) \pm 1$. Multiplying the above equation by $(-A^3)^{-\omega(L_0)}$ and using the definition of $F_L(A)$ gives: $A(-A^3)F_{L_+}(A) - A^{-1}(-A^3)F_{L_-}(A) = (A^2 - A^{-2})F_{L_0}(A)$. Letting $A \to t^{-1/4}$ one has the following equation: $t^{-1}F_{L_+}(t^{-1/4}) - tF_{L_-}(t^{-1/4}) + (t^{-1/2} - t^{1/2})F_{L_0}(t^{-1/4}) = 0$.

By one induction on the number of crossings in the diagram of the knot or link and by a second induction by the minimum number of crossings that must be changed in a given diagram to obtain a diagram for the trivial knot orlink, one shows that $F_L(t^{-\frac{1}{4}})$ is equal to the Jones polynomial. First, they agree for the diagrams of the trivial links without crossings. Second, they both satisfy the same recursion relationship and, therefore, if they are equal for the corresponding two of the three cases they are equal for the third. This occurs in the induction since L_0 has fewer crossings and one of the cases, + or −, requires fewer crossing changes to arrive at a diagram for the trivial knot or link, according to the standard algorithm.

An Example: Using the properties of the bracket polynomial we shall compute F for the right handed trefoil knot.

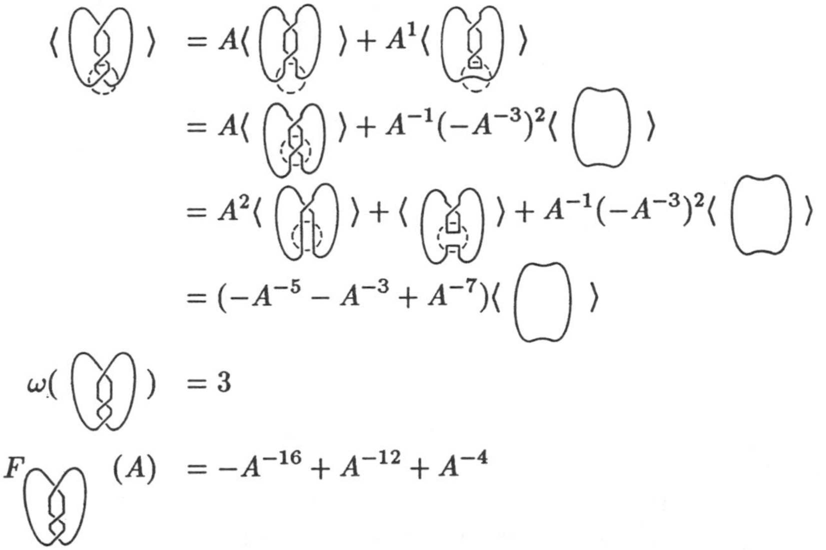

$$\langle\ \rangle = A\langle\ \rangle + A^{1}\langle\ \rangle$$

$$= A\langle\ \rangle + A^{-1}(-A^{-3})^2\langle\ \rangle$$

$$= A^2\langle\ \rangle + \langle\ \rangle + A^{-1}(-A^{-3})^2\langle\ \rangle$$

$$= (-A^{-5} - A^{-3} + A^{-7})\langle\ \rangle$$

$$\omega(\ ) = 3$$

$$F(A) = -A^{-16} + A^{-12} + A^{-4}$$

This example illustrates several important properties of the Jones polynomial. First, one can express the polynomial of the mirror reflection, $\overline{L}$, of the link L in terms of the polynomial of L since reflection exchanges the A and A^{-1} contributions to the crossing removal in the definition of the bracket polynomial. Since the writhe also changes sign, one has

Proposition: $F_{\overline{L}}(A) = F_L(A^{-1})$.

Because the polynomial of the right trefoil, computed above, is changed under the transformation $A \to A^{-1}$ to a different equation for the left trefoil, $A^4 + A^{12} - A^{16}$, we see that the two trefoils are distinct. Thus the trefoil is a chiral knot. Computation of the polynomial of the figure 8 knot will give an equation which is unchanged. This reflects the fact that the figure 8 knot is achiral. Unfortunately, there are knots whose equations are invariant under taking A to A^{-1} but which are, nevertheless, chiral knots.

A basic property of the polynomial is that the equation associated to the connected sum of two knots or links (Beware, this is not well defined for links!) is the product of the equations of the constituent knots or links, *i.e.*

PROPOSITION: $F_{K\#L}(A) = F_K(A)F_L(A)$.

Although the Jones polynomial contains some information about the relative orientations of the constituents of the link this dependence is reflected in the following proposition concerning the effect of changing the orientation of one component, K, of the link, L. If L^* is the link formed from L by changing the orientation of K and λ denotes sum of the linking numbers of K with all the other components of L, then

PROPOSITION: $F_L * (A) = A^{3\lambda/4}F_L(A)$

Another provocative observation is the fact that the Jones polynomial is not able to distinguish between mutants of a knot or link, even though other classical invariants such as the genus is able to do so.

The **breadth** of a Laurent polynomial, $\beta(F(A))$, is defined to be the difference between the largest and smallest exponents appearing with non-zero coefficients. Thus, the breadth of the polynomial associated with the trefoil is

$$-4 - (-16) = 12 = 4 \times 3.$$

Thus, by dividing by 4, one shows that the breadth of the Jones polynomial of the trefoil knot is equal to 3. Some experimental work quickly leads one to conjecture that there is a close connection between the breadth and the number of crossings in a diagram. Indeed, this is the case.

In the definition of the bracket polynomial each crossing is split according to the define rule and either A or A^{-1} is assigned to the resulting configuration. The assignment of such a split to each crossing of a diagram is called a state, s, of the bracket. Let $\{s\}$ be the number of circles which arise in the state s. A presentation is said to be + **adequate** if, at every crossing split, the two closed segments of the reconnection to which A is associated belong to distinct circles. Similarly, − **adequate** is defined and the presentation is said to be **adequate** if it is both + and − adequate.

Recall the definition of the k-parallel companion of a knot or link diagram. It is the result of drawing k-parallel copies of the knot or link with respect to the given diagram. The k-parallel of different diagrams for the same knot or link can give different links. The fact is exploited in the definition of 3-manifold invariants. The following diagram, Figure 2.2, shows the 2-parallel of the knot 13_{6714}. Rather than drawing the parallel diagram we shall use a natural number adjacent to the diagram of a component to designate the number of parallel copies. The 0-parallel denotes the empty diagram and the 1-parallel is the original knot.

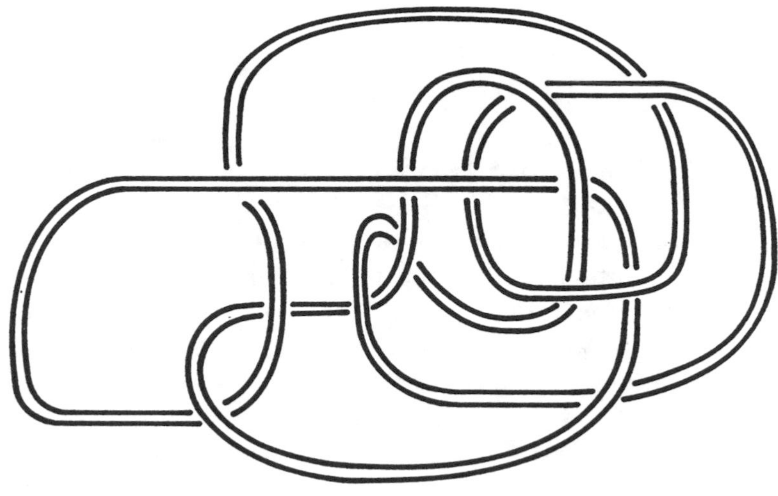

Fig. 2.2

The following proposition provides a useful connection between the diagram and the breadth of the Jones polynomial.

PROPOSITION (Lickorish-Thistlethwaite, [LT1]): An alternating presentation without nugatory crossings is adequate. Furthermore, the k-parallel companion of an adequate presentation is adequate.

PROPOSITION: Let K denote an n crossing presentation, then $\beta(F_K) \leq 2n + 2(\{s_+\} + \{s_-\}) - 4$. If the presentation is adequate, equality holds.

Some consequences of the Jones polynomial are the following:

(i) (Kauffman and Murasugi, [Kau3&5, Mur3&4]) The breadth of the Jones polynomial of an oriented link having a connected presentation with n crossings is no larger than n and is equal to n if the presentation is alternating. This proves the conjecture of Tait that all alternating presentations of a knot without nugatory crossing must have the same number of crossings. Furthermore, no presentation of the knot can have fewer crossings. Note, $\beta(F_K) = 4n$, since $\{s_+\} + \{s_-\} = n + 2$, by the Euler identity.

(ii) (Murasugi and Thistlethwaite, [Mur4&5, Th2–5]) The breadth of the Jones polynomial of a link having a prime non-alternating n crossing presentation is strictly less than n. Note $\beta(F_K) < 4n$.

REMARK: The breadth of the polynomial of the k-parallel of an n crossing alternating knot diagram without nugatory crossings, K, can be computed as follows:

$\{s_+\} + \{s_-\} = 2$ for K

For the k parallel, $n \to nk^2, \{s_+\} \to k\{s_+\}$, and $\{s_-\} \to k\{s_-\}$. Thus $\beta(F_{kK}) = 2nk^2 + 2(k\{s_+\} + k\{s_-\}) - 4 = 2nk^2 + 2k(n+2) - 4 = 2nk^2 + 2(n+2)k - 4$ which grows as a quadratic function of k when $n > 0$.

(iii) (Thistlethwaite, [Th2]) Suppose that K has a connected alternating presentation without nugatory crossings, then the signs of the coefficients of the Jones polynomial alternate, the coefficients of extreme degrees are ± 1, and no intervening term has a zero coefficient unless K is a $(2,n)$ torus knot or link.

(iv) (Murasugi, [Mur5]) The algebraic crossing number of an alternating presentation without nugatory crossings is an invariant of the link. This proves another old conjecture of Tait.

(v) (Menasco and Thistlethwaite, [MTh]) Any two alternating n crossing presentations without nugatory crossings of a knot are related by a sequence of flype moves.

3. The 3-Manifold Invariants

The goal of this section is to briefly provide a description of the $su(2)_q$ Witten; 3-manifold invariants, [Wi1&2], following the approach of Reshetekhin and Turaev, [Re1], as described by Lickorish, [Li11&12]. My purpose for doing this is to introduce a new method of developing invariants associated to diagrams in which the role of the k-parallel is central. I am hoping that this will suggest analogous methods involving braided satellites which will prove useful in the study of the topological aspects of fluid dynamics.

The first key result asserts that links can be used to construct any closed (*i.e.* compact and without boundary) oriented 3-manifold. One considers the standard solid torus with a specific meridian circle identified and a "framed" link, L, in the 3-sphere. The framing of each component of the link consists of an integer which will play a critical role in the construction. **Surgery** is performed on the framed link, *i.e.*: The interior of the solid torus tubular neighborhood of each component of L is evacuated and replaced by the interior of the standard solid torus by taking its meridian to the curve consisting of a longitude plus the appropriate integer number of meridians associated to that component of L. This is illustrated in Figure 3.1 for the case of "f"$= +5$ surgery on the figure 8 knot.

THEOREM (Lickorish-Wallace [Li1, Wall]): Any closed oriented 3-manifold can be obtained from the 3-sphere by surgery on a framed link.

Fig. 3.1

In order to develop a calculus which encodes both the framing and the possibility of k-parallels in a single diagram, Lickorish, [Li11&12], suggests using the writhe of each component to represent the framing associated with the component and using a natural number to label a component of a diagram to indicate the number of parallel copies which will be required in the diagrammatic definition of the 3-manifold invariants. Since the writhe of a diagram is unchanged under regular isotopy, one has the following:

PROPOSITION: Regular isotopy classes of link diagrams represent framed links and, therefore, a closed oriented 3-manifold is represented by link diagram.

Example: The diagram in Figure 3.2 represents some 3-manifold described by surgery on a three component link in the 3-sphere.

Just as two distinct knot or link diagrams or two braids may represent the same knot or link, two diagrams may represent the same closed oriented 3-manifold. The equivalence relation on the diagrammatic level determined by equivalence of the 3-manifold was determined in the 1978 paper of R. Kirby, [Kir1]. This was simplified by Fenn and Rourke in 1979, [Fe], to give the following:

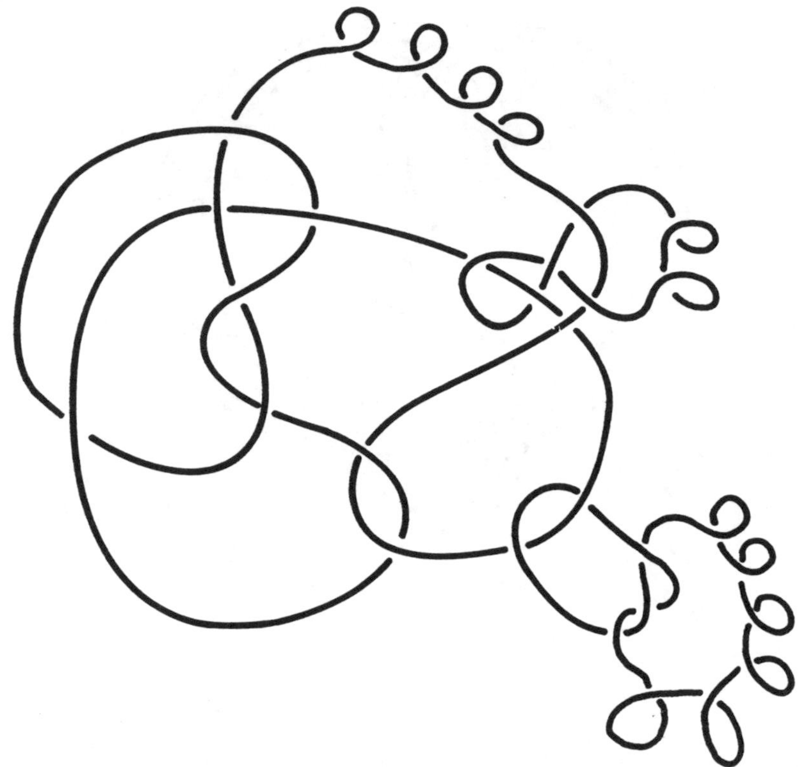

Fig. 3.2

PROPOSITION: Two diagrams represent the same 3-manifold if, and only if, they are related by the relation generated by regular isotopy of diagrams and the two elementary equivalences shown in Figure 3.3.

As was the case with the relationship between knot or link diagrams and their equivalence and with the braid descriptions of knots or links and the Markov relations, the description of 3-manifolds by virtue of link diagrams is not "solved" by the use of these relations. The complexity of the possible equivalent presentations and the lack of an effective procedure to determine whether or not two presentations are related by a sequence of these elementary equivalences make the problem too difficult.

Recall that a component of a link labeled with a natural number represents that number of planar parallels in the indicated portion of the diagram. If no number is given, then one is understood. The number 0 would indicate that that strand is eliminated.

In order to describe the 3-manifold invariants it is useful to apply a ver-

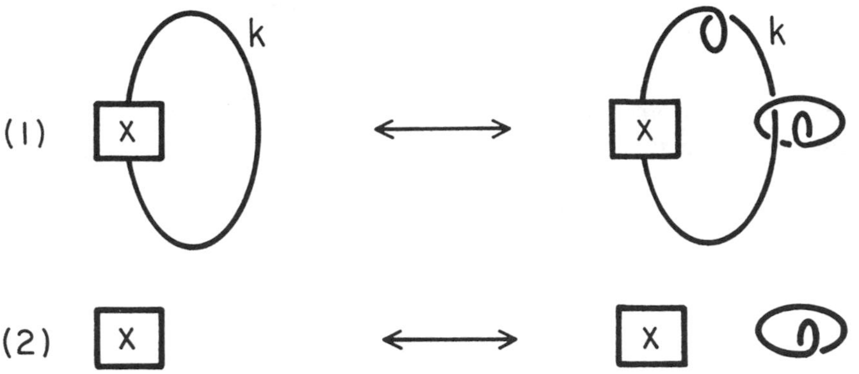

Fig. 3.3 3-manifold equivalences

sion of the bracket polynomial which is normalized by assigning the invariant 1 to the empty knot, ϕ, *i.e.*

Kauffman Bracket (renormalized):

$$\langle \quad \rangle : \ \{\text{Link Diagrams}\} \to Z[A^{\pm 1}]$$

(i) $\langle \phi \rangle = 1$

(ii) $\langle D \cup \mathbf{O} \rangle = -(A^{-2} + A^2)\langle D \rangle$

(iii) $\langle \times \rangle = A^{-1}\langle \asymp \rangle + A\langle\ )(\ \rangle$

is a regular isotopy invariant.

Let $r \geq 3$m be an integer and consider A a primitive $2r^{\text{th}}$ root of unity. Let D denote an n component link diagram representing a closed oriented 3-manifold, M. Let C denote the set of functions from the set $\{1, 2, ..., n\}$ into the set$\{0, 1, ..., r-2\}$ and, for each $c\varepsilon C$, let c^*D denote the link diagram by replacing the ith component of the link with $c(i)$ planar parallel copies of the component. Associated to the diagram D defining the framed oriented link one has the matrix of linking numbers of the components with the diagonal terms being the self linking or framing of the component. The signature, σ, and the nullity, ν, of this matrix is independent of the choices of orientations of the components of the link and are therefore defined by the unoriented diagram, D.

THEOREM (Witten, Reshetekhin-Turaev, Lickorish): If one defines

$$T_i = \langle \;\; _i \rangle$$

then (i) There is a unique set of complex numbers $\{\lambda_0, \lambda_1, ..., \lambda_{r-2}\}$ giving a solution to the equation

$$\sum_{i=0}^{r-2} \lambda_i T_{i+j} = (-A^{-2} - A^2)^j \qquad j = 0, 1, ..., r-2$$

and (ii)

$$I(D) = \left(\sum_{i=0}^{r-2} \lambda_i T_{i+j}\right)^{(\sigma+\nu-n)/2} \sum_{c \epsilon C} \lambda_{c(1)} ... \lambda_{c(n)} \langle c^* D \rangle$$

is an invariant of the corresponding 3-manifold determined by D.

This theorem describes all of the Witten $su(2)_q$ invariants. Blanchet, Habegger, Mausbaum and Vogel have shown that the invariants which can be represented by such sums exist only for these eigenvalues and are essentially unique, in the sense that any other such invariant is the same up to a factor depending only on standard algebraic topological invariants of the manifold. Kirby and Melvin [KM1&2], have developed another elementary approach in their construction. These quantum 3-manifold invariants are associated to the representation theory of the classical Lie groups and, thereby, give potentially rich interactions between the structure of these quantum invariants of the manifold, the classical algebraic topological invariants of the manifold, and deep aspects of number theory which are only now being discovered.

Just as is the situation with the invariants of knots and links, Lickorish has given examples of distinct 3-manifolds all of whose Witten $su(2)_q$ invariants are equal.

References

[A1] ALEXANDER, J.W., 1923 A lemma on systems of knotted curves. *Proc. Not. Acad. Sci. U.S.A.* **9**, 93–95.

[A2] ALEXANDER, J.W., 1928 Topological invariants of knots and links. *Trans. Amer. Math. Soc.* **20**, 275–306.

[AB] ALEXANDER, J.W., AND BRIGGS, G.B., 1927 On types of knotted curves. *Ann. of Math* **2**, 562–586.

[B] BANKEWITZ, C., 1930 Uber die torsionzahlen der alternierenden Knoten. *Math. Ann.* **103**, 145–161.

[B1] BLANCHET, C., HABEGGER, N., MASBAUM, G., AND VOGEL, P., 1990 Three manifold invariants derived from the Kauffman bracket.

[Bir1] BIRMAN, J., 1974 Braids, links and mapping class groups. *Ann. of Math. Studies* **82** (Princeton University Press).

[Bir2] BIRMAN, J., 1985 On the Jones polynomial of closed 3-braids. *Invent. Math.* **81**, 287–294.

[BirW] BIRMAN, J. AND WENZL, H., 1989 Link polynomials and a new algebra. *Trans. Amer. Math. Soc.* **313**, 249–273.

[BLM] BRANDT, R.D., LICKORISH, W.B.R., AND MILLETT, K.C., 1986 A polynomial invariant for nonoriented knots and links. *Invent. Math.* **84**, 563–573.

[BZ] BURDE, G., AND ZIESCHANG, H., 1986 *Knots* (De Greyter).

[Con] CONWAY, J.H., 1969 An enumeration of knots and links and some of their algebraic properties. *Computational Problems in Abstract Algebra*, John Leech, ed., 329–358 (Pergamon Press, Oxford and New York).

[Cr] CROWELL, R.H., AND FOX, R.H., 1963 *Introduction to Knot Theory.* (Blaisdell Pub. Co.)

[De] DEHN, M., 1914 Die Beiden Kleeblattschlinger. *Math. Ann* **102**, 402–414.

[EMS] EDWARDS, R., MILLETT, K., AND SULLIVAN, D., 1977 Foliations with all leaves compact. *Topology* **16**, 13–22.

[E1] EPSTEIN, D., 1972 Periodic flows on 3-manifolds. *Ann. of Math.* **95**, 68–82.

[E2] EPSTEIN, D., 1976 Foliations with all leaves compact. *Ann. Inst. Fourier* **26**, 265–282.

[EM] EWING, B., AND MILLETT, K., 1991 A load balance algorithm for the calculation of the polynomial knot and link invariants. *The Mathamtical Heritage of C.F. Gauss* (World Scientific Publishing Co., Singapore).

[EM2] EWING, B., AND MILLETT, K., 1990 Computational algorithms and the complexity of link polynomials.

[EMT] EPSTEIN, D., MILLETT, K., AND TISCHLER, D., 1977 Leaves without holonomy. *J. London Math. Soc.* **16**, 548–552.

[EV] EPSTEIN, D., AND VOGT, E., 1978 A counterexample to the periodic orbit conjecture in codimension 3. *Ann. of Math* **108**, 539–552.

[ES1] ERNST, C., AND SUMNERS, D.W, 1987 The growth number of prime knots. *Math. Proc. Camb. Phil. Soc.* **102**, 303–315.

[Fe] FENN, R., AND ROURKE, C., 1979 On Kirby's calculus of links. *Topology* **18**, 1–15.

[Fox] FOX, R.H., 1962 A quick trip through knot theory. *Topology of 3-Manifolds*, M.K. Fort Jr., ed., 120–167 (Prentice-Hall).

[FYHLMO] FREYD, P., YETTER, D., HOSTE, J., LICKORISH, W.B.R., MILLETT, K., AND OCEANU, A., 1985 A new polynomial invariant for knots and links. *Bulletin (New Series) of the Amer. Math. Soc.* Vol. 12, **2**, 239–246.

[Gab] GABAI, D., 1984 Foliations and genera of links. *Topology* **23**, 381–400.

[Ga] GAUSS, C.F., 1877 Zur mathematischen theorie der electrodynamischen Wirkungen, Werke. *Koniglichen Gesellschaft der Wissinchaften zu Gottingen*, vol. 5, 602–629.

[Gor] GORDON, C.MCA., 1978 Some aspects of classical knot theory. *Lecture Notes in Math.* **685**, 1–60.

[GorL] GORDON, C.MCA., AND LITHERLAND, R., 1978 On the signature of a link. *Invent. Math.* **47**, 53–69.

[GorLu] GORDON, C. MCA., AND LUEKE, J., 1989 Knots are determined by their complements. *J. Amer. Math. Soc* **3**, 371–415.

[HaKW] DE LA HARPE, P., KERVAIRE, M., AND WEBER, C., 1986 On the Jones polynomial. *L'Enseign. Math.* **32**, 271–335.

[Har] HARTLEY, R., 1983 Conway potential functions for links. *Comment. Math. Helv.* **58**, 365–378.

[Ho] HO, C.F., 1985 A new polynomial for knots and links – preliminary report. *Abstracts A.M.S.* **6**, 4, abstract 82-57-16.

[Hos] HOSTE, J., 1986 A polynomial invariant of knots and links. *Pacific J. Math.* **124**, 295–320.

[Ja] JAEGER, F., VERTIGAN, D.L., AND WELSH, D.J.A., 1990 On the computational complexity of the Jones and Tutte polynomials. *Math. Proc. Camb. Phil. Soc.* **108**, 35–53.

[Jon1] JONES, V.F.R, 1985 A polynomial invariant for knots *via* von Neumann algebras. *Bull. Amer. Math. Soc.* **12**, 103–111.

[Jon2] JONES, V.F.R, 1987 Hecke algebra representations of braid groups and link polynomials. *Ann. of Math.* **126**, 335–388.

[Jon3] JONES, V.F.R, 1986 Subfactors of type II_1 factors and related topics. *Proc. Int. Congress Math., Berkeley, California U.S.A.*, 939–947.

[Jon4] JONES, V.F.R, 1989 On knot invariants related to some statistical mechanical models. *Pacific J. Math.* **137**, 311-334.

[Jon5] JONES, V.F.R, 1988. On a certain value of the Kauffman polynomial. Preprint.

[Ka] KANENOBU, T., 1986 Infinitely many knots with the same polynomial invariant. *Proc. A.M.S.* **97**, 158–162.

[Kau1] KAUFFMAN, L., 1981 The Conway polynomial. *Topology* **20**, 101–108.

[Kau2] KAUFFMAN, L., 1983 Formal knot theory. *Math. Notes* **30** (Princeton University Press).

[Kau3] KAUFFMAN, L., 1987 On knots. *Annal of Math. Studies* **115** (Princeton University Press).

[Kau4] KAUFFMAN, L., 1990 An invariant of regular isotopy. *Trans. Amer. Math. Soc.* **318**, 417–471.

[Kau5] KAUFFMAN, L., 1987 State models and the Jones Polynomial. *Topology* **26**, 395–407.

[Ki] KIDWELL, M., 1987 On the degree of the Brandt-Lickorish-Millett-Ho polynomial of a link. *Proc. Amer. Math. Soc.* **100**, 755–762.

[Kir1] KIRBY, R., 1978 A calculus for framed links in S^3. *Invent. Math.* **45**, 35–56.

[KM1] KIRBY, R., AND MELVIN, P., 1989 Evaluations of the 3-manifold invariants of Witten and Reshetikhin-Turaev for $sl(2, C)$. *Geometry of Low-dimensional Manifolds* (Durham); 1990 *London Math. Soc. Lect. Notes* **151**, (Cambridge University Press).

[KM2] KIRBY, R., AND MELVIN, P., 1991 The 3-manifold invariants of Witten and Reshetikhin-Turaev for $sl(2, C)$. *Invent. Math.* **105**, 473–545.

[Li1] LICKORISH, W.B.R., 1962 A representation of orientable, combinatorial 3-manifolds. *Ann. Math.* **76**, 531–540.

[Li2] LICKORISH, W.B.R., 1986 A relationship between link polynomials. *Math. Proc Camb. Phil. Soc.* **100**, 109–112.

[Li3] LICKORISH, W.B.R., 1981 Prime knots and tangles. *Trans. Amer. Math. Soc.* **271**(1), 321-332.

[Li4] LICKORISH, W.B.R., 1985 The unknotting number of a classical knot. *Contemp. Math. A.M.S.* **44**, 117–119.

[Li5] LICKORISH, W.B.R., 1987 Linear skein theory and link polynomials. *Topology* **27**, 100, 109–112.

[Li6] LICKORISH, W.B.R., Unknotting by adding a twisted band. Preprint.

[Li7] LICKORISH, W.B.R., Some link-polynomial relations. *Math. Proc. Camb. Phil. Soc.*, to appear.

[Li8] LICKORISH, W.B.R., 1986 The panorama of polynomials for knots, links, and skeins. Proceedings of Artin's Braid Group Conference, Santa Cruz, California.

[Li9] LICKORISH, W.B.R., 1988 Polynomials for links. *Bull. London Math. Soc.* **20**, 558–588.

[Li10] LICKORISH, W.B.R., Link polynomials related. To appear.

[Li11] LICKORISH, W.B.R., Invariants for 3-manifolds from the combinatorics of the Jones polynomial. *Pacific J. Math.*, to appear.

[Li12] LICKORISH, W.B.R., 1990 3-manifolds and the Temperly-Lieb algebra.

[LiLi] LICKORISH, W.B.R., AND LIPSON, A.S., 1987 Polynomials of 2-cable-like links. *Proc. A.M.S.* **100**, 355-361.

[LM1] LICKORISH, W.B.R., AND MILLETT, K.C., 1987 A polynomial invariant for oriented links. *Topology* **26**, 107–141.

[LM2] LICKORISH, W.B.R., AND MILLETT, K.C., 1986 The Jones reversing result. *Pacific J. Math.* **124**, 173–176.

[LM3] LICKORISH, W.B.R., AND MILLETT, K.C., 1986 Some evaluations of link polynomials. *Comment. Math. Helv.* **61**, 349–359.

[LM4] LICKORISH, W.B.R., AND MILLETT, K.C., 1988 The new polynomials for knots and links. *Math. Magazine* **61**, 3–23.

[LM5] LICKORISH, W.B.R., AND MILLETT, K.C., An evaluation of the F-polynomial for links. Proceedings of the Siegen Topology Conference.

[LT] LICKORISH, W.B.R., AND THISTLETHWAITE, M.B., Some links with non-trivial polynomials and their crossing-numbers. *Comment. Math. Helv.*, to appear.

[Lit1] LITTLE, C.N., 1885 On knots, with a census for order ten. *Trans. Conn. Acad. Sci.* **18**, 374–378.

[Lit2] LITTLE, C.N., 1889 Non-alternate ± knots, of orders eight or nine. *Trans. Royals Soc. Edin.* **35**, 663–664.

[Lit3] LITTLE, C.N., 1890 Non-alternate ± knots, of orders 11. *Trans. Royals Soc. Edin.* **36**, 253–255.

[Lit4] LITTLE, C.N., 1900 Non-alternate ± knots. *Trans. Royals Soc. Edin.* **39**, 771–778.

[Men] MENASCO, W., 1984 Closed incompressible surfaces in alternating knot and link complements. *Topology* **23**, 37–44.

[MtH] MENASCO, W., AND THISTLETHWAITE, M., 1991 *Bull. Amer. Math. Soc.* **25**, 403–412.

[Mi1] MILLETT, K., 1976 Compact foliations. *Differential Topology, L.N.M.* 484 (Springer-Verlag).

[Mi2] MILLETT, K., 1986 Configuration census, topological chirality and the new combinatorial invariants. Proceedings of the International Symposium on Applications of Mathematical Concepts to Chemistry, *Croatia Chemica Acta* **59** (3), 669–684.

[MuJ1] MURAKAMI, J., 1987 The Kauffman polynomial of links and representation theory. *Osaka J. Math.* **24**, 745–758.

[MuJ2] MURAKAMI, J., The parallel version of link invariants. Preprint.

[Mur1] MURASUGI, K., 1965 On a certain numerical invariant of link types. *Trans. Amer. Math. Soc.* **114**, 377–383.

[Mur2] MURASUGI, K., 1958 On the genus of the alternating knot I, II. *J. Math. Soc. Japan* **10**, 94–105 and 235–248.

[Mur3] MURASUGI, K., 1986 Jones polynomials of alternating links. *Trans. Amer. Math. Soc.* **295**, 147–174.

[Mur4] MURASUGI, K., 1987 Jones polynomials and classical conjectures in knot theory. *Topology* **26**, 187–194.

[Mur5] MURASUGI, K., 1987 Jones polynomials and classical conjectures in knot theory II. *Math. Proc. Camb. Phil. Soc.* **102**, 317–318.

[O] OCNEANU, A., A polynomial invariant for knots: a combinatorial and an algebraic approach. Preprint.

[PT] PRZYTYCKI, J., AND TRACZYK, P., 1987 Invariants of the Conway type. *Kobe J. Math.* **4**, 115–139.

[Re] REEB, G., 1952 Sur Certaines proprietes topologiques des varietes feuilletees. *Act. Sci. Indust.* (Hermann, Paris).

[R] REIDEMEISTER, K., 1948 Knotentheorie. Reprint (Chelsea, New York).

[Re1] RESHETIKHIN, N.Y., AND TURAEV, V.G., 1991 Invariants of 3-manifolds *via* link polynomials and quantum groups. *Invent. Math.* **103**, 547–597.

[Ro1] ROLFSEN, D., 1990 Knots and Links (Publish or Perish Press).

[Sch] SCHARLEMANN, M., 1985 Unknotting number one knots at prime. *Invent. Math.* **82**, 387–422.

[Sul] SULLIVAN, D., 1976 A counterexample to the periodic orbit conjecture. *Publ. Math. IHES* **46**, 5–14.

[Su] SUMNERS, D.W., 1988 The knot enumeration problem. Proceedings of an international course and conference on the interfaces between mathematics, chemistry and computer science, Dubrovnik, Yugoslavia, *Studies in Physical and Theoretical Chemistry* **54**, 67–82.

[Ta] TAIT, P.G., On knots, I, II, III. *Scientific Paper* **1**, 273–347 (Cambridge University Press, London).

[Th1] THISTLETHWAITE, M., 1985 Knot tabulations and related topics. *Aspects of Topol-*

ogy, I.M. James and E.H. Kronheimer, eds. *L.M.S. Lecture Notes* **93**, 1–76.

[Th2] THISTLETHWAITE, M., 1987 A spanning tree expansion of the Jones polynomial. *Topology* **26**, 297–309.

[Th3] THISTLETHWAITE, M., Kauffman's polynomial and prime knots. Preprint.

[Th4] THISTLETHWAITE, M., Kauffman's polynomial and alternating links. *Topology*, to appear.

[Th5] THISTLETHWAITE, M., An upper bound for the breadth of the Jones polynomial. Preprint.

[Th6] THISTLETHWAITE, M., 1988 On the Kauffman polynomial of an adequate link. *Invent. Math.* **93**, 285–296.

[Th7] THISTLETHWAITE, M., On flypes and alternating tangles. Preprint.

[Tho] THOMPSON, W., 1867 On vortex atoms. *Philosophical Magazine*, 34 July, 15–24; *Mathematical and Physical papers* **4**, Cambridge (1910).

[Tr] TROTTER, H.F., 1987 Non-invertible knots exist. *Topology* **2**, 341–358.

[Tu1] TURAEV, V.G., 1987 A simple proof of Murasugi and Kauffman theorem on alternating links. *L'Ensign. Math.* **33**, 203–225.

[Tu2] TURAEV, V.G., 1988 The Yang-Baxter equations and invariants of links. LOMI preprint E-3-87, Steklov Institute, Leningrad; *Invent. Math.* **92**, 527–554.

[Tu3] TURAEV, V.G., The Conway and Kauffman modules of the solid torus with and appendix on the operator invariants of tangles. LOMI preprint E-6-88, Steklov Institute, Leningrad.

[Tu4] TURAEV, V.G., Algebras of loops on surfaces, algebras of knots, and quantization. LOMI preprint E-10-88, Steklov Institute, Leningrad.

[Tu5] TURAEV, V.G., 1990 Quantum invariants of links and 3-valent graphs in 3-manifolds.

[Tu6] TURAEV, V.G., 1990 Shadow links and face models of statistical mechanics.

[TV] TURAEV, V.G., AND VIRO, O., 1990 State sum invariants of 3-manifolds and quantum $6j$ symbols.

[V] VOGEL, P., 1990 Representation of links by braids: a new algorithm. *Comment. Math. Helv.* **65**, 104–113.

[Vo1] VOGT, E., Foliations of codimension 2 with all leaves compact. *Manuscripta Math.* **18**, 187–212.

[Vo2] VOGT, E., 1989 A foliation of R^3 and other punctured 3-manifolds by circles. *Publ. IHES* **69**, 215–232.

[Wall] WALLACE, A.H., 1960 Modifications and cobounding manifolds. *Can. J. Math.* **12**, 503–528.

[Wi1] WITTEN, E., Address to Centennial Meeting of the American Mathematical Society, Providence, Rhode Island, August, 1988.

[Wi2] WITTEN, E., 1989 Quantum field theory and Jones polynomial. *Comment.Math.Phys.* **121**, 351–399.

[Ya1] YAMADA, S., 1987 The minimal number of Seifert circles equals the braid index of a link. *Invent. Math.* **89**, 347–356.

TOPOLOGY OF KNOTS.

Martin Scharlemann*
Mathematics Department
University of California
Santa Barbara, California 93106-3080, U.S.A.

ABSTRACT.
This is meant to be a gentle introduction to the theory of knots, aimed specifically at physicists who want some background to Freedman and He's work (1991) relating asymptotic crossing number of knots to estimates of the energy in a knotted divergence-free flow.

1. Knotting and unknotting

DEFINITION 1. *A knot K in a smooth 3-manifold M^3 is the image of a smooth non-singular imbedding $f : S^1 \to M^3$.*

Here "non-singular" means that the derivative of f is non-zero and is included only to exclude the sort of pathological behavior shown in figure 1.

Two knots are regarded as equivalent, or isotopic, if you can move one to the other in M^3. Explicitly

DEFINITION 2. *Two knots K and K' are isotopic (write $K \sim K'$) if there is a smooth non-singular level-preserving embedding $F : S^1 \times I \to M \times I$ such that $F(S^1 \times \{0\}) = K$ and $F(S^1 \times \{1\}) = K'$.*

Once again non-singular is thrown in to eliminate pathological behavior. In particular, without it, all knots are unknotted! (See figure 2.)

Although these definitions are the most useful from the point of view of differential topology, the pathologies above often tempt topologists to think in a somewhat different category, called the PL category. In this category M^3 has a local linear structure, and in it knots are closed polygonal curves. From the point of view of the topology of the knot, these viewpoints are

* Research supported in part by a National Science Foundation Grant

H. K. Moffatt et al. (eds.), Topological Aspects of the Dynamics of Fluids and Plasmas, 65–82.

Fig. 1. Singular, or "wild" point.

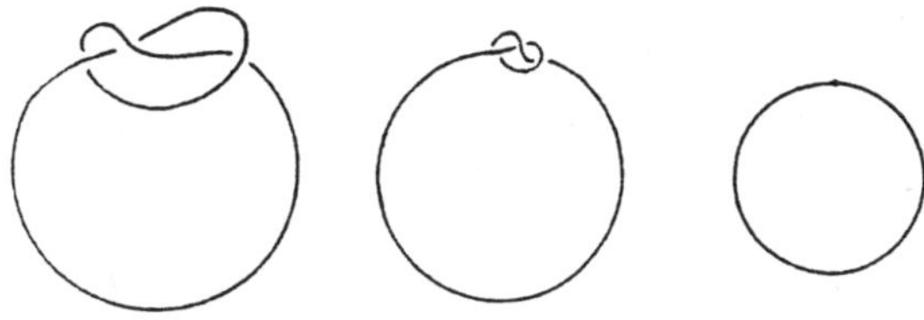

Fig. 2. A singular unknotting.

identical, but of course the corners may play havoc with any arguments involving the knot's differential structure (*e. g.* the curvature of the knot.)

Here we'll only be concerned with knots in 3-space. Since topologists have a preference for compact spaces, often they secretly regard the knots as lying in the 3-sphere. For most elementary purposes, the distinction between R^3 and S^3 is unimportant, for the difference is a single point at infinity. This point can be taken to be disjoint from the knot and its isotopies.

An advantage of working with knots in the 3-sphere (or 3-space) instead of an arbitrary 3-manifold, is the following theorem, called the Alexander trick:

THEOREM 1. *$K \sim K'$ in S^3 if and only if there is an orientation preserving diffeomorphism of S^3 carrying K to K'.*

DEFINITION 3. *K is unknotted (or the unknot) if it is isotopic to $S^1 \subset S^3$ (equivalently, to the round circle in the plane in 3-space).*

Here's an equivalent formulation:

THEOREM 2. *K is unknotted in S^3 if and only if K bounds a smooth disk in S^3.*

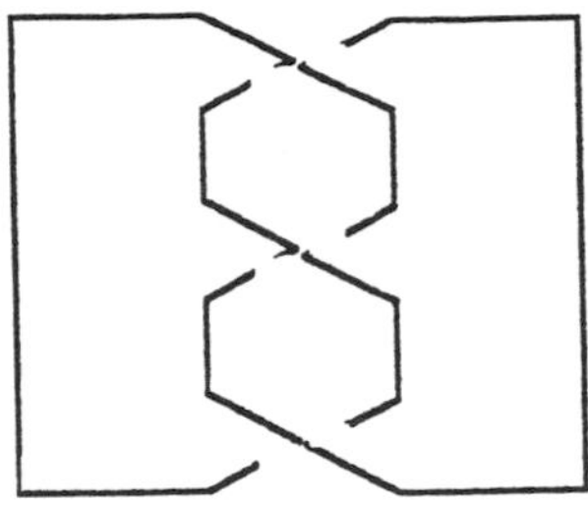

Fig. 3. Polyhedral picture of *trefoil* knot.

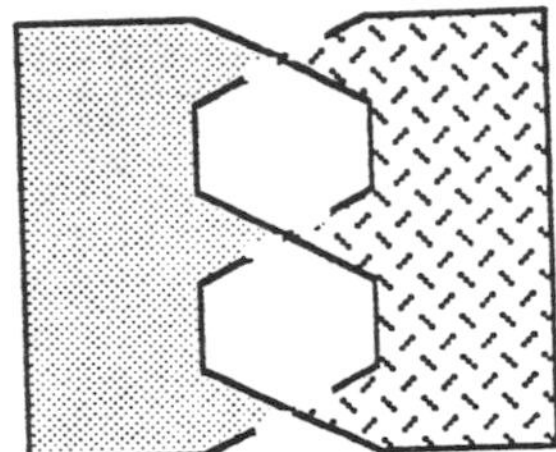

Fig. 4. Seifert surface for the trefoil knot.

Proof: Suppose K is unknotted in S^3, so there is a diffeomorphism of S^3 taking K to $S^1 \subset S^3$. Now S^1 bounds a hemisphere of $S^2 \subset S^3$; carry the hemisphere back by the inverse of the diffeomorphism.

Conversely, suppose K bounds a smooth disk D in S^3. A tiny neighborhood of the center of D is essentially a flat round disk D', whose boundary is a round circle. The annulus $D - D'$ then isotopes $K = \partial D$ to $\partial D'$. □

2. Seifert surfaces and knot genus

Although only the unknot bounds a disk, a general statement holds for all knots:

THEOREM 3. *(Seifert) Any knot bounds some orientable surface in* S^3.

Such a surface is called a Seifert surface of the knot. A Seifert surface of a knot isn't unique, even up to isotopy, for one can always attach tubes to a given surface and make it more complicated. Even if we restrict to the simplest possible Seifert surface, there are knots with infinitely many non-isotopic such surfaces. Nonetheless, the complexity of such surfaces is a useful indicator of the complexity of the knot.

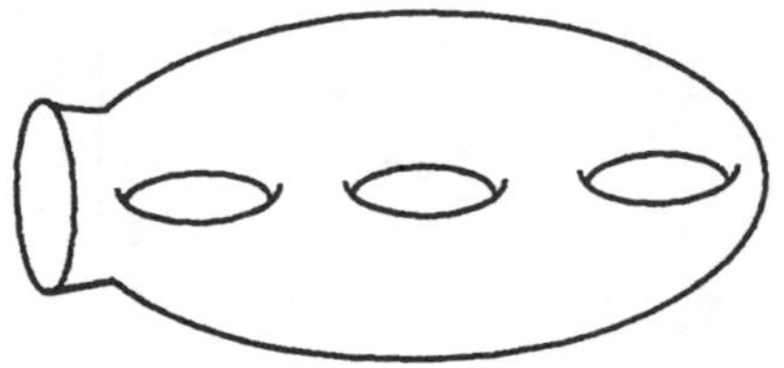

Fig. 5. A genus three surface.

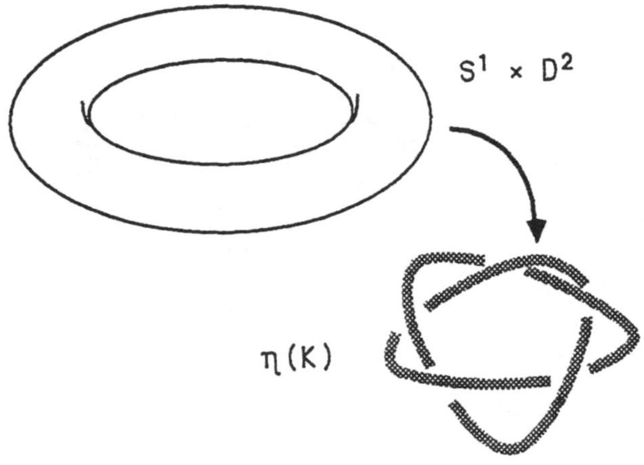

Fig. 6. A regular neighborhood η of a knot K.

DEFINITION 4. *A knot K in S^3 has genus n if it bounds an orientable surface of genus n, but not one of genus $n-1$.*

Note: The Seifert surface in figure 4 and the following corollary show that the trefoil knot has genus one.

COROLLARY 1. *A knot K has genus 0 if and only if it is unknotted.*

Proof: A genus 0 surface with one boundary component is a disk. □

THEOREM 4. *A smooth knot K has a tubular neighborhood. That is, there is a neighborhood $\eta(K)$ of K so that the pair $(\eta(K); K)$ is diffeomorphic to $S^1 \times (D^2, 0)$.*

DEFINITION 5. *The boundary $\partial\eta(K)$ of a tubular neighborhood of K is diffeomorphic to the torus $S^1 \times S^1$. It contains two special circles:*

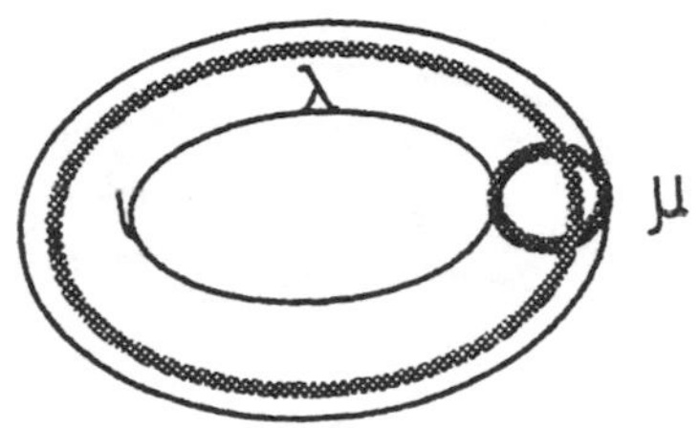

Fig. 7. Meridian μ and longitude λ.

The meridian is the boundary of a cross-sectional disk $\{*\} \times D^2 \subset S^1 \times D^2$ and is denoted μ.

The longitude is denoted λ and is the boundary of a Seifert surface.

THEOREM 5. *The meridian and longitude of a knot K are uniquely defined in $\partial\eta(K)$ up to isotopy.*

Proof: The harder part is showing the uniqueness of λ . This can be done by studying the arcs of intersection of two different Seifert surfaces. Alternately, there's an easy proof from algebraic topology. First note that two simple closed curves c and c' imbedded in $S^1 \times S^1$ are isotopic if and only if the algebraic intersection $c \cdot c'$ is trivial. If S and S' are two Seifert surfaces for K, then $\partial S \cdot \partial S' = \partial[S \cdot S']$. But $S \cdot S'$ is in $H_1(M, \partial M)$, which is trivial. □

Remark: The natural choice of circles λ and μ then allows us to describe any simple closed curve on $\partial\eta(K)$ by a number in the extended rationals $Q \cup (\infty)$. The curve homologous to $p\mu + q\lambda$ we associate with the rational number p/q. (*e. g.* $\mu \leftrightarrow \infty$).

3. Winding, (w)rapping, and satellites

DEFINITION 6. *Suppose k is a knot in $S^1 \times D^2$. Then the winding number of k, denoted $w(k)$, is the algebraic intersection $k \cdot D^2$ of k with a cross-sectional disk $\{*\} \times D^2$. Here algebraic intersection means that an intersection in one direction across the disk will cancel an intersection going in the other direction. In contrast, the wrapping number $r(k)$ of k is the number of times which k intersects D^2 (regardless of direction), minimized by isotopies of k in $S^1 \times D^2$.*

Figure 8 illustrates a knot which has winding number zero, but wrapping number 2.

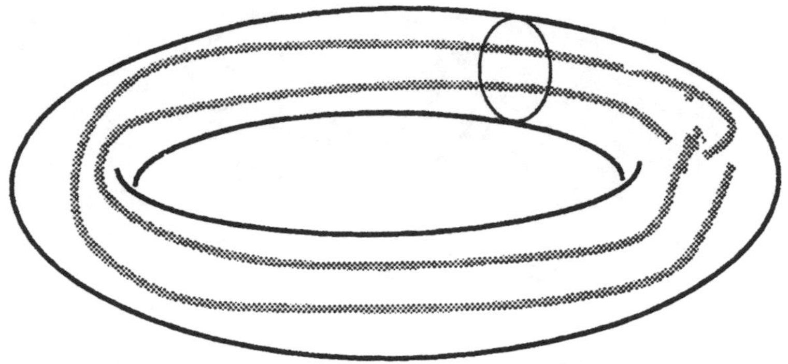

Fig. 8. How to "double" a knot.

One way of creating new knots is by imbedding simple closed curves in the tubular neighborhoods of other knots. Specifically, suppose K is a knot in S^3 . We would like to make the following *preliminary definition: k is a satellite* of K if $k \subset \eta(K)$.

But a moments reflection shows that some caution is in order. Here are some problems:

a) Any knot is a satellite of the unknot. Indeed, put a little loop l around a given knot k. Then the complement of l in S^3 is itself a (very large) regular neighborhood of an unnamed unknot. Since k lies in the complement of l, k would automatically be a satellite of the unknot.

b) k may be isotopic to K in $\eta(K)$, so in effect k is K.

c) If $r(k) = 0$ then k misses a cross-sectional disk of $\eta(K)$. The complement of such a disk in $\eta(K)$ is just a 3-ball. Any knot can be isotoped into any 3-ball (just shrink R^3), so if we were to allow $r(k) = 0$, then any knot would be a satellite of any other!

With these things in mind we make the final

DEFINITION 7. *a) $k \subset \eta(K)$ is a satellite of K if $r(k) \neq 0$ and k is not isotopic to K in $\eta(K)$.*

b) k is a satellite knot if it's a satellite of K, for K not the unknot.

(Note that a satellite of the unknot is not necessarily a satellite knot!)

DEFINITION 8. *Suppose k is a satellite of K. The pattern of k is $h(k)$, for $h : \eta(K) \to S^1 \times D^2$. In general, ambiguity in the choice of h from possible twisting of $S^1 \times D^2$ would result in ambiguity of the pattern. This can be circumvented by insisting that the longitude in $\eta(K)$ go to $S^1 \times \{*\}$, for some point $*$ in ∂D^2.*

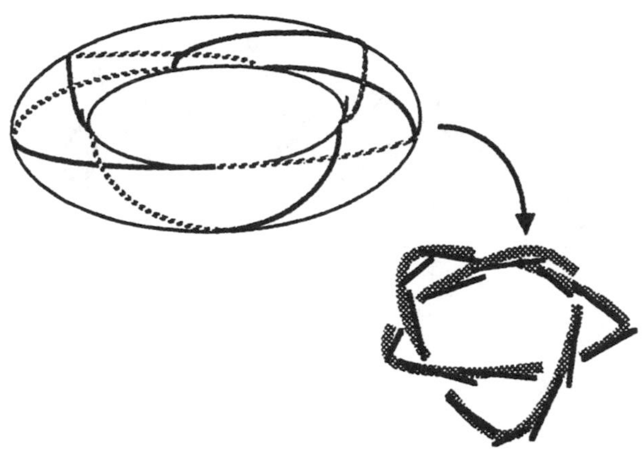

Fig. 9. How to cable a knot.

Some important examples of satellite knots are those obtained from the following patterns:

1) Double of a knot: Pick the simplest pattern with $w(k) = 0$, $r(k) = 2$. See figure 8.

2) Parallels of a knot: Pick for k several parallel copies of the core knot K. (Strictly speaking, this is a link, since there are several components).

3) Cables of a knot: Let the pattern be the torus knot on $\partial\eta(K)$. See figure 9.

In some sense, detecting satellite knots is fairly easy, thanks to the following theorem of Alexander.

THEOREM 6. *Any $T^2 \subset S^3$ bounds a solid torus on at least one side.*

COROLLARY 2. *If there's a torus $T^2 \subset S^3 - k$ which is not parallel to $\partial\eta(k)$ and doesn't bound a solid torus in $S^3 - k$, then k is a satellite knot.*

4. Band sums and connected sums of knots

DEFINITION 9. *A link is an embedded disjoint union of circles in S^3. A link is split if it's the distant union of two proper sublinks.*

DEFINITION 10. *The band-sum $K\#_b K'$ is obtained by banding together distant copies of K and K'.*

DEFINITION 11. *The connected sum $K\#K'$ is obtained by banding together distant copies of K and K' via a band crossing a splitting sphere once. $K\#K'$ is called a composite knot.*

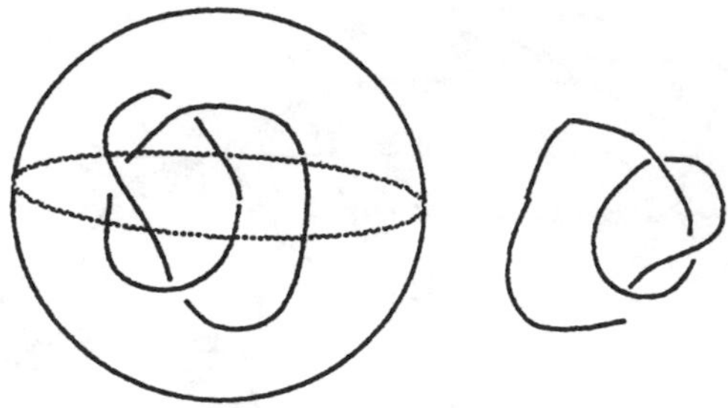

Fig. 10. A split link.

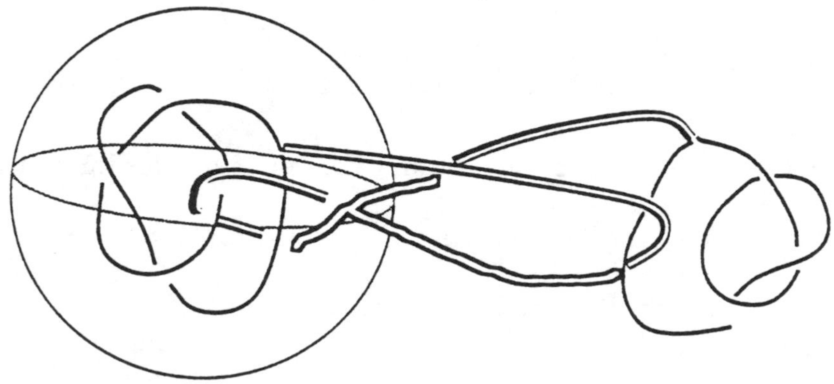

Fig. 11. The band sum $K\#_b K'$ of knots.

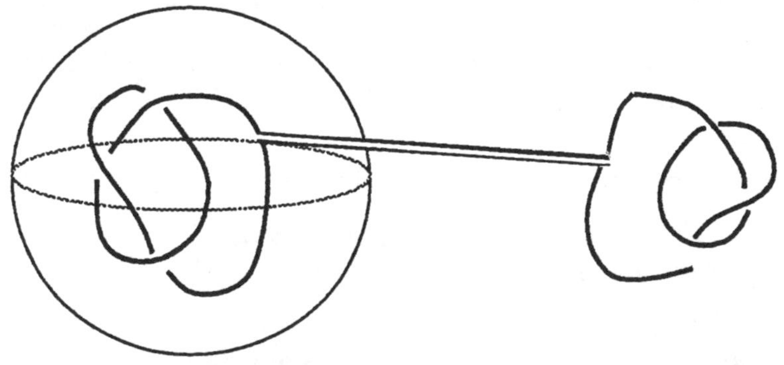

Fig. 12. The connected sum $K\#K'$ of knots.

In general, the knot obtained from $K\#_b K'$ depends heavily on the choice of b as well as that of K and K'. Not so, when we consider just the connected sum. Indeed

THEOREM 7. *$K\#K'$ is well-defined.*

Proof: : The proof is surprisingly simple: Suppose in figure 12 a band had been chosen so that it intersected the sphere once, but the part of the band outside the sphere was complicated. Just shrink the ball on the left, containing one summand, down to the size of a pea, and unravel the band. Then blow the ball up again. Now the band is straight outside the ball. Repeat the process using a ball around the other summand to straighten out the band inside the sphere. □

In light of this theorem, we can regard the connected sum as a binary operation on the set of knots. It's easy to see that the operation is commutative ($K_1\#K_2$ is isotopic to $K_2\#K_1$), associative ($(K_1\#K_2)\#K_3$ is isotopic to $K_1\#(K_2\#K_3)$) and that the unknot O is an identity element ($K\#O$ is isotopic to K). To a mathematician this means that # induces a "commutative monoid structure". To be a commutative group, all that's missing is the existence of an inverse. Hence we get the natural question: Given a knot K, is there a knot K' such that $K\#K' = O$?

As an applied problem, this can be translated to: If a garden hose is knotted, is there a way of tying another knot in an end of the hose, so that when slid down to the first knot, the two cancel and the hose is unknotted. Experienced gardeners know that this is impossible, but the mathematician requires proof. In fact, it's an elementary corollary of the following more general:

THEOREM 8. *For K and K' two knots in S^3, genus$(K\#K')$ = genus(K)+ genus(K').*

Proof: 1) The inequality genus$(K\#K')$ $\leq$genus(K)+genus(K') is immediate: If S and S' are minimal genus Seifert surfaces for K and K' respectively, just glue an arc on $\partial S = K$ to an arc on $\partial S' = K'$. (See figure 13.) The genus of the resulting surface T is genus(K)+genus(K'), and $\partial T = K\#K'$. T may not be of minimal genus among all Seifert surfaces of $K\#K'$, so we only get an inequality, not an equality.

2) The inequality genus$(K\#K')$ $\geq$genus(K)+genus(K') is more difficult. What we need to show is that given a Seifert surface T for $K\#K'$, it can be viewed, as above, as a Seifert surfaces for K and K' glued together along an arc. The idea is to start with T and modify it in a way that does not increase genus but so that it will eventually intersect a sphere S^2 separating K from K' in a single arc. Now $S^2-(K\#K')$ is an annulus A whose ends are

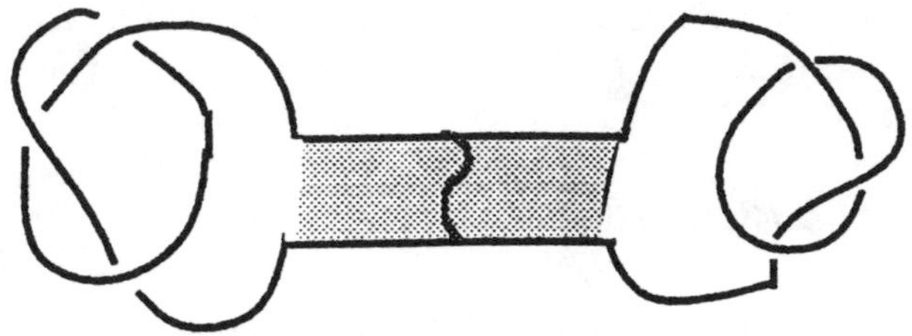

Fig. 13. Gluing together Seifert surfaces of K and K'.

Fig. 14. $T \cap A$

meridians of $K \# K'$. Since $\partial T = K \# K'$, $T \cap A$ is a 1-manifold with precisely two ends, one at each end of A. Viewed in A, this means that either $T \cap A$ is a single arc, and we're done, or $T \cap A$ contains also a bunch of circles (see figure 14).

In the latter case, at least one component of $T \cap A$ bounds a disk in A containing no other component of $T \cap A$. Figure 15 shows how to use this disk to modify T so that this component, and maybe more, is removed from $T \cap A$.

The operation may decrease, but won't increase, the genus of T. After sufficiently many such operations, all circles of intersection are removed. □

Remark: It is a much deeper result, proven only recently (Gabai (1987), Scharlemann (1989)), that genus($K \#_b K'$) $\geq$genus(K)+genus(K').

COROLLARY 3. *(Absence of inverses) If $K \# K' = O$ then $K = K' = O$.*

Proof: : Since genus(K)+genus(K') $= 0$, genus(K) $=$ genus(K') $= 0$, so K and K' are both the unknot. □

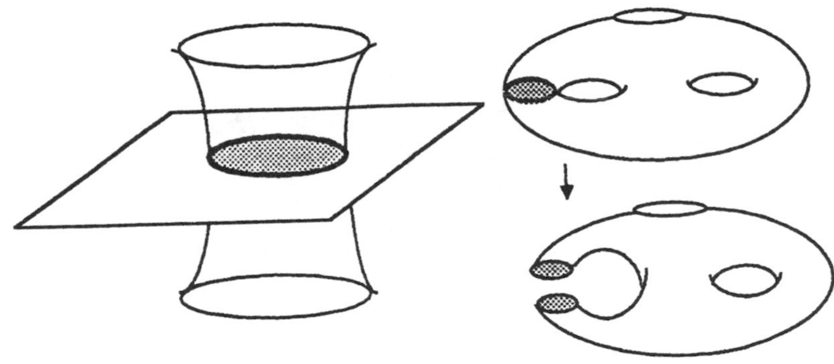

Fig. 15. "Compressing" T.

COROLLARY 4. *Genus one knots are prime (= not composite).*

Proof: : Suppose $\overline{K}$ is composite, so $\overline{K} = K \# K'$, with genus(K),genus(K') $\geq$ 1. Then genus($\overline{K}$) $\geq 1 + 1 = 2$. □

THEOREM 9. *A knot k is a satellite of a knot K_1 with $r(k) = 1$ if and only if k is the connected sum of K_1 with another knot K_0.*

Proof: The proof is in the figure 16, which shows how $K_0 \# K_1$ is imbedded in a tubular neighborhood of K_1. The illustrated torus T is called a "follow-swallow" torus, because it follows K_1 around, but swallows K_0. □

5. Bridge number and curvature

DEFINITION 12. *The bridge number β of a link is the minimum number of bridges needed to construct a freeway system supporting the link. The traffic rule is: you cross over each bridge exactly once, but can go under each bridge any number of times. By convention, $\beta(O) = 1$ (not 0).*

To see that this definition is symmetric, note that a path on the ground is travelled over once, though a bridge may pass over each path any number of times.

Figure 17 shows a two-bridge link, with the bridges appearing as horizontal segments.

There is an equivalent definition: If we look at the link in bridge position from the side, e. g. standing on the ground a good distance away, then the bridges can be viewed as maxima of a height function on K, and the paths

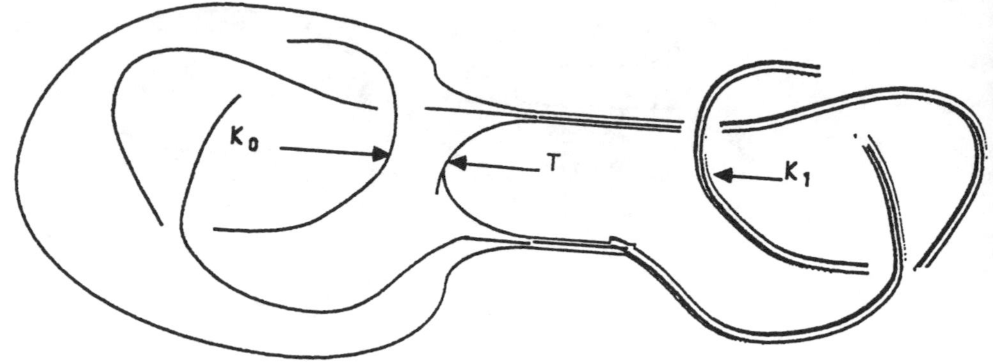

Fig. 16. The follow-swallow torus.

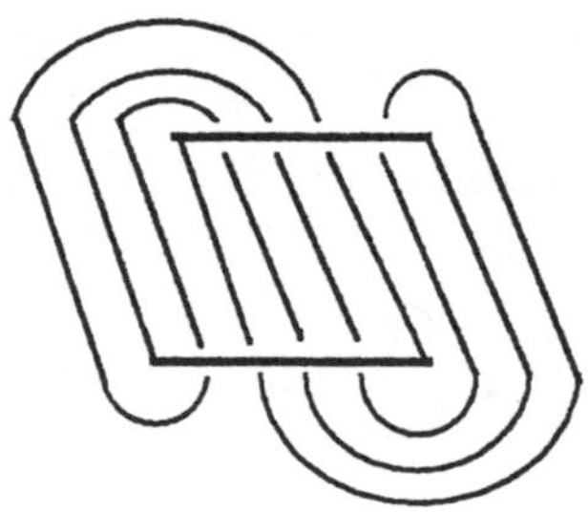

Fig. 17. A two-bridge link.

on the ground as minima. So we get: There's a height function on K with n maxima if and only if K has bridge number $\leq n$.

Here's yet a third picture. Imagine taking such a height function and exaggerating it, so the knot is actually hanging from its maxima, and its minima are pegged to a lower level. Then running between the top and the bottom is a braid of $2n$-strands. This is said to be a *$2n$-plat presentation* for the link (see figure 18).

For a fourth picture, imagine a plane which cuts through the plat presentation just above the minima. Then below the plane the minima just appear as a family of unknotted arcs. What's slightly more difficult to see is that if we ignore everything below the plane, and just look at the arcs lying above the plane, they too are isotopic (as proper arcs) to a family of unknotted arcs, by sliding their end points around on the plane. So k has bridge number $\leq n$ if and only if there is a plane cutting it into two families of n unknotted unlinked arcs.

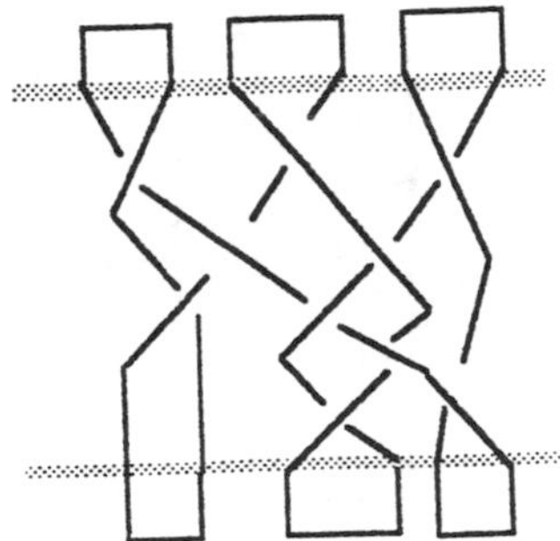

Fig. 18. A plat presentation for a three-bridge knot.

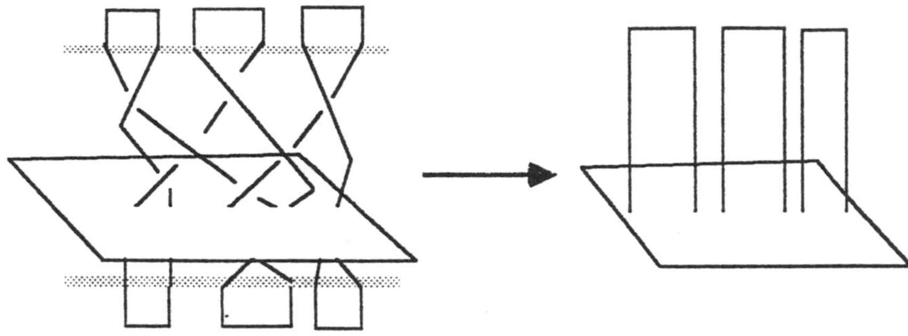

Fig. 19. Trivializing arcs above the plane.

THEOREM 10. *If K has bridge number one then K is the unknot.*

Proof: : There is a single maxima, hence a single minima. Between these two critical points K intersects each plane in two points. In each level plane, connect the two points of intersection with an arc, which can be taken to vary continuously with the level. The result is a disk whose boundary is the knot. □

Here is a deep theorem of Schubert (1954). For a modern proof, see Doll (1991), where it is also shown that it's possible to define a notion of bridge number for surfaces more complicated than planes (equivalently spheres).

THEOREM 11. *a) If k is a satellite of K then $\beta(k) \geq r(k) \cdot \beta(K)$.*
b) $\beta(K_1) + \beta(K_2) - 1 = \beta(K_1 \# K_2)$.

Remark: b) follows from a) by using the "follow-swallow" torus knot above.

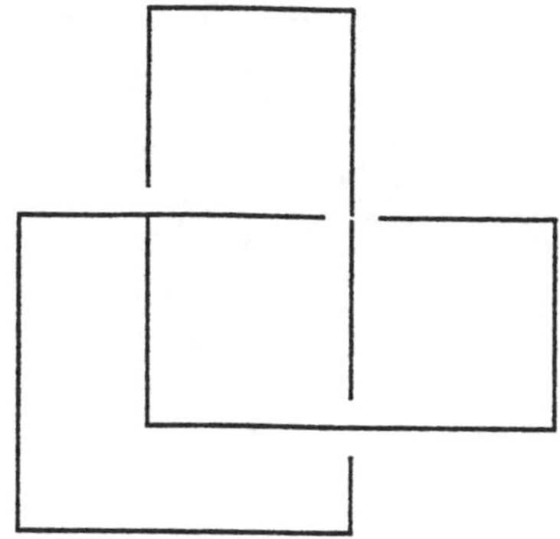

Fig. 20. A nearly flat trefoil knot, with κ near 4π

COROLLARY 5. *2-bridge knots are prime and not satellites.*

There is a remarkable connection between the curvature of a knot and its bridge number, discovered by John Milnor (1950) (when he was a teenager!). It is interesting in its own right, but is included here also because of its prescient connection to two notions of energy for a knot. Let $\kappa(k)$ denote the total curvature of a knot k in S^3 . It has the following polyhedral interpretation (indeed, this motivates the definition of curvature): For k a polyhedral knot in R^3 , let $\kappa(k)$ denote the sum of the exterior angles at corners of k.

THEOREM 12. $\kappa(k) \geq 2\pi\beta(k)$.

COROLLARY 6. *If* $\kappa(k) < 4\pi$ *then* k *is the unknot.*

Remark: In fact, $\kappa(k) \leq 4\pi$ implies that k is unknotted. Figure 20 shows that the estimate is sharp. The total exterior angle looks just like 4π, but this is an optical illusion. At least one of the straight lines must be bent or broken slightly, raising $\kappa(k)$ above 4π.

Proof of Milnor's theorem: We'll prove the polyhedral version; the smooth version can be gotten by polyhedral approximation. Let $p_1, ..., p_n$ be the n corners of the polyhedral approximation, and let α_i denote the exterior angle at p_i. Then $\kappa(k) = \sum \alpha_i$.

For any angle α in 3-space, there is a circular sector of directions σ_α in S^2 so that a height function with gradient in that direction will have a maxima at the angle's vertex (see figure 21).

We have area(σ_α)/area$(sphere) = \alpha/2\pi$, so area$(\sigma_\alpha) = 2 \cdot \alpha$. Let σ_i denote the angle corresponding to α_i. In particular the vector $\mathbf{v}$ is in σ_i if and only if α_i is a maximum for a height function whose gradient is $\mathbf{v}$. Put another way, if the number of maxima for k along such a height function is n, then $\mathbf{v}$ lies in n of the σ_i. The average number of maxima appearing over

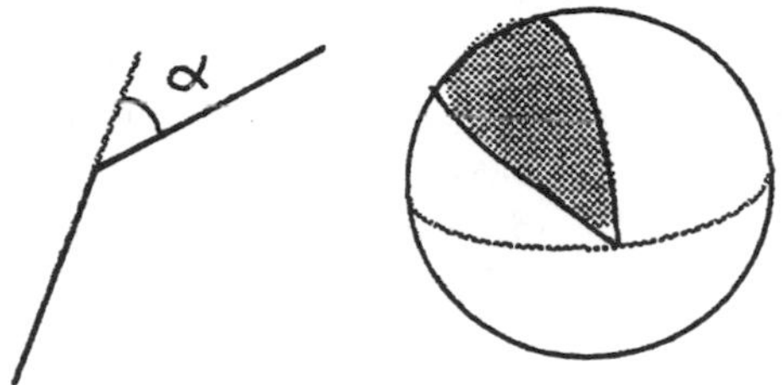

Fig. 21. A corner and its associated sphere segment.

all directions **v** is then simply $\sum$area(σ_i) divided by the area of the sphere, 4π. If we then let n_0 be the minimum number of maxima appearing at any direction **v**, it will clearly be no larger than the average, so we have

$$\beta(k) \leq n_0 \leq \sum area(\sigma_i)/4\pi = (\sum 2 \cdot \alpha_i)/4\pi = (\kappa(k))/2\pi \tag{5.1}$$

as required. □

6. Connections with knot energy

There are two connections here with notions of knot energy. The first is directly mathematical. If we imagine the knot made from an elastic straight wire, Bernoulli-Euler theory states that the amount of bending energy required to create the knot is the integral of κ^2 around the knot. A polyhedral approximation here is not allowed, for to create a corner requires an infinite amount of energy. This notion of energy for a knot is relatively accessible, and much is known (the definitive treatment is Langer and Singer (1985). For a knot theorist, however, the theory is ultimately disappointing. Langer and Singer show that an energy-reducing flow will always lead to self-intersections of a knot, changing its knot type. Eventually such a flow will terminate in a round circle.

The second connection is not so direct, but leads to the Freedman-He (1991) notion of average crossing number. We have seen how the total curvature of a knot describes an average bridge number for a specific position of the knot in R^3. Here the average is taken over all possible directions for a linear height functions on the knot in R^3.

Instead of the bridge number, consider another invariant of a knot k in R^3. For each direction in R^3 (i. e. unit vector $\mathbf{v}$ in S^2), project the knot onto the plane perpendicular to $\mathbf{v}$, and count the number of crossings in this projection. (Here $\mathbf{v}$ may have to be moved just a bit to ensure that the projection intersects itself transversally). Denote this number by $\zeta(\mathbf{v})$. If we isotope k and pick $\mathbf{v}$ so as to minimize $\zeta(\mathbf{v})$ the number we get is called the *crossing number* of k.

In analogy to the above discussion of bridge number and curvature, suppose we fix k in R^3 and don't allow it to change by isotopies. Regard $\zeta(\mathbf{v})$ as a function $S^2 \to \mathbf{Z}_+$ and integrate it over S^2. After dividing by the area of S^2 we obtain a number $\overline{\zeta} = (\int_{S^2} \zeta(\mathbf{v}) dS)/4\pi$, which it's natural to call the *average crossing number* of k.

Here is another way of describing $\overline{\zeta}$. Any pair of points $p \neq q$ on the knot k determine a line in R^3 whose direction may be regarded as a point in the projective plane RP^2 (= the space of lines in R^3 going through the origin, or the space derived from S^2 by identifying each pair of antipodal points). Thereby we have defined a map $\Gamma : k \times k \to RP^2$, where by convention $\Gamma(p,p)$ is defined to be the tangent line to k at p. A pair of points (p,q) in $k \times k$ is mapped to a point $\mathbf{v}$ in RP^2 if and only if when k is projected orthogonally to $\mathbf{v}$, p and q coincide. Generically, this means there is a crossing which identifies p and q. So (generically) $\mathbf{v}$ is covered n times by the image of Γ if and only if k, when viewed from the direction $\mathbf{v}$, has $n/2$ crossings (since there is a single crossing corresponding to both (p,q) and (q,p)). Hence $\overline{\zeta}$ can also be interpreted as the total variation (i. e. the area of the image, neglecting sign) of the map Γ, divided by twice the area 2π of RP^2. That is,

$$\overline{\zeta} = (\int_{k \times k} |det(d\Gamma)| dS)/4\pi \tag{6.1}$$

.

It's not hard to calculate $|det(d\Gamma)|$. The formula goes back to Gauss, who had to calculate $det(d\Gamma)$ in a slightly different context, that of the Gauss map $\tilde{\Gamma} : k \times k' \to S^2$ between a link of two components $k \cup k'$. In Gauss's situation tangent lines don't need to be considered, so there is always a natural orientation for the image line, say from p to q. Hence $\tilde{\Gamma}$ can be defined as a map into S^2 not RP^2. But his formula still applies: let $\gamma(p)$ and $\gamma(q)$ be unit vectors tangent to k at $p \neq q$. Then

$$|det(d\Gamma(p,q))| = |(\gamma(p), \gamma(q), p-q)|/|p-q|^3 \tag{6.2}$$

In the numerator is the vector triple product $(\gamma(p), \gamma(q), p-q) = (\gamma(p) \times \gamma(q)) \cdot (p-q)$. This then gives rise to the Freedman-He "crossing number version" of the Gauss formula.

It's appropriate here to mention a remarkable theorem of W. Pohl (1968). Let k be a knot, and consider the projection onto a plane perpendicular to a unit vector $\mathbf{v}$. Assign a sign $\pm$ to each crossing as follows: Pick either orientation for k. At the crossing examine whether the pair of oriented arcs obey the right hand rule $(+)$ or the left hand rule $(-)$. The answer is independent of the original orientation. If we add the signs of all the crossings, we get the signed crossing number $\zeta_{\pm}(\mathbf{v})$.

Just as in our development of $\overline{\zeta}$, it makes sense to define the average signed crossing number of k, $\overline{\zeta}_{\pm}$, where the average is taken over all directions of projection $\mathbf{v}$ in S^2. (Note that the sign of a crossing is also the same when viewed from the other side of the projection plane.) $\overline{\zeta}_{\pm}$, too, can be calculated via the Gauss integral, but to define $\tilde{\Gamma}$ here it's necessary to suppress the problem of orientation for tangent lines by removing the diagonal $\Delta = \{(p,p)|p \in k\}$ from $k \times k$. That is, take $X = (k \times k) - \Delta$, and define a map $\Gamma : X \to S^2$ as Gauss does for links. Then just as in the discussion of average crossing number we discover that

$$4\pi\overline{\zeta}_{\pm} = \int_X det(d\Gamma)dS \tag{6.3}$$

.

Now $\overline{\zeta}_{\pm}$ is rarely an integer. But what Pohl shows is that, as long as the curvature κ of k never vanishes, the sum of $\overline{\zeta}_{\pm}$ and the average torsion $\overline{\tau} = (\int_k \tau ds)/2\pi$ of k is an integer! This integer, called the self-linking number (or writhe) of k, is not an invariant of isotopy. For example, consider the unknot. A round circle clearly has trivial $\overline{\zeta}_{\pm}$ and $\overline{\tau}$, hence trivial writhe. On the other hand, given any integer n, the unknot can be isotoped so a projection to a plane has signed crossing number n. Use this projection to isotope the unknot near the projection plane, so $\overline{\tau}$ approaches zero and $\overline{\zeta}_{\pm}$ approaches n. Then this imbedding of the unknot must have writhe n. What Pohl's theorem does allow us to conclude is that one cannot perform such an isotopy through curves which have everywhere non-vanishing curvature.

7. Entry into the literature

This has just been the briefest of outlines of a rich and growing subject. Of the many books in the literature, two deserve special mention. For an informal treatment that begins at the beginning, but assumes some background in topology, *e. g.* homology theory, see Rolfsen (1976). For a more formal approach, with a truly astonishing bibliography, see Burde and Zieschang (1985).

References

BURDE, G. & ZIESCHANG, H., 1985 Knots. *Walter de Gruyter Press.*

DOLL, H., 1991 A generalized bridge number for knots in 3-manifolds. Ph. D. thesis, Santa Barbara, U. S. A.

FREEDMAN, M. & HE, Z.-X., 1991 Divergence-free fields: energy and asymptotic crossing number. *Annals of Math.*, **134**, pp. 189-229.

GABAI D., 1987 Genus is superadditive under band-connected sum. *Topology*, **26**, pp. 209-210.

LANGER, J. & SINGER, D., 1985 Curve straightening and a minimax argument for closed classic curves. *Topology*, **24**, pp. 75-88.

MILNOR, J., 1950 On the total curvature of knots. *Ann. of Math.*, **52**, pp. 248-257.

POHL, W., 1968 The self-linking number of a closed space curve. *Jour. of Math. Mech.*, **17**, pp. 975-985.

ROLFSEN, D., 1976 Knots and links. *Publish or Perish.*

SCHARLEMANN, M. 1989 Sutured manifolds and generalized Thurston norms. *Jour. Diff. Geom.*, **29**, pp. 557-614.

SCHUBERT, H., 1954 Über eine numerische Knoteninvariante. *Math. Zeit.*, **61**, pp. 245-288.

STRETCHING AND ALIGNMENT IN GENERAL FLOW FIELDS: CLASSICAL TRAJECTORIES FROM REYNOLDS NUMBER ZERO TO INFINITY.

M. TABOR*
Institute for Theoretical Physics
University of California
Santa-Barbara, California 93106-4030, U.S.A.

1. Introduction

The aim of these lectures is to describe the general properties of classical trajectories - be they associated with simple deterministic dynamical systems (Reynolds number very small) or the paths followed by fluid particles in turbulent flow fields (Reynolds number very large). The idea is to look for geometric characterizations of an orbit which are applicable to these two, very different, regimes and to investigate the way which they may change as the transition is made from deterministic flows to what are, effectively, stochastic flows. The reason that such an approach is possible is that in the case of fluid dynamics the description of the fluid in terms of fluid particles, namely the Lagrangian description, is governed by ordinary differential equations (as is the case, of course, with deterministic flows) - the complexity lies in the immensely complicated spatio-temporal behaviour of the velocity field. The processes that are a consequence of having a given velocity field, irrespective of its origin, is the subject matter of kinematics. The processes that govern the velocity field itself is the topic of dynamics - in the case of fluids this dynamics is governed by the Navier-Stokes equation.

The Lagrangian description of fluids provides a very natural bridge between fluid dynamics and dynamical systems. As we shall see some of the well established ideas of dynamical systems theory can be carried over to turbulent flows. However, this must be done with care: for example, the stretching of material elements in turbulence can be quantified by what

* Permanent address after January 1st 1992: Department of Mathematics, University of Arizona, Tucson, Arizona 85721

H. K. Moffatt et al. (eds.), Topological Aspects of the Dynamics of Fluids and Plasmas, 83–110.

looks very much like a Lyapunov exponent but, as we will see, this equivalence can only go so far. On the other hand, as we study ideas of stretching and alignment in turbulence, we start to see new ideas that might actually be useful in the study of dynamical systems. It should again be emphasized that these close links come about from the use of the Lagrangian description in which the phase space that the fluid particles explore is just the three dimensional configuration space (x, y, z) in which the fluid lives. This is very different from from those low order dynamical systems used to describe certain fluid processes (*e.g.* the famous Lorenz model) which govern the amplitudes of certain Fourier (or other) modes. In these cases some of the fascinating behaviour of the system, such as the appearance of strange attractors, may have very little to do with the real fluid dynamics that it is attempting to model.

The Lagrangian description of fluid dynamics has tended to be neglected in favour of the Eulerian description since the latter is often more tractable both analytically and computationally. However, in many cases, the Lagrangian frame is, in fact, the more natural one to use. Some developments in the Lagrangian picture of fluid dynamics are described in a recent monograph (Stuart and Tabor (1990)).

Apart from comparing and contrasting the behaviour and characterization of trajectories in deterministic and stochastic flows our aim is also to answer such basic questions as the rate of stretching of material elements in turbulence. In doing so we are naturally led to the all important issue of the stretching and alignment of vorticity itself and here we will see a fascinating interplay of kinematics and dynamics.

Before proceeding it is worth giving a little "translation table" for certain pieces of fluid dynamics terminology. Here $\mathbf{u} = (u, v, w)$ denotes the three component velocity field (vector field), and $\mathbf{x} = (x, y, z)$ the three dimensional configuration space (phase space). In the case of tensorial quantities we use the obvious notation u_i and x_i, $i = 1, 2, 3$, for the three components of velocity and space respectively.

2. Re → 0 : Simple Deterministic Dynamical Systems

Since we are interested in incompressible fluids we only consider volume preserving flows, *i.e.*

$$\nabla . \mathrm{u} = 0 \tag{2.1}$$

where the divergence operator is either two or three dimensional. A striking and important difference between these two cases is that the dynamics of fluid particles in two dimensional flows is Hamiltonian whereas the three dimensional case is not.

Fluid dynamical terminology	**Meaning**
Eulerian description	study of the velocity field $\mathbf{u} = \mathbf{u}(\mathbf{x}, t)$
Lagrangian description	study of fluid particle trajectories, *i.e.* $\frac{d\mathbf{x}}{dt} = \mathbf{u}(\mathbf{x}, t)$
Dynamics	processes that govern the form of $\mathbf{u}$
Kinematics	processes governed by a given $\mathbf{u}$
Steady flow	autonomous vector field, $\mathbf{u} = \mathbf{u}(\mathbf{x})$
Unsteady flow	nonautonomous vector field, $\mathbf{u} = \mathbf{u}(\mathbf{x}, t)$
Streamline	A curve $\Gamma(s) = (x(s), y(s), z(s))$, at a particular time t, which has the same direction as $\mathbf{u}$ at each point, *i.e.* $\frac{dx/ds}{u} = \frac{dy/ds}{v} = \frac{dz/ds}{w}$ For steady flows particle paths follow streamlines, for unsteady flows they do not.
Chaotic advection	chaotic trajectories
Velocity gradient tensor $A_{ij} = \frac{\partial u_i}{\partial x_j}$	tangent map
Rate of strain tensor $S_{ij} = \frac{1}{2}\left(\frac{\partial u_i}{\partial x_j} + \frac{\partial u_j}{\partial x_i}\right)$	symmetrized tangent map
Vorticity tensor $\Omega_{ij} = \frac{1}{2}\left(\frac{\partial u_i}{\partial x_j} - \frac{\partial u_j}{\partial x_i}\right)$	antisymmetrized tangent map
Euler equation	$\frac{\partial \mathbf{u}}{\partial t} + (\mathbf{u}.\boldsymbol{\nabla})\ \mathbf{u} = -\frac{1}{\rho}\boldsymbol{\nabla} p$
Navier-Stokes equation	$\frac{\partial \mathbf{u}}{\partial t} + (\mathbf{u}.\boldsymbol{\nabla})\ \mathbf{u} = -\frac{1}{\rho}\boldsymbol{\nabla} p + \nu \boldsymbol{\nabla}^2 \mathbf{u}$ ν is the kinematic viscosity p is the pressure field ρ is the fluid density

TABLE I
Jargon Translator

The underlying idea is, of course, that for two dimensional incompressible flows a stream function $\psi = \psi(x, y, t)$ can be introduced such that the two velocity components are given by the standard relations $u = \partial\psi/\partial y$ and $v = -\,\partial\psi/\partial x$. The stream function plays the role of a Hamiltonian and the fluid particle motions are governed by the dynamical system

$$\dot{x} = \frac{\partial\psi}{\partial y} \qquad \dot{y} = -\frac{\partial\psi}{\partial x} \tag{2.2}$$

in which x and y play the role of canonical variables. For steady flows the Hamiltonian is autonomous, *i.e.* $\psi = \psi(x, y)$, leading to the well known result that flow is integrable and all the trajectories lie on smooth curves (KAM tori in the Hamiltonian language and streamlines in the fluid dynamical language). For unsteady flows, *i.e.* explicitly time dependent ψ, the motion is typically nonintegrable and the particle dynamics is generically chaotic. By drawing on the extensive understanding of Hamiltonian dynamics the observed chaotic advection[1] phenomena can be analyzed (and predicted) in detail (see, for example, Aref (1984), Chaiken *et al.* (1986), Ottino (1989), Rom-Kedar *et al.* (1989)). Although the flows that have been analyzed in these references belong to a rather limited class (low Reynolds number, unsteady, two dimensional) they have provided some valuable insights into mixing and transport processes.

In three dimensions the fluid particle trajectories are governed by the non-Hamiltonian dynamics

$$\begin{aligned} \dot{x} &= u(x, y, z, t) \\ \dot{y} &= v(x, y, z, t) \\ \dot{z} &= w(x, y, z, t) \end{aligned} \tag{2.3}$$

Here chaotic motion is still possible for stationary flows, *i.e.* the right hand side of (2.3) is time independent. A valuable model of a stationary three dimensional flow field is the ABC system studied by Dombre *et al.* (1986) which takes the form

$$\begin{aligned} u &= A\sin z + C\cos y \\ v &= B\sin x + A\cos z \\ w &= C\sin y + B\cos x \end{aligned} \tag{2.4}$$

This is a solution to the (time-independent) Euler equations

[1] The term "chaotic advection", introduced by Aref (1984) in his pioneering paper is often used to describe chaotic fluid particle motion. However, since the chaos is purely kinematic in origin (the flow field is nonintegrable) and the term "advection" has its own fluid dynamical connotations we have proposed the alternative term "kinematic chaos" (Stuart and Tabor (1990)).

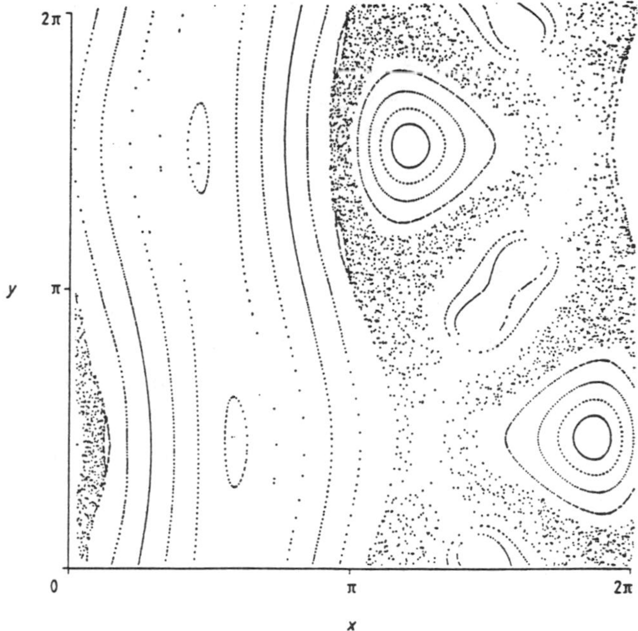

Fig. 1. Typical phase plane (at $z = 0$) for ABC flow with $A = 1$, $B = 1/\sqrt{2}$, $C = 1/\sqrt{3}$.

$$(\mathbf{u}.\boldsymbol{\nabla})\ \mathbf{u} = -\frac{1}{\rho}\boldsymbol{\nabla} p \tag{2.5}$$

and is an example of a Beltrami flow, namely a flow for which the vorticity is parallel to the velocity, that is $\boldsymbol{\omega} = \lambda \mathbf{u}$, where $\boldsymbol{\omega} = \boldsymbol{\nabla} \wedge \mathbf{u}$ is the vorticity vector[2]. Beltrami flows, or to be more precise, regions with the Beltrami property will be discussed frequently in this volume. The detailed study of (2.4) by Dombre *et al.* reveals the expected generic mixture of regular and chaotic orbits and a complicated stable/unstable fixed point structure. A typical phase plane is shown in figure 1. A discrete time mapping version of the ABC flow has been studied by Feingold *et al.* (1988). It is worth noting that although the dynamics of three dimensional, autonomous, volume pre-

[2] Since the ABC flow is a solution to the Euler equations it can, in a sense, be thought of as a flow corresponding to $Re = \infty$, in contradiction to the section heading $Re \to 0$! Here our sense of $Re \to 0$ is that of flows resolving few spatial scales: the ABC flow (2.4) resolves just one scale.

serving systems, such as (2.4), shows much of the phenomenology exhibited by nonintegrable Hamiltonian systems, namely a chaotic sea interspersed with invariant KAM like regions, far less in the way of rigorous results is available at this time; *e.g.* there is no equivalent of a KAM theorem on the preservation of invariant regions under perturbation.

In nature, as opposed to on a computer, we tend to observe distributions of points, such as a blob of dye, evolve under a given flow field. Thus we need to ask how such distributions evolve and, in particular, the morphologies of the evolving patterns that can be seen in an actual experiment. These distributions might correspond, as just mentioned, to a little blob of dye or hot fluid (a passive scalar distribution) or a piece of magnetic field line (a passive vector, to a first approximation) or a vorticity filament (a non passive vector). In the case of passive quantities in unsteady two dimensional flows these morphologies were predicted some time ago by Berry *et al.* (1979) to be either whorls, which are tight curling structures due to the presence in the phase plane of elliptic fixed points, or tendrils, which are flailing, exponentially growing structures due to hyperbolic fixed points (see figure 2). The latter, with their essential combination of stretching and folding characteristics, are responsible for efficient mixing in 2 dimensions. These morphologies are seen very clearly in the journal bearing experiment of Chaiken *et al.* (1986). Although these evolving shapes are associated with passive advection many studies of vorticity dynamics in two dimensions also show similar characteristic patterns (see, for example, Dritschel (1990)). The profound differences between vorticity evolution and passive scalar evolution will be discussed in detail later on. Another fundamental issue that we will investigate is the characterization of stretching and folding. The former is naturally quantified in terms of Lyapunov exponents whereas there has yet to be devised a useful quantification of folding. The folding process is, as discussed elsewhere in this volume, of particular significance in the dynamo problem. The possible morphologies that can emerge for evolving distributions in three dimensional flows is not as well understood as in the two dimensional case. This is because the structure and intersections of the stable and unstable manifolds in three dimensional phase space is extremely complicated (see, for example, the discussion of null-null lines in Dombre *et al.*). One approach to studying this structure will (inevitably) involve detailed numerical studies using three-dimensional graphics facilities. Nonetheless one can probably make the reasonable conjecture that the three dimensional analogue of whorls are likely to be "corkscrews" (see figure 3 in which a line of points evolving in the neighbourhood of an elliptic region (now tube-like in the three dimensional phase space) will wrap itself around the nested tubes each of which will have a smoothly varying winding number. It is not clear if there is an obvious analogue to tendrils in three dimensions.

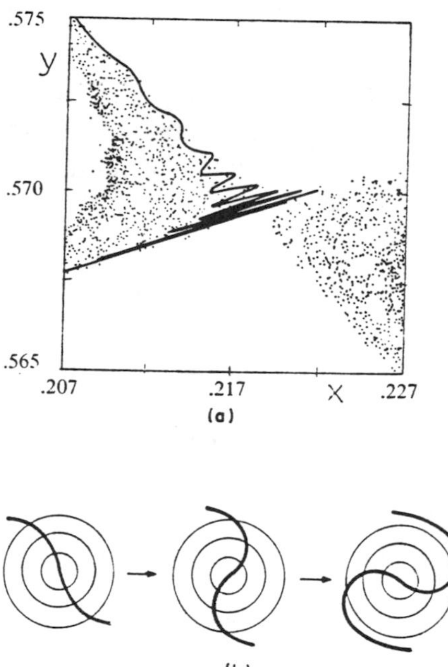

Fig. 2. (a) Line element stretching and folding in the neighbourhood of a hyperbolic fixed point forming a "tendril".

(b) Line element wrapping itself around an elliptic fixed point forming a "whorl".

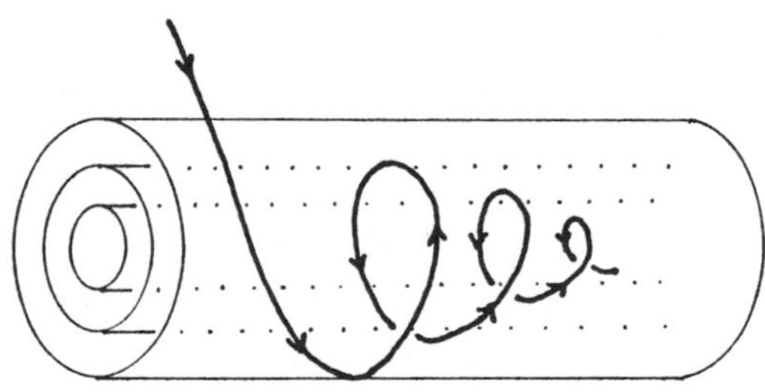

Fig. 3. Line element wrapping itself around a stable region of nested tori forming a "corkscrew".

3. The Geometry of Trajectories: Stretching, Rotation and Linking

Although the calculation of Lyapunov exponents, especially the largest one, is by now standard knowledge (see, for example, Tabor (1989)) we remind the reader of how it is done for two reasons: (*i*) to highlight the suprisingly many open questions that still remain and (*ii*) to set the stage for a comparison with similar looking quantities used to quantify line stretching in turbulent flows.

For an nth order flow of the form

$$\frac{dx_i}{dt} = F_i(x_1, \ldots, x_n) \qquad i = 1, 2, \ldots, n \tag{3.1}$$

linearization around some reference orbit $\mathbf{x} = (x_1, \ldots, x_n)$ gives the standard tangent mapping

$$\frac{d\delta x_i}{dt} = \sum_{j=1}^{n} \left(\delta x_j \frac{\partial F_i}{\partial x_j} \right)_{\mathbf{x} = \mathbf{x}(t)} \tag{3.2}$$

Introducing the norm

$$d(t) = \sqrt{\sum_i (\delta x_i(t))^2} \tag{3.3}$$

one can then define the mean rate of divergence, σ, of two initial points separated by a distance $d(0)$, namely

$$\sigma = \lim_{t \to \infty} \frac{1}{t} \ln \frac{d(t)}{d(0)} \tag{3.4}$$

This is the largest Lyapunov exponent, $\sigma_1(\mathbf{x}(0))$, which is sensitively dependent on the initial condition $\mathbf{x}(0)$. It should also be emphasized that there is no guarantee that the limit $t \to \infty$ in (3.4) exists for all $\mathbf{x}(0)$. When the limit does exist this suggests that there is also some asymptotic direction, termed the Lyapunov (eigen)vector, along which the stretching occurs.

It is well known that (3.4) is not the most practical way of computing σ_1 since exponential growth of $d(t)$ will lead to overflow. The standard way to proceed is to compute the separation for some small time interval, τ, and then rescale the separation and start again - this is illustrated in figure 4. The separation after the jth time step is defined as

$$d_j = |\delta \mathbf{x}^{(j-1)}(\tau)|$$

where

$$\delta \mathbf{x}^j(0) = \frac{\delta \mathbf{x}^{(j-1)}(\tau)}{d_j}$$

One then defines the "finite time" quantity $\sigma^{(N)}$,

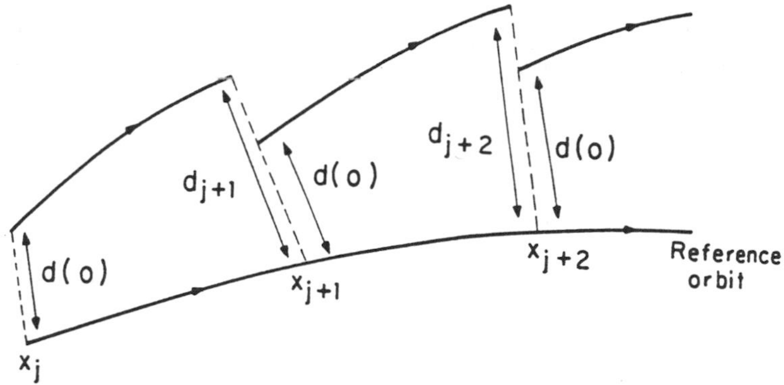

Fig. 4. Computation of the largest Lyapunov exponent: after each period τ the distance from the reference orbit is rescaled back to $d(0)$.

$$\sigma^{(N)} = \frac{1}{N\tau}\sum_{j=1}^{N}\ln(d_j) \tag{3.5}$$

and claims that

$$\sigma_1 = \lim_{N\to\infty} \sigma^{(N)} \tag{3.6}$$

The quantity (3.5) represents a time average over the d_j's, but each d_j in the sum is precisely correlated with its preceding one. Later we shall see a quantity representing a mean stretching rate in turbulence which looks very much like (3.5) but, as will be explained, the similarity is superficial.

So far we have only discussed the calculation of the largest Lyapunov exponent. For a n-dimensional system there are, in fact, a spectrum of them: $\sigma_1, \ldots, \sigma_n$ - each with an associated eigendirection ξ_i - and their calculation represents a formidable challenge. Here, following Benettin *et al.* (1980), the idea is to consider the deformation of a parallelepiped spanned by a k $(1 < k < n)$ dimensional vector space and follow the growth of the associated volume (of course for a volume-preserving flow the net change of volume for a n-dimensional parallelepiped is zero). The volume is defined (invariantly) in terms of determinant: for example, for a three dimensional parallelepiped spanned by the vectors $\mathbf{v}_i$, $i = 1, 2, 3$, one has the standard result that volume $= \mathbf{v}_1.(\mathbf{v}_2 \wedge \mathbf{v}_3) = \det |\{\mathbf{v}_1, \mathbf{v}_2, \mathbf{v}_3\}|$.

More formally, one can express the result as follows (here we essentially quote Benettin *et al.*): Let M be a compact connected Riemannian manifold of class C^1 and T a diffeomorphism onto itself. Let $\mathrm{Vol}^p[\ldots]$ denote the p-dimensional volume of the parallelepiped spanned by the vectors $\mathbf{v}_1, \ldots \mathbf{v}_p$, then

$$\lim_{t\to\infty} \frac{1}{t} \ln \left(\mathrm{Vol}^p\left[dT_x\mathbf{v}_1, \ldots, dT_x\mathbf{v}_p\right]\right) = \sigma_1(x) + \sigma_2(x) + \ldots + \sigma_p(x) \qquad (3.7)$$

where dT_x denotes the tangent map at time t associated with the (initial) point x. For an n-dimensional volume preserving flow one then has the result: $\sigma_1(x) + \sigma_2(x) + \ldots \sigma_n(x) = 0$. By studing the volume growth for all vector spaces, from 1 to n-dimensional, one can then deduce all the σ_i. These ideas are usefully illustrated by the case of a two dimensional flow (if volume preserving the two exponents will, of course, be equal and opposite). Assuming that the two exponents σ_1, σ_2 and their associated eigendirections $\boldsymbol{\xi}_1$, $\boldsymbol{\xi}_2$ exist, we can represent two independent vectors $\mathbf{v}_1$, $\mathbf{v}_2$ in the plane as

$$\mathbf{v}_1 = \alpha\boldsymbol{\xi}_1 + \beta\boldsymbol{\xi}_2 \qquad \text{and} \qquad \mathbf{v}_2 = \gamma\boldsymbol{\xi}_1 + \delta\boldsymbol{\xi}_2$$

In the limit $t \to \infty$ one can write

$$\begin{aligned} dT_x\mathbf{v}_1 &= \alpha e^{\sigma_1 t}\boldsymbol{\xi}_1 + \beta e^{\sigma_2 t}\boldsymbol{\xi}_2 \\ dT_x\mathbf{v}_2 &= \gamma e^{\sigma_1 t}\boldsymbol{\xi}_1 + \delta e^{\sigma_2 t}\boldsymbol{\xi}_2 \end{aligned}$$

If we consider the stretching of just the individual $\mathbf{v}_i$ it is easy to see that

$$\lim_{t\to\infty} \frac{1}{t} \ln |dT_x\mathbf{v}_i| = \sigma_1 \qquad \text{for} \quad i = 1, 2$$

Thus vectors spanning a one dimensional subspace (*i.e.* the lowest dimensional subspace) stretch by the largest Lyapunov exponent. On the other hand, considering the space spanned by $\mathbf{v}_1$ and $\mathbf{v}_2$ shows that

$$\begin{aligned} \lim_{t\to\infty} \frac{1}{t} \ln \left(\mathrm{Vol}^2[dT_x\mathbf{v}_1, dT_x\mathbf{v}_2]\right) &= \lim_{t\to\infty} \frac{1}{t} \ln \left(\det \begin{vmatrix} \alpha e^{\sigma_1 t} & \beta e^{\sigma_2 t} \\ \gamma e^{\sigma_1 t} & \delta e^{\sigma_2 t} \end{vmatrix}\right) \\ &= \sigma_1 + \sigma_2 \end{aligned}$$

Benettin *et al.* have proposed an elegant computaional procedure analogous to the "rescaling" method in the one dimensonal case. An initial p-dimensional vector space is spanned by a set of p orthogonal vectors. After they have evolved for a small time they will have stretched out into a long, thin parallelepiped. By using the Gram-Schmidt orthogonalization procedure these vectors are "straightened out" into a new (and renormalized) orthogonal set spanning the same vector space. This procedure is then repeated for many steps and an average of the short-time volume growth rates taken just as in the one dimensional case. Lest all this effort seem in vain we mention that one use of knowing all the Lyapunov exponents is to calculate the Kaplan-Yorke dimension which is given by the formula

$$d_{YK} = j + \frac{\sigma_1 + \ldots + \sigma_j}{|\sigma_{j+1}|}$$

where j is the largest dimension for which the sum of exponents is positive, *i.e.* $\sigma_1 + \ldots \sigma_j > 0$. The formula is essentially an interpolation between positive and negative exponents to determine the "dimension" which would have a zero exponent.

Before proceeding any further it is worth pausing to ask the following questions:

(i) In the definition of exponents does the limit $t \to \infty$ exist? This question has been answered by Oseledec (1968) who proved that there are initial conditions (so called regular points) for which the limit does indeed exist.

(ii) How well does the limit $t \to \infty$ (or, in practice, $N \to \infty$) converge? Little is known about this - often, as we shall see, $\sigma^{(N)}$ is a highly oscillatory function of N. Whether these oscillations contain any useful information is an interesting question.

(iii) What is the meaning of the eigenvectors $\boldsymbol{\xi}_i$? Although they must say something about the local geometry of the set on which the orbit in question lies, what this something is is not understood. A better understanding of this issue would be useful for the dynamo problem where the (asymptotic) direction in which magnetic field lines point is of fundamental importance.

In calculating Lyapunov exponents fundamantal information about rotation is discarded. This can be seen from the following discussion (following Goldhirsch et al (1987)). For the nth order flow (3.1) we define the propagator

$$\mathbf{M}(t,0) = T : \exp\left(\int_0^t \mathbf{A}\left(\mathbf{x}\left(t'\right)\right) dt'\right) \tag{3.8}$$

where here T : denotes the time ordering operator and $\mathbf{A}$ the (tangent map) matrix evaluated along the reference trajectory $\mathbf{x}(t)$, *i.e.* $\mathbf{A}_{ij} = (\partial F_i / \partial x_j)_{\mathbf{x}=\mathbf{x}(t)}$. $\mathbf{M}$ satisfies the equation

$$\frac{d\mathbf{M}}{dt} = \mathbf{A}\left(\mathbf{x}(t)\right)\ \mathbf{M}(t,0). \tag{3.9}$$

For every regular point $\mathbf{x}(0)$ there exists an orthonormal set of vectors $\boldsymbol{\xi}_i(t)$ such that

$$\sigma_i = \lim_{t\to\infty} \frac{1}{t} \ln ||\mathbf{M}(t,0)\ \boldsymbol{\xi}_i(t)|| \tag{3.10}$$

exists. By noting that

$$||\mathbf{M}(t,0)\ \boldsymbol{\xi}_i(t)||^2 = \boldsymbol{\xi}_i^\dagger(t)\ \mathbf{M}^\dagger(t,0)\ \mathbf{M}(t,0)\ \boldsymbol{\xi}_i(t) \tag{3.11}$$

we see that the σ_i^2 and $\boldsymbol{\xi}_i$ are the eigenvalues and eigenvectors, respectively, of $\mathbf{M}^\dagger\mathbf{M}$. Since this is a real symmetric matrix (hence real eigenvalues and orthogonal eigenvectors) one can write

$$\mathbf{M}^\dagger\mathbf{M} = \mathbf{S}^2 \tag{3.12}$$

where $\mathbf{S}$ is another real symmetric matrix. Thus we can write

$$\mathbf{M} = \mathbf{RS} \tag{3.13}$$

where $\mathbf{R}$ is a real unitary matrix. Clearly the rotational information contained in $\mathbf{R}$ is lost in the calculation of the eigenvalues and eigenvectors of $\mathbf{M}^\dagger\mathbf{M}$

We have begun an investigation of this problem. A typical behaviour of $\mathbf{R}$ is illustrated by a study of the standard map. In figures 5,6,7 we show, respectively, a typical chaotic phase plane (generated by one trajectory), the associated computation of the (positive) Lyapunov exponent $\sigma^{(N)}$ as a function of the number of time steps N, and the corresponding evolution of $R_{11} = \cos\theta$ (for this two dimensional problem $\mathbf{R}$ is just the standard 2×2 rotation matrix). We first note the non-trivial behaviour of $\sigma^{(N)}$: the large fluctuations during the first few thousand timesteps clearly correspond to the orbit staking out its "set" and although the worst of this transient behaviour is over by $N = 10\,000$, it is still not clear by $N = 20\,000$ to what $\sigma^{(N)}$ is converging. The plot of $\cos(\theta(N))$ *vs.* N shows an almost random splatter of points: however there is clearly a greatest probability of finding $\cos(\theta(N)) = \pm 1$, which says that the most likely orientation of the instantaneous stretching direction (the tangent vector) is parallel (but pointing either "up" or "down") to the Lyapunov direction. There also appears to be a slightly increased probability for finding $\cos(\theta(N)) \approx 0.65$: the significance of this has yet to be understood.

Clearly a complete picture of local rotation can be extracted from (3.13) by computing $\mathbf{R} = \mathbf{M}\ \mathbf{S}^{-1}$ which becomes nontrivial for higher dimensional flows. A simple way of obtaining at least some of the rotational information (Dresselhaus and Tabor(1989)) is to compute the quantity

$$\mu^2 = \frac{1}{2}\ \text{Trace}\ \mathbf{A}^2 \tag{3.14}$$

where $\mathbf{A}$ denotes the tangent map (velocity gradient tensor) along a given trajectory. It gives a simple and compact measure of the straining and rotational components of the motion since it is easily shown that

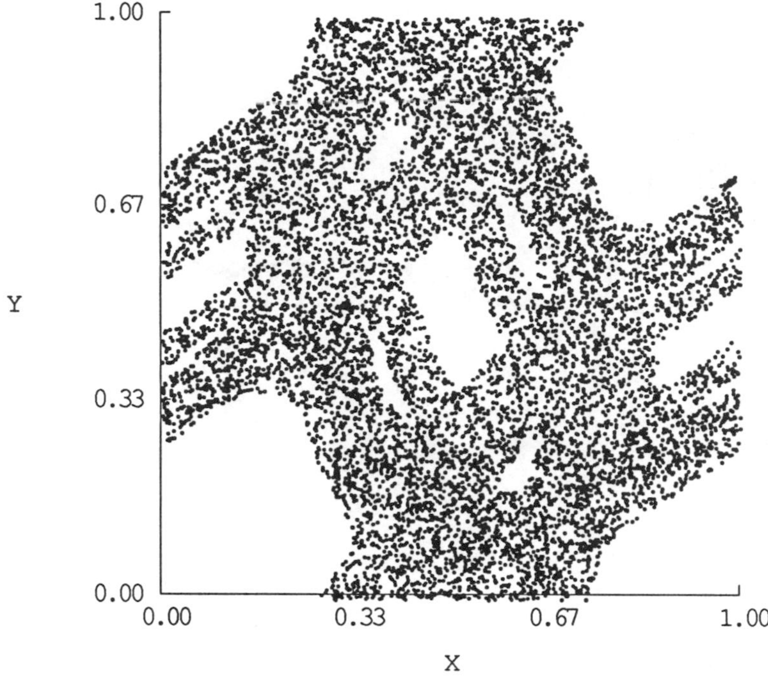

Fig. 5. Typical chaotic phase plane for the standard map: $\{x_{i+1} = x_i + y_{i+1}, y_{i+1} = y_i + (k/2\pi)\ \sin\ (2\pi x_i)\ ;\ \mathrm{mod}\{x, y\} = 1\}$ at $k = 1.25$. Initial conditions are $(x_0, y_0) = (0.2, 0.4)$.

$$\mu^2 = \sum_{i=1}^{3} \left(s_i^2 - \Omega_i^2\right) \tag{3.15}$$

where s_i and Ω_i are the real and imaginary parts, respectively, of the eigenvalues of the $\mathbf{A}$. Clearly if $\mu^2 < 0$ the motion is rotation dominated and the associated morphology will be whorl-like, whereas for $\mu^2 > 0$ the motion is strain dominated with a tendency for tendril formation. The study of μ^2 in the Lagrangian frame of fluid dynamics has a nice interpretation since one may easily show, by taking the divergence of the Navier-Stokes equation, that

$$\boldsymbol{\nabla}.(\mathbf{u}.\boldsymbol{\nabla})\mathbf{u} = -\frac{1}{\rho}\nabla^2 p = \text{Trace } \mathbf{A}^2 = \mu^2 \tag{3.16}$$

which shows the connection between local stretching and rotation and the pressure field p. (In the particular case of steady two dimensional flows μ^2

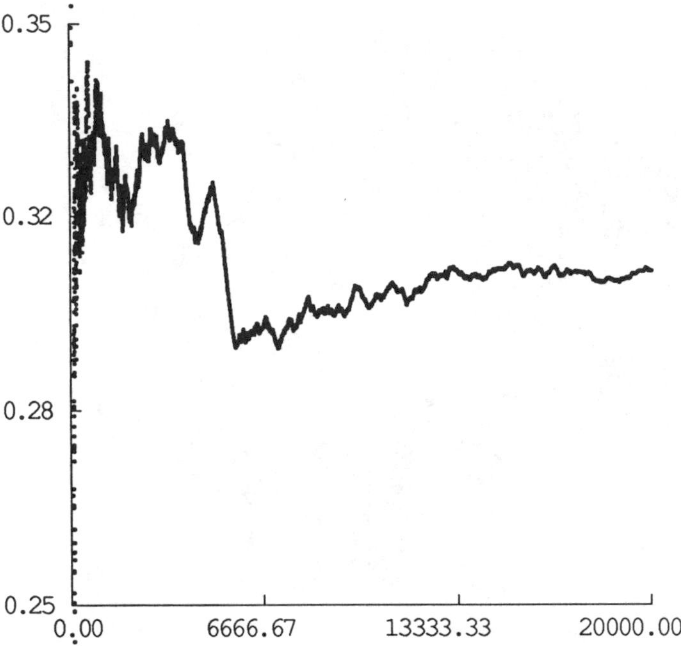

Fig. 6. Computation of finite time Lyapunov exponent $\sigma^{(N)}$ versus N for same orbit shown in 5.

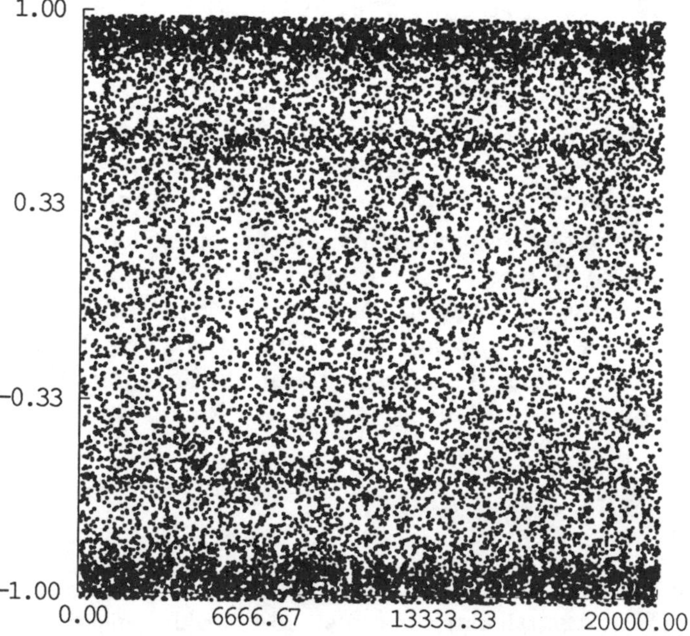

Fig. 7. Computation of alignment angle $\cos(\theta(N))$ versus N for the same orbit shown in 5.

has the additional geometric intepretation of being the Gaussian curvature of the stream lines.) For a given initial condition μ^2 shows rapid fluctuations as a function of time and a more useful quantity may be the long time average of the real and imaginary parts of μ itself (*i.e.* $\mu = (\text{Trace}\mathbf{A})^{1/2}$) along a trajectory. The ratio

$$\chi = \frac{\sigma_{\text{Re}}}{\sigma_{\text{Im}}} \tag{3.17}$$

gives a simple measure of the mean strain-rotation ratio experienced by particle orbits. (The inverse of this quantity was introduced long ago in fluid mechanics by Truesdell (1954)).

Additional geometric information about trajectories can also be obtained by studying their curvature and torsion (see Dresselhaus and Tabor (1990)).

Our discussion of the stretching and rotation of orbits involves consideration of a single trajectory. An important characterization of the topological relationship between different orbits is provided by consideration of their linking number. Here, as we shall discuss, this quantity, which can in principle be computed for any divergence free vector field, has its most useful interpretation in the fluid dynamical context (for a nice review see Arnold and Khesin (1992)). The basic idea is to consider two trajectories, emanating from two (different) initial conditions, $\mathbf{x}$ and $\mathbf{y}$ say. (The vector field, $\mathbf{u}$, is assumed to live on some manifold M, in R^3, with a Riemannian metric and measure $d\mu$). After some long time, T, these orbits are artificially closed on themselves by certain "shortest paths" thereby defining two closed paths γ and γ', say. The linking number $\ell(\gamma, \gamma')$ is defined as the number of (signed) intersection points of γ with the surface bounded by γ'. The asymptotic linking number, associated with the orbits emanating from $\mathbf{x}$ and $\mathbf{y}$ is then defined as

$$\lambda(\mathbf{x}, \mathbf{y}) = \lim_{T \to \infty} \frac{\ell_T(\gamma, \gamma')}{T^2}$$

where $\ell_T(\gamma, \gamma')$ is the linking number computed for orbits integrated for time T. The limit exists for almost every pair $\mathbf{x}$, $\mathbf{y}$ and $\lambda(\mathbf{x}, \mathbf{y})$ is of L^1 class. The average linking number is the average over the entire domain, *i.e.*

$$\lambda = \iint_{M \times M} \lambda(\mathbf{x}, \mathbf{y}) \, dx dy$$

The important result is that λ, which measures the mean linking of any two orbits of the vector field, is equal to the "Hopf invariant", $\mathcal{H}$, of that field. $\mathcal{H}$ itself is defined as

$$\mathcal{H} = \int_M \left(\text{curl}^{-1}(\mathbf{u}), \mathbf{u}\right) d\mu$$

where (,) denotes scalar product and curl^{-1} denotes the formal inverse of the curl operation (the 'Biot-Savart operator'). Thus, in the fluid mechanical context, since $\mathrm{curl}(\mathbf{u}) = \boldsymbol{\omega}$ and hence $\mathrm{curl}^{-1}(\boldsymbol{\omega}) = \mathbf{u}$, $\mathcal{H}$ is seen to be nothing more than the helicity of the flow (see Moffatt and Tsinober (1992) and Moffatt in this volume). Whereas helicity and its interpretation as a mean linking number is of fundamental importance in fluid mechanics, its usefulness, and computation, for dynamical systems has yet to be explored.

4. Re $\to \infty$: Stretching In Turbulent Flows

The study of the stretching of line and surface elements in turbulent flows has a long history (see for example Monin and Yaglom (1975)); although these investigations have tended to proceed independently of advances in the understanding of dynamical systems.

The first major treatment of material elements in turbulent flows was given by Batchelor (1951) who made two fundamental assumptions. The first was that a finite material element can be considered as a collection of infinitesimal elements since in a turbulent flow the stretching of elements separated by more than a few Kolomogorov length scales should be completely decorrelated. As long as one is averaging over many different material elements this appears to be a very reasonable assumption. The second assumption was that the infinitesimal elements $\boldsymbol{\ell}$ do, in fact, stretch exponentially. On the basis of this a local stretching rate was introduced

$$\overline{\zeta} = \overline{\frac{1}{|\,\boldsymbol{\ell}\,|}\;\frac{d\,|\,\boldsymbol{\ell}\,|}{d\,t}} = \overline{\frac{d}{dt}\ \ln\ |\,\boldsymbol{\ell}\,(t)\,|} \tag{4.1}$$

where the overbar represents an average over different realizations of the (turbulent) velocity field. In particular it was claimed that

$$\lim_{t\to\infty} \overline{\zeta} > 0 \tag{4.2}$$

At first sight this all looks very much like the discussion of the largest Lyapunov exponent given in the previous section with (4.1) and (4.2) being analogous to (3.5) and (3.6) respectively. However, in (4.1) the average is over different realizations of the flow field rather than along a single trajectory (this would be like removing the correlations between the succesive d_j's in (3.5)) and, furthermore, there is no analogue of Oseledec's theorem that guarantees that the limit $t \to \infty$ in (4.2) exists. Thus the idea of a stretching rate in turbulence being, somehow, the equivalent of a Lyapunov exponent has to be treated with caution.

At the time of Batchelor's investigations little was known about the validity of (4.2) and it was not until much later that Cocke (1969,1971) demonstrated, for Gaussian random fields, that *(i)* $\ln \mid \boldsymbol{\ell}(T) \mid \; > \; \ln \mid \boldsymbol{\ell}(0) \mid$ for $T > 0$, *i.e.* two points do separate on average for a finite time but that

(ii) $\lim_{t\to\infty} \overline{\zeta} = 0$, *i.e.* there is no long time exponential separation.

It is instructive to reproduce a simplified version (due to Orzsag (1971)) of the proof of *(i)*. The infinitesimal element $\boldsymbol{\ell}$ evolves according to the standard formula

$$\frac{d\,\boldsymbol{\ell}}{dt} = \mathbf{A}\,\boldsymbol{\ell} \tag{4.3}$$

where $\mathbf{A}$ is the velocity gradient tensor defined before. This equation can be formally integrated to give

$$\boldsymbol{\ell}(t) = \mathbf{M}(t,0)\,\boldsymbol{\ell}(0)$$

where the propagator $\mathbf{M}$ is defined in equation (3.8). Thus one can write

$$\mid \boldsymbol{\ell}(t) \mid^2 = \sum_{ij} W_{ij}\,\ell_i(0)\,\ell_j(0)$$

where W_{ij} are the matrix elements of $\mathbf{W} = \mathbf{M}^{\dagger}\mathbf{M}$. The matrix $\mathbf{W}$ is now assumed to be random with (random) eigenvalues w_1, w_2, w_3 which, by incompressibility, satisfy the constraints

$$w_1\, w_2\, w_3 = 1 \qquad \text{and} \qquad w_1 > w_2 > w_3 > 0$$

An important consequence of these two conditions is that

$$w_1 + w_2 + w_3 \geq 3 \tag{4.4}$$

If we imagine placing an infinitesimal sphere of radius δ at $\boldsymbol{\ell}_0 = \boldsymbol{\ell}(0)$, it will be deformed by the flow into an ellipse with principal axes of length $w_1\delta,\ w_2\delta,\ w_3\delta$. If one defines a polar coordinate system with respect to the orthogonal axes associated with eigenvectors of $\mathbf{W}$, one can show that

$$\mid \boldsymbol{\ell}(t) \mid^2 = \left[\sin^2\theta \left(w_1 \cos^2\phi + w_2 \sin^2\phi\right) + w_3 \cos^2\theta\right] \mid \boldsymbol{\ell}_0 \mid^2$$

If the polar angles $\theta,\ \phi$ and the eigenvalues w_i are taken to be random variables and that the former, for isotropic turbulence, are uniformly distributed over the unit sphere, one has the result that

$$\mid \boldsymbol{\ell}(t) \mid^2 = \frac{1}{3}\,(w_1 + w_2 + w_3) \mid \boldsymbol{\ell}_0 \mid^2$$

where we have used the averages $\overline{\sin^2\theta\cos^2\phi} = \overline{\sin^2\theta\sin^2\phi} = \overline{\cos^2\theta} = 1/3$. As a consequence of (4.4) one can then conclude that

$$\mid \boldsymbol{\ell}(t) \mid^2 \geq \mid \boldsymbol{\ell}_0 \mid^2$$

Although we must now bear in mind that $\overline{\zeta}$ is only a finite time quantity it is still a useful one to study. Noting that (4.3) can also be written as

$$\frac{d\,\boldsymbol{\ell}}{dt} = \mathbf{S}\,\boldsymbol{\ell} + \boldsymbol{\Omega} \wedge \boldsymbol{\ell} \tag{4.5}$$

where $\mathbf{S}$ and $\boldsymbol{\Omega}$ are the rate of strain and vorticity tensors respectively, the local stretching rate ζ can be expressed as

$$\zeta = \sum_{ij} S_{ij}\,\hat{\ell}_i\,\hat{\ell}_j \qquad i,j = 1,2,3 \tag{4.6}$$

where the $\hat{\ell}_i$ are the unit vector components associated with $\boldsymbol{\ell}$. The natural question to ask is for an estimate of $\overline{\zeta}$ in turbulent flows. Here Batchelor and Townsend (1956) suggested that

$$\overline{\zeta} = \overline{s_1} \tag{4.7}$$

i.e. the mean of the largest eigenvalue of the rate of strain tensor. Implicit in this estimate is the principle of persistence of strain by which is meant that the combined effect of the vorticity and the rotation of the principal axes of the straining field is sufficiently small such that the material element has time to align with the strain axis with largest eigenvalue and hence experience maximum stretching. This idea was consistent with an earlier experiment of Townsend (1951) on the decay of heat spots in grid turbulence. (He found that these spots experienced maximum cooling - which was consistent with his theoretical model in which the strain axes were stationary, *i.e.* the strain was persistent). Batchelor and Townsend went on to give the estimate

$$\overline{s_1} = 0.43/\tau_k$$

where τ_k is the Kolmogorov turn-over time given by $(\nu/\varepsilon)^{1/2}$, where ε is the rate of energy dissipation and ν the kinematic viscosity.

Pursuing, for a moment, the idea of estimates for stretching rates in turbulence it is clear that two, infinitesimally close particles will only separate significantly over regions of space for which the associated velocity gradient (*i.e.* straining field) is well correlated. The largest scale over which this can occur in (fully developed) turbulent flow is, almost certainly, the Kolmogorov micro-scale - this is the idea underlying the Batchelor-Townsend estimate. As a function of Reynolds number, Re, the micro-scale scales as $Re^{-3/4}$. Of course, as this scale becomes smaller, the associated gradient increases. This gradient (the micro-scale frequency) scales as $Re^{1/2}$. Considering that a mean stretching rate is a long-time average of particle separation the competition between these two effects as a function of Re is clearly very complicated. Just because a velocity field becomes more turbulent does not necessarily mean that the overall stretching will increase.

For many years the persistence of strain paradigm became firmly entrenched in the lore of fluid dynamics (see, for example, Monin and Yaglom). Some fairly recent pieces of evidence have now challenged this idea:

(i) A study of vorticity alignment by Ashurst *et al.* (1987) in large scale numerical simulations of the Navier-Stokes equations showed that the vorticity tended to align with the intermediate strain axis rather than the direction with the largest eigenvalue. In ideal fluids the stretching of vorticity (discussed in detail later) looks formally similar to that of line stretching and one would therefore expect vorticity alignment with the axis with largest eigenvalue if the strain were persistent. Other numerical experiments all now seem to confirm the observations of Ashurst *et al.*. We will eventually develop an alignment theory that explains these observations.

(ii) Girimaji and Pope (1990) studied Lagrangian orbits and line stretching in spectral code simulations of homogeneous isotropic turbulence at Taylor scale Reynolds numbers, R_λ, of up to 90 and found that the mean strain rate on a material line to be about one third of the Batchelor and Townsend result. We also note their finding that the average direction cosine between the (infinitesimal) line element(s) and the strain axis with largest eigenvalue was around 0.9 (for persistent strain it should be 1.0).These results, coupled with other numerical tests of various timescales, strongly indicated that the strain was not persistent.

(iii) An exact analytical solution to the Euler equations due to Vieillefosse (1982, 1984) showed that the vorticity actually blows up in finite time in the direction of the intermediate strain axis. Although his model makes some strong, and probably unrealistic assumptions, about the pressure field it is very suggestive. A similar analysis has also been carried out more recently by Majda (1990).

We conclude this section by briefly discussing one area of fluid dynamics in which the persistence of strain hypothesis plays an important role. This is the problem of polymeric drag reduction. Here, it is has long been known that the addition of minute quantities of polymer (typically, $10-50$ ppm) can lead to an enormous reduction in turbulent drag - an effect that has many important applications ranging from its use in oil pipe-lines to fire hoses. A satisfactory theoretical explanation has proved to be elusive. For many years it has been assumed that the effect was due to the enhancement of the so-called extensional viscosity: This non- Newtonian effect, which has long been known experimentally, corresponds to a significant increase in the viscosity of even very dilute polymer solutions under a purely extensional, steady, flow (*e.g.* vertical flow under gravity from a capillary). This effect has a satisfactory explanation in terms of the significant extension that the polymer molecules can undergo in such flow fields. However, the key is that the extensional flow is steady, *i.e.* the strain is persistent. This not only enables the polymer to stretch significantly but also have time to relax back against the imposed straining field. This means that there is relative motion between the polymer and the fluid and hence the possibility of viscous effects. Virtually nothing is known about this phenomenon in unsteady flows. In these cases, especially for the rapidly fluctuating straining fields that might occur in turbulence, it seems more likely that the polymer deforms "affinely". That is, it does not have time to respond to the local straining field and, instead, obediently follows the deformation of the surrounding fluid element. In this case any effects of the polymer on the flow are going to be elastic rather than viscous. This suggest a different mechanism for drag reduction (Tabor and deGennes (1986)).

5. Alignment: The Interplay of Kinematics and Dynamics

In order to understand exactly what is going in the stretching and alignment process it is necessary to perform an analysis in the orthonormal basis $(\boldsymbol{\xi}_1, \boldsymbol{\xi}_2, \boldsymbol{\xi}_3)$ in which the rate-of-strain tensor is diagonal. Denoting $\mathbf{X}$ as the matrix whose columns are the positively oriented orthonormal eigenvectors $\boldsymbol{\xi}_i$ (*i.e.* $\det \mathbf{X} = +1$ for all time) we can write

$$\mathbf{S}(t) = \mathbf{X}^{\dagger} \begin{pmatrix} s_1 & 0 & 0 \\ 0 & s_2 & 0 \\ 0 & 0 & s_3 \end{pmatrix} \mathbf{X}$$

where the eigenvalues of $\mathbf{S}$, s_1, s_2, s_3, are ordered so that $s_1 \geq s_2 \geq s_3$. This moving coordinate system will be referred to as the "strain coordinates". In addition we will assume that the flow is incompressible, *i.e.* $s_1 + s_2 + s_3 = 0$ so that $s_1 \geq 0$ and $s_3 \leq 0$ for all time. By transforming $\boldsymbol{\ell}$ into the strain coordinate system, the local separation rate (4.6) becomes

$$\zeta_{\text{line}} = \sum_{i=1}^{3} s_i \, \lambda_i^2 \tag{5.1}$$

where $\boldsymbol{\lambda} = \mathbf{X} \, \boldsymbol{\ell}$ is the material element in strain coordinates. The λ_i are the cosines of the angles between $\boldsymbol{\ell}$ and $\boldsymbol{\xi}_i$. Thus equation (5.1) shows the intimate connection between the local stretching rate of a material element and its alignment with the strain axes. Furthermore we note the largest value attained by (5.1) corresponds to $\boldsymbol{\lambda} = (\pm 1, 0, 0)$, *i.e.* $\zeta_{\text{line}} = s_1$ which is just the persistent strain estimate of Batchelor and Townsend.

The time evolution of the λ_i is determined from the relationship

$$\frac{d}{dt}\lambda_i = \frac{d}{dt}\langle \boldsymbol{\ell}, \boldsymbol{\xi}_i \rangle = \left\langle \frac{d}{dt}\boldsymbol{\ell}, \boldsymbol{\xi}_i \right\rangle + \left\langle \boldsymbol{\ell}, \frac{d}{dt}\boldsymbol{\xi}_i \right\rangle \tag{5.2}$$

The first of the two terms on the right hand side of (5.2) can be computed using equation (4.5). In order to evaluate the second term, expressions for the evolution of the $\boldsymbol{\xi}_i$ must be developed. Since the eigenvectors are orthonormal, *i.e.* $\langle \boldsymbol{\xi}_i, \boldsymbol{\xi}_j \rangle = \delta_{ij}$ it follows that $\langle d\boldsymbol{\xi}_i/dt, \boldsymbol{\xi}_i \rangle = 0$ for all time. Thus we can write

$$\frac{d}{dt}\boldsymbol{\xi}_i = \boldsymbol{\Omega}' \wedge \boldsymbol{\xi}_i \tag{5.3}$$

where the vector $\boldsymbol{\Omega}'$ is the instantaneous axis of rotation of strain axes and we shall refer to it as the strain rotation vector (we will occasionally use the more compact term "orienticity"). It will be seen to play a crucial role in the alignment equations. The components of $\boldsymbol{\Omega}'$, in the basis of strain coordinates, can be determined by differentiating with (respect to time) the characteristic equation $\mathbf{S}\ \boldsymbol{\xi}_j = s_j \boldsymbol{\xi}_j$, using equation (5.3) and taking the inner product of both sides with $\boldsymbol{\xi}_i$. The result is

$$\Omega'_k = \varepsilon_{ijk} \frac{\left\langle \boldsymbol{\xi}_i\, ,\ \dfrac{d}{dt}\mathbf{S}\ \boldsymbol{\xi}_j \right\rangle}{s_i - s_j} \tag{5.4}$$

where ε_{ijk} is the usual third order totally antisymmetric tensor. (For two dimensions one may easily verify that the Ω'_k remain finite should the eigenvalues become degenerate.) At this stage the definition of $\boldsymbol{\Omega}'$ is purely kinematical in origin - the dynamical content is in $d\mathbf{S}/dt$ which, in the fluid dynamical context, is governed by the Navier-Stokes equation.

Using the above results equation (5.2) becomes

$$\frac{d}{dt}\boldsymbol{\lambda} = \begin{pmatrix} s_1 & 0 & 0 \\ 0 & s_2 & 0 \\ 0 & 0 & s_3 \end{pmatrix} \boldsymbol{\lambda} - \zeta_{\text{line}}\, \boldsymbol{\lambda} + (\boldsymbol{\Omega} - \boldsymbol{\Omega}') \wedge \boldsymbol{\lambda} \tag{5.5}$$

where $\boldsymbol{\Omega} = \nabla \wedge \mathbf{u}$ is the local angular velocity (to be distinguished from the actual vorticity $\boldsymbol{\omega} = \frac{1}{2}\nabla \wedge \mathbf{u}$) and ζ_{line} is given by equation (5.1). Thus the second term in (5.5) is nonlinear. The rotational terms $\boldsymbol{\Omega} - \boldsymbol{\Omega}'$, which we will sometimes refer to as the effective rotation, $\mathbf{R}$, represent the net rotation seen by the element in the strain basis. The first two terms in (5.5) cause the material element to align with the principal strain axis with the largest strain rate, namely $\boldsymbol{\xi}_1$. This claim can be justified by the following argument. Setting $\mathbf{R} = 0$ and setting $\gamma_i = s_i - \zeta_{\text{line}}$, equation (5.5) becomes

$$\frac{d}{dt}\lambda_i = \gamma_i\ \lambda_i \tag{5.6}$$

Thus the components of the material element are instantaneously either growing or shrinking exponentially. Furthermore the incompressibility condition, the ordering $s_1 \leq s_2 \leq s_3$ and the condition $\zeta_{\text{line}} \leq s_1$ immediately gives the inequality

$$0 \leq \gamma_1 \geq \gamma_2 \geq \gamma_3 \leq 0 \tag{5.7}$$

Thus λ_1 always grows to 1 and λ_3 always shrinks to zero. The complexity lies in the behaviour of λ_2 - it will grow if s_2 is positive and shrink if s_2 is negative. This can be seen as follows. After transients have died out equation (5.7) will ensure that

$$\lambda_1^2 \leq \lambda_2^2 \leq \lambda_3^2 \tag{5.8}$$

For s_2 positive it then follows that

$$\begin{aligned} \gamma_2 &= s_2 - (s_1\lambda_1^2 + s_2\lambda_2^2 + s_3\lambda_3^2) \\ &\geq -s_1\lambda_1^2 - s_2\lambda_2^2 \\ &\geq -(s_1 + s_3)\lambda_3^2 = s_2\lambda_3^2 \geq 0 \quad . \end{aligned} \tag{5.9}$$

A similar argument carries through if s_2 is negative, namely

$$\begin{aligned} \gamma_2 &= s_2 - (s_1\lambda_1^2 + s_2\lambda_2^2 + s_3\lambda_3^2) \\ &\leq -s_1\lambda_1^2 - s_3\lambda_3^2 \\ &\leq -(s_1 + s_3)\lambda_1^2 = s_2\lambda_1^2 \leq 0 \quad . \end{aligned} \tag{5.10}$$

Thus in absence of an overall rotation $\mathbf{R}$ the instantaneous sign of $s_2(t)$ determines the extent to which $\boldsymbol{\lambda}$ attains perfect alignment with $\boldsymbol{\xi}_1$. Thus if s_2 is positive the alignment will be "poor" (since there is a tendency to align towards $\boldsymbol{\xi}_2$ as well) whereas if s_2 is negative the alignment will be "good". Although our model is purely kinematical and the above arguments ignore the effect of rotation, it seems to be consistent with the results of various dynamical calculations. Thus in the work of Ashurst *et al.* the average value $\overline{s_2}$ is positive and the alignment of vorticity with $\boldsymbol{\xi}_1$ is found to be relatively poor.

Alignment equations analogous to (5.5) can also be derived for the vorticity. By taking the curl of the Euler equations (here we work with $\boldsymbol{\Omega} = \nabla \wedge \mathbf{u}$ rather than $\boldsymbol{\omega}$) one finds that

$$\frac{d\boldsymbol{\Omega}}{dt} = \frac{\partial \boldsymbol{\Omega}}{\partial t} + (\mathbf{u}.\nabla)\ \boldsymbol{\Omega} = (\boldsymbol{\Omega}.\nabla)\ \mathbf{u} = \mathbf{A}\ \boldsymbol{\Omega} \tag{5.11}$$

which is identical in form to the evolution equation for material lines (4.3). For an infinitesimal vorticity element in the strain basis one can introduce the direction cosines $\boldsymbol{\mu} = (\mu_1, \mu_2, \mu_3)$ representing the angles between the unit vectors associated with $\boldsymbol{\Omega}$ and the strain axes and define a local stretching rate

$$\zeta_{\text{vort}} = \sum_{i=1}^{3} s_i\ \mu_i^2 \tag{5.12}$$

Following the same steps used in the derivation of (5.5) one obtains

$$\frac{d}{dt}\,\boldsymbol{\mu} = \begin{pmatrix} s_1 & 0 & 0 \\ 0 & s_2 & 0 \\ 0 & 0 & s_3 \end{pmatrix} \boldsymbol{\mu} - \zeta_{\text{line}}\,\boldsymbol{\mu} + \boldsymbol{\Omega}' \wedge \boldsymbol{\mu} \tag{5.13}$$

An important difference between (5.5) and (5.13) is, of course, the absence of the angular velocity term $\boldsymbol{\Omega}$ since, by definition, $\boldsymbol{\mu}$ and $\boldsymbol{\Omega}$ are parallel.

Comparison of the two alignment equations leads to some interesting conclusions. In the case of both line and vorticity stretching we suppose that the alignment is with some combination of $\boldsymbol{\xi}_1$ and $\boldsymbol{\xi}_2$, that is we can write

$$\boldsymbol{\mu} \approx \mu_1\,\boldsymbol{\xi}_1 + \mu_2\,\boldsymbol{\xi}_2 \tag{5.14a}$$

and

$$\boldsymbol{\lambda} \approx \lambda_1\,\boldsymbol{\xi}_1 + \lambda_2\,\boldsymbol{\xi}_2 \tag{5.14b}$$

where, on the basis of Ashurst *et al.* we expect $\mu_2 > \mu_1$ in (5.14a) and, from Girimaji and Pope, we expect $\lambda_1 > \lambda_2$ (in (5.14b). The difference between (5.5) and (5.13) lies in the term $\boldsymbol{\Omega} \wedge \boldsymbol{\lambda}$, which from (5.14) can be estimated as

$$\boldsymbol{\Omega} \wedge \boldsymbol{\lambda} \approx (\lambda_1\mu_2 - \lambda_2\mu_1)\,\boldsymbol{\xi}_3.$$

From the above comments about the relative sizes of the $\mu_{1,2}$ and $\lambda_{1,2}$ we can expect this term to be non negligible (whereas if both $\boldsymbol{\Omega}$ and $\boldsymbol{\ell}$ aligned with $\boldsymbol{\xi}_1$ we would expect it to be smaller). Since the contribution to $\boldsymbol{\Omega} \wedge \boldsymbol{\lambda}$ is through $\boldsymbol{\xi}_3$ this will result in fluctuations in λ_3 and hence, since $s_3\lambda_3^2$ is negative definite, a reduction in ζ_{line}. Thus we make the prediction

$$\zeta_{\text{line}} \leq \zeta_{\text{vort}}$$

A rigorous proof of this conjecture is a nice theoretical challenge.

The alignment equation (5.5) is purely kinematical in origin and hence may be of value in analyzing the alignment and rotation questions we raised earlier for dynamical systems in general. The vorticity alignment equation (5.13), however, has an essential piece of dynamics built in since the starting point of its derivation, equation (5.11), makes use of the Euler equations. In order to make further comparisons of line and vorticity stretching in turbulence it is necessary to introduce the Navier-Stokes dynamics which enters into the alignment equations through expression (5.4) for the strain rotation $\boldsymbol{\Omega}'$. By taking the antisymmetric and symmetric gradients, respectively, of the Navier-Stokes equations one can derive the general evolution equations for angular velocity (vorticity) $\boldsymbol{\Omega} = \boldsymbol{\Omega}(\mathbf{x}, t)$

$$\frac{d\boldsymbol{\Omega}}{dt} = \mathbf{S}\ \boldsymbol{\Omega} + \nu \nabla^2 \mathbf{u} \tag{5.15}$$

and strain $\mathbf{S} = \mathbf{S}(\mathbf{x}, t)$

$$\frac{d\mathbf{S}}{dt} = -\left(\mathbf{S}^2 + \boldsymbol{\Omega}^2\right) - \frac{1}{\rho}\mathbf{P} + \nu \nabla^2 \mathbf{S} \tag{5.16}$$

where $\mathbf{P}$ is the pressure Hessian, *i e.* $\mathbf{P}_{ij} = \partial^2 p/\partial x_i x_j$. Using (5.16) in (5.4) leads to

$$\boldsymbol{\Omega}'_k = \varepsilon_{ijk} \frac{-\boldsymbol{\Omega}_i \boldsymbol{\Omega}_j - \frac{1}{\rho} P_{ij} + \nu D_{ij}}{s_i - s_j} \tag{5.17}$$

where $\boldsymbol{\Omega}_i = \boldsymbol{\Omega}\ \boldsymbol{\xi}_i$, $P_{ij} = (\boldsymbol{\xi}_i, \mathbf{P}\boldsymbol{\xi}_j)$ and $D_{ij} = (\boldsymbol{\xi}_i, (\nabla^2\mathbf{S})\boldsymbol{\xi}_j)$. Thus overall we can write

$$\boldsymbol{\Omega}' = \boldsymbol{\Omega}'_\Omega + \boldsymbol{\Omega}'_P + \boldsymbol{\Omega}'_\nu \tag{5.18}$$

where $\boldsymbol{\Omega}'_\Omega$, $\boldsymbol{\Omega}'_P$, $\boldsymbol{\Omega}'_\nu$ represent the contributions of, respectively, vorticity, pressure and dissipation to the strain rotation. It is important to note that in $\boldsymbol{\Omega}'_\Omega$ the contribution is quadratic in the vorticity.

Before proceeding a few words about the nature of the pressure field, $p = p(\mathbf{x}, t)$ are in order. From (3.16) we recall that the Laplacian of the pressure is the difference of the local strains and angular velocity (vorticity), *i.e.* $\nabla^2 p = \sum_i(\boldsymbol{\Omega}_i^2 - s_i^2)$. Treating the right hand side as a "charge" the equation can be solved by standard techniques to give

$$p\ (\mathbf{x}, t) = \frac{1}{4\pi} \int \frac{\sum_i \left(\boldsymbol{\Omega}_i^2(\mathbf{x}', t) - s_i^2(\mathbf{x}', t)\right)}{\|\mathbf{x} - \mathbf{x}'\|}\, d\mathbf{x}' \tag{5.19}$$

emphasizing the long range nature of pressure (the integrand is slowly decaying) and its quadratic dependence on the strain and vorticity of nearby regions. The P_{ij} needed to evaluate the $\boldsymbol{\Omega}'_P$ contribution to $\boldsymbol{\Omega}'$ are found by suitable differentiation of (5.19). The interplay between the local quadratic contributions of vorticity to $\boldsymbol{\Omega}'$, through $\boldsymbol{\Omega}'_\Omega$, and the corresponding nonlocal ones, through $\boldsymbol{\Omega}'_P$, is a fascinating and poorly understood problem.

Recognition of the different contributions to $\boldsymbol{\Omega}'$ is important. In analyzing their results, Girimaji and Pope attempt to distinguish between vorticity and strain rotation contributions to non-persistent straining by modifying the basic line stretching equation in the following ways: *(i)* to isolate the effect of vorticity they change (4.3) to

$$\frac{d\,\boldsymbol{\ell}}{dt} = \overline{\mathbf{A}}(0)\ \boldsymbol{\ell} \tag{5.20}$$

where $\overline{\mathbf{A}}(0)$ is some mean stationary straining field estimated from the numerical data - the idea here is that by "freezing" $\mathbf{A}$, the only effect that can affect stretching is vorticity and *(ii)* to isolate the effects of strain rotation they eliminate the effect of vorticity by computing with

$$\frac{d\,\boldsymbol{\ell}}{dt} = \left(\mathbf{A} + \mathbf{A}^{\dagger}\right)\,\boldsymbol{\ell} \tag{5.21}$$

However, what (5.18) tells us is that strain rotation is itself quadratically dependent on vorticity (both locally and non-locally) and hence use of (5.20) does not eliminate the role of strain rotation. Considerations such as these and identification of the term $(\boldsymbol{\Omega} - \boldsymbol{\Omega}') \wedge \boldsymbol{\lambda}$ as the agent for misalignment in (5.5) suggests more precise ways of estimating the various contributions to the non-persistence of strain. We have proposed (Dresselhaus and Tabor (1991)) the following dimensionless quantities

$$\begin{aligned} Q_{\Omega}(t) &= \|\boldsymbol{\Omega} \quad\wedge\ \boldsymbol{\lambda}\ \|\ /S \\ Q'_{\Omega}(t) &= \|\boldsymbol{\Omega}'_{\Omega}\ \wedge\ \boldsymbol{\lambda}\ \|\ /S \\ Q'_{P}(t) &= \|\boldsymbol{\Omega}'_{P}\ \wedge\ \boldsymbol{\lambda}\ \|\ /S \\ Q'_{\nu}(t) &= \|\boldsymbol{\Omega}'_{\nu}\ \wedge\ \boldsymbol{\lambda}\ \|\ /S \end{aligned}$$

where $S = \|\ \mathbf{S}\,\boldsymbol{\lambda} \wedge \boldsymbol{\lambda}\ \|$. Thus $Q_{\Omega}(t)$ measures the contribution of "pure" vorticity, $Q'_{\Omega}(t)$ measures the effect of vorticity through strain rotation, $Q'_{P}(t)$ the effects of pressure and $Q'_{\nu}(t)$ the effects of viscous terms.

The equations for vorticity and strain evolution (5.15) and (5.16) can be rewritten in the strain basis and, with simplification, take the form

$$\frac{ds_j}{dt} = -s_j^2 + (|\boldsymbol{\Omega}|^2 - \boldsymbol{\Omega}_j) - P_{jj} \qquad j = 1, 2 \tag{5.22}$$

$$\frac{d\boldsymbol{\Omega}_i}{dt} = (s_i + \alpha_i)\ \boldsymbol{\Omega}_i + \beta_i \qquad i = 1, 2, 3 \tag{5.23}$$

where

$$\alpha_i = \sum_{j \neq i} \frac{\boldsymbol{\Omega}_j^2}{s_j - s_i} \tag{5.24}$$

and

$$\beta_i = -(\boldsymbol{\Omega}'_P \wedge \boldsymbol{\Omega}) = \sum_{j \neq i} \frac{\boldsymbol{\Omega}_j^2}{s_j - s_i} \tag{5.25}$$

In (5.22) we need only consider the evolution of s_1 and s_2 since $s_3 = -(s_1 + s_2)$. From (5.23) and (5.24) we a see a significant concept, namely that in vorticity stretching the rotation of the strain axes due to vorticity actually plays the role of a nonlinear stretching term, not as a rotation. Since this contribution is quadratic in the vorticity it plays a significant role.

Study of equations (5.22) and (5.23) leads to the following qualitative description of vortex stretching when viewed from the Lagrangian frame. Here I reproduce the argument given in Dresselhaus and Tabor (1991) (a similar scenario is briefly summarized in She et al. (1991)). Suppose that at some time t the intermediate strain is approximately zero *i.e.* $s_2(t) \approx 0$, and that the (local) vorticity is weak, *i.e.* $\boldsymbol{\Omega} \approx 0$. At this instant the nonlinear stretching rates α_i are small and the vorticity will begin to stretch in the $\boldsymbol{\xi}_1$ direction. From equation (5.22) it follows that since $s_2 \approx 0$ the stretched $\boldsymbol{\xi}_1$ component of vorticity will cause s_2 to grow and become positive resulting in an increase in $\boldsymbol{\Omega}_2$. As a consequence of this sequence of events α_1 will become negative and large and will quickly act to nullify the previous growth in $\boldsymbol{\Omega}_1$. Since s_2 is still small compared to s_1 the $\boldsymbol{\Omega}_2$ stretching will proceed for a longer time than the initial stretching of $\boldsymbol{\Omega}_1$ due to s_1. This relative long-time persistent alignment with $\boldsymbol{\xi}_2$ survives until $\boldsymbol{\Omega}_2$ itself becomes large at which point contraction of vorticity in the $\boldsymbol{\xi}_3$ direction becomes significant (due to the quadratic contribution of $\boldsymbol{\Omega}_2$ to the always positive non-linear stretching rate α_3). The net result is that now α_2 now becomes negative and large. This suppresses the previous stretching of the intermediate vorticity $\boldsymbol{\Omega}_2$. The alignment with the strong contracting direction causes a quick but not total destruction of the previous vorticity build up. It should be emphasized that the relatively small magnitude of s_2 implies that that the vortex stretching in the intermediate direction is the most long lived of the stretchings described in this scenario. This sequence of events is observed to be robust in a variety of simulations (Dresselhaus 1991) - even in the presence of random forms of the pressure terms and is consistent with vortex stretching results in turbulence simulations (Ashurst *et al.*, She *et al.* (1991)).

Acknowledgements

The author thanks Boris Khesin, Eliot Dresselhaus and Jean-Daniel Fournier for many valuable conversations and the AFOSR for support under grant AFOSR-90-0284.

This article was prepared at the Institute for Theoretical Physics, Santa Barbara which is supported by NSF Grant PHY89-04035.

References

AREF, H., 1984 Stirring by chaotic advection. *J. Fluid Mech.*, **143**, pp. 1-21.

ARNOLD, V.I. & KHESIN, B., 1992 Topological methods in hydrodynamics. *Ann. Rev. Fluid. Mech.*, **24**, pp. 145-166.

ASHURST, W.T. KERSTEIN, A.R. KERR, R.M. & GIBSON, C.H., 1986 Alignment of vorticity and scalar gradient with strain rate in simulated Navier-Stokes turbulence. *Phys. Fluids.*, **30**, pp. 2343-2353.

BATCHELOR, G.K., 1952 The effect of homogeneous turbulence in materials lines and surfaces. *Proc. Roy. Soc. (London)*, **A213**, pp. 349-366.

BATCHELOR, G.K. & TOWNSEND, A.A., 1956 Turbulent diffusion. Surveys in Dynamics, (G.K. Batchelor and R.M. Davies, eds.), 352-399. Cambridge University Press.

BENETTIN, G. GALGANI, L., GIORGILLI, A. & STRELCYN, J.-M., 1980 Lyapunov characterisitc exponents for smooth dynamical systems and for Hamiltonian systems; a method for computing all of them. *Meccanica*, **15**, pp. 9-21.

BERRY, M.V., BALAZC, N.L., TABOR, M. & VOROS, A., 1979 Quantum Maps. *Ann. Phys. N.Y.*, **122**, pp. 26-63.

CHAIKEN, J., CHEVRAY, R., TABOR, M. & TAN, Q.M., 1986 Experimental study of Lagrangian turbulence in a Stokes flow. *Proc. Roy. Soc. (London)*, **A** 408, pp. 165-174.

COCKE. W.J., 1969 Turbulent Hydrodynamic Line Stretching: Consequences of Isotropy. *Phys. Fluids*, **12**, pp. 2488-2492.

COCKE, W.J., 1971 Turbulent Hydrodynamic Line Stretching: The Random Walk Limit. *Phys. Fluids*, **14**, pp. 1624-1628.

DOMBRE, T., FRISCH, U., GREENE, J.M., HÉNON, M., MEHR, A. & SOWARD, A.M., 1986 Chaotic streamlines in the ABC flows. *J. Fluid Mech.*, **167**, pp. 353-391.

DRESSELHAUS, E. & TABOR, M., 1989 The persistence of strain in dynamical systems. *J. Phys. A: Math Gen.*, **22**, pp. 971-984.

DRESSELHAUS, E. & TABOR, M., 1990 The geometry of Lagrangian orbits. "Topological Fluid Mechanics", Proceedings IUTAM Symposium, Cambridge, August 13-18,1989, Eds: H.K. Moffatt & A. Tsinober, Cambridge University Press.

DRESSELHAUS, E. & TABOR, M., 1991 The stretching and alignment of material elements in general flow fields. *J. Fluid Mech.*, accepted for publication.

DRESSELHAUS, E., 1991 Ph.D. thesis. Columbia University, New York.

DRITSCHEL, D.G., 1990 The stability of elliptical vortices in an external straining flow. *J. Fluid Mech.*, **210**, pp. 233-261.

FEINGOLD, M., KADANOFF, L.P. & PIRO, O., 1988 Passive scalars, three dimensional volume-preserving maps and chaos. *J. Stat. Phys.*, **50**, pp. 529-565.

GIRIMAJI, S.S. & POPE, S.B., 1990 Material element deformation in isotropic turbulence. *J. Fluid Mech.*, **220**, pp. 427-458.

GOLDHIRSCH, I., SULEM, P.-L. & ORSZAG, S.A., 1987 Stability and Lyapunov stability of dynamical systems: a differential approach and a numerical method. *Physica*, **27D**, pp. 311-337.

MAJDA, A., 1991 Vorticity, turbulence, and acoustics in fluid flow. *SIAM Review, to appear.*

MOFFATT, H.K. & TSINOBER, A., 1992 Helicity in Laminar and Turbulent Flow. *Ann. Rev. Fluid. Mech.*, **24**, pp. 281-312.

MONIN, A.M. AND YAGLOM, I., 1975 Statistical Fluid Mechanics. *MIT press* **2**.

ORSZAG, S.A., 1970 Comments on: "Turbulent hydrodynamic line stretching: consequences of isotropy". *Phys. Fluids*, **13**, pp. 2203.

OSELDEC, V.I., 1968 A multaplicative ergodic theorem. Lyapunov characteristic numbers for dynamical systems. *Trans. Moscow Math. Soc.*, **19**, pp. 197.

OTTINO, J.M., 1989 The Kinematics of Mixing: Stretching, Chaos and Transport. *Cambridge University Press.*

ROM-KEDAR, V., LEONARD, A. & WIGGINS, S., 1990 An analytical study of transport, mixing and chaos in an unsteady vortical flow. *J. Fluid Mech.*, **214**, pp. 347-394.

SHE, Z.-S., JACKSON, E. & ORSZAG, S.A., 1991 Structure and Dynamics of Homogeneous Turbulence: Models and Simulations, *Proc. Roy. Soc.*, **A434**, pp. 101.

STUART, J.T. & TABOR, M., EDS., 1990 The Lagrangian Picture of Fluid Dynamics. *Phil. Trans. Roy. Soc.*, **A333**.

TABOR, M. & DE GENNES, P.G., 1986 A cascade theory of drag reduction. *Europhys. Lett.*, **2**, pp. 519-522.

TABOR, M., 1989 Chaos and Integrability in Nonlinear Dynamics: An Introduction. *John*

Wiley, New York.

TOWNSEND, A.A., 1951 The diffusion of heat spots in isotopic turbulence. *Proc. Roy. Soc. (London)*, **A209**, pp. 418-430.

TRUESDELL, C., 1954 The Kinematics of Vorticity. *Indian University Press.*

VIEILLEFOSSE, P., 1982 Local interaction between voricity and shear in a perfect incompressible fluid. *J. Physique*, **43**, pp. 837-842.

VIEILLEFOSSE, P., 1984 Internal motion of small element of fluid in an inviscid flow. *Physica* **125 A**, pp. 150.

FAST DYNAMO THEORY

Stephen Childress
New York University
Courant Institute of Mathematical Sciences
251 Mercer Street
New York, NY 10012

ABSTRACT.
Basic ideas of dynamo theory are reviewed, the emphasis being on kinematic induction in the limit of infinite magnetic Reynolds number. The distinction between a dynamo in a perfect conductor, and one in a "near perfect" conductor of large but finite conductivity, is emphasized. The analysis of the "perfect" fast dynamo is identified with the geometrical problem of average orientation of material lines in a stretching flow. Constructions based upon baker's maps and pulsed flows in the unsteady case, and stochastic webs in the steady case, are reviewed. Finally, analytical problems arising in the perfect case, mainly associated with the non-existence of smooth eigenfunctions, are discussed. It is concluded that, while fast dynamos are very likely ubiquitous objects among three-dimensional fluid motions, and the examples are convincing of accessible kinematic mechanisms, there is as yet no proper theoretical setting for rigorously establishing perfect fast dynamo action.

1. Kinematic Dynamos Slow and Fast

1.1. INTRODUCTION TO KINEMATIC DYNAMO THEORY

1.1.1. A Brief History. This paper will focus on the so-called *fast* branch of a subject, *dynamo theory*, which is itself a small part of magnetohydrodynamics or MHD. A classic introduction to dynamo theory is contained in a little book on MHD by Cowling (1957) which still provides a good discussion of the astrophysical problems which motivated the early work on models of the magnetic fields of the Earth and Sun. While we cannot here do justice to the numerous research directions that have grown out from these origins, and particularly to the explosion of work in the last forty years, there are fortunately two excellent books (Moffatt, 1978, and Krause and Rädler, 1981), treating primarily the slow dynamo, where the reader can find extensive surveys. Among the many shorter review papers, the most recent (Roberts and Soward, 1991) includes a discussion of fast dynamos and an up-to-date bibliography.

H. K. Moffatt et al. (eds.), Topological Aspects of the Dynamics of Fluids and Plasmas, 111–147.

The paper is organized as follows. First we define and review some basic aspects of the slow and fast dynamo problems. The second section is devoted to examples of constructions aimed at discovering how steady and unsteady fast dynamos might work. We conclude with some rudiments of a theory of fast dynamo action based upon the kinematics of a perfectly conducting fluid. If there is a logical thread to the arguments below, it is that a goal of fast dynamo theory should be to find reasonable models using the perfectly conducting limit. In other words, we should evolve the magnetic field in a perfect conductor in such a way that it closely approximates the perfectly conducting limit of a real fluid. By reducing the problem to the perfect case, fast dynamo theory is essentially made to coincide with the kinematics of material lines. The dynamo aspect of the problem brings both line folding (i.e. some measure of line orientation) as well as line stretching into consideration. In this respect fast dynamo theory is the study of the average properties of *oriented* material lines.

In physical terms, dynamo theory seeks to understand how magnetic fields survive in natural dissipative systems such as a planet or a star. In the case of the Earth, there is little doubt that natural fluid motions are indeed generating a magnetic field within the electrically conducting fluid core. For, if these processes were not operating, the observed magnetic field would die away in about 10^5 years, or about 10^{-3} times the age of the Earth. It is thought that Newtonian fluid mechanics, together with Maxwell's equations, provides enough physics to understand the essential features of the geodynamo. For the Earth's dynamo then, we have a problem of MHD in a rotating fluid body. The solar magnetic field provides another familiar instance of dynamo activity, with the active magnetic cycle of 22 years providing the evidence that magnetic field is repeatedly created and destroyed. The physics of the solar photosphere is of course quite different from that of the Earth's fluid core, but again MHD approximations (for a compressible fluid) give a reasonable first approximation.

Astrophysical systems are typically large, much larger than typical diffusion scales, and so in a sense they are nearly free of dissipation; that is, they are "nearly perfect" systems. Slow dynamo theory, we emphasize at once, embraces dissipation– at least what there is–as a basic ingredient of the dynamo cycle. The role of diffusion in slow dynamo action is complex, not only altering the evolution of the magnetic field but also removing small-scale "noise". Fast dynamo theory, on the other hand, takes the opposite view, embracing the dissipation-free environment as a proper setting for discussing dynamo action in astrophysics. New, interesting problems arise from this viewpoint. A common theme in applied fluid dynamics is the use of carefully chosen solutions of a perfect system to represent solutions of the corresponding system with slight dissipation. Applied to dynamos, this is essentially the problem we want to address here.

The development of the dynamo theory of natural magnetism has been largely determined by the mathematical tools available. Early efforts to build working fluid dynamos naturally adopted a kinematic model (given fluid motion) and the simplest geometry– namely fields with axial symmetry–to lower the dimension of the underlying space. However, Cowling (1934,1957) showed that a dynamo cannot maintain itself if the magnetic field has this symmetry. Cowling's theorem provides an early example where the issue under discussion is the geometry or topology of the desired solution. (A modern example of such a theorem, excluding a class of slow dynamos from the ranks of fast dynamos, will be considered in section 1.2.4). It turns out that it is rather easy to construct dynamos in the kinematic approximation, if the *velocity* is axisymmetric, provided the magnetic field is allowed to break the symmetry. But real, nearly perfect flows tend to be turbulent, so more complicated motions have been studied as representative of natural dynamos.

Notable among the explicit constructions was the proposal of Parker to model the solar and terrestrial fields explicitly using a multiscale model, and the later realization of Parker's model in the context of nearly axisymmetric, nearly perfect systems by Braginsky (see e.g. Moffatt 1978 for references). Multiscale methods subsequently provided a partial estimate of the "density" of slow kinematic dynamos in the space of flows. Roberts (1970) stated such a result for spatially-periodic, square integrable flows: almost all flows will work!

A second basic topic in dynamo research deals with the MHD dynamos, i.e. dynamos which are self-consistent in their dynamics. This problem is far more difficult than the kinematic problem, the latter being now a special case arising for sufficiently weak magnetic fields.[1] The construction of self-consistent models has as its object the explicit modeling of the main observed features of, say, the geodynamo, at the approximate descriptive level of atmospheric models of climate. Modern computers are close to making this possibility a reality, although a number of unanswered theoretical question remain. The present discussion will deal with only the kinematic theory, but it is important to keep in mind that self-consistent fast dynamos, which

[1] The most common criticism put to kinematic dynamo theorists is to justify a theory that takes no account of the back-reaction of the induced magnetic field! A word of support for the kinematic approximation is therefore called for, as it is in some respects the more basic problem. There are thermodynamic constraints on the magnetic energy which can be produced in any MHD system. Thus the growth of the field *must* be arrested, and the level at which this occurs, as well as the nature of the equilibrated field, are among the predictions of the MHD theory. However the cause of the dynamo is really the instability of the fluid system with zero magnetic field to the addition of small seed field. In the early phase of the instability the magnetic field remains too small to interact dynamically, and kinematic theory is sufficient for the analysis of dynamo action. The situation is analogous to the use of linear stability theory at fixed points of a dynamical system to gain information about the existence of a limit cycle.

have yet to receive any attention, remain the ultimate goal of this work.

Before turning to a more precise formulation, we note that the fascination of kinematic dynamo theory in the limit of a nearly perfect medium comes from its strong focus on the *geometry* of flows. The subject bears the same relation to three-dimensional vector fields that the problem of thermal transport in fluids bears to scalar fields. It is far simpler than the dynamics of vorticity according to Euler's equations, but provides similar challenges to the imagination.

1.1.2. The Induction Equation and the Perfectly Conducting Limit. A *kinematic dynamo* will be a vector velocity field $\mathbf{u}(\mathbf{x}, t)$ which allows dynamo-like solutions of the induction equation

$$Ł_R\mathbf{B} \equiv \mathbf{B}_t + \nabla \times (\mathbf{B} \times \mathbf{u}) - \frac{1}{R}\nabla^2\mathbf{B} = 0. \tag{1}$$

It is understood throughout that $\mathbf{B}$ is to be a divergence-free field. We have begun with an equation in dimensionless form; the important parameter R is called, in analogy with the central parameter of fluid mechanics, the *magnetic Reynolds number.*[2] It is defined by

$$R = \frac{UL}{\eta} \tag{2}$$

where U, L are scales of speed and length relevant to the flow $\mathbf{u} = (u, v, w) = (u_1, u_2, u_3)$, and η is a diffusivity for the magnetic field. We have taken $T = L/U$ as the characteristic time scale. If we assume that the magnetic field also varies on the scales L, T, then $\epsilon \equiv R^{-1}$ is an estimate of the size of the diffusion term $\nabla^2\mathbf{B}$ in (1) relative to the other terms.[3] We can also write

$$R = L^2/\eta T \tag{3}$$

which emphasizes the role of the time scale when the length scale is fixed. Either physically large systems or rapid movements can thus realize an almost dissipation-free environment.

The formal limit $R \to \infty$, or $\epsilon \to 0$, is called the *perfectly conducting limit.* When the fluid is a perfect conductor, (1) is replaced by

$$Ł_\infty\mathbf{B} \equiv \mathbf{B}_t + \nabla \times (\mathbf{B} \times \mathbf{u}) = 0. \tag{4}$$

[2] The induction equation (1) is derived in dimensional form from the laws of Ampère, Faraday, and Ohm. These are, respectively, $\nabla \times \mathbf{B} = \mu\mathbf{J}$, $\nabla \times \mathbf{E} = -\partial\mathbf{B}/\partial t$, and $\mathbf{J} = \sigma(\mathbf{E} + \mathbf{u} \times \mathbf{B})$. Here $\mathbf{E}$ is the electric field and $\mathbf{J}$ is the current density, with $\eta = 1/(\sigma\mu)$. The quantity $\sigma(\mathbf{u} \times \mathbf{B})$ is called the *induced current.*

[3] ϵ will also be used below as a general small parameter.

We then have an equation free of parameters. If dynamo-like solutions are obtained with (4), we say $\mathbf{u}$ is a *perfect fast dynamo.* A major issue below is the question of what constitutes dynamo-like solutions in the diffusive problem as well as in the perfectly-conducting limit.

If $\mathbf{l}(\mathbf{x},t)$ is an infinitesimal element of a material line for the flow $\mathbf{u}$ located at the point $\mathbf{x}$, then we have (as discussed in the article by Tabor elsewhere in this volume)

$$\mathbf{l}_t + \mathbf{u}\cdot\nabla\mathbf{l} \equiv d\mathbf{l}/dt = \mathbf{l}\cdot\nabla\mathbf{u}. \tag{5}$$

Using a vector identity, we see that for the case where $\nabla\cdot\mathbf{u} = 0$, i.e. the flow is divergence-free, $\mathbf{B}$ and $\mathbf{l}$ satisfy the same equation. For a general flow, a material density $\rho(\mathbf{x},t)$ can be introduced and the same result is true of $\mathbf{B}/\rho$. Thus the magnetic field vector normalized by density in the perfectly conducting limit satisfies the same equation as the material line element. Assuming that $\mathbf{u}$ is regular enough to insure the uniqueness of the solution to the initial-value problem for (4), we can express $\mathbf{B}$ in terms of the Lagrangian description of the flow. If $\mathbf{x}(\mathbf{a},t)$ is the Lagrangian variable, defined by

$$d\mathbf{x}/dt = u(\mathbf{x}(\mathbf{a},t),t), \quad \mathbf{x}(\mathbf{a},0) = \mathbf{a}, \tag{6}$$

then

$$l_i(\mathbf{a},t) = J_{ij}(\mathbf{a},t)l_j(\mathbf{a},0) \tag{7}$$

in terms of the Jacobian

$$J_{ij} = \partial x_i/\partial a_j(\mathbf{a},t). \tag{8}$$

Thus, in the divergence-free case

$$B_i(\mathbf{x}(\mathbf{a},t),t) = J_{ij}(\mathbf{a},t)B_j(\mathbf{a},0). \tag{9}$$

Often (9) is referred to as the *Cauchy* representation of the field. The flow lines of $\mathbf{B}$, called the *lines of force*, are thus material lines of the flow in the perfectly conducting limit. The lines of force may be visualized as carried about by the fluid, while undergoing stretching, twisting, and folding; magnetic field is thus said to be "frozen" into the deformable conductor.

One consequence of (9) is that the magnetic flux through a material surface is invariant under the flow. For the divergence-free case this can be understood as a consequence of the frozen-in field and the conservation of volume, i.e. line element (field) dotted into area element (flux surface) is conserved. This property will lead to simple maps of space which are clearly able to increase flux, see section 1.3.1.

The underlying equation (4) for $\mathbf{B}$ is identical in form to that satisfied by the *vorticity field* $\Omega \equiv \nabla\times\mathbf{u}$. The vorticity equation defines the perfect

Euler flows. This equation is nonlinear, as opposed to the linear equation determining the perfect fast dynamos. Of course Euler flows provide solutions of (4), but not necessarily the other way around.[4] Turbulence does not unfortunately give us ready examples of fast dynamos, since with finite viscosity a forced system (bounded work being done) will equilibrate with vorticity bounded from above. Thus the exponential growth of a dynamo solution is ruled out.

1.1.3. Slow and Fast Dynamos. We assume that R is finite and that (1) is to be solved for some $\mathbf{B}$ within some class of functions $\mathcal{B}$. Applications of kinematic dynamo theory can involve finite conductors, with appropriate boundary conditions added, but for the present discussion it is suitable to take the conductor to fill all of space and to choose $\mathcal{B}$ to be, for example, *steady* divergence-free square-integrable vector fields which are periodic functions of $\mathbf{x}$. We then consider solutions of (1) of the form $\mathbf{B} = e^{\lambda t}\mathbf{b}(\mathbf{x})$; thus

$$\nabla \times (\mathbf{u} \times \mathbf{b}) + \frac{1}{R}\nabla^2 \mathbf{b} = \lambda \mathbf{b} \tag{10}$$

determines the eigenfunction associated with the eigenvalue λ. The eigenparameter λ is just the *complex growth rate* of the magnetic field. We say that the solution is dynamo-like, and that $\mathbf{u}$ is a (kinematic) dynamo, provided that $Re(\lambda) > 0$. We are thus faced with a standard eigenvalue problem for an elliptic operator with variable coefficients determined by $\mathbf{u}$. A dynamo is a field $\mathbf{u}$ which causes the magnetic field to grow exponentially for any initial field with a nonzero projection onto the eigenfunction $\mathbf{b}$. Note that. along with the field, the magnetic energy E, defined over the domain periodicity domain V by

$$2E = \int_V B^2 dV, \tag{11}$$

also grows exponentially. The energy is being supplied by mechanical work against the electromagnetic Lorentz force $\mathbf{J} \times \mathbf{B}$. Even if $\mathbf{u}$ has the desirable property of having bounded kinetic energy for all time, within the kinematic theory the work which is done on the fluid grows exponentially when dynamo action occurs. In real systems this growth is arrested by the dynamical reaction of the magnetic field on the flow, which modifies the motion so as to remove it from the class of kinematic dynamos.[5]

[4] Assuming the uniqueness of the solution, whenever $\mathbf{u}$ is an Euler flow then $\mathbf{B} = \Omega$ for all time if it does so initially.

[5] The case of *stationary* dynamos $Re(\lambda) = 0$ is also acceptable in the slow theory. However it is not useful in the context of fast dynamos since analysis based upon the perfectly conducting limit, where all fields are frozen in, makes even a rigid conductor a stationary dynamo.

Let us suppose that, for some fixed $\mathbf{u}$, dynamo action occurs for some initial field $\mathbf{B}(\mathbf{x},0)$ (which may depend upon R), for all sufficiently large values of R. Let $\lambda(R)$ be the maximal growth rate of the field. We then say that $\mathbf{u}$ is a *fast dynamo*, provided that

$$\lim_{R\to\infty} Re(\lambda) > 0. \tag{12}$$

It is easily shown that $Re(\lambda) = O(1)$ in the limit, but many examples of dynamos are known for which the above limit is zero. In the latter case we say the dynamo is *slow*. The terminology thus refers to growth relative to the natural time scale L/U of the flow. If the limit is positive, we say the dynamo is *fast*. There is a subclass of fast dynamos which are less deserving of the title because of the manner in which diffusion continues to operate in the limit. These *intermediate fast dynamos* are discussed in section 1.2.2.

We should remark that the term *fast* is used here in a rather restrictive sense. In other problems, e.g. thermal convection and MHD reconnection, another terminology is used, wherein a process is fast if it occurs on a time scale which is much smaller than a molecular diffusion time (here L^2/η). In this setting some formally slow dynamos would qualify as fast. This larger class of fast processes is probably the interesting class in physical terms, but we prefer to adhere to a stricter interpretation in order to emphasize the critical nature of significant exponential stretching.

For general $\mathbf{u}$, steady or unsteady, we may assume that the kinetic energy is bounded for all time (to prevent spurious dynamo action) and replace the eigenparameter by

$$\lambda(R) = \lim_{t\to\infty} \frac{1}{2t} \log E. \tag{13}$$

in the above definitions of slow and fast dynamo action.[6]

1.2. HELICAL FLOWS AND GENERALIZATIONS

1.2.1. The Roberts Cell as the Prototype of a Diffusive Dynamo. We turn to a representative computation of dynamo action, using as our motion one of a family of two-dimensional flows studied by G.O. Roberts (1972). The flow is two-dimensional in the sense that the velocity field is independent of one coordinate (z), but the velocity field has three nonzero components. It is a divergence-free flow and has the form

$$\mathbf{u} = (\psi_y, -\psi_x, K\psi), \quad \psi = \sin x \sin y. \tag{14}$$

[6] In cases were the limit of $Re(\lambda)$ fails to exist, it is sufficient to define fast dynamos in terms of the supremum limit, as in (13). This insures that regardless of how large R may be, there are still "windows" of growth at a rate which is bounded from below by a positive number.

Here K is an arbitrary constant. Note that $\nabla \times \mathbf{u} = (K\psi_y, -K\psi_x, 2\psi)$ so that, if $K = \sqrt{2}$, vorticity is everywhere parallel to velocity, which defines a *Beltrami field.* Beltrami fields have some nice properties in the context of slow dynamo theory; their relevance to fast dynamo theory is less clear, although we will use them in section 2.2. Such fields are prototypes of flows with constant *helicity* (discussed elsewhere in this Workshop), reflected in (14) by the helical particle paths. Adjacent cells have opposite circulations in the $x - y$ plane, but also opposite *vertical* motions parallel to the z-axis, so all helical paths have the same orientation and thus the same sign of helicity.

This cellular flow, being independent of z, allows separation of variables and we may seek solutions of (1) in the form $\mathbf{B}(x, y, z, t) = e^{ijz+\lambda t}\mathbf{b}(x, y)$ where j is the wavenumber. For arbitrary R numerical methods must be used to find $\mathbf{b}$, and Roberts (1972) gave results for values of R up to 64. For any fixed j, the growth rate is negative, then rises to a positive maximum, then decreases to zero as R is increases. At any R there is a j of maximal growth, which increases with R; the maximal growth rate decreases slowly with R. These calculations suggest that as R increases the dynamo tends to produce field on smaller and smaller length scales (as determined by the j of maximal growth), but that ultimately the growth decays to zero at infinite R. It turns out that both of the conclusions are correct, but their verification involves considerable analysis and the decay of growth rate at large R is extremely weak. Soward (1987) studied the problem asymptotically for large R, using boundary layer theory (see also Childress 1979). His analysis yields a j of maximal growth of order $(R/\log R)^{1/2}$ and a maximal growth rate of order $\log(\log R)/\log R$. The latter estimate indicates that the dynamo is slow, but the decay is sufficiently weak to warrant a close look at the physical mechanisms involved.

Suppose that we assume a nonzero x-component of the mean field, $\bar{B}_x \neq 0$, with R large. Since the field tends to be carried by the flow, the only way for it to avoid being wound up in the cells and subsequently dissipated is to form magnetic boundary layers near the separatrices of the cells, the lines $x = m\pi$, $y = n\pi$ for integer m, n,. These layers, of thickness $O(R^{-1/2})$ relative to the cell, must carry the assumed x-directed flux. In a steady system, the lines of force must cross separatrices parallel to the y-axis, and the flow then tends to pull them out as tongues of flux in the y direction. Since B_y changes sign along with the vertical velocity w as the separatrix is crossed, the first term $(\mathbf{u} \times \mathbf{B})_x = u_z B_y - u_y B_z$ contributes to the mean induced x-current. In the second term, the z-field developed by twisting the tips of the tongues creates an induced current from the y-motion on the separatrices. A mean current is thus realized in the thin boundary layers because of the formation of twisted tongues of flux.

The mean induced current, which is approximately proportional to the mean magnetic field, is said to be created by the *alpha effect* (Moffatt 1978). The alpha effect has been of limited use in analyzing fast dynamos, for reasons not really understood, and we do not pursue this method further. It is more revealing to treat the formation of flux tongues as if the process were time-dependent, using frozen field arguments. As Soward (1987) points out, the lifting and stretching of a line of force into the tongue, followed by the action of the vertical shear, brings together enhanced field of like sign, thus reinforcing the the z-structure. Keep in mind, however , that the process is really diffusive. It is only in the diffusive layers that flux field can survive the "winding up" created by the rotation within the cells. This winding up of interior field in the cells until dissipation removes it and only the layers remain is called *flux expulsion.* The expulsion occurs on a time scale $O(R^{1/3})$ (see e.g. Moffatt and Kamkar, 1983).

Because of this restriction of field to layers, the dynamo mechanism can be understood from the flow structure in the immediate neighborhood of the separatrix streamline $\psi = 0$. The latter is actually a lattice surface consisting of the planes $x = m\pi, y = n\pi$ for integer m, n. In the vicinity of the stagnation points the velocity is small and the time of residence of a fluid particle is large, of order $-\log|\psi|$. The horizontal "corner" motion can be understood by approximating the streamfunction (near (0,0)) by $\psi \approx xy$. One sees easily that small material line elements aligned initially with a streamline contract exponentially on entering and expand exponentially on leaving the corner. If such an element initially makes an angle with a streamline, the endpoint with the greater value of $|\psi|$ will move faster through a corner and so the element will undergo stretching. By this mechanism stagnation points (here X points) introduce large streamwise field components into the magnetic structure. (See in particular section 2.3.2.) In regions of the helical cell away from the separatrices, this large field would develop also large shear because of the winding up of flux during the expulsion phase. What remains in the boundary layers is the large streamwise field (or order $R^{1/2}$ times the spatially averaged field) which diffusion cannot remove.

Thus, while the dynamo is slow the details of the excitation at large R is closely linked to the structure of the flow in the separatrix region. Soward found, in fact, that the flow could be converted to a truly fast dynamo by modifying it within a small cylinders with axes on the stagnation lines, in such a way that the sojourn time at the corners for all fluid particles was finite uniformly for arbitrarily small ψ. This can be done in such a way that the velocity is continuous, the vorticity however becoming infinite.

1.2.2. The Ponomarenko Dynamo. In both the fast and near-fast behavior, Roberts' cell is however special in its magnetic structure, in that the vertical wave number of the magnetic mode grows with R and so the length scale of dominant magnetic structure shrinks to zero. This allows diffusion to continue to operate in an essential way in the perfectly conducting limit, contrary to the view of fast dynamo action as creating field on length scales of the flow (here, the cell size), as we emphasize below in section 1.3. Dynamos with this property have been called *intermediate* (Molchanov *et al.* 1985).

The simplest such dynamo is probably the Ponomarenko dynamo (see Gilbert, 1988, and Ruzmaĭkin *et al.*). The flow has a discontinuity on a cylindrical surface but otherwise is a cylindrically symmetric helical cell:

$$(u_r, u_\theta, u_z) = \begin{cases} (0, r\omega, 1), & \text{if } r < 1; \\ 0, & \text{if } r > 1. \end{cases} \tag{15}$$

If the discontinuity is smoothed, using velocity $(0, r\omega(r), W(r))$, the dynamo is strictly slow, with a growth rate of order $R^{-1/3}$ occurring with s magnetic mode having a wavenumber of order $R^{1/3}$. This is quite different from the Roberts cell, and reflects the absence of stagnation points in the Ponomarenko case. The resonant layers where dynamo action occurs are similar to the tearing modes of plasma instability theory. A fast dynamo mechanism based on this similarity has been discussed by Strauss (1986).

The Ponomarenko dynamo otherwise is a prototypical slow dynamo on a par with the Roberts cell. If a spatially-periodic analog is sought, however, stagnation points must be introduced and the flow resembles helical cells or their cat's eye extensions (see below). In such flows we can expect to see a combination of dynamo excitation at resonant layers of the helical structure, as well as "flux tongue" excitation at critical points.

1.2.3. Flows With Channels. There are many elaborations of the helical cell obtained by adding small perturbing components to the streamfunction given by (14). A general steady family of interest has the form (Childress and Soward, 1989, Soward and Childress, 1990)

$$\mathbf{u} = (\psi_y, -\psi_x, K\psi) + \epsilon(\cos\theta, \sin\theta, 0), \qquad \psi = \sin x \sin y + \delta \cos x \cos y. \tag{16}$$

Here δ, ϵ, θ are fixed parameters. Dynamo studies were carried out for small δ and zero ϵ, and the other way around. The first case converts the cells to "cat's eyes" which are separated by "channels". Within the channels flux expulsion cannot occur and there is in general intense magnetic field. In the second case channels also form, but now they can be filled densely (modulo the periodic cell) by a single particle path, provided that $\tan\theta$ is irrational. The calculations do not offer much hope for true fast dynamo action in these flows, although the the results to date are incomplete.

So what are the characteristics of slow dynamos (including intermediate cases)? While very few flows have been studied systematically for large R, it is common to all that diffusion remains an essential part of the process of excitation. In essence the simplicity of slow dynamos is allowed because the magnetic field structure can be significantly modified by diffusion. Generally a specific magnetic mode arising in the eigenvalue problem will exhibit a "window" of growth as R increases, and a value of R where the growth rate is largest. In the intermediate case, the R of maximal growth can be made arbitrarily large by taking modes of arbitrarily small spatial scale, but the "fine tuning" of the excitation by the diffusion is always in evidence.

Mathematically, the slow dynamo is characterized by negligible effect of time dependence of the absolute value of the field at large R, i.e. it is in this sense "quasi-steady". We have to be careful here since only the real part of the complex growth rate is involved in the definition of a slow dynamo. If we assume that the complex part of the growth rate *also* becomes small, the slow dynamo should have a dominant magnetic field $\mathbf{B}_0$ which, for large R, satisfies $\nabla \times (\mathbf{u} \times \mathbf{B}_0) = 0$. For steady flows, an obvious choice is $\mathbf{B}_0 = f\mathbf{u}$ where $\mathbf{u} \cdot \nabla f = 0$. Such examples encompass the dynamos of Braginsky (see Moffatt, 1978 and Soward, 1972), and a more general class studied by Soward ((1990). Of course a full characterization of slow dynamos is complementary to classification of the fast dynamos and this is one the outstanding problems of the field at this time.

1.2.4. Stretching Flows and Vishik's Theorem. A key result, due to Vishik (1989) concerns the role of *stretching* of line elements in fast dynamo action. We have seen in the paper of Tabor in this volume that an average stretching rate for a Lagrangian line element is measured by the Liapunov exponent

$$\lambda_{liap}(\mathbf{a}, \mathbf{e}) = \limsup_{t \to \infty} \frac{1}{2t} \log(\Lambda_{ij} e_i e_j) \tag{17}$$

where

$$\Lambda_{ij} = \frac{\partial x_k}{\partial a_i} \frac{\partial x_k}{\partial a_j} = J_{ki} J_{kj}. \tag{18}$$

Thus $\Lambda_{ij} e_i e_j$ is the square of the stretching factor at time t for a line element initially at $\mathbf{a}$ with direction$\mathbf{e}$. Vishik shows that if a smooth steady divergence-free flow in three dimensions has the property that λ_{liap} as defined by (17) is invariably zero, then the flow cannot be a fast dynamo. We shall say that a flow is a stretching flow if for some $\mathbf{a}$,$\mathbf{e}$ the Liapunov exponent is positive. So Vishik shows that flows which are not stretching can be at most slow dynamos.

The number of possible distinct values of λ_{liap} associated with a given point $\mathbf{a}$ equals the dimension of the space. We shall usually consider divergence-free flows and for these the sum of the λ_{liap}'s is zero, so a flow has stretching if some λ_{liap} is non-zero.. It is likely that *almost all* three-dimensional flows are stretching flows over a set of $\mathbf{a}$ of positive measure. A stretching flow need not provide a fast dynamo, however, as the Roberts cell shows. For the helical flow the Liapunov exponent obtained from elements parallel to the z-axis is invariably zero. Two points on a streamline within a cell have a periodic motion and so the streamwise $\mathbf{e}$ also yields zero. Since the sum is zero the choice of $\mathbf{e}$ orthogonal to a streamline also gives zero within a cell. This leaves the separatrices and stagnation points. On this set of measure zero stretching occurs, as is easily seen by considering two nearby streamwise points initially on a separatrix but not at a stagnation point. These points will flow into a stagnation point, and the distance between them will decay exponentially yielding a Liapunov exponent of -1. Hence the cross-stream components stretch with exponent +1. Similar results apply to the stagnation points under a limiting procedure. The stretching here is insufficient to produce a fast dynamo. In the classification introduced above, the Roberts cell is thus a intermediate case. But it is suggestive of belonging to the "boundary" of the set of fast dynamos. We are thus led to the working hypothesis that for fast dynamo action to occur, the domain of stretching should be broadened onto a set of positive measure, where Lagrangian orbits are in some sense ergodic. Thus flows exhibiting Lagrangian chaos become attractive as sources of fast dynamo action (Vainshteĭn and Zeldovich, 1972).

1.3. ORIGINS OF THE FAST DYNAMO

1.3.1. The STF Map. The archetype of the unsteady fast dynamo is a volume-preserving mapping of space in three dimensions which operates on a toroidal region as shown in Figure 1. Given that flux is conserved, we see that the mechanism, which involves three distinct steps of stretching (S), twisting (T), and folding (F), essentially doubles the flux in one sequence. The basic idea appears in an early paper by Alfvén (1950), who did not however make a distinction between slow and fast processes. Later Vainshteĭn and Zel'dovich (1972) emphasized this distinction and the importance of the twist to a fast dynamo. This allows doubling to occur without splitting the original loop into two daughter loops, a process which is taken to proceed more slowly.

While the STF process is compelling, it is not clear when or how a flow field will produce it. In particular the process is time-dependent, with no obvious steady-state counterpart. It does suggest the role of maps of space in setting up the construction of a fast dynamo, and emphasizes the reorienting of stretched field as the key ingredient. Finally, it suggest that magnetic structure on ever finer scales will develop in perfect fast dynamo action.

1.3.2. The Moffatt-Proctor Analysis. The last property would imply that in the perfectly conducting limit the eigenvalue problem for a steady fast dynamo cannot be studied in the traditional manner since smooth eigenfunctions do not exist. A direct attack on this eigenvalue problem (Moffatt

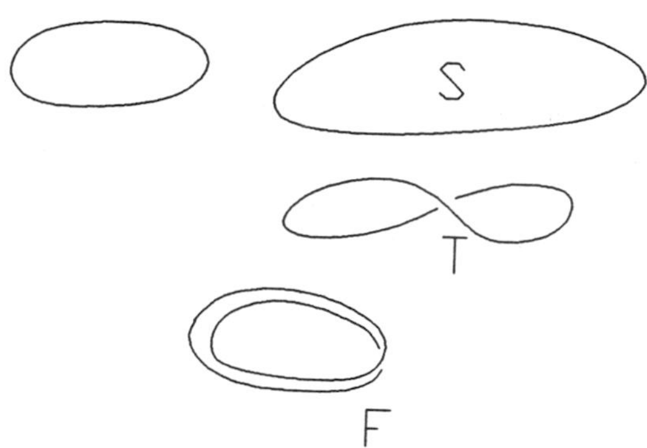

Fig. 1. The STF map. An oriented loop of flux is first stretched to twice its length, then twisted, and finally folded over to produce approximately the starting configuration, but with twice the flux in the double loop.

and Proctor (1985) encounters such a difficulty. The approach is to solve (10) using the vector potential $\mathbf{A}$, $\mathbf{b} = \nabla \times \mathbf{A}$. Upon integration, the arbitrary gradient can be absorbed and we have

$$\mathbf{u} \times (\nabla \times \mathbf{A}) = \lambda \mathbf{A}. \tag{19}$$

This implies that $\lambda \mathbf{A} \cdot \nabla \times \mathbf{A} = 0$. From this, Moffatt and Proctor deduce under various conditions that $Re(\lambda) = 0$. The conclusion is consistent with the fact that eigenfunctions do not in fact exist in a strong sense, and also that to treat the problem in classical terms one must introduce dissipation. To make the diffusive term essential, it was argued that $O(R^{-1/2})$ length-scales are introduced throughout the domain of magnetic activity. We shall see examples of this below.

1.3.3. Flows Which Stretch. The STF paradigm together with Vishik's theorem lead to the conjecture that flows where stretching occurs on finite volume should be good candidates for fast dynamo action. Arnold et al. (1981) (see also Arnold and Avez 1967) construct a steady stretching flow by redefining the problem on a special three-dimensional Riemannian manifold, so that a uniform *flow* on the manifold realizes a uniform *stretching* in Cartesian space. The model utilizes a torus map, and an identification of end tori for extension to all space. While the flow cannot be realized in physical space, it offers a novel approach to stationary fast dynamos with non-uniform stretching, and was the first construction to make use of an underlying mapping operation.

Stretching alone is not, however, sufficient for fast dynamo action. It is known that two-dimensional motion (particle paths all in parallel planes) cannot achieve dynamo action, fast or slow. Since two-dimensional stretching flows clearly exist, dynamo action comes from three-dimensional effects analogous to the "twist" step of the STF map. This suggests that in fast dynamo action we are dealing with a residual effect in which the stretching of field lines prevails over the partial cancellation coming from the folding of lines and dissipation on small spatial scales.

2. Examples of Fast Dynamos in Steady and Unsteady Flows

2.1. BAKER'S MAPS

2.1.1. Cut-and-Paste Models of the Perfect Fast Dynamo. The simplest kinematic models of unsteady perfect fast dynamo action utilize the conductor as a baker would work with pastry dough–stretching, cutting, and layering in a variety of ways. Finn and Ott have explored a number of examples, for steady as well as unsteady dynamo action (Finn and Ott, 1988a,b,1989). The models utilize virtual deformations in three-dimensional space but the final map is two-dimensional. Consider for example the unit cube with uniform field of magnitude 1 aligned with the x-axis. Let $(x_0 = 0, x_1, x_2, ..., x_{N-1} = 1)$ be a partition of unity which cuts the cube into N slices. Each slice is then stretched longitudinally until it has unit length, while preserving volume by compression transversely. The new slices are then stacked to restore the cube. The magnetic field in the slabs can be reversed during the stacking according to some fixed rule, determined by orientation factors $\epsilon_i = \pm 1$ (see Figure 2). If the magnetic field is initially unity everywhere, after one iteration the field strength (by conservation of flux in a perfect conductor) has strengths ϵ_i / p_i where $p_i = (x_i - x_{i-1})$. Continued iteration builds up a multilayered structure with field variations on arbitrarily small scales.

We shall use the fact that as a Lagrangian particle is tracked under iteration of this map, it will visit the N layers with probabilities p_i. [7] The Liapunov exponent can thus be computed as

$$\lambda_{liap} = -\sum_{i=1}^{N} p_i \log p_i \tag{20}$$

assuming that the time between applications of the map is normalized to unity. Because of the small-scale layering, it is natural to measure net field strength by the total flux in the cube. Since each of the N layers carry the

[7] This can be understood by noting that a particle picked at random will lie in slice m_1 with probability p_{m_1}. Depending upon its location within the m_1th slice, it will next visit slice m_2 with probability p_{m_2} etc.

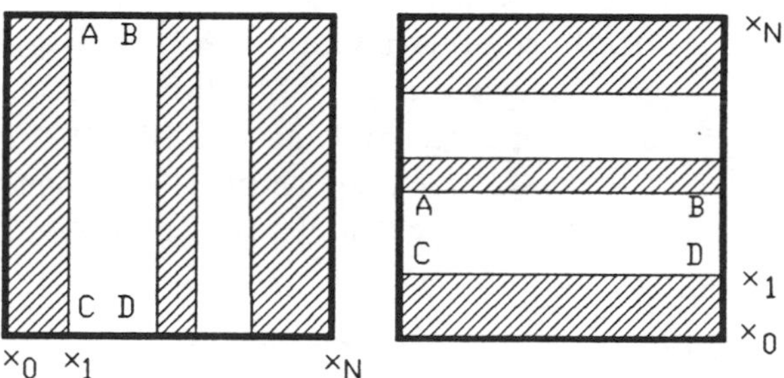

Fig. 2. A Baker's map. N slices are stretched tto the same length and stacked vertically with possible reversals of field direction.

same flux (up to an orientation), we see that flux increases at each step by the factor e^{Γ} where

$$\Gamma = \log |\sum_{i=1}^{N} \epsilon_i|. \tag{21}$$

Note that $\lambda_{liap} = \Gamma$ for $p_i = 1/N$ and ϵ_i of the same sign. Keeping like orientations, λ_{liap} will be smaller than Γ for any other choice of p_i. Thus nonuniform stretching makes average flux grow faster than the length of a *particular* line element.

On the other hand we may also have $\lambda_{liap} > \Gamma$ if not all e_i have the same sign; that is, the cancellation of flux coming from folding of the field can lower growth below the Liapunov exponent. It is thus possible, under conditions of both non-uniform stretching and folding, to obtain flux growth rates which are not too different from the Liapunov exponent.

Another measure of stretching focuses on the simultaneous action of the map on an extensive geometrical object, in this case a line. We observe that a line $y = y_0, z = z_0, 0 < x < 1$ is actually stretched to length N before being cut into N pieces of unit length. The growth rate of line length (as opposed to line element length) is a global measure of stretching which can (in this example) be identified with the *topological* or *KS entropy* of the map. (The use of the term entropy is suggested by (20).) Denoting this quantity by λ_{top}, we have

$$\lambda_{top} = \log N \tag{22}$$

in this example, where clearly $\lambda_{top} \geq \Gamma$.

2.1.2. Lagrangian Maps and Pulsed Flows. We consider briefly the analysis of maps in the perfectly conducting limit based on Lagrangian variables. Ideally we would like to start with a smooth flow $\mathbf{u}(\mathbf{x},t)$, then replace its effect on a magnetic field over some time interval $(0,t)$ by a map $M_t : \mathbf{a} \to \mathbf{x}$ on some volume of space onto itself, determined from Lagrangian variables as described in section 1.1.2. The cut and past maps are discontinuous Lagrangian maps, e.g. classic Baker's map with a single cut,

$$(x,y) \to \begin{cases} (2x, y/2), & \text{if } 0 \leq x < 1/2; \\ (2x-1, (y+1)/2), & \text{if } 1/2 \leq x < 1. \end{cases} \tag{23}$$

We have seen from the Lagrangian kinematics of the magnetic field that $\mathbf{B}(\mathbf{x},0)$ maps into $\mathbf{B}(\mathbf{x},t) \equiv G_t\mathbf{B}(\mathbf{x},0)$ where

$$G_t\mathbf{B}(\mathbf{x},0) = \mathbf{J}(M_t^{-1}\mathbf{x},t) \cdot \mathbf{B}(M_t^{-1}\mathbf{x},0), \tag{24}$$

This equation states that the magnetic field at a given point of the domain is, after application of the map, equal to the field at the pre-image of $\mathbf{x}$ over the flow for time T, multiplied by the Jacobian realized at time t for an element starting at $M_t^{-1}\mathbf{x}$.

The operator which is the formal *adjoint* of G_t will also be of interest to us. We define the formal adjoint L^+ of an operator L by $(L^+A, B) = (A, LB)$ for all A, B in the domain of L, where the complex inner product is used. In our case the latter will involve the dot product of two vector fields integrated over a volume V. Let us write $J_{ij}(\mathbf{a},t) = \tilde{J}_{ij}(\mathbf{a},\mathbf{x})$. Then we have

$$(\mathbf{A}, G_t\mathbf{B}) = \int A_i^*(\mathbf{x})\tilde{J}_{ij}(\mathbf{x},\mathbf{a})B_j(\mathbf{a})d\mathbf{x}. \tag{25}$$

We thus find that

$$G_t^+\mathbf{A}(\mathbf{a}) = \tilde{\mathbf{K}}(\mathbf{x},\mathbf{a}) \cdot \mathbf{A}(\mathbf{x}), K_{ij} = J_{ji}. \tag{26}$$

Thus the adjoint involves the transpose of the Jacobian, and the image point and pre-image are interchanged. But this is the same as letting the *inverse* of $\mathbf{K}$ evolve under the *time-reversed* flow which carries $\mathbf{x}$ to $\mathbf{a}$. To see this, note that $\tilde{K}_{ij}(\mathbf{a},\mathbf{x})\tilde{K}_{jk}(\mathbf{x},\mathbf{a}) = \delta_{ik}$ and therefore

$$G_t^+\mathbf{B}(\mathbf{a},0) = \mathbf{K}^{-1}(M_t\mathbf{a},t) \cdot \mathbf{B}(M_t\mathbf{a},0), \tag{27}$$

It can be shown (Bayly and Childress (1989)) that the adjoint is thus equivalent to evolving area rather than line elements under the time-reversed flow.

Lagrangian maps will be assumed to apply whenever a flow acts in a conductor for a finite time at very high values of R. A finite Lagrangian map (with length scale L say) can thus be realized in an arbitrarily short time interval T (since R is proportional to L^2/T). This allows diffusion to be split off from the motion and allowed to act only in time intervals of length T_D. This is what we shall refer to as a *pulsed flow*. The motion of the conductor occurs only time intervals of length T, separated by diffusion intervals of length T_D where the conductor is at rest. This is an old technique in dynamo theory which was applied by Backus (1958) in an early proof of diffusive dynamo action. The method is similar in spirit to modern techniques of operator splitting in numerical analysis of partial differential equations.[8]

Thus, to discuss perfect fast dynamo action in the unsteady case we can simply iterate a Lagrangian map. To study fast dynamo action in the perfectly conducting limit we can use pulsing and steadily reduce the value of $\eta T_D/L^2$,treating all maps as instantaneous.

2.1.3. The SFS Fast Dynamo. We consider now a model for unsteady fast dynamo action based upon a Baker's map, involving perfect cancellation of flux under a fold, but adding a shear in the otherwise ignorable z-direction. The magnetic field will, in addition, have a z-dependence, allowing the field to be modified by the shear. The motivation for the form of the map comes from the Roberts-Soward process operating in the magnetic boundary-layers of the steady helical cell, as discussed in section 1.2.1. Diffusion will be incorporated using the pulsing technique.

The stretch-fold-shear or SFS map is defined as follows (Bayly and Childress, 1988, 1989): it is a map of the unit cube into itself given by

$$(x,y,z) \to \begin{cases} (2x, y/2, z+\alpha(y-1)/2), & \text{if } 0 \le x < 1/2; \\ (2-2x, 1-y/2, z+\alpha(1-y)/2), & \text{if } 1/2 \le x < 1. \end{cases} \tag{28}$$

The action of this map on a piecewise constant field pointing in the x-direction in shown in Figure 3. Considered as a perfect dynamo the Lagrangian operator on the magnetic field under the SFS map will be denoted by G. (For pulsed flows the time subscript on G is extraneous.) We set $\mathbf{B} = (b(y), 0, 0)e^{2\pi i z}$ in which case

$$Gb = 2\,sgn(1/2-y)e^{-2\pi i\alpha(y-1/2)}b(\tau(y)) \tag{29}$$

where $\tau(y) = min(2y, 2-2y)$.

For large but finite R iterations of the map were separated by time intervals of unit duration, where $b(y)$ evolved according to the heat equation. To monitor fast dynamo action only the average of $b(y)$ was used. The numerical results indicated fast dynamo action above a critical threshold of

[8] Pulsing is easily seen to be consistent, in that the length scales which survive diffusion will behave as perfect modes during the pulses, provided only that $T \ll T_D$. For, the smallest magnetic scales remaining after diffusion are of size $L_D = \sqrt{\eta T_D}$. During the pulse, to use a Lagrangian map it is sufficient that $L_D^2/T\eta$ be large, which is the condition stated.

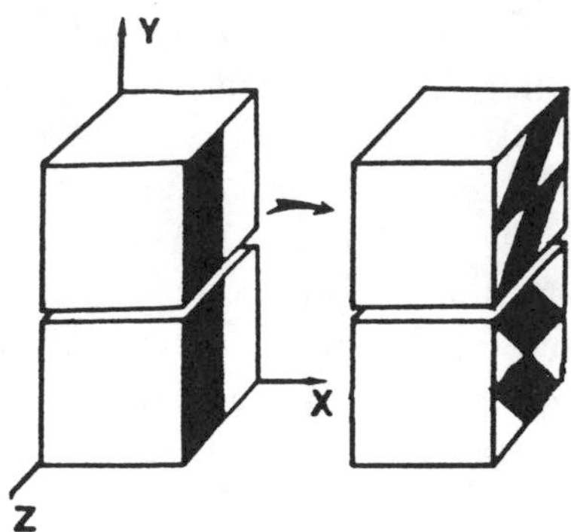

Fig. 3. The SFS map.Here white denotes a uniform field pointing in the direction of positive x, black in the direction of negative x. The field strength doubles with each iteration.

α of about .45, with clear indication of growth at a rate of about .3 for α between .9 and 1.6. The Liapunov exponent for the map is just $\log 2 = .693$ so this is an example where stretching is uniform and field cancellation by folding reduces the growth rate. (The topological entropy and the Liapunov exponent are identical because of the uniform stretching.) It should be emphasized that we are dealing with sub-dominant or exponentially small flux relative to field magnitude, and it is for this reason that the critical α is difficult to calculate, even in this simple one-dimensional setting.

The effect of increasing R is most clearly seen in the spectrum of $b(y)$, which developed, as expected, smaller scale structure and a cascade to high wavenumbers. At fixed R, the magnetic field converged under iteration to an eigenfunction, whose complexity increased with R. In the limit there is no eigenfunction of fixed structure.

The small-scale structure played a passive role in the excitation, however, as can be seen by studying a filtered map with a fixed cut-off wavenumber. It was found that only a few dozen Fourier modes are sufficient to compute reasonably well the growth rate at large R (above 10000). This is consistent with the smoothness of the eigenfunctions of the adjoint of G. The adjoint is defined by

$$G^{+}b(y) = e^{\pi i\alpha(y-1)}b(y/2) - e^{-\pi i\alpha(y-1)}b(1-y/2). \tag{30}$$

The averaging of two terms leads to a smoothing effect, compared with the introduction of small scales by G.

2.1.4. The Slide Map. The SFS map incorporates a constant shear which attempts to simulate an effect of a smooth flow. However it can be argued that the discontinuity introduced into the fold hardly warrants any more regular treatment of the shear. Klapper (1991c) thus replaces the shear of the SFS map by a *slide* in which the z-velocity component is a piecewise constant function of y. If we replace the map of z by, for example $z \to z+\alpha/2\, sgn(x-1/2)$, the problem becomes much easier to analyze. A related class of models for the steady flow problem was introduced by Finn *et al.*(1989b).

The magnetic operator is now

$$G_s b = 2\, sgn(1/2 - y) e^{-\pi i \alpha\, sgn(y-1/2)} b(\tau(y)) \tag{31}$$

It is possible to compute the growth rate of flux explicitly in the SFSl map:

$$F(G_s b) \equiv \int_0^1 Gb(y)dy = 2i \sin \pi\alpha F(b), \tag{32}$$

so $\Gamma = \log(2\sin \pi\alpha)$. As α is increased from 0, the critical value is thus 1/6.

The SFS, slide, and related models are very special in their clear separation of expanding and contracting directions (the x and y axes respectively). We have already noted in section 2.1.2 that the adjoint can be viewed in terms of a time-reversed flow operating on area elements. The distinctly different effects of a stretch-fold-shear and a unshear-unfold-compress accounts for the prominent role of the adjoint. As we shall indicate below (see section 2.2.1 other unsteady models are essentially the same under time reversal and for these the adjoint formulation probably offers no special advantage. This suggests that unsteady motions with significant asymmetry under time reversal may be efficient fast dynamos.

2.2. OTHER UNSTEADY MODELS

2.2.1. Pulsed Beltrami Waves. Baker's map models of the kind considered above oversimplify the process of stretching and folding which can occur in smooth flows. A magnetic field (1,0,0) acted upon by a flow $(0, \cos x, 0$ will indeed be stretched and folded as the lines of force are forced into a wavy shape, which is crudely accounted for by the stretch-fold sequence. Similarly, the flow $(0, 0, \sin x)$ will produce an analogous shear. We are thus led to introduce the *Beltrami wave* having the general form

$$\mathbf{u}_B = \mathbf{a}k \sin(\mathbf{k}\cdot\mathbf{x}+\phi) + (\mathbf{k}\times\mathbf{a})\cos(\mathbf{k}\cdot\mathbf{x}+\phi). \tag{33}$$

where $\mathbf{a}\cdot\mathbf{k} = 0$ and ϕ is an arbitrary phase. This flow has the dynamo property $\nabla\times\mathbf{u}_B = k\mathbf{u}_B$. Associated with any such flow, acting on a conductor for time T, we have the Lagrangian map

$$\mathbf{x} \to \mathbf{x} + T\mathbf{u}_B \tag{34}$$

which generates a very non-trivial Jacobian. In particular lines of force parallel to $\mathbf{k}$ are deformed into helices. Useful special cases are the three orthogonal waves with $k = -1$.

$$\mathbf{u}_x(x) = (0, \cos x, \sin x), \mathbf{u}_y(y) = (\sin y, 0, \cos y), \mathbf{u}_z(z) = (\cos z, \sin z, 0). \tag{35}$$

The steady flow $\mathbf{u}_x + \mathbf{u}_y$ is equivalent to the Roberts cell under rigid body motion and scaling. It is therefore of interest to break the integrability of the Roberts cell by considering time-dependent motions of the form

$$\mathbf{u}(x, y, z, t) = \alpha(t)\mathbf{u}_x + \beta(t)\mathbf{u}_y \tag{36}$$

where the functions of time are arbitrary. For modest amplitudes the resulting Lagrangian motion is chaotic in web-like regions. A pulsed flow can be generated by taking the functions to be piecewise constant and passing to a limit. For the perfect case we may simply alternate between $(\alpha, \beta) = (1, 0)$ and $(0, 1)$. The magnetic field will still have the separated form $\mathbf{B} = e^{2\pi i n z}\mathbf{b}(x, y, t)$, so the problem is two-dimensional (rather than one-dimensional as for the SFS model).

A number of calculations have been carried out giving evidence of fast dynamo action. Bayly and Childress (1988) formulated the problem slightly differently, taking $(a \cos x, 0, b \sin x)$ as the basic wave (which is Beltrami when $a = b$). They followed the pulsing of this flow by the rotation $x, y, z \to y, x, z$. Iteration of the composite map is equivalent to alternate pulsing of two generalized waves with $\alpha = \beta$. Spectral methods were used. Growth rates were observed to converge with increasing R, the Beltrami case being not far from optimal. When the amplitudes were chosen to approximate the SFS map, growth rates in the two models were indeed not too different. Otani (1989) considered a similar problem with $\alpha = \sin^2 t, \beta = \cos^2 t$, obtaining growth with $R = 1000$, again using spectral methods. Otani also developed a time-dependent graphical display of the dynamo action. The strongest excitation occurs in thin sheets, with continual reinforcement of the stretched field coming from motion in the vertical direction. Reasonable agreement with the pulsed cases studied by Bayly and Childress (1989) was obtained. More recently, Bayly (1991) has considered the pulsed perfect case, and studied the growth rate as a function of resolution and the flow parameters, using Lagrangian maps directly. His results offer further evidence of convergence with refinement of the partition of space for Lagrangian analysis, and also indicate that the dynamo excitation is turned on by the growth of the region of Lagrangian chaos.

2.2.2. A Perturbed Four-Corner Cell. A somewhat different approach has been taken by by Klapper (1991b,c). The basic model is a time perturbation of a Roberts cell, but this is simplified to make the Lagrangian analysis explicit. The streamfunction of the unperturbed motion is made up of four contiguous corner flows (one component being a streamfunction of the form xy for example). The perturbation velocity if of the form $\epsilon(0, \sin t, 0)$.

The velocity is thus continuous but the Jacobians are not, and jump conditions must be computed across the lines where the corner flows join.

The model is nevertheless well suited to accurate Lagrangian calculations. Klapper also treats diffusion using stochastic Lagrangian methods; these are effective at extremely large R and results are obtained for $R = 10^6$. Fast dynamo action was observed to occur and the growth rates obtained using (i) flux averaging, (ii) averaging of white noise, and (iii) an approximate (Gaussian) method for modelling the effect of large but finite R (Klapper 19991a) were compared.

2.2.3. Random Pulsed Flows. There are two ways stochastic methods can enter into fast dynamo theory. One of them, just mentioned, is the use of added noise and stochastic averaging to simulate the effects of diffusion. The theoretical underpinnings of the method are secure and it has the advantage as a numerical tool of improved accuracy with increasing R.

The second basic application of random methods is to pulsed flows. (Molchanov *et al.*1985, Zeldovich *et al.* 1988). A *renewing flow* is defined as a random flow where in each successive time interval of length T an independent choice is made from an ensemble of flows. The ensemble is equipped with given probability density function. In a diffusive conductor, each pulse is followed by a diffusive phase with fluid at rest. However, the role of diffusion is secondary and averages out, so that in effect fast dynamos and perfect fast dynamos coincide in a renewing flow. This fact has been exploited previously in dynamo theory, in the modelling of turbulent dynamos (Moffatt, 1978).

The ensemble of random Beltrami waves is the most natural set of flows for our purposes. Some recent calculations have been carried out by Gilbert and Bayly (1990), for two cases: (i) $\mathbf{k}$ uniformly distributed over the sphere of radius k, and (ii) $\mathbf{k}$ restricted to three orthogonal directions. Dynamo action was observed in both cases. As the wavenumber of the magnetic field relative to that of the flow is increased growth is observed in a sequence of intervals, depending upon the duration time of action of the flow relative to a turnover time. It is interesting to compare these results with the steady Beltrami calculation considered in section 2.3.3.

2.3. STEADY FAST DYNAMOS

2.3.1. General Remarks. We thus see that unsteady fast dynamos are not only easily visualized through the STF map and its variants, but also seem to be common among chaotic flows. In steady flow, on the other hand, the process is less accessible. We can think of a *perfect* steady fast dynamo as a Lagrangian structure which causes exponential growth of (average) field in some volume of space. The "assembly " of magnetic structure from the stretching and folding of initial field proceeds at a constant rate instead of at an intermittent or pulsed rate. We can make an analogy with a hand-built automobile, where a sequence of steps is carried to completion, then repeated, versus a production line with a steady output of finished product.

One way to approach the steady problem in three dimensions is through the unsteady problem in two dimensions, by relating the z coordinate to time. Finn et al. (1989a,b) have used this suspension technique to construct a class of steady perfect fast dynamos in a compressible conductor, using a flux condition for dynamo action. The flow is periodic in z with period 2L, and the motion consists of two parts. For $0 \leq z \leq L$ the fluid moves with unit speed in the z-direction and also with a transverse velocity independent of z. In the interval $L \leq z \leq 2L$, the fluid moves only in the z direction with a velocity depending upon x, y, z. The first step realizes a transverse mapping of the magnetic field, while the second step applies vertical shear. The z dependence of the velocity in the second step makes the divergence nonzero in general so a density variation must be introduced to be consistent with conservation of mass. The model thus has some similarity to SFS-type models in its division of steps.

2.3.2. Stochastic Webs. We focus on the so-called *stochastic webs* formed at the onset of Lagrangian chaos by perturbing an integrable steady-state flow (Lichtenberg and Lieberman 1983, Zaslavsky *et al.*, 1988). Here, we perturb a Roberts cell by a small multiple of the Beltrami wave $\mathbf{u}_z$, cf. (35). Scaling the z-coordinate to make the entire flow Beltrami with positive helicity, we assume

$$\mathbf{u} = (\psi_x, -\psi_y, \psi) + \epsilon(\cos\sqrt{2}z, -\sin\sqrt{2}z, 0) \tag{37}$$

where again $\psi = \sin x \sin y$ and ϵ is assumed small. We refer to (37) as the "11ϵ" flow. The perturbation will be the source of fast dynamo activity through its effect on the neighborhood of the separatrices. Lagrangian chaos is introduced by the small horizontal motions, which enable fluid particles to wander between the streamlines $\psi = constant$ of the unperturbed flow. Instead of simply moving around a cell in a helical fashion, a particle near the separatrix can now move to a different cell, and indeed travel on an erratic helical path taking it from cell to cell throughout space. At the same time, the vertical motion continuously changes the form of the horizontal perturbation. These two effects of streamline wandering and vertical motion lead to deterministic chaos in the particle paths near the separatrix.

Fast dynamo action by the flow (37) has been described recently by Gilbert (1991b) (see also Gilbert and Childress, 1990). His treatment makes use of two basic approximations. The first is a local approximation which essentially replaces ψ by its leading term in the vicinity of the separatrix. The second is a mutually exclusive separation of the effect of the perturbation from that of the vertical motion. The resulting approximation to the separatrix flow, which is termed the the 11ϵs (for separatrix) flow, has an extremely simple Lagrangian structure, allowing very accurate computation of a Poincaré map.

The structure of the 11ϵs flow thus divides naturally into 2 kinds of domains. Centered on each line of stagnation point of the unperturbed cell is a square cylinder of side κ. Projected onto the horizontal plane, these define the "X-regions". The "S-regions" are similar with projection onto rectangles spanning the separatrices and connecting the X-regions. Within an X-region, ψ is approximated by a corner flow and the perturbation does not act. Within the S-regions, ψ is approximated by a linear function in the direction normal to the separatrix and the vertical motion is assumed zero. These assumptions approximate the exact flow for small ϵ because the fluid particles spend a significant time in the X-regions (because of a factor $|\log \psi|$), while most of the streamline drift occurs during rapid passage between X-regions.

In this approximate geometry, it is the useful to use the Lagrangian time t, vertical position z, and unperturbed streamfunction ψ as coordinates. (The map $xyz \to \psi, z, t$ preserves volume.) A Poincaré map can be set up between the ψ, z plane bisecting a horizontal segment of the separatrix connecting two stagnation points, and the subsequent plane reached as the point flows through an X-region. This allows a very compact mapping problem, which can be exploited in numerical simulations.

The magnetic field structure is largely determined by the stretching at X-points, as is to be expected. Mathematically, this is now apparent from the structure of the Jacobian matrix,

$$\mathbf{J} = \frac{\partial(\hat{\psi}, \hat{z}, \hat{t})}{\partial(\psi, z, t)} = \begin{pmatrix} O(1) & O(1) & 0 \\ O(1) & O(1) & 0 \\ \gg 1 & \gg 1 & 1 \end{pmatrix} \tag{38}$$

This implies that large fields are developed parallel to the streamlines, from the stretching of cross-streamline and z-directed components. This is the effect of the X-regions and the critical points of the flow, which produce local exponential stretching as described in section 1.2.1. At the same time, the two zeros in the right-hand column of the Jacobian indicate that this large streamwise field is passive and not converted to cross-stream components. Since the large streamwise field is difficult to compute accurately, dynamo action is best monitored by a flux integral over a cross-stream component.

To study the evolution of the field numerically, a region is selected and partitioned. At each point the inverse map is performed and the Cauchy representation calculated along the path from the preimage. The resulting magnetic structure is then built up and the flux computed at the current time. Perfect fast dynamo action is thus observed in regions of Lagrangian chaos, and non-dynamo actionis found in a regular or integrable parts of the flow. The growth rates which are computed are insensitive to the region over which flux is computed; comparable values are obtained from a surface and a line. The structure of the initial field does however influence the growth rate. As in an eigenvalue problem, different modes of growth can be extracted with slightly different growth rates. The group of symmetries of the flow can be used to divide up the initial conditions into four symmetry classes, two of which lead to dynamo action. Within each class growth rates are essentially independent of initial condition, but the rates of the two classes are not the same. A fifth class has a mixed character.

This model offers the possibility of extracting some details of the dynamo mechanism and Gilbert finds that the pasting together of flux having different histories is more important than local shear, so the cut-and-paste models of section 2.1.1 may be appropriate. The symmetry properties of the flow may be used to advantage in deducing relevant models, but it is not yet understood how the stretch-fold operations are actually created in the flow.

2.3.3. General Beltrami Flows. The family of steady Beltrami flows

$$\mathbf{u} = A\mathbf{u}_x + B\mathbf{u}_y + C\mathbf{u}_z \tag{39}$$

has been studied in both slow and fast dynamo theory. The case $A = B = C = 1$ is exceptional in that straight streamlines connect an array stagnation points and there is maximal symmetry (Childress and Soward 1984, Dombre *et al.* 1986). The flow contains three families of helical structures analogous in topology to the Roberts cell, and exhibits a region of Lagrangian chaos. Numerical studies of the 111 flow as a possible fast dynamo were initiated by Arnold and Korkina (1983). They studied 2π-periodic magnetic fields obtaining a limited band of R where dynamo action occurred. Galloway and Frisch gf extended these computations and obtained dynamo action out to $R = 550$. Gilbert (1991a,b) has computed field growth in the perfectly conducting case for the 111 flow using Lagrangian methods. For (39) the Jacobian cannot be computed explicitly and is instead determined by solving

$$dJ_{ij}/dt = S_{ik}J_{kj} \tag{40}$$

where $S_{ij} = \partial u_i/\partial x_j$. Although numerical error limits the time of computation relative to the $11\epsilon s$ model, exponential growth of flux of cross-streamline components is again observed.

2.3.4. Piecewise Constant Flows. We mention an interesting class of flows studied by Avellaneda and Almgren (private communication). The idea is to replace the sines and cosines in (39) by their piecewise constant analogs with period 4:

$$\sin x \to \begin{cases} 1, & \text{if } 0 \le x < 2; \\ -1, & \text{if } 2 \le x < 4\,. \end{cases} \tag{41}$$

$$\cos x \to \begin{cases} 1, & \text{if } 0 \le x < 1 \text{ or } 3 \le x < 4; \\ -1, & \text{if } 1 \le x < 3\,. \end{cases} \tag{42}$$

The Lagrangian particle paths are now polygonal lines. The perfect dynamo problem can in principle be studied in such flows, but the discontinuities in the velocity field introduce singular flux sheets. The piecewise constant (PC) 111 flow, in contrast to its continuous counterpart, does not exhibit chaos, although the PC11ϵ flow does. Because of the simplicity of these flows for numerical simulation, their capacity for fast dynamo action warrants attention.

3. Analysis of the Perfectly Conducting Limit

We first summarize some features of fast dynamos based upon the above examples. We first note that it is the complexity of the flow which literally forces the field to amplify, as in the STF map. Second, the true fast dynamo is orchestrated on the scale of the velocity field itself. The cascade to small magnetic scales accompanies the dynamo action, but is secondary to it. By definition, the fast dynamo must become independent of dissipation, which means that the ever-smaller magnetic scales (which always eventually dissipate) will not be important once R is sufficiently large. Finally, it appears that, in addition to the necessary condition that the Liapunov exponent to be nonzero somewhere, smooth flows must also be stretch over a sufficiently large set of points, and fold in a sufficiently constructive way. A working hypothesis thus seems to be: *fast dynamo action occurs in many smooth three-dimensional flows in regions of Lagrangian chaos.*

We turn now to an attempt to devise an approach to the analysis of fast excitation. Little will actually be accomplished here toward a proof of fast dynamo action in a nontrivial setting, but we hope to raise some issues which might prove to be useful in formulating the perfect fast dynamo problem precisely.

3.1. SOME RESULTS OF SPECTRAL THEORY

The operator L will here be taken as some given linear operator, but we shall want it to be the Lagrangian time one map corresponding to some flow which is periodic in time with period one. We know that the classic eigenvalue problem $LB = \lambda B$ does not make sense for the perfect dynamo, so we should not be looking for point spectrum of L. We shall summarize the basic classification of the non-point spectrum of a linear operator (see e.g. Friedman, 1956). Let us assume that we are given the vector space $\mathcal{B}$ of magnetic fields B, equipped with complex inner product (A, B). The *point spectrum* (or discrete spectrum) of L consists of all complex numbers λ such that the equation $LB = \lambda B$ has a non-trivial solution. If this equation has

only the trivial solution we can distinguish three cases:
(i) The closure of the range of $L-\lambda$ is all of $\mathcal{B}$ and its inverse is bounded. In this case the inhomogeneous equation $(L-\lambda)B = F$ has a bounded solution, and λ is said to belong the the *resolvent set* of L.
(ii) The closure of the range of $L-\lambda$ is all of $\mathcal{B}$, but the inverse of $L-\lambda$ is unbounded. In this case λ is said to belong to the *continuous spectrum* of L. At a point of the continuous spectrum has the property that $(L-\lambda)B$ can be made as small as desired with fields of norm 1.
(iii) The last case is that the closure of the range of $L-\lambda$ is a proper subset of $\mathcal{B}$. Then λ is said to belong to the *residual spectrum* of $\mathcal{B}$.[9]

A useful result for us is:

THEOREM 1. *If λ is in the residual spectrum of L, then λ^* is in the point spectrum of the adjoint L^+ to L.*

The proof depends upon the fact that a null vectors of $L^+ - \lambda^*$ are orthogonal to every vector in the range of $L-\lambda$, that is, they make up the orthogonal complement of the range of $L-\lambda$. Since by assumption the latter set is not empty, λ^* belongs to the point spectrum of L^+.

3.1.1. The Slide Map Reconsidered. Let us consider the above classification for the slide map, see section 2.1.4. Using this notation, we recall

$$LB(y) = 2\,sgn(1/2-y)e^{-\pi i\alpha\, sgn(y-1/2)}B(\tau(y)) \tag{43}$$

where $\mathcal{B}$ will be the square-integrable complex-valued functions of y defined over [0,1]. We first note that 1 is not in the range of $L-\lambda_0$ where $\lambda_0 = 2i\sin\pi\alpha$, since the average over y is zero. Thus we recover λ_0^* as a point eigenvalue of L^+ corresponding to the eigenfunction 1. We now look for more eigenvalues of L^+. We have

$$L^+B(y) = e^{-i\pi\alpha}B(y/2) - e^{i\pi\alpha}B(1-y/2), \tag{44}$$

we next observe that any field of the form $B_0(y) = e^{-\pi i\alpha\, sgn(y-1/2)}f(\tau(y))$ for some f in $\mathcal{B}$ is in the null space of L^+. Thus 0 is in the point spectrum of L^+. In fact 0 is also in the residual spectrum of L.[10]

Are there any other point eigenvalues of L^+? If $L^+B = \lambda^*B$ we have, using the triangle inequality (with $\| B \| = \sqrt{(B,B)}$),

$$|\lambda|\,\| B \| \le \sqrt{2}[(\int_0^{1/2} B^2(y)dy)^{1/2} + (\int_{1/2}^1 B^2(y)dy)^{1/2}] \le 2\,\| B \| \tag{45}$$

[9] There are various alternatives to this classification, especially in the elimination of residual spectrum as a separate object (Kato, 1966).

[10] We verify that if $LB = B_0$ then $2B = f$ and $-2B = f$ which is impossible if f does not vanish. This has to be verified since we could in principle also have a point eigenvalue of L.

and so necessarily $|\lambda| \leq 2$. We next try a linear function. Trying $B = a + b(y - 1/2)$ yields immediately $\lambda^* = \cos \pi\alpha$ with eigenfunction

$$B = \cos \pi\alpha - (2\cos \pi\alpha + 4i \sin \pi\alpha)(y - 1/2). \tag{46}$$

In fact, it is not difficult to show that an eigenfunction of of the form

$$B_k(y) = \sum_{j=0}^{k} a_j (y - 1/2)^j \tag{47}$$

exists corresponding to the point eigenvalue

$$\lambda_k^* = 2^{-k}[e^{-\pi i\alpha} + (-1)^{1+k} e^{\pi i\alpha}]. \tag{48}$$

Indeed the eigenvalue problem then transforms (by taking coefficients of $(y-1)^k$ to a homogeneous upper-triangular matrix equation. The vanishing of the single element in the Nth row yields (48), with a_N arbitrary; the remaining diagonal elements are then nonzero. Backsubstitution then yields the remaining coefficients $a_k, k = 0, ..., N-1$ in terms of a_N. We thus obtain a sequence of point eigenvalues converging to the origin, where the eigenvalue of infinite degeneracy lies. Since these polynomial eigenfunctions are linearly independent they can be combined to construct Legendre polynomials and so are complete in $\mathcal{B}$.

Since we have easily $\| L^+ \| \leq 2$, the Neumann series for $(L^+ - \lambda^*)^{-1}$ exists for $|\lambda| > 2$. This domain is in the resolvent set of L^+.(The same is true for the L.) Since we have a complete set of eigenfunctions we can in fact define (corresponding to the nonzero eigenvalues given above), the inverse $(L^+ - \lambda^*)^{-1}$, whenever $\lambda^* \neq \lambda_k, 0$, by

$$(L^+ - \lambda^*)^{-1} f \sim \sum_{k=0}^{\infty} \frac{a_k B_k(y)}{(\lambda_k^* - \lambda^*}, f \sim \sum_{k=0}^{\infty} a_k B_k(y). \tag{49}$$

We now consider the operator L. It is of course this operator which causes difficulty with the eigenvalue problem. The source of the problem can be traced to the failure of L to be a *compact* operator. The last property means that for any bounded sequence $\{b_k\}$, $\{Lb_k\}$ contains a subsequence converging to some limit in $\mathcal{B}$. As a bounded sequence which does not have this property, we may take $b_k = T^k(1)/(2\sin \pi\alpha)^k$. One can track using the binary expansion the number of visits to the two half-intervals, with associated multiplication by $e^{\pm\pi i\alpha}$. The result is that we do not have convergence in the L^2 norm to any element of $\mathcal{B}$. The stretch-fold-slide sequence does not have the nice property of pointwise convergence.

We next show that L has no point eigenvalues in $\mathcal{B}$ if $|\lambda| < \sqrt{2}$. Indeed if $2e^{\pi i\alpha} B(2y) = \lambda B(y)$ for $0 \leq y < 1/2$ then

$$2 \int_0^1 |B|^2(y) dy = |\lambda|^2 \int_0^{1/2} |B|^2(y) dy \leq |\lambda|^2 \int_0^1 |B|^2(y) dy \tag{50}$$

which establishes the result. This still leaves open the possibility of point eigenvalues with $\sqrt{2} \leq |\lambda| \leq 2$.

However we can show that if λ^* is in the resolvent set of L^+, then λ is in the resolvent set of L. For $(L^+)^+ = L$ and therefore the inverse of $L - \lambda$ is given by

$$(L - \lambda)^{-1} = ((L^+ - \lambda^*)^{-1})^+. \tag{51}$$

Thus the remaining possibility is that $2i \sin \pi\alpha$ is a point eigenvalue of L when α makes its modulus greater than or equal to $\sqrt{2}$. But then we see that the eigenfunction would have to satisfy $B(y) = 1/2(B(y/2) + B(1 - y/2))$. Iterating this operation, the limit and only solution is $B = constant$. But this is possible only if $2e^{\pi i\alpha} = 2i \sin \pi\alpha$ or $\alpha = 1/2$. This is, of course, the case where the slide map does produce perfect doubling of the field structure. Thus we have the following result:

THEOREM 2. *If $0 < \alpha < 1/2$ the spectrum of the slide map consists of residual spectrum at the points $\lambda_k = 2^{-k}[e^{\pi i\alpha} + (-1)^{1+k}e^{-\pi i\alpha}], k = 0, 1, ...$ as well as at the point $\lambda = 0$. The λ_k^* are point eigenvalues of the adjoint of the slide map, with associated eigenfunctions complete in L^2. The origin is a point eigenvalue of L^+ of infinite degeneracy.*

No analogous results are known for the SFS map, other than that the the adjoint eigenfunction for any growing mode are smooth functions (Bayly and Childress, 1989). The eigenfunction for the largest eigenvalue is easily computed by iteration. However we conjecture that the same situation applies, with a countable set of smooth eigenfunctions of the adjoint forming a complete set, and the eigenvalues of the adjoint matched with residual spectrum of the map. The origin is again clearly an eigenvalue of the adjoint having infinite degeneracy.

3.1.2. The Spectrum of Pulsed Flows. More perplexing is the behavior of maps which are *not* of baker's type. The clean separation of the expanding and contracting directions has led in the baker's maps to the asymmetry between the operator and adjoint. If this asymmetry is absent, as it may be for pulsed Beltrami waves in some cases, we seem to be led toward point eigenvalues for both adjoint and operator, and this stands in contradistinction with the ever increasing complexity we expect to see in the magnetic structure of a perfect fast dynamo. We look now at this issue in the simplest possible setting.

The pure baker's map, the $\alpha = 0$ case in the SFS map, can be compared with a *sinusoid* map

$$(x, y) \rightarrow (y + \alpha \sin 2\pi x, x) \; mod \, 1. \tag{52}$$

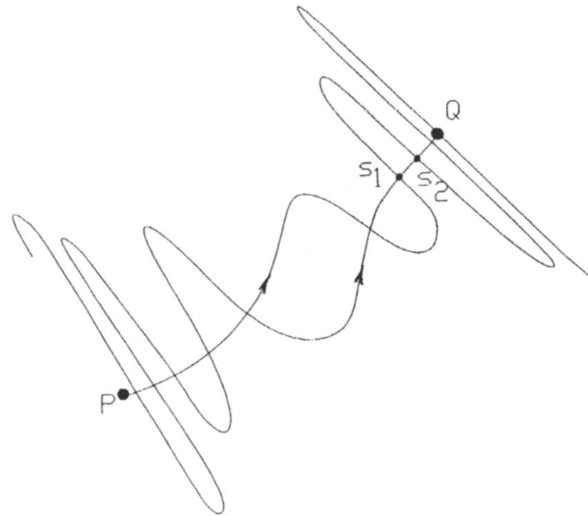

Fig. 4. The heteroclinic tangle for the sinusoid map, where $P(0,0)$ and $Q(1/2,1/2)$ are critical points. Near Q the unstable manifold from P intersects the stable manifold at Q at the infinite sequence of points $s_1, s_2, \ldots$. Net flux carried by this infinitely sheeted structure can model the flux introduced into the manifold at P.

To obtain the action of this map on a two-dimensional magnetic field it is convenient to work with a scalar magnetic potential $A(x,y)$, $\mathbf{B} = (A_y, -A_x)$, where for a perfect conductor A is a material invariant. Now the baker's map has no point eigenvalues, but for the sinusoid map we can look to concentrate field at a invariant curve. An invariant curve is either an unstable or a stable manifold associated with a given critical point of the sinusoid map. Restricting attention to the unit square the critical points of the map are at the corners and center of the square. The unstable manifold at P is an invariant curve containing (0,0) such that $(y + \alpha \sin 2\pi x, x)$ $mod\,1$ belongs to the curve whenever (x, y) does (see Figure 4). Near the point P, the slope m of the invariant manifolds is determined from the eigenvectors of the Jacobian of the map at (0,0):

$$m_P = -\pi\alpha \pm \sqrt{\pi^2\alpha^2 + 1}. \tag{53}$$

The plus sign yields the unstable manifold from P, the minus sign the stable. For positive α, the unstable manifold from point P, say, intersects the stable manifold from Q *transversely*, resulting in the deterministic chaos of the map. The homoclinic nature of the intersection is a result of indentification of equivalent critical points under the periodicity of the map. For small α the invariant manifolds are close to the curve $y = x - \alpha/2 \sin 2\pi x$ connecting P to Q, and the transversal intersection is very close to the terminal critical point, probably in an exponentially (relative to α) region containing the critical point.

We emphasize that we have here (at least for small α) an explicit example of a distribution eigenfunction of the sinusoid map for eigenvalue close to 1. That is to say, to the extent that a transversal intersection does *not* occur, as in the steady Roberts cell e.g., finite flux can in principle be confined to the separatrix. Being a delta function, this eigenfunction is not square integrable, and therefore in the spectral classification considered above, 1 is not in the point spectrum. It is also not in the residual spectrum, since if the operator L propagates A, we have $LA(x,y) = A(y, x - \alpha \sin 2\pi y)$ (involving the inverse $x, y \to y, x - \alpha \sin 2\pi y$ of the sinusoid map), and also $L^+ A(x,y) = A(x + \alpha \sin 2\pi x, x)$. Thus it is not possible that 1 is in the point spectrum of L^+. In fact the eigenvalue 1 is in the *continuous* spectrum of L.[11] The continuous spectrum thus arises directly from the distribution nature of the perfect eigenfunction in a case where residual spectrum cannot occur. We suspect this is generally going to be true for pulsed flow perfect fast dynamos, but know of no way to prove it. We could then expect the flux to be concentrated into Cantor sets once the heteroclinic tangle is established. In this picture fast dynamo action is realized by the stretching and folding inherent in transversal intersection of invariant manifolds. A qualitative model based upon this idea is given below.

3.1.3. Weak Convergence. The appearance of distributions as eigenfunctions sets the stage for a brief discussion of weak convergence. The slide maps are sufficiently simple to allow exploratory studies of the structure of the eigenfunction. Bayly (1991) has raised the question of weak convergence of iterates as a way of dealing with the non-existence of eigenfunctions in a perfect conductor. If L is a general Lagrangian map on a complex inner-product space of magnetic fields, Bayly proposes that a perfect fast dynamo be identified by the condition that there exist smooth f and g such that

$$\lim_{m \to \infty} \frac{1}{m} \log |(f, L^m g)| > 0. \tag{54}$$

This is similar to, but more general than, a flux estimate, which has the advantage of allowing use of the adjoint operator.

The slide map indeed has a much stronger property. We introduce a baker's slide map does away with the fold, utilizing the classic version of the baker's map (Arnold and Avez, 1967). Let the inverse of the map be

$$(x, y, z) \to \begin{cases} (x/2, 2y, z + \alpha/2), & \text{if } y < 1/2; \\ ((1 + x)/2, 2y - 1, z - \alpha(y - 1/2)), & \text{if } y > 1/2. \end{cases} \tag{55}$$

[11] The easiest way to see this is to define, following Friedman (1956), the *approximate spectrum* as follows: λ is in the approximate spectrum of L there is a sequence of vectors $\{B_k\}$ having norm 1 such that $\| (L - \lambda) B_k \| < 1/k$. The approximate spectrum contains the discrete and continuous spectra, and may or may not contain the residual spectrum. In our case if 1 is in the approximate spectrum it must be in the continuous spectrum. To show that it is in the approximate spectrum, we define the delta function as the limit of a sequence, producing unbounded L^2 norm. Normalization of the sequence then leads to the small estimate on $L - 1$.

The mapping of field is then

$$G_{s*}b(y) = 2e^{-\pi i\alpha\, sgn(y-1/2)}b(\beta(y)) \tag{56}$$

where

$$\beta(y) = \begin{cases} 2y, & \text{if } 0 \le y < 1/2; \\ 2y-1, & \text{if } 1/2 \le y < 1\,. \end{cases} \tag{57}$$

Here $f(G_{s*}b) = 2\cos\pi\alpha F(b)$. The adjoint of this slide operator is given by

$$G_{s*}^{+}b = e^{-i\pi\alpha}b(y/2) + e^{i\pi\alpha}b((1+y)/2). \tag{58}$$

We define $Q = G_{s*}^{+}/2\cos\pi\alpha$ we can write, with $p = e^{-i\pi\alpha}/2\cos\pi\alpha, q = e^{i\pi\alpha}/2\cos\pi\alpha$,

$$Q^m f = \sum_{k=0}^{m} \frac{m!}{(m-k)!k!}p^{m-k}q^k \bar{f}_k \tag{59}$$

where $\bar{f}_k$ is an average of the values of f at all points $y_i = y/2^m + c_i$ where the c_i are those members of the set $[i/2^m, i = 0, ..., 2^m - 1]$ whose binary representation contains exactly k 1's. As m becomes large, the dominant contribution to the sum comes from terms where $\bar{f}_k \approx \bar{f} \equiv \int_0^1 f dy = F(f)$ by the mixing property of the baker's map (Arnold and Avez, 1967). We thus see that

$$(f, G_{s*}^m g)/(2\cos\pi\alpha)^m = (Q^m f, g) \to F(f)F(g) \tag{60}$$

as $m \to \infty$. We can express this weak convergence to and exponentially growing mean value by

$$L^m/\lambda^m \overset{w}{\to} \int_0^1 (\cdot)dy \tag{61}$$

where $\lambda = 2\cos\pi\alpha$ is the dominant eigenvalue. In particular, we see that the total flux $F(b)$ is here equivalent to projection onto the adjoint eigenfunction 1 . Weak convergence is thus an attractive alternative to point spectrum when dealing with the perfect fast dynamo, and is more flexible than a flux average. Strictly speaking, any one estimate could fail to detect fast Lagrangian excitation in the unlikely event that it is orthogonal to the projection. In effect, weak convergence considers all trial estimates and so samples the local averages over all parts of the domain.

The criterion (54) may need to be modified in examples which are more complicated than the baker's maps. In the above weak test of growth there are two functions, the function f which in effect provides the averaging over small scales, and the initial condition g which leads to perfect dynamo action. In general the choice of initial condition will be important. From the point of view of the physics we are interested in the predominant magnetic structure, so we want to take a supremum over g in computing both the growth rate (maximal eigenvalue) and the weak magnetic structure. On the other hand the smoothing function f could also play a role in determining growth.[12] We can therefore modify (54) and say that the perfect dynamo has growth rate Γ provided that

$$\Gamma = \sup_g \sup_f \lim_{m\to\infty} \frac{1}{m} \log |(f, L^m g)| \tag{62}$$

where f, g are smooth square integrable functions. A perfect fast dynamo will have $\Gamma > 0$.

To discuss weak convergence of eigenfunctions, we first fix g and take the supremum over only f in (62). If the resulting growth rate is Γ, we would say that the magnetic field converges weakly to a distribution B provided that there is a complex number λ with modulus e^Γ such that

$$\lim_{m\to\infty} (f, \tilde{L}^m g) = (f, B), \tilde{L} = L/\lambda \tag{63}$$

for every smooth f. The symbol (f, B) means a linear functional on f determining the distribution B. The problem with this definition is that it B is here very unstable relative to g, since there may exist larger growth rates. Also, Gilbert (1991b) has found in steady Beltrami flows that several λ with the same modulus can be found as the initial condition is altered. So we now restore the supremum over g in the definition of Γ, and introduce a set $\mathcal{M}$ of distributions of magnetic structure associated with the dynamo. We say that B belongs to $\mathcal{M}$ provided that (63) holds for some smooth g and some λ whose modulus is e^Γ, in terms of the (maximal) growth rate Γ.

This last definition is stable in that it deals with the dominant magnetic signals. Indeed we now have

THEOREM 3. *If B corresponding to g belongs to $\mathcal{M}$, then $B = Pg$ where P is a projection onto a distribution in the weak sense that $(f, P^2 g) = (f, Pg)$ for all smooth f.*

It is clear that B and g are linearly related, so to prove this we need only demonstrate the projection property. Note that

$$(f, \tilde{L}^m Pg - Pg) = (f, \tilde{L}^{m+n} g - B) + (f, \tilde{L}^m (B - \tilde{L}^n g)). \tag{64}$$

For large m, n both terms on the right become small because of the fact that $\tilde{L}$ is a bounded operator (a property not available without the supremum over g). This shows that $(f, (P^2 - P)g) \to 0$ for this g and any smooth f, as required.

[12] For example a flux estimate would be inappropriate over a surface where symmetry insured that net flux always vanished.

It should be emphasized that we are dealing with very singular magnetic fields in these problems, and it is not obvious that this definition of magnetic structure is sufficiently weak to exist in interesting cases. It does encompass some known cases such as the baker's map dynamos however.

We finish this section by reconsidering topological entropy in comparison with the above definition of Γ. It has been conjectured that topological entropy λ_{top} is $\geq \Gamma$ in the sense of a flux average (Finn and Ott, 1988b). Let us first give a precise definition of λ_{top}. (See Guckenheimer and Holmes, 1983, and also section 2.1.1above). Let a map M have some invariant set S. A subset s of S is said to be (n, ϵ) *separated* provided that for any distinct x, y in s there is an i, $0 \leq i < n$ such that the distance from $M^i(x)$ to $M^i(y)$ exceeds ϵ. Let $\sigma(n, \epsilon)$ be the maximum number of points obtained over all (n, ϵ) separated subsets of S. Then we define

$$\lambda_{top} = \lim_{\epsilon \to \infty} \limsup_{n \to \infty} \frac{1}{n} \log \sigma(n, \epsilon). \tag{65}$$

In essence, for fixed ϵ, λ_{top} measures the rate of growth of distinct separated orbits of the map, and a function of the length of the orbit. Since our interest is in orbits which diverge, the number should increase exponentially, and so the greatest separation will tend to be achieved in the last iterates. This suggests a close identification of topological entropy with global growth of line length or area. Indeed Pesin (1977)has expressed topological entropy as a spatial Lebesque integral of the sum of the positive Liapunov exponents. λ_{top} is thus attractive as an upper bound on the growth of magnetic energy in a perfectly conducting fluid, whereas Γ will never exceed one-half of the growth rate of energy, because of cancellation in the flux average. (Added note: The paper by Vishik in this volume addresses this issue successfully and establishes such a bound.)

3.2. A MODEL FOR UNSTEADY FAST DYNAMOS AT THE ONSET OF CHAOS

The above picture of the invariant manifold emanating from critical points and carrying flux concentrations is of course similar to the *diffusive* structure of flux tongues in the helical slow dynamos of section 2. This suggests that there is a natural extension of that picture of dynamo action to the perfect fast dynamo.

Let us consider the following adaptation of the pulsed Beltrami flows of section 2.2.1. We perform the sinusoid map in time 1/2, then turn on the flow with velocity $(0, 0, \alpha \cos 2\pi x)$ for time 1/2. We set $\mathbf{B} = e^{2\pi i z}\mathbf{b}(x, y)$ and measure the flux of $\mathbf{b}$ through the *unstable* manifold emerging from Q (see Figure 4). We regard the flux as concentrated on the unstable manifold from P. Because of the transversal intersection property, the field will be a complicated object built up of folds upon folds; we can view the object as the tangled support for a delta function defined along a Jordan curve. For our purposes we simply concentrate the object on a curve and endow it with a flux though any plane orthogonal to the tangent vector of the curve. Because of three-dimensional effects, the flux will depend upon position s along the curve.

To see how flux can grow, we take α to be small to put the tangles close to the critical points. We shall compute flux on a short segment of the line with slope $\pi\alpha + \sqrt{\pi^2\alpha^2 + 1}$ and center at Q; this will, by symmetry, correspond to the flux introduced into the tangle at P. The unstable manifold from P will intersect this line segment close to Q, at points $s_1, s_2, ...$ say. Under conditions of exponential growth, folds of the manifolds are added to the structure at P at each iteration, so that flux will in fact tend to decrease at fixed time along the manifold from P to Q. The exponential growth of flux must arise in this model from the reinforcement of folded flux sheets which are shifted in the vertical relative to one another. The mechanism is similar to the SFS dynamo; Klapper (1991c) noted the probable existence of this dynamo effect in his unsteady model.

We may estimate the effect of molecular diffusion on the growth rate. If we assume that the points of intersection s_k are on an orbit carrying a Liapunov exponent of λ, then as the structure is stretched the lateral dimensions shrink exponentially and the molecular diffusion becomes important when $e^{-2\lambda t}R$ is of order unity; for a unit period we see that diffusion begins to affect the structure after N steps where $N \approx \log R/2\lambda$. Now after N steps the separation of intersection points s_k has shrunk by $1/m^N$ (with m corresponding to the stable manifold at Q). Consequently the diffusion scale is comparable to the separation of intersections (and therefore the distinction between adjacent folds is removed by diffusion) after N^* steps where $n^* \approx \log R/2\log m$. Depending upon the values of λ and m diffusion may or may not modify the invariant curve structure before the folds coalesce. The limits on our analysis are placed by the coalescence however. The fact that N^* grows with R is all we need to conclude that the series which yields the flux through the line element at Q, when restricted to terms unaffected by diffusion, will converge to the result for a perfect fluid. Thus the model establishes the heteroclinic tangle as a fast dynamo as well as a perfect fast dynamo. A general approach to small diffusivity scalar problems in chaotic flows, leading to a Gaussian approximation similar to that used here, has recently been developed by Klapper (1991a).

To summarize, we propose that unsteady fast dynamos be studied quantitatively by direct analysis of the flux contributed by the "heteroclinic tangles" where the field should tend to be concentrated. The key ingredient is the phase change contributed by the vertical flow. It is not unlike the SFS map except that shear effects are replaced by a time delay. That is, relative drift is effected here by the different times which the lobes of the heteroclinic tangle spend in the vertical flow. This model is also a direct extension of the flux-tongue picture of Roberts and Soward to the chaotic neighborhoods of critical points of pulsed flows. Together tearing and recombination which seems to operate in steady stochastic webs, we begin to see several direct routes to fast dynamo action from the geometry of chaotic flow.

We also mention the prospect that a precise calculation of flux growth rate might be possible if this method is applied to simpler unsteady models. We note in this regard Klapper's perturbed four-corner cell, see section 2.2.2, and the linked twist mappings studied by Devaney (1983).

Acknowledgements

Many of the ideas discussed in this paper, not already formally credited, are the product of collaborations with Bruce Bayly, Andrew Gilbert, Isaac Klapper, and Andrew Soward.

References

ALFVÉN, H., 1950 Discussion of the origin of the terrestrial and solar magnetic fields. *Tellus*, **2**, pp. 74-82.

ARNOLD, V.I. & KORKINA, E.I., 1983 The growth of magnetic field in steady incompressible flow. *Vest. Mosk. Un. Ta. Ser. 1. Mat. Mec.*, **3**, pp. 43-46.

ARNOLD, V.I. & AVEZ, A., 1967 *Problèms Ergodiques de la Mécanique Classique*. Gauthier-Villars, Paris.

ARNOLD, V.I., ZEL'DOVICH, YA. B., RUZMAIKIN, A.A., & SOKOLOFF, D.D. (1981) A magnetic field in stationary flow with stretching in Riemannian space. *Soviet Phys. JETP* **54**, pp. 1083-1086.

ARTUSO, ROBERTO, AURELL, ERIK, & CVITANOVIĆ, PREDRAG, 1990 Recycling of strange sets: I. Cycle expansions. *Nonlinearity*, **3**, pp. 325-359.

BACKUS, G., 1958 A class of self-sustaining dissipative dynamos. *Ann. Physics*, **4**, pp. 372-447.

BAYLY, B.J., 1986 Fast magnetic dynamos in chaotic flow. *Phys. Rev. Lett.*, **57**, pp. 2800-2803.

BAYLY, B.J., 1991 Infinitely conducting dynamos and other horrible eigenproblems. To appear in Proceedings of Workshop on Nonlinear Phenomena in Atmospheric and Ocean Sciences, IMA(Minnesota), June 1990.

BAYLY, B. & CHILDRESS, S., 1988 Construction of fast dynamos using unsteady flows and maps in three dimensions. *Geophys. Astrophys. Fluid Dyn.*, **44**, pp. 211-240.

BAYLY, B. & CHILDRESS, S., 1989 Unsteady dynamo effects at large magnetic Reynolds numbers. *Geophys. Astrophys. Fluid Dyn.*, **49**, pp. 23-43.

CHILDRESS, S., 1979 Alpha-effect in flux ropes and sheets. *Phys. Earth Planet. Int.*, **20**, pp. 172-180.

CHILDRESS, S. & SOWARD, A.M., 1984 On the rapid generation of magnetic fields. In *Chaos in Astrophysics*, NATO Advanced Research Workshop, Palm Coast, Florida, USA.

CHILDRESS, S. & SOWARD, A.M., 1989 Scalar transport and alpha-effect for a family of cat's-eye flows. *J. Fluid Mech.*, **205**, pp. 99-133.

CHILDRESS, S. & STRAUSS, H.R., 1989 *An Introduction to Solar MHD*, Lecture notes, Courant Institute.

COWLING, T.G., 1957 *Magnetohydrodynamics*, Interscience Publishers, Inc.

COWLING, T.G., 1934 The magnetic field of sunspots. *Mon. Not. Roy. Astr. Soc.*, **140**, pp. 39-48.

COWLING, T.G., 1957 The dynamo maintenance of steady magnetic fields. *Quart. J. Mech. and Appl. Math.*, **10**, pp. 129-137.

DEVANEY, ROBERT L., 1983 Linked twist mappings are almost Anosov. In Lecture Notes in Mathematics, **819**, *Global Theory of Dynamical Systems*, Springer-Verlag, New York.

DOMBRE, T. FRISCH, U., GREENE, J.M., HÉNON, M., MEHR, A., & SOWARD, A.M., 1986 Chaotic streamlines and Lagrangian turbulence: the ABC flows. *J. Fluid Mech.*, **167**, pp. 353-391.

FINN. J.M. & OTT, E., 1988a Chaotic flows and fast magnetic dynamos. *Phys. Rev. Lett.*, **60**, pp. 760-763.

FINN, J.M. & OTT, E., 1988b Chaotic flows and fast magnetic dynamos. *Phys. Fluids*, **31**, pp. 2992-3011.

FINN, JOHN M. & OTT, E., 1989 The fast kinematic magnetic dynamo and the dissipationless limit. University of Maryland Plasma Preprint UMLPR 90-003.

FINN, J.M., 1990 Chaotic generation of magnetic fields – the fast dynamo problem. University of Maryland Plasma Preprint UMLPR 90-031.

FINN, J.M., HANSEN, J.D., KAN, I. & OTT, E., 1989a Do steady fast dynamos exist? *Phys. Rev. Lett.*, **62**, pp. 2965-2968.

FINN, J.M., HANSEN, J.D., KAN, I. & OTT, E., 1989b Steady fast dynamo flows. University of Maryland Plasma Preprint UMLPR 90-015.

FRIEDMAN, BERNARD, 1956 *Principles and Techniques of Applied Mathematics*, John Wiley and Sons, Inc., New York.

GALLOWAY, D. & FRISCH, U., 1986 Dynamo action in a family of flows with chaotic streamlines. *Geophys. Astrophys. Fluid Dyn.*, **36**, pp. 53-83.

GILBERT, A.D., 1988 Fast dynamo action in the Ponomarenko dynamo. *Geophys. Astrophys. Fluid Dyn.*, **44**, pp. 241-258.

GILBERT, A.D., 1991a Fast dynamo action in a steady chaotic flow. *Nature*, **350**, pp. 483-485

GILBERT, ANDREW D., 1991b Magnetic field evolution in steady chaotic flows. Preprint.

GILBERT, A.D. AND BAYLY, B.J., 1990 Magnetic field intermittency and fast dynamo action in random helical flows, preprint.

GILBERT, A.D. & CHILDRESS, S., 1990 Evidence for fast dynamo action in a chaotic web. *Phys. Rev. Lett.*, **65**, pp. 2133-2136.

KATO, T., 1966 *Perturbation theory of linear operators*, Springer-Verlag New York, Inc., p. 515.

GUCKENHEIMER, JOHN, & HOLMES, PHILIP, 1983 *Nonlinear Oscillations, Dynamical Systems, and Bifurcations of Vector Fields.* Springer-Verlag, New York, p. 285.

KLAPPER, I., 1991a The role of small diffusivity in passive scalar problems, preprint.

KLAPPER. I., 1991b On fast dynamo action in chaotic helical cells, preprint.

KLAPPER, I., 1991c Chaotic Fast Magnetic Dynamos. PhD Thesis, New York University.

KRAUSE, F. & RÄDLER, K.-H., 1981 *Mean Field Magnetohydrodynamics and Dynamo Theory*, Pergamon Press, Oxford.

LICHTENBERG, A.J. & LIEBERMAN, M.A., 1983 *Regular and Stochastic Motion*, Springer-Verlag, New York.

MOLCHANOV, S.A., RUZMAĬKIN, A.A. , & SOKOLOV, D.D., 1985 Kinematic dynamo in random flow. *Usp. Fiz Nauk*, **145**, pp. 593-628 (*Sov. Phys. Usp.*, **28**, pp. 307-327).

MOFFATT, H.K., 1978 *Magnetic Field Generation in Electrically Conducting Liquids*, Cambridge University Press.

MOFFATT, H.K. & KAMKAR, H., 1983 On the time-scale associated with flux expulsion. In *Stellar and Planetary Magnetism*, A.M. Soward, Ed. Gordon and Breach Science Publishers.

MOFFATT, H.K. & PROCTOR, M.R.E., 1985 Topological constraints associated with fast dynamo action. *J. Fluid Mech.*, **154**, pp. 493-507.

OTANI, NIELS F., 1989 Computer simulation of fast kinematic dynamos. *Trans. Geophys.*

Union , **69**, No. 44, Nov. 1, Abstract No. HS51-15, p. 1366.

PESIN, J.B., 1977 Characteristic Lyapunpov exponents and smooth ergodic theory. *Russ. Math. Surv.*, **32**, pp. 55-114.

ROBERTS, P.H. & SOWARD, A.M., 1991 Dynamo Theory. To appear in *Ann. Rev. Fluid Mech.*.

ROBERTS, G.O., 1970 Spatially periodic dynamos. *Phil. Trans. Roy. Soc. Lond.* , **A 266**, pp. 535-558.

ROBERTS, G.O., 1972 Dynamo action of fluid motions with two-dimensional periodicity. *Phil. Trans. Roy. Soc. Lond.*, **A 271**, pp. 411-454.

RUZMAĬKIN, A., SOKOLOFF, D., & SHUKUROV, D.D., 1988 A hydromagnetic screw dynamo. *J. Fluid Mech.*, **197**, pp. 39-56.

SOWARD, A.M. 1972 A kinematic theory of large magnetic Reynolds number dynamos. *Phil. Trans. Roy. Soc. Lond.*, **A 272**, pp. 431-462.

SOWARD, A.M. , 1987 Fast dynamo action in steady flow. *J. Fluid Mech.*, **180**, pp. 267-295.

SOWARD, A.M., 1990 A unified approach to a class of slow dynamos. *Geophys Astrophys. Fluid Dyn.* in press.

SOWARD, A.M. & CHILDRESS, S., 1990 Large magnetic Reynolds number dynamo action in spatially periodic flow with mean motion. *Trans. Roy. Soc. Lond. A*, **331**, pp. 649-733.

STRAUSS, H.R. 1986 Resonant fast dynamo. *Phys. Rev Lett.*, **57**, pp. 2231-2233.

VAINSHTEĬN, S.I. & ZEL'DOVICH, YA. B., 1972 Origin of magnetic fields in astrophysics. *Usp.Fiz. Nauk.*, **106**, pp. 431-457 (*Sov. Phys. Usp.*, **15**, pp. 159-172).

VISHIK, M., 1989 Magnetic field generation by the motion of a highly conducting fluid. *Geophys. Astrophys. Fluid Dyn.*, **48**, pp. 151-167.

ZASLAVSKIĬ, G.M., SAGDEEV, R.Z., & CHERNIKOV, A.A., 1988 Stochastic nature of streamlines in steady-state flows. *Sov. Phys. JETP*, **67** (2), pp. 270-277.

ZELDOVICH, YA.B., MOLCHANOV, S.A., RUZMAĬKIN, A.A., & SOKOLOFF, D.D., 1988 Intermittency, diffusion, and generation in a nonstationary medium. *Sov. Sci. Rev. C. Math. Phys.*, **7**, pp. 1-110.

PART II

RELAXATION AND MINIMUM ENERGY STATES

RELAXATION AND TOPOLOGY IN PLASMA EXPERIMENTS

J. B. TAYLOR
Institute for Fusion Studies
University of Texas
Austin, TX 78712, USA

ABSTRACT. Topological factors control the behavior of magnetised turbulent plasma in two ways. One is through constraints on the magnetic field. This leads to the concept of the relaxed state. The other is through the topology of the mechanical and magnetic boundaries of the plasma. Consequently, while the constraints are universal, the relaxed states themselves differ greatly from one situation to another. In a toroidal system there are two classes of relaxed state, one exhibiting field reversal and another exhibiting current saturation. The relaxed states of spherical systems depend on whether both the mechanical and magnetic boundaries, or only the mechanical boundary, is topologically spherical. In the first case the relaxed state is unique, in the other it depends on the properties of the boundary. These differences are discussed in connection with a number of experiments and the theoretical predictions are compared with the experimental data.

1. Introduction

In this talk I'd like to discuss two ways in which topological factors control the behavior of turbulent plasmas in actual magnetic confinement experiments. The first is through constraints on the magnetic field, the other is through the topology of the plasma container itself. I shall also discuss the role of symmetry.

The importance of topological factors in plasmas first emerged (Taylor, 1974 and 1986) in connection with the simple pinch. This experiment involves only a toroidal vessel in which a field B is first created by external coils (Fig. 1). Then a toroidal current I is induced. This heats and compresses the plasma. Initially the plasma is unstable and highly turbulent, but it then settles into a more quiescent state. In this state the mean magnetic field profiles are universal and depend only on a single parameter $\Theta = 2I/aB$. More surprisingly, if Θ exceeds a critical value ≈ 1.2, the toroidal field created by the plasma is reversed near the wall. This behavior is a clear indication that the plasma seeks a preferred configuration – the "Relaxed State".

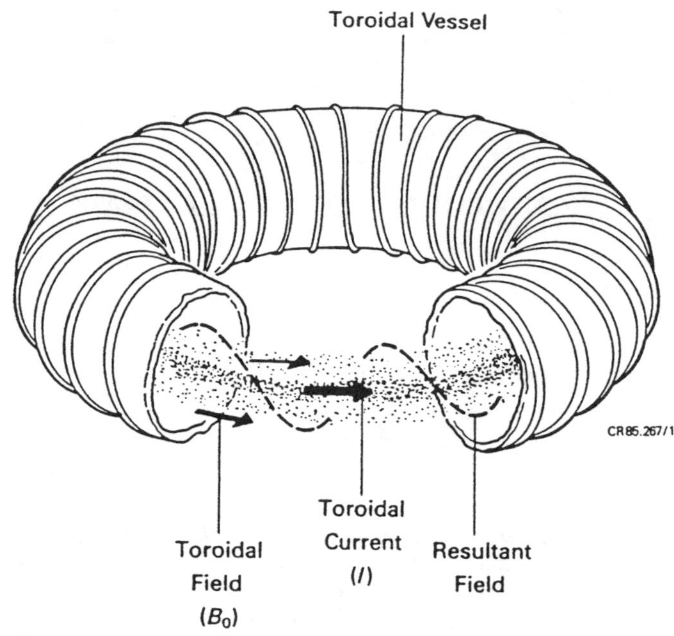

Fig. 1. The toroidal pinch

The Relaxed State is a configuration of minimum energy subject to appropriate constraints on the variation of the magnetic field. These constraints are due to the high conductivity of the plasma. If it were *perfectly* conducting, there would be an infinity of constraints which can be described by saying that for each closed field-line there is an invariant

$$K(\alpha, \beta) = \int \mathbf{A} \cdot \mathbf{B} d\tau \tag{1.1}$$

(where the integration volume is an infinitesimal flux tube surrounding the field-line). These invariants describe the linkages of field-lines and are *topological*.

However, the state of minimum energy under this infinity of constraints is simply any force-free equilibrium and is in no way a universal configuration.

To remedy this, we must recognise that the topological constraints (1.1) require the *identification* of individual field-lines. In a resistive plasma this is impossible – because field-lines break and rejoin at local concentrations of current – so that the individual field-line constraints are no longer valid. However, if the resistivity is small, the total magnetic helicity

$$K = \int_V \mathbf{A} \cdot \mathbf{B} d\tau \tag{1.2}$$

(where V is now the volume of the system) remains an effective invariant. Of course, K itself has a topological interpretation, but it differs from the individual $K(\alpha,\beta)$ in that it does not require identification of field-lines.

The state of minimum energy subject to constant total helicity satisfies

$$\nabla \times \mathbf{B} = \mu \mathbf{B} \tag{1.3}$$

The magnetic field in this state depends only on a single parameter μ, directly related to the pinch parameter by $\Theta = \mu a/2$.

Before I discuss the relaxed state itself I should remind you that the global helicity K is well-defined only if the boundary of V is a flux surface (*i.e.*, no field-lines penetrate the boundary). Otherwise it is not gauge invariant. This means that one can specify the total helicity inside a flux surface, but not where this helicity is located – any more than one can specify the location of the linkage between two closed loops! Nevertheless the helicity is not just an esoteric mathematical notion. Indeed it is closely related to the "Volt-seconds" stored in the discharge – a very solid electrical engineering concept.

To return to the relaxed state. Its full specification requires two independent quantities, one to fix μ, the other to fix the scale of B. In a toroidal system these two quantities are the global helicity K and the toroidal flux Ψ. In fact μa is a function of K/Ψ^2 and either K or Ψ suffices to set the scale. (Note that the relaxed state is completely determined from first principles without empirical parameters.)

2. Large Aspect Ratio Torus

In a large aspect ratio torus with circular cross section the field profiles in the relaxed state are given by

$$B_r = 0, \qquad B_\theta = B_0 J_1(\mu r), \qquad B_\phi = B_0 J_0(\mu r). \tag{2.1}$$

These agree well with the universal profiles observed in Pinch experiments (Fig. 2). The critical value for spontaneous field reversal is $\mu a = 2.4$, which also agrees well with experiment.

In fact, however, the determination of relaxed states is more complex than I have indicated. This is because Eq. (1.3) may have many solutions compatible with the given constraints. One must then select that solution which has the lowest energy.

One finds that there are just two possible solutions which may represent the relaxed state. One is the "primitive" solution already described. The other is a superposition of the primitive solution and the lowest eigenfunction – defined by $\nabla \times \mathbf{B} = \lambda \mathbf{B}$ with the supplementary condition that the toroidal flux be zero. (NB, without this condition Eq. (1.3) does not define an eigenvalue.)

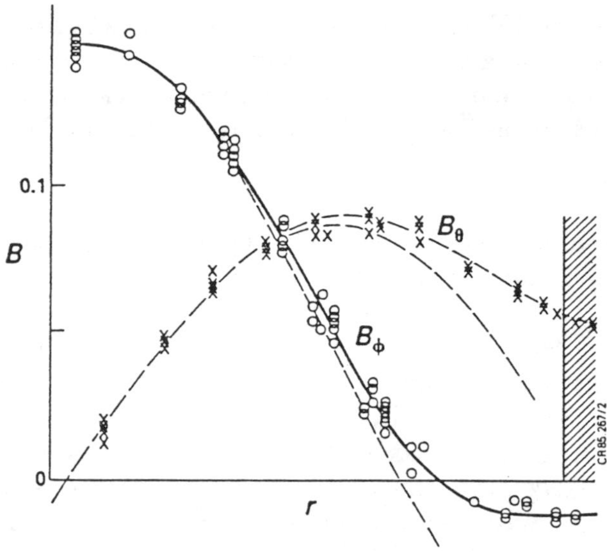

Fig. 2. Experimental and theoretical magnetic field profiles. HBTX1A. (From Bodin, 1984.)

When K/Ψ^2 is small the relaxed state is described by the primitive solution and μ is an increasing function of K/Ψ^2. However, when K/Ψ^2 is large, the relaxed state is described by the mixed solution. Then K/Ψ^2 no longer determines μ, which is "locked" at the eigenvalue ($\mu a = 3.2$): instead it determines the relative amplitudes of the primitive and eigenfunction contributions to the field. (Note that the relaxed state is again completely specified by the two invariants, but in a new way.)

Incidentally, the eigenfunction involved in the mixed state is not axisymmetric (it is helical); so the mixed state is an interesting example of spontaneous symmetry breaking!

A remarkable consequence of the fact that in the relaxed state μ never exceeds the lowest eigenvalue, is that there is a maximum plasma current (corresponding to $\Theta \approx 1.6$) in any toroidal pinch – regardless of how many volt-secs are applied. This current limitation has been confirmed in the HBTX experiment (Bodin and Newton, 1980). Fig. (3) shows the behaviour of Θ when a very high voltage was applied to create the discharge. Initially Θ rose to a large value but quickly dropped to ~ 1.6 and remained there for the duration of the discharge.

3. The Multipinch Experiment

The relaxed states in any axisymmetric torus have similar properties to those of the simple pinch described above – except in one respect. In that

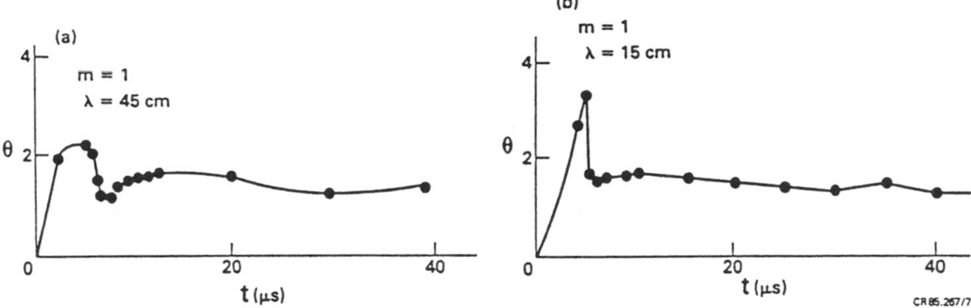

Fig. 3. Limitation of Θ. HBTX1. (From Bodin and Newton, 1980.)

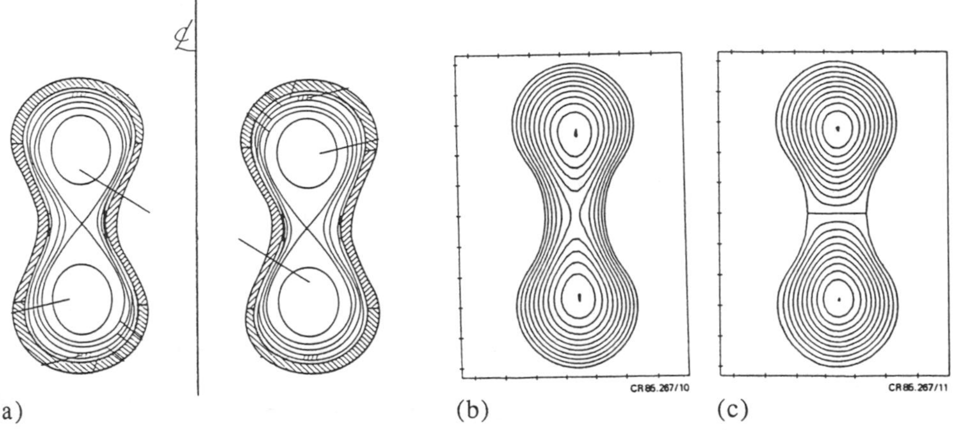

Fig. 4. (a) Multipinch experiment. (From La Haye, *et al.*, 1985); (b) Relaxed profile in Multipinch. $\mu a = 1.5$; (c) Eigenfunction for Multipinch. $\mu a = 2.21$.

example axial symmetry is broken by the eigenfunction in the mixed state. However in some systems the eigenfunction is itself also axisymmetric: but spontaneous symmetry breaking may still occur! Such behaviour occurs in the Multipinch.

The Multipinch (La Haye *et al.*, 1986) is an axisymmetric toroidal experiment with a non-circular cross-section having the shape of a figure-of-eight (Fig. 4). For this configuration the lowest eigenfunction is axisymmetric, but it is antisymmetric between the two lobes of the figure-eight (giving zero toroidal flux as required).

The experiment dramatically illustrates current saturation in the mixed state. Fig. (5) shows the plasma current I as a function of the driving transformer voltage V_{CB} (roughly proportional to Volt-secs). At low voltage the plasma current, which is equally shared between the two lobes of the cross-

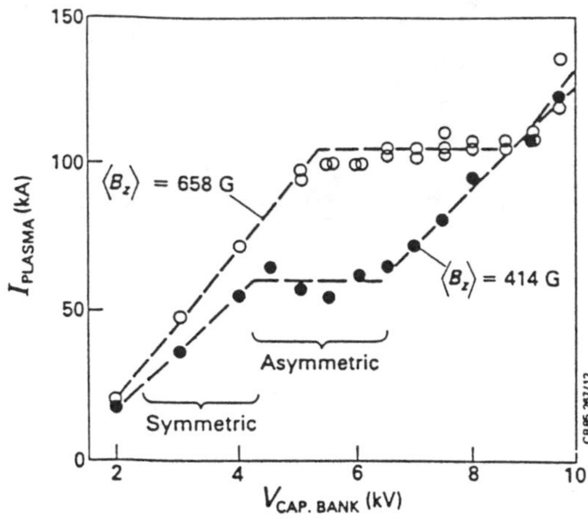

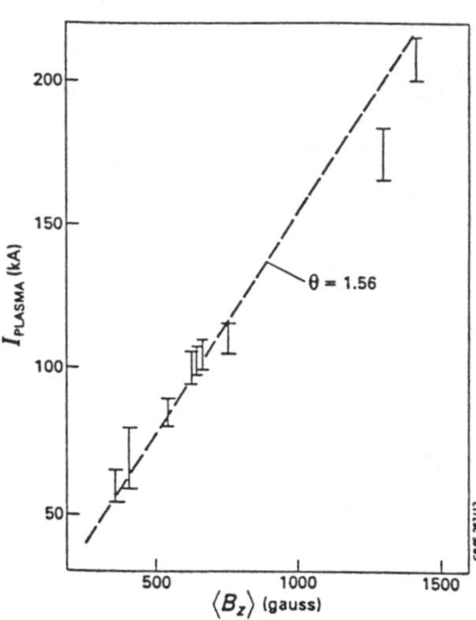

Fig. 5. Plasma current vs. driving voltage in Multipinch. (From La Haye, *et al.*, 1985.)

Fig. 6. Variation of I_{sat} with toroidal flux in Multipinch. (N.B. $\Theta = 1.56$ corresponds to $\mu a = 2.42$ in this configuration. (From La Haye, *et al.*, 1985)

section, increases with V_{CB}. At higher voltage the current saturates and more flows in one lobe than the other. This asymmetry increases as V_{CB} is further increased.

Fig. (6) shows that the saturation current is proportional to the toroidal flux, as the theory predicts. The slope of the line corresponds to $\mu a = 2.42$ – in remarkable agreement with the calculated eigenvalue $\mu a = 2.21$.

In theory, the asymmetry in the current-saturated state should increase indefinitely with V_{CB} until eventually the current in one lobe of the cross-section would reverse. In practice the asymmetry increases only until it falls to zero. Beyond this point the discharge is confined to the other lobe and behaves much as a simple pinch.

Before leaving the discussion of relaxed states in toroidal systems, it is useful to note that the theory has a particularly concise form in terms of the vector potential: the relaxed state is a solution of

$$\nabla \times \nabla \times \mathbf{A} = \mu \nabla \times \mathbf{A} \tag{3.1}$$

and is completely specified by the helicity K and the values of $\oint \mathbf{A} \cdot d\mathbf{l}$ and $\oint \mathbf{A} \cdot d\mathbf{s}$, the loop-integrals the long and short way around the toroidal boundary. The eigenfunction is specified by $\oint \mathbf{A} \cdot d\mathbf{s} = 0$.

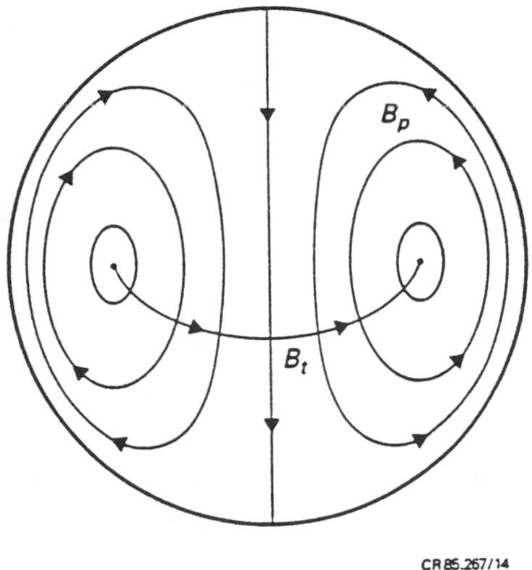

Fig. 7. Spheromak configuration. Schematic.

4. Spherical Systems

So far we have discussed only toroidal systems. However, relaxed states are equally important in spherical configurations, of which the prototype is the Spheromak (Rosenbluth and Bussac, 1979) (Fig. 7).

In this configuration the magnetic field has toroidal surfaces, as in the toroidal pinch, but it is topologically distinguished from it by the fact that there is no central aperture for toroidal field coils. (Consequently the toroidal field is zero everywhere on the boundary of a Spheromak.)

It is difficult to produce high temperature plasma in a Spheromak. A popular method (Turner *et al.*, 1983; Kitson and Browning, 1990) uses a co-axial gun to inject plasma into a confinement chamber (Fig. 8). Plasma formed in the gun carries both poloidal field, provided by coils within the gun, and toroidal field produced by plasma currents created when a high voltage is applied across the coaxial electrodes. The magnetic forces then eject the plasma into the container (known as the flux-conserver) where it relaxes to a Spheromak configuration.

As in the toroidal case, the helicity is conserved during relaxation and the relaxed state satisfies Eq. (1.3). Despite this similarity, however, the change in topology substantially alters the theory. As we have seen, in a torus there are two invariants, the helicity K and the toroidal flux Ψ, and these are just sufficient to specify the relaxed state. But in a Spheromak, toroidal flux is not an invariant quantity; it can be annihilated and created at the axis of symmetry. Consequently there is only a single invariant K for a spherical system.

On the other hand, because the interior of a Spheromak is a simply

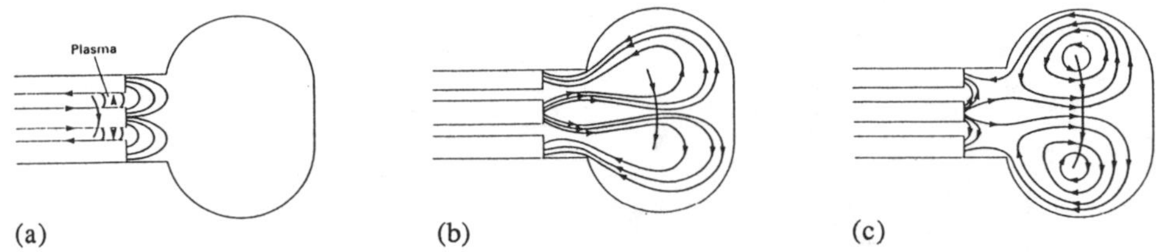

Fig. 8. Magentic field equilibria in gun-injected spheromak. SPHEX formation. The plasma is formed between two concentric annular electrodes (a) and is accelerated by its own $\mathbf{J} \times \mathbf{B}$ force into the flux conserver, taking with it the field from the solenoid inside the inner electrode (b), (c).

connected region, there is no distinction between $\oint \mathbf{A} \cdot d\mathbf{l}$ and $\oint \mathbf{A} \cdot d\mathbf{s}$ and both integrals vanish. But this is precisely the eigenvalue condition so the *only* solutions of Eq. (1.3) for a spherical system are the eigenfunctions and μ can only be an eigenvalue. Consequently, in a sphere, there is no primitive solution with a range of μ, and no mixed solution. The relaxed state is just the lowest eigenfunction and is determined by the shape of the container. The invariant K merely determines the magnitude of the magnetic field.

The magnetic fields observed in Spheromak experiments agree well with the calculated eigenfunctions. An example is shown in Fig. (9) taken from the BETA II experiment (Turner *et al.*, 1983). Further confirmation of relaxation in Spheromaks is illustrated in Fig. (10), taken from the S-1 experiment (Hart *et al.*, 1986). This shows the poloidal current against the poloidal flux across a diameter of the plasma. It confirms that current is proportional to magnetic field, as required of a relaxed state, and the slope of the line agrees remarkably well with the eigenvalue calculated for the (ellipsoidal) flux conserver used in the S-1 experiment.

A very striking demonstration of relaxation (Janos *et al.*), also from S-1, is provided by Fig. (11). This shows the *evolution* of the poloidal and toroidal magnetic flux *during* the relaxation phase. It can be seen that during relaxation poloidal flux is destroyed and toroidal flux is created, until their ratio (represented in Fig. (11c) by the pitch-length q of field lines near the magnetic axis) reaches the calculated theoretical value (corresponding to $\Theta = 0.65$). Thereafter the ratio remains constant.

5. Flux Core Spheromaks (FCS)

An interesting, and topologically distinct, development of the Spheromak is the configuration obtained from it by introducing a central core of externally produced magnetic flux along the axis of symmetry (Fig. 12). This externally linked flux enters through one polar-cap and leaves through the other. The resulting configuration may appear to resemble a Toroidal Pinch,

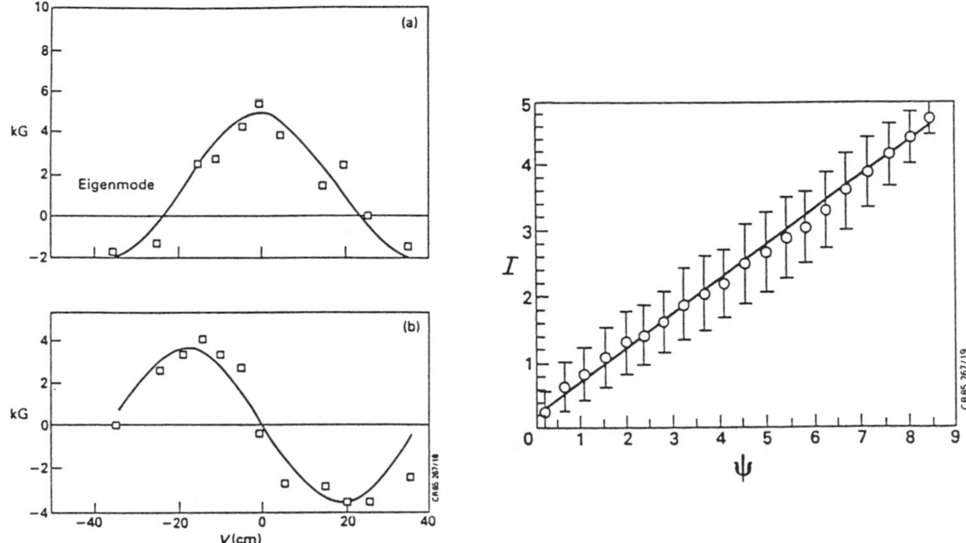

Fig. 9. Magnetic filed in BETA II Spheromak. Experiment and theory. (a) Poloidal field; (b) toroidal field. (From Turner, *et al.*, 1983.)

Fig. 10. Poloidal current vs. poloidal flux in Spheromak (S-1). (From Hart *al.*, 1985.)

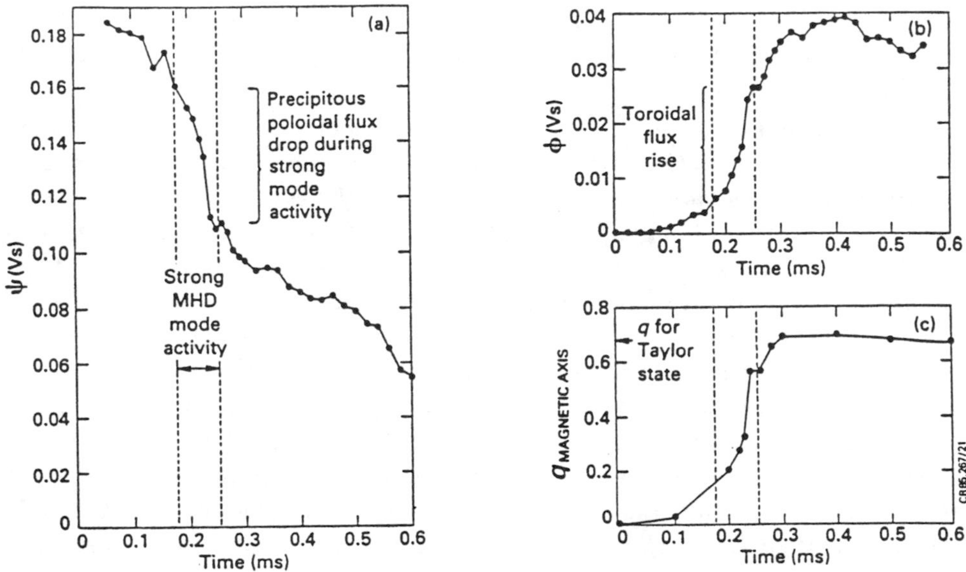

Fig. 11. Time dependence of magnetic fields in S-1 Spheromak. (From Janos *et al.*, 1985.)

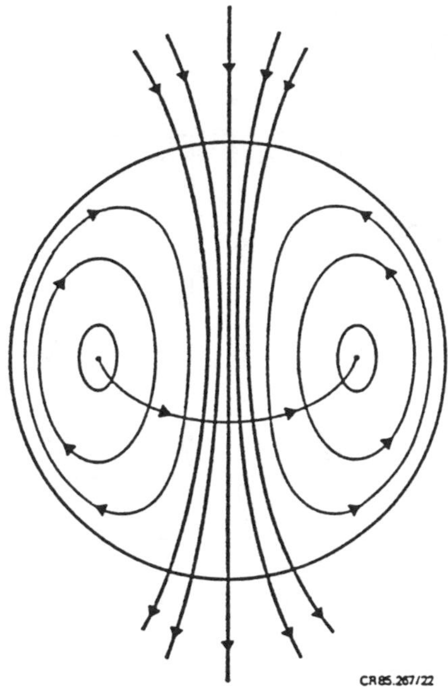

Fig. 12. Flux core Spheromak. Schematic.

but, as in the simple Spheromak, toroidal flux is not invariant and in that respect the FCS is topologically toroidal. However, because the magnetic field is linked through the polar caps it is also topologically distinct from the simple Spheromak.

As the boundary is not a flux surface, the helicity K is not immediately well-defined in a FCS. One way to rectify this is to imagine that the flux leaving and entering the boundary is extended through the exterior as a vacuum field. Then the total helicity $\int \mathbf{A} \cdot \mathbf{B}$ inside and outside the sphere is well defined. Furthermore, if the surface of the sphere is perfectly conducting, the normal component of $\mathbf{B}$ is "frozen in" and changes in the interior do not affect the hypothetical external field. The difference in helicity between two fields which differ inside the Spheromak but have identical normal components on the boundary (and hence identical hypothetical extensions outside) may be defined as their "relative helicity". This is well-defined and gauge invariant. (Note that it is necessary to include the contribution to $\int \mathbf{A} \cdot \mathbf{B}$ from both the interior and exterior regions even though the exterior field does not change. This reflects the fact that helicity is not a local quantity.)

Incidentally, the conceptual external extension of the internal field, brings out the unique topology of the FCS. The *plasma* is contained within a sphere, but the *field* is contained within a torus!

If the plasma boundary is an equipotential, the relative helicity is constant

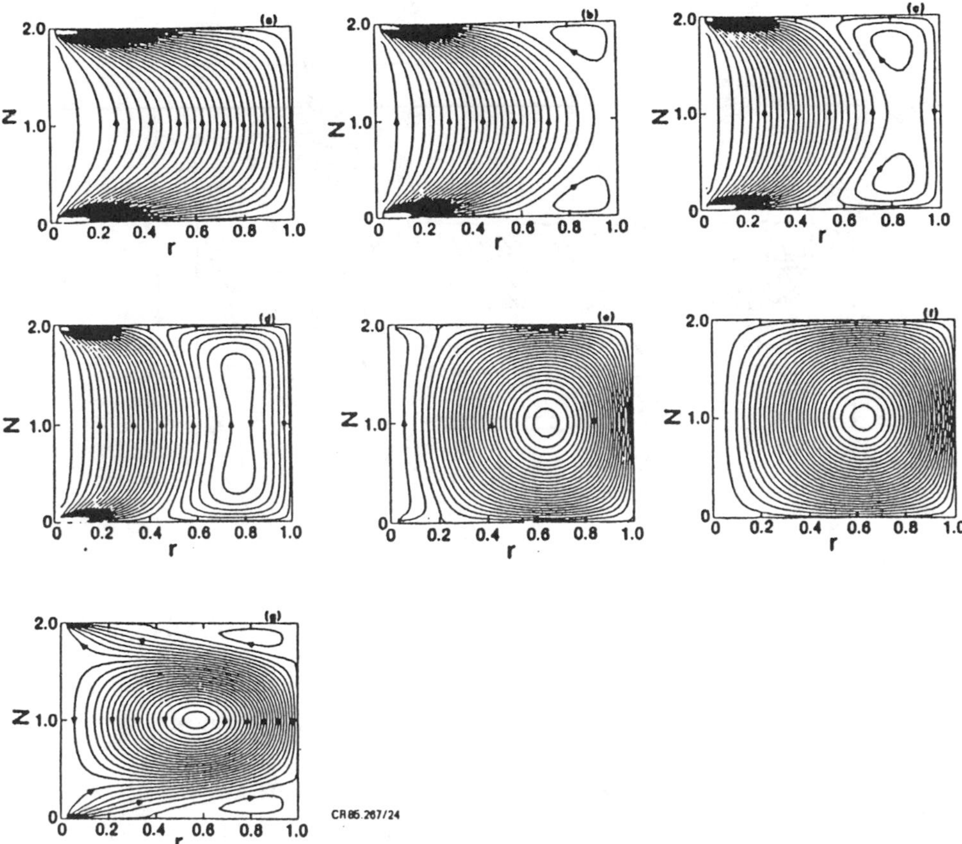

Fig. 13. Field in idealised FCS (vanishing polar cap radius) at various values of μa: (a) $\mu a = .001$, (b) $\mu a = 2.25$, (c) $\mu a = 2.70$, (d) $\mu a = 3.00$, (e) $\mu a = 4.00$, (f) $\mu a = 4.14$, (g) $\mu a = 5.00$. (From Turner, 1984.)

during relaxation. However, relative helicity may be created if one of the polar caps is maintained at a different potential to the other. The rate of change of helicity is

$$\frac{dK}{dt} = 2V_p\Psi_p \tag{5.1}$$

where V_p is the voltage between the polar caps and Ψ_p is the flux through them. Note again the connection with volt-secs.

The relaxed state of the FCS is obtained by minimising the energy subject to two invariants, the (relative) helicity and the flux Ψ_p through the polar caps. Once again this leads to equation (1.3), but there are two new ways in which the appropriate solution is to be selected. In the first μ is determined by the ratio K/Ψ_p^2 (much as the primitive state in a torus is determined by K/Ψ^2) and Ψ_p determines the scale. This is appropriate if the polar caps

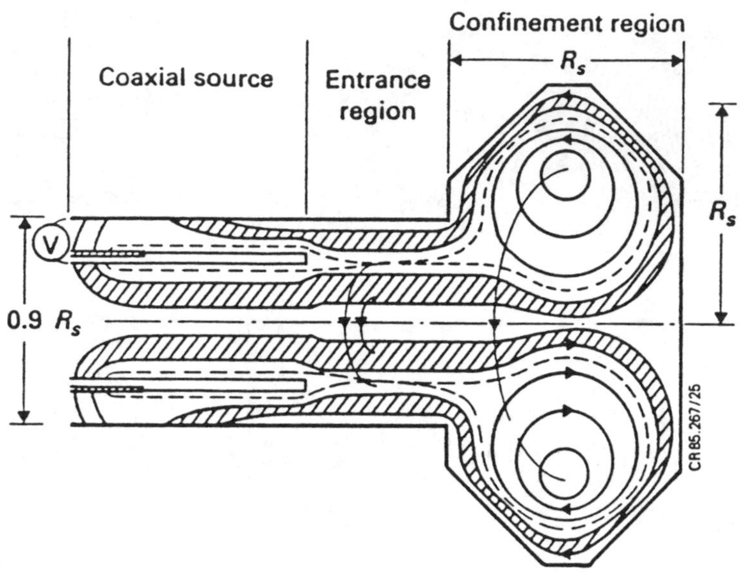

Fig. 14. Sustained configuration in CTX experiment. (From Jarboe, 1985.)

can supply the current required for the relaxed state without introducing large potentials.

Alternatively if one of the polar caps is electrically insulated from the rest of the container, μ can be selected by controlling the current I_p allowed through that polar cap ($\mu = I_p/\Psi_p$). In this case the helicity K is not independently specified but is forced to conform with μ by the voltage required to control the current I_p.

An interesting example of the field in a FCS is shown in Fig. (13). This shows an idealisation in which the polar-caps are shrunk to points (though retaining finite flux through them) and illustrates the the effect of increasing μ. When μ is much smaller than the lowest eigenvalue, the externally linked flux forms a large part of the total flux. As μ approaches the eigenvalue the self-generated flux increases indefinitely and the externally linked flux is reduced to a narrow pencil on the axis. As μ exceeds the eigenvalue the externally linked flux flips to the outside of the Spheromak. (However, one would not expect this to be a stable situation.)

A schematic diagram of a real Flux Core Spheromak experiment (Jarboe, 1985) is shown in Fig. (14). This may not immediately look like Fig. (12), but closer examination reveals a core of flux (hatched) passing from the inner electrode of the gun, along the axis of the Spheromak and returning around it to the outer electrode. From the topological point of view, the gun-voltage is applied across the flux-core, though the "polar caps" are somewhat remote!

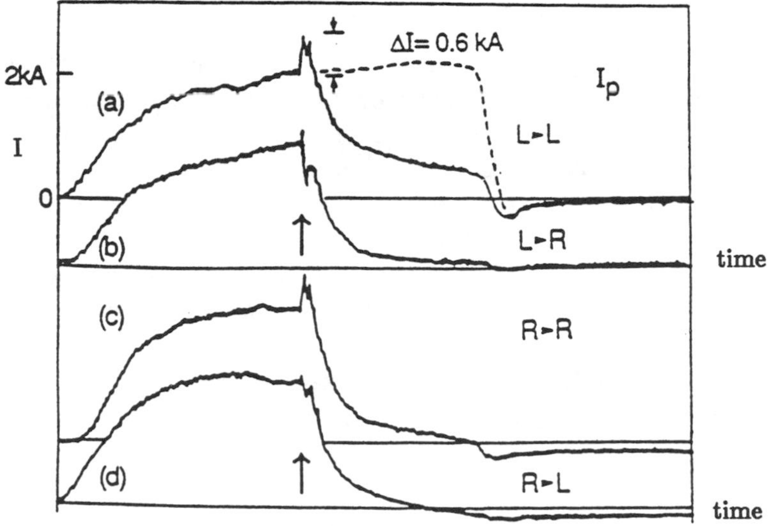

Fig. 15. Spheromak injection into Tokamak. (Arrow marks time of injection.)

6. Spheromak Injection

Finally, I would like to describe an experiment (Brown and Bellan, 1991) which involves both spherical and toroidal plasmas. In this a small Spheromak is injected into a toroidal (Tokamak) discharge from a coaxial plasma gun placed in a side-arm from the main toroidal chamber. In operation a normal Tokamak discharge is first formed in the main torus, then a Spheromak is produced in the gun and injected into the torus.

To understand the result we must recall that helicity is a pseudo-scalar, *i.e.*, it has a "handedness" (which *e.g.* is reversed if the relative directions of the initial magnetic field and the initial current are reversed in the Tokamak). Thus there are four permutations of the experiment; a left- or right-hand Spheromak injected into a left- or right-hand Tokamak.

The effect of injection on the toroidal current in the four cases is shown in Fig. (15). (The broken line is the behavior if there is no injection.) It can be seen that when injection is "favourable", *i.e.* when the Spheromak and Tokamak have the same handedness, as in cases (a) and (c), it produces a rise in the toroidal current. Conversely, when the injection is unfavourable, as in cases (b) and (d), it produces a drop in the current. (In all cases the initial change is followed by a collapse of the current; this is due to the influx of cold plasma and gas from the gun.)

7. Summary and Conclusions

We have seen that topological factors control the behavior of turbulent plasmas in magnetic containment experiments. One way they do so is through topological constraints on the magnetic field. The other is through the topology of the mechanical and magnetic boundaries of the plasma. As a result, although the constraints are universal, their consequencies are quite different in different experimental situations.

In any toroidal system the relaxed state is defined by two invariants, the helicity K and the toroidal flux Ψ. However it has different properties according as K/Ψ^2 is less or greater than a critical value. In the first case it exhibits field reversal; in the second it exhibits current saturation.

In a spherical system, if both the magnetic and mechanical boundaries are spherical, there is only one invariant, but this alone determines the unique relaxed state.

If the mechanical boundary is spherical but the magnetic field penetrates it, and is therefore topologically toroidal, the relaxed state depends on the electrical properties of the boundary. If this is equipotential and emits current freely, the relaxed state is determined by two invariants, the relative helicity and the flux through the boundary. Alternatively the relaxed state may be selected externally by controlling the current through the boundary. (In this case the helicity "adiabatically" adjusts to conform to the externally controlled current.)

The behaviour predicted by the theory in all these different topological situations has been *quantitatively* confirmed in many different experiments. It seems clear that magnetised plasmas do indeed recognise the importance of topological concepts and obey them.

References

N.B., the results covered here have been selected from a wide range of investigations. A comprehensive list of references can be found in Taylor (1986), below.

Bodin, H.A.B. and Newton, A.A., 1980 Reversed field pinch research. *Nucl. Fusion* **20**, 1255–1324.

Brown, M.R., and Bellan, P.R., 1991 Spheromak injection into Tokamak. *Phys. Fluids* **B2**, 1306–1310.

Hart, G.W., et.al., 1986 Verification of Taylor state in spheromak. *Phys. Fluids* **29**, 1994–1997.

Janos, A., Hare, G.W., Nam, C.H. and Yamada, M., 1985 Global magnetic fluctuations in spheromak plasmas and relaxation to minimum energy state. *Phys. Fluids* **28**, 3667–3673.

Jarboe, T.R., 1985 A kinked Z-pitch as helicity source for Spheromak. *Comments Plasma Physics* **9**, 161–168.

Kitson, D.A., and Browning, P.K., 1990 Partially relaxed magnetic field equilibria. *Plasma Physics and Controlled Fusion* **32**, 1265–1287.

La Haye, R.J., et.al., 1986 Multipinch – a reversed field pinch with magnetic well. *Nucl. Fusion* **26**, 255-265.

ROSENBLUTH, M. N., AND BUSSAC, M. N., 1979 MHD Stability of spheromak. *Nucl. Fusion* **19**, 489-498.

TAYLOR, J.B., 1974 Relaxation of toroidal plasma and generation of reversed magnetic field. *Phys. Rev. Lett.* **33**, 1139 -1141.

TAYLOR, J.B., 1986 Relaxation and magnetic reconnection in plasmas. *Rev. Mod. Phys.* **58**, 741-763.

TURNER, W.C., ET.AL., 1983 Investigation of magnetic structure and decay of compact torus. *Phys. Fluids* **26**, 1965-1986.

TAYLOR'S RELAXATION IN AN UNBOUNDED DOMAIN OF SPACE

J.J. ALY
Service d'Astrophysique
C.E. Saclay
91191 Gif sur Yvette Cedex, France

ABSTRACT. We discuss the problem of the existence of a minimum energy state among all the magnetic fields: i) occupying some unbounded domain of space D (which may be either the exterior of a bounded region, or a half-space, or a half-cylinder); ii) having a given normal component on the boundary of D; iii) having a prescribed relative helicity. We show that, depending on the shape of D and on the values of the parameters, this problem has either a unique solution (which is the unique constant-α force-free field satisfying the constraints), or no solutions at all ("minimizing sequences" losing, when passing to the limit, a part of their energy and helicity, which gets diluted in the infinite volume of D). This result is used in particular to discuss some recent attempts to extend to astrophysical systems (*e.g.* the coronae of stars and of accretion disks) Taylor's theory describing the turbulent relaxation of a plasma in some laboratory devices.

1. Introduction

The aim of this *Communication* is to discuss the possibility to extend to unbounded domains D of some particular types (exterior of a bounded region; half-space; half-cylinder) the following result: *The set of all the magnetic fields occupying some bounded domain D and having given relative helicity and normal component on the boundary (and satisfying also some "period" conditions if D is multiply connected),contains a minimum energy state which is a linear force-free field with the lowest possible value of* $|\alpha|$ (α is a number measuring the ratio of the electric current density to the field at each point) (Taylor (1986) and references therein; Berger (1985); Laurence and Avallaneda (1991)). This statement is well known in plasma physics where it constitutes in particular one of the building blocks of the theory proposed by Taylor (see *e.g.* Taylor (1986)) for explaining a remarkable phenomenon occurring in several laboratory devices: whatever the state in which the plasma is created inside these machines, it is observed to relax through a short highly turbulent phase to a quiescent state with a magnetic

H. K. Moffatt et al. (eds.), Topological Aspects of the Dynamics of Fluids and Plasmas, 167–175.

structure well approximated by a linear force-free field. According to Taylor, this can be understood by combining the mathematical result above with the physically quite reasonable assumption that the plasma is seeking the minimum energy state compatible with the value of the helicity initially injected into the field—helicity being actually the only "ideal" invariant which should be also conserved in the presence of strong turbulence, when all the magnetic lines suffer repeated reconnections.

Our motivation for looking at this problem comes essentially from astrophysics. In many cosmic situations, one has to consider the structure of an unbounded region D (*e.g.* the corona of a star or of an accretion disk) which contains a magnetic field embedded in a low-beta highly conducting plasma. The bulk of current research on this type of system is largely based on a general picture which has progressively emerged during the last two decades, and describes its evolution as follows: most of the time, the field evolves quasi-statically through a sequence of nonlinear force-free configurations because of the motions which are *imposed* on its footpoints on the boundary ∂D by the dense slowly moving plasma in which they are firmly anchored—helicity being thus continuously injected into that field (or withdrawn from it) at a given rate; sporadically, however, the field suffers short phases of dynamical evolution because of the development of instabilities which lead to fast reconnection processes *during which helicity must be (approximately) conserved,*—these events being for instance at the origin of big flare-like phenomena in which there is a global rearrangement of the magnetic structure (Aly (1990a)) or of a general heating of the plasma when magnetic energy is released and converted in a more diffuse way (Heyvaerts and Priest (1984), Browning (1988)). Thus—and this is the important point here—the magnetic helicity can be considered at each time as a given quantity, and the statement we want to discuss would be clearly quite useful (if true!), as it would allow us to compute at each time a lower bound (actually independent of the particular structure of the field!) on the magnetic energy stored in D, as well as a upper bound on the energy which could be spontaneously released in a fast transition starting from a force-free state.

2. Statement of the Problem

Let D be a simply connected unbounded domain of space and $\mathcal{H}$ be the set of all the magnetic fields $\mathbf{B}$ satisfying the conditions (in addition of the usual constraint: $\nabla.\mathbf{B} = 0$):

i) The normal component B_n of $\mathbf{B}$ on the boundary ∂D of D is equal to a given function q (with zero flux through ∂D).

ii) $\mathbf{B}$ has a finite energy, *i.e.*

$$C[\mathbf{B}] := \int_D |\mathbf{B}|^2 d\mathbf{r} < \infty. \tag{2.1}$$

iii) **B** has a relative helicity (Berger and Field (1984), Finn and Antonsen (1986)) equal to a prescribed value H, *i.e.*

$$H[\mathbf{B}] := \int_D (\mathbf{A} + \mathbf{A}_0).(\mathbf{B} - \mathbf{B}_0)\, d\mathbf{r} = H, \tag{2.2}$$

where $\mathbf{B}_0$ is the current-free field in D ($\nabla \times \mathbf{B}_0 = 0$) satisfying (i) and (ii) above, and $\mathbf{A}$ and $\mathbf{A}_0$ are potential vectors for $\mathbf{B}$ and $\mathbf{B}_0$, respectively.

We want to discuss for some particular domains D (exterior of a bounded region; half-space; half-cylinder) the existence and the nature of a minimum energy state in $\mathcal{H}$, *i.e.* of a field $\mathbf{B}^- \in \mathcal{H}$ such that

$$C[\mathbf{B}^-] = \inf_{\mathcal{H}} C[\mathbf{B}]. \tag{2.3}$$

Our strategy for solving this problem is straightforward:

i) We first note that, if $\mathbf{B}^-$ does exist, then it must satisfy the Euler-Lagrange equation associated with the minimization problem (3). This equation is readily found to be

$$\nabla \times \mathbf{B}_\alpha = \alpha \mathbf{B}_\alpha \;\; in \; D, \tag{2.4}$$

where α is some constant (to be determined *a posteriori*) which introduces itself as a Lagrange multiplier relative to the constraint (2).

ii) Thus we determine all the solutions of (4) which belong, for some value of α, to our set $\mathcal{H}$. For the types of domain we consider here, it turns out that there is either no such solution or only one.

iii) If there is none, we are done: problem (3) has no solution. If there is one —call it $\mathbf{B}_\alpha$—we try to compare its energy with that of all the other fields of $\mathcal{H}$.

iv) For that, we use in particular a formula for $\delta C[\mathbf{B}] := C[\mathbf{B}] - C[\mathbf{B}_\alpha]$ due originally to Berger (1985). We combine it with some general mathematical inequalities to get a lower bound on $\delta C[\mathbf{B}]$. If this bound is positive, we can thus conclude that $\mathbf{B}_\alpha$ is a minimum energy state in $\mathcal{H}$ indeed. If it is not, we apply Berger's formula to some test functions $\mathbf{B}^{(h)}$ depending on the parameters h, that we try to adjust to make $\delta C[\mathbf{B}^{(h)}] < 0$. If we can do it, we conclude that our minimization problem does not admit any solution.

3. The Minimization Problem in an External Domain and in a Half-Space

3.1. EXISTENCE OF A FIELD $\mathbf{B}_\alpha$ IN $\mathcal{H}$

Let us first assume that D is the exterior of some bounded region V and suppose that the corresponding set $\mathcal{H}$ contains a constant-α force-free field $\mathbf{B}_\alpha$, with $\alpha \neq 0$.Then a simple application to each of its cartesian components $B_{\alpha i}$ (which are analytic functions satisfying in D the Helmholtz's equation: $(\nabla^2 + \alpha^2)B_{\alpha i} = 0$ (Aly,1991a)) of a theorem due to Rellich (see *e.g.* Colton and Kress (1983), Section 3.3) shows that $\mathbf{B}_\alpha$, being of finite energy, must vanish outside a sphere of radius large enough to enclose V—and thus in the whole domain D by analyticity (this point was first made by Seehafer (1978) for the case when V is a sphere). For domains of the exterior type, we can thus conclude at once that the only field which could be an energy minimizer is the potential field $\mathbf{B}_0$, Rellich's theorem not applying when $\alpha = 0$. But $\mathbf{B}_0$ belongs to $\mathcal{H}$ only if $H = 0$—in which case it is a minimum energy state indeed by the well known Thompson's principle. Thus we can conclude that our minimization problem has a (unique) solution in $\mathcal{H}$ if and only if $H = 0$.

When D is the half-space $\{z > 0\}$, Rellich's theorem, which concerns the behaviour of a solution of Helmholtz's equation outside a sphere, clearly does not apply, and the situation is a bit more complicated, but if $H = 0$, in which case $\mathbf{B}_0$ is as usual the unique energy minimizer in $\mathcal{H}$. For $|H| > 0$, the existence of a field $\mathbf{B}_\alpha$ in $\mathcal{H}$ turns out to depend on the imposed data. It may be shown indeed that (Alissandrakis (1981), Berger (1985), Aly (1991a)):

There exists a critical value $H_\kappa \geq 0$ (depending on the boundary function q) such that:

- *For $0 \leq |H| \leq H_\kappa$, $\mathcal{H}$ contains one and only one field $\mathbf{B}_\alpha$.*
- *For $H_\kappa < |H| < \infty$, there is no field $\mathbf{B}_\alpha$ in $\mathcal{H}$.*

The value of H_κ is strictly positive if $\kappa > 0$, where $\kappa = \sup\{s|\hat{q}(\mathbf{k}) = 0, \forall\mathbf{k}$ such that $|\mathbf{k}| \leq s\}$, with $\hat{q}(\mathbf{k})$ being the Fourier transform of q.

Therefore, when $|H| > 0$ and $|H| > H_\kappa$ (a relation which should hold in the general case, as for "most" boundary functions q, $\kappa = 0$), we are in a situation similar to that we faced when D is an exterior domain, and we can conclude that $\mathcal{H}$ contains no minimum energy state. For $\kappa > 0$ and $0 < |H| \leq H_\kappa$, however, $\mathcal{H}$ contains a field $\mathbf{B}_\alpha$, and we need to go ahead one more step to conclude.

3.2. IS $\mathbf{B}_\alpha$ A MINIMUM ENERGY STATE IN $\mathcal{H}$ WHEN D IS A HALF-SPACE AND $= |H| \leq H_\kappa$?

Let us then consider the case $D = \{z > 0\}$ and $0 < |H| \leq H_\kappa$, for which the arguments above have not allowed us to reach a firm conclusion, just indicating that there is in $\mathcal{H}$ one serious candidate for the job of energy minimizer. It turns out, however, that this field $\mathbf{B}_\alpha$ is not good enough, and is unable to go successfully through the second test. It can be proved indeed (Aly (1991a)) that it is possible to construct in $\mathcal{H}$ a sequence of fields $\mathbf{B}^{(k)}$ such that

$$\lim_{k\to\infty} C[\mathbf{B}^{(k)}] = C[\mathbf{B}_0]. \tag{3.1}$$

As (by the principle already quoted) $C[\mathbf{B}_0] < C[\mathbf{B}]$ for any $\mathbf{B}(\neq \mathbf{B}_0) \in \mathcal{H}$, it thus turns out that $\inf_{\mathcal{H}} C[\mathbf{B}] = C[\mathbf{B}_0]$, and that the lower bound on the energy could be reached in $\mathcal{H}$ only by $\mathbf{B}_0$, would that field belong to that set. As this last property holds true only when $H = 0$, we need to conclude that, although the energy is bounded from below and the associated Euler-Lagrange equation has a solution in $\mathcal{H}$, our minimization problem has no solution if $|H| > 0$. It is worth noticing that the possibility of finding fields in $\mathcal{H}$ with an energy as close as we want to that of the potential field $\mathbf{B}_0$, also holds in all the situations where we have found in Subsection 3.1 that $\mathcal{H}$ does not contain a constant-α force-free field $\mathbf{B}_\alpha$ (see the discussion in (Berger, 1985) for the case when D is the exterior of a sphere, and in (Aly, 1991a) for the case when D is a half-space).

To summarize this Section: we have proved that, except in the simple case $H = 0$, there is no minimum energy state having given H and B_n when D is an external domain or a half-space.

4. The Minimization Problem in a Half-Cylinder

4.1. ASSUMPTIONS AND EXISTENCE OF A FIELD $\mathbf{B}_\alpha$ IN $\mathcal{H}$

Let us now consider the case when $D = \{r < R,\ z > 0\}$, ((r, ϕ, z) being a standard cylindrical coordinate system and R a given constant). We choose a boundary function which vanishes on the vertical part of ∂D (*i.e.* on $\{r = R,\ z > 0\}$) and is ϕ-independant on its horizontal part $\{z = 0\}$, where we thus have

$$q(r, 0) = B_0 \sum_{i=N}^{\infty} c_i J_0(\nu_i r/R), \tag{4.1}$$

with B_0 and c_i being constants (with $c_N \neq 0$), and ν_i being a zero of the Bessel's function J_1.

With these assumptions, it may be shown that (Aly (1991b)), *whichever be the value of H, $\mathcal{H}$ contains one and only one constant-α force-free field* $\mathbf{B}_\alpha$ (which is then axisymmetric, and can actually be computed explicitly). The corresponding value of α is found to satisfy the inequality: $|\alpha| < \nu_N/R$.

4.2. IS $\mathbf{B}_\alpha$ A MINIMUM ENERGY STATE IN $\mathcal{H}_{AX}$?

Let us first compare the energy of $\mathbf{B}_\alpha$ with that of all the axisymmetric fields of $\mathcal{H}$ (we shall denote by $\mathcal{H}_{ax}$ that subset of $\mathcal{H}$, to which $\mathbf{B}_\alpha$ belongs, as noted above). Following the general strategy outlined in Section 2 above, it is possible to prove the following result (Aly (1991b)):

– *If* $N = 1$ *(* $c_1 \neq 0$ *and* $R|\alpha| < \nu_1 = 3.82$*) and:* $-\infty < H < +\infty$*, then* $\mathbf{B}_\alpha$ *is the unique energy minimizer in* $\mathcal{H}_{ax}$*.*

– *If* $N > 1$ *(* $c_1 = 0$ *):*
  - *For* $|H| \leq H_c^{ax} = H[\mathbf{B}_{\nu_1/R}]$*,* $\mathbf{B}_\alpha$ *is the unique energy minimizer in* $\mathcal{H}_{ax}$*.*
  - *For* $H_c^{ax} < |H|$*, there is no minimum energy state in* $\mathcal{H}_{ax}$*.*

In this last case, "minimizing sequences" (*i.e.* sequences $\{\mathbf{B}^{(k)}\}$ with $\mathbf{B}^{(k)} \in \mathcal{H}_{ax}$ and $\lim_{k\to\infty} C[\mathbf{B}^{(k)}] = \inf_{\mathcal{H}_{ax}} C[\mathbf{B}] = C[\mathbf{B}_{\nu_1/R}] + (\nu_1/R).(H - H_c^{ax})$) converge to $\mathbf{B}_{\nu_1/R} \notin \mathcal{H}_{ax}$, a part of their energy and helicity getting diluted in the unbounded domain D when $k \to \infty$.

4.3. IS $\mathbf{B}_\alpha$ A MINIMUM ENERGY STATE AMONG THE AXISYMMETRIC FIELDS OF $\mathcal{H}$?

Let us now consider the possibility for $\mathbf{B}_\alpha$ to minimize the energy in the whole set $\mathcal{H}$. It is possible to show that (Aly (1991b)):

There exists a critical value H_c *of* $|H|$ *(with* $H_c < H_c^{ax}$ *depending on* q*) such that:*

– *If* $|H| \leq H_c$*, then* $\mathbf{B}_\alpha$ *is the unique minimum energy state in* $\mathcal{H}$*.*

– *If* $H_c < |H|$*, then there in no field in* $\mathcal{H}$ *minimizing the energy.*

In fact, it turns out that, for large enough values of $|H|$, there do exist in $\mathcal{H}$ fully three-dimensional fields with an energy smaller than that of $\mathbf{B}_\alpha$—a property which is related to the "spontaneous breaking of symmetry" which occurs in Taylor's theory set in an infinite cylinder (Taylor(1986)): in that case, the energy minimizer (per unit of z-length!) is axisymmetric for low values of $|H|$, but is helicoidal when this parameter is taken large enough. In our case, however, there is no available three-dimensional equilibria in $\mathcal{H}$ to which minimizing sequences could converge, and one faces again the phenomenon of energy and helicity loss at infinity quoted above. The exact value of H_c has not yet been computed, but it is known that the associated value $\alpha_c R$ of αR is between 2.01 and 3.11.

The results of Subsections 4.2–4.3 can be summarized in the following table (where MES stands for "minimum energy state" and H_c^{ax} is allowed to take the value $+\infty$:

—Helicity—	$0 \leq \|H\| \leq H_c$	$H_c < \|H\| \leq H_c^{ax}$	$H_c^{ax} < \|H\| < \infty$
$\mathbf{B}_\alpha$ MES in $\mathcal{H}_{ax}$	yes	yes	no
$\mathbf{B}_\alpha$ MES in $\mathcal{H}$	yes	no	no

5. Discussion and Conclusion

The existence of a minimum energy state among all the magnetic fields occupying an unbounded domain of space D and having prescribed helicity

H and normal component B_n on ∂D, has been shown in this paper to depend on the nature of D and on the values which are chosen for the parameters. Although the energy is bounded from below in $\mathcal{H}$, we have presented explicit examples in which this quantity does not reach its lower bound when $|H|$ exceeds a critical value H_c (which may possibly be zero). As this fact may look strange to some people, it may be useful to give an example of a very simple system which is shown by elementary arguments to exhibit a similar behaviour. Consider an incompressible liquid drop of volume $\mathcal{V}$ which is constrained to stay above the horizontal plane $\{z = 0\}$ in which there is a golf hole of volume $\mathcal{V}_c$, and address the question of the minimum potential energy state of this system in the presence of a uniform gravitational field $\mathbf{g} = -|g|\mathbf{z}$. It is quite obvious that, if $\mathcal{V} \leq \mathcal{V}_c$, the energy is minimized when the drop completely fills the hole up to some horizontal level. If on the contrary $\mathcal{V}_c < \mathcal{V}$, the liquid excess of volume $\mathcal{V} - \mathcal{V}_c$ which cannot fit inside the hole gets dispersed at infinity when trying to minimize its energy: in that case, there cannot exist an energy minimizer satisfying the volume constraint! Analogy with our magnetic problem is obtained at once by making the correspondence: – plane with hole $\rightarrow$ domain D (with D being a half-cylinder (*resp.* a half-space or an external domain) when $\mathcal{V}_c > 0$ (*resp.* $= 0$);—$\mathcal{V} \rightarrow |H|$ and $\mathcal{V}_c \rightarrow H_c$.

To conclude, let us discuss briefly the possible consequences of our results for a theory of coronal heating of the type originally proposed by Heyvaerts and Priest (1984). In the most satisfying statement of this theory—which fits in the general paradigm outlined in the Introduction—the evolution of an isolated coronal magnetic structure containing a low-beta plasma and occupying some unbounded region D, is described by a repetitive series of two-steps processes (Dixon *et al.* (1989), Aly (1990b)): in the first step, the field evolves quasi-statically through a sequence of nonlinear force-free configurations along which dissipative effects play a negligible role. In the second step, a process of fast maximal dissipation develops in a bounded part D' of D limited by some magnetic surface (*e.g.* the interior of a highly twisted flux tube) in which there is a high concentration of magnetohydrodynamic turbulence, and the field Taylor-relaxes to a minimum energy state compatible with the conservation of the only relative helicity in D' and with the ideal constraints in $D \setminus D'$. Clearly, in that final state, the field is a linear force-free field in D' and still a nonlinear one in $D \setminus D'$, with the *free boundary* between these two regions being self-consistently fixed by pressure balance. Unfortunately, solving this highly nonlinear problem is a quite difficult task and nothing is yet known on the existence of a solution. Therefore, it is interesting to look at simplified formulations of the problem. The difficulty arising from the fact that D' is a part of D which is not known *a priori*, the first natural thing to be done is to consider the case when D' coincides with D, in which case one comes back to Heyvaerts and Priest's

(1984) model (or to one of its numerous avatars; see *e.g.* Browning (1988), and references therein, Dixon *et al.* (1988)) and one meets the existence problem we have discussed here (and which had not been addressed before, all the workers in that field taking for granted that it should be solved by a constant-α force-free field). But then the results we have reported above imply that the existence of a minimum energy state depends in a crucial way on the choice which is made for $D = D'$, and it appears difficult to guess from them what could be the nature of the minimum energy state in the more realistic model described above. Quite tentatively, however, we may boldly conjecture that it should behave more or less like the solution of the minimization problem in a half-cylinder—a choice of domain with which one keeps a certain realistic amount of boundedness for the turbulent region in which relaxation is taking place, as well as the physical openness of the domain, what is certainly required for any model of corona, such a system being well known to exhibit numerous ejection phenomena). Thus one should expect that there exists a solution for small values of the helicity in D', but no solution at all when this quantity becomes too large, the field losing a part of its energy and helicity at infinity when trying to relax to the lowest possible energy state —an interesting mathematical phenomenon which could be actually given a straightforward physical interpretation: in some circumstances (*e.g.* when the field is too much stressed), the turbulent system expels during a relaxation event what could be called "plasmoids", which carry away an excess of energy (as well as helicity) which cannot be dissipated locally into heat.

References

ALISSANDRAKIS, C.E., 1981 On the computation of constant-α force-free magnetic field *Astr. Ap.*, **100**, pp.197-200.

ALY, J.J., 1990a Quasi-static evolution of a force-free magnetic field. *Comp. Phys. Comm.*, **59**, pp.13-20.

ALY, J.J., 1990b Flux constraints on magnetic energy release in a highly conducting plasma. *Topological Fluid Mechanics*, Proceedings IUTAM Symposium, Cambridge, August 13-18,1989, Eds: H.K. Moffat & A. Tsinober, Cambridge University Press.

ALY, J.J., 1991a Some properties of finite energy constant-α force-free magnetic fields in a half-space. *Solar Phys.* (in press).

ALY, J.J., 1991b A model for magnetic energy storage and Taylor's relaxation in the solar corona. I. Helicity constrained minimum energy state in a half-cylinder. *Phys.Fluids B* (submitted).

BERGER, M.A., 1985 Structure and stability of constant-α force-free fields. *Ap. J. Suppl.*, **59**, pp. 433-444.

BERGER, M.A. & FIELD, G.B., 1984 The topological properties of magnetic helicity. *J. Fluid Mech.*, **147**, pp. 133-148.

BROWNING, P.K., 1988 Magnetohydrodynamics in solar coronal and laboratory plasmas: a comparative study. *Phys. Rep.*, **169**, pp. 329-384.

COLTON, D. & KRESS, R., 1983 Integral Equation Methods in Scattering Theory. *John Wiley, New York.*

DIXON, A.M.,BERGER, M.A., BROWNING, P.K. & PRIEST, E.R., 1989 A generalization of the Woltjer minimum-energy principle. *Astr. Ap.*, **225**, pp. 156-166.

DIXON, A.M., BROWNING, P.K. & PRIEST, E.R., 1988 Coronal heating by relaxation in a sunspot magnetic field. *Geophys. Astrophys. Fluid Dynamics*, **40**, pp. 293-327.

FINN, J.M. & ANTONSEN, T.M., 1985 Magnetic helicity: what is it and what is it good for? *Comments Plasma Phys. Controlled Fusion*, **9**, pp. 111-126.

HEYVAERTS, J. & PRIEST, E.R., 1984 Coronal heating by reconnection in d.c. current systems. A theory based on Taylor's hypothesis. *Astr. Ap.*, **137**, pp.63-80.

LAURENCE, P. & AVALLANEDA, M., 1991 On Woltjer's variational principle for force-free fields. *J. Math. Phys.*, **32**, pp. 1240-1253.

SEEHAFER, N., 1978 Determination of constant-α force-free solar magnetic fields from magnetograph data. *Solar Phys.*, **58**, pp. 215-223.

TAYLOR, J.B., 1986 Relaxation and magnetic reconnection in plasmas. *Rev. Mod. Phys.*, **58**, pp. 741-763.

FORCE-FREE MAGNETIC FIELDS WITH CONSTANT ALPHA

S.I. Vainshtein*
Institute for Theoretical Physics
University of California
Santa Barbara, California 93106-4030, U.S.A.

ABSTRACT. According to Taylor's conjecture (1986), the force-free field with constant α is a final state of a magnetic configuration after reconnection and subsequent relaxation.This conjecture is based on the assumption of existence and completeness of the set of eigenfunctions of the operator $\nabla\times$, *i.e.*, the set of force-free solutions with constant α. It turns out, however, that the existence of solutions is very sensitive to the geometry of the plasma configuration. Actually, only for a boundary with high degree of symmetry the solutions are found. If the shape of the boundary possesses only axial symmetry, the force-free fields are presumably axisymmetric as well. Therefore, the set of eigenfunctions is not complete in that case. For a body with a "bad" shape force-free fields with constant α may not exist at all. As a result, the question of the final state of a magnetic configuration inside the body remains open. As pointed out by Moffatt (1985), there is an analogy between magnetostatic equilibria and steady Euler flows. According to this analogy, constant-alpha force-free fields are equivalent to Beltrami flows. Additionally, only this latter class of fields may be stochastic, so that magnetic surfaces do not exist (Arnold, 1973). According to Moffatt (1985), the relaxation of an initially ergodic field (in some domain) under topological constraints has to end up as a constant-alpha force-free field. In addition to Moffatt's conclusion, we would suggest that the relaxation of the field under topological constraints is accompanied by a change of the shape towards a more symmetric geometry of the domain.

1. Introduction

Force-free fields have attracted attention for many years because of their importance in astrophysics. They play a conspicuous role in stellar atmospheres, in particular in the solar corona. The prevalence of force-free fields in stellar atmospheres lies in the inability of the pressure, p, of a low density plasma to balance magnetic forces. To be more specific, a low density atmosphere corresponds to low β,

* Permanent address: Department of Astronomy and Astrophysics University of Chicago Chicago, Ill 60637

H. K. Moffatt et al. (eds.), Topological Aspects of the Dynamics of Fluids and Plasmas, 177–193.

$$\beta = \frac{p}{(B^2/8\pi)} \ll 1 \tag{1.1}$$

B being magnetic field strength. Thus if β is small the field tends to settle down to force-free state.

Much attention has been paid to so called linear force-free fields, or fields with constant alpha. The reason is obvious: they are relatively simple to describe; however, more importantly, the constant-alpha field is regarded as a final state after reconnection of the lines of force and subsequent reconstruction of the field, as suggested by Taylor (1986). This hypothesis has been used more than once to account for corona heating.

Another aspect of force-free fields with constant alpha was introduced by Arnold (1973) and Moffatt (1985). They found only this kind of equilibrium field has the potential to stochastic field lines. As pointed out by Moffatt, an analogy exists between magnetic equilibrium and Euler flow of arbitrarily complex topology; the constant alpha field is equivalent to Beltrami flow. Thus studying the force-free fields gives an insight into fluid dynamics.

Much work has been devoted to the construction of force-free configurations. On one hand, one can construct the field lines of a configuration given an analytic expression for the field. It is in this way that stochastic lines of force have been found. On the other hand, one of the outstanding problems of the solar corona is to construct the field configuration from observations of the photospheric field. It should be kept in mind that existence and uniqueness of the solution to this problem are still open questions.

For the constant-alpha field, in a finite domain, this problem can be formulated as a eigenvalue and eigenfunction problem. In this paper we will analyze and discuss the existence of the constant-alpha, force-free solutions.

2. Formulation of the Problem

The momentum equation for a plasma in magnetostatic equilibrium takes the form

$$0 = \mathbf{g}\rho - \nabla p - \frac{1}{4\pi}\mathbf{B} \times \nabla \times \mathbf{B} \tag{2.1}$$

In the case of a rarefied atmosphere, both the pressure and the density are negligible, leaving only the last term on the right side of equation (2.1). The resulting equation has the following general solution:

$$\nabla \times \mathbf{B} = \alpha \mathbf{B}, \tag{2.2a}$$

$$\mathbf{B} \cdot \nabla \alpha = 0 \tag{2.2b}$$

Generally α is space-dependent, and is constrained by (2.2b) to be constant along field lines and magnetic surfaces. The special case

$$\alpha = \text{constant} \tag{2.3}$$

everywhere in space has been studied in detail. In fact, if $\alpha = \text{const}$, it follows from (2.2b) that the magnetic field should satisfy:

$$\nabla \times \nabla \times \mathbf{B} = \alpha^2 \mathbf{B}. \tag{2.4}$$

Equation (2.4) looks like eigenvalue problem, but generally with inhomogeneous boundary conditions. For force-free magnetic fields these are typically encountered in the following two forms:

i) The field (or at least the normal component $\mathbf{B}_n$) is given on the boundary;

ii) $\mathbf{B}_n = 0$ on the boundary. As a general rule, condition ii) is applied to a closed boundary.

Condition i) is used to construct coronal force-free fields with given field strength on the photosphere. In the case of $\alpha = \text{const}$, it is enough to know only $\mathbf{B}_n$, and the value of α. Generally, however, $\alpha \neq \text{const}$ and one needs to know all three components of the magnetic field on the photosphere.

Condition ii) appears to be homogeneous; however, this is not quite the case (see *e.g.* equation (3.4) below). Note, condition ii) does not correspond to classical electrodynamics. Recall that for classical electrodynamics there are two relevant cases:

1) An ideal conductor is outside the body under consideration so that on the boundary

$$\mathbf{B}_n = 0, \tag{2.5a}$$

$$\{\nabla \times \mathbf{B}\}_t = 0, \tag{2.5b}$$

where $\{...\}_t$ means two tangential components.

2) An ideal insulator (vacuum) is outside the body. Then, on the surface,

$$\{\nabla \times \mathbf{B}\}_n = 0 \tag{2.6a}$$

while $\mathbf{B}$ is continuous across the surface. And outside the body

$$\nabla \times \mathbf{B} = 0 \tag{2.6b}$$

$$\mathbf{B} = 0(r^3) \text{ as } \mathbf{r} \to \infty \tag{2.6c}$$

The asymptotic behavior of the magnetic field results from the absence of external sources.

Operator of equation (2.4) with either (2.5) or (2.6) imposed is self-adjoint, (Landau & Lifshitz, 1982), resulting in all eigenvalues α^2 being positive.

Whereas condition (2.5a) coincides with ii), imposing additional condition (2.5b) on equations (2.2), only a trivial solution $\mathbf{B} \equiv 0$ is possible. To prove this, note that it immediately follows from (2.2a) and (2.5b) that $\mathbf{B}_t = 0$ on the boundary. Combining this result with (2.5a) we obtain $\mathbf{B} = 0$ on the boundary. To deduce the field inside the body, let us rewrite (2.1) with $\rho = 0$ and $p = 0$ and the final term in the more useful form

$$(\partial_j x_i - \delta_{ij})(\delta_{ij} B^2/2 - B_i B_j) = 0, \tag{2.7}$$

with summation performed over repeating indices. Multiplying (2.7) by x_i, we get

$$(\partial_j x_i - \delta_{ij})(\delta_{ij} B^2/2 - B_i B_j) = 0 \tag{2.8}$$

Integrating (2.8) over the region inside the boundary, applying Stokes theorem and utilizing the boundary conditions just obtained, we get

$$\frac{1}{2} \int B^2 d\mathbf{r} = 0. \tag{2.9}$$

This implies $\mathbf{B} \equiv 0$.

Likewise, it can be shown that condition (2.6) also admits only the trivial solution. As proof, it is obvious that if ii) is satisfied, then for a force-free field (for which (2.2a) holds) condition (2.6a) is automatically realized. In addition, if we impose constraints (2.6b) and (2.6c), we again get the trivial solution $B \equiv 0$. The remainder of the proof is essentially the same as above, only now equation (2.7) holds in the whole space so that (2.8) is integrated over the entire space. In fact, both cases are representations of the virial theorem. In summary, the "classical" boundary conditions (2.5) and (2.6) are irrelevant to a force-free field.

For the remainder of this paper we will focus our attention on condition ii) for a finite body with a closed boundary. It should be noted, first of all, that in spite of the resemblance of equation (2.4) to a conventional eigenvalue problem, we should keep in mind that the field under consideration satisfies not only (2.4), but also (2.2). Therefore it is enough to consider (2.2). Thus the search for solutions for force-free fields with constant α is interpreted as an eigenvalue problem of the operator $\nabla\times$.

The main difficulty of this interpretation is that the operator is not self-adjoint when considered with boundary condition ii). As a result it is impossible to use the theorems of existence and completeness of eigenfunction systems. Indeed, the scalar product of (2.2a) and the solution of the adjoint problem, $\tilde{\mathbf{B}}$, results in the following integrals over all space:

$$\int \tilde{\mathbf{B}} \cdot \nabla \times \mathbf{B} d\mathbf{r} = \int \mathbf{B} \cdot \nabla \times \tilde{\mathbf{B}} d\mathbf{r} + \int \left\{ \mathbf{B} \times \tilde{\mathbf{B}} \right\}_n dS \tag{2.10}$$

The last integral is taken over the boundary and appears due to Gauss' theorem. This integral does not vanish under condition ii) (for the field of the adjoint problem we should impose the same boundary condition $\tilde{\mathbf{B}}_n = 0$). Therefore, the adjoint problem is not described by equation (2.2a).

3. Construction of Force-Fields

Force-free fields with constant α can be constructed as follows :

$$\mathbf{B} = \alpha \mathbf{B}_T + \nabla \times \mathbf{B}_T \tag{3.1a}$$

where $\mathbf{B}_T$ satisfies the equation:

$$\nabla \times \nabla \times \mathbf{B}_T = \lambda \mathbf{B}_T \tag{3.1b}$$

$$\alpha = \pm\sqrt{\lambda}$$

The field $\mathbf{B}_T$ is sometimes called toroidal component of the field, and $\nabla \times \mathbf{B}_T$, the poloidal component. The terminology is appropriate for a spherical geometry (for a cylinder, however, $\mathbf{B}_T$ is the poloidal component). Equation (3.1b) can be considered as an eigenvalue problem with boundary condition ii), *i.e.*,

$$\alpha \{\mathbf{B}_T\}_n + \{\nabla \times \mathbf{B}_T\}_n = 0 \tag{3.2}$$

Representation (3.1) proves to be useful only for simple geometries, such as spherical and cylindrical, for which the form of $\mathbf{B}_T$ can be just guessed. For example, for the torus, $\mathbf{B}_T$ is known only for axisymmetric fields.

3.1. CYLINDER-LIKE GEOMETRY

Consider a body with a boundary parallel to Z axis. Its projection onto the XY-plane may be of arbitrary shape. If it is a circle, then the body itself is a circular cylinder. In general case $\mathbf{B}_T$ in (3.1) may be represented as

$$\mathbf{B}_T = \nabla \times \mathbf{A} \tag{3.3a}$$

$$\mathbf{A} = \psi \mathbf{e}_z \tag{3.3b}$$

Then (3.1b) reduces to

$$\nabla^2 \psi = -\lambda \psi + h \tag{3.4}$$

where h is a constant. Force-free fields in cylindrical coordinates $\{\varpi, \phi, z\}$ are represented as follows:

$$\mathbf{B} = \alpha\left\{\frac{1}{\varpi}\frac{\partial\psi}{\partial\phi}, -\frac{\partial\psi}{\partial\varpi}, 0\right\} +$$

$$\left\{\frac{\partial^2\psi}{\partial z\partial\varpi}, \frac{1}{\varpi}\frac{\partial^2\psi}{\partial z\partial\phi}, -\frac{1}{\varpi}\left(\frac{\partial}{\partial\varpi}\varpi\frac{\partial\psi}{\partial\varpi} + \frac{1}{\varpi}\frac{\partial^2\psi}{\partial\phi^2}\right)\right\} \tag{3.5}$$

In Cartesian coordinates:

$$\mathbf{B} = \alpha\left\{\frac{\partial\psi}{\partial y}, -\frac{\partial\psi}{\partial x}, 0\right\} +$$

$$\left\{\frac{\partial^2\psi}{\partial z\partial x}, \frac{\partial^2\psi}{\partial z\partial y}, -\left(\frac{\partial^2}{\partial x^2} + \frac{\partial^2}{\partial y^2}\right)\psi\right\}. \tag{3.6}$$

3.1.1. Translationally Invariant Solutions ($\partial_z = 0$). For solutions invariant to translations along the Z axis, boundary condition (3.2) reduces to

$$\psi = 0. \tag{3.7}$$

Eigenvalue problem (3.4) possesses now a *continuous spectrum* because of the constant on the right side. Denoting ψ_m and λ_m as the eigenfunctions and eigenvalues of problem (3.4), (3.7), the general solution has the form:

$$\psi = \psi_m \tag{3.8a}$$

for $h = 0$ and $\lambda = \lambda_m$, *i.e.*, λ is an eigenvalue; and

$$\psi = \sum_m \frac{h_m}{\lambda - \lambda_m} \tag{3.8b}$$

$$h_m = \int h\psi_m dxdy$$

for $\lambda \neq \lambda_m$, for any m.

In the particular case of a cylinder with radius a, the solution is simple:

$$\psi = J_n(\sqrt{\lambda}\varpi)e^{in\phi} \quad \text{for } n > 0 \tag{3.9a}$$

$$\psi = J_0(\sqrt{\lambda}\varpi) - J_0(\sqrt{\lambda}a) \tag{3.9b}$$

Here J_n is Bessel function. Note that if $n > 0$, then $J_n(\sqrt{\lambda}a) = 0$. For $n = 0$ expression (3.9b) replaces (3.8).

For another particular case of a rectangular region $0 \leq x \leq a$, $0 \leq y \leq b$, we have the following eigenfunctions:

$$\psi = \sin\frac{nx\pi}{a}\sin\frac{my\pi}{b} \tag{3.9c}$$

where n and m are integers.

3.1.2. General Solutions. Actually, the general solution is known only for a cylinder:

$$\psi = J_n(\beta\varpi)\left[a_{nk}\cos\{n\phi + kz\} + b_{nk}\sin\{n\phi + kz\}\right], \tag{3.10}$$

where $\beta = \sqrt{\lambda - k^2}$ and λ, the eigenvalue, is determined from

$$\alpha n J_n(\beta a) + ka\beta J_n'(\beta a) = 0, \tag{3.11}$$

where the prime ($'$) denotes differentiation.

For the rectangular geometry, we seek solutions of the form:

$$\psi = \psi_1 \cos kz + \psi_2 \sin kz \tag{3.12}$$

Then according to (3.2) and (3.6), on the boundary the following conditions should be satisfied:

$$\alpha\frac{\partial\psi_1}{\partial y} + k\frac{\partial\psi_2}{\partial x} = 0 \tag{3.13a}$$

$$k\frac{\partial\psi_1}{\partial x} - \alpha\frac{\partial\psi_2}{\partial y} = 0 \tag{3.13b}$$

at $x = 0, a$ and

$$\alpha\frac{\partial\psi_1}{\partial x} - k\frac{\partial\psi_2}{\partial y} = 0 \tag{3.13c}$$

$$k\frac{\partial\psi_1}{\partial y} + \alpha\frac{\partial\psi_2}{\partial x} = 0 \tag{3.13d}$$

at $y = 0, b$. We seek for solution of the form

$$\psi_1 = a_1\cos k_x x\cos k_y y + b_1\cos k_x x\sin k_y y + c_1\sin k_x x\cos k_y y + d_1\sin k_x x\sin k_y y$$

$$\psi_2 = a_2\cos k_x x\cos k_y y + b_2\cos k_x x\sin k_y y + c_2\sin k_x x\cos k_y y + d_2\sin k_x x\sin k_y y.$$

Substituting these expressions into (3.13), we find that in order to satisfy the boundary conditions we have to set all the coefficients equal to zero (unless, of course, $k = 0$, which returns us back to solution (3.9c))!

The same situation holds for elliptic cylindrical geometry $\{u, v, z\}$ with the following components of the metric tensor:

$$g_{uu} = g_{vv} = a^2(\cosh 2u - \cos 2v)$$

$$0 \le u \le u_0 \qquad 0 \le v \le \pi.$$

For this case, boundary condition (3.2) can be written as

$$\alpha \frac{\partial \psi}{\partial v} + \frac{\partial^2 \psi}{\partial z \partial u} = 0. \tag{3.14}$$

Naturally, we look for solutions of the form (3.12), and then we can rewrite (3.14) as

$$\alpha \frac{\partial \psi_1}{\partial v} + \frac{\partial \psi_2}{\partial u} = 0 \tag{3.15a}$$

$$k \frac{\partial \psi_1}{\partial u} - \frac{\partial \psi_2}{\partial v} = 0 \tag{3.15b}$$

(cf. (3.13)). Both ψ_1 and ψ_2 satisfy (3.4), or, to be more specific

$$\left(\frac{\partial^2}{\partial u^2} + \frac{\partial^2}{\partial v^2} \right) \psi_{1,2} = a^2(k^2 - \lambda)(\cosh 2u - \cos 2v)\psi_{1,2}. \tag{3.16}$$

For translationally invariant solutions, $k = 0$, only one function remains in (3.12), namely ψ_1, and condition (3.15) reduces to (3.7). This is a classical eigenvalue problem and solutions exist, of course. If $k \neq 0$ the situation is different: there is no solution at all.

In order to see this, consider at first a limiting case of a strongly elongated ellipse, $u_0 \ll 1$. Then $\cosh 2u \approx 1$ and variables are separable in (3.16):

$$\psi_1 = \psi_1^{(u)}(u)\psi_1^{(v)}(v) \qquad \psi_2 = \psi_2^{(u)}(u)\psi_2^{(v)}(v) \tag{3.17}$$

where $\psi_{1,2}^{(v)}$ are solutions of the equation

$$\frac{d^2}{dv^2}\psi_{1,2}^{(v)} = (\kappa_1 - \kappa_2 \cos 2v)\psi_{1,2}^{(v)} \tag{3.18}$$

with κ_1, κ_2 constant. Substituting (3.17) into (3.15), we get another equation for $\psi_{1,2}^{(v)}$:

$$\left(\frac{d^2}{dv^2} + \omega^2 \right) \psi_{1,2}^{(v)} = 0, \tag{3.19}$$

where

$$\omega^2 = \frac{k^2}{\alpha^2} \frac{d\psi_1^{(u)}(u_0)/du)(d\psi_2^{(u)}(u_0)/du)}{\psi_1^{(u)}(u_0)\psi_2^{(u)}(u_0)} = \text{const.}$$

Obviously the solution of (3.19) ($\sim \sin nv$ or $\sim \cos nv$) does not coincide with that of (3.18). In other words, there is no nontrivial solution of (3.16) with boundary conditions (3.15). When $k = 0$, equation (3.19) does not make sense anymore because now only one function remains in (3.12), in the first place. In that case the boundary condition can be written down as follows :

$$\psi_1^{(u)}(u_0) = 0$$

and a solution does exist.

Another way to separate variable is to consider the case $\lambda = \alpha^2 = k^2$. Then according to (3.16)

$$\psi_1 = c_1 \sinh 2nu \cos 2nv \qquad \psi_2 = c_2 \sinh 2nu \sin 2nv$$

Substituting this solution into (3.15), we get $c_1 = c_2 = 0$.

The proof of nonexistence of solutions for a more general case $u_0 \leq f, f = O(1)$ is more cumbersome. Due to the periodicity requirement, the solution should be represented in a Fourier series:

$$\psi_{1,2} = \sum_m {}^1\psi_{1,2}^{(m)} \cos 2mv +{}^2 \psi_{1,2}^{(m)} \sin 2mv. \tag{3.20}$$

Both ${}^1\psi$ and ${}^2\psi$ satisfy the Fourier-representation of (3.16):

$$\left(\frac{d^2}{du^2} - (2m)^2\right) \psi^{(m)} = a^2(k^2 - \lambda)(\psi^{(m)} \cosh 2u - \frac{1}{2}[\psi^{(m+1)} + \psi^{(m-1)}]) \tag{3.21}$$

(where all unnecessary induces are left out). Now instead of boundary conditions (3.15), we have:

$$\alpha 2m {}^1\psi_1^{(m)} - k\frac{d}{du}\ {}^2\psi_2^{(m)} = 0 \tag{3.22a}$$

$$\alpha 2m {}^2\psi_1^{(m)} + k\frac{d}{du}\ {}^1\psi_2^{(m)} = 0 \tag{3.22b}$$

$$k\frac{d}{du}\ {}^1\psi_1^{(m)} - \alpha 2m {}^2\psi_2^{(m)} = 0 \tag{3.22c}$$

$$k\frac{d}{du}\ {}^2\psi_1^{(m)} + \alpha 2m {}^1\psi_2^{(m)} = 0 \tag{3.22d}$$

at $u = u_0$. Note, similar conditions are applied for the case of cylindrical geometry to get (3.10), (3.11). We need only make the following replacements:

$$m \Rightarrow n$$
$$\frac{d}{du} \Rightarrow a\frac{d}{d\varpi}.$$

Then

$${}^k\psi_l^{(n)} = J_n(\beta\varpi)^k T_l$$

where

$$\hat{T} = \begin{pmatrix} 1 & 1 \\ 1 & -1 \end{pmatrix},$$

$T_{ij} =^i T_j$. With this notation, all conditions (3.22) reduce to (3.11).

Now, however, the situation is quite different from cylindrical case. The variables are not separable, so that each single term of the series (3.20) is *not* a solution of (3.16). Therefore $\hat{\psi}^{(m)}$ should satisfy (3.22) for *all* m: this is plainly the Fourier-representation of condition (3.2). λ (and therefore α) is fixed for all harmonics with number m, that quantity being an eigenvalue.

In order to see that equation (3.21) is inconsistent with boundary conditions (3.22), we investigate the asymptotic behavior of the harmonics at large m. The second term on the left-hand side of (3.21) can be balanced only by the last one on the right-hand side, *i.e.*,

$$\psi^{(m)} = p\frac{1}{2(2m)^2}\psi^{(m-1)} + O\left(\frac{1}{(2m)^4}\right), \tag{3.23}$$

$p = a^2(k^2 - \lambda)$. In this case, the series (3.20) converges. Supposing the (m - 1)th harmonic satisfies (3.22), we substitute (3.23) into (3.22) to find that $\psi^{(m)}$ does not satisfy (3.22) to within $O(\frac{1}{2m})$. Note, however, that if $k = 0$, $\psi_1^{(m)}(u = u_0) = 0$, which is consistent with (3.23).

As for the limiting case $u_0 \gg 1$, the previous asymptotic analysis holds for $2m \gg \cosh 2u_0$.

3.2. ASYMMETRIC BOUNDARY

If the boundary does not have some kind of symmetry, force free solutions with constant α do not appear to exist. This statement is very difficult, if not impossible, to prove. We can only give some examples illustrating the link between symmetry and existence of solution.

3.2.1. Rectangular Geometry. Consider a parallelepiped $0 \leq x \leq a$, $0 \leq y \leq b$, $0 \leq z \leq c$. One may try to construct a solution with help of the vector-potential (3.3b); but, as we saw above in Sec. 3.1.1, 3.1.2, only translationally invariant solutions exist, namely (3.9c), for which, according to (3.6), $B_z \neq 0$ and is independent of z. Therefore B_z-component can not be zero at horizontal boundaries $z = 0, c$.

The nonexistence of eigenvalues of (2.2a) is very surprising because problems (2.5), (2.6) of classical electrodynamics are, of course, solvable for a parallelepiped. We have to use different approaches in order to make certain of this statement. First, we may look for solutions of the general form:

$$\begin{aligned}\mathbf{B} = \mathbf{a}e^{i(k_x x+k+_y y+k_z z)} + \mathbf{b}e^{i(k_x x+k_y y-k_z z)} \\ +\mathbf{c}e^{i(k_x x-k_y y+k_z z)} + \mathbf{d}e^{i(-k_x x+k_y y+k_z z)} + c.c.\end{aligned} \tag{3.24}$$

However, substituting (3.24) into (2.2a), we find that all coefficients in (3.24) must be zero.

Still, there may be left a suspicion that the solution cannot be expressed through known functions. We may seek for it in terms of its Taylor series at the vicinity of the origin:

$$B_i = a_{ij}x_j + a_{ijk}x_j x_k + a_{ijkl}x_j x_k x_l + \dots \tag{3.25}$$

($\mathbf{B} = 0$ at the origin). The requirement that $\mathbf{B}_n = 0$ reduces (3.25) to

$$\begin{aligned}B_i = a_{ii}x_i + (a_{iij} + a_{iji})x_i x_j + \\ (a_{iijk} + a_{ijik} + a_{ijki})x_i x_j x_k + \dots\end{aligned} \tag{3.26}$$

where summation is carried out for even numbers of identical induces only (!). Substitution of expression (3.26) into (2.2a) and equating terms of the same order on the left and right sides results in all coefficients in (3.26) being zero in consecutive order.

3.2.2. Cylinder-like Body Restricted by Horizontal Planes. Consider a cylinder-like geometry between planes $z = 0$ and $z = c$. Naturally, we look for solution of the form (3.12). Due the boundary conditions on both planes we have : $\psi_1 = 0$ and $k = n\pi/c$ with n, an integer. Considering the lateral boundary (parallel to the Z-axis), for general (orthogonal) coordinates $\{x_1, x_2, z\}$ the boundary condition gives:

$$\alpha\frac{1}{\sqrt{g_{22}}}\frac{\partial\psi}{\partial x_2} + \frac{1}{\sqrt{g_{11}}}\frac{\partial^2\psi}{\partial z\partial x_1} = 0 \qquad \text{at } x_1 = x_1^{(0)} \tag{3.27}$$

(cf. (3.14)). This expression results, as before, in *two* conditions (like (3.15a)) and (3.15b)), but with only *one* function ψ_2 this time:

$$\frac{\partial \psi_2}{\partial x_1}(x_1^{(0)}) = 0 \tag{3.28a}$$

and

$$\frac{\partial \psi_2(x_1^{(0)})}{\partial x_2} = 0 \quad \text{or} \quad \psi_2(x_1^{(0)}) = 0. \tag{3.28b}$$

The second equality in (3.28b) is essentially due to the constant h in the main equation (3.4). It should be noted that the classical eigenfunction problem is formulated with only *one* condition: (3.28a) – Neumann problem, or (3.28b) – Dirichlet problem. It is clear (see Vainshtein, 1984) that nontrivial solutions are possible only for some degenerative case of high symmetry. And indeed, solutions exist if the projection of the lateral boundary on the XY-plane is a circle. For a circle with radius a, *i.e.*, the region itself is part of a cylinder,

$$\psi_2 = J_0(\beta\varpi) - J_0(\beta a), \tag{3.29}$$

(*e.g.*, Taylor (1986)). Note that solution (3.29) is independent of ϕ (cf. (3.10)).

The situation is different when the projection of the boundary on the XY-plane is rectangular. For this case, as we saw above in Sec. 3.2.1, there is no solution. Also, if the projection is an ellipse, then only translationally invariant, $\partial/\partial z = 0$, solutions exist (see Sec. 3.1.2), which cannot be adjusted to restrictions on the planes $z = 0, c$: here B_z cannot vanish.

3.2.3. Modified Axial Symmetry. Let us begin our discussion with axisymmetric boundary expressed in spherical coordinates $\{r, \theta, \phi\}$. In this case axisymmetric solutions should satisfy the following equations:

$$\left(\frac{\partial}{\partial r}\frac{\partial}{\partial r} + \frac{\sin\theta}{r^2}\frac{\partial}{\partial\theta}\frac{1}{\sin\theta}\frac{\partial}{\partial\theta}\right)\tilde{A}_\phi = f(\tilde{A}_\phi), \qquad f(\tilde{A}_\phi) = -(1/2)\frac{d\tilde{B}_\phi^2}{d\tilde{A}_\phi} \tag{3.30}$$

$$\tilde{B}_\phi = \tilde{B}_\phi(\tilde{A}_\phi) \quad \frac{d\tilde{B}_\phi}{d\tilde{A}_\phi} = \alpha$$

$$\tilde{A}_\phi = A_\phi r\sin\theta \qquad \tilde{B}_\phi = B_\phi r\sin\theta$$

and

$$B_r = \frac{1}{r\sin\theta}\frac{\partial}{\partial\theta}A_\phi\sin\theta \qquad B_\theta = -\frac{1}{r}\frac{\partial}{\partial r}A_\phi r$$

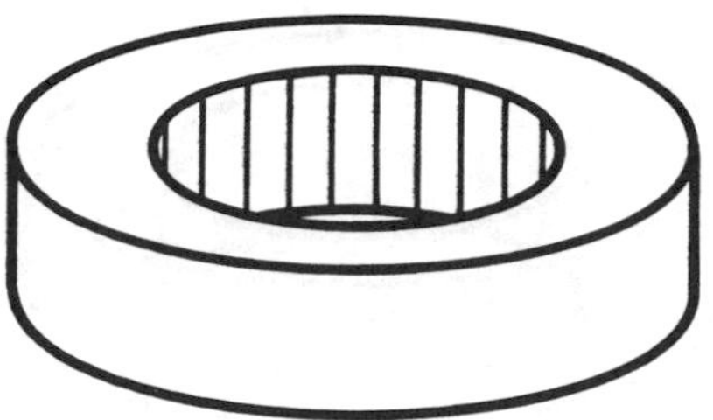

Fig. 1.

For $\alpha =$ const equation (3.30) is linear and for boundary condition $\tilde{A}_\phi = 0$, which corresponds to $\mathbf{B}_n = 0$, we have a classical eigenvalue problem. Solutions, of course, exist.

The question arises: are there solutions which do not possess axial symmetry? Looking for solution of the form (3.1), $\mathbf{A}$ (see (3.3a)) may be represented as

$$\mathbf{A} = \psi \mathbf{e}_r.$$

Then

$$\mathbf{B} = \alpha \left\{ 0, \frac{1}{r\sin\theta}\frac{\partial\psi}{\partial\phi}, -\frac{1}{r}\frac{\partial\psi}{\partial\theta} \right\} + \left\{ -\frac{1}{r\sin\theta}\left(\frac{\partial}{\partial\theta}\sin\theta\frac{\partial}{\partial\theta} + \frac{1}{\sin\theta}\frac{\partial^2}{\partial\theta^2}\right)\psi, \right.$$
$$\left. \frac{1}{r}\frac{\partial^2\psi}{\partial r\partial\theta}, \frac{1}{r\sin\theta}\frac{\partial^2\psi}{\partial r\partial\phi} \right\}. \tag{3.31}$$

The solution is well known for a sphere with radius a:

$$\psi = r^{-1/2} J_{n+1/2}(\alpha r) P_n^m(\cos\theta)(a_m \cos m\phi + b_m \sin m\phi), \tag{3.32}$$

with α satisfying $J_{n+1/2}(\alpha a) = 0$ (Chandrasekhar & Kendall, 1957). But if we impose a superconductive "partition" at the meridional plane $\phi = 0$, *i.e.*, demand $B_\phi = 0$ at $\phi = 0$, then the solution disappears. Indeed, for this case, nontrivial solutions for a_m, b_m exist only if

$$\begin{vmatrix} \{P_m^n\}' J_{n+1/2} & -\frac{m}{\sin\theta} P_n^m J'_{n+1/2} \\ \frac{m}{\sin\theta} P_n^m J'_{n+1/2} & \{P_m^n\}' J_{n+1/2} \end{vmatrix} = 0. \tag{3.33}$$

Since (3.33) cannot be satisfied for *arbitrary* r and θ, $a_m, b_m \equiv 0$.

Consider now a torus in cylindrical coordinates $\varpi_0 \leq \varpi \leq \varpi_1$ and $0 \leq z \leq c$, Fig. 1. We seek solutions of the form (3.12), where

$$\psi_{1,2} = {}^1\psi_{1,2} \cos m\phi + {}^2\psi_{1,2} \sin m\phi \tag{3.34}$$

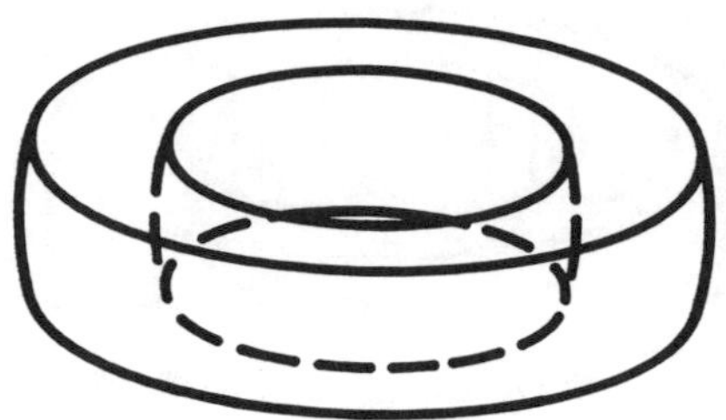

Fig. 2.

(cf. (3.20)). Now each ${}^k\phi$ can be represented as

$$ {}^k\psi_{1,2} = {}^k a_{1,2} J_n(\beta\varpi) + {}^k b_{1,2} Y_n(\beta\varpi). \tag{3.35}$$

On the bottom and top, *i.e.*, at $z = 0, c$, we have $B_z = 0$, so that $\psi_1 = 0$ and $k = n\pi/c$. Making use of the condition $B_\varpi = 0$ at $\varpi = \varpi_0$, taking into account (3.5), we immediately get that both ${}^k\psi(\varpi_0)$ and $\frac{d}{d\varpi}{}^k\psi(\varpi_0)$ are zero. Using (3.35), we see that all ${}^k a = {}^k b = 0$.

Thus, there are no solutions lacking axial symmetry. Note that for the axisymmetric case, the second condition remains, $\frac{d}{d\varpi}{}^k\psi(\varpi_0) = 0$, which, after combining with analogous condition at $\varpi = \varpi_1$, results in a dispersion relation for α.

It is clear that if we introduce a "partition" at $\phi = 0$, we lose the solution completely because B_ϕ cannot be zero at $\phi = 0$ for an axisymmetric field.

Finally, we note that the analogous situation holds for a torus which is bounded by $r = r_0, r = r_1$ and $\theta = \theta_0, \theta = \theta_1$, Fig. 2.

3.3. SYMMETRY CONSIDERATIONS

All of the examples constructed above suggest the link between the existence of solutions and the symmetry of the boundary. We saw for cylinder-like geometries that it is necessary to have *four* functions ${}^k\psi_l, k; l = 1, 2$ (see (3.20) and (3.22)) with the same eigenvalue λ in order to construct the general solution; in other words, one needs a degeneracy of fourth order, which actually appears due to the symmetry of the problem. The cylindrical geometry possesses two symmetries: translational invariance and axial symmetry. Because of the first one, there is a natural second order degeneracy: $\sin kz$ and $\cos kz$ are equivalent (see (3.12)). The second symmetry also gives rise to a second order degeneracy: now $\cos n\phi$ and $\sin n\phi$ are equivalent. As a result, the general solution (3.10) can be constructed. One may expect that the system of all solutions (3.10) with boundary condition (3.11) forms a complete set, as suggested by Taylor (1986).

As soon as the problem is deprived of translational symmetry – by imposing two planes $z = 0, c$ – it is left only one function in (3.12), and consequently the set of solutions (3.12), (3.29) (with different k's) is no longer complete; the point is all functions in (3.29) are independent of ϕ. Still, solutions exist due to the axial symmetry.

In another approach we deprive the geometry of axial symmetry by considering boundaries with rectangular or elliptic projections onto the XY-plane, then the ϕ-degeneracy is lifted, and again the set of eigenfunctions (3.8), lacking z-dependence, is incomplete. Depriving the geometry of both symmetries, we lose the solutions altogether.

The same concept can be maintained considering modified axial symmetric boundaries, Sec. 3.2.3. A sphere possesses two symmetries: the boundary $r = a$ is independent of both ϕ and θ. This is why there exists a solution (3.32), which depends on these two variables and, of course, on r. This circumstance suggests the completeness of the set (3.32) along the lines of Taylor's (1986) conjecture. Imposing a boundary at a meridional plane, we deprive the geometry of both of the symmetries and, as a result, we end up with no force-free field with constant α.

Finally, a torus possesses only one kind of symmetry, namely, axial. As a result, there exist only solutions independent of ϕ; *i.e.*, the set of eigenfunctions is definitely incomplete. Now, again, the "partition" deprives the geometry of the only symmetry, and solutions no longer exist. All these considerations will receive further support in the next section devoted to the perturbation approach. Still, we should admit there is no proof of nonexistence of force-free fields with constant α for an asymmetric geometry. Moreover, we may expect that one might find solutions by finding some closed magnetic surfaces, which may serve as a boundary, in known solutions. Nevertheless, it looks like force-free fields with constant α "prefer" simple geometries.

4. Perturbation Theory

4.1. PERTURBATION OF A SPHERE

Suppose the sphere is displaced by

$$\xi = \delta \cos^2 \theta \tag{4.1}$$

so that the boundary is described by $r = S$, where

$$S = a + \delta \cos^2 \theta. \tag{4.2}$$

Now the condition $\mathbf{B}_n = 0$, *i.e.*, $\mathbf{B}\nabla(r - S) = 0$, can be written down in an explicit way:

$$B_r(S) + B_\theta(S)\frac{\delta}{S}2\cos\theta\sin\theta = 0 \tag{4.3}$$

At zeroth approximation $\delta = 0$ and (4.3) reduces to $B_r(a) = 0$ with solution (3.32), $J_{n+1/2}(\alpha a) = 0$. At first approximation

$$B_r(a) + B_r'(a)\delta\cos^2\theta + B_\theta(a)\frac{\delta}{a}2\cos\theta\sin\theta = 0 \tag{4.4}$$

(Rosenbluth & Bussac (1979)). The solution of (3.4) is (3.32), which we substitute into (4.4). It exists, *i.e.*, one of the coefficients, a_m or b_m, is unequal to zero, if and only if

$$\begin{vmatrix} a_{11} & a_{12} \\ a_{21} & a_{22} \end{vmatrix} = 0, \tag{4.5}$$

where

$$\begin{aligned} a_{11} &= a_{22} = n(n+1)P_n^m\{a^{-5/2}J_{n+1/2}(\alpha a) + (r^{-5/2}J_{n+1/2})'|_{r=a}\delta\cos^2\theta \\ &\quad + \frac{1}{a}\{P_n^m\}'(r^{-5/2}J_{n+1/2}'|_{r=a}\frac{\delta}{a}2\cos\theta\sin\theta \\ a_{12} &= -a_{21} = \alpha\frac{m}{a\sin\theta}a^{-1/2}J_{n+1/2}(\alpha a)P_n^m\frac{\delta}{a}2\cos\theta sin\theta. \end{aligned}$$

Determinant (4.5) should be equal to zero at *arbitrary* θ (and $r = a$). It is possible if $a_{12} = a_{21} = 0$, *i.e.*, $J_{n+1/2}(\alpha a) = 0$, and $a_{11} = a_{22} = 0$, *i.e.*, $J_{n+1/2}'(\alpha a) = 0$. These two requirements are incompatible, and therefore $a_m = b_m = 0$. Nontrivial solution does not exist.

Note that displacement (49) deprives the boundary of one of the symmetries, still, it remains axisymmetric. That is why a solution possessing axial symmetry exists: in this case $m = 0, a_{12} = a_{21} \equiv 0$, and only one requirement remains: $a_{11} = a_{22} = 0$. But asymmetric solution no longer exists.

5. Discussion

First, let us focus our attention on Taylor's conjecture that a force-free field with constant α is a final state of a magnetic configuration after reconnection and subsequent reconstruction. It is clear from the above that this process proves to be very sensitive to the geometry. As we have seen, for some cases, asymmetric solutions are not allowed.

Therefore, the kind of force-free fields predicted to appear for strongly helical fields may not be realized if the boundary is not symmetric enough.

For a body with a "bad" shape force-free field with constant α may not exist at all. As a result, the question of a final state of a magnetic configuration inside the body remains open. The theory should be revised, anyway, because Taylor's conjecture is based on an assumption of existence and completeness of the set of eigenfunctions of the operator $\nabla\times$.

Another aspect of force-free solutions with constant α is that only this class of fields may be stochastic, so that magnetic surfaces do not exist, Arnold (1973). Additionally, as pointed out by Moffatt (1985), magnetic equilibrium is described by an equation analogous to that of stationary Euler flow, and it turns out that Beltrami flow is equivalent to the constant-alpha field.

Moffatt (1985) put forth an idea concerning the relaxation of magnetic fields (or correspondingly, Euler flow) under topological constraints. The point is the topology of the field cannot easily change under the frozen-in conditions. And, if the field is initially ergodic in some domain, it has to relax to a constant-alpha force-free field; there is no escaping it. As we have seen above, the geometry of the domain is also important, in addition to the topology of the field. Namely, alpha-constant solutions are not possible for any shape of the boundary of the domain.

Therefore, in addition to Moffatt's conclusion, we would suggest that relaxation of the field under topological constraints is accompanied by the changing and *simplifying* of the geometry of the domain. It is not clear what will happen if the ergodic region is tangled with other domains, imposing constraints on the geometry change. Perhaps the relaxation will induce the appearance of discontinuities or current sheets to account for the complexity of the evolving geometry.

Acknowledgements

This article was prepared at the Institute for Theoretical Physics, Santa Barbara which is supported by NSF Grant PHY89-04035.

References

ARNOLD, V.I., 1974 The asymptotic Hopf invariant and its applications. *Proc. Summer School in Differential Equations, Erevan, Armenian SSR Acad. Sci.* Transl. 1986, in *Sel. Math. Sov.*,**5**, 327-345 (From Russian).

LANDAU, L.D. & LIFSHITZ, E.M., 1982 Electrodynamics of continuous media. *Nauka.*

MOFFATT, H.K., 1985 Magnetostatic equilibrium and analagous Euler flows of arbitrarily complex topology. *J. Fluid Mech.*, **159**, 359-378.

ROSENBLUTH, M.N. & BUSSAC, M.N., 1979 MHD stability of sheromak *Nuclear Fusion*, **19**, 489-498.

TAYLOR, J.B., 1986 Relaxation and magnetic reconnection in plasmas. *Rev. Mod. Phys.*, **58**, 741-763

VAINSHTEIN, S.I., 1984 Formation of current layers in nonequilibrium magnetic configurations. *Sov. Phys. JETP*, **59**, 262-268

MINIMUM ENERGY MAGNETIC FIELDS WITH TOROIDAL TOPOLOGY

A. Y. K. CHUI * and H. K. MOFFATT *
Institute for Theoretical Physics
University of California
Santa-Barbara, California 93106-4030, U.S.A.

ABSTRACT. Consider a flux tube initially constructed in an incompressible perfectly conducting fluid around an arbitrary knot K in a standard way such that the total helicity is prescribed; the minimum energy of the flux tube is then determined (in principle) solely by the topology of K. In this paper we restrict attention to axisymmetric configurations, so that K is a circle. The functional relationship between minimum energy (in axisymmetric relaxed states) and helicity is determined via a variational principle approach. The result is confirmed by considering the scaling properties of the Grad-Shafranov equation. We discuss briefly how this result is modified when non-axisymmetric equilibrium states are allowed.

1. Introduction

It has been shown (Moffatt 1990) that, if a tube of volume V and carrying magnetic flux Φ is constructed around an arbitrary knot K, in a standard way such that the helicity of the field is given by $\mathcal{H} = h\Phi^2$, then the minimum magnetic energy of the field under the group of volume-preserving distortions (realizable by incompressible flow) is given by

$$\mathcal{M}_{min} = m(h)\Phi^2 V^{-1/3} \tag{1.1}$$

where $m(h)$ is a function determined (in principle) solely by the topology of K. If K is the 'unknot', then h is the twist (or 'rotational transform') of the magnetic field $\mathbf{B}$ in the toroidal tube surrounding K. If we restrict attention to axisymmetric configurations, so that K is a circle, then we may seek to determine the function $m_a(h)$ corresponding to the minimum magnetic energy of axisymmetric field configurations with toroidal topology. These

* Permanent address: DAMTP, University of Cambridge, Silver Street, Cambridge CB3 9EW, England

H. K. Moffatt et al. (eds.), Topological Aspects of the Dynamics of Fluids and Plasmas, 195–218.

axisymmetric configurations may be unstable to non-axisymmetric perturbations leading to lower energy equilibrium states. We may be confident however that $m_a(h)$ provides an upper bound for $m(h)$:

$$m(h) \leq m_a(h) \ . \tag{1.2}$$

Determination of $m_a(h)$, which is the objective of the present paper, is an essential preliminary to the determination of $m(h)$; this in turn is an essential preliminary to determination of $m(h)$ for any knot K of nontrival topology. We see this therefore as a first step in a program of increasing complexity.

There are however independent reasons for the study of axisymmetric configurations. Magnetic containment devices like the TOKAMAK are axisymmetric, and, in ideal circumstances, the magnetic field and the contained plasma adopt an axisymmetric magnetostatic configuration of minimum magnetic energy. Determination of such states is a classical problem of plasma physics (see, for example, Thompson 1962). These states are of course constrained by the particular geometry of the containment device and of the surrounding coils. Here we allow the toroidal flux tube to expand or contract in radius without any such constraint, the equilibrium state being determined solely by Φ,V and h; this procedure gives some useful insight concerning configurations that may be regarded as 'natural' and therefore most easily containable.

The minimum energy states are magnetostatic equilibria satisfying the equations

$$\mathbf{j} \wedge \mathbf{B} = \nabla p \ , \qquad \mu_0 \mathbf{j} = \nabla \wedge \mathbf{B} \quad \text{and} \quad \nabla \cdot \mathbf{B} = 0 \ , \tag{1.3}$$

where $\mathbf{j}$ is the current density and p is the pressure field. To each solution of these equations, there corresponds via the analogy $\mathbf{B} \leftrightarrow \mathbf{u}$, $\mu_0 \mathbf{j} \leftrightarrow \omega$ and $p_0 - p \leftrightarrow \mu_0 h$, a solution of the steady Euler equations for incompressible flow:

$$\mathbf{u} \wedge \omega = \nabla h \ , \qquad \omega = \nabla \wedge \mathbf{u} \quad \text{and} \quad \nabla \cdot \mathbf{u} = 0 \ . \tag{1.4}$$

The states that we shall describe correspond to toroidal vortex configurations which are stationary relative to the surrounding fluid, in fact $\mathbf{u} \equiv \mathbf{0}$ outside the torus to which the vorticity is confined. The total flux of vorticity around the tube (including a surface contribution) is zero. They are thus a degenerate form of 'vortex tube with swirl'. There is no guarantee that these flows are stable, even to axisymmetric disturbances. They are nevertheless of peculiar interest as examples of steady axisymmetric Euler flow of finite energy in unbounded fluid.

In section 2, we recapitulate the construction of a standard twisted flux tube, as described by Moffatt (1990), and we give an explicit calculation of the associated helicity. In section 3, we reformulate the minimum energy variational principle, incorporating the helicity and incompressibility constraints. In sections 4 and 5, we obtain asymptotic expressions for $m_a(h)$ for $h \gg 1$ and $h \ll 1$ respectively, and in section 6, we present the results of numerical calculation based on the variational principle, which confirm the asymptotic results, and we determine $m_a(h)$ in the transitional regime when $h = O(1)$. In section 7, the asymptotic results are interpreted in terms of appropriate scalings of the Grad-Shafranov equation; and in section 8, the results are summarised, and we discuss qualitatively the nature of the non-axisymmetric instabilities to which the equilibrium states may be subject.

2. Construction of Flux Tube of Prescribed Helicity

Let $\mathcal{T}$ be the torus defined by

$$\mathcal{T} = \{(r,\theta,z) : (r-R)^2 + z^2 \le (\varepsilon R)^2\} \tag{2.1}$$

where $\varepsilon \le 1$. The cross-section $\mathcal{A}$ of this torus is circular with area $A = \pi\varepsilon^2 R^2$, and its volume is

$$V = \int_{\mathcal{A}} 2\pi r\, dr\, dz = 2\pi A R = 2\pi^2\varepsilon^2 R^3 . \tag{2.2}$$

We now define a toroidal field $\mathbf{B}_T = (0, B, 0)$ given by

$$B = \begin{cases} 2\pi r\Phi/V & \text{inside } \mathcal{T}; \\ 0 & \text{otherwise.} \end{cases} \tag{2.3}$$

The toroidal flux of this field using (2.2) is

$$\int_{\mathcal{A}} 2\pi r\Phi/V\, dr\, dz = \Phi . \tag{2.4}$$

Note that $B/r = 2\pi\Phi/V$ is constant; under any volume-preserving axisymmetric frozen-field distortion, B/r is constant following a given field line; hence the field (2.3) is invariant in form within the tube $\mathcal{T}$ if the tube is radially expanded or contracted. It is moreover the unique field (of flux Φ) with this property.

We now wish to construct an associated poloidal field $\mathbf{B}_P$ by simple surgery as follows: imagine that we cut the flux tube at the section $\theta = 0$, twist the tube as if to make a twisted rope, the twist being uniformly distributed with respect to the angle θ, and then reconnect the flux tube (figure 1). Let us calculate the field $\mathbf{B}_P$ generated by this operation.

The velocity field that provides the required twisting effect is

$$\mathbf{v}(r,\theta,z) = \frac{\theta}{2\pi}\left(-\frac{1}{r}\frac{\partial\psi}{\partial z}, 0, \frac{1}{r}\frac{\partial\psi}{\partial r}\right) \stackrel{\text{def}}{=} \frac{\theta}{2\pi}\hat{\mathbf{v}} , \qquad (0 < \theta \le 2\pi) ; \tag{2.5}$$

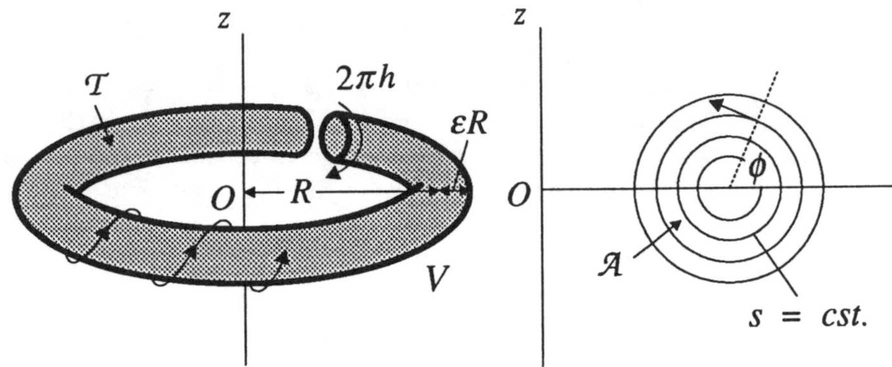

Fig. 1. The simple surgery consists of three steps: cutting the tube, twisting it uniformly and reconnecting it.

where $\psi = \psi(r,z)$ is the Stokes stream function of $\hat{\mathbf{v}}$. Note that $\mathbf{v}$ satisfies the incompressibility condition $\nabla \cdot \mathbf{v} = 0$, and is discontinuous across the cut at $\theta = 0$. We choose

$$\psi(r,z) = \begin{cases} \frac{1}{2}\psi_0\Big((r-R)^2 + z^2\Big) & \text{inside } \mathcal{T}, \\ 0 & \text{otherwise;} \end{cases} \tag{2.6}$$

so that the streamlines in any plane $\theta = cst.$ are circles centred on $(r,z) = (R,0)$. We will find it useful (see figure 1) to define the polar coordinates (s,ϕ) such that

$$r - R = s\cos\phi, \qquad z = s\sin\phi\ ; \tag{2.7}$$

then the velocity of the fluid particle on the circle of radius s at the plane $\theta = 2\pi$ is $v_\phi = \psi_0 s/r$, and the time to describe a complete circle is

$$T = \oint \frac{s\,d\phi}{v_\phi} = \frac{1}{\psi_0}\oint (s\cos\phi + R)\,d\phi = \frac{2\pi R}{\psi_0}\ , \tag{2.8}$$

independent of s.

Consider now the action of the velocity field (2.5) on the field (2.3). Let $\mathbf{B}(\mathbf{x},t) = \mathbf{B}_T(\mathbf{x},t) + \mathbf{B}_P(\mathbf{x},t)$ be the magnetic field at time t. From the frozen-field equation,

$$\frac{\partial \mathbf{B}}{\partial t} = \nabla \wedge (\mathbf{v} \wedge \mathbf{B}) = (\Phi/V)\,\hat{\mathbf{v}}\ , \tag{2.9}$$

and this field is axisymmetric and poloidal: only poloidal field is generated by the surgery, and the toroidal field is left invariant. At time $t = hT = 2\pi hR/\psi_0$ (when the tube is twisted through h turns),

$$\mathbf{B}_P = \frac{2\pi h R\Phi}{\psi_0 V}\hat{\mathbf{v}} \stackrel{\text{def}}{=} \nabla \wedge \mathbf{A}_T\ , \tag{2.10}$$

where

$$\mathbf{A}_T = \begin{cases} \left(0\,, \frac{\pi R h\Phi}{rV}((r-R)^2+z^2)\,, 0\right) & \text{inside } \mathcal{T}, \\ \mathbf{0} & \text{otherwise;} \end{cases} \tag{2.11}$$

and the total poloidal flux generated between the magnetic axis $s = 0$ and the tube boundary $s = \varepsilon R$ is then $\Phi_P = h\Phi$.

The helicity of the field can now be easily calculated. Note first that, with $\mathbf{B}_T = \nabla \wedge \mathbf{A}_P$,

$$\int \mathbf{B}_T \cdot \mathbf{A}_P\, dV = \int \mathbf{B}_P \cdot \mathbf{A}_T\, dV = 0 \tag{2.12}$$

and

$$\int \mathbf{B}_T \cdot \mathbf{A}_T\, dV = \int \mathbf{B}_P \cdot \mathbf{A}_P\, dV\ ; \tag{2.13}$$

hence

$$\begin{aligned} \mathcal{H} &= \int \mathbf{A}\cdot\mathbf{B}\, dV = 2\int_{\mathcal{T}} \mathbf{A}_T \cdot \mathbf{B}_T\, dV \\ &= \frac{2\pi h R\Phi^2}{V^2}\int_{\mathcal{T}} \left((r-R)^2+z^2\right) r\, dr\, dz \quad \text{using (2.3) and (2.11)} \\ &= \frac{2\pi h R\Phi^2}{(\pi\varepsilon^2 R^3)^2}\int_0^{2\pi}\int_0^{\varepsilon R} s^2(s\cos\phi + R)\, s\, ds\, d\phi \quad \text{using (2.7)} \\ &= h\Phi^2 = \Phi\Phi_P\ . \end{aligned} \tag{2.14}$$

When $h = 1$, then every pair of **B**-lines are linked with linking number $+1$, and the result $\mathcal{H} = \Phi^2$ may be obtained in this case from the formula [Moffatt 1990]

$$\mathcal{H} = 2\int_0^{\Phi} \varphi\, d\varphi = \Phi^2\ . \tag{2.15}$$

Hence (2.14) merely confirms what may be physically obvious. Note that if $h = p/q$ is rational (where p and q are co-prime integers), then each **B**-line (except the 'magnetic axis' $s = 0$) is a torus knot $K_{p,q}$. If h is irrational, then (with the same exception) each **B**-line covers a torus $s = \text{cst}$. In the context of toroidal containment devices, h is known as the 'rotational transform' and the nested tori defined by **B**-lines are called magnetic surfaces.

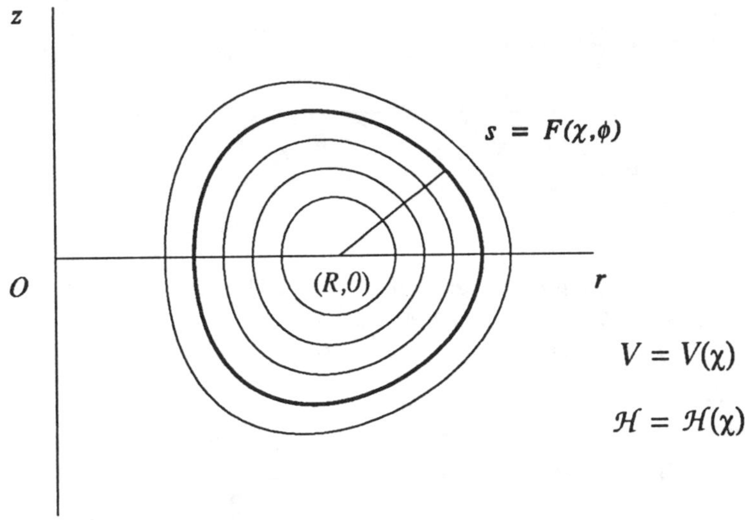

Fig. 2. The magnetic energy is minimised subject to two families of constraints which fix the volume $V(\chi)$ and the helicity $\mathcal{H}(\chi)$ inside each magnetic surface $\chi = cst.$

3. The Variational Principle

In this section, we shall develop the necessary tool for the problem. There are two approaches for obtaining relaxed states: we can follow the dynamical process directly using the MHD equations and including viscous (or equivalent) dissipation of energy; or we can make use of the fact that the relaxed state has minimum energy, which leads to a variational principle. In this paper we will adopt the second approach. It is known (Moffatt 1969) that the helicity is conserved inside each magnetic surface, and the incompressibility condition implies that the volume is also conserved inside each magnetic surface. Expressing **B** in the standard way

$$\mathbf{B} = \mathbf{B}_T + \mathbf{B}_P = \Big(0\,,B(r,z)\,,0\Big) + \Big(-\frac{1}{r}\frac{\partial\chi(r,z)}{\partial z}\,,0\,,\frac{1}{r}\frac{\partial\chi(r,z)}{\partial z}\Big) \tag{3.1}$$

where $\chi = cst.$ on the magnetic surfaces, we may formulate the variational principle as follows. Let $(R,0)$ be the position of the elliptic point (at which $\chi = 0$) in the $(r-z)$ plane at equilibrium. Define the polar coordinates (s,ϕ) with the origin at $(R,0)$ as in (2.7). Let

$$s = F(\chi,\phi) \tag{3.2}$$

be the contours of constant χ (figure 2). It is required to

Minimise

$$\mathcal{M} = \frac{1}{2}\int \mathbf{B}^2\, dV = \frac{1}{2}\int \left[B^2 + \frac{1}{r^2}\left(\left(\frac{\partial \chi}{\partial r}\right)^2 + \left(\frac{\partial \chi}{\partial z}\right)^2 \right) \right] dV \tag{3.3}$$

subject to

$$\int_{s \le F(\chi,\psi)} dV = V(\chi) \tag{3.4}$$

and

$$\int_{s \le F(\chi,\phi)} \mathbf{B}\cdot\mathbf{A}\, dV = \int_{s \le F(\chi,\phi)} 2\,\mathbf{B}_T\cdot\mathbf{A}_T\, dV = \int_{s \le F(\chi,\phi)} \frac{2B\chi}{r}\, dV = \mathcal{H}(\chi) \tag{3.5}$$

where $V(\chi)$ and $\mathcal{H}(\chi)$ are prescribed. Here $B = B(s,\phi)$, $F = F(\chi,\phi)$ and $R = cst.$ are to be determined. Note that (3.4) and (3.5) are two functional constraints which represent the infinite number of constraints on magnetic surfaces $\chi = cst.$, $0 \le \chi \le \chi_{max}$. This formulation of the problem is equivalent to that of Kruskal & Kulsrud (1958) who however expressed the constraints entirely in terms of flux and mass conservation.

Let us examine (3.3) in more detail. First note that

$$\begin{aligned} \frac{\partial \chi}{\partial r} &= \frac{\partial \chi}{\partial s}\frac{\partial s}{\partial r} + \frac{\partial \chi}{\partial \phi}\frac{\partial \phi}{\partial r} = \frac{\cos\phi}{F_\chi} + \frac{F_\phi}{F_\chi}\frac{\sin\phi}{s}\,, \\ \frac{\partial \chi}{\partial z} &= \frac{\partial \chi}{\partial s}\frac{\partial s}{\partial z} + \frac{\partial \chi}{\partial \phi}\frac{\partial \phi}{\partial z} = \frac{\sin\phi}{F_\chi} - \frac{F_\phi}{F_\chi}\frac{\cos\phi}{s}\,; \end{aligned} \tag{3.6}$$

hence, with (2.7) and (3.2), we simplify (3.3) to

$$\mathcal{M} = \pi \int_0^{2\pi}\int_0^{\chi_{max}} \left(G^2(R + F\cos\phi)FF_\chi + \frac{F^2 + F_\phi^2}{(R + F\cos\phi)FF_\chi} \right) d\chi\, d\phi \tag{3.7}$$

where

$$G(\chi,\phi) \equiv B(F(\chi,\phi),\phi)\,. \tag{3.8}$$

Now we manipulate (3.4) in the same way to obtain

$$2\pi \int_0^{2\pi}\int_0^{\chi} (R + F(\chi',\phi)\cos\phi)F(\chi',\phi)F_{\chi'}(\chi',\phi)\, d\chi'\, d\phi = V(\chi) \tag{3.9}$$

and differentiate to obtain

$$\frac{1}{2\pi}V'(\chi) = \int_0^{2\pi} (R + F\cos\phi) F F_\chi \, d\phi \,. \tag{3.10}$$

Similarly, (3.5) leads to

$$\int_0^{2\pi} G F F_\chi \, d\phi = \frac{1}{4\pi\chi}\mathcal{H}'(\chi) \,. \tag{3.11}$$

The constraints (3.9) and (3.11) mean simply that the volume and helicity between any two neighbouring magnetic surfaces are prescribed and fixed.

The function G can be determined by the usual Lagrange multiplier technique: if $\lambda(\chi)$ is the Lagrange multiplier for the constraint (3.11), we immediately find that

$$G = \frac{\lambda(\chi)}{2\,(R + F\cos\phi)} \tag{3.12}$$

where, from (3.11), $\lambda(\chi)$ satisfies

$$\lambda \int_0^{2\pi} \frac{F F_\chi \, d\phi}{2\,(R + F\cos\phi)} = \frac{\mathcal{H}'(\chi)}{4\pi\chi} \,. \tag{3.13}$$

Note that the first term in (3.3) becomes

$$\begin{aligned} \mathcal{M}_1 &= \pi \int_0^{\chi_{max}} \frac{\lambda}{2} \int_0^{2\pi} \frac{\lambda F F_\chi}{2(R + F\cos\phi)} \, d\phi \, d\chi = \pi \int_0^{\chi_{max}} \frac{\lambda}{2} \frac{\mathcal{H}(\chi)}{4\pi\chi} \, d\chi \\ &= \pi \int_0^{\chi_{max}} \left(\frac{\mathcal{H}(\chi)}{4\pi\chi}\right)^2 \left(\int_0^{2\pi} \frac{F F_\chi \, d\phi}{(R + F\cos\phi)}\right)^{-1} d\chi \,. \end{aligned} \tag{3.14}$$

The analysis so far is applicable to all axisymmetric fields with magnetic surfaces in the form of nested tori. For the particular case of the flux tube defined in section 2 we may calculate $V(\chi)$, $\mathcal{H}(\chi)$ and χ_{max} explicitly. From (2.11), we see that then (using $\mathbf{A}_T = (0, \chi/r, 0)$),

$$s = \left(\frac{\chi V}{h\pi R\Phi}\right)^{\frac{1}{2}} \stackrel{\text{def}}{=} S(\chi) \,. \tag{3.15}$$

Note that s attains its maximum εR when $\chi = \chi_{max}$, so that

$$\chi_{max} = h\pi R\,(\varepsilon R)^2 \frac{\Phi}{V} = \frac{h\Phi}{2\pi} \,. \tag{3.16}$$

Using (3.15), we now obtain

$$\begin{aligned} V(\chi) &= 2\pi \iint\limits_{s \le S(\chi)} r \, dr \, dz = 2\pi \iint\limits_{s \le S(\chi)} (R + s\cos\phi)\, s \, ds \, d\phi \\ &= 4\pi^2 R \left[\frac{s^2}{2}\right]_0^{S(\chi)} = \frac{2\pi\chi V}{h\Phi} \end{aligned} \tag{3.17}$$

and similarly

$$\mathcal{H}(\chi) = \frac{4\pi^2}{h}\chi^2 \,. \tag{3.18}$$

Note that $V(\chi_{max}) = V$ and the total helicity $\mathcal{H}(\chi_{max}) = h\Phi^2$ as expected. The variational problem now becomes:

Find the function $F(\chi, \phi)$ and the constant R that minimise

$$\mathcal{M} = \pi \int_0^{\chi_{max}} \left[\frac{4\pi^2}{h^2} \left(\int_0^{2\pi} \frac{F F_\chi \, d\phi}{R + F\cos\phi} \right)^{-1} + \int_0^{2\pi} \frac{(F^2 + F_\phi^2)\, d\phi}{(R + F\cos\phi) F F_\chi} \right] d\chi \tag{3.19}$$

subject to the constraint

$$\int_0^{2\pi} (R + F\cos\phi)\, F F_\chi \, d\phi = \frac{V}{h\Phi} \,. \tag{3.20}$$

This variational problem cannot be solved analytically. However, by restricting $F(\chi, \phi)$ to be within a certain class of functions, it is possible to obtain an upper bound for the true minimum. In the following sections, we will study, in turn, the cases of (i) large h, (ii) small h and (iii) $h = O(1)$.

4. The Case of Large h

In this section the minimum energy of the flux tube in the large h limit will be examined. In the order of complexity, we use three different methods to provide successively better estimates of the true minimum from above. A lower bound for the true minimum energy will also be estimated using the Poincaré inequality.

4.1. THE SCALING ARGUMENT

Consider the flux tube confined in the torus $\mathcal{T}$ as described in (2.1), carrying a magnetic field $\mathbf{B} = \mathbf{B}_T + \mathbf{B}_P$. Recall that R is the mean radius of the flux tube and A is the area of the cross-section $\mathcal{A}$. Let $\Phi = \int_{\mathcal{A}} |\mathbf{B}_T(r, z)|\, dr\, dz$ be the toroidal flux of $\mathbf{B}$ around Oz. Under any volume-preserving axisymmetric frozen-field distortion, both V (the volume of the tube) and Φ are invariant. If, for example, the tube is stretched so that R increases, then A decreases so that $V = 2\pi R A = cst$.

Under such a distortion, $|\mathbf{B}_T|$ increases in proportion to R; if the distortion is such that $\mathcal{A}$ contracts in self-similar manner, then $|\mathbf{B}_P|$ decreases in proportion to $A^{\frac{1}{2}}$; also if the poloidal field is generated by the cut-and-twist surgery as described in section 2, then $|\mathbf{B}_P|$ is proportional to h. On dimensional ground we therefore have

$$|\mathbf{B}_T| \sim \Phi V^{-1} R \,, \qquad |\mathbf{B}_P| \sim h\Phi (VR)^{-1/2}; \tag{4.1}$$

and the magnetic energy may be written in the form

$$\mathcal{M} = k_T \frac{\Phi^2}{V} R^2 + k_P \Phi^2 \frac{h^2}{R} \tag{4.2}$$

where k_T and k_P are dimensionless constants. Hence $\mathcal{M}$ is minimal when

$$R = (k_P/2k_T)^{\frac{1}{3}} V^{\frac{1}{3}} h^{\frac{2}{3}} \tag{4.3}$$

and then

$$\mathcal{M}_{min} \sim h^{\frac{4}{3}} \Phi^2 / V^{\frac{1}{3}} \tag{4.4}$$

for some constant k. Since h, Φ and V are invariant under all frozen field distortions of the flux-tube, this provides an important estimate (to within a constant of order unity) of the minimum magnetic energy that may be attained.

4.2. EVALUATING THE ENERGY INTEGRALS

With the magnetic field explicitly defined in section 2, we now calculate the energy

$$\mathcal{M} = \mathcal{M}_T + \mathcal{M}_P = \frac{1}{2} \int \mathbf{B}_T^2 \, dV + \frac{1}{2} \int \mathbf{B}_P^2 \, dV \tag{4.5}$$

of the twisted flux tube constructed as above. Firstly,

$$\begin{aligned} \mathcal{M}_T &= \frac{1}{2} \int_A (2\pi r)^2 \frac{\Phi^2}{V^2} 2\pi r \, dr \, dz \\ &= 4\pi^3 \frac{\Phi^2}{V^2} \int_0^{2\pi} \int_0^{\varepsilon R} (R + s\cos\phi)^3 s \, ds \, d\phi \quad \text{using (2.7)} \\ &= \frac{2\pi^2 R^2 \Phi^2}{V} \left(1 + \frac{3}{4}\varepsilon^2\right) . \end{aligned} \tag{4.6}$$

The first term dominates when $\varepsilon \ll 1$ (so that the tube is thin and long), and corresponds to the energy of a uniform flux tube of length $2\pi R$; the second term is the correction associated with the fact that the axis of the tube is curved, with radius of curvature R.

Secondly, since from (2.10) and (2.11),

$$\mathbf{B}_P^2 = \left(2\pi R h \frac{\Phi}{V} \frac{s}{r}\right)^2 , \tag{4.7}$$

the poloidal energy is given by

$$\begin{aligned} \mathcal{M}_P &= \frac{\pi \Phi^2}{V^2} \int_0^{2\pi} \int_0^{\varepsilon R} \frac{4\pi^2 R^2 s^2 h^2}{R + s\cos\phi} s \, ds \, d\phi \\ &= \frac{2^{\frac{4}{3}} \pi^{\frac{2}{3}} \Phi^2 h^2}{3 V^{\frac{1}{3}}} \varepsilon^{-\frac{10}{3}} \left(2 - (1 - \varepsilon^2)^{\frac{1}{2}} (2 + \varepsilon^2)\right) . \end{aligned} \tag{4.8}$$

Note here the limiting form for small ε

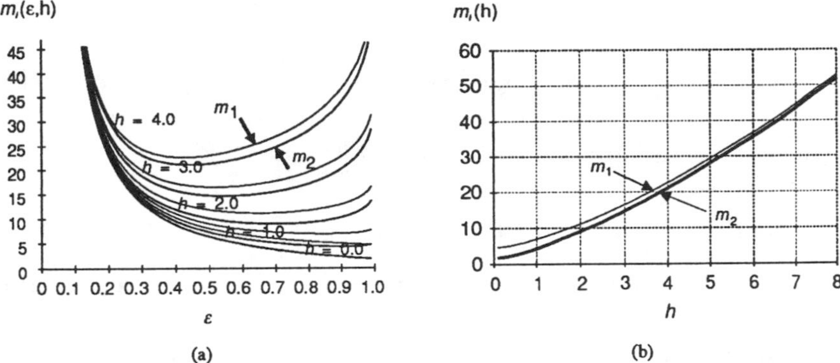

Fig. 3. (a) magnetic energy $m_1(\varepsilon, h)$ and $m_2(\varepsilon, h)$ are plotted against ε for various values of h; (b) the minimum energy $\min_\varepsilon m_1(\varepsilon, h)$ and $\min_\varepsilon m_2(\varepsilon, h)$ are plotted against h.

$$\mathcal{M}_P \approx \frac{2^{\frac{4}{3}}\pi^{\frac{2}{3}}\Phi^2 h^2}{3V^{\frac{1}{3}}}\varepsilon^{-\frac{10}{3}} \times \frac{3}{8}\varepsilon^4 = \frac{h^2\Phi^2}{2R} \quad \text{as} \quad \varepsilon \to 0 \tag{4.9}$$

(cf. 4.2). We now have

$$\mathcal{M} = \frac{\Phi^2}{V^{\frac{1}{3}}}\, m_1(\varepsilon, h) \tag{4.10}$$

where

$$m_1(\varepsilon, h) = \frac{2^{\frac{1}{3}}\pi^{\frac{2}{3}}}{\varepsilon^{\frac{4}{3}}}\left(1 + \frac{3}{4}\varepsilon^2 + \frac{2h^2}{3\varepsilon^2}\left(2 - (1-\varepsilon^2)^{\frac{1}{2}}(2+\varepsilon^2)\right)\right) . \tag{4.11}$$

This function is sketched in figure 3 for $0 < \varepsilon \leq 1$ and various values of h. Note the presence of a minimum value for all $h \neq 0$ (since $\partial m_1/\partial\varepsilon < 0$ as $\varepsilon \to 0$, but > 0 as $\varepsilon \to 1$).

For $h \gg 1$, m_1 attains its minimum when $\varepsilon \ll 1$, so $\mathcal{M}_{min}$ may be expanded in an asymptotic series (for small ε)

$$\mathcal{M} \approx \frac{2^{\frac{1}{3}}\pi^{\frac{2}{3}}\Phi^2}{V^{\frac{1}{3}}}\,\varepsilon^{-\frac{4}{3}}\left(1 + \left(\frac{3}{4} + \frac{1}{2}h^2\right)\varepsilon^2 + \cdots\right) \tag{4.12}$$

which has a minimum when

$$\varepsilon^2 = \frac{8}{3 + 2h^2} \approx 4/h^2 \,, \tag{4.13}$$

or equivalently when

$$R = R_{min} \approx \frac{1}{2}\left(\frac{Vh^2}{\pi}\right)^{\frac{1}{3}} \tag{4.14}$$

(cf. 4.3), and then

$$\mathcal{M} = \mathcal{M}_{min} \approx \frac{3\pi^{\frac{2}{3}}}{2} \frac{h^{\frac{4}{3}}\Phi^2}{V^{\frac{1}{3}}} \approx 3.22\, h^{\frac{4}{3}} \frac{\Phi^2}{V^{\frac{1}{3}}}\,, \tag{4.15}$$

in confirmation of the previous estimates (4.3) and (4.4).

4.3. THE ENERGY PRINCIPLE

In the above calculation, we have assumed that the tube cross-section is circular, and the field structure is invariant. We ignored the fact that the toroidal field may be re-distributed during the relaxation, even if each magnetic surface is constrained to have circular cross-section. We now calculate the effect of this redistribution.

Consider the variational principle (3.19)–(3.20) in the case of circular cross-sections. For large h we assume that F is independent of ϕ, so each contour with constant χ is circular; then (3.19) yields

$$FF_{\chi} = \frac{V}{2\pi h \Phi R} \tag{4.16}$$

and hence

$$F^2 = \frac{V\chi}{\pi h \Phi R}\,. \tag{4.17}$$

Now we have to minimise

$$\mathcal{M} = 4\pi^2 \int_0^{\chi_{max}} \left(\frac{\pi R \Phi}{hV}(R^2 - F^2) + \chi \right) \frac{d\chi}{\sqrt{R^2 - F^2}} \tag{4.18}$$

where $\chi_{max} = h\Phi/2\pi$ and F is given by (4.17). The integral can be evaluated to give

$$\mathcal{M} = \frac{\Phi^2}{V^{\frac{1}{3}}}\, m_2(\varepsilon, h) \tag{4.19}$$

where

$$m_2(\varepsilon, h) = 2^{\frac{4}{3}} \pi^{\frac{2}{3}} \varepsilon^{-\frac{10}{3}} \left(\frac{1-h^2}{3}(1 - (1-\varepsilon^2)^{\frac{3}{2}}) + h^2 (1 - (1-\varepsilon^2)^{\frac{1}{2}}) \right) \tag{4.20}$$

is to be minimised. This function (figure 3) provides an estimate of minimum energy less than that given in (4.11): in that calculation, the structure of the toroidal field was prescribed and fixed, so that the class of equilibrium states to which the tube could relax was narrowed. We also note that now the relaxed toroidal field is $(0, B_E(r,z), 0)$, where

$$B_E(r,z) = 2\pi R \frac{\Phi}{V} \frac{\sqrt{2rR - r^2 - z^2}}{r} \tag{4.21}$$

using (3.8),(3.12) and (3.13). However, the asymptotic form of $m_2(\varepsilon, h)$ for small ε is

$$m_2(\varepsilon, h) = 2^{\frac{4}{3}}\pi^{\frac{2}{3}}\varepsilon^{-\frac{10}{3}}\left(\frac{1}{2}\varepsilon^2 + \frac{2h^2-1}{8}\varepsilon^4 + \cdots\right) \tag{4.22}$$

which has a minimum when

$$\varepsilon^2 \approx 4/h^2 , \tag{4.23}$$

and then $\mathcal{M}_{min} = kh^{\frac{4}{3}}$ (with same constant k as before). These results are the same as given in (4.13)–(4.15): when $h \gg 1$, the area of cross-section is $O(R^{-1})$, so the details of the toroidal field distribution become unimportant.

4.4. THE LOWER BOUND

There is also a lower bound on the magnetic energy when $h \neq 0$ [Arnol'd 1974, Moffatt 1985, Freedman 1988]. This results from the Schwarz inequality

$$\int \mathbf{B}^2 dV \cdot \int \mathbf{A}^2 dV \geq \left(\int \mathbf{A}\cdot\mathbf{B}\, dV\right)^2 \tag{4.24}$$

and the Poincaré inequality

$$\frac{\int \mathbf{B}^2 dV}{\int \mathbf{A}^2 dV} \geq q_0^2 \tag{4.25}$$

where q_0 is a constant (with dimensions (length)$^{-1}$) depending on the geometry of the fluid domain. Hence

$$\int \mathbf{B}^2 dV \geq q_0 \left|\int \mathbf{A}\cdot\mathbf{B}\, dV\right| . \tag{4.26}$$

Writing

$$\mathcal{M} = \frac{1}{2}\int \mathbf{B}^2\, dV = m_a(h)\frac{\Phi^2}{V^{\frac{1}{3}}} \tag{4.27}$$

and

$$\mathcal{H} = \int \mathbf{A}\cdot\mathbf{B}\, dV = h\Phi^2 , \tag{4.28}$$

we then have

$$m_a(h) \geq \frac{1}{2}q_0 V^{\frac{1}{3}}|h| . \tag{4.29}$$

For example [1], the fluid domain may be chosen to be a sphere of radius

$$R' = R(1+\varepsilon) = \left(\frac{V}{2\pi^2\varepsilon^2}\right)^{\frac{1}{3}}(1+\varepsilon)\,, \tag{4.30}$$

which just contains the flux tube. Then $q_0 = \pi/R'$, so that (4.29) becomes

$$m_a(h) \geq 2^{-\frac{2}{3}}\pi^{\frac{5}{3}}|h|\frac{\varepsilon^{\frac{2}{3}}(h)}{1+\varepsilon(h)}\,. \tag{4.31}$$

In the limit $h \to \infty$, $\varepsilon \to 2/h$ so that

$$m_a(h) \geq \pi^{\frac{5}{3}}|h^{\frac{1}{3}}| \tag{4.32}$$

approximately. Combining this with the previous result, we have

$$\pi^{\frac{5}{3}}h^{\frac{1}{3}} \leq m_a(h) \leq \frac{3\pi^{\frac{2}{3}}}{2}h^{\frac{4}{3}} \tag{4.33}$$

in the limit $h \to \infty$.

5. THE CASE OF SMALL h

In the case of small h the poloidal field is weak; the toroidal field then tends to contract and give up energy, and hence to squeeze the flux tube onto the axis of symmetry. The cross-section is then long and thin, and the flux tube may be approximated by a hollow cylinder

$$\mathcal{C} = \{(r,\theta,z) : R - \varepsilon R < r < R + \varepsilon R\,,\ |z| < L\} \tag{5.1}$$

(see figure 4a), of volume $V = 8\pi\varepsilon R^2 L$, $\varepsilon \leq 1$, with $R \ll L$. We shall assume that $\varepsilon = O(1)$ as $h \to 0$.

5.1. THE SCALING ARGUMENT

Let

$$\mathbf{B} = \left(B_r, 2\pi r\frac{\Phi}{V}, B_z\right) \tag{5.2}$$

so that the total toroidal flux is invariant. The toroidal energy has the same scaling property

$$\mathcal{M}_T = \frac{1}{2}\int |\mathbf{B}_T|^2 dV \sim \frac{R^2\Phi^2}{V} \tag{5.3}$$

[1] This provides a rather poor lower bound when $h \gg 1$; a better lower bound could be obtained by, for example, taking the fluid domain to be the spherical annulus $R(1-\varepsilon) < r < R(1+\varepsilon)$. The important point, however, is that there is *always* a positive lower bound for the energy when $h \neq 0$

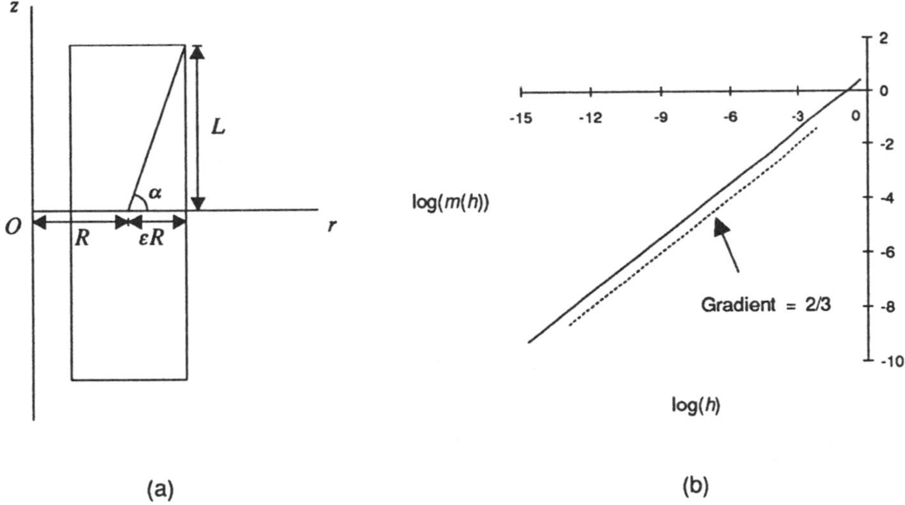

Fig. 4. (a) when h is small, the flux tube is approximated by a hollow cylinder, hence it has retangular cross-section; (b) $\log m_a(h)$ is plotted against $\log h$, showing that $m_a(h) = O(h^{2/3})$ in the limit $h \to 0$.

as before; the poloidal energy, however, scales differently. First note that the poloidal flux Φ_P about the magnetic axis can be estimated in two ways: it is the total flux across the ring $\{(r, \theta, z) : R < r < (1+\varepsilon)R, z = 0\}$ so that

$$|\Phi_P| \approx 2\pi R^2 |B_z| \tag{5.4}$$

or it is the total flux across the surface $\{(r, \theta, z) : r = R, 0 < z < L\}$so that

$$|\Phi_P| \approx 2\pi R L |B_r| \tag{5.5}$$

where $|B_z|$ and $|B_r|$ are the average values. It follows that

$$\left|\frac{B_r}{B_z}\right| \sim \frac{R}{L}\,, \tag{5.6}$$

i.e. $|B_r| \ll |B_z|$. Using $\Phi_P = h\Phi$, the poloidal magnetic energy is then

$$\mathcal{M}_P = \frac{1}{2}\int (B_r^2 + B_z^2)\, dV \sim k_P \frac{h^2 \Phi^2 V}{R^4}\,. \tag{5.7}$$

Combining (5.3) and (5.7), the total magnetic energy is then

$$\mathcal{M} = k_T \frac{R^2 \Phi^2}{V} + k_P \frac{h^2 \Phi^2 V}{R^4} \tag{5.8}$$

where k_T and k_P are constants. This has a minimum when

$$R = \left(\frac{2k_P}{k_T}\right)^{\frac{1}{6}} h^{\frac{1}{3}} V^{\frac{1}{3}} \tag{5.9}$$

and then

$$\mathcal{M}_{min} \sim h^{\frac{2}{3}} \Phi^2 / V^{\frac{1}{3}} . \tag{5.10}$$

Note the different scalings of the minimum energy

$$\mathcal{M}_{min} = \begin{cases} O\left(h^{\frac{2}{3}}\right) , & \text{when } h \text{ is small;} \\ O\left(h^{\frac{4}{3}}\right) , & \text{when } h \text{ is large.} \end{cases} \tag{5.11}$$

5.2. THE VARIATIONAL PRINCIPLE

We assume that the magnetic surfaces are similar in shape, and that each takes the form of a hollow cylinder as described above. Hence we define

$$F(\chi, \phi) = \begin{cases} +\dfrac{f(\chi)}{\cos\phi} , & \phi \leq \alpha ; \\ \dfrac{f(\chi)}{\sin\phi} \tan\alpha , & \alpha \leq \phi \leq \pi - \alpha; \\ -\dfrac{f(\chi)}{\cos\phi} , & \pi - \alpha < \phi \leq \pi ; \end{cases} \tag{5.12}$$

where $\tan\alpha = L(\varepsilon R)^{-1}$ (figure 4a), and $f(\chi)$ is an arbitrary function of χ. The integral in the constraint (3.20) can now be evaluated, and is simplified to

$$f f_\chi = \frac{V}{8Rh\Phi \tan\alpha} = \frac{\pi\varepsilon^2 R^2}{h\Phi} \tag{5.13}$$

so that

$$f^2 = \frac{2\pi\varepsilon^2 R^2 \chi}{h\Phi} . \tag{5.14}$$

Hence the arbitrary function $f(\chi)$ is completely determined by the (volume) constraint. Using this result, and the normalizations

$$\chi \rightarrow \frac{h\Phi}{2\pi}\chi , \qquad R \rightarrow V^{\frac{1}{3}} R ; \tag{5.15}$$

we simplify (3.19) and obtain

$$\mathcal{M} = \frac{\Phi^2}{V^{\frac{1}{3}}} \int_0^1 \left(4\pi^2 R^2 J_1^{-1} + h^2 J_2\right) d\chi \tag{5.16}$$

where

$$
\begin{aligned}
J_1 &= \frac{1}{1-\varepsilon^2\chi} + \frac{1}{\varepsilon\chi^{\frac{1}{2}}}\tanh^{-1}(\varepsilon\chi^{\frac{1}{2}}) \,, \\
J_2 &= \frac{\chi}{4\pi^2\varepsilon^2(1-\varepsilon^2\chi)R^4} + 16\varepsilon\chi^{\frac{1}{2}}R^2\tanh^{-1}(\varepsilon\chi^{\frac{1}{2}}) \,;
\end{aligned}
\tag{5.17}
$$

and (5.22) is minimised subject to

$$
R > 0 \,, \qquad 0 < \varepsilon < 1 \,. \tag{5.18}
$$

This has been done numerically, and $\log m_a(h)$ is plotted against $\log h$ in figure 4b. The slope is 2/3, confirming the estimate (5.11); we also found that $\varepsilon \not\to 1$ as $h \to 0$ (see below). However, if it were assumed that the contours are elliptic at the equilibrium state, then the closest distance between the outermost contour and the central axis should tend to zero.

In fact we can now prove the assertion that $\mathcal{M}_{min} = O(h^{2/3})$ for small h. Defining

$$
\begin{aligned}
A(\varepsilon) &= \int_0^1 4\pi^2\left(\frac{1}{1-\varepsilon^2\chi} + \frac{1}{\varepsilon\chi^{\frac{1}{2}}}\tanh^{-1}(\varepsilon\chi^{\frac{1}{2}})\right)^{-1} d\chi \,, \\
B(\varepsilon) &= \int_0^1 \frac{\chi\, d\chi}{4\pi^2\varepsilon^2(1-\varepsilon^2\chi)} \,, \\
C(\varepsilon) &= \int_0^1 16\varepsilon\chi^{\frac{1}{2}}\tanh^{-1}(\varepsilon\chi^{\frac{1}{2}})\, d\chi \,,
\end{aligned}
\tag{5.19}
$$

then (5.22) becomes

$$
m_a(R,\varepsilon,h) = R^2(A(\varepsilon) + h^2C(\varepsilon)) + \frac{h^2}{R^4}B(\varepsilon) \tag{5.20}
$$

where $\mathcal{M} = \Phi^2 V^{-1/3} m_a$ as before. For any fixed ε, assuming that $A(\varepsilon)$, $B(\varepsilon)$ and $C(\varepsilon)$ are $O(1)$, then $m_a(R,\varepsilon,h)$ is minimal when $R = R^*$, where

$$
R^{*2} = h^{\frac{2}{3}}\left(\frac{2B(\varepsilon)}{A(\varepsilon) + h^2C(\varepsilon)}\right)^{\frac{1}{3}} \tag{5.21}
$$

and then

$$
m_a(\varepsilon,h) = \frac{3}{2^{\frac{2}{3}}}h^{\frac{2}{3}}\left(A(\varepsilon) + h^2C(\varepsilon)\right)^{\frac{2}{3}} B^{\frac{1}{3}}(\varepsilon) \,. \tag{5.22}
$$

Assuming $C(\varepsilon) = O(1)$ as $h \to 0$, we then have

$$
m_a \approx \frac{3}{2^{\frac{2}{3}}}h^{\frac{2}{3}}A^{\frac{2}{3}}B^{\frac{1}{3}} \tag{5.23}
$$

and we find by computation of (5.19) that this is minimised when

$$\varepsilon = \varepsilon^* \approx 0.9059\ . \tag{5.24}$$

Hence $A(\varepsilon^*) = 13.35$, $B(\varepsilon^*) = 4.115 \times 10^{-2}$ and $C(\varepsilon^*) = 8.700$ are $O(1)$. We also note that

$$R^* \to \left(\frac{2B(\varepsilon^*)}{A(\varepsilon^*)}\right)^{\frac{1}{6}} h^{\frac{1}{3}} \approx 0.4282\, h^{\frac{1}{3}}\ , \quad m_a(h) \approx 3.6718\, h^{\frac{2}{3}} \tag{5.25}$$

as $h \to 0$.

5.3. THE LOWER BOUND

As in the previous section, we imagine the flux tube is bounded by a fluid domain in the form of a sphere. The optimal choice of the radius R' of the sphere is

$$R' = \sqrt{L^2 + R^2(1+\varepsilon)^2} \approx L \tag{5.26}$$

and therefore, from (4.29),

$$m_a(h) \geq \frac{\pi V^{\frac{1}{3}}|h|}{\sqrt{L^2 + R^2(1+\varepsilon)^2}} = \frac{\pi|h|}{\sqrt{(8\pi\varepsilon R^2)^{-2} + R^2(1+\varepsilon)^2}}\ . \tag{5.27}$$

In the limit $h \to 0$, $\varepsilon \to \varepsilon^*$ and $R \to R^* h^{2/3}$ where ε^* and R^* are constants; hence we have, approximately, that

$$m_a(h) \geq \frac{\pi h^{\frac{5}{3}}}{\sqrt{0.0574 + 0.6660 h^2}} \tag{5.28}$$

for small h. Combining with (5.25), we obtain

$$13.11\, h^{\frac{5}{3}} \leq m_a(h) \leq 3.6718\, h^{\frac{2}{3}} \tag{5.29}$$

as $h \to 0$.

6. THE CASE OF h=O(1)

In this case numerical methods must be used. The numerical solution of the variational problem (3.19)- (3.20) can be obtained by discretizing the variable $F(\chi, \phi)$ both in χ and in ϕ. For simplicity, however, we restrict $F(\chi, \phi)$ to take the form

$$F(\chi, \phi) = f(\chi)\, g(\phi) \tag{6.1}$$

where

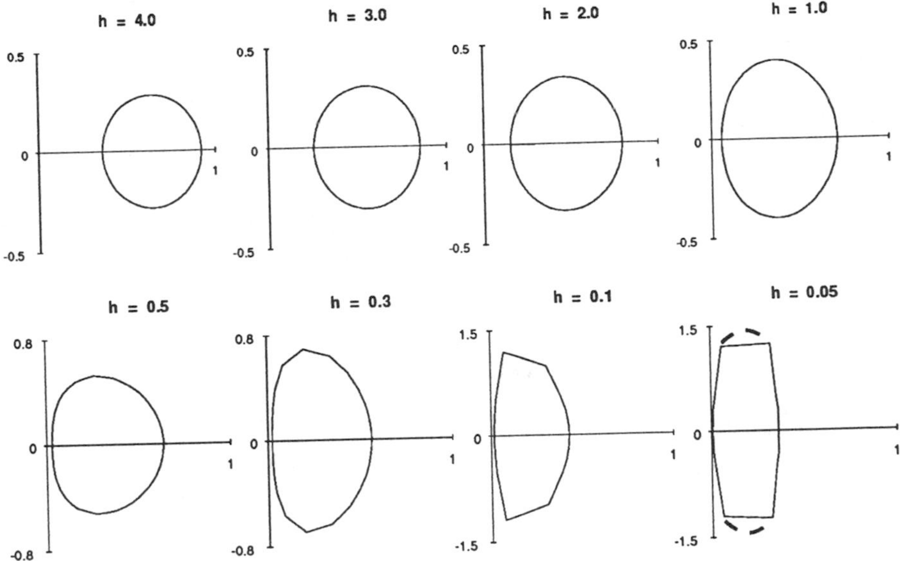

Fig. 5. The cross-section of minimum energy states for various values of h when the boundary is approximated by (16+1) points. Note that the numerical solutions are poor when h is small.

$$f(\chi) = \left(\frac{V\chi}{\pi h \Phi R}\right)^{\frac{1}{2}} . \tag{6.2}$$

The contours of constant χ are then similar in shape; and $f(\chi)$ is chosen to be compatible with the large h results given in section 4. We expect that $g(\phi) \to 1$ as $h \to 0$ since in the limit the shape of the cross-section at equilibrium is a circle.

The constraint (3.20) becomes

$$\int_0^{2\pi} (R + fg\cos\phi)\, g^2 d\phi = 2\pi R \tag{6.3}$$

which is decomposed into

$$\int_0^{2\pi} g^2 d\phi = 2\pi , \qquad \int_0^{2\pi} g^3 \cos\phi \, d\phi = 0 . \tag{6.4}$$

The objective function $\mathcal{M}$ can then be expressed in terms of $g(\phi)$ and R, and is minimized subject to the 'realistic conditions'

$$R + f^* g(\phi)\cos\phi > 0 , \qquad R > 0 , \qquad g(\phi) > 0 \tag{6.5}$$

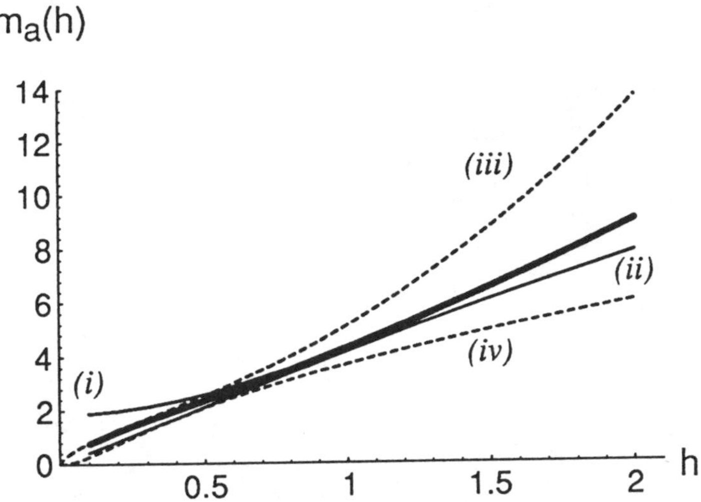

Fig. 6. The minimum energy $m_a(h)$ against h, calculated numerically (thick curve). This plot also shows the estimated (i) upper bound for large h, obtained by numerical minimisation of (4.20); (ii) lower bound for large h, (4.32); (iii) upper bound for small h, obtained by numerical minimisation of(5.22), and (iv) lower bound for small h, (5.28).

where

$$f^* = (2\pi^2 R)^{-\frac{1}{2}} \tag{6.6}$$

(since otherwise the surface is not defined in a meaningful way). The optimization problem is solved by standard numerical routines. Without going into details, we simply state the result in figure 5, which shows the cross-sections for different values of h at relaxed states. Note that the cross-sections deform as h decreases in such a way that magnetic field lines are contracting to the central axis as expected. Also the minimum energy calculated numerically confirms the estimates from previous sections, as summarised in figure 6. Note that $m_a(h)$ has an inflexion point near $h = 1$.

7. THE GRAD–SHAFRANOV EQUATION

Magnetostatic equilibrium with flux function $\chi(r, z)$, pressure field $p = P(\chi)$ and toroidal field $B_\theta(r, z) = r^{-1}F(\chi)$ is governed by the Grad-Shafranov equation

$$D^2\chi = -r^2 P'(\chi) - F(\chi)F'(\chi) \tag{7.1}$$

where

$$D^2 = \frac{\partial^2}{\partial r^2} - \frac{1}{r}\frac{\partial}{\partial r} + \frac{\partial^2}{\partial z^2} \,. \tag{7.2}$$

When the equilibrium is a circular flux tube of mean radius R and cross-section $A \ll R^2$, the operator D^2 may be approximated by

$$D^2 \approx \frac{\partial^2}{\partial r^2} + \frac{\partial^2}{\partial z^2} \tag{7.3}$$

and the Grad-Shafranov equation may be approximated by

$$\left(\frac{\partial^2}{\partial r^2} + \frac{\partial^2}{\partial z^2}\right)\chi = -R^2 P'(\chi) - F(\chi)F'(\chi) \,. \tag{7.4}$$

Here, (r, z) may be regarded as cartesian coordinates in the meridian plane, centred on the axis of the flux tube.

Consider now the scale transformation

$$R \to \lambda R \,, \qquad (r, z) \to \lambda^{-\frac{1}{2}}(r, z) \tag{7.5}$$

which corresponds to stretching the flux tube by a factor λ and shrinking its cross-section by the same factor to conserve the volume. It may easily be verified that (7.4) is then invariant under the transformation

$$F \to \lambda^2 F \,, \quad \chi \to \lambda^{\frac{3}{2}}\chi \quad \text{and} \quad P \to \lambda^2 P \,. \tag{7.6}$$

The toroidal flux Φ_T is invariant under this transformation, but the total poloidal flux, equal to $2\pi\Delta\chi$ where $\Delta\chi$ is the jump in χ from the centre of the flux tube to its boundary, scales as $\Phi_P \sim \lambda^{\frac{3}{2}}$. Hence the helicity scales as

$$\mathcal{H} \sim \Phi_T \Phi_P \sim \lambda^{\frac{3}{2}} \tag{7.7}$$

or equivalently $\lambda \sim \mathcal{H}^{\frac{2}{3}}$. This is the counterpart of the result (4.14), and provides confirmation that the results (4.13)–(4.15) are characteristic of toroidal magnetostatic equilibrium (when $R \gg A^{1/2}$), even when the cross-section of the flux-tube is not circular.

Similarly, in the case of small h, the operator D is approximated by

$$D^2 \approx \frac{\partial^2}{\partial r^2} - \frac{1}{r}\frac{\partial}{\partial r} \tag{7.8}$$

and (7.1) is invariant under the scale transformation

$$r \to \lambda r \,, \quad z \to \lambda^{-2} z \,; \qquad F \to \lambda^2 F \,, \quad \chi \to \lambda^3 \chi \quad \text{and} \quad P \to \lambda P \,. \tag{7.9}$$

This corresponds to stretching the flux tube by a factor λ in the r–direction and compressing it by a factor λ^2 in the z–direction to conserve the volume. The toroidal flux is again invariant under this transformation, but poloidal flux scales as λ^3. Hence the helicity scales as

$$\mathcal{H} \sim \Phi_T \Phi_P \sim \lambda^3 \tag{7.10}$$

or $\lambda \sim \mathcal{H}^{\frac{1}{3}}$. This is the counterpart of the result (5.25).

8. THE EFFECT OF NON-AXISYMMETRIC INSTABILITY

In this paper we have studied the minimum energy $m_a(h)$ of the twisted flux tube as a function of h. The main result,

$$m_a(h) \sim \begin{cases} 3.67\, h^{\frac{2}{3}} & \text{when } h \ll 1, \\ 3.22\, h^{\frac{4}{3}} & \text{when } h \gg 1, \end{cases} \tag{8.1}$$

provides a close estimate from above for the minimum energy of the flux tube. We also estimate a lower bound and the results are summarised in figure 6. All calculations assume that the minimum energy state is axisymmetric. However, preliminary calculation suggests that this configuration is, in fact, unstable to kink mode perturbations when h is larger than a critical value of order unity. The flux tube may then relax further to a state with lower energy, and the result (8.1) should therefore be understood as an upper bound of the true minimum energy $m(h)$.

Although we are unable to calculate the exact form of $m(h)$, there is a simple argument suggesting that $m(h) \sim Ch^{4/3}$ when $h \gg 1$, like $m_a(h)$, but with smaller proportionality constant C. Figure 7a shows a generic situation when a twisted flux tube is deformed by kink mode displacement. The total helicity is unchanged, but the twist ingredient is transformed into a torsion ingredient (associated with the torsion of the axis, see Ricca & Moffatt 1992). Figure 7b shows how a flux tube with a large value of h may relax. A possible model for this relaxed state is a twisted cylindrical flux tube, shown in figure 7c, carrying equal and opposite magnetic flux (of magnitude Φ); the corrections at the two ends are negligible when h is large. In cylindrical polar coordinates, let $\mathbf{B} = (0, B_\theta, B_z)$ be the uniform twisted field confined in the cylinder $\mathcal{C} = \{(r, \theta, z) : 0 \leq r \leq a,\ 0 \leq z \leq L\}$ of volume V and area of cross-section $A = \pi a^2$, then

$$|B_z| = 2\Phi/A\ , \qquad B_\theta = 2\pi b r \Phi/V \tag{8.2}$$

where b is to be determined. The 'net' axial flux is $\Phi_z = 2\Phi$, and the flux associated with the azimuthal component of the field is

$$|\Phi_\theta| = \frac{\Phi}{V} \iint 2\pi b r\, dr\, dz = b\Phi\ ; \tag{8.3}$$

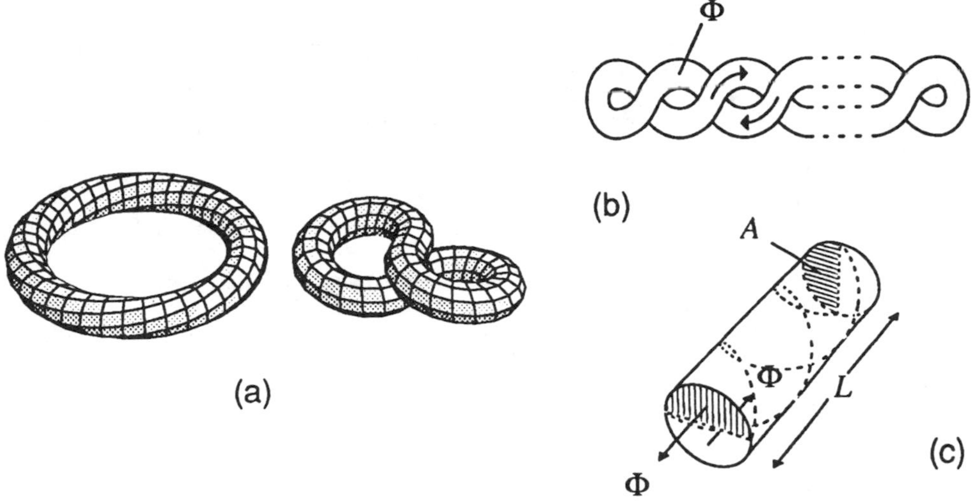

Fig. 7. (a) a torus with twist (of angle 2π) can be deformed into a figure 8 without twist; (b) the non-axisymmetric equilibrium state; (c) the flux tube is approximated by a cylinder with twisted magnetic field.

hence the helicity of the field (8.2) is

$$\mathcal{H} = \Phi_z \Phi_\theta = 2b\Phi^2. \tag{8.4}$$

Since the helicity is an invariant and has the value $h\Phi^2$, therefore $b = h/2$. The magnetic energy is then

$$\mathcal{M} = \frac{1}{2}\int \left(B_z^2 + B_\theta^2\right) dV = \Phi^2 \left(\frac{2V}{A^2} + \frac{h^2\pi}{4V}A\right) , \tag{8.5}$$

which has a minimum when $A^3 = 16V^2(h^2\pi)^{-1}$, and then, using (4.15),

$$\mathcal{M} = \mathcal{M}_{min} = \frac{3\pi^{\frac{2}{3}}}{2^{\frac{5}{3}}} h^{\frac{4}{3}} \frac{\Phi^2}{V^{\frac{1}{3}}} = \frac{1}{2^{\frac{2}{3}}} m_a(h) \frac{\Phi^2}{V^{\frac{1}{3}}} = m(h)\frac{\Phi^2}{V^{\frac{1}{3}}}, \tag{8.6}$$

where

$$m(h) = 2^{-2/3} m_a(h) \approx 0.63 m_a(h) , \tag{8.7}$$

consistent with the conjecture that the relaxed state with lowest energy is non-axisymmetric.

Acknowledgements

A.Y.K.C. would like to thank M.J.D. Powell for his advice on numerical computations; he is also grateful to J. Greene and A. Bhattacharjee for discussions. A.Y.K.C. was supported by the Croucher Foundation.

References

ARNOL'D, V.I. 1974 The asymptotic Hopf invariant and its applications (in Russian). In *Proc. Summer School in Differential Equations, Erevan.* Armenian SSR Acad. Sci., Transl. 1986 in *Sel. Math. Sov.*, **5**, pp. 327-345.

FREEDMAN, M.H. 1988 A note on topology and magnetic energy in incompressible perfectly conducting fluids. *J. Fluid Mech.*, **194**, pp. 549-551.

KRUSKAL, M.D. & KULSRUD, R.M. 1958 Equilibrium of a magnetically confined plasma in a toroid. *Phys. Fluids*, **1**, pp. 265-274.

MOFFATT, H.K. 1969 The degree of knottedness of tangled vortex lines. *J. Fluid Mech.*, **35**, pp. 117-129.

MOFFATT, H.K. 1985 Magnetostatic equilibria and analogous Euler flows of arbitrary complex topology. Part 1. Fundamentals. *J. Fluid Mech.*, **159**, pp. 359-378.

MOFFATT, H.K. 1990 The energy spectrum of knots and links. *Nature*, **347**, pp. 367-369.

RICCA, R.L. & MOFFATT, H.K. 1992 The helicity of a knotted vortex filament. *This volume.*

THOMPSON, W.B. 1962 *Introduction to plasma physics.* Pergamon.

RESEARCH ANNOUNCEMENT ON THE "ENERGY" OF KNOTS

MICHAEL H. FREEDMAN
Department of Mathematics
University of California at San Diego
9500 Gilman Drive
La Jolla, California 92093-0112, USA

and

ZHENG-XU HE
Department of Mathematics
Princeton University
Fine Hall, Washington Street
Princeton, New Jersey 08544-1000, USA

It is a natural idea that a loop of string might somehow be charged and placed in a bottle of honey, there to assume its least (potential) energy or "best" shape. In the late 1970's Alan Hatcher announced a remarkable geometric theorem (Hatcher, 1983) (the "Smale conjecture": $\mathit{Diff}(S^3) \simeq O(4)$). Among its many consequences is the fact that the space of all smooth unknots – that is, simple closed curves in three space ($\mathbf{R}^3$) which can be deformed to a round circle – actually can be continuously deformed to the subspace of round circles in three space. This has heightened interests in the honey jar experiment. For the theorem tells us that there is some method – continuous in initial conditions – which untangles arbitrary unknots and makes them round. The key point is continuity; there should be no intelligent intervention. No choices should be made. A person untangling string makes lots of judgements but perhaps the charged unknotted loop in honey would always make its way to a round circle without incident. Perhaps not. Computer experiments may be able to suggest the correct answer to this interesting mathematical problem. Actually the Newtonian potential $\frac{1}{r}$ is too weak at small scales to create an infinite energy barrier to crossings. This requires at least the $\frac{1}{r^2}$ potential described below. Whether or not weaker po-

H. K. Moffatt et al. (eds.), Topological Aspects of the Dynamics of Fluids and Plasmas, 219–222.

tentials $\frac{1}{r^p}$, $p < 2$ might yet determine gradient flows which respect topology (*i.e.*, admit no crossings) is an open question.

Let $\gamma = \gamma(u)$ be a rectifiable curve in $\mathbf{R}^3$ parameterized so that the derivative $\dot{\gamma}(u)$ exists a.e.; where u belongs to $\mathbf{R}$, an interval of $\mathbf{R}$ or the circle S^1. For any pair of points $\gamma(u)$, $\gamma(v)$ on γ, denote by $D(\gamma(u),\gamma(v))$ the distance between them on the curve; *i.e.*, the minimum of the lengths of sub-arcs of γ with one endpoint at $\gamma(u)$ and the other at $\gamma(v)$. We define the *energy of the curve* γ *relative to the point* $\gamma(u)$ to be the following integral:

$$E(\gamma,\gamma(u)) = \int \left\{ \frac{1}{|\gamma(v)-\gamma(u)|^2} - \frac{1}{D(\gamma(v),\gamma(u))^2} \right\} |\dot{\gamma}(v)|\, dv. \tag{1.1}$$

Note that the function $u \mapsto E(\gamma,\gamma(u)) \in [0,\infty]$ is measurable. The following integral

$$E(\gamma) = \int E(\gamma,\gamma(u))|\dot{\gamma}(u)|\, du \tag{1.2}$$

will be called the *energy* of the curve γ.

By (1) and (2), we obtain

$$E(\gamma) = \int\int \left\{ \frac{1}{|\gamma(v)-\gamma(u)|^2} - \frac{1}{D(\gamma(v),\gamma(u))^2} \right\} |\dot{\gamma}(u)|\, |\dot{\gamma}(v)|\, dudv. \tag{1.3}$$

The following lemma is immediate.

LEMMA 1. (i) $E(\gamma,\gamma(u))$ and $E(\gamma)$ do not depend on the parametrizations or orientations of the curve. (ii) Let $T\colon \mathbf{R}^3 \to \mathbf{R}^3$ be an affine similarity. Then $E(T\circ\gamma, T\circ\gamma(u)) = E(\gamma,\gamma(u))$ and $E(T\circ\gamma) = E(\gamma)$.

If γ is a simple closed curve in $\mathbf{R}^3$ whose curvature is defined and uniformly bounded, then the energy of γ is finite.

O'Hara (1991) proved that given simultaneous upper bounds on several geometric quantities: energy, length, and the L^2-norm of the curvature, only finitely many knot types can occur. We drop the hypotheses on length and the L^2-norm of the curvature to prove:

THEOREM 2. If a knot K is represented by an imbedding $\gamma : S^1 \to \mathbf{R}^3$ then

$$\text{crossing } \# \ (K) \le \frac{11}{12\pi}\ E(\gamma) + \frac{1}{\pi}.$$

Recall that (see Freedman and He, 1991, pp. 196-197) the *average crossing number* $c(\gamma)$ of a (not necessarily closed) curve γ over itself is

$$c(\gamma) = c(\gamma,\gamma) = \frac{1}{4\pi}\int_{S^1\times}\int_{S^1} \frac{|(\dot{\gamma}(x),\ \dot{\gamma}(y),\ \gamma(y)-\gamma(x))|}{|\gamma(y)-\gamma(x)|^3}\, dxdy. \tag{1.4}$$

In fact, Theorem 2 is a corollary of the following.

THEOREM 3. For any simple rectifiable curve $\gamma: S^1 \to \mathbf{R}^3$, we have

$$c(\gamma) \leq \frac{11}{12\pi} E(\gamma) + \frac{1}{\pi}. \tag{1.5}$$

It may be estimated (Sumners, 1987; Tutte, 1963; Welsh) that the number $K_c(n)$ of distinct knots of at most n crossings satisfies:

$$2^n \leq K_c(n) \leq 2 \cdot 24^n.$$

So the number of knot types with representatives below a given energy threshold can also be bounded by an exponential.

COROLLARY 4. $K_e(E) \leq 6C^E$ where $C = 24^{11/12\pi}$.

In our normalization all round circles have energy E (round) $= 4$ which, presumably, is the smallest possible value for the energy of closed curves in $\mathbf{R}^3$. On the other hand, if $E(\gamma) < Const. \approx 9.38526$, then E is unknotted.

It is interesting to compare E with the total curvature

$$T(\gamma) = \int \left| \left(\frac{\gamma'(u)}{|\gamma'(u)|} \right)' \right| \, du.$$

Clearly T (round) $= 2\pi$ and according to Milnor (1950), $T(\gamma) \leq 4\pi$ implies γ is unknotted. The functional T is less coersive than E since unlike Theorem 2 there are infinitely many 2-bridge knots all having representatives with $T = 4\pi + \epsilon$.

We also consider the existence and regularity of extremals.

THEOREM 5. There exists a simple closed rectifiable curve $\gamma_0 \subset \mathbf{R}^3$ whose energy realizes the infimum in the class of closed rectifiable curves in $\mathbf{R}^3$. Any such extremal curve is a C^1, convex, plane curve.

As we remarked, we conjecture that any extremal curve in the above theorem must be a round circle.

Given a knot K, it is interesting to find the position of the knot with minimal energy. However, in order to establish existence of an extremal curve representing K, we have to work with proper rectifiable imbeddings of the real line $\mathbf{R}$ in $\mathbf{R}^3$. Such proper rectifiable imbeddings of $\mathbf{R}$, also termed as *proper rectifiable lines* in $\mathbf{R}^3$, can be considered as simple closed curves in $\overline{\mathbf{R}}^3 = \mathbf{R}^3 \cup \{\infty\}$ which pass through the point at infinity. Analogous to Theorem.2, any proper rectifiable line γ in $\mathbf{R}^3$ with finite energy represents a knot, say, K, with

$$\text{crossing } \# \, (K) \leq \frac{1}{2\pi} \, E(\gamma). \tag{1.6}$$

A smooth knot is called *irreducible* (or *prime*) if it is not equal to a connected sum of two nontrivial knots.

THEOREM 6. Let K be an irreducible knot type. There exists a proper rectifiable line γ_K with knot type K such that $E(\gamma_K) \leq E(\gamma)$ for any other proper rectifiable line γ of knot type K.

We believe that the extremal curves must be smooth when parametrized by arc length. However, we have not yet proved that they are even continuously differentiable.

References

FREEDMAN, M. H. AND Z.-X. HE 1991 Divergence-free fields. Energy and asymptotical crossing number. *Annals of Math* **134**, 189-229.

HATCHER, A., 1983 A proof of the Smale conjecture, *Annals of Math.* **117**, 553-607.

O'HARA, J., 1991 Energy of a knot. *Topology* **30**, 241-247.

MILNOR, J., 1950 On the total curvature of knots. *Annals of Math.* **52**, 248-257; Zbl. 389; M. R. 373.

SUMNERS, DEWITT, 1987 The growth of the number of prime knots. *Math. Proc. Comb. Phil. Soc.* **102**, 303-315.

TUTTE, W. T., 1963 A census of planar maps. *Canadian J. Math.*, **15**, 249-271.

WELSH, D.J.A., On the number of knots. Preprint.

PART III

HELICITY, LINKAGE, AND FLOW TOPOLOGY

THE HELICITY OF A KNOTTED VORTEX FILAMENT

R.L. RICCA* and H.K. MOFFATT*
Institute for Theoretical Physics
University of California
Santa Barbara, California 93106-4030
U.S.A.

ABSTRACT. The helicity $\mathcal{H}$ associated with a knotted vortex filament is considered. The filament is first constructed starting from a circular tube, in three stages involving injection of (integer) twist, deformation and switching of crossings. This produces a vortex tube in the form of an arbitrary knot K; each vortex line in the tube is a (trivial) satellite of K, and the linking number of any pair of vortex lines in the tube is the same integer n. It is shown that in these circumstances the helicity is given by $\mathcal{H} = n\kappa^2$ where κ is the circulation associated with the tube. This result is discussed in relation to earlier works, in particular the work of Călugăreanu (1959, 1961) which establishes that, for a twisted ribbon with axis $\mathcal{C}$ the number n is the sum of three ingredients:

$$\frac{\mathcal{H}}{\kappa^2} = n = W(\mathcal{C}) + \mathcal{T}(\mathcal{C}) + \frac{1}{2\pi}[\Delta\Theta]_{\mathcal{C}}$$

where $W(\mathcal{C})$ is the writhing number and $\mathcal{T}(\mathcal{C})$ is the total torsion. The quantity $[\Delta\Theta]_{\mathcal{C}}$ represents the net angle of rotation of the spanwise vector on the ribbon relative to the Frenet triad in one passage round $\mathcal{C}$. Both $\mathcal{T}(\mathcal{C})$ and $[\Delta\Theta]_{\mathcal{C}}$ are discontinuous in deformations that take $\mathcal{C}$ through an inflexion point. The generic behaviour in such passage through an inflexion point is analysed and clarified in §6.

1. Introduction

Let $\mathbf{u}(\mathbf{x}, t)$ be the velocity field in an inviscid incompressible fluid, evolving under the Euler equations, and let $\boldsymbol{\omega}(\mathbf{x}, t) = \nabla \times \mathbf{u}$ be the corresponding vorticity field. Let $\mathcal{S}$ be any closed orientable surface moving with the fluid on which $\boldsymbol{\omega} \cdot \mathbf{n} = 0$. Then it is well-known (Moffatt, 1969) that the helicity integral

$$\mathcal{H} = \int_{\mathcal{V}} \mathbf{u} \cdot \boldsymbol{\omega} \, d\mathcal{V} \tag{1}$$

* Permanent address: Department of Applied Mathematics and Theoretical Physics, Silver Street, Cambridge CB3 9EW, U.K.

H. K. Moffatt et al. (eds.), Topological Aspects of the Dynamics of Fluids and Plasmas, 225–236.

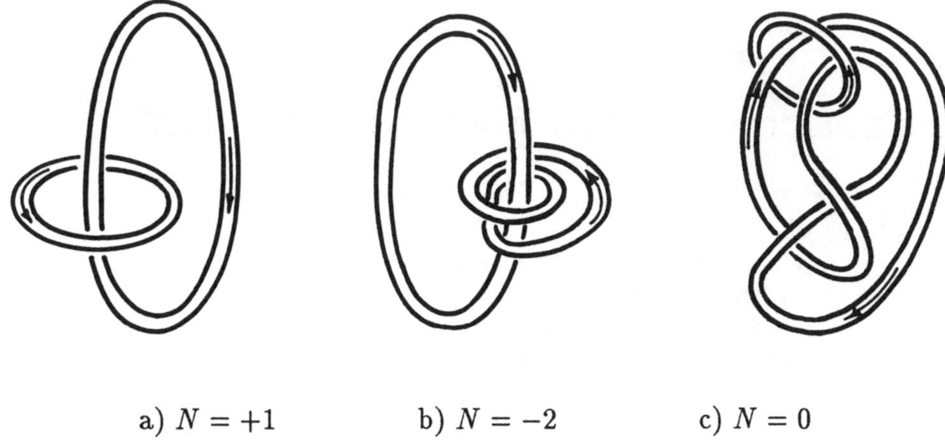

a) $N = +1$ b) $N = -2$ c) $N = 0$

Fig. 1. Linking of oriented vortex tubes.

where $\mathcal{V}$ is the volume inside $\mathcal{S}$, is an invariant under this Euler evolution, this invariance being associated with the fact that the vortex lines are frozen in the fluid and the topology of the vorticity field is therefore conserved.

The topological interpretation is most transparent for the simple situation in which $\boldsymbol{\omega} \equiv 0$ except in two linked vortex filaments of vanishingly small cross-sections and of circulations κ_1, κ_2; then, provided each vortex tube is unknotted and the vorticity field has no internal twist within each tube, it is easily shown that

$$\mathcal{H} = 2N\kappa_1\kappa_2 \tag{2}$$

where N is the Gauss (linking) number of the axes C_1, C_2 of the tubes, positive or negative according as the orientation of the linkage is right-handed or left-handed (for examples, see Fig. 1).

This interpretation has been extended by Arnol'd (1974) to situations in which linked vortex lines are not closed curves but wind around each other infinitely often. The integral (1) is still invariant in this situation and has been described by Arnol'd as the "asymptotic Hopf invariant".

For a single knotted vortex filament, the situation is not so simple. If the axis of the tube is in the form of a knot K, then each vortex line is (if closed) a satellite of K, and the helicity invariant may be expected to bear the imprint of K in the limit as the tube cross-section shrinks to zero. However there is now an unavoidable twist of the field $\boldsymbol{\omega}$ within the tube, partly associated with torsion of the axis C of the tube, and evaluation of $\mathcal{H}$ presents consequential difficulties.

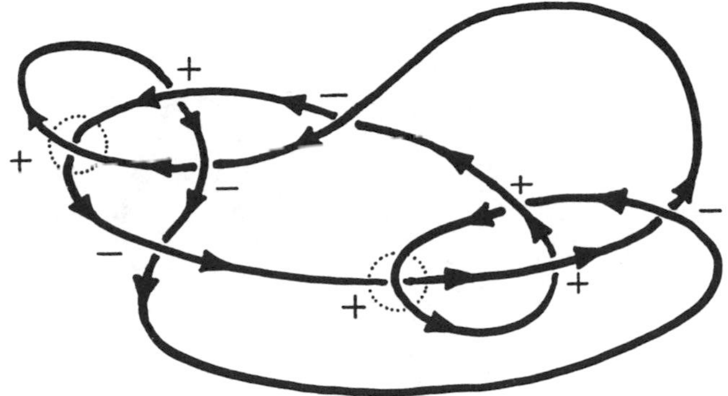

Fig. 2. Alternating knot (overcrossings alternate with undercrossings) with 9 crossings, 5 positive and 4 negative (at a positive crossing, the overstrand must be rotated anticlockwise to come into coincidence with the understrand with arrows pointing the same way). Switching of two positive crossings converts this to the unknot.

It has been conjectured (Moffatt, 1981) that $\mathcal{H}$ must be asymptotically identifiable with the invariant of Călugăreanu (1959, 1961) and this is indeed implicit in later works (*e.g.* Berger & Field, 1984) in which the helicity of a knotted vortex tube (or equivalently magnetic flux tube) is expressed as the sum of writhe and twist components. However a direct evaluation of $\mathcal{H}$ for a knotted vortex filament has never been given. We aim to provide this in the present contribution.

2. Construction of a knotted vortex tube of prescribed helicity

We recall first the construction of a knotted vortex tube described by Moffatt (1990). Let K be an arbitrary (tame) knot, and suppose that we view it in standard plane projection with a finite number of crossings, each of which is either positive or negative (for an example, see Fig. 2). By a finite number of crossing "switches", K may be converted to the unknot, which may be continuously deformed to a circle $\mathcal{C}_0$. By reversing these steps, $\mathcal{C}_0$ may be reconverted to K.

Now let T_0 be a tubular neighbourhood of $\mathcal{C}_0$, and let ω_0 be a vorticity field in T_0, uniform over each (small) cross-section of T_0, each vortex line being a circle parallel to $\mathcal{C}_0$. Let κ be the circulation of the vortex tube. The helicity of ω_0 is zero. We may inject helicity $\mathcal{H}_0 = h\kappa^2$ into this vorticity field by *Dehn surgery*: cut the tube at some section, twist through an angle $2\pi h$, and reconnect (Fig. 3). If h is an integer n_0 (as we shall suppose), then each

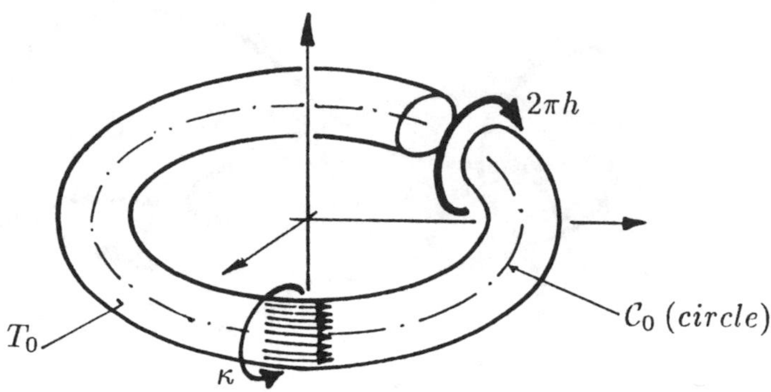

Fig. 3. Dehn surgery (cut, twist, reconnect) on a circular vortex tube.

ω line is a closed curve which closes on itself after one passage around T_0 (in knot terminology, it has winding number 1). Now convert $\mathcal{C}_0$ (carrying T_0 with it) to the knot K by the steps indicated above. $\mathcal{H}_0$ is conserved during deformation but changes by an amount $\pm 2\kappa^2$ with each switch creating a positive or negative crossing (Fig. 4). Hence the helicity of the knotted tube becomes

$$\mathcal{H} = (n_0 + 2N_+ - 2N_-)\kappa^2 = n\kappa^2, \tag{3}$$

say, where $N_\pm$ are (respectively) the numbers of positive and negative switches used to create K. Thus, by this construction, $\mathcal{H}/\kappa^2$ is an integer.

Note the special choice $n_0 = -2(N_+ - N_-)$ which makes $\mathcal{H} = 0$. In this situation, each ω-line is still a replica of K so that the topology of the ω-field is decidedly non-trivial. However the linking number of each pair of ω-lines is zero (this does not necessarily mean that they are unlinked! — see the example of Fig. 1c).

3. Helicity and the self-linking number of a framed knot

The number n in (3) is in fact the linking number of any pair of ω-lines in the knotted tube filament. This may be proved as follows.

Let us divide the tube up into m "sub-tubes" each with the same circulation (flux of vorticity) κ/m. Suppose that the linking number of each pair of vortex lines is N; this is then also the linking number of each pair of sub-tubes.

If $\mathcal{H}$ is the total helicity, then the helicity associated with the vorticity in a sub-tube is

$$\mathcal{H}_m = \mathcal{H}/m^2 \tag{4}$$

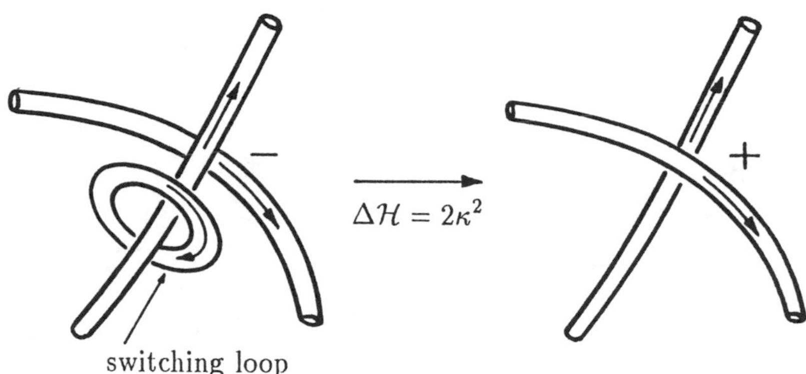

Fig. 4. Switching a negative crossing to become a positive crossing; this is equivalent to the insertion of a "switching loop" which cancels the field of the understrand below the crossing and recreates it above the crossing. This corresponds to increasing the helicity of the knot by $2\kappa^2$. Similarly the reverse switch changes the helicity by $-2\kappa^2$.

(since helicity is a quadratic functional of vorticity). $\mathcal{H}_m$ may be thought of as the "self-helicity" of a sub-tube T_m associated with the linkage of vortex lines within T_m.

The total helicity $\mathcal{H}$ is the sum of these self-helicities plus the sum of the interactive helicities (cf. eq. 2) arising from linkage of flux tubes, *i.e.*

$$\mathcal{H} = m\mathcal{H}_m + 2\sum_{\substack{i,j\\ i\neq j}} N\kappa_i\kappa_j \tag{5}$$

with $\kappa_i = \kappa/m$ ($i = 1, 2, \ldots, m$). Hence

$$\mathcal{H} = \frac{\mathcal{H}}{m} + 2N\frac{1}{2}m(m-1)\left(\frac{\kappa}{m}\right)^2 \tag{6}$$

i.e.

$$\mathcal{H} = N\kappa^2, \tag{7}$$

a result that is independent of the degree m of subdivision of the tube. Hence, comparing with (3), $N = n$ as asserted.

Any two vortex lines C_1 and C_2 in the tube are the boundaries of a ribbon R_{12} contained within the tube. A frame of reference $(\mathbf{e}_1, \mathbf{n}_{12}, \mathbf{e}_1 \times \mathbf{n}_{12})$ may be constructed on this ribbon, where $\mathbf{e}_1$ is the tangent vector to C_1 (a function of arc-length s_1 on C_1), and $\mathbf{n}_{12}$ is the spanwise vector on R_{12} from C_1 to C_2 (also a function of s_1). Choice of such a frame constitutes a "framing" of

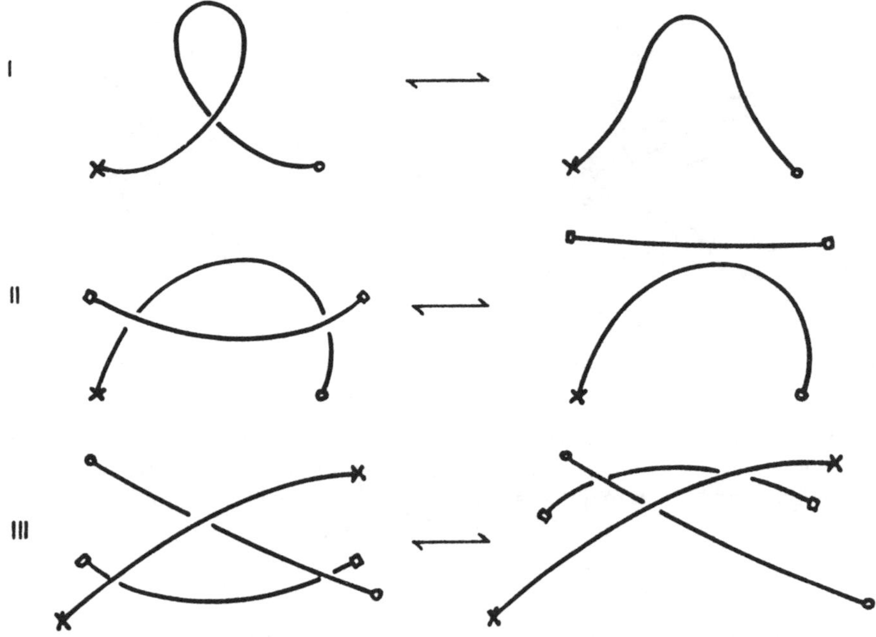

Fig. 5. The three Reidemeister moves (Kauffman, p. 9, 1987).

the knot K. The number n may then be described (Pohl, 1968) as the self-linking number of the framed knot. The self-linking number has no meaning unless a frame is specified.

4. Helicity and Reidemeister's first move

One of the classical results of knot theory is that two knots K and K' are isotopic (*i.e.* one may be continuously deformed into the other) if and only if this can be achieved by a succession of Reidemeister moves (Fig. 5) acting on (any) plane projection of K (or K'). A distinction is made between "ambient isotopy" which allows moves of types I, II and III, and "regular isotopy" which allows only moves of types II and III. The description "ambient isotopic" is synonymous (for knots) with "topologically equivalent" (Kauffman, p. 9, 1987).

If we consider the Reidemeister moves applied to a vortex filament, then we can think of these as being localised deformations of the vorticity field that conserve global helicity. There is no difficulty whatsoever in relation to the moves II and III in this respect. There is however a subtlety in relation to move I, as may be seen from consideration of figure 6 which represents the twisting of a loop through an angle 2π, thus converting a negative crossing

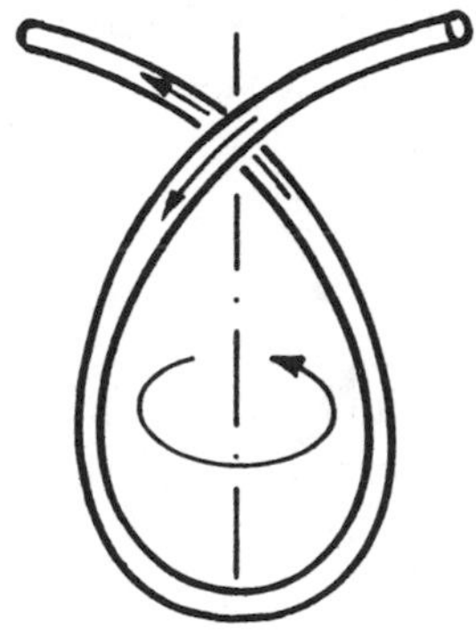
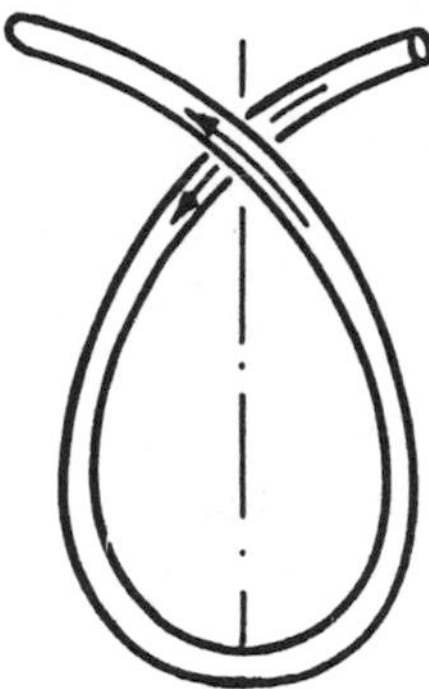

Fig. 6. A right-handed twist of the loop through 2π (equivalent to two type-I Reidemeister moves) converts a negative crossing to a positive crossing.

to a positive crossing. If the same effect were achieved by a crossing switch as described in §2 above, then the helicity would increase by $2\kappa^2$. Since the Reidemeister twisting move conserves helicity, this increase must be compensated by an equal decrease $-2\kappa^2$ arising from the twist of ω-lines within the tube, *i.e.* a combination of switching and Dehn surgery is needed to maintain helicity:

$$2 \times \text{Reidemeister move I} \equiv \text{switch } (- \rightarrow +) + \text{Dehn surgery } (-4\pi) \qquad (8)$$

The essence of this equivalence may be appreciated by playing with a belt or a paper ribbon!

5. Helicity and the Călugăreanu invariant

Now let C be the central axis of the vortex tube (itself a vortex line); let s be the arc-length from a point O of C, and suppose that the curvature $c(s)$ is everywhere positive on C (*i.e.* there are no points of inflexion). Let $(\mathbf{e}, \mathbf{n}, \mathbf{b})$ be the orthogonal triad of unit vectors ($\mathbf{e}$ = tangent vector, $\mathbf{n}$ = principal normal, $\mathbf{b}$ = binormal) which satisfy the Frenet-Serret equations

$$\frac{d\mathbf{e}}{ds} = c\mathbf{n}, \qquad \frac{d\mathbf{n}}{ds} = -c\mathbf{e} + \tau\mathbf{b}, \qquad \frac{d\mathbf{b}}{ds} = -\tau\mathbf{n}, \qquad (9)$$

where $\tau(s)$ is the torsion of C.

If $\mathbf{x} = \mathbf{x}(s)$ is a point on C, then a neighbouring curve C' may be defined by $\mathbf{x}' = \mathbf{x}(s) + \varepsilon\mathbf{n}(s)$ where ε is a small positive parameter. The linking number of C and C' is given by Gauss's formula

$$G(C, C') = \frac{1}{4\pi} \oint_C \oint_{C'} \frac{d\mathbf{x} \times d\mathbf{x}' \cdot (\mathbf{x} - \mathbf{x}')}{|\mathbf{x} - \mathbf{x}'|^3} \qquad (10)$$

and this is of course an integer. Călugăreanu (1959) considered the limit of this integral as $\varepsilon \to 0$. One (obvious) contribution in the limit is what is known (Fuller, 1971) as the writhing number of $\mathcal{C}$:

$$W(\mathcal{C}) = \frac{1}{4\pi} \oint_{\mathcal{C}} \oint_{\mathcal{C}} \frac{d\mathbf{x} \times d\mathbf{x}' \cdot (\mathbf{x} - \mathbf{x}')}{|\mathbf{x} - \mathbf{x}'|^3}. \tag{11}$$

There is however a second contribution to the limit arising from pairs of points $\mathbf{x}$, $\mathbf{x}'$ such that $|\mathbf{x} - \mathbf{x}'| = O(\varepsilon)$. Călugăreanu showed that this second contribution is given by

$$\mathcal{T}(\mathcal{C}) = \frac{1}{2\pi} \oint_{\mathcal{C}} \tau(s)\, ds \tag{12}$$

i.e. the total torsion divided by 2π, and therefore that

$$\lim_{\varepsilon \to 0} G(\mathcal{C}, \mathcal{C}') = W(\mathcal{C}) + \mathcal{T}(\mathcal{C}) \tag{13}$$

is an invariant under distortions of $\mathcal{C}$ *which do not introduce any inflexion point*. This is a very severe restriction, because as will be shown in the following section it excludes twisting deformations, *i.e.* Reidemeister moves of type I.

6. Generic behaviour associated with inflexion points

At an inflexion point s_c on a curve $\mathbf{x} = \mathbf{x}(s)$ (where s is arc-length), $d\mathbf{e}/ds = d^2\mathbf{x}/ds^2 = 0$, so that near $s = s_c$,

$$\mathbf{e}(s) = \mathbf{e}_c + \frac{1}{2}(s - s_c)^2 \mathbf{e}_c'' + \ldots \tag{14}$$

and

$$\mathbf{x}(s) = \mathbf{x}_c + (s - s_c)\mathbf{e}_c + \frac{1}{6}(s - s_c)^3 \mathbf{e}_c'' + \ldots \tag{15}$$

Moreover $\mathbf{e}_c''$ is perpendicular to $\mathbf{e}_c$ since

$$(\mathbf{e}'' \cdot \mathbf{e})_{s=s_c} = \frac{d^2}{ds^2}(\mathbf{e}^2)\bigg|_{s=s_c} = 0. \tag{16}$$

Choosing origin at the inflexion point ($\mathbf{x}_c = 0$, $s_c = 0$), and axes $Oxyz$ with Ox parallel to $\mathbf{e}_c$ and Oz parallel to $\mathbf{e}_c''$, the form of the curve near the inflexion is

$$\mathbf{x}(s) = (s,\, 0,\, \alpha s^3) \tag{17}$$

where $\alpha = \frac{1}{6}|\mathbf{e}_c''|$, *i.e.* $y = 0$, $z = \alpha x^3$. Without loss of generality we may take $\alpha = 1$. We consider a time-dependent twisted cubic curve

$$\mathcal{C}: \quad \mathbf{x}(s,t) = \left(s - \frac{2}{3}t^2 s^3,\, ts^2,\, s^3\right) \tag{18}$$

which passes through the plane inflexional configuration (17) at time $t = 0$ (Fig. 7). We shall suppose that $|t|$ and $|s|$ are small, and we calculate the torsion $\tau(s,t)$ to leading order near $t = s = 0$. First note that

$$\mathbf{e}(s,t) = \frac{\partial \mathbf{x}}{\partial s} = (1 - 2t^2 s^2,\, 2ts,\, 3s^2) \tag{19}$$

and that

$$|\mathbf{e}| = 1 + O(s^4) \tag{20}$$

so that, neglecting terms of order s^4, $\mathbf{e}(s,t)$ is indeed the unit tangent vector near $s = 0$.

We now have

$$\frac{\partial \mathbf{e}}{\partial s} = (-4t^2 s,\, 2t,\, 6s) \approx (0,\, 2t,\, 6s) \tag{21}$$

so that the curvature is

$$c(s,t) = \left|\frac{\partial \mathbf{e}}{\partial s}\right| \approx 2(t^2 + 9s^2)^{\frac{1}{2}} \tag{22}$$

near $t = s = 0$. As expected, c vanishes at $s = 0$, $t = 0$, but there is no inflexion point when $|t| \neq 0$; thus $\mathcal{C}(t)$ contains an inflexion point at $s = 0$ at the single instant $t = 0$. The principal normal is

$$\mathbf{n}(s,t) = \frac{1}{c}\frac{\partial \mathbf{e}}{\partial s} = (t^2 + 9s^2)^{-\frac{1}{2}}(0,\, t,\, 3s) \tag{23}$$

and the binormal is then, to leading order,

$$\mathbf{b}(s,t) = \mathbf{e} \times \mathbf{n} = (t^2 + 9s^2)^{-\frac{1}{2}}(0,\, -3s,\, t). \tag{24}$$

From the third Frenet-Serret equation $\partial \mathbf{b}/\partial s = -\tau \mathbf{n}$, we now easily find that to leading order near $t = s = 0$,

$$\tau(s,t) = \frac{3t}{t^2 + 9s^2} \tag{25}$$

This result (see Fig. 8) reveals the nature of the singularity of τ at $t = s = 0$; indeed the total torsion between $s = -s_0$ and $s = +s_0$ for $t \neq 0$ is

$$\int_{-s_0}^{s_0} \tau(s,t)\, ds = 2\int_0^{s_0} \frac{3t}{t^2 + 9s^2}\, ds = 2\tan^{-1}\left(\frac{3s_0}{t}\right). \tag{26}$$

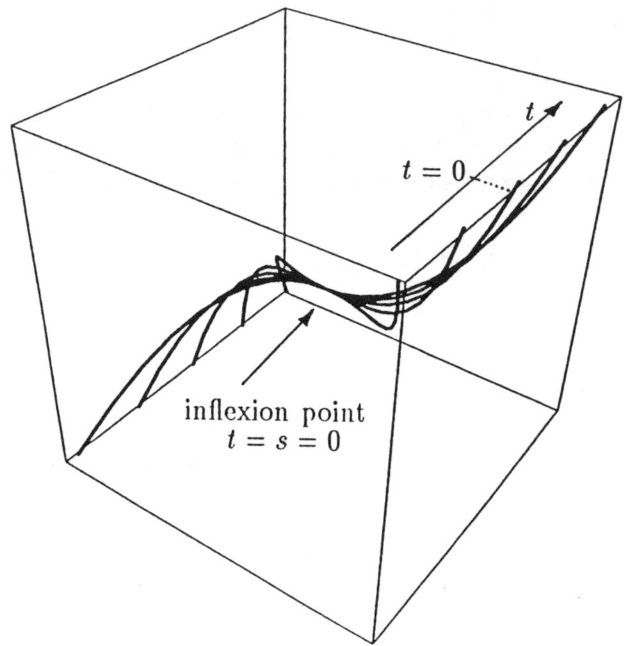

Fig. 7. The twisted cubic (18) for $-1 < s < 1$ and for various values of t. The curve contains an inflexion point at $s = 0$ when $t = 0$.

This jumps from $-\pi$ to $+\pi$ as t increases through zero, irrespective of the value of s_0.

This type of behaviour was recognised by Călugăreanu (1961) who devised a particular example of a *closed* curve deforming through an inflexional configuration. However, Călugăreanu misleadingly refers to "a discontinuity of τ" at the point of inflexion[1], whereas in fact it is the integral with respect to arc-length of $\tau(s,t)$ that is dicontinuous (by an amount 2π) at $t = 0$. This discontinuity is just the amount expected for a Reidemeister move I, by the argument of §4.

Since $\mathcal{W}(\mathcal{C})$ is continuous and $\mathcal{T}(\mathcal{C})$ is discontinuous for such distortions, the sum is no longer invariant when distortions through an inflexion point take place. However, the situation is rectified by framing the curve as in §3 and by including the total angle of twist $[\Delta\Theta]_{\mathcal{C}}$ of the spanwise vector from $\mathcal{C}$ to $\mathcal{C}'$ relative to the Frenet vectors $(\mathbf{n}, \mathbf{b})$ in one passage round $\mathcal{C}$. The "modified" Călugăreanu invariant is then

$$\frac{\mathcal{H}}{\kappa^2} = \mathcal{W}(\mathcal{C}) + \mathcal{T}(\mathcal{C}) + \frac{1}{2\pi}[\Delta\Theta]_{\mathcal{C}} \tag{27}$$

[1] Călugăreanu's paper is in French; he writes "...la torsion passe nécessairement par un point de discontinuité ..." (Călugăreanu, 1961, p. 616).

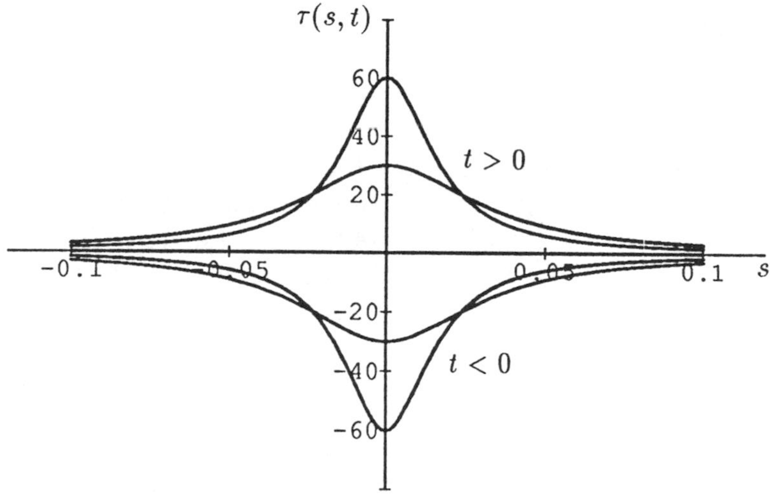

Fig. 8. The torsion function $\tau(s,t)$ as given by (25) for various values of t.

and we write it in this way to emphasise that it is none other than the helicity invariant divided by κ^2. When $\mathcal{C}$ contains an inflexion point, $\mathcal{T}(\mathcal{C})$ and $\frac{1}{2\pi}[\Delta\Theta]_{\mathcal{C}}$ are both indeterminate. In deformation through a configuration containing an inflexion point, both $\mathcal{T}(\mathcal{C})$ and $\frac{1}{2\pi}[\Delta\Theta]_{\mathcal{C}}$ are discontinuous but by equal and opposite amount so that $\mathcal{T}(\mathcal{C})+\frac{1}{2\pi}[\Delta\Theta]_{\mathcal{C}}$ is continuous; we may then adopt the common limit (as $t \to t_c$ from above or below) to give meaning to $\mathcal{T}(\mathcal{C})+\frac{1}{2\pi}[\Delta\Theta]_{\mathcal{C}}$ at the moment of discontinuity.

Of course all these subtleties are avoided if we simply adopt the helicity $\mathcal{H} = n\kappa^2$ of the (framed) vortex tube as the fundamental invariant which is insensitive to the presence or absence of inflexion points!

7. Summary and discussion

We have shown that, if the vortex lines in a knotted vortex filament are twisted in such a way that each vortex line is a closed curve which closes after one passage around the tube, and each pair of vortex lines in the tube has linking number n, then the helicity of the vorticity field is given by

$$\mathcal{H} = \int_{\mathcal{V}} \mathbf{u}\cdot\boldsymbol{\omega}\, d\mathcal{V} = n\kappa^2 \tag{28}$$

where κ is the circulation associated with the tube. The integer n is an invariant under frozen field distorsion of the tube, and is identified with the

Călugăreanu (1961) invariant:

$$n = W(C) + \mathcal{T}(C) + \frac{1}{2\pi}[\Delta\Theta]_C \tag{29}$$

where the writhing number $W(C)$ and the total torsion $\mathcal{T}(C)$ are defined by eqs. (11) and (12). The angle $[\Delta\Theta]_C$ represents the total angle of rotation of a neighbouring curve C' (which we here take to be a vortex line) relative to the Frenet pair $(\mathbf{n}, \mathbf{b})$ in one passage around the tube; $\frac{1}{2\pi}[\Delta\Theta]_C$ is an integer and so therefore is $W(C) + \mathcal{T}(C)$.

If C is deformed continuously, then it may pass through configurations containing one or more inflexion points. When C contains an inflexion point (or an odd number of inflexion points), the torsion is singular. This behaviour is analysed in §6 and it is shown that (generically) $\mathcal{T}(C)$ jumps discontinuously through ± 1 as C passes through the inflexional configuration. By virtue of the invariance of $\mathcal{H}$, there is then a compensating jump of $\mp 2\pi$ in $[\Delta\Theta]_C$. This behaviour is associated with the classical Reidemeister move of type I.

Călugăreanu's proof of the result (29) is long and complicated. It should be possible to derive the result directly from consideration of the limiting form of the helicity integral. We hope to present such a derivation in a future paper.

Acknowledgements

One of us (R.L.R.) wishes to thank the Associazione per lo Sviluppo Scientifico e Tecnologico del Piemonte (Turin, Italy) for financial support.

References

Arnol'd, V.I. 1974 The Asymptotic Hopf Invariant and Its Applications (in Russian). In *Proc. Summer School in Differential Equations, Erevan*, Armenian Acad. Sci. (English translation: *Sel. Math. Sov.* **5**, 327 (1986)).

Berger, M.A. & Field, G.B. 1984 The topological properties of magnetic helicity. *J. Fluid Mech.* **147**, 133.

Călugăreanu, G. 1959 L'intégrale de Gauss et l'analyse des nœuds tridimensionnels. *Rev. Math. Pures Appl.* **4**, 5.

Călugăreanu, G. 1961 Sur les classes d'isotopie des nœuds tridimensionnels et leurs invariants. *Czechoslovak Math. J.* T **11**, 588.

Fuller, F.B. 1971 The Writhing Number of a Space Curve. *Proc. Natl. Acad. Sci. USA* **68**, 815.

Kauffman, L. 1987 *On Knots*. Princeton University Press, Princeton NJ.

Moffatt, H.K. 1969 The degree of knottedness of tangled vortex lines. *J. Fluid Mech.* **35**, 117.

Moffatt, H.K. 1981 Some developments in the theory of turbulence. *J. Fluid Mech.* **106**, 27.

Moffatt, H.K. 1990 The energy spectrum of knots and links. *Nature* **347**, 367.

Pohl, W.F. 1968 The Self-Linking Number of a Closed Space Curve. *J. Math. Mech.* **17**, 975.

A HIERARCHY OF LINKING INTEGRALS

N.W. EVANS
Department of Applied Mathematics and Theoretical Physics
Silver St
Cambridge CB3 9EW , U.K.

and

M.A. BERGER
Department of Mathematics
University College
Gower Street
London WC1 , U.K.

ABSTRACT. Since the days of Maxwell and Tait, it has been known that the Gauss linking integral fails to detect entanglements with equal numbers of oppositely signed crossings, such as the Borromean rings. As the helicity or Gauss integral is quadratic in the fluxes or vortex strengths, it measures second order linking. It is the lowest member of a hierarchy of linking integrals. A third order linking integral (based on the Massey product) describes the subtle interlocking of the Borromean rings. An entanglement of four rings – any pair or triple of which is unlinked – is presented, together with a fourth order linking integral which distinguishes the delicate tangling from four unlinked rings. Links are close relatives of braids and the hierarchy of linking integrals now becomes a hierarchy of winding numbers. Detailed examples are given for $n = 3$ and 4 and the extension to n orders is sketched in the conclusions.

1. Introduction

Gauss (1867) writes in a letter dated January 22, 1833 that "a major task on the boundary of topology (Geometria Situs) and differential geometry (Geometria Magnitudinis) is to enumerate the linkages of two paths which are closed or extend to infinity". He adds that the double line integral (written in modern notation)

$$L_{12} = \frac{-1}{4\pi} \oint_{C_1} \oint_{C_2} \left[\frac{1}{|\mathbf{x}_2 - \mathbf{x}_1|^3} \frac{d\mathbf{x}_1}{ds} . \frac{d\mathbf{x}_2}{dt} \times (\mathbf{x}_2 - \mathbf{x}_1) \right] dsdt, \qquad (1.1)$$

where $\mathbf{x}_1(s)$ and $\mathbf{x}_2(t)$ are the parametric equations of curves C_1 and C_2, has a value equal to the signed number of crossings. This is the famous

H. K. Moffatt et al. (eds.), Topological Aspects of the Dynamics of Fluids and Plasmas, 237–248.

Gauss linking integral. The first scientist to realise that the result is not a comprehensive description of the linkage between two curves seems to have been Clerk Maxwell (1873). In his *Electricity and Magnetism*, he gives a very clear investigation (but without vectors) of the Gauss integral, together with an explicit statement that it may vanish even though the curves are inseparably linked. What is now called the Whitehead link is drawn out as an example of two intertwined curves with an equal number of oppositely signed crossings so that the integral (1.1) is zero.

The physical importance of the Gauss integral arises on interpreting the curves C_1 and C_2 as closed circuits running along the axes of two vortex or magnetic flux tubes. Pictorially, the curves are enclosed in thin toroidal volumes U_1 and U_2 and each torus is filled with unit longitudinal flux. Using Ampère's theorem, the vector potential $\mathbf{A_2}$ due to U_2 is

$$\mathbf{A_2}(\mathbf{x_1}) = \frac{-1}{4\pi} \oint_{C_2} \left[\frac{1}{|\mathbf{x_2} - \mathbf{x_1}|^3} \frac{d\mathbf{x_2}}{dt} \times (\mathbf{x_2} - \mathbf{x_1}) \right] dt, \tag{1.2}$$

where $\mathbf{x_1} \in C_1$ and $\mathbf{x_2} \in C_2$. Now, (1.1) can be simplified to yield a single line integral of the form

$$L_{12} = \oint_{C_1} \mathbf{A_2}.\mathbf{dl} = \oint_{C_2} \mathbf{A_1}.\mathbf{dl}, \tag{1.3}$$

or reformulated as a volume integral over the helicity density (see *e.g.*, Moffatt (1978))

$$L_{12} = \int_{U_1} \mathbf{A_2}.\mathbf{B_1} dV = \int_{U_2} \mathbf{A_1}.\mathbf{B_2} dV. \tag{1.4}$$

The significance of the helicity L_{12} (or any linking integral) is that it is an exact invariant of ideal MHD, where the magnetic field is (Lie) dragged under the evolving velocity field and so the topology of magnetic field lines is preserved. The helicity is quadratic in the magnetic fluxes and hence describes *second order linking*. For future comparison, we note that the Gauss integral can be written as a surface integral by defining a new field $\mathbf{M_{12}} = \mathbf{A_1} \times \mathbf{A_2}$ which is divergence-free everywhere outside U_1 and U_2. Now exploiting the divergence theorem gives

$$L_{12} = \int_{\partial U_1} \mathbf{M_{12}}.\hat{\mathbf{n}} dS = -\int_{\partial U_2} \mathbf{M_{12}}.\hat{\mathbf{n}} dS, \tag{1.5}$$

where ∂U_1 means the boundary of U_1 and $\hat{\mathbf{n}}$ is a unit normal. Full proofs of these formulae can be found for example in Moffatt (1978) or Berger & Field (1984).

The problem of finding higher order integral invariants analogous to (1.1) which distinguish between unlinked circuits and linkages for which there is an equal number of oppositely signed crossings is emphasised in Moffatt (1981) who draws attention to the striking example of the Borromean rings. This very old heraldic device is used on the coat of arms of the Borromeo family of Italian nob ility. It comprises three linked rings, no two of which are linked (see figure 1a). The Gauss integral taken over any two rings vanishes and so the helicity of the flux tube configuration is zero. The subtle intertwining of the Borromean rings may however be described by a *third order linking integral* as shown in Berger (1990, 1991) and reviewed in section 2.

This though is just the start of a hierarchy. Section 3 presents four interlocked rings, any two and any three of which are unlinked. For any pair of rings, the Gauss integral vanishes and for any triple of rings, the third order linking integral vanishes. The entanglement can be captured by a *fourth order linking integral.* An extension to nth order is suggested in the concluding section 4, which identifies further problems for study.

2. Third Order Linking

A full account of the higher order linking integrals is eased by the use of differential forms. We use mainly the exterior calculus (see Schutz (1980) for an accessible introduction) but translate the important results into vectorial language (see eqs. (2.10–11) and (3.8–9)). Henceforth, tildes above bold-face letters distinguish differential forms from vectors. All indices label distinct flux tubes.

The three closed curves C_i $(i = 1, 2, 3)$ of the Borromean rings are encased in toroidal volumes U_i with associated one-forms $\tilde{\mathbf{A}}_\mathbf{i}$. These are the one-forms with the same components as the vector potentials $\mathbf{A}_\mathbf{i}$. Let us introduce the two-forms

$$\tilde{\mathbf{M}}_{\mathbf{ij}} = \tilde{\mathbf{A}}_\mathbf{i} \wedge \tilde{\mathbf{A}}_\mathbf{j} = -\tilde{\mathbf{A}}_\mathbf{j} \wedge \tilde{\mathbf{A}}_\mathbf{i} \qquad i \neq j, \tag{2.1}$$

so that the Gauss integral is just [c.f., eq. (1.5)]

$$L_{ij} = \int_{\partial U_i} \tilde{\mathbf{M}}_{\mathbf{ij}} = -\int_{\partial U_j} \tilde{\mathbf{M}}_{\mathbf{ij}}. \tag{2.2}$$

Applying Stokes' theorem gives [c.f., eq. (1.4)]

$$L_{ij} = \int_{U_i} \tilde{\mathbf{d}}\tilde{\mathbf{M}}_{\mathbf{ij}} = \int_{U_i} \tilde{\mathbf{B}}_\mathbf{i} \wedge \tilde{\mathbf{A}}_\mathbf{j} = -\int_{U_j} \tilde{\mathbf{A}}_\mathbf{i} \wedge \tilde{\mathbf{B}}_\mathbf{j}, \tag{2.3}$$

where $\tilde{\mathbf{d}}$ denotes the exterior derivative operator and the two-form $\tilde{\mathbf{B}}_\mathbf{i} = \tilde{\mathbf{d}}\tilde{\mathbf{A}}_\mathbf{i}$ is the dual (*e.g.*, Schutz, 1980) of the ordinary magnetic field vector $\mathbf{B}_\mathbf{i}$. For the Borromean rings, L_{12}, L_{23} and L_{31} all vanish. Therefore, the Poincaré lemma guarantees everywhere the local existence of a one-form $\tilde{\mathbf{N}}_{\mathbf{ij}}$ such that

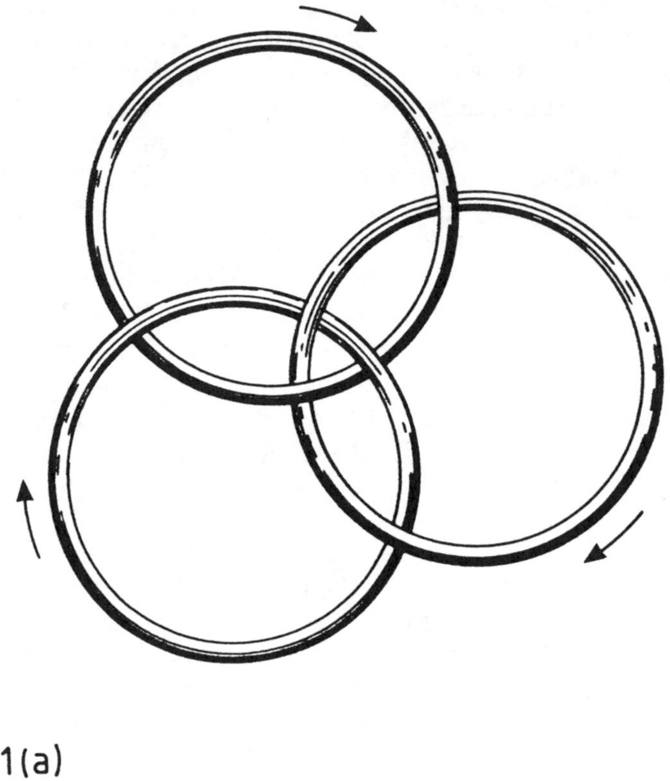

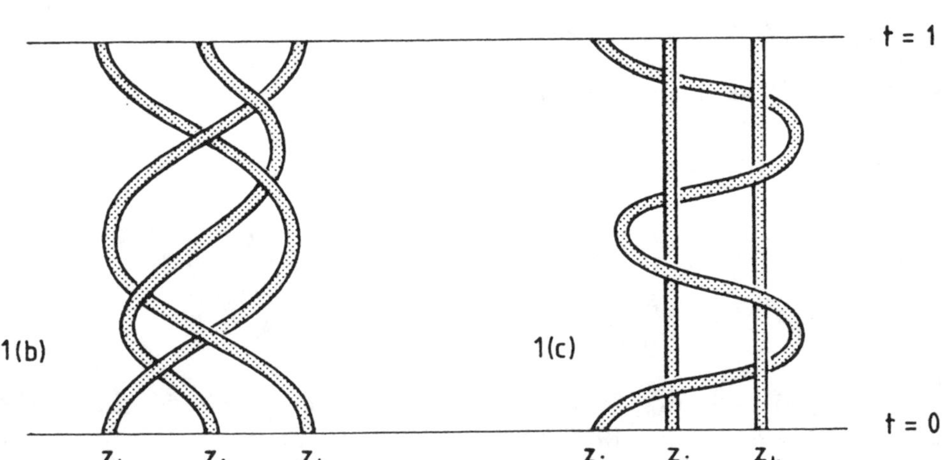

Fig. 1. (a) the Borromean rings, (b) the pigtail braid, (c) the combed pigtail braid.

$$\tilde{\mathbf{M}}_{\mathbf{ij}} = \tilde{\mathbf{d}}\tilde{\mathbf{N}}_{\mathbf{ij}}, \tag{2.4}$$

provided $\tilde{\mathbf{d}}\tilde{\mathbf{M}}_{\mathbf{ij}}$ is everywhere zero. This is possible on modifying (2.1) to read

$$\tilde{\mathbf{M}}_{\mathbf{ij}} = \begin{cases} \tilde{\mathbf{A}}_{\mathbf{i}} \wedge \tilde{\mathbf{A}}_{\mathbf{j}} - \phi_{(i)j}\tilde{\mathbf{B}}_{\mathbf{i}}, & \text{if } x \in U_i, \\ \tilde{\mathbf{A}}_{\mathbf{i}} \wedge \tilde{\mathbf{A}}_{\mathbf{j}} + \phi_{(j)i}\tilde{\mathbf{B}}_{\mathbf{j}}, & \text{if } x \in U_j, \\ \tilde{\mathbf{A}}_{\mathbf{i}} \wedge \tilde{\mathbf{A}}_{\mathbf{j}}, & \text{elsewhere,} \end{cases} \tag{2.5}$$

where $\phi_{(i)j}$ is a scalar potential satisfying $\tilde{\mathbf{A}}_{\mathbf{j}} = \tilde{\mathbf{d}}\phi_{(i)j}$. Note that $\phi_{(i)j}$ exists in U_i as the magnetic field $\tilde{\mathbf{B}}_{\mathbf{j}} = \tilde{\mathbf{d}}\tilde{\mathbf{A}}_{\mathbf{j}}$ is there zero, but is not globally defined. The third order linking integral can be written as

$$L_{ijk} = \int_{\partial U_i} \tilde{\mathbf{M}}_{\mathbf{ijk}} = -\int_{\partial U_k} \tilde{\mathbf{M}}_{\mathbf{ijk}} \qquad i \neq j \neq k, \tag{2.6}$$

where $\tilde{\mathbf{M}}_{\mathbf{ijk}}$ is the Massey triple product (really a mapping between cohomology classes (Massey, 1958; Fenn, 1983))

$$\tilde{\mathbf{M}}_{\mathbf{ijk}} = \tilde{\mathbf{A}}_{\mathbf{i}} \wedge \tilde{\mathbf{N}}_{\mathbf{jk}} + \tilde{\mathbf{N}}_{\mathbf{ij}} \wedge \tilde{\mathbf{A}}_{\mathbf{k}}. \tag{2.7}$$

Using Stokes' theorem, L_{ijk} can be formulated as a volume integral [c.f., eq. (1.4)]

$$L_{ijk} = \int_{U_i} \tilde{\mathbf{d}}\tilde{\mathbf{M}}_{\mathbf{ijk}} = -\int_{U_k} \tilde{\mathbf{d}}\tilde{\mathbf{M}}_{\mathbf{ijk}}, \tag{2.8}$$

or – most conveniently of all – as a line integral [c.f., eq. (1.3)]

$$L_{ijk} = \oint_{C_i} \tilde{\mathbf{N}}_{\mathbf{jk}} - \phi_{(i)j}\tilde{\mathbf{A}}_{\mathbf{k}} = \oint_{C_k} \tilde{\mathbf{N}}_{\mathbf{ij}} + \phi_{(k)j}\tilde{\mathbf{A}}_{\mathbf{i}}. \tag{2.9}$$

Berger (1990) shows that the third order linking integral $L_{123} = L_{231} = L_{312}$ is both gauge invariant and unchanged by deformations of the three curves. It is equal to ± 1 for the Borromean rings but vanishes for three unlinked circuits.

The vectorial equivalent of the Poincaré lemma (2.4) is that $\text{curl}^{-1}\mathbf{M}_{\mathbf{ij}} = \text{curl}^{-1}(\mathbf{A}_{\mathbf{i}} \times \mathbf{A}_{\mathbf{j}})$ is well-defined if and only if L_{ij} vanishes. The Massey triple product in terms of vector calculus then becomes

$$\mathbf{M}_{\mathbf{ijk}} = \mathbf{A}_{\mathbf{i}} \times \text{curl}^{-1}\mathbf{M}_{\mathbf{jk}} + \text{curl}^{-1}\mathbf{M}_{\mathbf{ij}} \times \mathbf{A}_{\mathbf{k}}. \tag{2.10}$$

which integrated over the surface of the ith flux tube gives the third order linking integral. This can be recast as a line integral

$$L_{ijk} = \oint_{C_i} \left[\text{curl}^{-1}\mathbf{M}_{\mathbf{jk}} - \phi_{(i)j}\mathbf{A}_{\mathbf{k}} \right] .\mathbf{dl}, \tag{2.11}$$

where $\nabla\phi_{(i)j} = \mathbf{A_j}$. This potential exists in U_i but not globally.

Braids (*e.g.*, see figure 1b) are close relatives of knots and links. Berger (1991) shows how to define a *third order winding number* for braids on three strings. It is exactly the third order linking number for those braids corresponding to three component links with $L_{12} = L_{23} = L_{31} = 0$. A braid on three strings is specified by three curves $z_i(t)$ $(i = 1, 2, 3)$ in the complex plane $\mathcal{C}$. The planes of constant t are cross-sections of the braid, which stretches from $t = 0$ to $t = 1$. It is sometimes helpful to visualise the braid as a single curve in $\mathcal{C}^3$ by writing $\gamma(t) = (z_1(t), z_2(t), z_3(t))$. Links may be constructed from braids by identifying endpoints. Joining the ends of the pigtail braid (see figure 1b) gives the Borromean rings, for example. The *second order winding numbers* are integrals of closed one-forms $\tilde{\mathbf{d}}\theta_{ij}$ over the curve $\beta = (z_i(t), z_j(t))$, viz.,

$$\Theta_{ij}(t) = \int_\beta \tilde{\mathbf{d}}\theta_{ij} = \mathcal{I}m\ \log\left[\frac{z_j(t) - z_i(t)}{z_j(0) - z_i(0)}\right]. \tag{2.12}$$

For the pigtail braid, all the net second order winding numbers $\Theta_{12}(1) = \Theta_{23}(1) = \Theta_{31}(1)$ are zero but the strings may not be untangled and the braid is not equivalent to the identity.

The generators of the cohomology ring of the braid group are the one-forms $\tilde{\omega}_{\mathbf{ij}}$

$$\tilde{\omega}_{\mathbf{ij}} = \frac{1}{2\pi i}\frac{\tilde{\mathbf{d}}z_j - \tilde{\mathbf{d}}z_i}{z_j - z_i}. \tag{2.13}$$

The structure of the cohomology ring is determined by the identity (Arnol'd, 1969)

$$\tilde{\omega}_{\mathbf{ij}} \wedge \tilde{\omega}_{\mathbf{jk}} + \tilde{\omega}_{\mathbf{jk}} \wedge \tilde{\omega}_{\mathbf{ki}} + \tilde{\omega}_{\mathbf{ki}} \wedge \tilde{\omega}_{\mathbf{ij}} = 0. \tag{2.14}$$

Let us define

$$\lambda_{ij} = \frac{1}{2\pi i}\log\left[\frac{z_j(t) - z_i(t)}{z_j(0) - z_i(0)}\right], \tag{2.15}$$

so that $\tilde{\omega}_{\mathbf{ij}} = \tilde{\mathbf{d}}\lambda_{ij}$. An immediate consequence of (2.14) is that the one-form

$$\tilde{\psi}_{\mathbf{ijk}} = \lambda_{ij}\tilde{\mathbf{d}}\lambda_{jk} + \lambda_{jk}\tilde{\mathbf{d}}\lambda_{ki} + \lambda_{ki}\tilde{\mathbf{d}}\lambda_{ij}, \tag{2.16}$$

is closed and so [c.f., eq. (2.12)]

$$\Psi_{ijk} = \mathcal{R}e\int_\gamma \tilde{\psi}_{\mathbf{ijk}} = \mathcal{R}e\int_0^1\left[\lambda_{ij}\frac{d\lambda_{jk}}{dt} + \lambda_{jk}\frac{d\lambda_{ki}}{dt} + \lambda_{ki}\frac{d\lambda_{ij}}{dt}\right]dt, \tag{2.17}$$

is the same for any two homotopic curves and thus topologically invariant. This is the third order winding number. We emphasise that this is a well-defined braid invariant irrespective of whether the net second order winding numbers vanish or not. If they do, it is equivalent to (2.9), which we shall now show. A simple way to calculate (2.17) is to comb the braid. When the net second order winding numbers all vanish, a combed 3-braid (Artin, 1950) has two strings $z_j(t)$ and $z_k(t)$ fixed and always vertical, while the remaining string $z_i(t)$ weaves around them. For the pigtail braid, this is drawn out in figure 1c. The third order winding number simplifies to

$$\Psi_{ijk} = \mathcal{R}e \int_{z_i} \lambda_{ki} \tilde{\mathbf{d}} \lambda_{ij} = -\frac{1}{4\pi^2} \int_0^1 \left[\log r_{ki} \frac{d \log r_{ij}}{dt} - \theta_{ki} \frac{d\theta_{ij}}{dt} \right] dt, \qquad (2.18)$$

where r_{ij} and θ_{ij} are the modulus and argument of $z_i - z_j$. As the strings z_j and z_k are vertical, we can choose (see Berger (1991) for more details)

$$\tilde{\mathbf{A}}_{\mathbf{k}} = \frac{1}{2\pi} \tilde{\mathbf{d}} \theta_{ik}, \qquad \phi_{(i)j} = \frac{\theta_{ij}}{2\pi}, \qquad \tilde{\mathbf{N}}_{\mathbf{jk}} = -\frac{1}{4\pi^2} \log r_{ik} \tilde{\mathbf{d}} \log r_{ij}. \qquad (2.19)$$

Inserting these in (2.9) and integrating by parts recovers (2.18) exactly.

3. Fourth Order Linking

Figure 2a gives an example of a four component link with the unusual property that the removal of any one component enables the remaining ones to be pulled free (there are several ways to do this – see Kauffman (1991)). All sets of second and third order linking numbers L_{ij} and L_{ijk} are zero. Can we find a fourth order linking integral to describe the intricate tangling?

The vanishing of the L_{ijk} is suggestive of the local exactness of the two-forms $\tilde{\mathbf{M}}_{\mathbf{ijk}}$. The Poincaré lemma ensures that everywhere there locally exists a one-form $\tilde{\mathbf{N}}_{\mathbf{ijk}}$ such that

$$\tilde{\mathbf{M}}_{\mathbf{ijk}} = \tilde{\mathbf{d}} \tilde{\mathbf{N}}_{\mathbf{ijk}}, \qquad (3.1)$$

when $\tilde{\mathbf{M}}_{\mathbf{ijk}}$ is taken as

$$\tilde{\mathbf{M}}_{\mathbf{ijk}} = \begin{cases} \tilde{\mathbf{A}}_{\mathbf{i}} \wedge \tilde{\mathbf{N}}_{\mathbf{jk}} + \tilde{\mathbf{N}}_{\mathbf{ij}} \wedge \tilde{\mathbf{A}}_{\mathbf{k}} - \eta_{(i)jk} \tilde{\mathbf{B}}_{\mathbf{i}}, & \text{if } x \in U_i, \\ \tilde{\mathbf{A}}_{\mathbf{i}} \wedge \tilde{\mathbf{N}}_{\mathbf{jk}} + \tilde{\mathbf{N}}_{\mathbf{ij}} \wedge \tilde{\mathbf{A}}_{\mathbf{k}} - \phi_{(j)i} \phi_{(j)k} \tilde{\mathbf{B}}_{\mathbf{j}}, & \text{if } x \in U_j, \\ \tilde{\mathbf{A}}_{\mathbf{i}} \wedge \tilde{\mathbf{N}}_{\mathbf{jk}} + \tilde{\mathbf{N}}_{\mathbf{ij}} \wedge \tilde{\mathbf{A}}_{\mathbf{k}} + \eta_{(k)ij} \tilde{\mathbf{B}}_{\mathbf{k}}, & \text{if } x \in U_k, \\ \tilde{\mathbf{A}}_{\mathbf{i}} \wedge \tilde{\mathbf{N}}_{\mathbf{jk}} + \tilde{\mathbf{N}}_{\mathbf{ij}} \wedge \tilde{\mathbf{A}}_{\mathbf{k}}, & \text{elsewhere.} \end{cases} \qquad (3.2)$$

so that $\tilde{\mathbf{d}}^2 \tilde{\mathbf{N}}_{\mathbf{ijk}} = 0$, as it should. Here, we have introduced the scalar potentials $\eta_{(i)jk}$ and $\eta_{(k)ij}$ such that

$$\tilde{\mathbf{d}} \eta_{(i)jk} = \tilde{\mathbf{N}}_{\mathbf{jk}} - \phi_{(i)j} \tilde{\mathbf{A}}_{\mathbf{k}}, \qquad \tilde{\mathbf{d}} \eta_{(k)ij} = \tilde{\mathbf{N}}_{\mathbf{ij}} + \phi_{(k)j} \tilde{\mathbf{A}}_{\mathbf{i}}, \qquad (3.3)$$

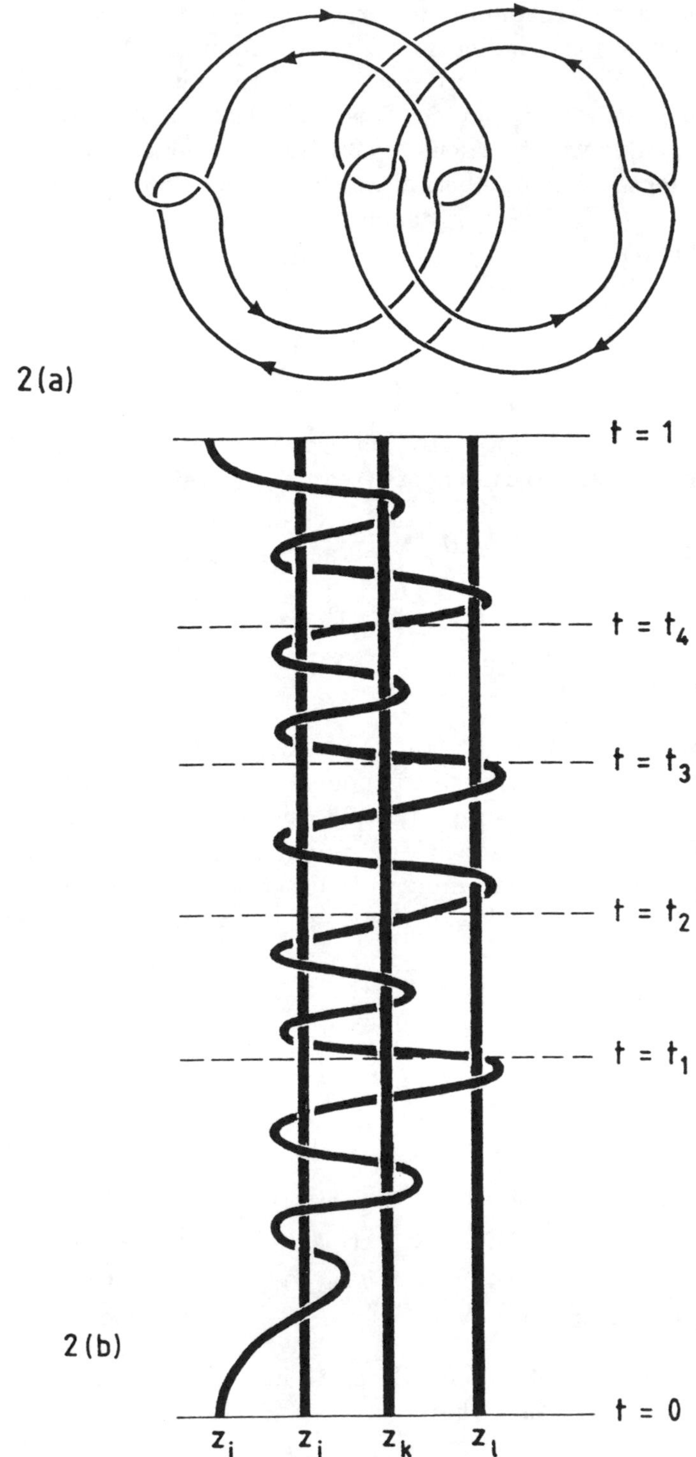

Fig. 2. (a) the tangle of four rings with all $L_{ij} = L_{ijk} = 0$, (b) the combed 4-braid.

which is fine if and only if all the $L_{ij} = L_{ijk} = 0$. By analogy with (2.2) and (2.3), the fourth order linking integral is

$$L_{ijkl} = \int_{\partial U_i} \tilde{\mathbf{M}}_{\mathbf{ijkl}} = -\int_{\partial U_l} \tilde{\mathbf{M}}_{\mathbf{ijkl}} \qquad i \neq j \neq k \neq l, \tag{3.4}$$

where the Massey quadruple product is

$$\tilde{\mathbf{M}}_{\mathbf{ijkl}} = \tilde{\mathbf{A}}_{\mathbf{i}} \wedge \tilde{\mathbf{N}}_{\mathbf{jkl}} + \tilde{\mathbf{N}}_{\mathbf{ij}} \wedge \tilde{\mathbf{N}}_{\mathbf{kl}} + \tilde{\mathbf{N}}_{\mathbf{ijk}} \wedge \tilde{\mathbf{A}}_{\mathbf{l}}. \tag{3.5}$$

Again, Stokes' theorem converts (3.4) into a volume integral [c.f., eq. (2.8)]

$$L_{ijkl} = \int_{U_i} \tilde{\mathbf{d}}\tilde{\mathbf{M}}_{\mathbf{ijkl}} = -\int_{U_l} \tilde{\mathbf{d}}\tilde{\mathbf{M}}_{\mathbf{ijkl}}, \tag{3.6}$$

or a line integral [c.f., eq. (2.9)]

$$L_{ijkl} = \int_{C_i} \tilde{\mathbf{N}}_{\mathbf{jkl}} - \phi_{(i)j}\tilde{\mathbf{N}}_{\mathbf{kl}} - \eta_{(i)jk}\tilde{\mathbf{A}}_{\mathbf{l}} = \int_{C_l} \tilde{\mathbf{N}}_{\mathbf{ijk}} + \phi_{(l)k}\tilde{\mathbf{N}}_{\mathbf{ij}} + \eta_{(l)jk}\tilde{\mathbf{A}}_{\mathbf{i}}. \tag{3.7}$$

The fourth order linking integral can be restated in vector algebra as the surface integral over the ith flux tube of the vector field [c.f., eq. (2.10)]

$$\mathbf{M}_{\mathbf{ijkl}} = \mathbf{A}_{\mathbf{i}} \times \mathrm{curl}^{-1}\mathbf{M}_{\mathbf{jkl}} + \mathrm{curl}^{-1}\mathbf{M}_{\mathbf{ij}} \times \mathrm{curl}^{-1}\mathbf{M}_{\mathbf{kl}} + \mathrm{curl}^{-1}\mathbf{M}_{\mathbf{ijk}} \times \mathbf{A}_{\mathbf{l}}, \tag{3.8}$$

where the inverse curls are well-defined only if all the $L_{ij} = L_{ijk} = 0$. As a line integral, it becomes [c.f., eq. (2.11)]

$$L_{ijkl} = \oint_{C_i} \left[\mathrm{curl}^{-1}\mathbf{M}_{\mathbf{jkl}} - \phi_{(i)j}\mathrm{curl}^{-1}\mathbf{M}_{\mathbf{kl}} - \eta_{(i)jk}\mathbf{A}_{\mathbf{l}} \right] . \mathbf{dl}, \tag{3.9}$$

with $\nabla\phi_{(i)j} = \mathbf{A}_{\mathbf{j}}$ and $\nabla\eta_{(i)jk} = \mathrm{curl}^{-1}\mathbf{M}_{\mathbf{jk}} - \phi_{(i)j}\mathbf{A}_{\mathbf{k}}$. These potentials exist in U_i.

Evans & Berger (in preparation) demonstrate that the fourth order linking integral is unaffected by changes of gauge and by motions of the flux tubes. We shall show here that it vanishes for four unlinked rings but equals ± 1 for the tangle of figure 2a. If the rings are unlinked, then each ring U_i may be separately enclosed by a surface S. Integrating throughout the volume V bounded by S and ∂U_i, we have by Stokes' theorem

$$\int_V \tilde{\mathbf{d}}\tilde{\mathbf{M}}_{\mathbf{ijkl}} = \int_S \tilde{\mathbf{M}}_{\mathbf{ijkl}} - \int_{\partial U_i} \tilde{\mathbf{M}}_{\mathbf{ijkl}}. \tag{3.10}$$

The first integral on the rhs of (3.10) is

$$\int_S \tilde{\mathbf{M}}_{\mathbf{ijkl}} = \int_S \tilde{\mathbf{d}}(\phi_i\tilde{\mathbf{N}}_{\mathbf{jkl}} + \phi_i\phi_l\tilde{\mathbf{N}}_{\mathbf{jk}} - \phi_l\tilde{\mathbf{N}}_{\mathbf{ijk}}) = 0, \tag{3.11}$$

where $\tilde{\mathbf{A}}_{\mathbf{i}} = \tilde{\mathbf{d}}\phi_i$ (permissable as the magnetic field vanishes on S) and we have made a further application of Stokes' theorem to get the last equality. The second integral on the rhs of (3.10) is just the fourth order linking number L_{ijkl}. However, the lhs of (3.10) is also zero as $\tilde{\mathbf{d}}\tilde{\mathbf{M}}_{\mathbf{ijkl}} = 0$ everywhere external to the toroids U_i (because the magnetic field is zero). So, L_{ijkl} vanishes as claimed.

Now let us sketch the calculation for the tangle of figure 2a. Figure 2b is a drawing of a 4-braid – joining the ends of this braid gives exactly figure 2a. The flux tubes labelled j, k and l are fixed and vertical. All the complication of the tangle is in the weaving and winding of the ith flux tube. We have already quoted the formulae for $\tilde{\mathbf{N}}_{\mathbf{kl}}$, $\tilde{\mathbf{A}}_{\mathbf{l}}$ and $\phi_{(i)j}$ in (2.19), which hold whenever the j, k and l flux tubes puncture the surfaces of constant t at right angles. It remains to find the $\tilde{\mathbf{N}}_{\mathbf{jkl}}$. This can be done quite easily as

$$\tilde{\mathbf{M}}_{\mathbf{jkl}} = \frac{1}{8\pi^3}\left[-\log r_{il}\tilde{\mathbf{d}}\theta_{ij} \wedge \tilde{\mathbf{d}}\log r_{ik} + \log r_{ij}\tilde{\mathbf{d}}\log r_{ik} \wedge \tilde{\mathbf{d}}\theta_{il}\right], \qquad (3.12)$$

where, as before, r_{ij} and θ_{ij} are the modulus and argument of $z_j - z_i$. Now we use the simple identity $\tilde{\mathbf{d}}\theta_{ij} \wedge \tilde{\mathbf{d}}\log r_{ik} = \tilde{\mathbf{d}}\theta_{ik} \wedge \tilde{\mathbf{d}}\log r_{ij}$ to obtain

$$\tilde{\mathbf{N}}_{\mathbf{jkl}} = \frac{1}{8\pi^3}\log r_{ij}\log r_{il}\tilde{\mathbf{d}}\theta_{ik}. \qquad (3.13)$$

So, the fourth order linking integral is

$$L_{ijkl} = \frac{1}{8\pi^3}\int_0^1 \left[\log r_{ij}\log r_{il}\frac{d\theta_{ik}}{dt} - \theta_{ij}\log r_{ik}\frac{d\log r_{il}}{dt} - \Psi(t)\frac{d\theta_{il}}{dt}\right] dt, (3.14)$$

where

$$\Psi(t) = 4\pi^2 \int \tilde{\mathbf{N}}_{\mathbf{jk}} - \phi_{(i)j}\tilde{\mathbf{A}}_{\mathbf{k}} = \int_0^t \left[\log r_{ij}\frac{d\log r_{ik}}{dt} - \theta_{ij}\frac{d\theta_{ik}}{dt}\right] dt. \qquad (3.15)$$

To evaluate the fourth order linking integral is possible by direct calculation in the complex plane. Let us choose the fixed positions of strings j, k and l to be the points $(1,0), (3,0)$ and $(5,0)$ and let string i weave between the others in semi-circular arcs of radius unity. This choice we are entitled to make, as the fourth order linking integral is homotopy invariant. The advantage is that now only the angular terms contribute to the expressions (3.14) and (3.15). For example, the first move of the combed braid (see figure 2b) is string i passing in front of string j and is described in the complex plane by

$$z_i = 1 - \cos\phi - i\sin\phi, \quad z_j = 1, \quad z_k = 3, \quad z_l = 5, \qquad (3.16)$$

where the parameter ϕ runs from 0 to π. The combed braid is composed of 40 such moves and it is straightforward (but lengthy) to add up all the contributions and show

$$L_{ijkl} = \frac{1}{4\pi^2}(\Psi(t_2)-\Psi(t_1)+\Psi(t_4)-\Psi(t_3)) = -\frac{1}{4\pi^2}(2\pi^2+2\pi^2) = -1. \quad (3.17)$$

Here, $\Psi(t_i)$ refers to the third order linking integral (3.15) calculated at t_i (marked on figure 2b). Remarkably, the fourth order linking integral depends only on the value of the third order linking integral at the four points t_i – and these add to give -1. (Reversing the orientation of one of the components gives +1). This validates our assertion that it distinguishes four unlinked rings from the entanglement of figure 2a.

4. Conclusions

The Gauss integral L_{ij} is the first member of a hierarchy of linking integrals. For three entangled rings like the Borromean with $L_{ij} = L_{jk} = L_{ki} = 0$, there exists a third order linking integral L_{ijk}. Likewise, four linked rings with vanishing L_{ij} and L_{ijk} are captured by a fourth order linking integral. And so on.

The nth member of the hierarchy describes an interlocking of n circuits, which have the property that the removal of any one of the circuits enables all the others to be untangled. The nth linking integral is built from the $n-1$th as follows. Suppose the $n-1$th linking integral is given by the integral of the two-form $\tilde{\mathbf{M}}_{\mathbf{ij..k}}$ over the surface ∂U_i and suppose it vanishes, *i.e.*,

$$\int_{\partial U_i} \tilde{\mathbf{M}}_{\mathbf{ij...k}} = 0. \quad (4.1)$$

The important point is that ∂U_i has no boundary. By Stokes' theorem, this is the necessary condition for $\tilde{\mathbf{M}}_{\mathbf{ij...k}}$ to be exact (*i.e.*, the exterior derivative of another form) on ∂U_i. If it is to be locally exact everywhere then we must impose the Poincaré lemma, $\tilde{\mathbf{d}}\tilde{\mathbf{M}}_{\mathbf{ij...k}} = 0$. This can be satisfied by modifying the definition of $\tilde{\mathbf{M}}_{\mathbf{ij...k}}$ within the $n-1$ flux tubes themselves. So, the vanishing $n-1$th linking integral is used via Stokes' theorem to give an exact one-form with which the nth order linking integral is built. This algorithm is illustrated above by the construction of third and fourth order linking integrals. There is an analogous hierarchy of winding numbers for braids.

For isolated flux tubes, the Gauss integral is the helicity. Helicity can also be interpreted as the Hopf invariant or asymptotic linking number of divergence-free vector fields in $\mathcal{R}^3$. It is natural to ask whether the higher-order linking integrals define analogous invariants. The answer is not known for certain, but is probably no – it is impossible to define the higher order linking numbers unless the second order linkage vanishes and it is difficult to see how this condition applies to a space-filling magnetic field. We emphasise that this limitation does not apply to the hierarchy of winding numbers.

Consider a tangle of many magnetic field lines between two plates at $t = 0$ and $t = 1$, say. The sum of the absolute value of the third order winding number over all the triangles of footpoints z_i, z_j, z_k on $t = 0$ is a well-defined invariant which measures the complexity of the intertwining. (The absolute value is needed because each triangle of footpoints counts six times, three times with positive and three times with negative signs). Such invariants should have many applications in physics.

Another point of interest is to trace the connection between the differential geometric invariants described in this review and the combinatorially defined invariants of the Conway-Alexander polynomials (see Kauffman, 1987). The Witten theory (summarised in Kauffman, 1991) provides functional integrals as interpretations of the combinatorics. It is natural to ask what are the interrelationships between the hierarchy of linking integrals and the Witten invariants.

Acknowledgements

Both MAB and NWE are grateful to the Institute for Theoretical Physics for financial support. NWE is funded by a reasearch fellowship from King's College, Cambridge.

References

ARNOL'D, V.I., 1969 The cohomology ring of the coloured braid group. *Math. Notes Acad. Sci. USSR*, **5**, pp. 227-232 (Engl. trans. 1974 *Math. Notes*, **5**, pp. 138-140)

ARTIN, E., 1950 The theory of braids. *American Scientist*,**38**, pp. 112-119

BERGER, M.A., & FIELD, G.B., 1984 The topological properties of magnetic helicity. *J. Fluid Mech.*, **147**, pp. 133-148

BERGER, M.A., 1990 Third order link invariants. *J. Phys. A: Math. Gen.*, **23**, pp. 2787-2793

BERGER, M.A., 1991 Third order braid invariants. *J. Phys. A: Math. Gen.*,**24**, pp. 4027-4036

FENN, R.A., 1983 Techniques of geometric topology, *London Math. Soc. Lecture Note Series*, **57**, Cambridge University Press

GAUSS, C.F., 1867 *Werke*, vol. V, Gottingen, p. 605

KAUFFMAN, L.H., 1987 *On Knots*, Princeton University Press, Princeton, chap. 2

KAUFFMAN, L.H., 1991 *Knots and Physics*, World Scientific Publishing, Singapore, p. 38

MASSEY, W.S., 1958 Some higher order cohomology operations. "Symp. International Topologia Algebraica", Mexico, UNESCO.

MAXWELL, J.C., 1877 *A Treatise on Electricity and Magnetism*, vol. II, Clarendon Press, Oxford, p. 41

MOFFATT, H.K., 1978 *Magnetic Field Generation in Electrically Conducting Fluids*, Cambridge University Press, chap. 2

MOFFATT, H.K., 1981 Some developments in the theory of turbulence. *J. Fluid Mech.*, **106**, pp. 27-47

SCHUTZ, B., 1980 *Geometrical Methods of Mathematical Physics*, Cambridge University Press, chap. 4

BORROMEANISM AND BORDISM

PETER AKHMET'EV
Institute of Terrestrial Magnetism
Ionosphere and Radio Wave Propagation
Academy of Sciences of USSR
Troitsk
Moscow Region
142092
USSR

and

ALEXANDER RUZMAIKIN
Institute for Theoretical Physics
University of California
Santa Barbara
CA 94106-4300
USA

ABSTRACT. A fourth-order topological invariant determined by the geometrical position of flux tubes linked with an even number of linkings is considered in ideal MHD. The topological invariant is determined by the properties of a surface bounded by flux tubes. For the Seifert surface it coincides with the Arf invariant. In particular, this topological invariant distinguishes "the Borromean rings" (three linked rings no two of which link each other) from three unlinked and unknotted rings.

Topological evolution of magnetic (or vortex) flux tubes in the limit of small diffusivity within a time such that helicity is conserved can be formalized as "a bordism", an orientable surface without self-intersections in 4-dimensional space (3-space plus time) bounded by the set of flux tubes initially and at any subsequent time. The "births" and "deaths" of flux tubes correspond to the maxima and minima of the bordism, the reconnections are the saddle points of the bordism. Two sets of flux tubes belong to the same framed bordism if and only if they have the same helicities.

1. Introduction

The basic element of ideal magnetohydrodynamics is a magnetic field line which has no endpoints. For this reason studies of topology of the closed or ergodic curves have proven very useful in this area. Among the best known topological invariants is the helicity which measures the linkage of magnetic field or vortex lines (Moffatt, 1969). Berger and Field (1984) pointed out

H. K. Moffatt et al. (eds.), Topological Aspects of the Dynamics of Fluids and Plasmas, 249–264.

the important contribution to the helicity coming from the twist and writhe of the flux tubes. [Writhe: to twist into coils or folds.]

This paper develops the study of topological properties of magnetic flux tubes. (The discussion, of course, is valid for vorticity as for magnetic field.) It uses a concept of "a frame" for the magnetic flux tube which allows in particular to define the self-linking number of the flux tube. A fourth-order topological invariant, an invariant on a surface bounded by evenly linked flux tubes with the help of the framed vectors, is introduced. In particular, this topological invariant distinguishes "the Borromean rings" (a configuration with zero helicity) from three unlinked and unknotted rings. Another known example of configuration with zero helicity is the Whitehead link (see Lickorish and Millett, 1988). The topological complexity inherent to this type of configurations is called "borromeanism".

Berger (1990), by use of Massey triple products, constructed a third-order linking number for configurations where all the second-order linking numbers (hence the helicity) vanish. The fourth-order topological invariant under consideration here however does not imply necessarily that all low-order invariants vanish. It is valid for any even linkages.

In real magnetohydrodynamics with diffusion there is in general no concept of magnetic (or vortex) lines. However, when the magnetic diffusivity is small (more correctly, the dimensionless magnetic Reynolds number is large) it is still reasonable to consider magnetic lines, taking into account diffusion effects only in regions of strong field gradients. In particular, strong gradients exist when oppositely directed magnetic lines meet each other. As a result the two magnetic lines reconnect in a region small compared to the size of the whole magnetic configuration (see for example Parker, 1979).

The evolution of the magnetic lines due to reconnections can be studied by assuming reconnections occur instantly at meeting points of two magnetic lines. Some justification for this assumption is given by the Taylor conjecture that the helicity is approximately conserved during magnetic relaxation (Taylor, 1974). In this case the reconnection process can be formalized as a bordism (see section 3) and some useful mathematical theorems proven for bordisms are available.

2. Linking Matrix

To simplify the study, only closed lines concentrated into magnetic flux tubes of unit magnetic flux will be considered. A magnetic flux tube is assumed oriented i.e. a direction of magnetic field along the tube is specified. More complicated morphologies of the magnetic fields can be studied by separating the space into regions bounded by magnetic surfaces, $\mathbf{B}.\mathbf{n}\,|_s = 0$, (Berger and Field, 1984). The analysis of ergodic lines can be reduced to the case of closed tubes (Arnold, 1974; Arnold and Khesin, 1992).

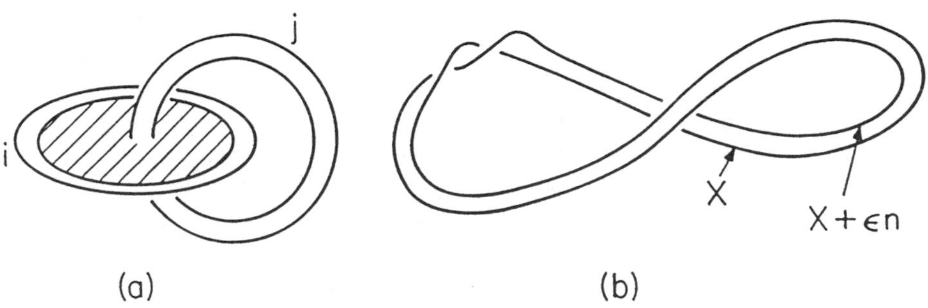

Fig. 1. Definitions of (a) the linking number L_{ij}, and (b) self-linking number L_{ii}. The self-linking number can be represented as a sum of the twist number and writhing number, $L_{ii} \equiv L(\mathbf{X}, \mathbf{X} + \varepsilon \mathbf{n}) = Tw + Wr$.

Second-order topological properties can be characterized by the linking matrix L_{ij} where $i, j = 1, 2, ..., N$ and N is the number of flux tubes in the configuration. A linking number L_{ij} of two different flux tubes i and j is defined as follows: Consider a disk having the flux tube i as a boundary (see figure 1a). Then L_{ij}, $i \neq j$ is the algebraic sum of intersections of this disk by the flux tube $j \neq i$. Intersections are counted as positive or negative in accordance with the mutual orientation of the magnetic fields in i, j tubes (right hand screw rule). It is evident that $L_{ij} = L_{ji}$. If the disk is a singular surface the intersections are assumed to avoid the singularities. In addition, intersection points are not permitted to coincide with one another.

It is a more delicate matter to define the diagonal elements L_{ii} of the matrix (self-linking numbers). An idea (Fuller, 1978) is to consider a central magnetic line of flux tube as a smooth curve $\mathbf{X}(s)$ where s is a parameter along the curve. Shift the curve at each point along normals $\mathbf{n}(\mathbf{s})$ to the curve by small value ε to get a new curve $\mathbf{X}(s) + \varepsilon \mathbf{n}(s)$ (figure 1b). Since $\mathbf{X}$ and $\mathbf{X} + \varepsilon \mathbf{n}$ are disjoint closed curves their linking number can be defined. This linking number is independent of ε for sufficiently small ε and can be identified with the diagonal element L_{ii}. However it is dependent on the choice of $\mathbf{n}(s)$, i.e. on a frame which includes the normal, tangent and a third vector orthogonal to them. It is natural to use the frame for which the set of normals to the axis is constructed in such a way that the end of each normal lies on the magnetic line determined by a solution of the induction equation without dissipation. Note that this frame is generally different from the triad determined by the well known Frenet equations for the tangent, normal, and binormal vectors. In particular the frame used in this study is smooth, and there are no in flex ion points where the normal jumps.

Geometrically the self-linking of a flux tube originates from writhing the axis of the tube and twisting of the magnetic lines over the axis (Călugăreanu, 1961; White, 1969; Fuller, 1978). Let ω denote the angular rate at which the vector $\mathbf{n}(s)$ revolves about the central axis. The twist number is defined as the integral of ω over the closed magnetic line: $Tw = (1/2\pi)\oint \omega ds$. The linking number L_{ii} and the twist number Tw are frame dependent. Only their difference, the writhing number $Wr = L_{ii} - Tw$, is frame independent.

For a given frame the self-linking number is a topological invariant. However the decomposition of the self-linking into twists and writhes is not topologically invariant because twist and writhe can convert into each other.

Note that an untwisted flux tube contains only the toroidal component of the magnetic field while in the twisted flux tube both the toroidal and poloidal components are present.

The sum of all elements of the linkage matrix L_{ij} for a set of flux tubes is called the helicity (see Ricca and Moffatt, this volume):

$$H = \sum_i \sum_{j \neq i} L_{ij} + \sum_i L_{ii} \tag{1}$$

Configurations for which all off-diagonal elements of the linking matrix are even numbers are called even linkages. The other linkages are called odd. Thus all self-linking configurations are even in so far as they have only diagonal elements.

Some examples of flux tube linkages for which the linking matrix and the helicity can be easily calculated are presented in figure 2. The configurations of the last two types in figure 2, the Borromean rings and the Whitehead link present a critical challenge. Each is a non-trivial configuration with a zero linking matrix. Thus one can expect that there may be a topological invariant of a higher order.

3. Seifert's Frame

For a knotted flux tube there is a topological invariant (Robertello, 1965) belonging to the two-component group $\{0, 1\}$ and independent of the choice of frame. The invariant is determined by the geometrical position of the flux tubes in the space. Some extension of the invariant have been found by Libgober (1988).

In the next section the invariant is used for even links to distinguish the Borromean rings from the three unlinked and unknotted rings so it is natural to call it "the Borromean invariant", Bo. Note that the Borromean rings each ring has even linkages with the other two rings. One can easily check this fact by constructing the Borromean rings with a help of three ropes (see figure 3). To define this invariant it is useful to use the concept of "the Seifert surface" (see for instance Browder, 1972).

The Seifert surface for a set of flux tubes can be constructed as a help of the vector-potential outside the tubes (figure 4). Let $\mathbf{A} = \mathbf{curl}^{-1}\mathbf{B}$ be a vector-potential of the magnetic field concentrated in the flux tubes. Outside

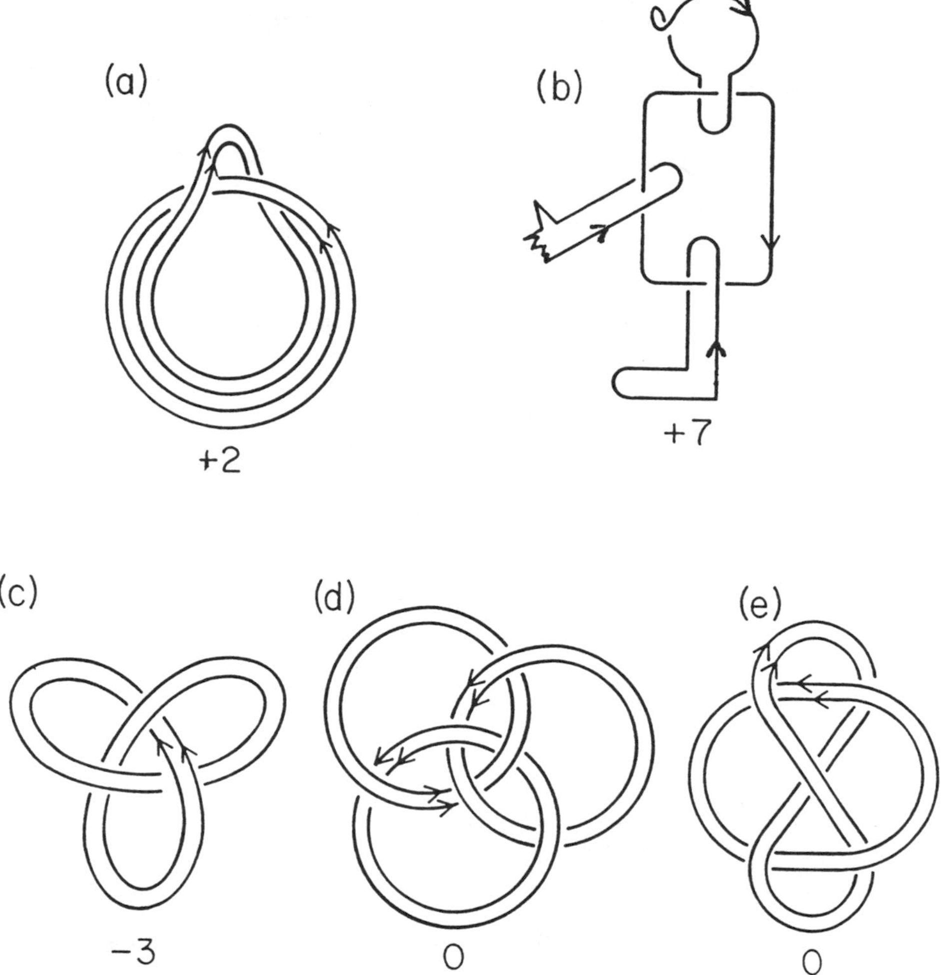

Fig. 2. Examples of odd (a,b) and even (c,d,e) linkages. The number below each figure gives the value of helicity. Figure 2c is the trefoil knot, 2d is the Borromean rings, and 2e is the Whitehead link.

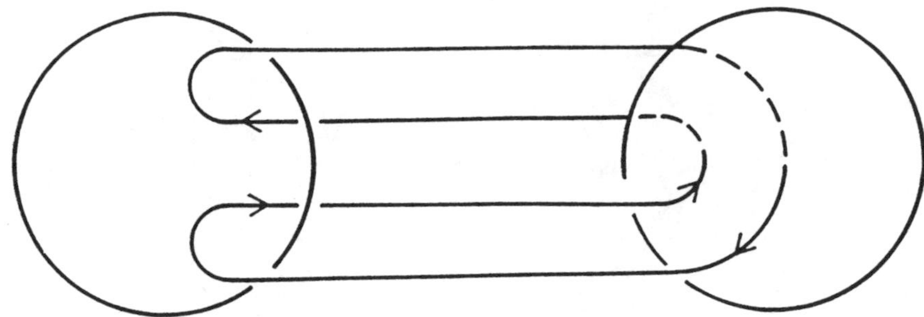

Fig. 3. Another presentation of the Borromean rings clarifies the nature of this configuration

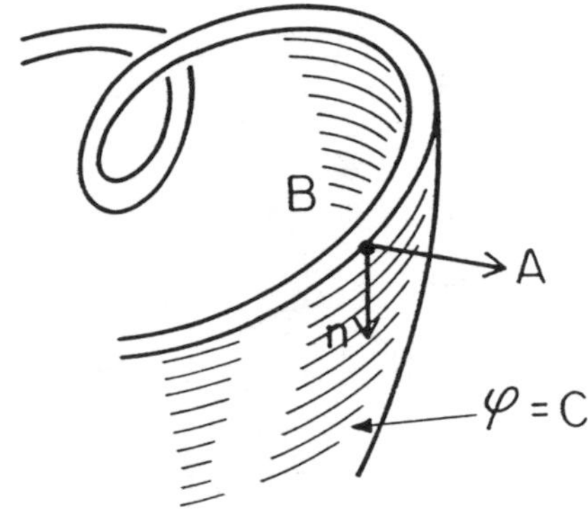

Fig. 4. Seifert's frame is a set of normals from the axis of the tube to the **A**–lines. It defines a boundary of zero helicity

the tubes $\mathbf{B} = 0$, so that the vector-potential is a gradient of a scalar function of the space coordinates, $\mathbf{A} = \nabla\varphi$. It is assumed also that **A** vanishes at infinity. Because the vector-potential does not return to its value after a turn around a flux tube, the function φ is multi-valued ($\int \mathbf{A} d\mathbf{l}$ over a contour around a cross-section of the flux tube is the magnetic flux of the tube. The magnetic flux is assumed to be the same for all tubes, so that φ is changed by integer number times the magnetic flux.). However φ can be considered as a simple mapping of the three-dimensional space outside the tubes onto a circle parametrized by some parameter "c".

Let us take a regular point c on the circle i.e. a point the inverse image of which has $\nabla\varphi \neq 0$ at every point of three-dimensional space. The inverse image $\varphi^{-1}(c)$ determines a surface (the Seifert surface) without self-intersections in the space around the set of the flux tubes under consideration. To provide the compactness of the surface it is sufficient to avoid the zero point on the circle corresponding to infinity in three-dimensional space. Of course, there are the singular points on the circle. However according to the theorem of Sard (see Milnor,1965) the set of singular points has a zero measure on the circle. [In fact, consider all singular points in the three-dimensional space and construct a small ball around each of the point. Because of the condition $\nabla\varphi = 0$ these balls are projected only into a set of zero measure on the circle.]

To define the orientation of the surface it is enough to take a regular point on the circle and a second close regular point in the clockwise direction. The second point defines a new closed surface which determines the normal to the (Seifert) surface.

It is clear that the Seifert surfaces constructed from the regular points on the circle are equivalent in the sense that they may be deformed into one another without changing topology. However, a transition from one regular point to another through a singular point changes the topology of the corresponding Seifert surfaces.

According to the general topological classification of oriented surfaces (the Jordan-Möbius theorem, see for example Hirsch, 1976) the Seifert surface has the form of a sphere with holes and handles. The number of holes is evidently equal to the number of flux tubes in the configuration. For one flux tube this surface is a disk with handles. The number of handles is related to the twist and writhe in the tubes and their mutual geometrical positions.

A set of normals to the tubes along the Seifert surface forms "the Seifert frame" (see figure 4). Note that in the frame used in Section 2 the normals are perpendicular to the magnetic, **B**-lines. In the Seifert frame they are perpendicular to the vector-potential, **A**-lines.

The Seifert frame has a remarkable property: The total helicity of a set of flux tubes in the this frame, i.e. the boundary of the Seifert surface, is zero.

To show it let us represent each flux tube i of the set as a torus labeled by the coordinates μ_i (the meridian) and λ_i (the longitude). Consider a cycle ξ constructed on this tube in the form of a spiral having a step determined by the self-linking number $L_{ii} : \xi_i = \mu_i + L_{ii}\lambda_i$. Assume there is a frame for which $L_{ii} = -\sum_j L_{ij}$. Then the sum of the cycles, i.e. the border of the Seifert surface, is

$$\xi = \sum_i \xi_i = \sum_i (\mu_i + L_{ii}\lambda_i) =$$

$$= \sum_i \mu_i - \sum_i \sum_j L_{ij}\lambda_i = \sum_i \mu_i - \sum_j \sum_i L_{ji}\lambda_j =$$

$$= \sum_i (\mu_i - \sum_j L_{ij}\lambda_j) = 0$$

due to $\mu_i - \sum_j L_{ij}\lambda_j = 0$ (The cycle μ_i is homological to the sum of the cycles taken over the meridians of j-tubes piercing the i-tube).

Assume now that there is another Seifert surface coming to the flux tubes along a different frame in which $L_{ii} \neq -\sum_j L_{ij}$. Then a difference of these surfaces must be a surface having some number of the meridian cycles μ_i at its boundary. But this is a contradiction because the cycles must be homological to zero outside the flux tubes. Thus any two surfaces must come to the flux tubes along the same frame.

The result that the total helicity along the Seifert frame is zero may be understood in a more simple way. Consider a boundary of the Seifert surface with the flux tube configuration and shift the boundary curve along the surface. This curve can not become twisted or knotted because the Seifert surface has no intersections with the flux tubes.

4. An Invariant of Even Linkages

Consider an even linkage i.e. a set of flux tubes which have an even number of the linkings for every pair $i \neq j$. The simplest example of even linkage is one flux tube knotted or unknotted. There are simply no off-diagonal elements in this case. A non-trivial example of the even linkages is the Borromean rings, see figures 2e and 3.

For even linkages there is an invariant which can be constructed with the help of the Seifert surface as follows. Each handle p on the surface is a torus. Consider, as it has been done for the flux tubes, the cyclic coordinates μ_p and λ_p on a p-torus and find the self-linking numbers $l(\mu_p)$ and $l(\lambda_p)$ for the contours defined by these coordinates by using the shift of these contours along the normals to the surface. Note that the coordinates on the handles can be chosen in many ways depending how the coordinate lines go over the holes on the Seifert surface. However the choice does not affect $l(\mu_p)$ and $L(\lambda_p)$ for even linkages. Indeed, the self-linking number of the boundary curve around a hole is $L_{ii} = -\sum_j L_{ij}$ (see Section 4) which is zero modulo 2. Then the value

$$Arf = \sum_{p=1}^{h} l(\mu_p) l(\lambda_p) \tag{2}$$

(the so called Arf invariant, Arf, 1941; Robertello, 1965; Browder, 1972) is a characteristics of the set of flux tubes independent of the choice of the Seifert surface. The quadratic form in (2) has a property

$$l(x+y) = l(x) + l(y) + < x, y >,$$

where $< x, y >$ means intersection number: $< \mu_i \lambda_i >= 1$, $< \mu_i \mu_i >=< \lambda_i \lambda_i >= 0$. Thus the Arf invariant is a fourth-order invariant.

The meaning of the Arf invariant can be clarified in terms of the theory of closed surfaces immersed into three-dimensional space. Each Seifert surface

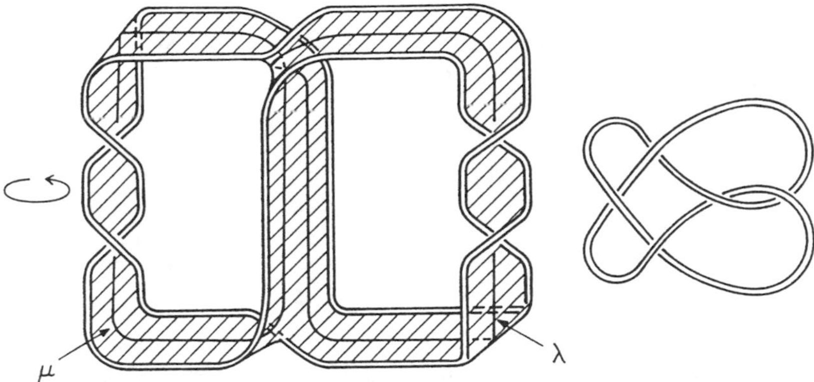

Fig. 5. The Seifert surface for the "eight knot" adapted from Francis (1987). The cyclic coordinate on the surface are labeled μ and λ. Each cycle is self-linked once so that $l(\lambda) = l(\mu) = 1$. Thus $Arf = 1$.

for even linkages can be extended to a closed surface (probably with self-intersections) by pasting some self-intersecting disks onto the holes. For example the pasting of the eight knot transforms (after deformations without singularities) its Seifert surface into the twisted torus (figure 6). Note that the deformation of the Seifert surface for a configuration with a zero Arf invariant results in an embedded closed surface, for example, an ordinary untwisted torus. For further details see the theory of deformation of the closed immersed surfaces (Guillou and Marin, 1986).

5. Roads to the Borromean Invariant

In order to find the invariant described in the previous section the Seifert surface must first be constructed. This construction is an art. See this art in Francis (1987). Figure 5 is adapted from this book. However after some surgery (pasting of the hole and deformation) conserving the Arf invariant this complicated surface can be transformed into a closed surface, so called twisted torus ("double-eight knot") on which the linkages of the coordinate lines are evident (figure 6).

In practice some other ways of finding the Borromean invariant may be easier to get. The standard topological method is to use the Alexander polynomials. It is known (Robertello, 1965) that the Arf invariant for the knot coincides modulo 2 with a coefficient of t^2 in the corresponding Alexander polynomial (see for example Marin, 1988). Knot theory (Crowell and Fox, 1963) gives for the trefoil knot (figure 2d) and "the eight knot" (figure 5), the following Alexander polynomials: $t^2 - t + 1$ and $t^2 - 3t + 1$, respectively.

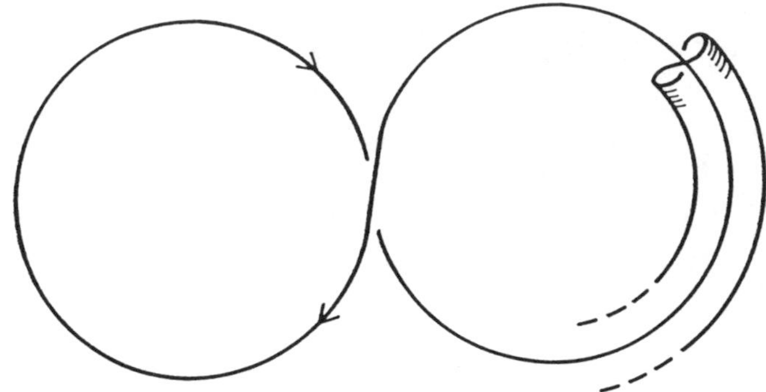

Fig. 6. The twisted torus is a pasted and deformed Seifert surface of figure 5. This closed surface can be viewed as a product of two "8"s . Because each "8" can be characterized by a second order topological invariant, Arf is a fourth-order topological invariant.

Thus in both cases the invariant equals 1. Because this invariant is the same on even bordisms (see section 6) the invariant for the Borromean rings must also be 1.

Another way to find the invariant is to construct, instead of the Seifert surface, a more primitive surface which can have self-intersections. However it must not intersect the flux tubes. The Arf invariant can also be defined for such surface. In many cases the Arf invariant on such surfaces is zero. However there are generally self-intersections and triple points on these surfaces. Let T be the number (mod 2) of the triple points on the surface and Hs be the total helicity (mod 2) of the curves defined by the self-intersections. Then

$$Bo = Arf + Hs + T \tag{3}$$

(see Appendix).

Such a surface for the Borromean rings may be constructed, for example, in the form of three tori with holes covering the rings. Two components of that surface are shown in figure 7.

6. Topological Evolution of Magnetic Flux Tubes

Now consider the dissipative situation. As is well known, the dissipation time is proportional to the squared scale of the magnetic field divided by the diffusivity η. So for time intervals much shorter than R^2/η, where R is the radius of the flux tube, one can neglect the dissipation everywhere except in regions where the flux tubes (or parts of a single flux tube) approach each other so that the small characteristic scales form (in particular when the approaching magnetic lines have opposite direction). In these region the

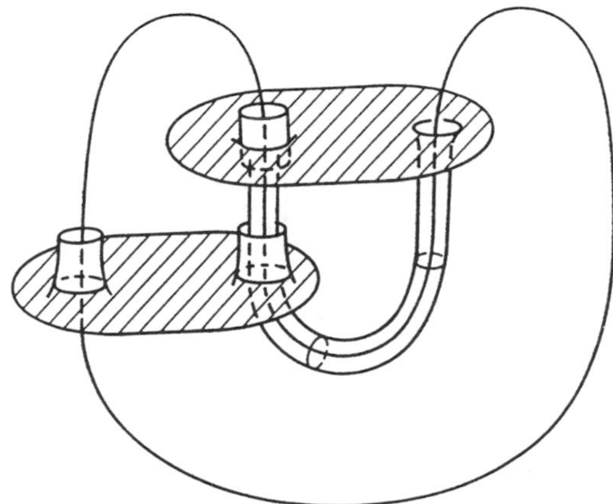

Fig. 7. A surface pulled onto the Borromean rings has the form of three tori with holes covering the rings. Two components of the surface are shown. For this surface $Arf = 0$, $Hs = 0$ and $T = 1$.

dissipation becomes very effective, the magnetic lines reconnect, and the configuration of the flux tubes changes.

Assume that in the limit of small diffusivity (very large Reynolds numbers) reconnections take place locally in space and time (instantly and pointwise). This assumption is justified by the Taylor hypothesis that in reconnections the helicity is changed more slowly than the energy (Taylor, 1974), and by some numerical results on reconnections between two vortices presented at this workshop by F. Hussain, and N. Zabusky.

With this assumption the reconnection process can be formalized mathematically as "a bordism" (see Milnor, 1966, this term is considered as more exact in the case under consideration than the term "cobordism" used in the book). The bordism β_t of a set of flux tubes is a surface without self-intersections in the 4-dimensional space (3-space plus time) that is bounded (with the same orientation) by the set of flux tubes initially and at any subsequent time t. For example, in the trivial case of one circular flux tube at rest the bordism is a surface of the cylinder with an axis directed along the time axis.

The "births" and "deaths" of flux tubes correspond to the maxima and minima of the bordism, reconnections are the saddle points of β_t. Figure 8 demonstrates a simple example of the non-trivial evolution of two flux tubes and the corresponding bordism: the flux tubes are created at time $t < t_{rec}$ and they reconnect at time $t = t_{rec}$. A bordism equipped with a smooth vector field of normals on it forms "a framed bordism" $(\beta_t, \mathbf{n})$. This bordism describes the evolution which evidently does not admit a change of helicity at the points of reconnections, creation or disappearance of flux tubes. Such an evolution of a set of flux tubes can be described by the following proposition:

Two sets of flux tubes can be converted into each other (i.e. they belong

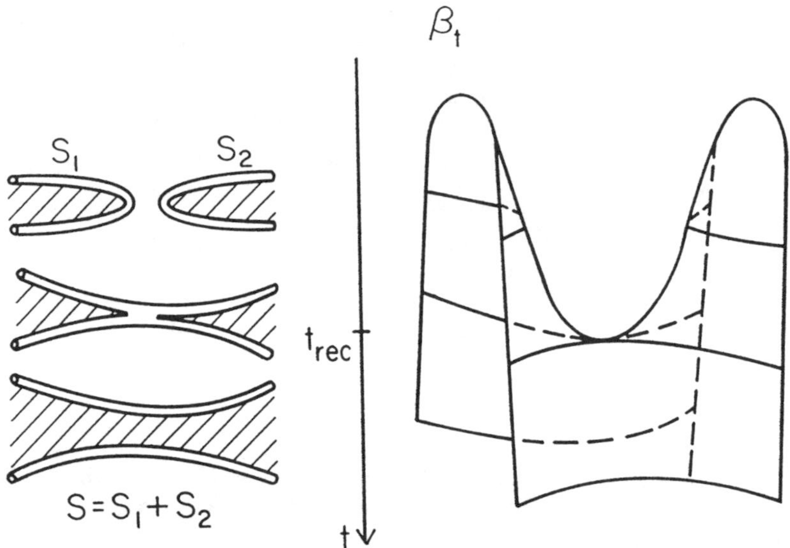

Fig. 8. The reconnection process for two flux tubes (the left side) and the corresponding bordism (the right side) are the characteristics of the topological evolution of two flux tubes. The time increases downwards.

to the same framed bordism) if and only if the sets have the same helicities. Differently speaking, the framed bordism is completely determined by the helicity, and a given helicity determines a class of equivalent bordisms.

First, note that the helicity does not change on the framed bordism. Cover each of two flux tubes i and j (figure 8) with the surfaces S_i and S_j the borders of which attach to the tubes along the frame vectors. The two surfaces join through the reconnection point into one surface $S = S_i + S_j$ covering the newly formed flux tube (see the left part of Figure 8). Because no new crossings appear in this process, the helicity remains unchanged.

Now consider two sets of flux tubes having the same helicities. Reduce each set by reconnections conserving the helicity to one flux tube with a self-linking number equal to the helicity. Now it is evident that the two sets of flux tubes belong to the same bordisms.

A corollary to the proposition is that transformations of flux tubes with different topological complexity by reconnections must be accompanied by twisting and/or writhing.

Examples demonstrating such topological evolution are shown in figure 9.

The proposition guarantees the conservation of helicity on bordisms. In terms of the continuous fluid it implies the conservation of the magnetic helicity $\int \mathbf{AB}dV$, where $\mathbf{B}$ is the magnetic field and $\mathbf{A}$ is its vector-potential.

On even bordisms, i.e. when reconnecting from an even linkage to another even one, the Borromean (Arf) invariant also is conserved, as one can conclude from the results of Robertello(1965) and Libgober(1988).

Consider a case when also $\int (\mathbf{AB})^2 dV = 0$, i.e. the vectors $\mathbf{A}$ and $\mathbf{B}$ are

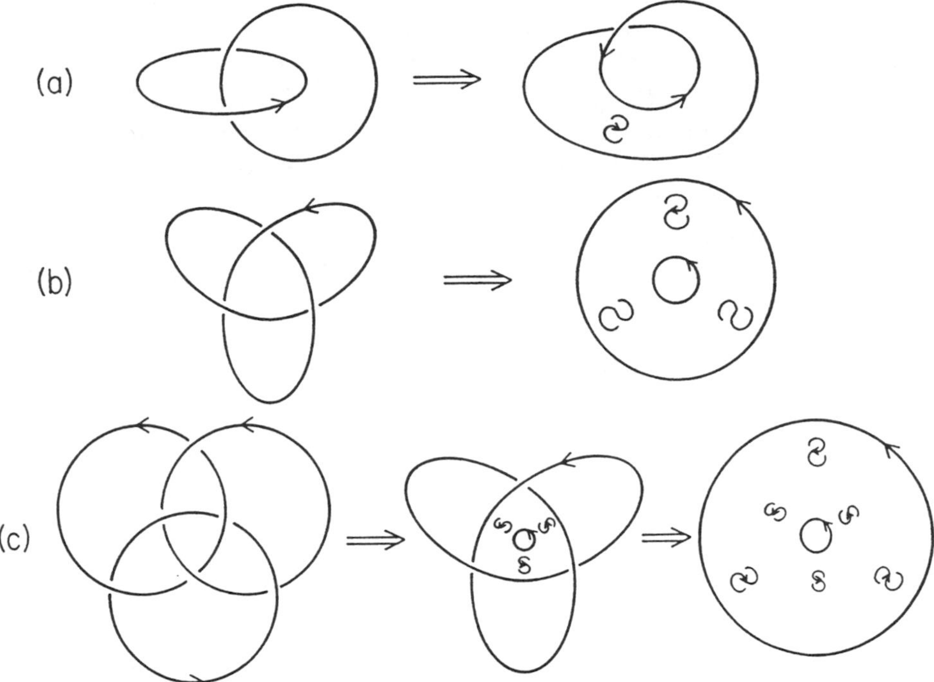

Fig. 9. The evolution of (a) simple odd linkage ($H = -2$), (b) the trefoil knot ($H = -3$), and (c) the Borromean rings ($H = 0$) under reconnections. Due to the helicity conservation a twist or writhe are created in the vicinity of the point of reconnection. To make situation more clear from the quantitative point of view in 9b,c instead of the twist the equivalent writhe is shown (the figure-8 has $H = \pm 1$ depending on its parity)

perpendicular at each point. A discreet analog of this integral is

$$Q = \sum_i \sum_{j \neq i} (L_{ij})^2 + \sum_i (L_{ii})^2.$$

Examples are three unknotted rings (a trivial case) and the Borromean rings. It appears that the sum must increase and then return to zero during transitions through the reconnections from the trivial configuration to the topologically non-trivial one.

In fact, consider two sets of flux tubes M_1 and M_2. Assume that the linking and self-linking numbers for each of the two sets are zero, i.e. $Q = 0$. Assume also that the two sets are distinguished by topological complexity so that $Arf(M_1) = 0, Arf(M_2) = 1$. For example, the set M_1 may consist of three unknotted flux tubes, the set M_2 is the Borromean flux tubes. These two sets put on one framed bordism conserving helicity. However the numbers Q , and $P = \sum_i L_{ii}{}^2$ must change during the reconnections. More precisely at some moment a system with $Q \geq 4, P \geq 2$ will exist.

Indeed, the reconnections cannot preserve the even linking numbers of the flux tubes (the Arf invariant must change). From this one concludes that two flux tubes with odd (non-zero) linking coefficients appear. At that moment $L_{ij} = L_{ji} \neq 0$ for a pair of flux tubes i and j. The helicity H of the system is conserved under the reconnection process and hence $L_{ii}+L_{jj} = -(L_{ij}+L_{ji})$, so that $L_{ii}^2 + L_{jj}^2 \geq 2$, and $Q \geq 4, P \geq 2$.

Acknowledgements

One of us (A.R.) would like to express best thanks to Mitchel Berger, Kenneth Millett, Keith Moffatt and Renzo Ricca for helpful discussions and critical comments.

References

Arf, C. 1941 Untersuchungen über quadratische Formen in Körpern der Charakteristik 2. (Teil 1). *J. Reine Angew. Math.*, **183**, 148–167.

Arnold, V.I. 1974 The asymptotic Hopf invariant and its applications. In *Proc. Summer School in Differential Equations*, Armenian SSR Acad. Sci.

Arnold, V.I., and Khesin, B.A. 1992 Topological methods in hydrodynamics. *Ann. Rev. Fluid Mech.* **24**, 145–166.

Berger, M.A., and Field, G.B. 1984 The topological properties of magnetic helicity. *J.Fluid Mech.* **147**, 133–148.

Berger, M.A. 1990 Third-order link integrals.*J. Phys. A: Math. Gen.* **23**, 2787–2793.

Browder, W. 1972 *Surgery on Simply-Connected Manifolds*, Springer-Verlag.

Crowell, R.H., and Fox, R.H. 1963 *Introduction to Knot Theory*, Ginn, Boston.

Fuller, F.B. 1978 Decomposition of the linking number of a closed ribbon: A problem from molecular biology. *Proc. Natl. Acad. Sci. USA* **75**, 3557–3561.

Francis, G.K. 1987 *A Topological Picturebook*, Springer-Verlag.

Guillou,L. and Marin, A. 1986 A la recherche de la topologie perdue. *Progress in Mathematics*, **62**, Birkhäuser, Boston-Basel-Stutgart.

Hirsch, M. 1976 *Differential Topology*, Springer-Verlag.

Lickorish, W.B.R., and Millett, K.C. The new polynomial invariants of knots and links. *Mathematics Magazine* **61**, 3–23.

Libgober, A. 1988 Theta characteristics on singular curves. Spin structures and Rohlin theorem. *Ann. Scient. Ec. Norm. Sup.*, **21**, 623.

Marin, A. 1988 Un nouvel invariant pour les spheres d'homologie de dimension trois. *Seminare Bourbaki, 1987-88, Asterisque* N **693**, 151.

Milnor,J. 1966 *Topology From the Differentiable Viewpoint*, University of Virginia Press, Charlottesville.

Moffatt, H.K. 1969 The degree of knottedness of tangled vortex lines. *J.Fluid Mech.* **159**, 359–378.

Robertello R. 1965 An invariant of knot cobordism. *Comm.Pure Appl. Math.***18**, 543–555.

Parker, E.N. 1979 *Cosmic Magnetic Fields*, Oxford Univ.Press.
Taylor, J.B. 1974 Relaxation of toroidal plasma and generation of reverse magnetic fields.*Phys. Rev. Lett.* **33**, 1139.
White, J.H. 1969 Self-linking and the Gauss integral in higher dimensions. *Am.J. Math.* **91**, 693.

7. Appendix. Proof of Formula (3).

Let S be a surface without singularities in the three-dimensional space having as its boundary the flux tubes and oriented in accordance with the direction of the magnetic field in the flux tubes. The surface does not necessary have be connected. Denote by Hs a sum of the self-linking numbers (mod 2) of all parts of a framed line of the self-intersection of the surface. The frame is determined by the normals to the surface. And let T be the number of triple points (mod 2) on the surface. It will be shown that the number

$$Bo = Arf + Hs + T,$$

is independent on the choice of the surface S. Here the Arf invariant on the surface is determined in the same way as in Section 4. When S is the Seifert surface, i.e. it has no self-intersections, the Bo invariant coincides with the Arf invariant for even linkages introduced in Section 4. From this it follows, in particular, that the invariant Bo does not change on even bordisms.

Consider another surface S_1 having the same Bo. The idea of the proof is to modify S_1 in such a way that, keeping Bo unchanged, S_1 will coincide with the surface S. The modification includes: (a) a surgery (cutting from, or adding to unknotted handles on the surface), (b) a stick-out operation for a part of the surface along a curve of self-intersection, (c) deformation without singularities. The operations (a) and (c) do not change any numbers (Arf, Hs or T). In operation (b) all the numbers can change, however the operation conserves the sum of the numbers.

The surfaces S_1 and S may be chosen to coincide near the boundary with flux tubes because all surfaces which are not intersected by the flux tubes come to the flux tubes along the Seifert frame (see Section 3).

By use of operation (a) the surface S_1 can be converted into some surface S_2 which is homotopic to S i.e. it has the same number of handles as S and can be deformed into S without changing the position of the boundary. The only difference between S_2 and S is the difference in the number of the handles having the form of the twisted torus shown in figure 6. This difference can be characterized by quadratic forms defined on these self-intersecting surfaces in the same way as in Section 4. The problem now is to make the quadratic forms equal. In this case, according to the theorem by Smale and Hirsch (Giullou and Marin, 1987), the surface S_2 can be deformed

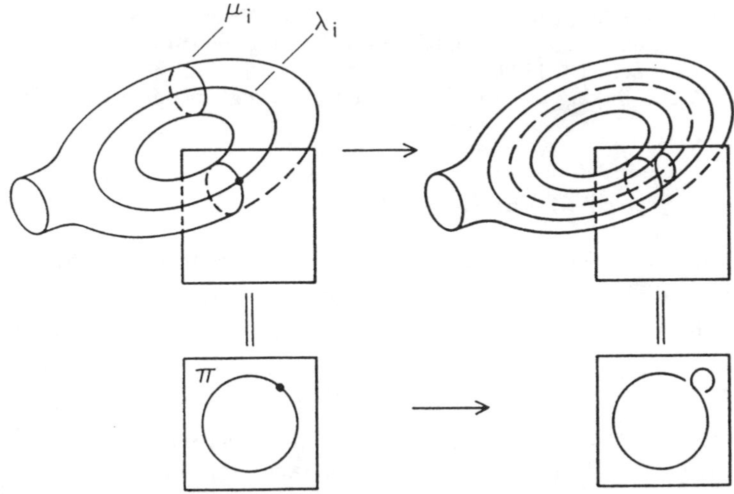

Fig. 10. The stick-out operation with along the longitude of a handle. A piece π, also shown separately, is a part of the surface S_2 intersecting the handle. The appearance of triple point and of a self-linkage of a curve of the surface self-intersection are shown.

into the surface S without singularities. The problem can be resolved as follows:

Let $\{\mu_i, \lambda_i\}$, where the index i means an i-handle, be cyclic coordinates on the surface S_2. The stick-out operation (b) along the closed coordinate line λ_i, see figure 10, results in

$$l(\lambda_i) \to l(\lambda_i), \qquad l(\mu_i) \to l(\mu_i) + 1.$$

The same operation can be done along the other coordinate line μ_i. By applying these operations many times one can make the quadratic form on the surface S_2 equal to the quadratic form on S. Thus it is sufficient to know how the number Bo is changing in one of the operations. It is clear that Arf (the product of l's) does not change (mod 2) when $l(\lambda_i) = 0$, and it changes when $l(\lambda_i) = 1$. The number Hs under this operation is changed by $l(\lambda_i) + T(\lambda_i)$, where $T(\lambda_i)$ is the number of triple points appearing when the sticked-out tube crosses pieces of the surface. As a result $Arf + Hs$ changes by $2l(\lambda_i)$, which is zero modulo 2, plus $T(\lambda_i)$. And because T also changes by $T(\lambda_i)$, and $2T(\lambda_i)$ is zero modulo 2, the number Bo does not change. Finally, regular deformations (operation (c) transforming the surface S_2 into S does not change any of the three numbers. Hence, formula (3) is justified.

TOPOLOGY OF STEADY FLUID FLOWS

VIKTOR L. GINZBURG
Department of Mathematics
Stanford University
Stanford, California 94305, USA

and

BORIS KHESIN
Department of Mathematics
University of California
Berkeley, California 94720, USA

ABSTRACT. Steady flows of an ideal fluid have a very special topological structure. For three dimensional analytic flows the region is fibered into tori and annuli invariant under the flow (V. Arnold's theorem). It turns out that the analogous picture holds in four dimensional case as well: four dimensional steady flows have invariant tori and are similar to integrable Hamiltonian systems with two degrees of freedom. We also established a strong relation between existence conditions of a smooth steady two-(and any even-) dimensional flow and the topology of the corresponding vorticity function. For instance, there is a well known example of a steady flow on the disk with vorticity monotonically decreasing along the radius. It turns out that on the disk there exists no smooth flow whose vorticity function is constant on the boundary and has local maximum and minimum inside this disk.

1. Introduction

Steady (or stationary) flows of compressible and incompressible fluids have a very special topology and existence conditions. These flows are often "attractors" in the phase space, and so their structure is an "approximate picture" of an arbitrary fluid motion after a long period of time.

So far, all the results on stationary flows concerned, to the best of our knowledge, the two- and three-dimensional case (see Arnold, 1965; Arnold, 1966; Moffatt, 1990; Kop'ev and Leont'ev, 1988; Troshkin, 1990). We discuss here higher dimensional analogs of these results, and, in particular, we describe the structure of generic four-dimensional steady flows (see section 3). We show that four-dimensional steady flows are quite similar to the integrable systems with two degrees of freedom.

H. K. Moffatt et al. (eds.), Topological Aspects of the Dynamics of Fluids and Plasmas, 265–272.

It turns out to be very useful to introduce a notion of an extra symmetry of a system, which is, roughly speaking, an analog of an additional first integral for Hamiltonian equations.

To have an extra symmetry is a very restrictive condition on the system: one may expect that a generic system does not admit extra symmetries. We also establish a strong relation between existence conditions of a smooth steady two- (and any even-) dimensional flow and the topology of the vorticity function (see section 4). We formulate certain conditions on this function which guarantee that the corresponding dynamical system has no smooth steady flows. For instance, there is a well known axisymmetric steady flow on the disc with the vorticity monotonically decreasing along radius. On the other hand, it turns out that on the disk there exists no smooth flow whose vorticity function is constant on the boundary and has a local maximum and minimum inside the disk. We emphasize that steady solutions with singularities may exist for a larger class of vorticity functions (see Hegna and Battacharjee, 1991; Moffatt, 1986).

In section 2 we recall the mathematical model of ideal hydrodynamics (see, *e.g.*, Arnold, 1966, or Arnold and Khesin, 1992) and the main concepts related to the fluid motion equations.

2. Mathematical model of ideal hydrodynamics

In this section we briefly discuss the mathematical model describing the motion of an ideal fluid.

The starting object in inviscid incompressible hydrodynamics is a manifold M equipped with a volume form μ. The phase space is the dual space $\mathcal{G}^*$ to the Lie algebra $\mathcal{G}$ of all divergence-free vector fields on M. (If $\partial M \neq \emptyset$, the vector fields must be tangent to the boundary of M.) The dual space $\mathcal{G}^*$ carries the natural linear Poisson structure, and Euler equation of the ideal fluid is Hamiltonian with respect to this structure (Arnold, 1989).

The Hamiltonian H is determined by an extra structure on the manifold M, namely, by a Riemannian metric g. In fact, the metric g gives rise to the nondegenerate quadratic form $\langle v, v\rangle = \int_M g(v,v)\mu$ on $\mathcal{G}$ and so, to the identification of $\mathcal{G}$ and $\mathcal{G}^*$. By definition, H is the corresponding quadratic form on $\mathcal{G}^*$.

The dual space $\mathcal{G}^*$ can be naturally viewed as the vector space of exact 2-forms on M (see Marsden and Weinstein,1983). The pairing of a 2-form ω with a divergence-free vector field v is given by $\langle \omega, v\rangle = \int \alpha(v)\mu$, where α is a primitive 1-form for $\omega : d\alpha = \omega$.

After this identification the Euler equation of ideal hydrodynamics has the following form

$$\dot{\omega} = L_v \omega \tag{2.1}$$

(called the Helmholtz equation), where the divergence-free vector field v (*i.e.*, v satisfying the equation $L_v\mu = 0$) is uniquely defined by the condition $d\alpha = \omega$ for $\alpha(*) = g(v, *)$.

REMARK: The Euler-Helmholtz equation means that the 2-form ω is frozen into the fluid (or is transported by the flow of v). This implies the following duality of incompressible flows on odd- and even-dimensional manifolds. For M^{2n+1} let ξ be the vorticity vector field of $\omega : i_\xi\mu = \omega^n$. For M^{2n} let l be the vorticity function of $\omega : l = \omega^n/\mu$. Then the vector field ξ for M^{2n+1} and the covector field dl (*i.e.*, the function l) for M^{2n} are both frozen, as well as ω itself, into the ideal fluid.

Our main goal is to describe the topology of smooth steady (or stationary) flows. By definition such a flow is a time independent smooth solution of the Euler-Helmholtz equation: $L_v\omega = 0$. The solutions are precisely critical points of H on the coadjoint orbits in $\mathcal{G}^*$ (Arnold, 1989). In other words, stationary flows are exactly the extremals of the energy functional among all isovorticed divergence-free vector fields.

For the three-dimensional case the complete description of analytic stationary flows is given by the following theorem.

THEOREM 2.1: (Arnold, 1965). *Assume that the region D is bounded by a compact analytic surface and that the field of velocities is analytic and not everywhere collinear with its curl. Then the region of the flow can be partitioned by an analytic submanifold into a finite number of cells, in each of which the flow is constructed in a standard way. Namely, the cells are of two types: those fibered into tori invariant under the flow and those fibered into surfaces invariant under the flow, diffeomorphic to the annulus $\mathbf{R} \times S^1$. On each of these tori the flow lines are either all closed or all dense, and on each annulus all flow lines are closed.*

REMARK: In this theorem it is important that the fields of velocity v and vorticity ξ are not collinear. Since $[\xi, v] = 0$, this means that the field ξ admits an "extra symmetry" and, therefore, so does every element of the coadjoint orbit through ξ. Note that the hypothesis of the theorem seems to hold very rarely: we rather expect that neither generic vector field nor "generic orbit" admits an extra symmetry.

In the next section we formulate an analog of this theorem for a four-dimensional region.

REMARK: The analogous description holds for steady flows of a compressible fluid (or gas dynamics). Indeed, the equations of this fluid include the equation of continuity: $\dot{\rho} + \mathrm{div}(v\rho) = 0$ for density ρ instead of the volume preserving condition $\mathrm{div} v = 0$. For a steady flow $\dot{\rho} = 0$ as well as $\dot{v} = 0$, and so a given stationary distribution of density ρ can be considered as a new volume form μ preserved by v.

3. The structure of steady four-dimensional flows

The main result of this section shows that steady flows of a four-dimensional fluid are very similar to integrable Hamiltonian systems with two degrees of freedom.

Recall that the equation of stationary flow has the form

$$L_v \omega = 0 \tag{3.1}$$

or, equivalently, $d\,(i_v\omega) = 0$. Thus for a simply connected M there exists a function h, called the Bernoulli function, such that $i_v\omega = dh$, and, as a consequence, the velocity field v is tangent to the levels of h, *i.e.*, $L_v h = 0$. (In the three-dimensional case this observation alone implies the existence of tori and annuli in the Arnold theorem (Arnold, 1965).

Further we will mainly work with an even-dimensional manifold M^{2n}. In this case, in addition to h, there is one more invariant function on M: $l(x) = \frac{\omega^n}{\mu}$, called the vorticity function. The function l is invariant, since $L_v\omega = 0$ and $L_v\mu = 0$. Actually, this means that l, as well as h, is the first integral of the dynamical system on M given by the stationary vector field v. Using this fact one can prove following

THEOREM 3.1: *Let M^4 be a closed four-dimensional manifold. Suppose that l and h are functionally independent almost everywhere on M, i.e., $\mathrm{rk}\pi_* = 2$ on an open dense subset of M. Then there exists a (singular) hypersurface Γ^3 such that the complement $M \setminus \Gamma$ is fibered into two-dimensional tori invariant under the flow. On each of these tori the flow lines are either all closed or all dense.*

REMARK: The hypersurface Γ is formed by the critical points the "momentum map" $\pi = (h, l) : M \to R^2$ and the zero level of l. The tori above are precisely regular levels of π for $l \neq 0$. Note also that for a four-dimensional manifold with boundary the corresponding analog of the Arnold theorem holds, provided that all the data are analytic.

Thus we have a more or less complete description of some four-dimensional smooth steady flows. The description is quite analogous to the one for the three-dimensional case (Arnold, 1965) and, in fact, resembles the structure of completely integrable systems with two degree of freedoms.

REMARK: For an arbitrary even dimensional manifold M^{2n} we can assert that M is a union of $(2n-2)$- (or less) dimensional submanifolds, such that the steady vector field v is tangent to them. These surfaces are the intersection of the levels $h = \text{const}$ and $l = \text{const}$.

REMARK: For an arbitrary odd dimensional M^{2n+1} instead of the function $l = \omega^n/\mu$ (and, hence, covector field dl) we can define the vorticity vector field ξ by $i_\xi\mu = \omega^n$. The field ξ and v commute and, thus, determine an R^2-action on M^{2n+1}. So in this case a steady flow gives rise to a foliation of dimension 2, unlike the foliation of codimension 2 in the even-dimensional case described above.

4. Topology of the vorticity function

In this section we formulate some very general topological properties of the vorticity function of a steady flow. As we mentioned in the introduction, these properties may be used in order to obtain important results on the existence of smooth steady flows. Note that even the application of our results to the simplest two-dimensional case seems to be new and nontrivial.

One can easily construct an example of an axisymmetric steady flow on the disk with the vorticity decreasing monotonically along radius (*e.g.* considering the flow with angular velocity of fluid particles decreasing along radius, see Kop'ev and Leont'ev, 1988; Moffatt, 1990). This vorticity function is constant on the boundary and has one maximum at the center.

Let us now consider a flow with a positive vorticity function which is also constant on the boundary and has nondegenerate critical points inside the (even-dimensional) disk. The formulated below general theorems have the following corollary: there is no smooth steady flow whose vorticity function satisfies the conditions above and has a maximum and a minimum inside the disk. An example of such a "prohibited" vorticity function is shown on Fig. 1. This picture is closely related to the problems of plasma physics, and, in particular, to the Kadomtsev conjecture on a stable state of plasma in tokamaks. These bans on the vorticity function presumably prohibit the analogous steady smooth magnetic field in toroidal tokamaks.

Now we describe this phenomenon in precise mathematical terms.

Let ω be a smooth stationary solution of Helmholtz equation (*i.e.*, a smooth steady flow) on an even-dimensional manifold M, and $\mathcal{O}$ be the set of all flows isovorticed with ω (in other words, 2-form ω belongs to the coadjoint orbit $\mathcal{O}$). In this section we investigate the topology of the vorticity function $l = \omega^n/\mu$ of the steady flow ω from a given orbit $\mathcal{O}$. It is obvious that "topological" invariants of the function l, such as a number of its critical points, their indices *etc.*, depend only on the orbit. We use these invariants in order to formulate certain properties which hold for the

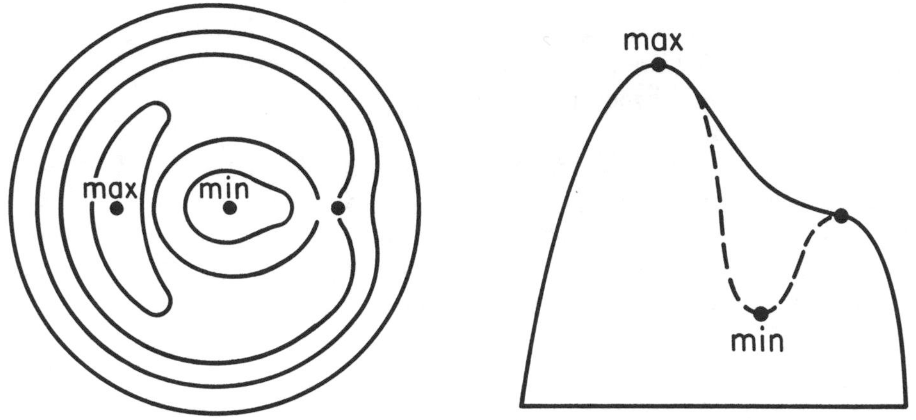

Fig. 1. Level surfaces and a profile of the vorticity function having no smooth steady flow.

vorticity function of a steady solution and so, for the orbit $\mathcal{O}$ which contains it, provided that the vorticity function l and the forms in $\mathcal{O}$ do not admit "too many symmetries". This will allow us to specify the topology of the steady flow (see Theorem 4.4 and 4.5), and to find certain orbits with no smooth steady solutions at all.

DEFINITION 4.1: A function f on a compact symplectic manifold (W, w) does not admit extra symmetries if an arbitrary function g, such that the Poisson bracket $\{f, g\} = 0$, is constant on connected components of the levels of the function f (*i.e.*, the differentials df and dg are proportional everywhere).

EXAMPLE: On a two-dimensional symplectic manifold any function does not admit extra symmetries. One can prove that functions without extra symmetries on a closed symplectic manifold form a C^k-dense subset of the space of smooth functions. Apparently, a generic function on such a manifold does not admit extra symmetries, however this, to the best of our knowledge, has not been proved yet.

DEFINITION 4.2: A coadjoint orbit $\mathcal{O} \subset \mathcal{G}^*$ does not admit extra symmetries if for any (or, equivalently, for all) $\omega \in \mathcal{O}$ and for any two regular values $0 < a < b$ (or $a < b < 0$) of the vorticity function l, this function does not admit extra symmetries on $l^{-1}([a, b])$.

Our definitions are consistent: a function f on a compact symplectic manifold does not admit extra symmetries, if and only if neither does its restriction to the inverse image of any segment with regular endpoints.

DEFINITION 4.3: An orbit $\mathcal{O} \subset \mathcal{G}^*$ has the Morse type if for any (or, equivalently, for all) $\omega \in \mathcal{O}$ the function l is a Morse function constant on connected component of ∂M and, moreover, 0 is not a critical value of l. The orbit is called positive if $l > 0$.

A smooth steady solution ω of the Euler-Helmholtz equation ($L_v\omega = 0$) is said to be exact if it has a single-valued Bernoulli function h, *i.e.* $dh = i_v\omega$, or, in other words the 1-form $i_v\omega$ is exact.

THEOREM 4.4: *Let* $\dim M \geq 4$ *and* $\mathcal{O}$ *be a Morse type orbit without extra symmetries. Assume that* $\mathcal{O}$ *contains a smooth exact steady solution. Then for every* $\omega \in \mathcal{O}$ *all the critical points of the vorticity function* l *have indices all either less than* $n+1$ *or more than* $n-1$ *on every connected component of* $M \backslash \{l = 0\}$.

We recall that the index of a nondegenerate critical point of a smooth function is the number of negative squares in its Morse normal form.

EXAMPLE: If $\mathcal{O}$ is as above and $l > 0$ on $M \backslash \partial M$, then l cannot have both a local maximum (index $2n$) and a minimum (index 0) on $M \backslash \partial M$.

THEOREM 4.5: *Let* $\dim M = 2$ *and* ∂M *be connected,* $\mathcal{O} \subset \mathcal{G}^*$ *a positive orbit of Morse type. If* $\mathcal{O}$ *contains an exact stationary solution, then for any* $\omega \in \mathcal{O}$ *the vorticity function* l *cannot simultaneously have a local maximum and a local minimum in* int$M$.

Theorems 4.4 and 4.5 provide us with a way to find coadjoint orbits and, thus vorticity functions without smooth steady solutions. The case of two-dimensional M is especially simple. Consider a disk $M = D^2 \subset \mathbf{R}^2(x, y)$, $\mu = dx \wedge dy$ and $\omega = l \cdot \mu$, where l is a positive Morse function on D, $l|_{\partial D} = \text{const}$. Assume also that l has both a local maximum and a local minimum in int$D$. Then the coadjoint orbit through ω does not contain a steady solution. In other words, there is no vorticity function isovorticed to ω which corresponds to a smooth steady solution.

The corollary can be generalized to higher dimensions. Note, however that this generalization is much less explicit, since it is based on the existence of orbits without extra symmetries. The details of the proofs, the discussion and further results on higher-dimensional case will appear elsewhere.

Acknowledgements

We are deeply grateful to V.I. Arnold, A.B. Givental, H.K. Moffatt, S.I. Vainshtein and A. Weinstein for numerous useful discussions. The second author would like to thank the Institute for Theoretical Physics at Santa Barbara and the organizers of the program "Topological Fluid Dynamics" for their hospitality and fruitful atmosphere stimulating this work. This research was partially supported by the NSF Grants PHY 89-04035 and DMS 85-05-550.

References

ARNOLD, V. I., 1965 Sur la topologie des écoulements stationaires des fluides parfaits. *C.R. Acad. Sci. Paris* **261**, 17–20.

ARNOLD, V. I., 1966 Sur la géométrie differentielle des groupes de Lie de dimension infinite et ses applications á hydrodynamique des fluides parfaits. *Ann. Inst. Fourier* **16**, 316–361

ARNOLD, V. I., 1989 *Mathematical Methods of Classical Mechanics* (Springer-Verlag).

ARNOLD, V. I. AND KHESIN, B. A., 1992 Topological methods in hydrodynamics. *Annual Rev. in Fluid Mechanics* **24**, 145–166.

HEGNA, C.C. AND BHATTACHARJEE, A., 1991 Islands in three-dimensional steady flows. *J. Fluid Mech* **227**, 527–542.

KOP'EV, V. F. AND LEONT'EV, E. A., 1988 Acoustic instability of plane vortex flows with circled flow. *Acousticheskii Zhurnal* **34**, No. 3, 475–580 (in Russian).

MARSDEN, J. AND WEINSTEIN, A., 1983 Coadjoint orbits, vortices and Clebsh variables for incompressible fluids. *Physica D* **7**, 305–323.

MOFFATT, H. K., 1986 On the existence of localized rotational disturbances which propagate without change of structure in an inviscid fluid. *J. Fluid Mech.* **173**, 289–302.

MOFFATT, H. K., 1990 Structure and stability of solutions of the Euler equations: a lagrangian approach. *Phil. Trans. R. Soc. London* **A333**, 321–342.

TROSHKIN, O. V., 1990 On topological analysis of structure of hydrodynamical flows. *Russian Math. Surveys* **43**, No. 4, 153–190.

PART IV

THE EULER EQUATIONS: EXTREMAL PROPERTIES AND FINITE TIME SINGULARITIES

EXTREMAL PROPERTIES AND HAMILTONIAN STRUCTURE OF THE EULER EQUATIONS

T.G. SHEPHERD
Department of Physics
University of Toronto
Toronto M5S 1A7 Canada

ABSTRACT. The Euler equations describing perfect-fluid motion represent a Hamiltonian dynamical system. The Hamiltonian framework implies, via Noether's theorem, that symmetries are connected with extremal properties. For example, steady solutions are conditional extrema of the energy under smooth perturbations that preserve vortex topology, as pointed out originally by Arnol'd (1966*a*). In the three-dimensional context such *isovortical* perturbations consist of those that maintain the strength and topology of vortex tubes; in the two-dimensional context they consist of those that maintain the vorticity of each fluid element, which is a considerably stronger constraint. Just as with temporal symmetry, solutions that are independent of a spatial coordinate are conditional extrema of the associated momentum invariant under isovortical perturbations; and steadily-translating solutions are conditional extrema of a linear combination of the energy and the momentum. This connection between symmetries and extremal properties is exploited to particular effect in Arnol'd's (1966*c*) nonlinear stability theorems and their recent extensions and applications. Finally, a class of algorithms is described that can find true energy and energy-momenta extrema (under isovortical perturbations) of the two- and three-dimensional Euler equations, when they exist.

1. Introduction

Hamiltonian structure provides the foundation upon which much of theoretical physics rests. This is true in such diverse fields as quantum mechanics, quantum field theory, statistical mechanics, relativity, optics, and celestial mechanics. Hamiltonian structure provides an important unifying framework for different theories, within which symmetry properties are readily apparent which can be connected to conservation laws through Noether's theorem.

Over the last few decades it has become apparent (see e.g. Morrison 1982; Marsden 1984) that fluid and plasma physics also fit into a Hamiltonian framework, though in a manner that requires a generalization of the notion of what "Hamiltonian" means. Rather than requiring canonical (q, p)

H. K. Moffatt et al. (eds.), Topological Aspects of the Dynamics of Fluids and Plasmas, 275–292.

variables and the canonical bracket (which generates Hamilton's equations), as in Goldstein (1980) for example, the modern approach is to take a more geometric perspective wherein Hamiltonian dynamics corresponds to symplectic structure on a phase-space manifold (Abraham & Marsden 1978; Arnol'd 1989; Tabor 1989, Appendix 2.2). The Poisson bracket associated with this symplectic structure may well turn out to be degenerate (or singular), in which case the phase space is foliated by so-called symplectic leaves and the Hamiltonian description is said to be *non-canonical*. The symplectic leaves are the level surfaces defined by the Casimir invariants (see e.g. Littlejohn 1982), which reflect the degeneracy of the bracket. A ramification of this degeneracy is that extremal or critical properties of the Hamiltonian can be expected to hold only for variations that are constrained to lie on the same symplectic leaf. For example, while steady solutions of a canonical Hamiltonian system are necessarily conditional extrema (critical points) of the Hamiltonian, for a non-canonical system they are rather *constrained* conditional extrema.

It turns out that virtually all Eulerian descriptions of fluid and plasma systems, including the Euler equations of ideal incompressible homogeneous fluid flow, correspond to such non-canonical Hamiltonian representations. (For examples, see e.g. Morrison 1982; Benjamin 1984; Holm *et al.* 1985; Shepherd 1990*a*). This is especially interesting from a geometric perspective because the Casimir invariants which foliate the phase space frequently correspond to *topological* invariants. In the case of the Euler equations, for example, specification of a symplectic leaf amounts to specification of the vortex topology (Kuznetsov & Mikhailov 1980; Marsden & Weinstein 1983; Arnol'd & Khesin 1992). This means that steady solutions will be extremal states of the Hamiltonian subject to vortex-topological constraints, a fact first pointed out by Arnol'd (1966*a*). This statement may be extended to any solution having a continuous symmetry, for example spatially-invariant or spatially-translating solutions.

2. Hamiltonian Structure

For fluids (and plasmas) we must consider not only non-canonical but also infinite-dimensional Hamiltonian systems, where the dynamical variables are fields. In this context a Hamiltonian representation may be considered in the abstract form

$$\frac{\partial \boldsymbol{u}}{\partial t} = J\frac{\delta \mathcal{H}}{\delta \boldsymbol{u}}, \tag{1}$$

where $\boldsymbol{u}(\boldsymbol{x},t)$ is the set of field variables, $\mathcal{H}[\boldsymbol{u}]$ is the Hamiltonian functional, $\delta/\delta\boldsymbol{u}$ denotes the functional or variational derivative, and J is a transformation (known as the *co-symplectic form*). In order to be Hamiltonian the

Poisson bracket induced by J, namely

$$[\mathcal{F},\mathcal{G}] = \Big(\frac{\delta\mathcal{F}}{\delta\boldsymbol{u}}, J\frac{\delta\mathcal{G}}{\delta\boldsymbol{u}}\Big), \tag{2}$$

must be bilinear, skew-symmetric, and satisfy the product rule as well as Jacobi's identity (e.g. Morrison 1982). In the above $\mathcal{F}[\boldsymbol{u}]$ and $\mathcal{G}[\boldsymbol{u}]$ are any two admissible functionals, and $(\cdot,\cdot)$ represents the relevant inner product on the function space $\{\boldsymbol{u}\}$.

Hamilton's equations follow from (1) upon choosing canonical variables $\boldsymbol{u} = (q_1(t),\ldots,q_N(t),p_1(t),\ldots,p_N(t))^T$ and the canonical bracket induced by

$$J = \begin{pmatrix} 0 & \mathrm{I} \\ -\mathrm{I} & 0 \end{pmatrix}, \tag{3}$$

where I is the $N \times N$ identity matrix. According to Darboux's theorem, any non-singular J may be (locally) transformed to (3) by a suitable change of variables, so we may regard any Hamiltonian system of the form (1) with a non-singular J as canonical. On the other hand, if the equation

$$J\frac{\delta\mathcal{C}}{\delta\boldsymbol{u}} = 0 \tag{4}$$

has non-trivial solutions, then J is singular and the Hamiltonian representation is said to be non-canonical; equivalently, there exist functionals $\mathcal{C}$ for which the Poisson bracket is degenerate:

$$[\mathcal{C},\mathcal{G}] = 0 \quad \forall\mathcal{G}. \tag{5}$$

Functionals satisfying (4) (equivalently (5)) are referred to as *Casimir* functionals. They are necessarily invariants (first integrals) of the dynamics (1) since

$$\frac{d\mathcal{C}}{dt} = \Big(\frac{\delta\mathcal{C}}{\delta\boldsymbol{u}}, \frac{\partial\boldsymbol{u}}{\partial t}\Big) = \Big(\frac{\delta\mathcal{C}}{\delta\boldsymbol{u}}, J\frac{\delta\mathcal{H}}{\delta\boldsymbol{u}}\Big) = -\Big(\frac{\delta\mathcal{H}}{\delta\boldsymbol{u}}, J\frac{\delta\mathcal{C}}{\delta\boldsymbol{u}}\Big) = 0, \tag{6}$$

after using (4) together with the skew-symmetry of J.

3. Euler Equations

The governing equations of three-dimensional ideal incompressible homogeneous flow in a domain D are

$$\frac{\partial\boldsymbol{v}}{\partial t} + \boldsymbol{v}\cdot\nabla\boldsymbol{v} = -\nabla p, \tag{7}$$

where $\boldsymbol{v}$ is the velocity and p the pressure, subject to

$$\nabla\cdot\boldsymbol{v} = 0 \quad \text{in } D, \qquad \boldsymbol{v}\cdot\hat{\boldsymbol{n}} = 0 \quad \text{on } \partial D. \tag{8}$$

They may be alternatively written in the vorticity form

$$\frac{\partial \boldsymbol{\omega}}{\partial t} = \nabla \times (\boldsymbol{v} \times \boldsymbol{\omega}), \qquad \boldsymbol{\omega} = \nabla \times \boldsymbol{v}, \tag{9}$$

with $\boldsymbol{v}$ constrained by (8). It is now well known that the above dynamics may be represented as an infinite-dimensional, non-canonical Hamiltonian dynamical system (Arnol'd 1966*b*, 1969; Ebin & Marsden 1970; Olver 1982; Arnol'd & Khesin 1992). This may be seen in a number of ways. For example, the vorticity equation (9) may be directly cast in the symplectic form

$$\frac{\partial \boldsymbol{\omega}}{\partial t} = J \frac{\delta \mathcal{H}}{\delta \boldsymbol{\omega}}. \tag{10}$$

Here $\boldsymbol{\omega}(\boldsymbol{x}, t)$ is the dynamical variable,

$$J = -\nabla \times (\boldsymbol{\omega} \times \nabla \times (\cdot)) \tag{11}$$

is the co-symplectic form, and the inner product is just

$$(\boldsymbol{f}, \boldsymbol{g}) = \iiint_D \boldsymbol{f} \cdot \boldsymbol{g} \, dxdydz.$$

The Hamiltonian $\mathcal{H}$ is just the usual energy

$$\mathcal{H} = \iiint_D \frac{1}{2} |\boldsymbol{v}|^2 dxdydz. \tag{12}$$

If we introduce the vector streamfunction $\boldsymbol{\psi}$, defined uniquely by

$$\boldsymbol{v} = -\nabla \times \boldsymbol{\psi}, \qquad \nabla \cdot \boldsymbol{\psi} = 0 \quad \text{in } D, \qquad |\boldsymbol{\psi}| = 0 \quad \text{on } \partial D, \tag{13}$$

then the functional derivative of $\mathcal{H}$ with respect to $\boldsymbol{\omega}$ follows:

$$\begin{aligned} \delta \mathcal{H} &= \iiint_D \boldsymbol{v} \cdot \delta \boldsymbol{v} \, dxdydz = - \iiint_D (\nabla \times \boldsymbol{\psi}) \cdot \delta \boldsymbol{v} \, dxdydz \\ &= - \iiint_D \boldsymbol{\psi} \cdot (\nabla \times \delta \boldsymbol{v}) \, dxdydz = - \iiint_D \boldsymbol{\psi} \cdot \delta \boldsymbol{\omega} \, dxdydz, \end{aligned}$$

whence

$$\frac{\delta \mathcal{H}}{\delta \boldsymbol{\omega}} = -\boldsymbol{\psi}. \tag{14}$$

Among other things, (14) confirms that the vorticity $\boldsymbol{\omega}$ *does* contain all necessary dynamical information! Substituting (11) and (14) into (10) produces the vorticity equation (9), as may easily be verified.

The transformation J induces the Poisson bracket

$$\begin{aligned} [\mathcal{F}, \mathcal{G}] = \left(\frac{\delta \mathcal{F}}{\delta \boldsymbol{\omega}}, J \frac{\delta \mathcal{G}}{\delta \boldsymbol{\omega}} \right) &= - \iiint_D \frac{\delta \mathcal{F}}{\delta \boldsymbol{\omega}} \cdot \nabla \times \left[\boldsymbol{\omega} \times \left(\nabla \times \frac{\delta \mathcal{G}}{\delta \boldsymbol{\omega}} \right) \right] dxdydz \\ &= - \iiint_D \left(\nabla \times \frac{\delta \mathcal{F}}{\delta \boldsymbol{\omega}} \right) \cdot \left[\boldsymbol{\omega} \times \left(\nabla \times \frac{\delta \mathcal{G}}{\delta \boldsymbol{\omega}} \right) \right] dxdydz \end{aligned}$$

$$\iff \quad [\mathcal{F},\mathcal{G}] = \iiint_D \boldsymbol{\omega}\cdot\left[\left(\nabla\times\frac{\delta\mathcal{F}}{\delta\boldsymbol{\omega}}\right)\times\left(\nabla\times\frac{\delta\mathcal{G}}{\delta\boldsymbol{\omega}}\right)\right]\,dx\,dy\,dz, \tag{15}$$

defined on admissible functionals $\mathcal{F}[\boldsymbol{\omega}]$, $\mathcal{G}[\boldsymbol{\omega}]$, so satisfying

$$\nabla\cdot\frac{\delta\mathcal{F}}{\delta\boldsymbol{\omega}} = 0 \quad \text{in } D, \qquad \left|\frac{\delta\mathcal{F}}{\delta\boldsymbol{\omega}}\right| = 0 \quad \text{on } \partial D \qquad (\text{ditto for } \mathcal{G}).$$

The symplectic representation of the dynamics can then be written

$$\frac{\partial\mathcal{F}}{\partial t} = [\mathcal{F},\mathcal{H}] \tag{16}$$

for admissible functionals $\mathcal{F}[\boldsymbol{\omega}]$, which is equivalent to (10).

In the case of two-dimensional Euler flow in the x-y plane,

$$\boldsymbol{\omega} = \omega\,\hat{\boldsymbol{z}}, \qquad \boldsymbol{\psi} = \psi\,\hat{\boldsymbol{z}},$$

and the governing equation (9) takes the form

$$\frac{\partial\omega}{\partial t} = -\partial(\psi,\omega) \equiv -\frac{\partial\psi}{\partial x}\frac{\partial\omega}{\partial y} + \frac{\partial\psi}{\partial y}\frac{\partial\omega}{\partial x}. \tag{17}$$

Equation (17) may be cast in symplectic form with

$$J = -\partial(\omega,\cdot), \tag{18}$$

(the two-dimensional version of (11)), which induces the bracket

$$[\mathcal{F},\mathcal{G}] = \iint_D \omega\,\partial\left(\frac{\delta\mathcal{F}}{\delta\omega},\frac{\delta\mathcal{G}}{\delta\omega}\right)\,dx\,dy \tag{19}$$

defined on admissible functionals $\mathcal{F}[\omega]$, $\mathcal{G}[\omega]$. The bracket (19), like (15), may be checked to satisfy the Jacobi condition and other requisite algebraic properties of Poisson brackets.

4. Energy Extrema and Steady States

Let $\boldsymbol{u} = \boldsymbol{U}$ be a steady solution of the Hamiltonian system (1). First suppose the Hamiltonian representation is canonical, in which case the transformation J is invertible (non-singular). Then

$$\frac{\partial\boldsymbol{U}}{\partial t} = 0 \implies J\left.\frac{\delta\mathcal{H}}{\delta\boldsymbol{u}}\right|_{\boldsymbol{u}=\boldsymbol{U}} = 0 \implies \left.\frac{\delta\mathcal{H}}{\delta\boldsymbol{u}}\right|_{\boldsymbol{u}=\boldsymbol{U}} = 0. \tag{20}$$

Hence steady solutions are *conditional extrema* (critical points) of the Hamiltonian.

In contrast, for a *non-canonical* system the second implication of (20) does not follow, since J is (by definition) non-invertible. However, steady states of non-canonical Hamiltonian systems still have extremal properties. This is because

$$J\frac{\delta\mathcal{H}}{\delta\boldsymbol{u}}\bigg|_{\boldsymbol{u}=\boldsymbol{U}}=0 \quad\Longrightarrow\quad \frac{\delta\mathcal{H}}{\delta\boldsymbol{u}}\bigg|_{\boldsymbol{u}=\boldsymbol{U}}=\frac{\delta\mathcal{C}}{\delta\boldsymbol{u}}\bigg|_{\boldsymbol{u}=\boldsymbol{U}} \tag{21}$$

for some Casimir $\mathcal{C}$. That is, there is (generically) some $\mathcal{C}$ locally "tangent" to $\mathcal{H}$. However, it must be mentioned that one may not always be able to find explicit expressions for these Casimirs in terms of the field variables of the problem; an important example is the case of three-dimensional Euler flow, as will be discussed further below.

It follows from (21) that steady solutions of non-canonical Hamiltonian systems are conditional extrema not of the Hamiltonian $\mathcal{H}$, but of the combined invariant $\mathcal{H}-\mathcal{C}$, for some $\mathcal{C}$ satisfying (21): that is,

$$\frac{\partial\boldsymbol{U}}{\partial t}=0 \quad\Longrightarrow\quad J\frac{\delta\mathcal{H}}{\delta\boldsymbol{u}}\bigg|_{\boldsymbol{u}=\boldsymbol{U}}=0 \quad\Longrightarrow\quad \frac{\delta(\mathcal{H}-\mathcal{C})}{\delta\boldsymbol{u}}\bigg|_{\boldsymbol{u}=\boldsymbol{U}}=0$$

for some Casimir $\mathcal{C}$. Equivalently, we see that steady solutions of non-canonical Hamiltonian systems are *constrained* conditional extrema of the Hamiltonian, where the perturbations are constrained to lie in the hypersurface defined by constancy of $\mathcal{C}$. This hypersurface lies within the symplectic leaf corresponding to the full class of Casimir invariants. The functional $\mathcal{C}$ is seen to play the role of a Lagrange multiplier.

It is instructive to consider the classical example of a rigid body (Goldstein 1980, Chapter 5; Arnol'd 1989, Chapter 6). The canonical representation involves a 12-dimensional phase space consisting of the three Cartesian positions of the centre-of-mass, the three Euler angles, and their associated momenta. However, it is well known that the force-free rigid-body dynamics may be expressed in a *reduced* three-dimensional phase space and described by *Euler's equations*:

$$\frac{dm_1}{dt}=\left(\frac{\mathrm{I}_2-\mathrm{I}_3}{\mathrm{I}_2\,\mathrm{I}_3}\right)m_2m_3, \quad \text{etc.}, \tag{22}$$

where m_i are the components of the angular momentum about principal axes, and I_i are the moments of inertia. The dynamics (22) can be represented in the symplectic form (1) if we take

$$\boldsymbol{u}=(m_1,m_2,m_3)^T, \qquad J=\begin{pmatrix} 0 & m_3 & -m_2 \\ -m_3 & 0 & m_1 \\ m_2 & -m_1 & 0 \end{pmatrix}, \tag{23}$$

together with the Hamiltonian

$$\mathcal{H}(\boldsymbol{m}) = \frac{1}{2}\left(\frac{m_1^2}{\mathrm{I}_1} + \frac{m_2^2}{\mathrm{I}_2} + \frac{m_3^2}{\mathrm{I}_3}\right), \tag{24}$$

after using the fact that

$$\frac{\delta\mathcal{H}}{\delta m_i} = \frac{m_i}{\mathrm{I}_i}. \tag{25}$$

This reduced representation is evidently non-canonical, because J is non-invertible (i.e. $\det J = 0$).

For the rigid body, the angular momentum $\mathcal{L} = \boldsymbol{m}\cdot\boldsymbol{m}$ is a constant of the motion. In the canonical representation, it would be associated with an explicit (rotational) symmetry in the usual way. But in the reduced representation, angles have been eliminated and the rotational symmetry is *invisible*; so where has the invariant gone? The answer is that it has become a Casimir invariant. This may be immediately verified by noting that

$$\frac{\delta\mathcal{L}}{\delta\boldsymbol{m}} = 2\boldsymbol{m}, \qquad J\frac{\delta\mathcal{L}}{\delta\boldsymbol{m}} = 0. \tag{26}$$

From the earlier discussion we expect that steady states of the reduced rigid-body dynamics will be conditional extrema not of $\mathcal{H}$, but rather of some combination $\mathcal{H}-\mathcal{C}$. This may be verified explicitly. For example, if the moments of inertia I_i are all distinct, then it is easy to see from (22) that steady solutions $\boldsymbol{m} = \boldsymbol{M}$ are of the form

$$\boldsymbol{M} = (\pm\ell, 0, 0)^T, \quad (0, \pm\ell, 0)^T, \quad (0, 0, \pm\ell)^T,$$

where $\ell = \sqrt{\mathcal{L}}$. Taking $\boldsymbol{M} = (\ell, 0, 0)^T$, note from (25) that

$$\left.\frac{\delta\mathcal{H}}{\delta\boldsymbol{m}}\right|_{\boldsymbol{m}=\boldsymbol{M}} = (\frac{\ell}{\mathrm{I}_1}, 0, 0)^T \neq 0;$$

but if we choose the Casimir

$$\mathcal{C} = \frac{\mathcal{L}}{2\mathrm{I}_1} = \frac{\boldsymbol{m}\cdot\boldsymbol{m}}{2\mathrm{I}_1}$$

then

$$\left.\frac{\delta(\mathcal{H}-\mathcal{C})}{\delta\boldsymbol{m}}\right|_{\boldsymbol{m}=\boldsymbol{M}} = 0.$$

That is, steady solutions of Euler's equations are indeed constrained conditional extrema of the Hamiltonian, as expected.

5. Steady Solutions of the Euler Equations

For the three-dimensional Euler equations, the *helicity*

$$\mathcal{C} = \iiint_D \boldsymbol{v} \cdot \boldsymbol{\omega} \, dxdydz, \tag{27}$$

which characterizes the knottedness of the vortex topology (Moffatt 1969; Arnol'd 1974; Kuznetsov & Mikhailov 1980), may be easily verified to be a Casimir invariant from (11) together with the fact that

$$\frac{\delta \mathcal{C}}{\delta \boldsymbol{\omega}} = 2\boldsymbol{v}. \tag{28}$$

It is sometimes referred to as a *Hopf invariant* (Arnol'd 1974). The invariance of the helicity is intimately related to the "frozen-in" character of vortex lines (the three-dimensional form of Kelvin's circulation theorem).

Arnol'd (1966*a*) defined "isovortical" perturbations as those preserving the circulation around all material contours, and showed that steady solutions of the Euler equations were conditional extrema of the Hamiltonian under such isovortical perturbations. To see this, note from (9) that a steady solution must satisfy

$$\nabla \times (\boldsymbol{v} \times \boldsymbol{\omega}) = 0 \qquad \Longleftrightarrow \qquad \boldsymbol{v} \times \boldsymbol{\omega} = \nabla \beta \tag{29}$$

for some scalar field β, while a general isovortical perturbation may be written in the form

$$\delta \boldsymbol{\omega} = \varepsilon \nabla \times (\boldsymbol{f} \times \boldsymbol{\omega}), \tag{30}$$

where ε is an infinitesimal parameter and $\boldsymbol{f}$ is some vector field satisfying

$$\nabla \cdot \boldsymbol{f} = 0 \quad \text{in } D, \qquad \boldsymbol{f} \cdot \hat{\boldsymbol{n}} = 0 \quad \text{on } \partial D, \tag{31}$$

but otherwise arbitrary. By considering the true evolution equation (9) under the constraints (8), $\boldsymbol{f}$ may be seen as some "pseudoadvective" vector field that is guaranteed to maintain the frozen-in character of the vortex lines. Under a perturbation of the form (30), the Hamiltonian varies according to

$$\begin{aligned} \delta \mathcal{H} &= -\iiint_D \boldsymbol{\psi} \cdot \delta \boldsymbol{\omega} \, dxdydz = -\varepsilon \iiint_D \boldsymbol{\psi} \cdot \nabla \times (\boldsymbol{f} \times \boldsymbol{\omega}) \, dxdydz \\ &= -\varepsilon \iiint_D \boldsymbol{f} \cdot (\boldsymbol{v} \times \boldsymbol{\omega}) \, dxdydz = -\varepsilon \iiint_D \boldsymbol{f} \cdot \nabla \beta \, dxdydz = 0, \end{aligned}$$

using (29) and (31). This proves the result.

Now in principle we should be able, for any steady solution of the Euler equations, to find a Casimir $\mathcal{C}$ such that (21) is satisfied (as in the rigid-body example). However, in the three-dimensional case this turns out to be less than straightforward. Taking the helicity (27), which is – as of the present time – the only known Casimir in the Eulerian representation, then substituting (14) and (28) into (21) (allowing a multiplicative factor in front of the Casimir) yields

$$\boldsymbol{\psi} = -\mu\,\boldsymbol{v} \qquad \Longrightarrow \qquad \boldsymbol{v} = \mu\boldsymbol{\omega} \tag{32}$$

for some constant μ; this condition describes *Beltrami flows*, including the well-known *ABC flows*. However, the condition (32) certainly does not describe all possible steady flows of the Euler equations. Abarbanel & Holm (1987) have shown that by extending the phase space to include Lagrangian (particle) labels, the complete set of Casimir invariants may be explicitly represented and condition (21) may be shown to hold for all steady Euler flows. This can be seen as follows. Let the field $\lambda(\boldsymbol{x}, t)$ represent particle labels, so

$$\frac{\partial\lambda}{\partial t} = -\boldsymbol{v}\cdot\boldsymbol{\nabla}\lambda. \tag{33}$$

The evolution equations (9) and (33) may be represented together in the symplectic form (1) if we take

$$\boldsymbol{u} = (\boldsymbol{\omega}, \lambda)^T, \qquad J = \begin{pmatrix} -\boldsymbol{\nabla}\times(\boldsymbol{\omega}\times\boldsymbol{\nabla}\times(\cdot)) & \boldsymbol{\nabla}\times((\cdot)\boldsymbol{\nabla}\lambda) \\ -\boldsymbol{\nabla}\lambda\cdot\boldsymbol{\nabla}\times(\cdot) & 0 \end{pmatrix}, \tag{34}$$

noting that

$$\frac{\delta\mathcal{H}}{\delta\lambda} = 0. \tag{35}$$

It follows from (9) and (33) that the quantity $q = \boldsymbol{\omega}\cdot\boldsymbol{\nabla}\lambda$ is a Lagrangian invariant:

$$\frac{\partial q}{\partial t} = -\boldsymbol{v}\cdot\boldsymbol{\nabla}q. \tag{36}$$

If we introduce the class of invariant functionals

$$\mathcal{C} = \iiint_D C(q)\, dxdydz, \tag{37}$$

where $C(\cdot)$ is an arbitrary function, then we have

$$\begin{aligned} \delta\mathcal{C} &= \iiint_D C'(q)\,\delta q\, dxdydz = \iiint_D C'(q)(\delta\boldsymbol{\omega}\cdot\boldsymbol{\nabla}\lambda + \boldsymbol{\omega}\cdot\boldsymbol{\nabla}\delta\lambda)\, dxdydz \\ &= \iiint_D \left[C'(q)\boldsymbol{\nabla}\lambda\cdot\delta\boldsymbol{\omega} - (\boldsymbol{\omega}\cdot\boldsymbol{\nabla}C'(q))\delta\lambda\right] dxdydz, \end{aligned}$$

whence

$$\frac{\delta\mathcal{C}}{\delta\boldsymbol{\omega}} = C'(q)\,\boldsymbol{\nabla}\lambda, \qquad \frac{\delta\mathcal{C}}{\delta\lambda} = -\boldsymbol{\omega}\cdot\boldsymbol{\nabla}C'(q). \tag{38}$$

Using (38) and (34), it may be verified that the functionals (37) satisfy (4) and thus are indeed Casimir invariants.

If we now impose the condition (21) using the class of Casimirs described by (37), then using (14), (35) and (38) we have the two conditions

$$-\boldsymbol{\psi} = C'(q)\,\nabla\lambda, \qquad \boldsymbol{\omega}\cdot\nabla C'(q) = 0. \tag{39,40}$$

It is convenient to set $g(q) = C'(q)$. Taking the curl of (39) yields

$$\boldsymbol{v} = \nabla\times(g\nabla\lambda) = \nabla g\times\nabla\lambda, \tag{41}$$

whence

$$\boldsymbol{v}\times\boldsymbol{\omega} = (\boldsymbol{\omega}\cdot\nabla g)\,\nabla\lambda - (\boldsymbol{\omega}\cdot\nabla\lambda)\,\nabla g. \tag{42}$$

But (40) implies that the first term on the right-hand side of (42) vanishes, while we may use the definition of q to simplify the second term, leaving

$$\boldsymbol{v}\times\boldsymbol{\omega} = -q\,\nabla g(q) = \nabla\beta(q), \qquad \text{with} \qquad \beta = -\int q\,g'(q)\,dq, \tag{43}$$

which is identical to the condition (29). This then demonstrates the point that all steady flows of the three-dimensional Euler equations are constrained conditional extrema of the Hamiltonian, *provided* the class of Casimir invariants is given by (37). However these Casimirs involve Lagrangian information. The issue of whether these Casimirs may, like the helicity, be expressible in some other way in terms of volume integrals involving the Eulerian field variables alone, remains an interesting open question.

For the two-dimensional Euler equations, the situation is far more satisfactory. The Casimir invariants is this case are of the form

$$\mathcal{C} = \iint_D C(\omega)\,dxdy, \tag{44}$$

where $C(\cdot)$ is an arbitrary function. Their invariance is intimately related to the material-conservation property of vorticity elements (the two-dimensional form of Kelvin's circulation theorem). Once again, steady solutions of the Euler equations are conditional extrema of the Hamiltonian under isovortical perturbations (Arnol'd 1966*a*) – which here are smooth rearrangements of the vorticity field.

To see this directly from the symplectic representation, note that

$$\frac{\delta\mathcal{H}}{\delta\omega} = -\psi, \qquad \frac{\delta\mathcal{C}}{\delta\omega} = C'(\omega), \tag{45}$$

so a flow will be an energy extremum under isovortical perturbations provided $\psi = -C'(\omega)$, whence $\partial(\psi,\omega) = 0$. It is evident from the evolution equation (17) that this condition encompasses all two-dimensional steady Euler flows, so the set of Casimirs (44) is seen to be complete.

Note that the family of Casimirs (44) imposes a far more powerful constraint on the dynamics of two-dimensional flow than does the helicity Casimir (27), or even the family of Casimirs (37), on three-dimensional flow; examples of this include the inverse energy cascade in two-dimensional turbulence (e.g. Lesieur 1990), and powerful nonlinear stability theorems (Arnol'd 1966*c*), to be discussed further below.

6. Symmetries and Conservation Laws

One of the most important results in theoretical mechanics is *Noether's theorem*, which links symmetries and conservation laws. It may be understood as follows. Define a one-parameter family of infinitesimal variations $\delta_{\mathcal{F}}\boldsymbol{u}$ generated by a functional $\mathcal{F}$ (e.g. Goldstein 1980) by

$$\delta_{\mathcal{F}}\boldsymbol{u} \equiv \varepsilon J \frac{\delta \mathcal{F}}{\delta \boldsymbol{u}} \tag{46}$$

(ε being the infinitesimal parameter). Then the change in another functional $\mathcal{G}$ induced by this variation is given by

$$\Delta_{\mathcal{F}}\mathcal{G} \equiv \mathcal{G}[\boldsymbol{u} + \delta_{\mathcal{F}}\boldsymbol{u}] - \mathcal{G}[\boldsymbol{u}] = \left(\frac{\delta \mathcal{G}}{\delta \boldsymbol{u}}, \delta_{\mathcal{F}}\boldsymbol{u}\right) + O((\delta_{\mathcal{F}}\boldsymbol{u})^2) = \varepsilon[\mathcal{G}, \mathcal{F}] + O(\varepsilon^2),$$

by the definition of the functional derivative. This yields *Noether's theorem*: The Hamiltonian is invariant under the infinitesimal variations generated by a functional $\mathcal{F}$, in the sense that $\Delta_{\mathcal{F}}\mathcal{H} = 0$, if and only if $\mathcal{F}$ is a constant of the motion.

From (4) and (46) it is evident that Casimir invariants correspond to an "invisible symmetry":

$$\delta_{\mathcal{C}}\boldsymbol{u} \equiv 0. \tag{47}$$

In the case of the Euler equations of fluid motion, the symmetry is invisible within the Eulerian (as opposed to Lagrangian) representation, and is sometimes referred to as the *particle-relabelling symmetry* (Ripa 1981; Salmon 1982; Marsden & Weinstein 1983).

It is straightforward to verify that the conserved functional associated with invariance of the Hamiltonian under translations in time is, as usual, just (the negative of) the Hamiltonian itself. In the same way, if the Hamiltonian is invariant under translations in the spatial coordinate x_j, say, then the associated conserved functional $\mathcal{M}$ is defined by

$$\delta_{\mathcal{M}}\boldsymbol{u} = -\varepsilon \frac{\partial \boldsymbol{u}}{\partial x_j} \qquad \Longleftrightarrow \qquad J\frac{\delta \mathcal{M}}{\delta \boldsymbol{u}} = -\frac{\partial \boldsymbol{u}}{\partial x_j}. \tag{48}$$

For example, in the Euler equations the x-component of momentum (or impulse), $\mathcal{M}$, satisfies

$$\frac{\delta\mathcal{M}}{\delta\boldsymbol{\omega}} = -\tfrac{1}{2}\boldsymbol{x}\times\hat{\boldsymbol{x}}, \qquad \nabla\times\frac{\delta\mathcal{M}}{\delta\boldsymbol{\omega}} = \hat{\boldsymbol{x}} \tag{49}$$

(note that this works for the extended representation (34) as well as for (11)), whence

$$\mathcal{M} = \iiint_D -\tfrac{1}{2}(\boldsymbol{x}\times\hat{\boldsymbol{x}})\cdot\boldsymbol{\omega}\,dxdydz = \iiint_D \boldsymbol{v}\cdot\hat{\boldsymbol{x}}\,dxdydz, \tag{50}$$

as expected. Similar expressions hold for the other components. In the two-dimensional case we obtain the well-known Kelvin's-impulse invariants

$$\iint_D y\omega\,dxdy = \iint_D \boldsymbol{v}\cdot\hat{\boldsymbol{x}}\,dxdy, \tag{51}$$

$$-\iint_D x\omega\,dxdy = \iint_D \boldsymbol{v}\cdot\hat{\boldsymbol{y}}\,dxdy, \tag{52}$$

associated respectively with symmetry in x and y, as may be verified using (18) and (48). In the same fashion, the angular momentum invariant

$$\iint_D -\tfrac{1}{2}(x^2+y^2)\omega\,dxdy = \iint_D \hat{\boldsymbol{z}}\cdot(\boldsymbol{x}\times\boldsymbol{v})\,dxdy \tag{53}$$

is associated with rotational symmetry. Note that the impulse form (i.e. that involving the vorticity) of the above invariants is more fundamental; the momentum form (i.e. that involving the velocity) is only valid when contributions at infinity vanish.

Just as steady solutions are constrained conditional extrema of $\mathcal{H}$, solutions that are invariant under translations in x_j are necessarily constrained conditional extrema of $\mathcal{M}$. This is immediately evident from (48), which leads to the causal chain

$$\frac{\partial\boldsymbol{U}}{\partial x_j} = 0 \quad\Longrightarrow\quad J\frac{\delta\mathcal{M}}{\delta\boldsymbol{u}}\bigg|_{\boldsymbol{u}=\boldsymbol{U}} = 0 \quad\Longrightarrow\quad \frac{\delta(\mathcal{M}-\mathcal{C})}{\delta\boldsymbol{u}}\bigg|_{\boldsymbol{u}=\boldsymbol{U}} = 0 \tag{54}$$

for some Casimir $\mathcal{C}$. In the case of three-dimensional Euler flow, for example, substituting (38) and (49) into (54) yields the two conditions

$$-\tfrac{1}{2}\boldsymbol{x}\times\hat{\boldsymbol{x}} = C'(q)\,\nabla\lambda, \qquad \boldsymbol{\omega}\cdot\nabla C'(q) = 0. \tag{55, 56}$$

Proceeding as with the steady-flow example, we set $g = C'(q)$ and take the curl of (55). This gives

$$\hat{\boldsymbol{x}} = \nabla\times(g\nabla\lambda) = \nabla g\times\nabla\lambda, \tag{57}$$

whence

$$\boldsymbol{\omega} \times \hat{\boldsymbol{x}} = (\boldsymbol{\omega} \cdot \nabla\lambda)\,\nabla g - (\boldsymbol{\omega} \cdot \nabla g)\,\nabla\lambda = q\,\nabla g(q) = -\nabla\beta(q), \tag{58}$$

after using (56) as well as the definition of g, where β is given by (43). It follows from (58) that $\boldsymbol{\omega}$ must be of the form

$$\boldsymbol{\omega} = \Big(\Omega, \frac{\partial\beta}{\partial z}, -\frac{\partial\beta}{\partial y}\Big),$$

with $\partial\beta/\partial x = 0$ since $\hat{\boldsymbol{x}} \cdot \nabla\beta = 0$, and $\partial\Omega/\partial x = 0$ since $\nabla \cdot \boldsymbol{\omega} = 0$. We may identify β with the x-component of velocity and Ω with the x-component of vorticity, and conclude that all fields are independent of x.

For two-dimensional Euler flow, (51) implies $\delta\mathcal{M}/\delta\omega = y$; using this together with (45) in (54) yields $C'(\omega) = y$, whence $\omega = \omega(y)$, as expected.

In exactly the same way, *steadily-translating* solutions may be classified as conditional extrema (cf. Benjamin 1984). To see this, let $\boldsymbol{u} = \boldsymbol{U}(x - ct, y, z)$, namely a solution translating in the x-direction at a speed c. We then have

$$\frac{\partial \boldsymbol{U}}{\partial t} = -c\frac{\partial \boldsymbol{U}}{\partial x} \qquad \Longleftrightarrow \qquad J\frac{\delta\mathcal{H}}{\delta\boldsymbol{u}}\bigg|_{\boldsymbol{u}=\boldsymbol{U}} = cJ\frac{\delta\mathcal{M}}{\delta\boldsymbol{u}}\bigg|_{\boldsymbol{u}=\boldsymbol{U}}$$

after using (1) and (48), whence

$$J\frac{\delta(\mathcal{H} - c\mathcal{M})}{\delta\boldsymbol{u}}\bigg|_{\boldsymbol{u}=\boldsymbol{U}} = 0 \qquad \Longleftrightarrow \qquad \frac{\delta(\mathcal{H} - c\mathcal{M} - \mathcal{C})}{\delta\boldsymbol{u}}\bigg|_{\boldsymbol{u}=\boldsymbol{U}} = 0$$

for some Casimir $\mathcal{C}$. Thus steadily-translating solutions are seen to be constrained conditional extrema of $\mathcal{H} - c\mathcal{M}$, where c is the translation speed. Applying this to the case of two-dimensional Euler flow, for example, we obtain

$$\psi + c\,y = -C'(\omega) \qquad \Longleftrightarrow \qquad \partial(\psi + c\,y, \omega) = 0. \tag{59}$$

Condition (59) describes flow patterns brought to rest by a constant flow $\boldsymbol{V} = -c\,\hat{\boldsymbol{x}}$, equivalently flow patterns translating with velocity $c\,\hat{\boldsymbol{x}}$.

7. Nonlinear Stability

We noted that a steady solution $\boldsymbol{U}$ of a *non-canonical* Hamiltonian system is a *constrained* conditional extremum of $\mathcal{H}$:

$$\frac{\partial \boldsymbol{U}}{\partial t} = 0 \quad \Longrightarrow \quad J\frac{\delta\mathcal{H}}{\delta\boldsymbol{u}}\bigg|_{\boldsymbol{u}=\boldsymbol{U}} = 0 \quad \Longrightarrow \quad \frac{\delta(\mathcal{H} - \mathcal{C})}{\delta\boldsymbol{u}}\bigg|_{\boldsymbol{u}=\boldsymbol{U}} = 0$$

for some Casimir $\mathcal{C}$. One may then investigate stability of the equilibrium by testing the sign-definiteness of $\delta^2(\mathcal{H} - \mathcal{C})$.

For finite-dimensional dynamical systems, where all norms are equivalent, this so-called *formal stability*, viz.

$$\delta(\mathcal{H}-\mathcal{C})=0 \qquad \text{and} \qquad \delta^2(\mathcal{H}-\mathcal{C})\neq 0 \qquad \forall\,\delta\boldsymbol{u} \tag{60}$$

implies normed stability to finite-amplitude disturbances, also known as *Liapunov stability*, viz.

$$\forall\,\varepsilon>0,\ \exists\,\delta>0 \ \text{ such that } \ \|\delta\boldsymbol{u}(0)\|<\delta \implies \|\delta\boldsymbol{u}(t)\|<\varepsilon \ \ \forall\,t,$$

where $\delta\boldsymbol{u}$ is the (finite-amplitude) disturbance. That is, small disturbances stay small for all time.

For infinite-dimensional systems (such as fluids), on the other hand, more work is required. In addition to (60), one must be able to find some disturbance norm $\|\cdot\|$ and *a priori* constants $c_1>0$, $c_2>0$ such that

$$c_1\|\delta\boldsymbol{u}\|<(\mathcal{H}-\mathcal{C})[\boldsymbol{u}]<c_2\|\delta\boldsymbol{u}\| \tag{61}$$

for arbitrary disturbances $\delta\boldsymbol{u}$. It then follows from (61) that

$$\|\delta\boldsymbol{u}(t)\|<c_1^{-1}(\mathcal{H}-\mathcal{C})(t)=c_1^{-1}(\mathcal{H}-\mathcal{C})(0)<c_1^{-1}c_2\|\delta\boldsymbol{u}(0)\|, \tag{62}$$

which establishes Liapunov stability in the norm $\|\cdot\|$. This is what is referred to as *Arnol'd's stability method* (Arnol'd 1966*c*; Holm *et al.* 1985).

From the considerations of the previous section, it is evident that Arnol'd's method need not be restricted to using the invariant $(\mathcal{H}-\mathcal{C})$. If the steady solution is independent of the coordinate x_j, for example, then stability may be provable using the combined invariant $(\mathcal{H}-\lambda\mathcal{M}-\mathcal{C})$ for some constant λ (Ripa 1983), or even just $(\mathcal{M}-\mathcal{C})$. In fact the most celebrated hydrodynamical stability theorems are usually of this latter type.

Many of the classical hydrodynamical stability theorems turn out to be but linear versions of fully nonlinear stability theorems which are obtainable using Arnol'd's method. These include Fjørtoft's theorem (Arnol'd 1966*c*) and Rayleigh's inflection-point theorem (Vladimirov 1986; McIntyre & Shepherd 1987; Carnevale & Shepherd 1990) for two-dimensional flow; Rayleigh's centrifugal-stability theorem for axisymmetric flow (Vladimirov 1986; Szeri & Holmes 1988; Shepherd 1991*a*); the Charney-Stern theorem for quasi-geostrophic baroclinic flow (Swaters 1986; McIntyre & Shepherd 1987; Shepherd 1989); stability of stably stratified flow at rest (Vladimirov 1987; Shepherd 1991*b*); and symmetric stability of rotating, stratified flow (Cho, Shepherd & Vladimirov 1991). A critical discussion of some of the successes and failures of Arnol'd's method is presented in Shepherd (1991*b*).

It has to be said that the power of Arnol'd's stability method depends on the richness of the Casimir invariants (cf. Holm *et al.* 1985). In the case of three-dimensional Euler flow, no stability theorem has yet been found

(Abarbanel & Holm 1987). But in the case of two-dimensional flow, the Casimir invariants exert a sufficiently strong constraint on the dynamics that Arnol'd's stability theorems can be used to determine rigorous constraints on the statistical-dynamical evolution of fully nonlinear flow. Examples include the development of an Eulerian statistical theory of fluctuations about stable states (Carnevale & Frederiksen 1987); the demonstration of the non-ergodicity of β-plane turbulence (Shepherd 1987); and rigorous saturation bounds on instabilities (Shepherd 1988, 1991*a*).

8. Extremization Algorithms

Given a Hamiltonian dynamical system in the symplectic form (1), consider the new dynamical system defined by (Shepherd 1990*b*)

$$\frac{\partial \boldsymbol{u}}{\partial t} = J\frac{\delta \mathcal{H}}{\delta \boldsymbol{u}} + J\alpha J\frac{\delta \mathcal{H}}{\delta \boldsymbol{u}}, \tag{63}$$

where α is a symmetric transformation with the inner product $(\boldsymbol{u}, \alpha\boldsymbol{u})$ of definite sign for all $\boldsymbol{u}$. The system (63) has the same Casimir invariants as the original system (1):

$$\frac{dC}{dt} = \left(\frac{\delta C}{\delta \boldsymbol{u}}, \frac{\partial \boldsymbol{u}}{\partial t}\right) = \left(\frac{\delta C}{\delta \boldsymbol{u}}, J\frac{\delta \mathcal{H}}{\delta \boldsymbol{u}} + J\alpha J\frac{\delta \mathcal{H}}{\delta \boldsymbol{u}}\right) = -\left(J\frac{\delta C}{\delta \boldsymbol{u}}, \frac{\delta \mathcal{H}}{\delta \boldsymbol{u}} + \alpha J\frac{\delta \mathcal{H}}{\delta \boldsymbol{u}}\right) = 0.$$

However the Hamiltonian functional $\mathcal{H}$ is no longer invariant, but will monotonically increase or decrease according to the sign of α:

$$\frac{d\mathcal{H}}{dt} = \left(\frac{\delta \mathcal{H}}{\delta \boldsymbol{u}}, \frac{\partial \boldsymbol{u}}{\partial t}\right) = \left(\frac{\delta \mathcal{H}}{\delta \boldsymbol{u}}, J\frac{\delta \mathcal{H}}{\delta \boldsymbol{u}} + J\alpha J\frac{\delta \mathcal{H}}{\delta \boldsymbol{u}}\right) = -\left(J\frac{\delta \mathcal{H}}{\delta \boldsymbol{u}}, \alpha J\frac{\delta \mathcal{H}}{\delta \boldsymbol{u}}\right). \tag{64}$$

The right-hand side of (64) is of definite sign, and is non-zero unless

$$J\frac{\delta \mathcal{H}}{\delta \boldsymbol{u}} = 0.$$

Hence steady solutions of the original dynamics (1) are also steady solutions of the "modified dynamics" (63). Furthermore, they will (by construction) be constrained *true* extrema of $\mathcal{H}$, and will thus be stable. The algorithm (63) is a generalization of the method proposed by Vallis, Carnevale & Young (1989) for certain fluid equations, but may be applied to *any* Hamiltonian system.

In the same way, the modified system

$$\frac{\partial \boldsymbol{u}}{\partial t} = J\frac{\delta(\mathcal{H} - c\mathcal{M})}{\delta \boldsymbol{u}} + J\alpha J\frac{\delta(\mathcal{H} - c\mathcal{M})}{\delta \boldsymbol{u}} \tag{65}$$

can be expected to approach stable, steadily-translating solutions.

In the case of the three-dimensional Euler equations, for example, such a prescription gives a modified dynamics of the form

$$\frac{\partial \boldsymbol{\omega}}{\partial t} = \boldsymbol{\nabla} \times (\tilde{\boldsymbol{v}} \times \boldsymbol{\omega}), \tag{66}$$

with

$$\tilde{\boldsymbol{v}} \equiv \boldsymbol{V} + \alpha \boldsymbol{\nabla} \times \boldsymbol{\nabla} \times (\boldsymbol{V} \times \boldsymbol{\omega}), \qquad \boldsymbol{V} \equiv \boldsymbol{v} - c\hat{\boldsymbol{x}}.$$

The dynamics is seen to be "pseudo-advective". In principle this scheme with $\alpha > 0$ would enable one to find energy minima for a given vortex topology. It bears a certain resemblance to the extremization algorithm of Moffatt (1985), which approaches energy minima for a given *streamline* topology. Unfortunately, there seems to be no way of knowing *a priori* that non-trivial constrained energy minima exist. To this point the algorithm (66) has yet to be used successfully.

As in so many other ways, the situation with two-dimensional flow is more satisfactory. In this case one can easily show, using a Poincaré inequality, that non-trivial energy *maxima* will exist for bounded or periodic domains. Carnevale & Vallis (1990) have applied modified dynamics in this case, and have demonstrated that non-trivial stable steady states appear that have strongly nonlinear $\psi(\omega)$ functional relationships. This indicates that the so-called "higher-order" vorticity invariants associated with the full family of Casimirs (44), and not just the enstrophy (the special case $C(\omega) = \omega^2$), matter in the imposition of the constraint. This is important because the numerical model used to evolve the modified dynamics is only guaranteed to conserve the enstrophy; evidently the quasi-conservation of vorticity following its pseudo-advective evolution is adequate for this purpose.

Acknowledgements

The author's research is supported by the Natural Sciences and Engineering Research Council and the Atmospheric Environment Service of Canada.

References

ABARBANEL, H.D.I. & HOLM, D.D., 1987 Nonlinear stability analysis of inviscid flows in three dimensions: incompressible fluids and barotropic fluids. *Phys.Fluids*, **30**, pp. 3369–3382.

ABRAHAM, R. & MARSDEN, J.E., 1978 *Foundations of Mechanics*, 2nd edn. Benjamin/Cummings, 806 pp.

ARNOL'D, V.I., 1966*a* Sur un principe variationnel pour les écoulements stationnaires des liquides parfaits et ses applications aux problèmes de stabilité non linéaires. *J.Méc.*, **5**, pp. 29–43.

ARNOL'D, V.I., 1966*b* Sur la géométrie différentielle des groupes de Lie de dimension infinie et ses applications à l'hydrodynamique des fluides parfaits. *Ann.Inst.Fourier (Grenoble)*, **16**, pp. 319–361.

ARNOL'D, V.I., 1966*c* On an a priori estimate in the theory of hydrodynamical stability, *Izv. Vyssh. Uchebn. Zaved. Matematika*, **54**, no.5, pp. 3–5. (English transl.: *Amer. Math. Soc. Transl., Series 2*, **79**, pp. 267–269 (1969).)

ARNOL'D, V.I., 1969 The Hamiltonian nature of the Euler equations in the dynamics of a rigid body and of a perfect fluid. *Usp. Mat. Nauk.*, **24**, no. 3, pp. 225–226. (In Russian.)

ARNOL'D, V.I., 1974 The asymptotic Hopf invariant and its applications. *Proceedings of Summer School in Differential Equations, 1973, Erevan.* (English transl.: *Sel. Math. Sov.*, **5**, pp. 327–345 (1986).)

ARNOL'D, V.I., 1989 *Mathematical Methods of Classical Mechanics*, 2nd edn. Springer-Verlag, 511 pp.

ARNOLD, V.I. & KHESIN, B., 1992 Topological methods in hydrodynamics. *Ann. Rev. Fluid Mech.*, **24**, pp. 145–166.

BENJAMIN, T.B., 1984 Impulse, flow force and variational principles. *IMA J. Appl. Math.*, **32**, pp. 3-68.

CARNEVALE, G.F. & FREDERIKSEN, J.S., 1987 Nonlinear stability and statistical mechanics of flow over topography. *J. Fluid Mech.*, **175**, pp. 157–181.

CARNEVALE, G.F. & SHEPHERD, T.G., 1990 On the interpretation of Andrews' theorem. *Geophys. Astrophys. Fluid Dyn.*, **51**, pp. 1–17.

CARNEVALE, G.F. & VALLIS, G.K., 1990 Pseudo-advective relaxation to stable states of inviscid two-dimensional fluids. *J. Fluid Mech.*, **213**, pp. 549–571.

CHO, H.-R., SHEPHERD, T.G. & VLADIMIROV, V.A., 1991 Application of the direct Liapunov method to the problem of symmetric stability in the atmosphere. *J. Atmos. Sci.*, submitted.

EBIN, D.G. & MARSDEN, J.E., 1970 Groups of diffeomorphisms and the motion of an incompressible fluid. *Ann. Math.*, **92**, pp. 102–163.

GOLDSTEIN, H., 1980 *Classical Mechanics*, 2nd edn. Addison-Wesley, 672 pp.

HOLM, D.D., MARSDEN, J.E., RATIU, T. & WEINSTEIN, A., 1985 Nonlinear stability of fluid and plasma equilibria. *Phys. Rep.*, **123**, pp. 1–116.

KUZNETSOV, E.A. & MIKHAILOV, A.V., 1980 On the topological meaning of canonical Clebsch variables. *Phys. Lett.*, **77A**, pp. 37–38.

LESIEUR, M., 1990 *Turbulence in Fluids*, 2nd revised edn. Kluwer Academic, 412 pp.

LITTLEJOHN, R.G., 1982 Singular Poisson tensors. In *Mathematical Methods in Hydrodynamics and Integrability in Dynamical Systems* (ed. M. Tabor & Y.M. Treve), Amer.Inst.Phys.Conf.Proc., vol.88, pp. 47-66.

MARSDEN, J.E. (ED.), 1984 *Fluids and Plasmas: Geometry and Dynamics.* Contemporary Mathematics, vol.28, American Mathematical Society.

MARSDEN, J.E. & WEINSTEIN, A., 1983 Coadjoint orbits, vortices, and Clebsch variables for incompressible fluids. *Physica*, **7D**, pp. 305–323.

MCINTYRE, M.E. & SHEPHERD, T.G., 1987 An exact local conservation theorem for finite-amplitude disturbances to non-parallel shear flows, with remarks on Hamiltonian structure and on Arnol'd's stability theorems. *J. Fluid Mech.*, **181**, pp. 527–565.

MOFFATT, H.K., 1969 The degree of knottedness of tangled vortex lines. *J. Fluid Mech.*, **35**, pp. 117-129.

MOFFATT, H.K., 1985 Magnetostatic equilibria and analogous Euler flows of arbitrarily complex topology. Part I. Fundamentals. *J. Fluid Mech.*, **159**, pp. 359–378.

MORRISON, P.J., 1982 Poisson brackets for fluids and plasmas. In *Mathematical Methods in Hydrodynamics and Integrability in Dynamical Systems* (ed. M. Tabor & Y.M. Treve), Amer.Inst.Phys.Conf.Proc., vol.88, pp.13-46.

OLVER, P.J., 1982 A nonlinear Hamiltonian structure for the Euler equations. *J. Math. Anal. Appl.*, **89**, pp. 233–250.

RIPA, P., 1981 Symmetries and conservation laws for internal gravity waves. In *Nonlinear Properties of Internal Waves* (ed. B.J.West), Amer.Inst.Phys.Conf.Proc., vol.76, pp. 281–306.

RIPA, P., 1983 General stability conditions for zonal flows in a one-layer model on the β-plane or the sphere. *J. Fluid Mech.*, **126**, pp. 463–489.

SALMON, R., 1982 Hamilton's principle and Ertel's theorem. In *Mathematical Methods in Hydrodynamics and Integrability in Dynamical Systems* (ed. M. Tabor & Y.M. Treve), Amer.Inst.Phys.Conf.Proc., vol.88, pp. 127–135.

SHEPHERD, T.G., 1987 Non-ergodicity of inviscid two-dimensional flow on a beta-plane and on the surface of a rotating sphere. *J.Fluid Mech.*, **184**, pp. 289–302.

SHEPHERD, T.G., 1988 Rigorous bounds on the nonlinear saturation of instabilities to parallel shear flows. *J.Fluid Mech.*, **196**, pp. 291–322.

SHEPHERD, T.G., 1989 Nonlinear saturation of baroclinic instability. Part II: Continuously-stratified fluid. *J.Atmos.Sci.*, **46**, pp. 888-907.

SHEPHERD, T.G., 1990*a* Symmetries, conservation laws, and Hamiltonian structure in geophysical fluid dynamics. *Adv.Geophys.*, **32**, pp. 287–338.

SHEPHERD, T.G., 1990*b* A general method for finding extremal states of Hamiltonian dynamical systems, with applications to perfect fluids. *J.Fluid Mech.*, **213**, pp. 573–587.

SHEPHERD, T.G., 1991*a* Nonlinear stability and the saturation of instabilities to axisymmetric vortices. *Eur.J.Mech.B/Fluids*, **10**, N° 2–Suppl., pp. 93–98.

SHEPHERD, T.G., 1991*b* Arnol'd stability applied to fluid flow: successes and failures. In *Nonlinear Phenomena in Atmospheric and Oceanic Sciences* (ed. G.F. Carnevale & R.T. Pierrehumbert), to appear. Springer-Verlag.

SWATERS, G.E., 1986 A nonlinear stability theorem for baroclinic quasi-geostrophic flow. *Phys.Fluids*, **29**, pp. 5–6.

SZERI, A. & HOLMES, P., 1988 Nonlinear stability of axisymmetric swirling flows. *Phil.Trans.R.Soc.Lond.A*, **326**, pp. 327–354.

TABOR, M., 1989 *Chaos and Integrability in Nonlinear Dynamics*. Wiley, 364 pp.

VALLIS, G.K., CARNEVALE, G.F. & YOUNG, W.R., 1989 Extremal energy properties and construction of stable solutions of the Euler equations. *J.Fluid Mech.*, **207**, pp. 133–152.

VLADIMIROV, V.A., 1986 Conditions for nonlinear stability of flows of an ideal incompressible liquid. *Zhurn.Prikl.Mekh.Tekh.Fiz.*, no.3, pp. 70–78. (English transl.: *J.Appl.Mech.Tech.Phys.*, **27**, no.3, pp. 382–389.)

VLADIMIROV, V.A., 1987 Integrals of two-dimensional motions of a perfect incompressible fluid of nonuniform density. *Izv.Akad.Nauk.SSSR, Mekh.Zhid.Gaza*, no.3, pp. 16–20. (English transl.: *Fluid Dynamics*, **22**, no.3, pp. 340–343.)

BLOW UP IN AXISYMMETRIC EULER FLOWS

A. PUMIR *, and E. D. SIGGIA
LASSP, Cornell University, Ithaca NY 14853-2501

ABSTRACT. The problem of development of singular solutions of the 3d Euler equations is considered in the particular case of flows with an axis of symmetry. There is a strong analogy between the physics of such flows, and Boussinesq convection in 2 dimension, the buoyancy been replaced by the centripetal acceleration. A hot bubble, initially at rest in cold fluid tends to rise. During the evolution, strong gradients of temperature develop on the cap of the bubble, while on the sides, vortex sheets tend to roll up. When the cap of the bubble starts folding, a rapid growth of the gradients was observed, suggesting a singularity of the vorticity in the 3d axisymmetric flows like $|\omega| \sim 1/(t^* - t)^2$. We have followed numerically this growth over an additional 5 orders of magnitude after this first instability, with no obvious impediment to further integration. Smaller and smaller scales are generated by repeated folding. Analytic estimates for the rate of stretching, consistent with our numerical observations are provided.

Generation of small scales of motion by a turbulent flow is a very characteristic feature of incompressible hydrodynamics at high Reynolds numbers. Understanding this important phenomenon has proven surprisingly difficult, and even very simple questions are still unanswered. As an example, it has been realized for years that vortex lines can be stretched by the flow. Because the stretching results from a nonlinear term, it could conceivably lead to finite time singularities, at least in the inviscid case.

Careful numerical studies have suggested that the vorticity grows exponentially, rather than algebraically (*i.e.*, catastrophically). This is the case for the Taylor-Green flow (Brachet *et al.*, 1983), and for certain class of antiparallel vortex tubes (Pumir and Siggia, 1990 and Meiron and Shelley, 1990), as well as for random initial conditions (Brachet *et al.*, 1991 and Kerr, 1991). These results show that the nonlinear term plays a rather subtle role. Of course, they clearly do not settle the problem of existence of finite time singularities for the 3d fluid equations (see also R. Kerr, these proceedings).

* Permanent address: LPS, Ecole Normale Supérieure, 75231 Paris, France

H. K. Moffatt et al. (eds.), Topological Aspects of the Dynamics of Fluids and Plasmas, 293–302.

In this paper we present some of our recent results, strongly suggesting the existence of axisymmetric solutions, blowing up in a finite time. This possibility has been suggested by Grauer and Sideris (1990). Our arguments in favor of a singularity are mainly numerical. It is also possible to obtain (nonrigorous) bounds for the rate of growth of the vorticity, useful to understand our numerical results. Technically, in the case of axisymmetric flows, one has to deal with an essentially 2 dimensional problem, which leads to obvious simplifications, and allow us to push our calculation much further than in the case of general 3d flows. Physically, a strong analogy exists between the 3d axisymmetric Euler equations and the 2d Boussinesq equations, as we will elaborate below. Although the two systems of equations have very similar structures, it has proven easier technically to study numerically the 2d Boussinesq equations. Our evidence for a finite singularity comes from our results on the Boussinesq problem. We expect them to hold also for a localized solution of the axisymmetric problem well away from the symmetry axis. A more detailed presentation of our results can be found elsewhere (Pumir and Siggia, 1991a).

An axisymmetric flow is defined by a velocity field independent of the azimuthal angle, ϕ around the symmetry axis, Oz. The velocity field, $\mathbf{u}(r,z,t)$ satisfies the 3d Euler equations :

$$\partial_t(ru_\phi) + (\mathbf{u}_{||}\cdot\nabla_{||})(ru_\phi) = 0 \tag{1}$$

$$\partial_t(\omega_\phi/r) + (\mathbf{u}_{||}\cdot\nabla_{||})(\omega_\phi/r) = -\frac{1}{r^4}\partial_z(ru_\phi)^2 \tag{2}$$

$$\frac{1}{r}\partial_r(ru_r) + \partial_z u_z = 0 \tag{3}$$

$$\omega_\phi = \partial_z u_r - \partial_r u_z \tag{4}$$

where $\mathbf{u}_{||} = (u_r, u_z)$ and $\nabla_{||} = (\partial_r, \partial_z)$.

Equation (1) expresses the conservation of circulation around a loop ($r = cst, z = cst.$), and equation (3) is the usual incompressibility condition. The mechanism of generation of azimuthal vorticity, as described from Eq. (2) is a consequence of the centripetal acceleration. It is very instructive to compare the system (1-4) with the 2-d Boussinesq equations with constant gravity, $\mathbf{g} = -g\hat{e}_y$:

$$\partial_t\theta + (\mathbf{u}\cdot\nabla)\theta = 0 \tag{5}$$

$$\partial_t\omega + (\mathbf{u}\cdot\nabla)\omega = \partial_x\theta \tag{6}$$

$$\partial_x u_x + \partial_y u_y = 0 \tag{7}$$

$$\omega = \partial_x u_y - \partial_y u_x \tag{8}$$

A comparison of the two systems of equations shows that the quantity $(ru_\phi)^2$, which is advected by the flow, (Eq(1)), plays the same role as the temperature as the source of vorticity (see Eq. (2) and (5)). The 'gravity' in the axisymmetric problem is directed radially outwards, and varies with the distance to the axis, r, as $\mathbf{g} = \hat{e}_r/r^4$. The velocity field (u_x, u_y) corresponds to (u_z, u_r). Rigorously, this analogy is valid only when the typical length scale of the solution is very much smaller than the distance to the axis, $\bar{r}$. In this case, one has the correspondence :

$$(\bar{r}\omega_r, \omega_\phi, \bar{r}\omega_z) \leftrightarrow \left(-\partial_x\theta^{1/2}, \omega, \partial_y\theta^{1/2}\right) \tag{9}$$

The analogy between 2d Boussinesq and the 3d axisymmetric problem is not useful to understand the development of singular solutions right at $r = 0$, a possibility we do not consider here.

We have simulated numerically both systems of equations. Finite difference techniques with adaptive mesh have been used. Because of the spontaneous appearance of very sharp gradients, it has been necessary to switch from centered differences to Total Variation Diminishing algorithms to control numerical instabilities (Osher and Chakravarthy, 1984 and for a more recent discussion, see Laney and Caughey (1991)). Our adaptive mesh technique is analogous to the one used in our previous 3d simulations [2]. To time step, say the Boussinesq equations, the streamfunction ψ, defined by $u_x = \partial_y\psi$ and $u_y = -\partial_x\psi$, was computed at each time step by solving the 2d Poisson equations : $\nabla^2\psi = \omega$. Efficient cyclic reductions methods (Swartzrauber, 1977) have been used. The time integration was done by a standard Runge-Kutta-Fehlberg algorithm.

Two extra simplifications are available for studying the 2d Boussinesq equation. First, one can continuously rescale space, time and vorticity in the following way : $X = x/a$, $\Omega = \omega a^{1/2}$, $\Psi = \psi/a^{3/2}$, $\theta = \Theta$ and $\partial_T t = a^{1/2}$. The equations then read :

$$\partial_T\Theta + \alpha\mathbf{X}\cdot\nabla_X\Theta + \frac{\partial(\Psi,\Omega)}{\partial(X,Y)} = 0 \tag{10}$$

$$\partial_T\Omega + \alpha\cdot\nabla_X\Omega + \frac{1}{2}\alpha\Omega + \frac{\partial(\Psi,\Omega)}{\partial(X,Y)} = \partial_X\Theta \tag{11}$$

with the definitions :

$$\nabla_X^2\Psi = \Omega,\ \alpha = -\partial_T lna \tag{12}$$

As it is the case for the nonlinear Schrodinger equation, continuously rescaling the solution is a very efficient way to maintain appropriate resolution while simulating a solution collapsing down to very small scales (LeMesurier *et al.*, 1988).

Also, in the Boussimesq case, it is possible to maintain the maximum gradient in the region where the resolution is highest by simply subtracting off the velocity of the point where the gradient is largest.

In the 3d axisymmetric case, the Galilean invariance and the possibility to continuously rescale the solution are lost. These properties make the numerical study of the Boussinesq equations much easier. In the following, we will restrict ourselves to the Boussinesq equations. Our numerical results on the axisymmetric problem can be found elsewhere (Pumir and Siggia, 1991a).

Although we have simulated a whole range of initial conditions, we will discuss here only the problem of a single bubble, initially at rest ($\omega = 0$ everywhere). More specifically, our initial condition was : $\theta(x, y) = (1 + 0.2y)/(1 + x^2 + y^2)^2$. This configuration is symmetric with respect to $x = 0$.

Qualitatively, hot regions tend to rise into colder regions. This situation is somewhat reminiscent of the 1d Burgers equation, where a similar mechanism leads to shocks in a finite time. Here of course, the incompressibility makes the problem much more subtle [1] The initial stages of the evolution lead to a region of sharp gradients ahead of the bubble, because of the tendency of hot regions to rise in colder regions. After awhile, a front separating a zone of high temperature underneath a zone of low temperature develops. In the process, fluid of intermediate temperature is expelled, leading to a jet on each side of the cap. This leads to the formation of arms of warm fluid, that eventually tend to roll-up.

It is well known that a front at rest separating hot fluid, underneath cold fluid is intrinsically unstable (Rayleigh-Taylor instability). In the case of a curved, rising front, the stretching induced by the motion of the fluid tends to strongly restabilize the instability, as it has been discovered in the case of curved flame fronts (Zel'dovitch *et al.*, 1980). The main physical consequence is that the interface separating hot and cold fluid remains stable for surprisingly long times. Quantitatively, an amplification of the gradient at the tip by a factor of 100 was observed before a destabilisation occurred. In this regime, a growth of the gradient slower than exponential was observed.

Once the instability develops, the growth of the gradients becomes very much faster. It can be shown in this regime, and over 5 orders of magnitude, that the gradient grows like $1/(t^* - t)^2$ (see Fig. 1). In the physical space, the evolution of the interface is characterized by a series of repeated foldings while the width of the interface is rapidly decreasing. These foldings keep reducing the radius of curvature, r_c, so as to have $r_c \sim \sigma$, where σ is the width of the interface. There is an obvious similarity with the physical picture that was found in the study of vortex filaments, described by the Biot-Savart model (Pumir and Siggia, 1987). We have maintained adequate

[1] The effect of incompressibility constraint on the nonlinear mechanism leading to shock formation in the Burgers equations can be studied in the 2d convection in porous media. The equations of motion are : $\mathbf{u} = \theta\hat{e}_y + \nabla p$, $\nabla \cdot \mathbf{u} = 0$, and $\partial_t\theta + (\mathbf{u} \cdot \nabla)\theta = 0$. These equations lead to the formation of steep interfaces, and to finite time singularities. The gradients of θ grows like $\frac{1}{(t^*-t)}$ (Pumir and Siggia, 1991b).

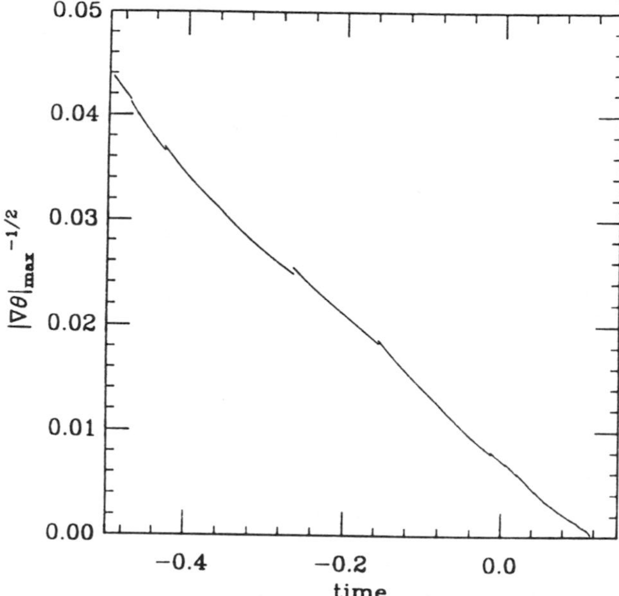

Fig. 1. The inverse square root of the maximum gradient $|\nabla\theta|_{max}^{-1/2}$ as a function of time in the last stage of the run. Time has been shifted by 10. The asymptotic regime started at $|\nabla\theta| \sim 100$, and the final value of $|\nabla\theta| = 1.810^7$

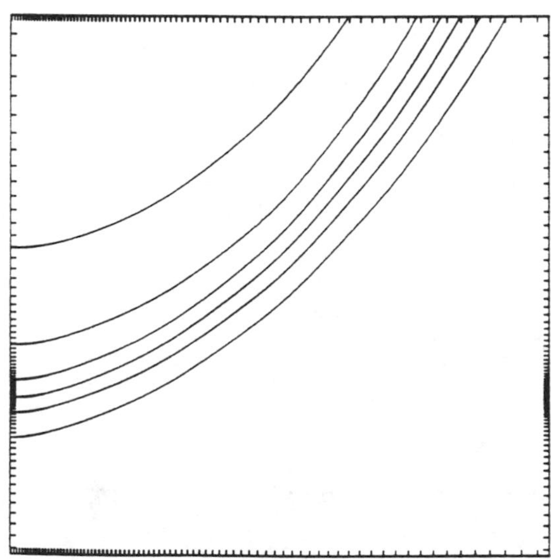

Fig. 2. Iso-θ lines near the cap of the bubble for $x \geq 0$, at time $t = 10.059$ ($|\nabla\theta|_{max} = 9.610^4$). The contour interval is 0.1. The x-coordinate range is : $0 \leq x \leq 1.810^{-4}$, and the y coordinate range is 4 times smaller.

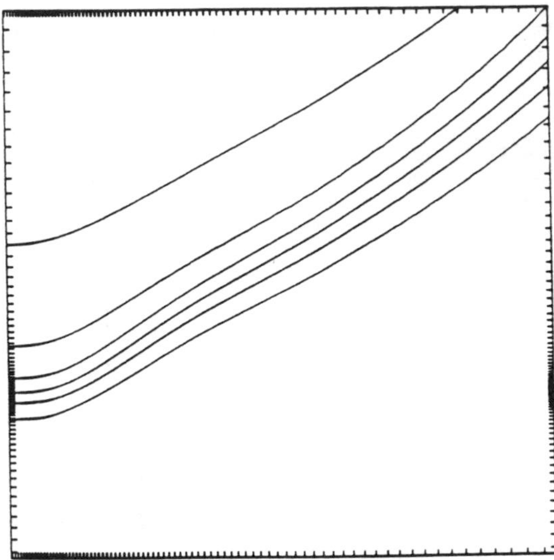

Fig. 3. Iso-θ lines near the cap of the bubble for $x \geq 0$, at time $t = 10.077$ ($|\nabla\theta|_{max} = 1.910^5$). The contour interval is the same as for Fig. 2 (0.1). The x-coordinate range is : $0 \leq x \leq 1.110^{-4}$, and the y coordinate range is 4 times smaller.

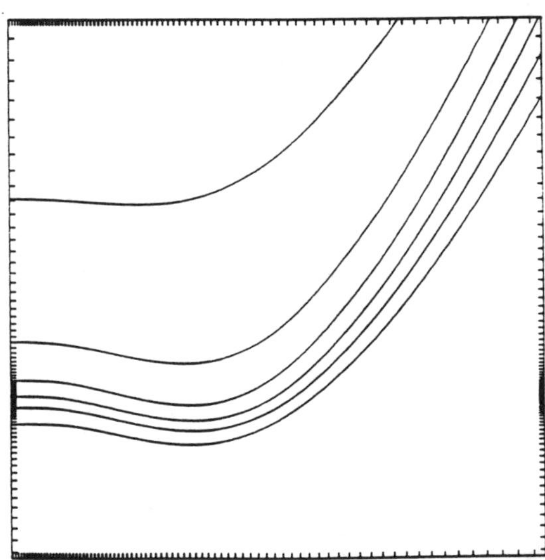

Fig. 4. Iso-θ lines near the cap of the bubble for $x \geq 0$, at time $t = 10.095$ ($|\nabla\theta|_{max} = 4.510^5$). The contour interval is the same as for Fig. 2 (0.1). The x-coordinate range is : $0 \leq x \leq 4.610^{-5}$, and the y coordinate range is 4 times smaller.

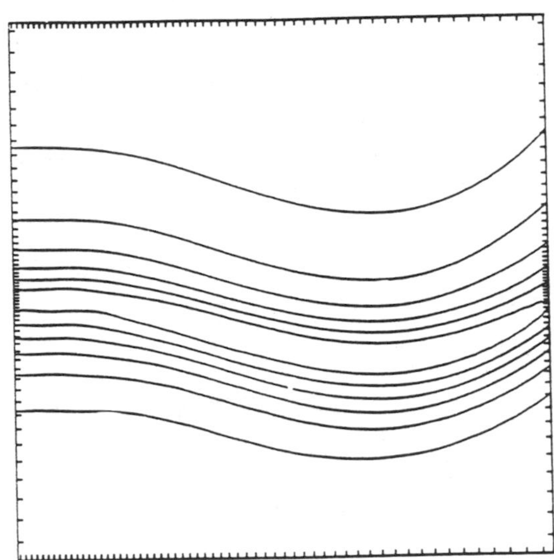

Fig. 5. Iso $|\nabla\theta|$ line at time $t = 10.095$ (same as Figure 4). The contour interval is 7.210^4. Note that the scales are more blown up than for Figure 4 : the coordinate range is 2.310^{-5} in x and 5 times smaller in y. The tick marks along the perimeter show the numerical grid points.

resolution in the region where the gradients are largest all the way to the end.

An example of folding of the interface is shown in Fig. (2,3,4), where iso-θ lines are shown near the cap of the bubble, on the right of the symmetry line ($x \geq 0$). Cold fluid is on top of hot fluid. The spatial scales are continuously blown up (see the figure captions). Note that while the front is becoming steeper, the upper contour line, corresponding to the lowest temperature, is drifting away from the location where the gradient is highest. The appearance of new foldings was essential to make $|\nabla\theta|$ grow, and to maintain the ratio r_c/σ between two constants : $0 < c_1 \leq r_c/\sigma \leq c_2$ where c_1 and c_2 are two numbers. After the interface has folded, maxima of $|\nabla\theta|$ start evolving, almost independently from one another in the regions where the interface is almost flat. To be able to fully resolve numerically one region where the gradient is maximum, it was necessary to pick one maximum, and to drop other parts of the interface that would also lead to a blow up. There is more than one point where the flow becomes singular. Figure 5 shows the norm of the gradient of θ at the same time as Figure 4, on a more blown up scale and shows the typical resolution we maintained in our runs. In the late stages of the evolution, the 2d vorticity in the Boussinesq problem grows approximately like : $|\omega| \sim 1/(t^* - t)$. The relative drift of the coldest iso-θ contour line away from the region where the front is steepest (Fig. 2-4) resulted in a steady decrease of the temperature jump across the front during the evolution. When we stopped our simulation the maximum gradient had grown by a factor 10^7, whereas the total jump in θ across the interface decreased by a factor $3-4$ only. There was no impediment to further integration. At the time where we stopped the integration, the overall shape of the interface was strikingly similar to what it was when $|\nabla\theta|$ was about ten times smaller, once the obvious scale factors were put in to really compare the two solutions. It is possible to fit our solutions with the following functional forms :

$$\theta = \theta_0 + (t^* - t)^\eta f(\mathbf{x}/(t^* - t)^{2+\eta}, \tau) \tag{13}$$

$$\omega = 1/(t^* - t) g(\mathbf{x}/(t^* - t)^{2+\eta}, \tau) \tag{14}$$

where $\tau = -ln(t^* - t)$ and η is a small number (roughly 0.2 ± 0.2), and the overall magnitude of the *singular* part of the velocity from (14) scales as $(t^* - t)^{1+\eta}$. Note that the Boussinesq equations are consistent with the scalings of equations (13,14). It is much harder to actually find solutions for f and g. In fact, our data suggest that the scaled functions f and g have a chaotic dynamics, in the sense that two nearby trajectories seem to diverge very fast as a function of τ. It would be interesting to make these ideas more precise.

Useful analytic estimates can be obtained by ignoring the actual shape of the interface, and using a contour dynamics limit. In this approximation, one describes the front by a temperature discontinuity, located on a curve $\mathbf{x}(\lambda, t)$, where λ is a Lagrangian parameter. Assuming $\omega = 0$ initially, vorticity is generated by buoyancy, and is nonzero only on the curve $\mathbf{x}(\lambda, t)$. The velocity $\mathbf{u}(\lambda, t)$ and the circulation $\gamma(\lambda, t)$ are related in the contour dynamics limit by the following equations :

$$u_i(x,t) = g\epsilon_{ij}\partial_j \int ln|(\mathbf{x} - \mathbf{x}(\lambda, t)|\gamma(\lambda)\frac{d\lambda}{2\pi} \tag{15}$$

$$\gamma(\lambda, t) = \int_0^t \partial_\lambda x_1(\lambda, t')dt' \tag{16}$$

The stretching, responsible for the growth of the gradients is given by

$$D_t|\nabla\theta|^2 = -2\nabla\theta \cdot e \cdot \nabla\theta \tag{17}$$

where D_t denotes the Lagrangian derivative and $e_{ij} = (\partial_i u_j + \partial_j u_i)/2$. From Eqns. (15-17), the following equation for the stretching can be derived :

$$\partial_t|\nabla\theta|^2 \sim c\int_0^t |\nabla\theta|^2 \kappa(t')dt' \tag{18}$$

Eq. (18) is useful to understand qualitatively the dynamics in the two regimes we found numerically. Before the cap of the bubble starts breaking up, $\kappa \sim cst.$, and Eq. (18) predicts that $\nabla\theta$ grows exponentially. When the instability starts, our numerical results suggest that $\kappa(t) \sim |\nabla\theta|$. This has to be understood in a statistical sense, since we never reached a truly self-similar state. In this case, it is easy to show that Eq. (18) leads to a growth asymptotically like : $|\nabla\theta| \sim 1/(t^* - t)^2$.

In the case of 3d axisymmetric Euler flows, the analogy with 2d Boussinesq convection suggests a divergence of the vorticity like $|\omega| \sim 1/(t^* - t)^2$ (see Eq. 9). The enstrophy itself, $\int |\omega|^2 d^3x$ does not diverge, but the quantity : $\int |\omega|^{2+\epsilon} d^3x$ diverges if $\epsilon > \eta$. The velocity itself remains finite. Because of the bounds found by Leray (1934), which show that the velocity has to diverge if a solution of the Navier Stokes equations is to diverge, we expect that our singular solutions will not survive in the viscous case. It is also expected that the axisymmetric solutions we found are completely unstable with respect to non axisymmetric perturbations. Some of the quantitative aspects of our solutions may very well apply to more general initial conditions. The salient features of our simulations, are the formation of vortex sheets and their repeated destabilisation, leading to the roll-up of the interface. Indeed, formation of vortex sheets has been observed in all the simulations of the 3d Euler equartions we are aware of. The roll-up of vortex sheets will lead to the formation of vortex tubes, a structure that has been reported many times in the study of 3d incompressible turbulence. It would be interesting to understand whether the physical mechanisms involved in the case of 3d axisymmetric flows have anything to do in the general 3d Euler initial value problem.

Acknowledgements

We are grateful to R. Grauer and T. Sideris for having communicated their results to us. One of us (A.P.) has benefitted from the hospitality of the Institute for Theoretical Physics, at the University of California, Santa Barbara. Our computations were done on RISC 6000 workstations donated by IBM. Support from the Air Force Office of Scientific Research under Grant number 91-0011 and the National Science Foundation under Grant number DMR-9012974 and Grant number PHY89-04035 are gratefully acknowledged.

References

BRACHET M., MEIRON D., NICKEL B., ORSZAG S. & FRISCH U. (1983), *J. Fluid Mech.*, **130**, 411 (1983)

BRACHET M., MENEGUZZI M., POLITANO H., SULEM P.L. & VINCENT A. (1991), private communication.

GRAUER R. & SIDERIS T. (1990), preprint.

KERR R. (1991), private communication.

C.B. LANEY & D.A. CAUGHEY (1991), *A.I.A.A.* paper 91-0632, 29th Aerospace Sciences

LEMESURIER B., PAPANICOLAOU G., SULEM C. & SULEM P.L. (1988), *Physica* **D31**, 78.

LERAY J. (1934), *Acta Mathematica* **63**, 193
MEIRON D. & SHELLEY M. (1991), preprint.
OSHER S. & CHAKRAVARTHY S. (1984) *SIAM J. Numer. Anal.*, **21**, 955
PUMIR A. & SIGGIA E.D. (1987), *Phys. Fluids* **30**, 1606
PUMIR A. & SIGGIA E.D. (1990), *Phys. Fluids* **A2**, 220 (1990)
PUMIR A. & SIGGIA E.D. (1991a), preprint.
PUMIR A. & SIGGIA E.D. (1991b) in preparation.
SWARTZRAUBER P.N. (1977), *SIAM Review* **19**, 490
ZELDOVICH YA. B., ISTRATOV A.G., KIDIN N.I. & LIBROVITCH V.B. (1980), *Combust, Sci. & Technology* **24**, 1

IS THERE A FINITE-TIME SINGULARITY IN AXISYMMETRIC EULER FLOWS?

XIAOGANG WANG and A. BHATTACHARJEE*
Institute for Theoretical Physics
University of California
Santa Barbara, California 93106-4030, U.S.A.

ABSTRACT. Recent numerical studies have identified a finite-time singularity in axisymmetric Euler flows with swirl. It is suggested here that what is being interpreted in these numerical studies as evidence for a finite-time singularity may, in fact, be evidence for an exponential growth of vorticity.

1. Introduction

Recently, Grauer and Sideris (1991) [hereafter, GS] have reported numerical evidence for the development of a finite-time singularity in axisymmetric incompressible Euler flows. This finding has been corroborated by Pumir and Siggia (1991) [hereafter, PS] who differ, however, from GS on the issue of how rapidly the divergence does occur. Whereas Grauer and Sideris report a singularity in the vorticity occurring as $(t_c - t)^{-1}$, Pumir and Siggia claim that the singularity occurs as $(t_c - t)^{-2}$ (where t_c is a constant).

Greene (1991) has given, in our opinion, persuasive arguments which put to doubt the occurrence of a finite-time singularity in GS. In this note we show, by arguments different from those given by Greene, that what is being interpreted in these numerical studies as evidence for a finite-time singularity may, in fact, be evidence for an exponential growth of vorticity.

In cylindrical coordinates (r, ϕ, z), the Euler equations for axisymmetric incompressible flow reduce to

$$\frac{dV}{dt} = 0, \tag{1}$$

$$\frac{d}{dt}\left(\frac{\omega_\phi}{r}\right) = \frac{1}{r^4}\frac{\partial}{\partial z}V^2, \tag{2}$$

* Permanent address: Department of Applied Physics, Columbia University, New York, NY 10027

H. K. Moffatt et al. (eds.), Topological Aspects of the Dynamics of Fluids and Plasmas, 303–308.

where $V \equiv rv_\phi$, and other symbols have the same meanings as in PS. Also,

$$\frac{d}{dt} \equiv \frac{\partial}{\partial t} + v_r\frac{\partial}{\partial r} + v_z\frac{\partial}{\partial z} = \frac{\partial}{\partial t} + \frac{1}{r}\frac{\partial(\psi, \ )}{\partial(r,z)}, \tag{3}$$

where ψ is a stream function such that

$$v_r = -\frac{1}{r}\frac{\partial\psi}{\partial z}, \quad v_z = \frac{1}{r}\frac{\partial\psi}{\partial r}. \tag{4}$$

ω_ϕ can be determined from the Poisson equation

$$r\frac{\partial}{\partial r}\left(\frac{1}{r}\frac{\partial\psi}{\partial r}\right) + \frac{\partial^2\psi}{\partial z^2} = -r\omega_\phi. \tag{5}$$

Equations (1), (2) and (5) constitute a closed set, solved numerically with smooth initial data in GS and PS. However, there are significant differences in the methodology and results given in these two papers. In GS equations (1), (2) and (5) are integrated numerically till blow-up occurs. The methods used in PS are more elaborate: the finite-time singularity is not seen in the axisymmetric Euler simulations, but in the analogous Boussinesq simulations. The argument we give relies on the useful analogy of axisymmetric Euler flows with two-dimensional Boussinesq convection, elucidated in PS.

Pumir and Siggia point out that away from the symmetry axis, axisymmetric incompressible Euler flows are isomorphic to two-dimensional Boussinesq convection. They exploit this analogy in their numerical study in order to maintain the resolution of the spatial mesh around the growing singularity. They work in an annular region $r_1 \leq r \leq r_2$, $(r_2 - r_1)/r_1 \gg 1$ which is mapped to rectangular coordinates $x \equiv z - z_0$, $y \equiv r - r_0$, where $z_0 \equiv (z_1 + z_2)/2$ and $r_0 \equiv (r_1 + r_2)/2$. Then equations (1), (2) and (5) can be approximated by the equivalent set

$$\frac{\partial\theta}{\partial t} + v_x\frac{\partial\theta}{\partial x} + v_y\frac{\partial\theta}{\partial y} = 0, \tag{6}$$

$$\frac{\partial\omega}{\partial t} + v_x\frac{\partial\omega}{\partial x} + v_y\frac{\partial\omega}{\partial y} - \frac{v_y}{r_0}\omega = g\frac{\partial\theta}{\partial x}, \tag{7}$$

and

$$\frac{\partial^2\psi}{\partial x^2} + \frac{\partial^2\psi}{\partial y^2} = \omega, \tag{8}$$

where $\theta \leftrightarrow V^2$, $\omega \leftrightarrow \omega_\phi$, $\psi \leftrightarrow -\psi/r_0$, and $g \equiv r_0^{-3}$. The "two-dimensional" velocity is given by

$$(v_x, v_y) = \left(-\frac{\partial\psi}{\partial y}, \frac{\partial\psi}{\partial x}\right). \tag{9}$$

The analogy with Boussinesq convection is now clear: g acts as a "gravitational acceleration" and θ is the "temperature" convected by the fluid.

Pumir and Siggia define scaled variables

$$\Omega(\mathbf{x}/a, T) = \Omega(\mathbf{X}, T) = \frac{\omega(\mathbf{x}, t)}{b}, \tag{10}$$

$$\Psi(\mathbf{x}/a, T) = \Psi(\mathbf{X}, T) = \frac{\psi(\mathbf{x}, t)}{ba^2}, \tag{11}$$

$$\Theta(\mathbf{x}/a, T) = \Theta(\mathbf{X}, T) = \theta(\mathbf{x}, t), \tag{12}$$

where T is a rescaled time variable, and the limit $a \to 0$, $b \to \infty$ is taken as the singularity is approached holding the gravitational acceleration fixed, that is, $g = ab^2 = 1$. Equations (6)–(8) then transform to

$$\frac{\partial \Theta}{\partial T} + \alpha \mathbf{X} \cdot \frac{\partial \Theta}{\partial \mathbf{X}} + \frac{\partial(\Psi, \Theta)}{\partial(X, Y)} = 0, \tag{13}$$

$$\frac{\partial \Omega}{\partial T} + \alpha \mathbf{X} \cdot \frac{\partial \Omega}{\partial \mathbf{X}} + \frac{\partial(\Psi, \Omega)}{\partial(X, Y)} = \frac{\partial \Theta}{\partial X}, \tag{14}$$

and

$$\frac{\partial}{\partial \mathbf{X}} \cdot \frac{\partial \Psi}{\partial \mathbf{X}} = \Omega, \tag{15}$$

with

$$\alpha = -\frac{\partial}{\partial T} \ln a, \tag{16}$$

and

$$\frac{\partial t}{\partial T} = a^{1/2}. \tag{17}$$

Here $\mathbf{X} = (X, Y)$. There is an additional term $\alpha\Omega/2$ in equation (10a) of PS. However, the imposed ordering implies that the corresponding term $-v_y \omega / r_0$ in our equation (7) is small, and has been neglected in equation (14).

The free positive parameter $\alpha(T)$ is adjusted by hand to maintain the required spatial resolution. If

$$\lim_{T \to 0} \int_o^T \alpha(T)\, dT = \bar{\alpha} > 0, \tag{18}$$

then, by converting back to the unscaled variables, Pumir and Siggia prove that there is a finite-time singularity. This is because, by (16) and (17),

$$\bar{\alpha} \sim -\frac{\partial t}{\partial T}\frac{\partial}{\partial t}\ln a = -\frac{1}{a^{1/2}}\frac{\partial a}{\partial t}, \tag{19}$$

whence

$$a \sim \frac{\bar{\alpha}^2}{4}(t_c - t)^2. \tag{20}$$

Consequently,

$$\left|\frac{\partial \theta}{\partial \mathbf{x}}\right| = \frac{1}{ga}\left|\frac{\partial \Theta}{\partial \mathbf{X}}\right| \sim \frac{4|\partial\Theta/\partial\mathbf{X}|}{g\bar{\alpha}^2(t_c - t)^2}. \tag{21}$$

If $|\partial\Theta/\partial\mathbf{X}|$ is a well-behaved function, $|\partial\theta/\partial\mathbf{x}|$ blows up as $(t_c - t)^{-2}$. Note that this implies that

$$\omega = b\Omega = \frac{\Omega}{a^{1/2}} \sim \frac{2\Omega}{\bar{\alpha}(t_c - t)}. \tag{22}$$

It follows then that if one accepts the scalings (10)–(12), ω diverges as $(t_c - t)^{-1}$. As pointed out in PS, the r- and z-components of the vorticity are even more singular, and diverge as $(t_c - t)^{-2}$.

We remark that the numerical integration of equations (13)–(15), by itself, says nothing about the occurrence of a finite-time singularity. The singularity is an inference, based on the scaling hypothesis (10)–(12) and the subsidiary relations (16) and (17). We now consider additional implications of this scaling hypothesis. Defining $\mathbf{v}_p \equiv (v_x, v_y)$, we have, as $a \to 0$, $b \to \infty$,

$$|\mathbf{v}_p| = \left|\frac{\partial \psi}{\partial \mathbf{x}}\right| = ab\left|\frac{\partial \Psi}{\partial \mathbf{X}}\right| \to 0. \tag{23}$$

This suggests that the singularity must be an x-point for the two-dimensional flow field $\mathbf{v}_p$. As far as we can determine, there are no numerical results reported in GS or PS which explicitly demonstrate the formation of such an x-point. One expects that the tendency to form an x-point will be evident before the actual occurrence of the singularity, and the demonstration of such a tendency in the flow field $\mathbf{v}_p$, without relying on scaling assumptions, is an important test for the numerical experiments.

A different possibility, and this appears to be supported by the numerical evidence presented in GS, is that $\mathbf{v}_p$ remains of $O(1)$ as the "singularity" is formed. Level surfaces of V ("isothermals") approach each other; contraction along the x-direction is accompanied by flattening along y. The definition $\mathbf{X} = \mathbf{x}/a$ in the scaling (10)–(12) assumes implicitly that there is only one length scale a, and cannot describe such dynamics. We now derive a different scaling to describe such dynamics, but rather remarkably, the new scaling also leads to equations (13)–(15). However, instead of a finite-time singularity, the new scaling predicts that the vorticity grows exponentially with time in the numerical simulations.

For the sake of generality, we write

$$\Omega(x/a, y/c, T) = \Omega(X, Y, T) = \omega(x, y, t)/b, \tag{24}$$

$$\Psi(x/a, y/c, T) = \Psi(X, Y, T) = \psi(x, y, t)/d, \tag{25}$$

$$\Theta(x/a, y/c, T) = \Theta(X, Y, T) = \theta(x, y, t), \tag{26}$$

where there are two length scales a and c, and b and d is to be determined. The case $a = c$ has already been considered by Pumir and Siggia. There are two other possibilities: $a/c \to 0$ or $a/c \to \infty$ as $a \to 0$. Then from the constraint

$$|\mathbf{v}_p| = \left|\frac{\partial \psi}{\partial \mathbf{x}}\right| = ab\left|\frac{\partial \Psi}{\partial \mathbf{X}}\right| \sim O(1), \tag{27}$$

as $a \to 0$ and $g = 1$, we can determine the other scaling parameters. Clearly, from (8) and (27), if $a/c \to \infty$, we have $d = c$ and $bc = 1$; on the other hand, if $a/c \to 0$, then $d = a$ and $ab = 1$. Consider the different terms in equation (7). For both cases,

$$v_x \frac{\partial \omega}{\partial x} + v_y \frac{\partial \omega}{\partial y} \sim \frac{1}{ac}, \tag{28}$$

and

$$g \frac{\partial \theta}{\partial x} \sim \frac{1}{a}. \tag{29}$$

It is now clear that the only consistent choice is $c = 1$, and hence $a/c \to 0$. The physical picture is clear for this case: vorticity intensifies on the small scale when two constant-θ surfaces come very close in the x-direction.

The limit $a \to 0$, $b \to \infty$ is now taken with $d = a$ and $g = ab = 1$. Then equations (13)–(15) continue to hold, except that the term $\alpha\Omega/2$ should now be restored on the left-hand-side of equation (14). Furthermore, the definition (17) must be changed to

$$\frac{\partial t}{\partial T} = 1. \tag{30}$$

When condition (18) holds, it follows that

$$\bar{\alpha} \sim -\frac{\partial}{\partial T} \ln a = -\frac{\partial}{\partial t} \ln a, \tag{31}$$

which gives

$$a \approx a_0 \exp(-\bar{\alpha} t), \qquad b \approx b_0 \exp(\bar{\alpha} t), \tag{32}$$

with $a_0 b_0 = 1$. The vorticity

$$\omega = b\Omega \sim b_0 \Omega \exp(\bar{\alpha} t), \tag{33}$$

then grows exponentially in time.

After the completion of this work, Dr. R. Kerr kindly brought to our attention that Grabowski and Clark (1991) have solved the problem of two-dimensional thermal convection numerically, and have reported evidence for the exponential growth of vorticity.

Acknowledgements

We are deeply grateful to Dr. J. M. Greene for sharing generously his insights on this problem, and to Dr. R. Grauer and Dr. T. Sideris for showing us their numerical results prior to publication. We also thank Dr. A. Pumir and Dr. E. Siggia for detailed discussions of their work, and for their critical comments on a first draft of this paper. This research was supported by the National Science foundation Grant No. PHY 98-04035 and ATM 91-00513.

References

GRABOWSKI, W. W. & CLARK, T. L., 1991 Cloud-environment interface instability: rising thermal calculations in two spatial dimensions. *J. Atmos. Sci*, **48**, pp. 527-546.

GRAUER, R. & SIDERIS, T., 1991 To appear in *Phys. Rev. Lett.*

GREENE, J. M., 1991 Private communication.

PUMIR, A. & SIGGIA E. D., 1991 Development of singular solutions to the axisymmetric Euler equations. Submitted to *Phys. Fluids A*.

EVIDENCE FOR A SINGULARITY OF THE THREE DIMENSIONAL, INCOMPRESSIBLE EULER EQUATIONS.

R. M. Kerr
Geophysical Turbulence Program
National Center for Atmospheric Research
Boulder, CO 80307-3000, U.S.A.

ABSTRACT. Three-dimensional, incompressible Euler calculations of the interaction of perturbed anti-parallel vortex tubes using a variety of smooth initial profiles in bounded domains with bounded initial vorticity is discussed. It will be shown that trends towards either exponential, non-singular growth of the peak vorticity or power law, singular behavior can be strongly dependent on details of the initial conditions. A numerical method that uses symmetries and additional resolution in the direction and location of maximum compression is used to simulate periodic boundary conditions in all directions. For the initial condition that yields singular type behavior the growth of the peak vorticity roughly obeys $(t_c - t)^{-1}$ and the growth of the strain along the vorticity at this point obeys $(t_c - t)^{-\gamma}$ where $\gamma \approx 1$.

1. Introduction

One of the most difficult problems in computational fluid dynamics that has been addressed is whether there is evidence for or against the existence of a singularity of the three-dimensional, incompressible Euler equations. Since two publications in 1987 in Physical Review Letters (Ashurst & Meiron 1987; Pumir & Kerr 1987) the favored approach to this problem has been direct, Eulerian simulation of an initial value problem with two anti-parallel vortices. One of the major points of these letters was that while Lagrangian vortex methods (Siggia 1985) are adequate for representing the initial phase of development, due to vortex core deformation they are inadequate for representing the late phases when the question of singularity arises. Since then, there has developed a growing body of evidence from a variety of inviscid calculations, including simulations of anti-parallel vortices (Pumir & Siggia 1990) and recent extensions (Brachet, private communication) of earlier simulations of the Taylor-Green problem (Brachet et. al. 1983) that suggest there is not a singularity and that at late times the growth in the peak vorticity is at most exponential. In addition, recent cal-

H. K. Moffatt et al. (eds.), Topological Aspects of the Dynamics of Fluids and Plasmas, 309–336.

culations by this author, to be discussed in a later paper, that use random initial conditions in a periodic box with 256^3 resolution, could also be interpreted as supporting exponential growth. To date, all the calculations that have suggested a singularity are inadequate either because they were viscous (Ashurst & Meiron 1987; Kerr & Hussain 1989) or in addition lacked resolution (Ashurst & Meiron 1987). The major difficulty with using a series of viscous calculations to infer trends towards the limit of zero viscosity is that the range of Reynolds numbers that can be used is too small. In addition, for at least one similar problem, whether there is a singularity of the inviscid, two-dimensional, incompressible, magnetohydrodynamic equations, it has been demonstrated that the trend towards a singularity indicated by viscous calculations (Orszag & Tang 1979) could not be reproduced by inviscid calculations (Frisch et. al. 1983). Therefore, due to the current weight of evidence against a singularity, any claim for a singularity must match the computational power of the calculations finding exponential growth, be able to reproduce the exponential behavior for similar conditions, and from this hopefully suggest a reason for at least two types of behavior, exponential and singular. In addition, the growth of various quantities such as the peak vorticity and the strain along this vorticity should have power law growth with time with constraints consistent with recent theoretical work (Caffarelle et. al. 1982; Beale et. al. 1984). In this paper, evidence for a singularity will be presented that attempts to meet these standards. It will be shown that the conclusion of Kerr & Hussain (1989) that their initial conditions gave signs of singular behavior was essentially correct.

The numerical method used in the calculations discussed here will use symmetries to reduce the computational domain and the fields will be represented in the different directions by Fourier transforms, sines, cosines, and combinations of Chebyshev polynomials. Nonlinear terms are calculated by a pseudospectral method using the 2/3 rule for dealiasing. All mesh sizes will refer to the number of mesh points in physical space. All the calculations will be inviscid. Only the highest resolution calculations will be presented. Complementary calculations at different resolutions that support the conclusion that there is a singularity will not be discussed. Neither will calculations with the Fourier method of Kerr & Hussain (1989) that treat more general initial conditions be discussed.

The inspiration for the particular initial vorticity profiles used here comes from two papers, Kerr & Hussain (1989) and Melander & Hussain (1989). Following these papers free-slip boundary conditions were imposed at the "dividing" plane between the vortices and at the "symmetry" plane of maximum perturbation of the vortices. In Kerr & Hussain (1989), the first step of initialization was a rather crude Gaussian profile that terminated at a given radius. In the second step a $\exp(-s(k_x{}^4 + k_y{}^4 + k_z{}^4))$ filter was imposed to smooth the rough edges. Due to the crudeness of the first step, the coefficient s was chosen to be substantial and reduced the initial peak vorticity by a factor of 6. This reduction was compensated for by choosing a small initial radius prior to filtering. The filtering expanded the vortex such that its edge was just above the dividing plane. Melander & Hussain (1989) devised a more sophisticated initial profile that went to zero smoothly at a given radius, but was still Gaussian in the center. In the calculations here the profile of Melander & Hussain (1989) is used as the first step of initialization procedure of Kerr & Hussain (1989), then different numerical experiments with and without a high wavenumber filter similar to that used by Kerr & Hussain (1989) are discussed. It will be shown that if a filter is used, that singular behavior is obtained, whereas without the filter the growth of the peak vorticity is exponential for most of the duration of the calculations. It will be suggested that the source of the differences in the dynamics are small regions of vorticity opposite in sign to the dominant vortices that appear in the initial conditions when the filter is not used.

2. Chebyshev

The major difficulty in addressing the singularity question is obtaining sufficient resolution. The approach of Pumir & Siggia (1990) uses a finite-difference nested mesh scheme to obtain more resolution in the region of maximum vorticity. In Kerr & Hussain (1989) a Fourier method with extra resolution in the direction of maximum compression was used. In the method used for the present calculations both mesh refinement and changing the resolution in the different directions are used in order to isolate the resolution in the direction of maximum compression to near the dividing plane. This will hopefully free computing resources that can be used to increase the resolution in the other directions, thereby increasing the overall resolution and the ability to extend the calculations further into the singular regime.

To accomplish these numerical objectives a split Chebyshev method seems ideal. The colocation points for Chebyshev methods are concentrated near the boundaries and high resolution can be achieved, since the inversion matrices needed to solve the Poisson equation can be done efficiently and there is a fast transform, the cosine transform, to physical space. Furthermore, if only half the Chebyshev polynomials are used for a given variable, then a free-slip symmetry forms in the center of the box that can be used to provide an upper boundary condition with less resolution. The split means that in addition to requiring a cosine transform, what is known as a quarter wave transform is needed. Like the cosine transform, the quarter wave transform can be related to Fourier transforms, for which there is a fast transform.

The numerical method for solving the Poisson equation with Chebyshev polynomials is based upon divergence-free vector functions that inherently satisfy the free-slip boundary conditions. The divergence-free vector functions decompose the velocity by a poloidal-toroidal decomposition. In this respect it is similar to the numerical method proposed by Moser et. al. (1983) for use with no-slip boundary conditions, except in this case the functions of Chebyshev polynomials needed to satisfy the boundary conditions are more complicated. The method requires the inversion of two narrow banded matrices to do implicit time advancement. The modification of Crank-Nicholson suggested by Spalart is used with the Wray's low storage third-order Runge-Kutta calculation of the nonlinear terms.

The largest meshes used for a $L_x : L_y : L_z$ domain of $4\pi \times 4\pi \times 2\pi$ are $256\times256\times192$ and $512\times256\times128$. The $256\times256\times192$ calculation achieves a factor of 30 increase in the overall peak vorticity, and a factor of 60 increase in the peak vorticity in the symmetry plane and shows that the trend towards a singularity is not a transient. But currently only the $512 \times 256 \times 128$ calculation, up to the time that a factor of 10 increase in the overall peak vorticity is seen, is considered good enough to estimate the singular time and exponents. Calculations with $512 \times 256 \times 192$ and $1024 \times 256 \times 128$ points are currently being conducted to support this conclusion and possibly increase the time over which exponents can be reliably estimated. (Note that a factor of 2 increase in z resolution is achieved by a factor of $\sqrt{2}$ increase in the number of z points due to the compression in colocation points near the boundary.) Word packing was not used when the number of colocation points in z exceeded 128, possibly due to the reasons discussed recently by Breuer & Everson (1991).

One drawback with this Chebyshev method that affected the initialization was that the order of Chebyshev polynomial in the poloidal and toroidal components is different. So, while within each component there is consistency in the application of incompressibility, if the initial vortex requires both components the difference in order between the components led to some anomalous energy in the higher order Chebyshev polynomials, which in turn led to numerical difficulties. The simple solution is to define the initial vorticity such that it is only toroidal, defined by $\delta_z = 0$ below, which creates initial vortices that are a constant distance from each other. Calculations with a variety of perturbations done with the Fourier code described by Kerr & Hussain (1989) show that this is one of the most effective initial conditions in obtaining maximum growth in the peak vorticity.

3. Calculation without initial filter

The calculation in this section is meant to show that if the high wavenumber filter is not applied to the initial conditions that for a period the strain saturates, during which period the growth of the peak vorticity is exponential. Only enough resolution to demonstrate this point is used. There are some weak signs that at the last time calculated that the singular type of of behavior might be beginning, that is the strain begins to increase, but given the success of the filtered initial conditions there does not seem to be any reason to follow this further.

Following Melander & Hussain (1989), for a vortex of radius r in xz around an initial trajectory of the form

$$(X, Y, Z) = (x_o + x(s), y, z_o + z(s)),$$

where s is a function of y, the following initial profile for the vortex cores was used:

$$\omega(r) = \exp(f(r)), \quad \text{where}$$
$$f(r) = (-r^2)/(1 - r^2) + r^2(1 + r^2 + r^4) \quad \text{and}$$
$$r = |(x, y, z) - (X, Y, Z)|/R \quad \text{for} \quad r \leq 1$$

This modification of the vortex core profile of Melander & Hussain (1989) was suggested by Melander (private communication).

Following Kerr & Hussain (1989) the initial trajectory is defined as

$$x(s) = x_o + \delta_x \cos(s)$$

$$z(s) = z_o + \delta_z \cos(s), \quad \text{where}$$
$$s(y) = y_2 + L_y \delta_{y1} \sin(\pi y_2 / L_y) \quad \text{and}$$
$$y_2 = y + L_y \delta_{y2} \sin(\pi y / L_y)$$

rather than a simple sinusoidal trajectory. Following Melander & Hussain (1989), the incompressible vorticity vector for this trajectory is proportional to:

$$\omega_x = -\pi\delta_x/L_x(1 + \pi\delta_{y2}\cos(\pi y/L_y))(1 + \pi\delta_{y1}\cos(\pi y_2/L_y))sin(\pi s(y))$$

$$\omega_y = 1$$

$$\omega_x = -\pi\delta_x/L_x(1 + \pi\delta_{y2}\cos(\pi y/L_y))(1 + \pi\delta_{y1}\cos(\pi y_2/L_y))sin(\pi s(y))$$

For $\delta_{y1} = \delta_{y2} = 0$ the trajectory is sinusoidal, which is one of the cases studied by Kerr & Hussain (1989) and is what is used by Melander & Hussain (1989). It was shown by Kerr & Hussain (1989) that if δ_{y1} and δ_{y2} are not zero, so that the perturbation is localized far from the periodic boundary conditions, that significantly larger growth in the peak vorticity can be obtained. Here the values of δ are: $\delta_{y1} = 0.5, \delta_{y2} = 0.4, \delta_x = 0.8$and $\delta_z = 0$ and $z_o = 1.57$ and $R = 0.75$. Figure 1 shows the y vorticity perpendicular to the symmetry plane in the symmetry xz plane initially and at one developed time.

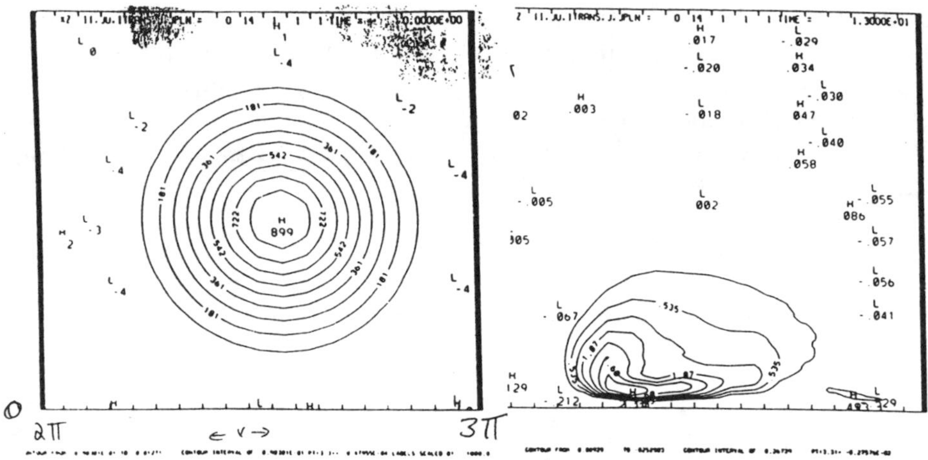

Fig. 1. y vorticity perpendicular in the symmetry plane. a: $t = 0$, b: $t = 13$.

This is the simplest initial condition attempted without filtering. More complicated initial conditions that overlayed vortices and reduced the magnitude of the regions of small negative vorticity that appear in figure 1a were also investigated, but did not yield substantially different results. Such small negative regions cannot be seen the contour plots unless either zero contours are plotted or highs and lows are given, so it cannot to determined whether they are present in other published calculations of vortex reconnection such as Pumir & Siggia (1990). A common feature of all the initial conditions without filtering is that the initial z spectra, or equivalently the initial distribution of Chebyshev polynomials, went to zero with increasing wavenumber in a sawtooth power law fashion. The Fourier spectrum is shown in figure 2. Increasing the resolution used with the initial conditions did not alter either the low wavenumber components of the spectra or the magnitude of the regions of negative vorticity. Regions of negative vorticity in roughly the same location and of the same magnitude appear for both the Chebyshev code and for the Fourier code. This type of spectral behavior is reminiscent of Gibbs phenomena, that is high wavenumber oscillations that occur when a Galerkin method tries to represent physical space discontinuities. This suggests that the origin of the sawtooth spectra and negative regions of vorticity is the sharp cutoff in the vorticity profile used by Melander & Hussain (1989), although possible inconsistencies between the initial perturbation and periodic boundary conditions might also have an influence.

Figure 1b shows that the vorticity that develops from initial condition 1a

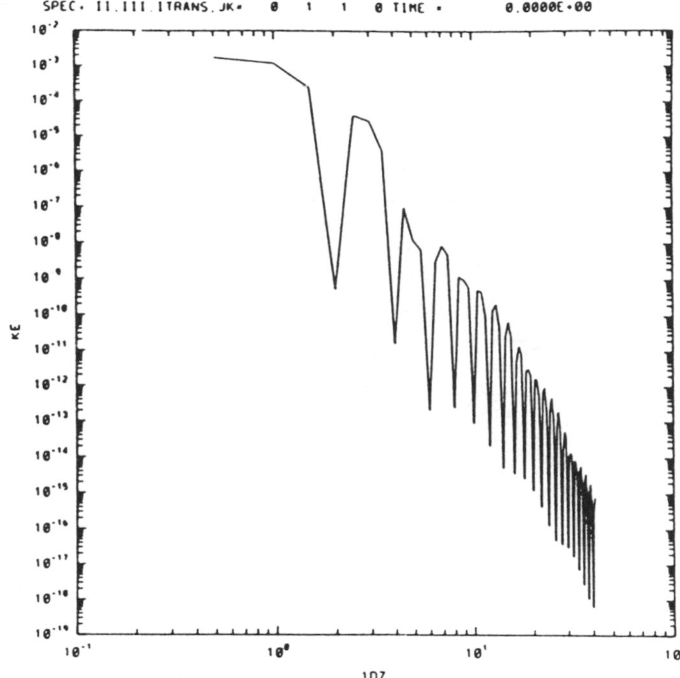

Fig. 2. z-spectra for unfiltered initial conditions.

is not any significantly different than the type of head-tail structure found in earlier calculations of vortex reconnection, except that when the regions of negative vorticity are shown it should be noted that even though they are initially very small, the order of 4×10^{-3} compared to the initial peak vorticity of 1, by the end of the calculation the negative peak is the same order of magnitude as the primary vortex. While difficult to see in figure 1b, these the negative regions have been sucked into the region between the primary vortices, that is along the dividing plane, and could have a major impact upon the dynamics. But detailed examination of graphics does not reveal any notable property to distinguish these calculations from the singular calculations discussed below, except that the peak of the strain along the vorticity in the symmetry plane is the same distance from the dividing plane (in z) as the peak vorticity, whereas in the singular calculations the strain peak is always somewhat further from the dividing plane. In particular, two-dimensionalization of the vortices, one of the properties observed by Pumir & Siggia (1990), is not seen, as shown by the three-dimensional isosurface plot in figure 3.

Figure 4 plots the peak strain along the vorticity, the peak vorticity and the peak negative vorticity as functions of time. This demonstrates that for this initial condition the strain saturates for an extended period, dur-

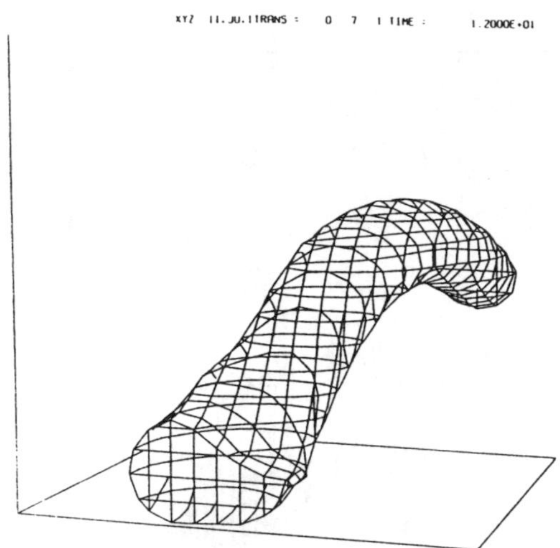

Fig. 3. Three-dimensional isosurface plot of the vorticity squared for (x,y,z) = (4.9:8.8,0:3.9,0:3.9) at t=12 for unfiltered initial conditions.

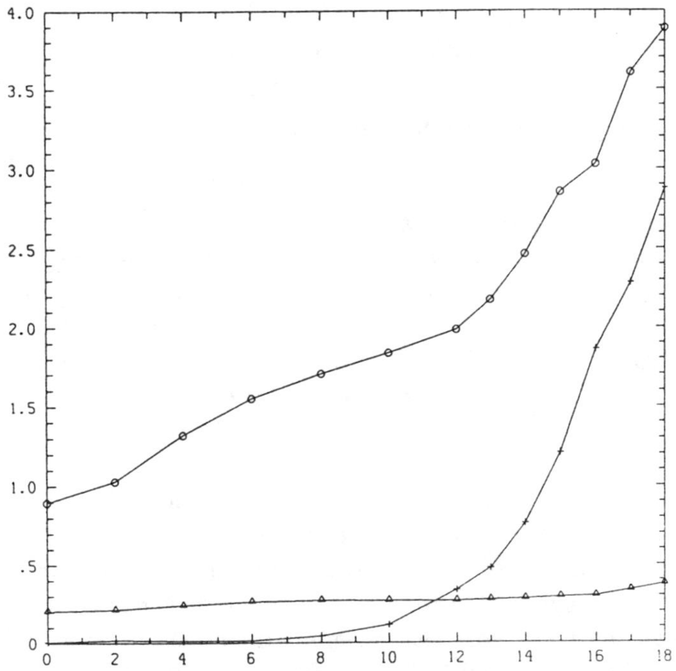

Fig. 4. Peak strain along the vorticity (triangle), peak vorticity (circle), peak negative vorticity (plus) for unfiltered initial conditions vs. time.

ing which period the growth of peak vorticity is exponential. In Pumir & Siggia (1990) the strain was observed to saturate in a similar manner, with exponential growth of the peak vorticity. There are significant differences between the trajectories of their initial vortices, they are hyperbolic, and the periodic trajectories used here, so definitive conclusions are out of the question, but based on these calculations the work of Pumir & Siggia (1990) should be checked to see if there are similar regions of negative vorticity playing a major role, and if so to determine whether these regions might then explain the saturation of the strain in their calculations.

4. Calculation with initial filter

Following upon the calculations of Kerr & Hussain (1989) a variety of initial conditions with filtering have been studied inviscidly with the Fourier code of Kerr & Hussain (1989). These include the same initial conditions with filtering used in their calculations, the initial conditions just discussed with filtering, as well as other initial perturbations with filtering. These calculations showed that by applying a high wavenumber filter that removed the regions of negative vorticity and gave smooth initial spectra that went to zero faster than a power law, that substantially different behavior than either the unfiltered results discussed above or Pumir & Siggia (1990) could be obtained. But the general reaction has been that the regime of singular-like behavior generated by these calculations was not sufficient to make a convincing case for a singularity. This is what directly inspired an attempt to use a Chebyshev method to obtain a longer regime of singular behavior. But while application of a wavenumber filter of the type used by Kerr & Hussain (1989) with a purely Fourier method is trivial, when one of the directions is not represented by a Fourier transform the application of a high wavenumber filter is not as easy. Various types of higher order smoothings were attempted with the Chebyshev method, but failed primarily due to the fact that with a sharp boundary to the vorticity profile there are significant higher order components of the Chebyshev distribution. When attempting to invert a matrix to do a higher order smoothing, these higher order components destroyed the inversion process. Using a lower order smoothing, like a Newtonian viscosity, worked better and gave the first clear indications that the Chebyshev code would yield singular results, but the final product was not any better than the results from the Fourier code.

A effective procedure for applying a high wavenumber filter was eventually found. It is to initialize with the Fourier code, with up to 1000 colocation points in z, filter, then map the results onto the Chebyshev code. The mapping procedure is not perfect, it introduces new regions of potentially dangerous negative vorticity, so more filtering than was necessary with the Fourier code is applied. The extra filtering smoothed the perturbation more

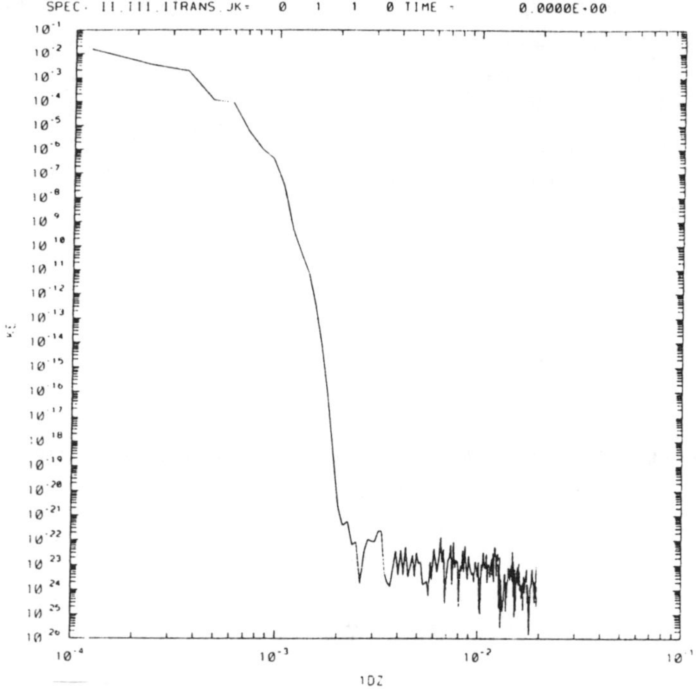

Fig. 5. Initial z-Chebyshev distribution (spectrum) for filtered intial conditions.

than desirable, so in addition δ_x was increased to 1.6 and the peak vorticity before smoothing was increased to 8. But the end result is a smooth initial condition that significantly extends the regime of singular behavior above what was obtained with the Fourier code, and for a shorter period yields results with sufficient resolution that singular times and exponents can be estimated. The initial distribution of Chebyshev polynomials (essentially the z-Chebyshev spectrum) is shown in figure 5. A spherically symmetric filter was used, $\exp(-s({k_x}^2+{k_y}^2+{k_z}^2)^2)$ with $s=0.05$. The Fourier results with filtered initial conditions are still significant because they show that for essentially the same initial vorticity profiles as used without filtering above, that singular behavior can be obtained.

Figure 6 shows the vorticity perpendicular to the symmetry plane ω_y in the xz symmetry plane at four early times, figure 7 shows ω_y at four late times. A complete description of these figures is given in the section on figures. The long initial tail is a remnant of the initialization procedure that required a larger perturbation and greater smoothing. Eventually the true head-tail structure develops. First the vorticity profile is dominated by a tilted flattened region that develops a dimple, represented by figure 6d at t=12.0. At later times this dimple becomes an corner separating the head and tail. The peak vorticity is located in this corner, not in the head as in

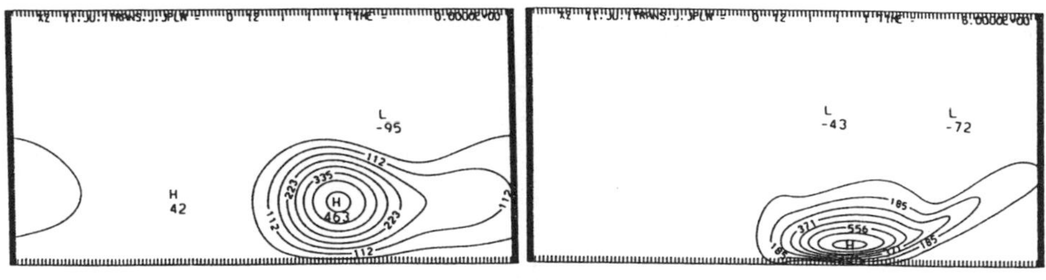

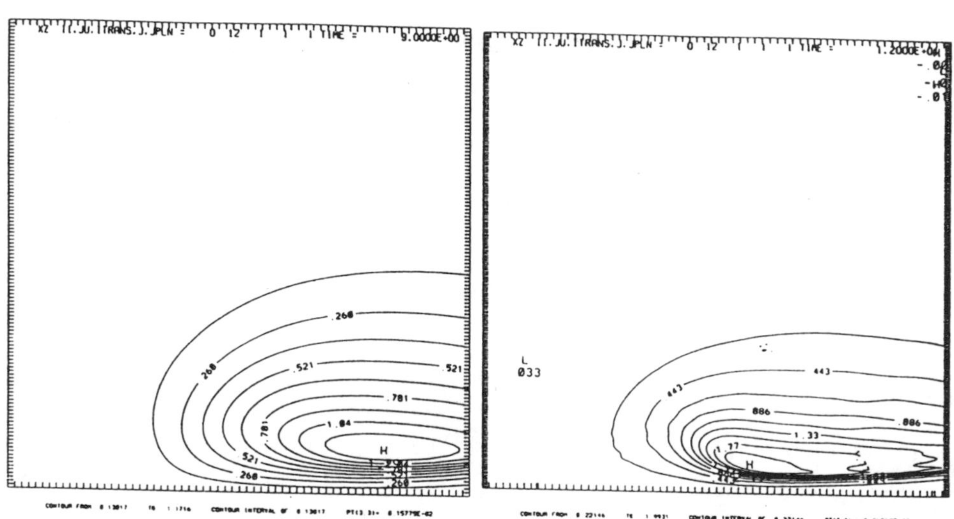

Fig. 6. Vorticity (ω_y) in the symmetry plane. a:t=0, b:t=6, c:t=9, d:t=12

earlier viscous calculations. In Pumir & Siggia (1990) at late times the region of peak vorticity was spread over more of the head and tail, although the vorticity structure just before the strain saturates has more of the features observed here. Figure 10 below expands the z by a factor of 5 both for direct comparison with the scaling used in the figures of Pumir & Siggia (1990) and for a better demonstration of how the head and tail are both vortex sheets meeting at a corner which contains the peak vorticity. Another feature of these calculations is that the signs of the components of the strain change rapidly in the region surrounding the peak vorticity, while in Pumir & Siggia

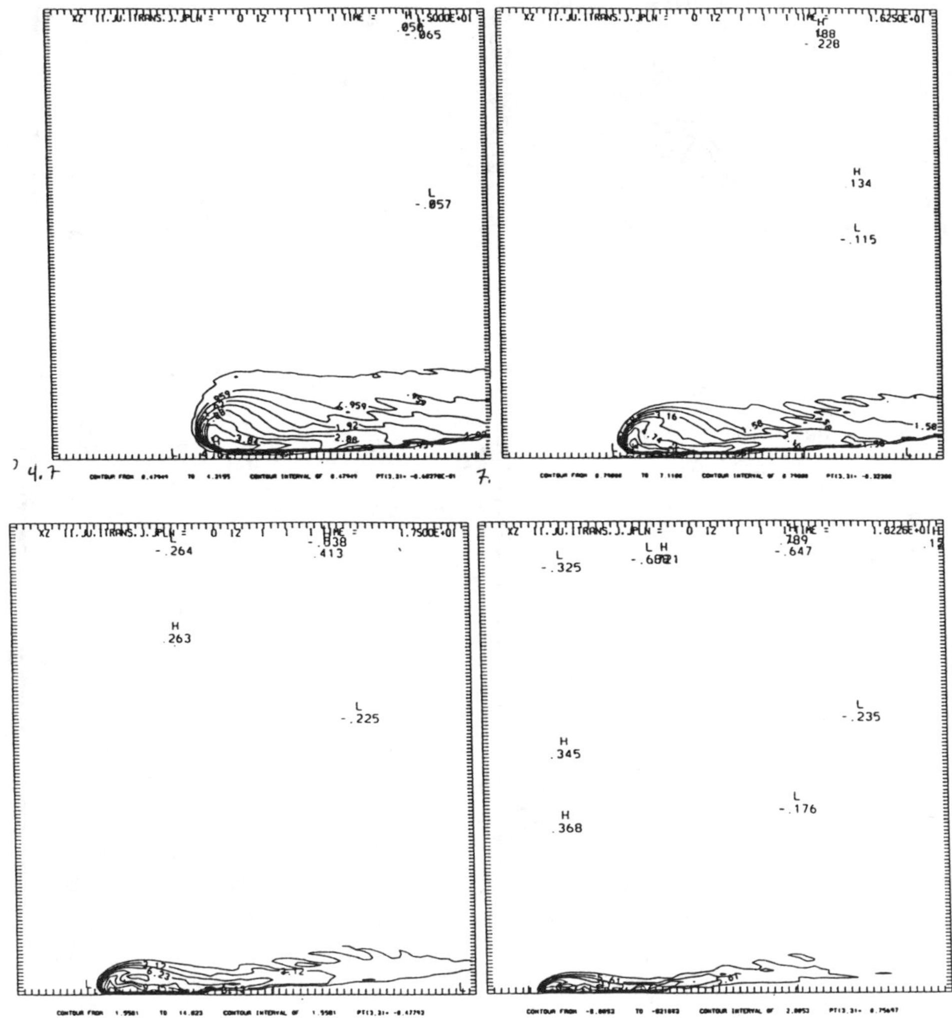

Fig. 7. Vorticity (ω_y) in the symmetry plane. a:t=15, b:t=16.25, c:t=17.5, d:t=18.2

(1990) the components of the strain varied slowly. This is shown in figures 12 and 13 below. Additional figures that are essential for demonstrating what mechanisms might maintain singular behavior and not allow the flow to slip into the configuration found by Pumir & Siggia (1990) are discussed the next section.

The small waves in the outer vorticity contours are the effects of resolution. Extensive resolution tests, to be discussed more fully in another paper, show that signs of singularity to be discussed are resolution independent up until t=17 for a resolution of $512 \times 256 \times 128$.

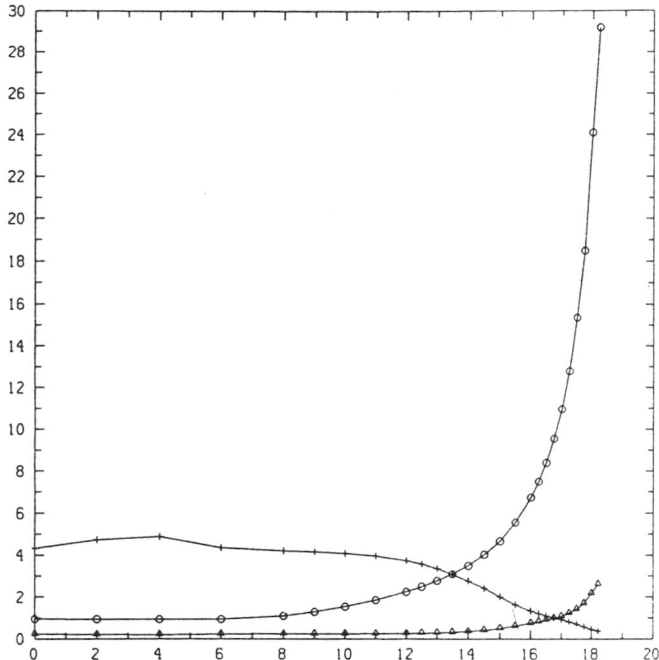

Fig. 8. Peak vorticity (circle), peak strain along the vorticity (triangle), inverse of that strain (plus) vs. time.

Figure 8 plots the peak vorticity, peak strain along the vorticity in the symmetry plane, and the inverse of that strain as functions of time for the $256 \times 256 \times 192$. The peak strain is plotted instead of the strain at the point of peak vorticity because its values were better behaved. For late times the peak strain is consistently about 50% larger than the strain at the point of peak vorticity. The linear behavior of the inverse strain is a clear sign of singular behavior. Theoretical predictions (Caffarelle et. al., 1982) are that the strain should go as $1/(t_c - t)$ and the peak vorticity should go as $1/(t_c - t)^\gamma$ where $\gamma \geq 1$. When $1/(t_c - t)$ is fitted to figure 7 between t=15.5 and 18.225, t_c is approximately 19.06. Using this value for t_c figure 8 shows a log-log plot of ω_p versus $1/(t_c - t)$. The best power law fit to figure 8 between t=12.0 and 18.225 gives $\gamma = 1.2$, consistent with theory. The slight divergence from a simple power law in figure 8 could be due to an error in the estimate of t_c or resolution. The $512 \times 256 \times 128$ results up to t=17 show the strain increasing slightly faster than in the $256 \times 256 \times 192$ results and the vorticity increasing slightly slower, both consistent with $\gamma = 1$ instead. This is the singular prediction of Pumir & Siggia (1987).

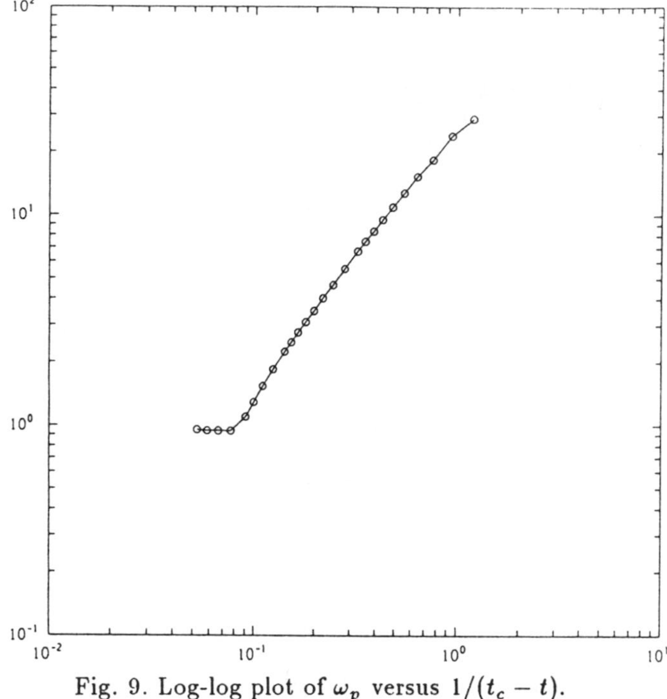

Fig. 9. Log-log plot of ω_p versus $1/(t_c - t)$.

5. Figures

The figures just presented are the minimum needed to show evidence for a singularity. But it has become clear that a better case can be made by studying the configuration of vorticity and strain that leads to singular behavior. In this section two-dimensional slices in a variety of directions and locations are used to show the strong curvature in the vortical direction y and how strain prevents the vorticity from flattening in the manner of Pumir & Siggia (1990). The graphics are taken from a variety of resolutions, some not the most resolved at a given time, and the descriptions are rather cryptic, due to the deadline for submission of this paper.

Probably the most informative figures are 12, 13, and 14. Figures 12 and 13 are two-dimensional vorticity plots in the symmetry plane and three-dimensional isosurface plots, respectively, with the axes stretched in the same manner as in Pumir & Siggia (1990). In figure 12 the vorticity is more localized in the corner between two vortex sheets than in Pumir & Siggia (1990), where one sheet is the "head" stretched in z and the other sheet is the "tail". The strong three-dimensional curvature in figure 13 contrasts strongly with the flattened structures of Pumir & Siggia (1990). In figure 14 the strain is more localized than in Pumir & Siggia (1990).

Figures 6 and 7 have contour plots of the vorticity (ω_y) in the symmetry plane above the dividing plane for different $x - z$ domains. 6a) t=0. Entire $x - z$ domain, 12.56×6.28. The initial peak vorticity in this plane (0.463) is roughly 1/2 the overall peak vorticity due to the smoothing. The initial tail is also an effect of the smoothing. 6b) t=6. Entire $x - z$ domain, 12.56×6.28. Some flattening. The tail is a remnant of the initial tail. 6c) t=9. For (x,z) = (4.71:7.85,3.14). 6d) t=12. For (x,z) = (4.71:7.85,3.14). Continued flattening. Profile is beginning to tip and peak vorticity is locating towards leading edge. Peak vorticity = 2.22. 7a) t=15. For (x,z) = (4.71:7.85,3.14). Peak vorticity is concentrating where vorticity profile is making a transition from parallel to the x-axis to perpendicular to it. Peak vorticity is 23 mesh points above the lower boundary. Peak vorticity = 4.7. 7b) t=16.25. For (x,z) = (4.71:7.85,3.14). Continuation of trends. Note hook where vorticity profile curves back above the peak vorticity. Flattening dominates the tail region, but around the location of peak vorticity there is concentration in both x and z. Peak vorticity (=7.0) is 15 mesh points above the lower boundary. 7c) t=17.5. For (x,z) = (4.71:7.85,3.14). Continuation of trends. More flattening in tail, more concentration near peak vorticity, peak vorticity still located where vorticity profiles change from being parallel to the x-axis to perpendicular to it, hook above vorticity remains. Some of the jaggedness of contours results from the poor resolution of the interpolation mesh (resolution shown by the vertical has marks) used for the graphics. See blowups in z in figure 3. Peak vorticity (15.5) is 15 mesh points above the lower boundary. 7d) t=18.2. For (x,z) = (4.71:7.85,3.14). Continuation of trends. Peak vorticity (=29.2) is 6 mesh points above the lower boundary.

Figures 10 and 11 have three-dimensional isosurface plots of the vorticity squared for a series of times. Different domains and different thresholds are used at the different times. All the plots have a perspective from the position (1.7,-3.5,0.9) if the box shown had unit sides. In figures 10a-c the domain plotted has dimensions $6.28 \times 6.28 \times 6.28$. In figures 10d,11a-d the domain has dimensions $3.9 \times 3.9 \times 3.9$ and covers locations (x,y,z) = (4.9:8.7,3.9,3.9)

10a) t=0. cutoff=0.05*max(ω^2). max(ω^2)=0.90 or max($||omega||$)=0.95. Domain is for (x,y,z) = (6.28:12.56,0:6.28,0:6.28) where the periodic dimensions are $12.56 \times 12.56 \times 6.28$. The initial peak vorticity in the symmetry plane is smaller than the overall maximum due to the initial smoothing. There is a secondary leg on the right, also due to the initial smoothing, that persists throughout the calculation. Based on this figure the maximum vorticity in this leg is about 0.2. At later times contour plots show about the same value and that its position does not change, hopefully implying that its effect on the late time behavior is negligible. This will be checked further. 10b) t=6. cutoff = 0.15*max(ω^2). max(ω^2) = 0.88 or max($||omega||$) = 0.94. and is still not in the symmetry plane. Domain is for (x,y,z) = (4.9:11.2,0:6.28,0:6.28). The vorticity in the symmetry plane is flat-

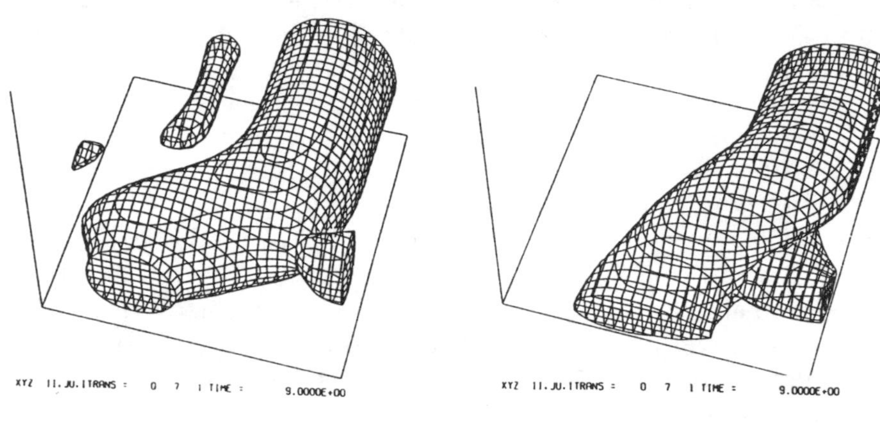

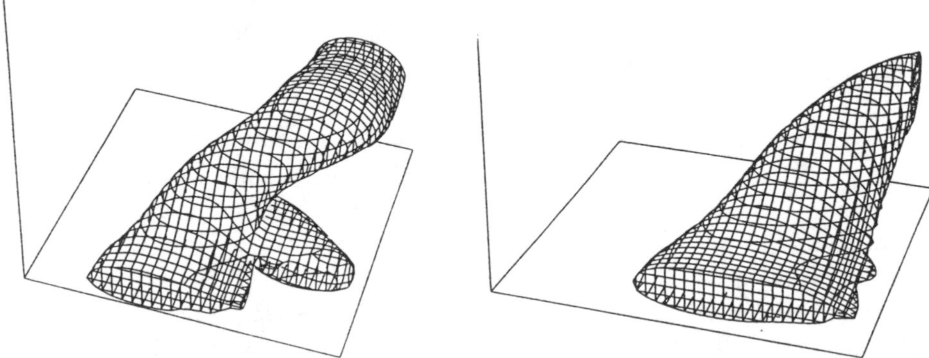

Fig. 10. Isosurface plots of the vorticity. a:t=0, b:t=6, c:t=9, d:t=9.

tening and approaching the dividing plane. 10c) t=9. cutoff = 0.15*max(ω^2). max(ω^2) = 1.66 or max($\|omega\|$) = 1.23. and is now in the symmetry plane. Domain is for (x,y,z) = (4.9:11.2,0:6.28,0:6.28). The tail is beginning to form in the symmetry plane. 10d) t=9. cutoff = 0.35*max(ω^2). max(ω^2) = 0.88 Domain is for (x,y,z) = (4.9:8.7,3.9,3.9). Cutoff chosen so that as peak vorticity increases in the subsequent figures roughly the same volume is covered. 11a) t=12. cutoff = 0.15*max(ω^2). max(ω^2) = 4.93 or max($\|omega\|$) = 2.22. Domain is for (x,y,z) = (4.9:8.7,3.9,3.9). 11b) t=15. cutoff = 0.15*max(ω^2). max(ω^2) = 22.1 or max($\|omega\|$) = 4.70.

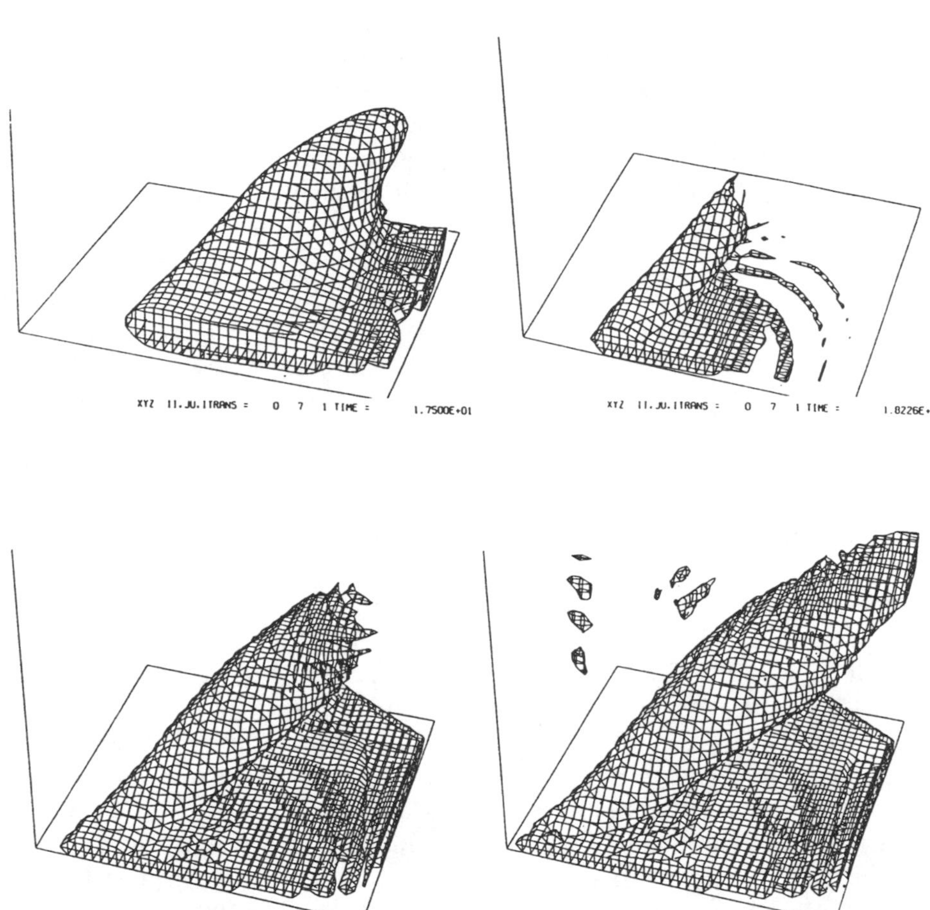

Fig. 11. Isosurface plots of the vorticity. a:t=12, b:t=15, c:t=17.5, d:t=18.2.

Domain is for (x,y,z) = (4.9:8.7,3.9,3.9). First clear evidence of a head tail structure. How head angles back as y increases should be noted. Filaments coming off the back of the tail have no significance. 11c) t=17.5 cutoff = 0.01*max(ω^2). max(ω^2) = 239 or max($\|omega\|$) = 15.5. Domain is for (x,y,z) = (4.9:8.7,3.9,3.9). Besides angling back as as y increases it seems the "head" also increases in radius. The head in the symmetry plane can no longer be resolved due to the resolution of the graphics. 11d) t=18.2 cutoff = 0.003*max(ω^2). max(ω^2) = 851 or max($\|omega\|$) = 29.2. Domain is for (x,y,z) = (4.9:8.7,3.9,3.9). Last time calculated. Continuation of trends in figure 11c.

Figure 12 has contour plots of the vorticity (ω_y) in the symmetry plane above the dividing plane for one $x-z$ domain with the z direction expanded by a factor of 5. The domain is for (x,z) = (4.7:8.9,0:0.63). (a,b,c,d) are for times (12,15,17.0,17.5) respectively. This emphasizes how the peak vorticity is in the corner where two vortex sheets, one at a slight angle to the x-axis and one almost perpendicular to the x-axis. As the vorticity increases it concentrates further into this corner. This is the same coordinates expansion as used by Pumir & Siggia (1990). The general picture is similar to an early time from their calculation, afterwards their profiles are much flatter, that is parallel to the x-axis with an insignificant vortex sheet perpendicular to the x-axis.

Figure 13 has several three-dimensional isosurface plots of the vorticity squared for t=15. The (x,y,z) axes have the ratio (6.5,3.25,1) for the domain (4.9:8.7,0:1.9,0:0.6). This is similar to the (7,3,1) ratio of axes used by Pumir & Siggia (1990). The perspectives all look down on the dividing plane and begin with a view of the symmetry plane for comparison with the two-dimensional contours, then rotate around in front of (downstream of) the reconnection region. cutoff=0.15*max(ω^2), the same as in figure 11b. a) A perspective from the position (1.7,-3.5,0.9) if the box shown had unit sides. This is the same view as figure 11b. b) Perspective (-1.7,-3.5,0.9). c) Perspective (-1.7,3.5,0.9). d) Perspective (1.7,3.5,0.9).

Figure 14 has contour plots in the symmetry plane of the out-of-plane strain along vorticity for (a,b,c,d), t=(12,15,17.5,18.2) respectively. The domain is for (x,z) = (4.71:7.85,3.14), the same as figure 7. This component of the strain is responsible for vorticity production in the symmetry plane. Its maximum is always approximately 50% larger than the the value at the peak vorticity. Both the maximum and the value go as $1/(t_c-t)$ Note how the peak of this component of the strain moves and concentrates with the peak vorticity, always remaining a bit above and behind the peak vorticity. This is similar to an early time from Pumir & Siggia (1990), but at later times their strain contours were more diffuse.

Figure 15 has contour plots in the symmetry plane of the in-plane stresses at t=15. The domain is for (x,z) = (4.71:7.85,3.14), the same as figures 7 and 14. The significance of these components is how they hold the vorticity in place and do not allow the vorticity to flatten and two-dimensionalize the way it does in Pumir & Siggia (1990). Of particular significance could be the sharp gradients of the components of the stress where they change sign near the peak vorticity. These should be compared with figure 7a.

a) $e_{xx}=\partial u_x/\partial x$. Just in front of the peak vorticity this is compressive. Only in the tail is it positive, with a magnitude so low that it no longer appears in this figure.

b) $\partial u_x/\partial z$. This is the dominant contribution to the vorticity ω_y and the cross strain e_{xz}. Due to this e_{xz} dominates the evaluation of the eigenvalues of the in-plane strain α, γ and their magnitude is much greater than the out-of-plane component $\beta = e_{yy}$ in figure 14. This implies that the normalized value of $\sqrt{6}\frac{\beta}{|e|}$ introduced by Kerr (1987) is nearly zero. It also implies that the γ, the negative principle component or compression is along the (1,1) direction, while the dominant positive component or expansion α is in the (-1,1) direction. This would squeeze the vorticity into the corner where the vortex sheets meet and pull it out into the two nearly perpendicular sheets. The rapid gradients of the other components of the strain are what turn the vorticity as it is pulled out and maintain it in the two nearly perpendicular vortex sheets.

c) $\partial u_x/\partial z$. The minor contribution to the vorticity ω_y and the cross strain e_{xz}.

d) $e_{zz} = \partial u_z/\partial z$. As with e_{xx} it is greatest just in front of the peak vorticity and changes sign rapidly near the peak vorticity. e_{xx} and e_{zz} just in front of the peak vorticity seem to be responsible for turning the vorticity that would otherwise be pulled out of the region of peak vorticity and maintaining it in the vortex sheet perpendicular to the x-axis. e_{zz} is negative in the tail, where is is responsible for the flattening of the tail.

Figure 16 has contour plots in the xy plane parallel to the dividing plane through the peak vorticity at $z = 0.11$. (x,y)=(4.7:7.8,0:3.14) Note the spatial dependence in y as the contours curve backwards. For the results of Pumir & Siggia (1990) these contours would be be much more extensive in y. a) ω_y. Note the sharp contours on the leading edge. b) Out-of-plane strain e_{yy}. Note how while it is stretching in the symmetry plane, it quickly changes to compressive to prevent the peak vorticity from flattening too much in y. c) e_{xx} Compressing. Sharp contours are coincident with the sharp contours of vorticity in 16a. d) e_{zz} Expanding into perpendicular vortex sheet. Sharp contours are again coincident with the sharp contours of vorticity in 16a.

Figures 17 and 18 have contour plots in the yz plane perpendicular to both the dividing plane and the symmetry plane through the peak vorticity at $x = 5.89$ (with one exception). (y,z) = (0:3.14,0:3.14). 17a) ω_y. 17b) Out-of-plane strain e_{yy}. It is positive, stretching, only right and below the peak vorticity. Around this it is compressing to help focus the peak vorticity. 17c) e_{xx} Compressing at the peak vorticity. Sharp contours are coincident with the sharp contours of vorticity in 17a. 17d) $\partial u_x/\partial z$. Essentially the same as the vorticity ω_y. 18a-c) $\partial u_z/\partial x$, the other component of ω_y, for two x locations, $x = 5.79$ just in front of the peak vorticity and $x = 5.89$ through the peak vorticity. Although this component of the velocity derivative does not contribute to the components of the vorticity, these two slices are used to demonstrate how the vorticity curves up from the dividing plane, out from the symmetry plane and back from the peak vorticity. 18d) e_{zz} At the point of peak vorticity there is upwards expansion into the vortex sheet perpendicular to the x-axis, but above that the behavior is different. Sharp contours are again coincident with the sharp contours of vorticity in 17a.

6. Conclusion

Evidence has been presented that there is a finite-time singularity of the incompressible Euler equations from smooth initial conditions in a finite geometry. The evidence suggests that the singularity occurs at a point, although since the energy spectrum does not become infinitely steep the singularity would not be on a set of measure zero. But it also is not the singularity that is predicted by spectral closures, which predict that the enstrophy, or mean square vorticity, blows up. In all these calculations the enstrophy remains finite. Viscous calculations in a periodic geometry currently suggest that the growth of enstrophy is exponential up until a viscous timescale.

Which of the two classes of solutions is relevant to understanding turbulence? The evidence is that the singular solution is relevant to understanding fully-developed turbulence. There are signs from calculations with fewer symmetries in a periodic box that some properties of the singularity, such as the exponent γ and the slope of the energy spectrum at the time of singularity, are universal. On the other hand the first solution might be more relevant to viscous boundary layers, where a region of negative vorticity between the primary vortex and the wall, similar to those appearing from the unfiltered initial conditions, does form and move in front of the primary vortex, leading to what is known as the bursting phenomena.

The energy spectrum associated with the singularity in this calculation and the periodic calculations is very steep, much too steep for the singularity being able to account directly for the -5/3 Kolmogorov spectrum. The spectra go roughly as k^{-4}, which is the spectrum for a vorticity jump. But there are some indications that the singularity might be able to account for roughly the value of the velocity derivative skewness, or normalized vortex stretching, which based on numerical simulations (Kerr, 1985) and moderate Reynolds number simulations might have a universal value of about -0.5. The singularity might also be closely associated with the distribution of strains observed by Kerr (1987) and Ashurst et. al. (1987) and the scatter plots of invariants of the strain by Chen et. al. (1990). In order to determine a relationship to the Kolmogorov cascade new simulations that relax the symmetries imposed here are needed. In particular simulations where helicity plays an active role are essential. Some work in this direction has been begun and suggests that the singularity might provide a seed for the viscous dynamics that produces the cascade.

Acknowledgements

The author wishes to thank R.D. Moser for suggesting the numerical method, M. Brachet for communicating results prior to publication, J. Herring for helpful suggestions and the participants in the workshop on Topological Fluid Dynamics at the Institute for Theoretical Physics at the University of California, Santa Barbara, which is supported by NSF Grant PHY89-04035. Support of AFOSR is acknowledged. NCAR is supported by the National Science Foundation.

References

ASHURST, W. T., KERSTEIN, A. R., KERR, R. M. AND GIBSON, C. H. 1987 Alignment of vorticity and scalar gradient with strain rate in simulated Navier-Stokes turbulence. *Phys. Fluids* **30**, 2343–2353.

ASHURST, W. & MEIRON, D. 1987 Numerical study of vortex reconnection. *Phys. Rev. Lett.* **58**, 1632–1635.

BEALE, J.T., KATO, T. & MAJDA, A. 1984 *Comm. Math. Phys.* **94**, 61.

BRACHET, M. E., MEIRON, D. I., ORSZAG, S. A., NICKEL, B. G., MORF, R. H., & FRISCH, U. 1983 Small-scale structure of the Taylor-Green vortex. *J. Fluid Mech.* **130**, 411–452.

BREUER, K.S. & EVERSON, R.M. 1991 On the errors incurred calculating derivatives using Chebyshev polynomials. Preprint.

CAFFARELLE, L., KOHN, R. & NIRENBERG, L. 1982 *Commun. Pure Appl. Math.* **35**, 771.

CHEN, J.H., CANTWELL, B., AND OTHERS 1990 in CTR-Summer '90 report. *CTR-S90*, Stanford U.

FRISCH, U., POUQUET, A., SULEM, P.-L. & MENEGUZZI, M. 1983 The dynamics of two-dimensional ideal MHD. *J. Mécanique Théor. Appliquée.* **2D**, 191–216.

KERR, R. M. 1985 Higher-order derivative correlations and the alignment of small-scale structures in isotropic numerical turbulence. *J. Fluid Mech.* **153**, 31–58.

KERR, R. M. 1987 Histograms of helicity and strain in numerical turbulence. *Phys. Rev. Lett.* **59**, 783–786.

KERR, R. & HUSSAIN, F. 1989 Simulation of vortex reconnection. *Physica D* **37**, 474–484.

MELANDER, M. V. & HUSSAIN, F. 1989 Cross-linking of two antiparallel vortex tubes. *Phys. Fluids.*, **A1**, 633–636.

MOSER, R. D., MOIN, P. & LEONARD, A. 1983 A spectral numerical method for the Navier-Stokes equations with application to Taylor-Couette flow. *J. Comp. Phys.* **52**, 524–544.

ORSZAG, S.A. & TANG, C.M. 1979 Small-scale structure of MHD turbulence. *J. Fluid Mech.* **90**, 129–143.

PUMIR, A. & KERR, R. M. 1987 Numerical simulation of interacting vortex tubes. *Phys. Rev. Lett.* **58**, 1636–1639.

PUMIR, A. & SIGGIA, E. D. 1987 Vortex dynamics and the existence of solutions of the Navier-Stokes equations. *Phys. Fluids.* **30**, 1606–1626.

PUMIR, A. & SIGGIA, E. D. 1990 Collapsing solutions to the 3-D Euler equations. *Phys. Fluids.* **A2**, 220–241.

SIGGIA, E. D. 1985 Collapse and amplification of a vortex filament. *Phys. Fluids.* **28**, 794–805.

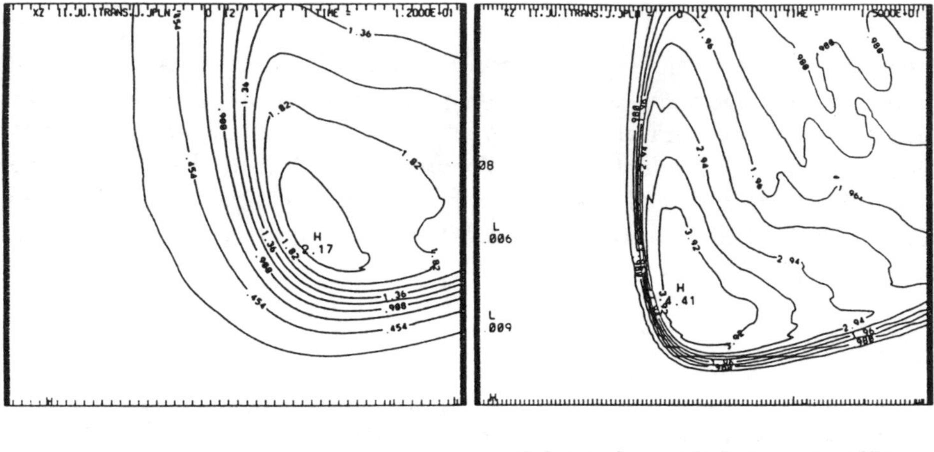

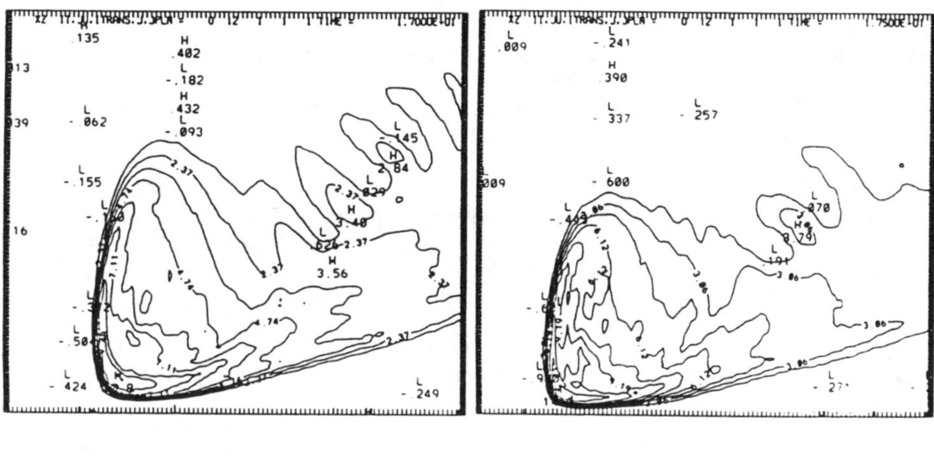

Fig. 12. Vorticity (ω_y) in the symmetry plane, z stretched by 5. a:t=12, b:t=15, c:t=17.0, d:t=17.5.

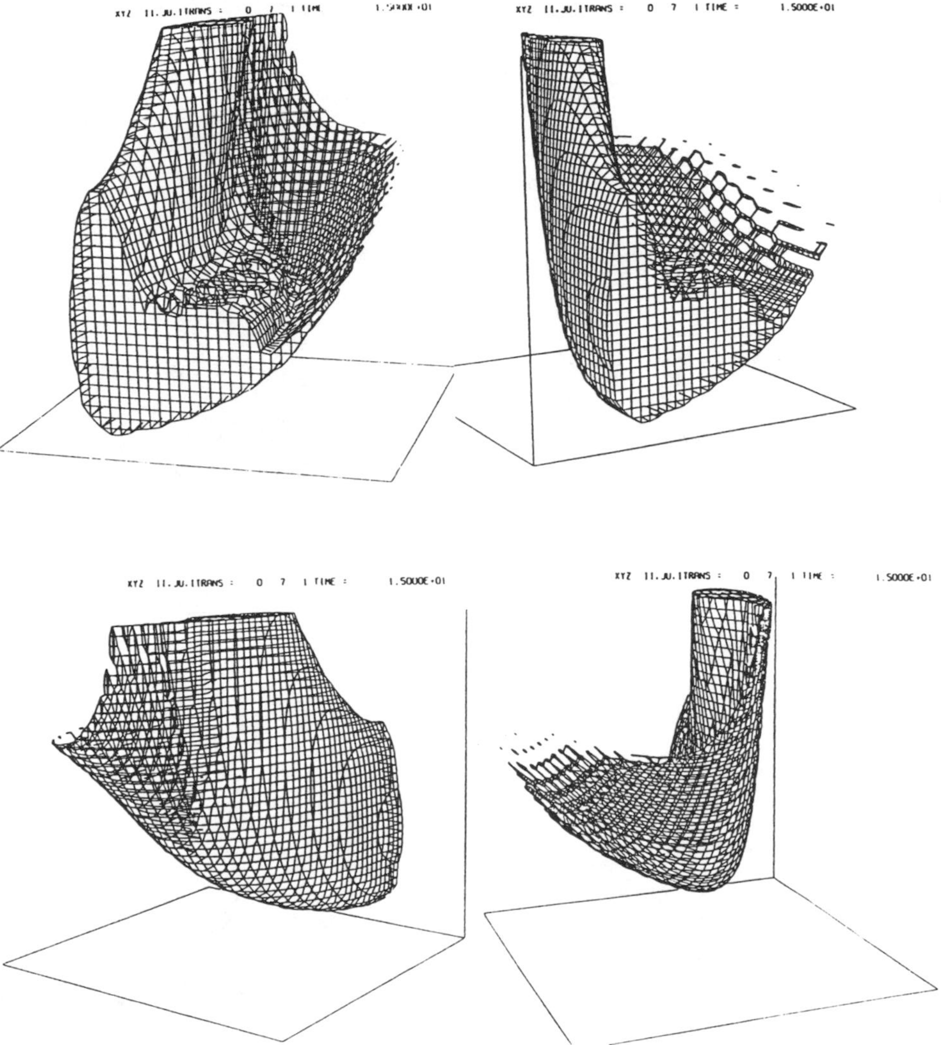

Fig. 13. Isosurface plots of the vorticity squared for t=15 from several perspectives.

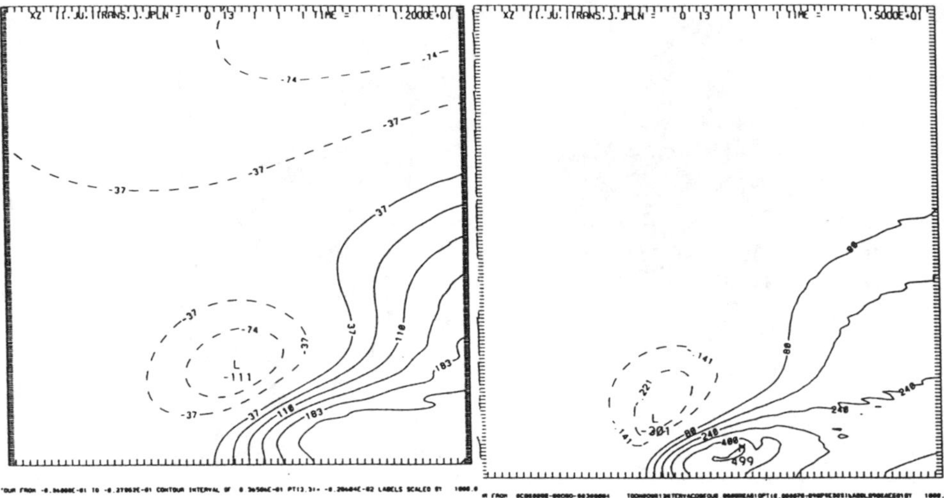

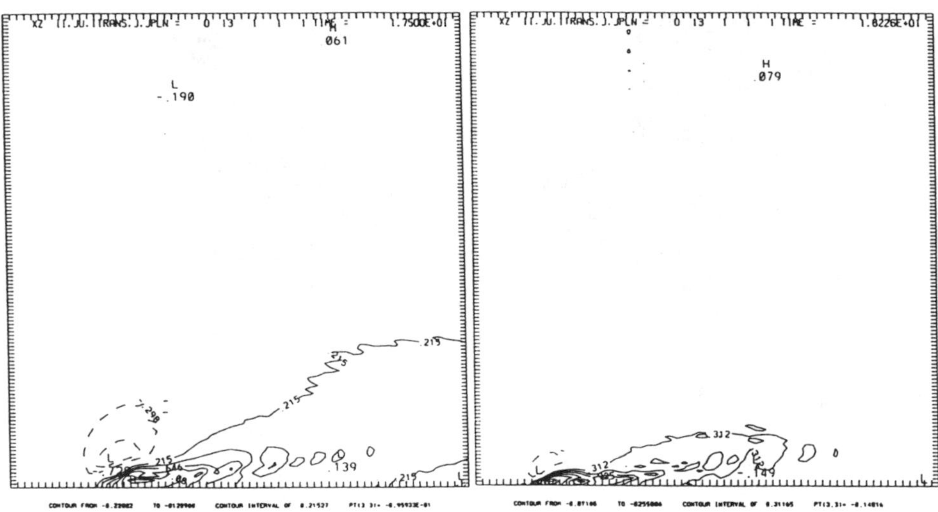

Fig. 14. Out-of-plane strain e_{yy} in the symmetry plane. a:t=12, b:t=15, c:t=17.5, d:t=18.2.

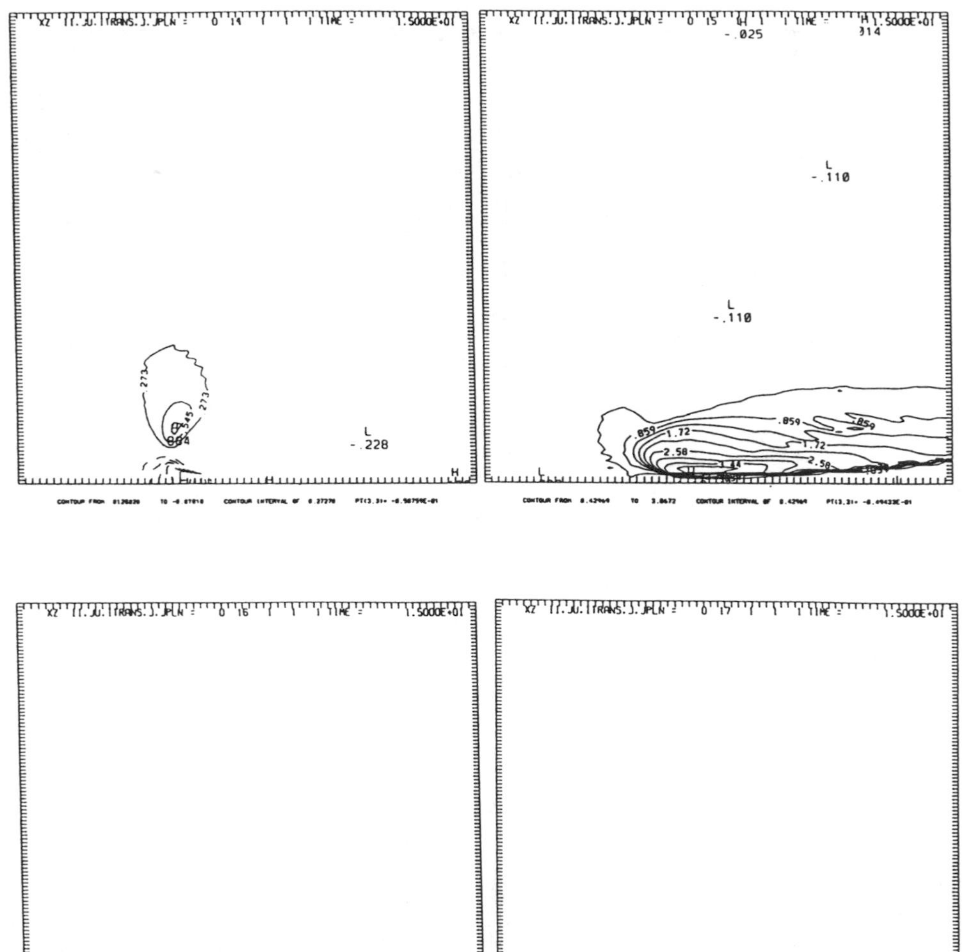

Fig. 15. In-plane stresses in the symmetry plane at t=15. a:$u_{x,x}$, b:$u_{x,z}$, c:$u_{z,x}$, d:$u_{z,z}$.

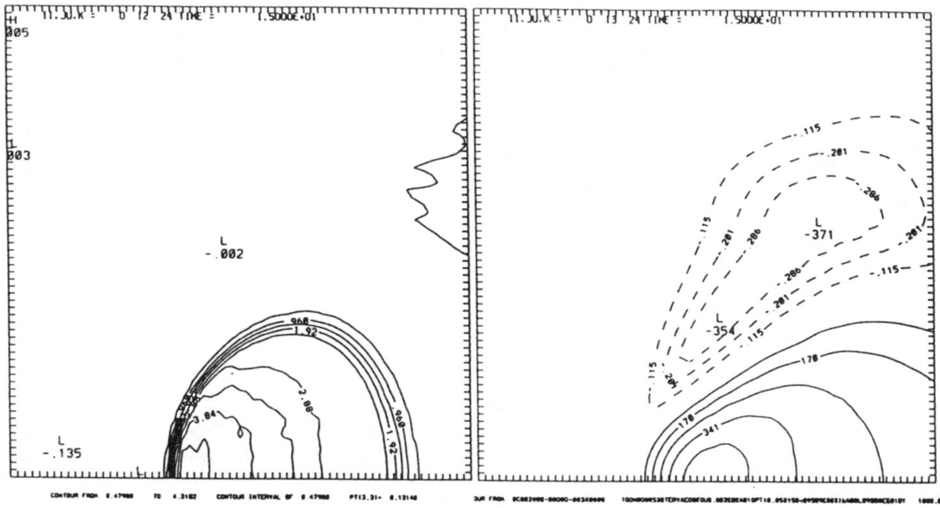

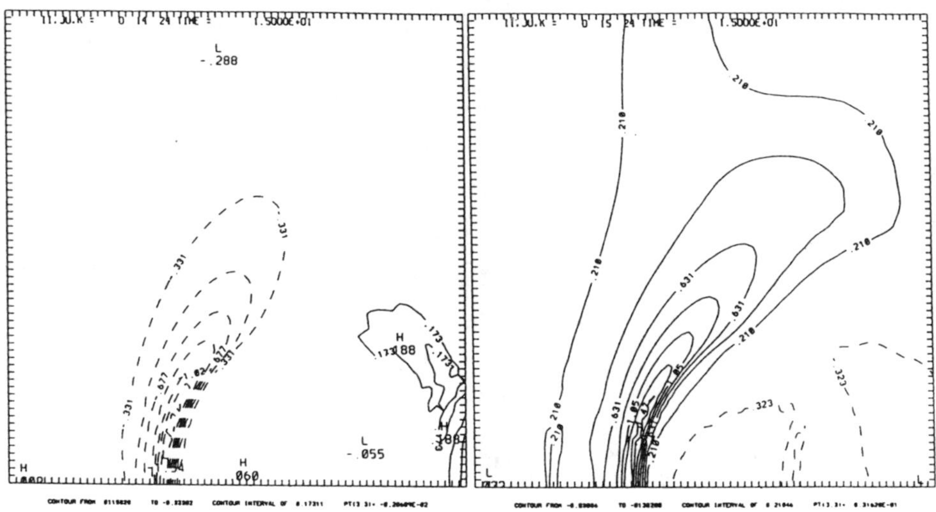

Fig. 16. xy contours at t=15. a:ω_y , b:e_{yy} , c:e_{xx} , d:e_{zz}.

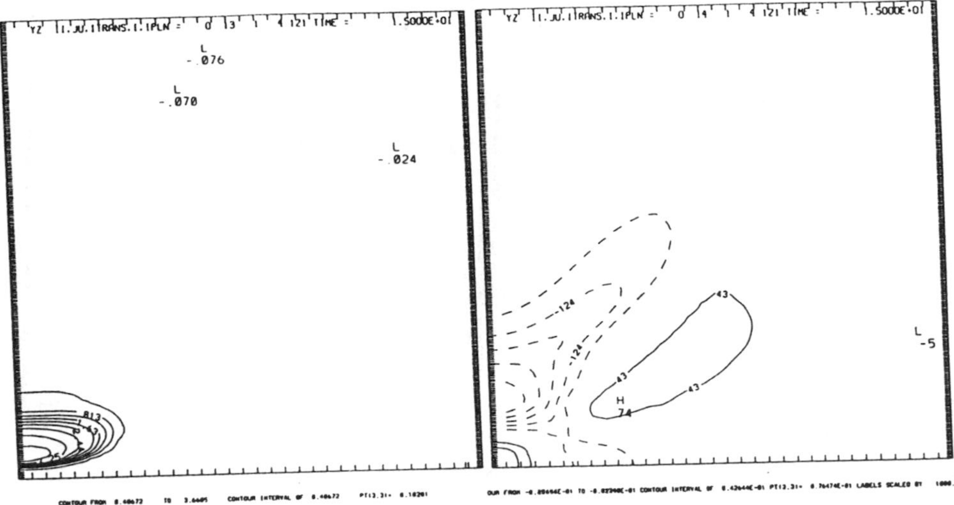

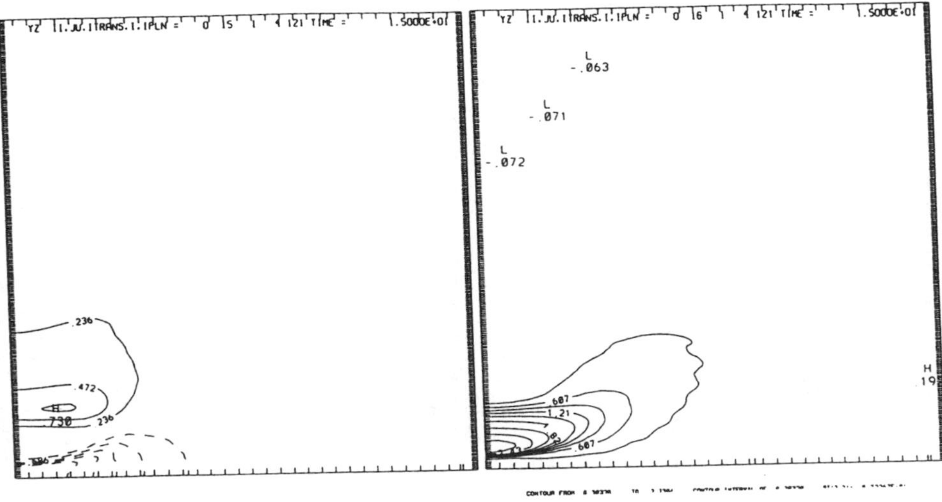

Fig. 17. yz contours at t=15. a:ω_y , b:e_{yy} , c:e_{xx} , d:$u_{x,z}$.

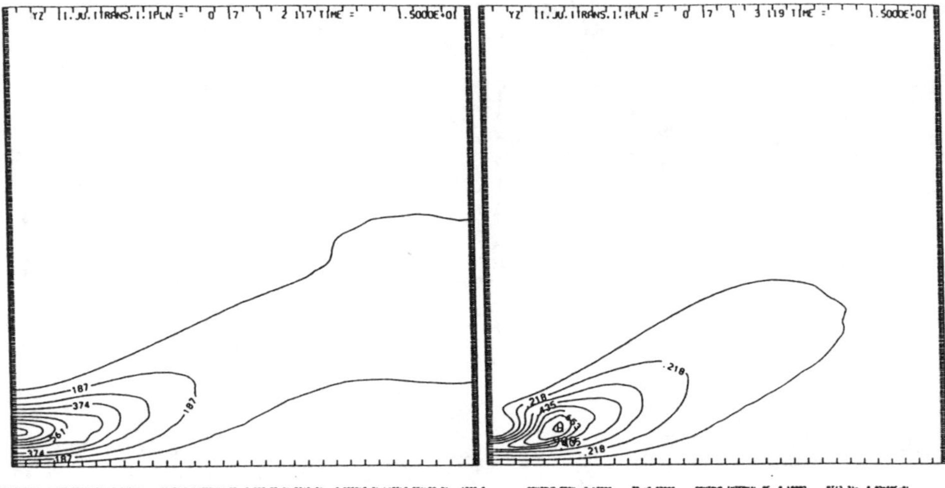

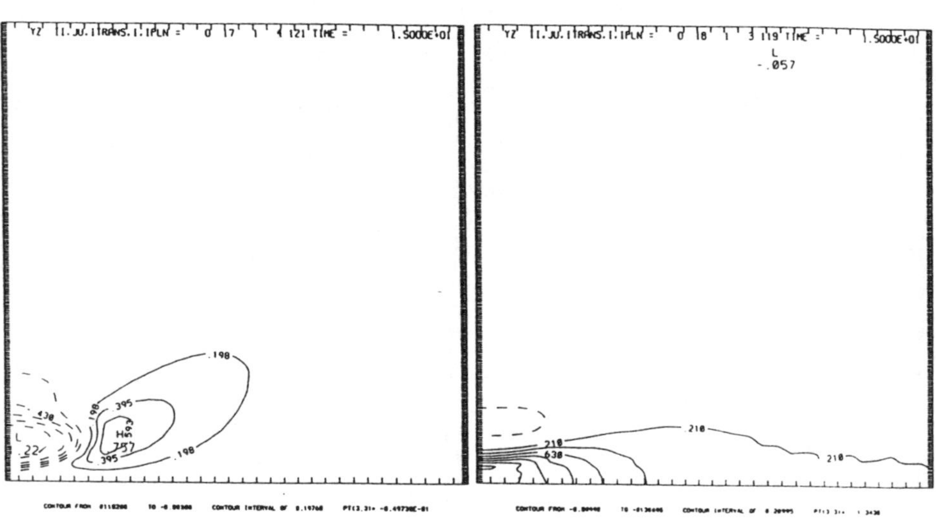

Fig. 18. yz contours at t=15. a:$u_{z,x}$, $x = 5.79$, b:$u_{z,x}$, $x = 5.89$, c:$u_{z,x}$, $x = 5.99$, d:e_{zz}.

SINGULARITY FORMATION ON VORTEX SHEETS: THE RAYLEIGH-TAYLOR PROBLEM

M. S. SIEGEL
Department of Mathematics
Ohio State University
231 W. 18 St.
Columbus, Ohio 43210

ABSTRACT. Evidence is now quite strong that vortex sheets with mean vorticity density develop a finite time curvature singularity when viscosity and surface tension are neglected. This singularity is due to a global Kelvin-Helmholtz instability, which in turn results from the presence of a mean sheet vorticity. We examine the possibility of singularity formation in the related Rayleigh-Taylor problem. In this problem density stratification causes vorticity to be barotropically generated on the sheet, although there is no mean sheet vorticity. A simple approximate theory is developed as a generalization of Moore's approximation for vortex sheets in the Kelvin-Helmholtz problem. For the approximate theory, a family of exact solutions is found in the form of traveling waves in the complex plane for which singularities develop on the interface. These exact solutions eliminate the need for a second asymptotic analysis like that employed by Moore. The resulting predictions for the time, type and location of singularity formation are directly verified by numerical computation for the full Rayleigh-Taylor problem. Excellent agreement between the theoretical predicitons and the numerical results is exhibited over a wide range of Atwood numbers. For A near 1, however, some aspects of the agreement break down. An additional traveling wave solution to the approximate equations is found which exhibits a new and stronger singularity.

1. Introduction

The Rayleigh-Taylor instability for 2-D incompressible, inviscid flows is the fingering instability which occurs when a heavy fluid falls into a light fluid under the influence of gravity. If the fluids are assumed to be irrotational, then all of the vorticity in the flow is confined to the interface. Consequently, the interface can be regarded as a vortex sheet whose strength changes in time due to the baroclinic generation of vorticity. The fluid velocity is determined by the location and strength of the sheet. This is one example of a flow which can be based on a vortex sheet representation; other examples are the Kelvin-Helmholtz flow in which the vorticity is present initially and

H. K. Moffatt et al. (eds.), Topological Aspects of the Dynamics of Fluids and Plasmas, 337–349.

Hele-Shaw flow for which vorticity is produced by a jump in viscosity across the interface.

Numerical vortex methods which utilize the simplicity inherent in the vortex sheet representation have been employed in several studies of the Rayleigh-Taylor problem. These provide us with the basic non-linear behavior of the vortex sheet, as well as elucidating the formation of certain structures on the sheet. The behavior of a single wavelength initial disturbance at large amplitudes and various density ratios can be found in Kerr (1986) , Trygvasson (1988) , and Baker, Meiron, & Orszag (1980) . Initially, the sheet deforms into a pattern containing rising bubbles of light fluid and falling spikes of heavy fluid. Local shear flows are produced on the sides of a penetrating finger. This region is then susceptible to the Kelvin-Helmholtz instability. After a sufficient amount of time this causes the interface to roll up into two counter rotating spirals. For larger density differences, the penetrating finger is narrower and the spirals appear further down on the sides. When the lower fluid has zero density (*i.e.* the heavy fluid is penetrating into a vacuum) there is no roll up. This is expected since a second fluid is necessary to produce the local Kelvin-Helmholtz instability which leads to roll up.

Attempts to numerically track the evolution of the vortex sheet fail at the onset of roll up. It has been presumed that the formation of a weak singularity in the vortex sheet contributes to the difficulties. Trygvasson (1988) points out that just before roll up in the pure Kelvin-Helmholtz problem the vortex sheet has a sharp spike, hence attempts to follow the roll up numerically by modeling the sheet as a row of point vortices are unreliable. Routines which have had success following the interface well into the roll up stage have employed some type of numerical regularization. Baker, Meiron & Orszag (1980) report only limited success with a non-regularizing boundary integral method, since in order to track the roll up it was necessary to keep the number of discretizing points low. The situation is different in the case of a fluid penetrating into a vacuum, where no roll up (and presumably no singularity) occurs and Baker *et al.* were able to follow the evolution of an interface perturbed by a single wave for as long as they wished.

The formation of singularities in some situations is also suggested by a linear stability analysis of small sinusoidal disturbances to a flat vortex sheet (Moore 1979) . Such disturbances grow at a rate given by

$$\omega(k) = \sqrt{\frac{1-A^2}{4}k^2\gamma_0{}^2 - Agk} \tag{1.1}$$

where A is the Atwood number (defined later), γ_0 is the mean sheet strength, and $-g$ is the downward acceleration due to gravity. Let the disturbance be given by $s(\alpha, t)$, where α is a variable which parameterizes the disturbance, and write s in a Fourier series as $s(\alpha,t) = \Sigma_k \hat{s}_k(t)e^{ik\alpha}$. If the initial disturbance is such that only the growing modes occur, then the kth Fourier coefficient of s has the form

$$\hat{s}_k(t) = \hat{s}_k(0)e^{\omega(k)t} \tag{1.2}$$

with $\omega(k)$ given in (1.1). Consider an initially analytic disturbance with $\hat{s}_k(0) = e^{-|k|}|k|^{-p}$ $(p > 0)$. For the Kelvin-Helmholtz problem, where $\gamma_0 \neq 0$, it is clear that a critical time exists where the exponential growth in k of $e^{\omega(k)t}$ overcomes the exponential decay of $\hat{s}_k(0)$ and the analyticity of the disturbance is lost. Alternatively, if one allows for complex α, then there exists a critical value of of $Im\ \alpha$ where the growth of $e^{ik\alpha}$ in k overcomes the decay of $\hat{s}_k(0)$ and again analyticity is lost. This motivates the idea that singularities in the Kelvin- Helmholtz problem form out in the complex plane and move toward the real α axis as time is increased. The singularities are physically realized when they intersect this axis.

Unfortunately, in the pure Rayleigh-Taylor problem (where $\gamma_0 = 0$) this argument breaks down as the growth of e^{wt} for $w(k) \sim \sqrt{k}$ and finite t is not strong enough to overcome the decay of $\hat{s}_k(0)$. On the other hand, baroclinic generation of vorticity on the interface creates regions along the sheet where there is a mean level to the vortex sheet strength locally. Moreover, the region along the sides of the spikes contains vorticity advecting downwards from the bubble and upwards from the spike. The conditions are right for the nonlinear development of singularities. In this paper we present a simple nonlinear theory and computations which show that singularities do occur.

2. Governing Equations

We consider the two dimensional motion due to gravity of a fluid of density ρ_1 lying under a fluid of density ρ_2 with $\rho_1 \leq \rho_2$. The fluids are assumed to be incompressible, inviscid and irrotational. We also assume that there is no surface tension at the interface between the fluids (the effects of surface tension are considered in Siegel (1991)). The interface is expressed as a function $z(\alpha,t) = x(\alpha,t) + iy(\alpha,t)$ in the complex plane, where t is time and α is a real Lagrangian variable used to parametrize the interface. As mentioned in section 1, the interface is a vortex sheet. The vorticity density (per unit arclength) is defined by

$$\Gamma = (\mathbf{u}_1 - \mathbf{u}_2) \cdot \hat{\boldsymbol{s}} \tag{2.1}$$

where $\hat{\boldsymbol{s}}$ is the unit tangent vector. The direction of positive $\hat{\boldsymbol{s}}$ is chosen so that Γ is positive for counterclockwise vorticity. It is also convenient to define the unnormalized vortex sheet strength γ by

$$\gamma = \Gamma \mid z_\alpha \mid . \tag{2.2}$$

Both Γ and γ change in time due to the presence of the density discontinuity.

The evolution equations for the position and strength of the sheet were derived by Baker *et al.* (1982) and are given by

$$\frac{\partial \bar{z}}{\partial t} = \bar{q} + \frac{\beta\gamma}{2z_\alpha} \tag{2.3}$$

$$\frac{\partial \gamma}{\partial t} = 2A\left[Re\left\{\frac{\partial \bar{q}}{\partial t}z_\alpha\right\} - \frac{1}{2}\beta\gamma Re\left\{\frac{q_\alpha}{z_\alpha}\right\} + \frac{1}{8}\frac{\partial}{\partial\alpha}\frac{\gamma^2}{z_\alpha\bar{z}_\alpha} + gy_\alpha\right] + \frac{\beta}{2}\frac{\partial}{\partial\alpha}\frac{\gamma^2}{z_\alpha\bar{z}_\alpha} \tag{2.4}$$

where

$$\bar{q} = \frac{1}{2\pi i}\int_{-\infty}^{\infty}\frac{\gamma(\alpha',t)}{z(\alpha,t)-z(\alpha',t)}d\alpha' \tag{2.5}$$

is the Birkhoff-Rott integral, $A = \dfrac{\rho_2-\rho_1}{\rho_1+\rho_2}$ is the Atwood number and g is the acceleration due to gravity directed along the -y axis. The slash indicates Cauchy principal value integration. The terms in (2.3) and (2.4) which contain the parameter β are a consequence of the freedom which is allowed in defining the tangential velocity of the sheet points. We define the velocity of a Lagrangian point labeled by α as a weighted average of the limiting velocities on either side of the interface:

$$\mathbf{U}_{sheet} = \frac{1+\beta}{2}\mathbf{U}_1 + \frac{1-\beta}{2}\mathbf{U}_2 \tag{2.6}$$

where $\mathbf{U}_1$ and $\mathbf{U}_2$ are the limiting velocities of the upper and lower fluids respectively as z approaches the sheet. Note that by choosing $\beta = 1$ the Lagrangian markers follow the upper fluid, and for $\beta = 0$ they move at the average velocity of the two fluids. A sheet velocity other than the average of the two limiting velocities was first used in Baker *et al.* (1982) in order to improve the accuracy of numerical simulations. We have also found some theoretical advantages in using this definition.

3. Localized Approximation

In this section we sketch the derivation of a system of local equations which approximate the Birkhoff-Rott and vorticity evolution equations (2.3, 2.4). We also discuss some properties of this system.

A simple solution to equations (2.3, 2.4) is

$$z(\alpha,t) = \alpha + \frac{\beta\gamma_0}{2}t \quad \gamma(\alpha,t) = \gamma_0$$

for any constant γ_0. This corresponds to a flat interface with uniform density whose points translate horizontally with speed $\beta\gamma_0/2$. This motion would be an equilibrium solution in an Eulerian frame of reference. We consider periodic perturbations of this basic state, so that

$$z(\alpha,t) = \alpha + \frac{\beta\gamma_0}{2}t + s(\alpha,t) \tag{3.1}$$

with $s(\alpha+2\pi,t) = s(\alpha,t)$. Define an operation * by

$$s^*(\alpha) = \bar{s}(\bar{\alpha}). \tag{3.2}$$

Note that s^* is the unique analytic function which equals $\bar{s}(\alpha)$ for α real. With this definition equation (2.3) can be continued analytically as

$$\frac{\partial s^*}{\partial t}(\alpha,t) = q^*[s,\alpha](\alpha,t) + \frac{\beta\gamma}{2(1+s_\alpha)} \tag{3.3}$$

where

$$q^*[s,\gamma](\alpha,t) = \frac{1}{2\pi i}\int_{-\infty}^{\infty}\frac{\gamma(\alpha',t)}{\alpha-\alpha'+s(\alpha,t)-s(\alpha',t)}d\alpha' \tag{3.4}$$

is the analytic continuation of the Birkhoff-Integral and it is implicitly understood that the integration is along the line $Im\ \alpha' = Im\ \alpha$. The equation for $\partial\gamma/\partial t$ can be analytically continued in a similar fashion.

It is clear that knowledge of the behavior in the complex plane of the kernel in the principal value integral (*i.e.* positions and types of singularities, behavior at infinity) would enable its evaluation via integration over a suitably deformed contour. This motivates a decomposition of the kernal into a sum of terms with known analytic behavior in the complex plane, plus an error term. Specifically, if s and γ are analytic in the strip $|Im\ \alpha| < \rho$, then we can write

$$\begin{aligned} s &= s_+ + s_- + s_0(t) \\ \gamma &= \gamma_+ + \gamma_- + \gamma_0 \end{aligned} \tag{3.5}$$

where

$$\begin{aligned} s_+(\alpha,t) &= \Sigma_{k>0}\hat{s}_k(t)e^{ik\alpha} \\ s_-(\alpha,t) &= \Sigma_{k<0}\hat{s}_k(t)e^{-ik\alpha} \\ \gamma_+(\alpha,t) &= \Sigma_{k>0}\hat{\gamma}_k(t)e^{ik\alpha} \\ \gamma_-(\alpha,t) &= \Sigma_{k<0}\hat{\gamma}_k(t)e^{ik\alpha} \end{aligned} \tag{3.6}$$

and γ_0 is a constant. For our purposes $s_0(t)$ may be ignored (see Baker, Caflisch & Siegel (1991)). Since s_+, γ_+ contain the positive wavenumber components of s, γ respectively, they are analytic in the upper half plane $Im\ \alpha > -\rho$. Similarly s_-, γ_- contain the negative wavenumber components and are analytic in the lower half plane $Im\alpha < \rho$. Now, for any function or operation $A[s,\gamma]$, we can write

$$A[s,\gamma] = A[s_+,\gamma_+] + A[s_-,\gamma_-] + E[s_+,s_-,\gamma_+,\gamma_-] \tag{3.7}$$

This is just a definition of the term E. The approximation takes the form

$$A[s,\gamma] \sim A[s_+,\gamma_+] + A[s_-,\gamma_-] \tag{3.8}$$

It can be deduced from (3.7) that E contains only terms with cross products of upper and lower analytic functions, such as s_+s_- or γ_-s_+. Pure products such as s_+^2 or γ_-s_- are contained in $A[s_+,\gamma_+]$ and $A[s_-,\gamma_-]$. Thus the approximation (3.8) amounts to neglecting all cross products of upper and lower analytic functions.

For $A[s,\gamma] = q^*[s,\gamma]$, *i.e.* the Birkhoff-Rott integral, the advantage of the approximation (3.8) over the original operator (3.7) is that the two integral operators on the right of (3.7) involve an integrand that is analytic in the upper and lower half planes. Therefore, as discussed, we can directly evaluate them by contour integration. This transforms the integral operators into local differential operators which come from the residues.

The approximation (3.8) is performed on each term of the evolution equations (2.3, 2.4). After evaluating the residues, projecting onto upper analytic function space, and differentiating with respect to α, we obtain the following system of equations for $\phi = 1 + s_{+\alpha}, \psi = 1 + s^*{}_{-\alpha}$ and $\omega = \gamma_+$;

$$\begin{aligned}
\frac{\partial\phi}{\partial t} &= \frac{\beta+1}{2}\frac{\partial}{\partial\alpha}\left(\frac{\omega}{\psi}\right)\\
\frac{\partial\psi}{\partial t} &= \frac{\beta-1}{2}\frac{\partial}{\partial\alpha}\left(\frac{\omega}{\phi}\right)\\
\frac{\partial\omega}{\partial t} &= \frac{\beta+A}{2}\frac{\partial}{\partial\alpha}\left(\frac{\omega}{\phi\psi}\right) - iAg(\phi-\psi)
\end{aligned} \tag{3.9}$$

This is the system of localized approximations equations to (2.3, 2.4) and is the main result of this section. The localized approximation equations for the pure Kelvin-Helmholtz problem are derived in Moore (1985) and Caflisch & Orellana (1986).

For the approximation (3.8) to be reasonable it is necessary that the error term E be negligible compared with the other terms in (3.7). That this is so can be seen from the following physical argument: Nonlinear interactions cause wavenumbers to add and result in energy flowing back and forth among Fourier components. However, the process of singularity development primarily involves the flow of energy from low wavenumbers to high wavenumbers. Our approximation is to keep this outflow of energy to high wave numbers, but to ignore all backflow. This is accomplished by neglecting the interaction of two wavenumbers k_1 and k_2 if the sign of k_1 differs from the sign of k_2. Mathematically this is the same as keeping interactions of s_+ with s_+ and of s_- with s_-, but neglecting interactions of s_+ with s_-.

A more detailed discussion of the error term E is given in Caflisch & Orellana (1986) and Caflisch, Orellana & Siegel (1990). A rigorous upper bound on the size of E can be obtained through an analysis similar to that employed by Caflisch and Orellana for the pure Kelvin Helmoltz problem, where $\gamma(\alpha, t) = \text{constant}$ (Siegel 1989).

Along with its local character, the system of equations (3.9) has several features which make it useful for prediction of singularity formation on the sheet and for analysis of the intial value problem prior to singularity formation. The first order system has characteristics which are in general complex, and so information can move towards the real α axis and reach it in finite time. In particular, singularities are formed out in the complex plane as envelopes of breaking characteristics and move toward the real α axis. The singularities are physically realized when the envelopes intersect that axis. In the next section we will present explicit solutions to the localized approximation equations that have singularities which move toward and reach the real α axis at some time t_c, at which time a curvature singularity appears on the interface.

4. Exact Traveling Wave Solutions

The localized approximation system (3.9) is upper analytic in the sense that it contains only upper analytic variables, namely ϕ, ψ and ω. The only other coefficients in the system are constants, *i.e.*, there are no coefficients or terms which are known functions of α and t. An unusual property of such systems is that they often exhibit periodic traveling wave solutions with speed determined by the linearized equations. This contrasts the behavior of traveling wave solutons to general non-linear PDE's, which have speeds that typically depend on the amplitude of the wave. Traveling wave solutions with imaginary speed are particulary significant since they provide explicit information about singularity formation in the localized approximation equations.

To illustrate the behavior outlined above, let $L[f_+] = 0$ be a PDE in α and t for the upper analytic variable f_+. Assume L has the properties listed above, and that $L[0] = 0$. We then look for solutions of the form

$$f_+ = a_1 e^{\xi} + a_2 e^{2\xi} + \ldots \text{ with } \xi = i\alpha + \sigma t \tag{4.1}$$

This is just a Fourier expansion of a periodic traveling wave with speed σ. The absence of a term a_0 is justified by the assumption $L[0] = 0$. Plugging this form into L and equating coefficients to like powers of e^{ξ}, we obtain the following heirarchy of equations

$$\begin{aligned} a_1 L_1(\sigma) &= 0 \\ L_2(\sigma; a_1, a_2) &= 0 \\ &\vdots \\ L_j(\sigma; a_1, a_2, \ldots, a_j) &= 0 \\ &\vdots \end{aligned} \tag{4.2}$$

The $L_i's$ are functions of the variables indicated. For the first equation to have non-trivial solutions with $a_1 \neq 0$, σ must satisfy $L_1(\sigma) = 0$. Thus σ is determined in the first (linear) order of the heirarchy. The remaining equations successively determine a_j as a function of $a_{j-1}, \ldots, a_1; \sigma$. The fact that the localized approximation models outflow of energy to higher wavenumbers and ignores inward flow is crucial in producing the structure of this heirarchy and the resulting traveling wave. We also remark that the existence of these special solutions makes the localized approximation a useful procedure, even for operators which are already local in form.

The above remarks imply that the localized approximation system (3.9) has periodic traveling wave solutions $\phi(\xi), \psi(\xi)$ and $\omega(\xi)$ for $\xi = i\alpha + \sigma t$ and with σ given by

$$\sigma = \frac{i(A+\beta)\gamma_0}{2} \pm \sqrt{\frac{(1-A^2)}{4}\gamma_0^2 - Ag} \tag{4.3}$$

Note the imaginary part of σ corresponds to propagation of the wave along the interface, and the real part corresponds to growth or decay of the wave. We are most interested in the positive real part of σ, which is due to the Rayleigh-Taylor instability .

The equation for ω is simplified considerable when $\beta = -A$. When (3.9) is recast as an ODE in the variable ξ and β is set to $-A$, it is easy to integrate and obtain a class of solutions given by

$$2\phi = \Phi_0 - \frac{(\eta+\gamma_1)}{\Psi_0} + \Phi_0\sqrt{1 + \frac{2A(\eta+\gamma_1)}{\Phi_0\Psi_0} + \left[\frac{\eta+\gamma_1}{\Phi_0\Psi_0}\right]^2} \tag{4.4}$$

$$2\psi = \Psi_0 + \frac{(\eta+\gamma_1)}{\Phi_0} + \Psi_0\sqrt{1 + \frac{2A(\eta+\gamma_1)}{\Phi_0\Psi_0} + \left[\frac{\eta+\gamma_1}{\Phi_0\Psi_0}\right]^2} \tag{4.5}$$

$$\omega = i\sigma\eta \tag{4.6}$$

where $\eta = \epsilon e^{\sigma t} e^{i\alpha}$, $\Phi_0 = 1 - \dfrac{i\gamma_0(1-A)}{2\sigma}$, $\Psi_0 = 1 + \dfrac{i\gamma_0(1+A)}{2\sigma}$ and $\gamma_1 = \gamma_0/i\sigma$. Here ϵ is a measure of the initial amplitude of the disturbance.

Several properties of the traveling wave solutions (4.4–4.6) are immediately apparent. For one, these solutions contain exclusively growing or decaying modes depending on the sign of σ. In the remainder we shall be interested in the behavior of the growing modes for $\gamma_0 = 0$ (*i.e.* pure Rayleigh-Taylor instability), so $\sigma = +\sqrt{Ag}$. These solutions also correspond to a special choice of initial condition for the sheet position and sheet strength obtained by setting $t = 0$ in (4.4–4.6).

The analysis of singularities in (4.4–4.6) is straightforward. While ω remains analytic, we see that ϕ and ψ have square root singularities whenever $\eta = -A \pm \sqrt{A^2 - 1}$. This implies that the singularities move with constant speed along the straight lines.

$$\alpha = i\sqrt{Ag}t - i\ln(-A \pm \sqrt{1 - A^2}) + i\ln\epsilon \tag{4.7}$$

in the complex α plane. They reach the real axis at time

$$t_c = \frac{1}{\sqrt{Ag}} \ln 1/\epsilon \tag{4.8}$$

and at positions given by

$$\alpha_c = \pi \pm \tan^{-1}\left(\frac{\sqrt{1 - A^2}}{A}\right) \tag{4.9}$$

The positions of the singularities on the interface $z(\alpha, t)$ are easily analyzed using (4.9). At $A = 0$, the two singularities occur at $\alpha = \pi/2$ and $\alpha = 3\pi/2$, which in terms of the physical variables x and y corresponds to singularities in x lying on the $y = 0$ axis at $x = \pi/2$ and $x = 3\pi/2$. There are no singularities in y or γ. This places the singularities at the edges of a bubble-spike pattern which as odd symmetry about $x = \pi/2$ and $x = 3\pi/2$. As A is increased, this symmetry is broken and the downward spike narrows while the rising bubble widens. For increasing A, the two singularities move symmetrically downward toward the tip of the penetrating spike, until at $A = 1$ they coalesce and cancel. The mathematical nature of the singularity cancellation is clearly seen from (4.4, 4.5). At $A = 1$ (and $\beta = -A$) the equations for ϕ and ψ become

$$\begin{aligned} 2\phi &= 1 + \sqrt{(1+\eta)^2} - \eta = 2 \\ 2\psi &= 1 + \sqrt{(1+\eta)^2} + \eta = 2(1+\eta) \end{aligned} \tag{4.10}$$

and the square root singularity is removed.

In fact, an examination of the original localized approximation system (3.9) shows that at $A = 1$ and $\beta = -1$ the first equation reduces to $\phi(\alpha, t) = \phi(\alpha, 0)$ and the remaining equations for ψ and ω are linear, with coefficients that depend on α. The absence of a singularity then follows from the fact that a linear system with entire coefficients and with entire intial data possesses a global solution (Garabedian 1986). We stress that this conclusion is not based on the traveling wave solutions and holds for all solutions to inital value problems for (3.9) with entire initial data. The traveling wave solution corresponds to the special case $\phi(\alpha,t) = \phi(\alpha,0) = 1$. This solution is "linear" in the sense that it satisfies a linear system with constant coefficients and therefore contains only wavenumbers present in the initial perturbation. However, this solution is non-linear in the Eulerian sense since a mode in the Lagrangian parameter α does not correspond to a perturbation which is sinusoidal in x, y space.

Since the traveling wave solutions are expressed in closed form, it is a simple matter to deduce the behavior of essentially any quantity of physical interest. For example, the fact that x_α has a square root singularity at t_c for $A \neq 1$ indicates that the curvature will blow up. A simple calculation of the curvature $\kappa = (x_\alpha y_{\alpha\alpha} - y_\alpha x_{\alpha\alpha})/(x_\alpha^2 + y_\alpha^2)^{3/2}$ shows that $\kappa \sim\mid t - t_c \mid^{-1/2}$ for $\alpha = \alpha_c$ and $\kappa \sim\mid \alpha - \alpha_c \mid^{-1/2}$ for $t = t_c$.

At $A = 1$, both x_α and y_α vanish for $\alpha = \pi, t = t_c$ indicating a possible geometric singularity . In fact a calculation shows that $y \sim \sqrt[3]{9/2}(x-\pi)^{2/3}$ locally near the tip of the Rayleigh-Taylor spike. Thus the interface exhibits a curvature singularity at $x = \pi$, although one of a different type than the $A \neq 1$ curvature singularity , where $x_{\alpha\alpha} \to \infty$. For $A = 1$ the derivatives of x (and y) with respect to α remain bounded. This geometric singularity does not affect the ability of a numerical code to procede, whereas the $A \neq 1$ singularity with $x_{\alpha\alpha} \to \infty$ leads to problems in numerics.

There is an additional class of traveling wave solutions to (3.9) for $\beta = -A$. One of these solutions, corresponding to $\beta = -.9, A = 1$ is discussed in Baker, Caflisch & Siegel (1991). In this solution, there is a new type of singularity where $\omega \to \infty$. More precisely, there is a point ξ_0 near which ω behaves like $\omega(\xi) \sim c(\xi - \xi_0)^{-1/2}$. This corresponds to an inverse square root singularity in γ and z_α.

A discussion of the behavior versus Atwood number of other quantities of physical interest, such as the vorticity density $\sigma = \Gamma/z_\alpha$, is given in Baker *et al.* (1991).

5. Numerical Results

Numerical computations on the exact equations (2.3, 2.4) were performed with the aim of resolving the form of the singularities and their motion in the complex plane, and to compare these with the predictions of the approximate analytic theory. The results are reported in Baker, Caflisch & Siegel (1991). Here we provide a very brief sketch of the numerical procedures and summarize the results.

In the method an equivalent form of the governing equations for the interface are solved using the point vortex method. An alternate point trapezoidal rule is used to approximate the Birkhoff-Rott integral, and the evolution equation for the vortex sheet strength is solved by iteration. A spectral filtering technique of Krasny is employed to prevent unstable perturbations

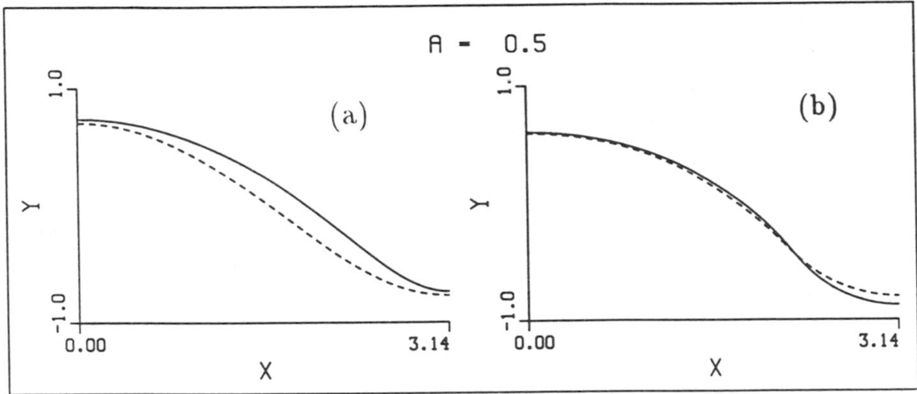

Fig. 1. Interface profiles at A=.5, t=$2/\sqrt{Ag}$. Only half of a symmetric wave profile is shown. (a) Linear profiles: The solid line is from Eulerian linear theory, and the dashed line is from Lagrangian linear theory. (b) The solid line is the numerical solution; the dashed line is calculated from the asymptotic theory.

which are excited by round-off from swamping the calculation.

The form of the singularities in $x(\alpha)$ as well as their positions are determined by asymptotic decay of Fourier coefficients. The details of this procedure are rather lengthy, and we refer the interested reader to Baker *et al.* (1991).

Figure 1 compares the shape of the interface at A=.5 as calculated from linear theory (both Eulerian and Lagrangian linear profiles are shown), the nonlinear theory of section 4, and the numerical solution. All the profiles use the traveling wave initial data obtained by evaluating (4.4, 4.6) at $t = 0$, and $\epsilon = .1$. They are plotted just before the singularity time. The linear theories show the overall trend in the profile, but the agreement is not quantitative nor do the linear theories predict singularity formation. The asymptotic theory is quite good for the range $0 < A < .9$ but deteriorates somewhat as $A \to 1$. The numerical results confirm the presence of square root singularities moving at constant speed along straight lines in the direction of the imaginary axis, as predicted by the nonlinear theory. An example of the correspondence between theory and numerics is given is figure 2(a), where we plot the imaginary part of the singularity position versus time for $A = .5$ (a second singularity exists with the same imaginary component). The asymptotic prediction is shown as a solid line, while the circles represent the numerical result. As A increases past .9 the two slopes begin to diverge. The degree of divergence can be seen in figure 2(b), which plots the speed of the singularity for various A. The agreement of theory and numerics is within 8% for Atwood numbers up to .7. The asymptotic theory predicts that the speed is a constant for fixed A and that is also observed by the numerical calculations over the full range of Atwood numbers. However, for $.9 < A < 1$ the constant speeds differ between the numerical calculations and the asymptotic theory. Indeed, the numerical results suggest that the speed of approach to the real α axis by the singularity may be vanishing as $A \to 1$. The time of singularity formation is of course related to the speed, implying that there is good agreement between the numerical and theoretical t_c for $0 < A < .9$, with the correspondence deteriorating as $A \to 1$. At

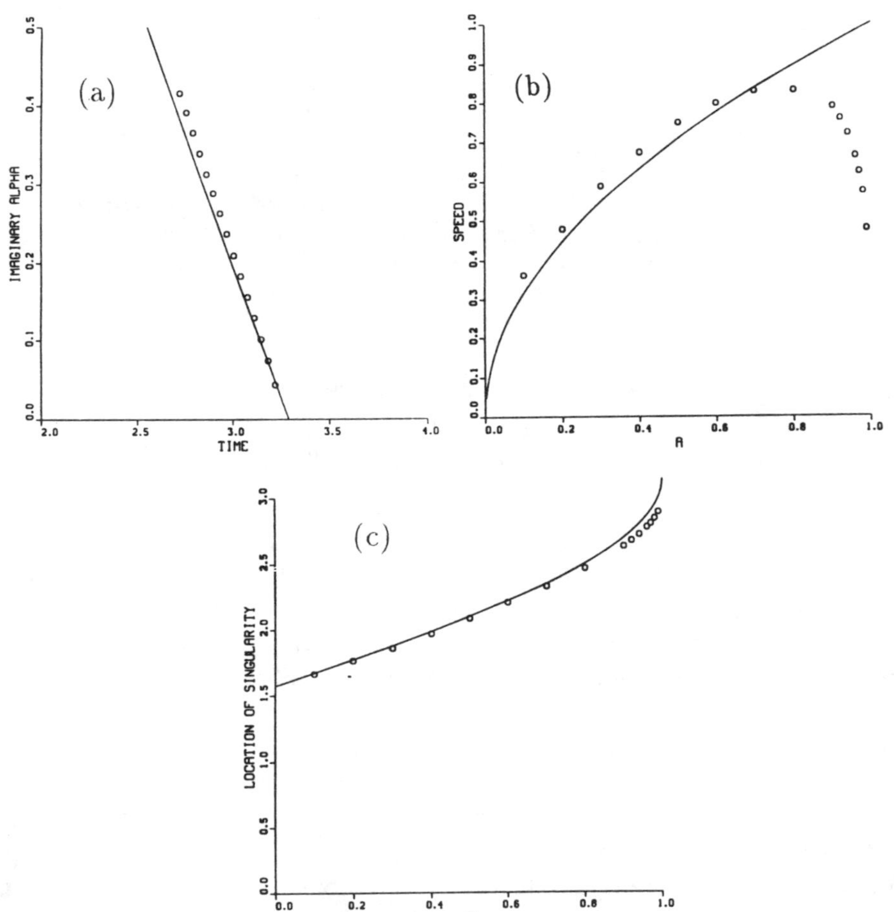

Fig. 2. Comparison of the predictions of asymptotic theory (solid line) with the numerical solution (circles) for various quantities. (a) The imaginary part of the singularity position versus time at A=.5. (b) The speed of the singularity versus A. (c) The point at which the complex singularity intersects the real α axis versus A.

$A = 1$, the numerical calculations for initial data generated by the traveling wave solution with $\beta = -1$ agree with the asymptotic theory in that no singularity is detected near $\alpha = \pi$. However, the curvature does not exhibit the geometric singularity discussed in section 4. For the initial data generated by the $A = 1, \beta = -.9$ traveling wave solution, the numerics confirm the prediction of a new singularity at $\alpha = \pi$ with $\omega = \infty$. However we see some evidence that the singularity is slowing down as it approaches the real axis.

Finally, we remark that the location where the singularity hits the real α axis is accurately predicted by the asymptotic theory for all Atwood numbers.

6. Conclusions

We have seen that the predictions of singularity formation obtained from the localized approximation equations are accurate over a wide range of Atwood numbers. For the range $.9 < A < 1$ some aspects of the agreement break down. However, the numerical calculations are exceedingly difficult in this regime, and it is possible that agreement will improve with further refinement of the numerics. Even if this is not the case, the success of the approximation for the range $0 \leq A < .9$ shows its potential and suggests its application in other problems.

Acknowledgements

The author acknowledges the valuable contributions of Greg Baker and Russ Caflisch in this work.

References

BAKER, G., CAFLISCH, R. & SIEGEL, M., 1991 Singularity Formation in the Rayleigh-Taylor Problem. Submitted to *J. Fluid Mech.*

BAKER, G.R., MEIRON, D.I. & ORSZAG, S. A,. 1980 Vortex simulations of the Rayleigh-Taylor instability. *Phys. Fluids* **23**(8), 1485-1490.

BAKER, G.R., MEIRON, D.I. & ORSZAG, S. A., 1982 Generalized vortex methods for free-surface flow problems. *J. Fluid Mech.* **123**, 477–501.

CAFLISCH, R. & ORELLANA, O., 1986 Long time existence for a slightly perturbed vortex sheet. *Comm. Pure Appl. Math.* **39**, 807–838.

CAFLISCH, R., ORELLANA, O. & SIEGEL, M., 1990 A localized approximation method for vortical flows. *SIAM J. Appl. Math.*, **50**(6), 1517–1532.

GARABEDIAN, P.R., 1986 *Partial Differential Equations* (Chelsea Publishing, New York).

KERR, R.M., 1986 Analysis of Rayleigh-Taylor flows using vortex blobs. Report, Lawrence Livermore National Laboratory, Report UCID-20915.

MOORE, D.W., 1979 The spontaneous appearance of a singularity in the shape of an evolving vortex sheet. *Proc. R. Soc. Lond. A* **365**, 105–119.

MOORE, D.W., 1985 Numerical and analytical aspects of Helmholtz instabilty. *Theoretical and Applied Mechanics*, F.I. Niordson and N. Olhoff, eds. (Elsevier, New York).

SIEGEL, M., 1989 An analytical and numerical study of singularity formation in the Rayleigh-Taylor problem. Ph.D. Thesis.

SIEGEL, M., 1991 The effects of surface tension in the Kelvin-Helmholtz problem. Preprint.

TRYGGVASON, G., 1988 Numerical simulations of the Rayleigh-Taylor instability. *J. Comp. Phys.* **75**, 253-282.

PART V

VORTEX INTERACTIONS AND THE STRUCTURE OF TURBULENCE

2D TURBULENCE: NEW RESULTS FOR $Re \to \infty$

David G. DRITSCHEL
Department of Applied Mathematics and Theoretical Physics
University of Cambridge
Silver Street
Cambridge CB3 9EW
ENGLAND

ABSTRACT. Novel physical-space vortex properties of nearly-inviscid, unforced two-dimensional turbulence are presented. They are obtained from the analysis of a large ensemble of turbulence calculations all beginning with a random distribution of vortex patches on a spherical surface. The numerical method is based on Contour Dynamics/Surgery (CD/CS) but includes a new way for calculating separated vortex interactions that results in a speed-up factor of 100 or more. The Lagrangian based numerical method is free from the vortex edge erosion effects brought on by fixed resolution in standard methods, and as a consequence reaches much higher effective Reynolds numbers. Moreover, the results it produces differ significantly from those obtained previously with fixed-resolution numerical methods, suggesting that previous results say nothing about the long-time behavior of a nearly inviscid fluid.

A theory of 2D turbulence requires an understanding of the essential characteristics of inelastic vortex interactions. Possible inroads into this huge task can be made in the limit of dilute turbulence. In this limit, curiously the famous collapse problem of 3 point vortices regains relevance.

1. Introduction

This note reports new results for two-dimensional turbulence in the nearly-inviscid limit generated using an accelerated contour dynamics/surgery algorithm (Dritschel 1991a & b). The algorithm makes use of a vast simplification of the interaction between separated vortices, a simplification equal to a hundred-fold increase in computational efficiency over the original contour surgery algorithm (Dritschel 1989a & refs.) in typical calculations of turbulence. For dilute turbulence, when vortices are nearly always widely separated, the computational cost reduces to calculating the interaction of point vortices. The computational overhead associated with resolving each vortex's shape, and occasional close-range vortex interactions, becomes insignificant compared to calculating the remaining vortex interactions. Since

H. K. Moffatt et al. (eds.), Topological Aspects of the Dynamics of Fluids and Plasmas, 353–362.

these interactions are almost all over a long range, one can approximate nearly all of them by equivalent point vortex interactions, for a prescribed maximum error tolerance.

This new algorithm reaches much higher effective Reynolds numbers than conventional, fixed-grid methods (e.g. pseudo-spectral; Dritschel 1991b, Dritschel & Legras 1991, Legras & Dritschel 1991). Its form of dissipation is selective, in that it removes only thin vorticity filaments (at a scale 10-100 times smaller than in typical fixed-grid methods). All evidence to date suggests that the removal of thin filaments should not produce serious intermediate to long time effects (Basdevant *et al.* 1981, Dritschel 1989b, Dritschel *et al.* 1991, Waugh & Dritschel 1991). [1] On the other hand, the erosion of sharp vorticity gradients, something not possible in contour surgery but typical of fixed-grid methods, would appear to have a big impact only after a certain number of eddy turn-around times on the order of the number of grid lengths spanning a typical vortex (Dritschel & Legras 1991). In simulations of turbulence, therefore, where each vortex receives only a small part of the total resolution, it does not take long before the effects of dissipation become dominant. It is not sufficient to quote the large-scale Reynolds number and expect nearly-inviscid evolution for a proportional number of eddy turn-around times.

In fact, the results discussed here depart significantly from those previously obtained using traditional computational methods (Benzi *et al.* 1989 & 1991, McWilliams 1990, Carnevale *et al.* 1991). The next section describes the calculations and presents several diagnostics of the typical vortex properties. §3 concludes with an optimistic possibility for developing a robust, simple model of dilute turbulence, that is when vortices are almost always well separated and rarely interact. In that case, one can argue that the only way of bringing two vortices close enough together for merger is to have a third, oppositely-signed vortex near enough to the corresponding collapse trajectory for *point* vortices.

2. Results

Ten calculations w ere performed differing only in a random number seed used to set up the initial conditions. The initial conditions consisted of 200 variable-sized vortex patches (uniform vorticity disks) distributed randomly over a spherical surface (spherical geometry is ideal owing to the simple Green function and the consequent accelerative possibilities). Half the patches had vorticity $+2\pi$, and the other half had vorticity -2π. The ratio of the largest to smallest vortex area was set to 100, and the specific areas were chosen to ensure that initially 10% of the surface was occupied

[1] We are not interested in the final state of the flow, as examined by Robert & Sommeria 1991.

by vorticity. First, the largest positive and negative vortices were placed on the surface, then successively smaller vortices were placed such that no two were closer than three times their average radii (this was done to avoid immediate merger).

All the calculations were run until $t = 100$ (50 eddy turnaround times; $t_e = 4\pi/|\omega| = 2$) except one which was taken to $t = 139$. This longer calculation is shown in Fig. 1. This figure does not do justice to the high resolution employed, since each pixel is 10 times larger than the surgical cutoff scale. Also, the few frames shown reveal only the partial elimination of intermediate to large-scale vortices and the slow growth of largest vortex size, giving the impression of an apparent reduction in vortex number. In fact, the number of vortices at the end of the calculation is greater than at the beginning because of the formation of a large population of small-scale vortices at or below the pixel size. This is the first fundamental difference between contour surgery and fixed-grid simulations of turbulence.

What is called a "vortex" here is any patch of uniform vorticity that is sufficiently coherent, in the sense displayed in Fig. 2. Using this criterion, Fig. 3 shows the total number of contours (coherent vortices plus filaments; labeled $c = 0$) versus time, averaged over the 10 calculations, as well as the number of coherent vortices selected for $c = 0.1$ and $c = 0.2$. The number of coherent vortices reaches a minimum between $t = 8$ and 10 because many of the vortices are at this stage in the process of merger or being strained out, and hence severely distorted. The total number of contours peaks later, at around $t = 24$, after a period in which many filaments have been detached from vortices by surgery. The decline in the total number of contours afterwards is almost entirely due to the elimination of extremely thin filaments. From $t = 12$, the number of coherent vortices grows and soon exceeds the initial number, finally leveling off at a value appreciably greater than the initial value.

What has happened is that the earlier inelastic vortex interactions have created a large population of small-scale vortices. This is most clearly seen in the number density distribution (for $c = 0.2$) in Fig. 4. It is also worth noting that this distribution changes very little after the initial surge of activity, and it is clearly *not* self-similar ($N(\hat{A})$ is not a power law). This is the second fundamental difference compared with fixed-grid calculations (Benzi *et al.* 1989 & 1991).

Another telling diagnostic is the average (inverse) aspect ratio versus vortex area, and its variance — see Fig. 5. Not only does this show that smaller vortices are more deformed on average than larger vortices (another indication of the lack of self-similarity), it also shows that vortices have an average aspect ratio of approximately 1.5. This value is appreciably greater than observed in fixed-grid calculations (in fact four times greater than found by McWilliams 1990). The increase in deformation with decreasing vortex

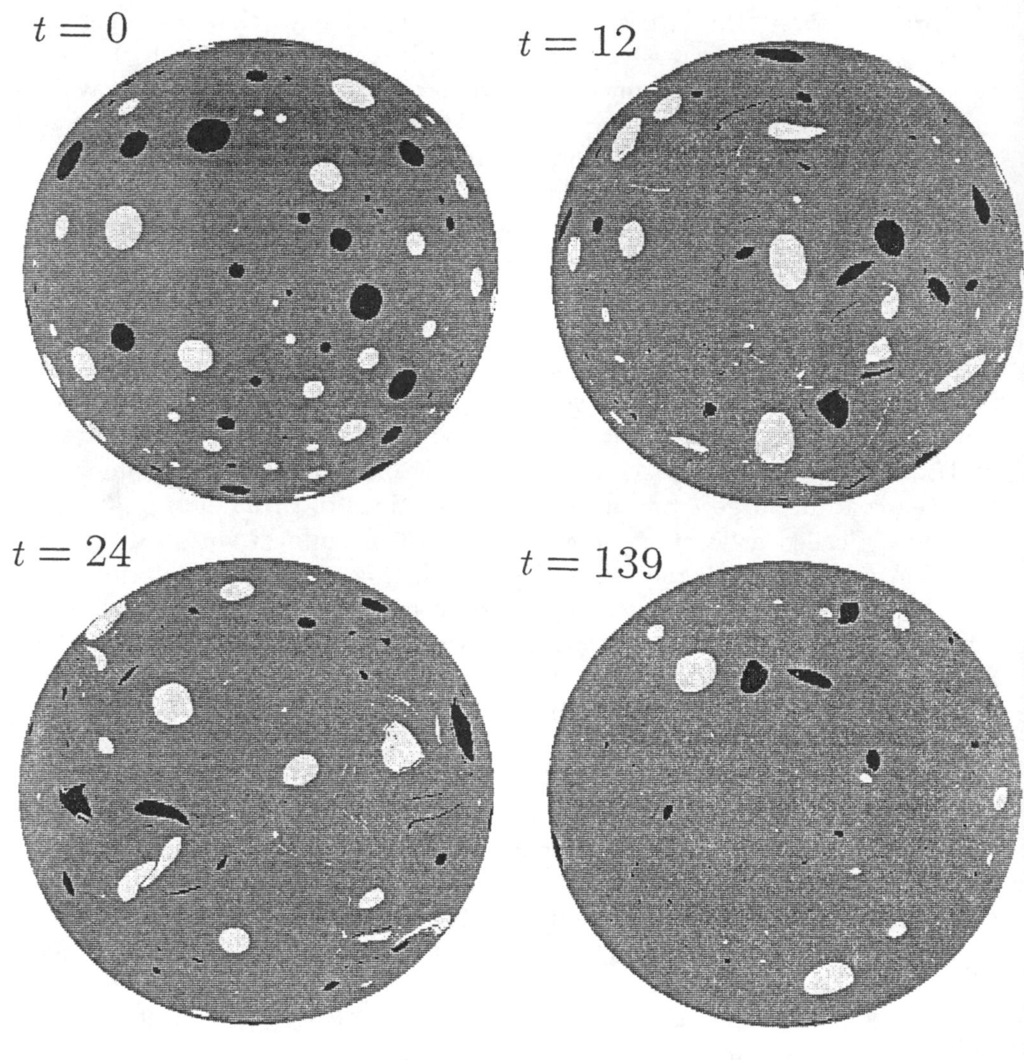

Fig. 1. Stages in the evolution of turbulence on a sphere. Only one side of the sphere is shown, and the projection is orthographic. The numerical algorithm parameters (see Dritschel 1989a & 1991a) used in this and the 9 other calculations are $\mu = 0.01$, $L = 10^{-1/2}$, $\delta = \mu^2 L/8 \approx 8.89 \times 10^{-4}$, $\Delta t = 0.1$, and $\varepsilon = 2\pi \times 10^{-4}$.

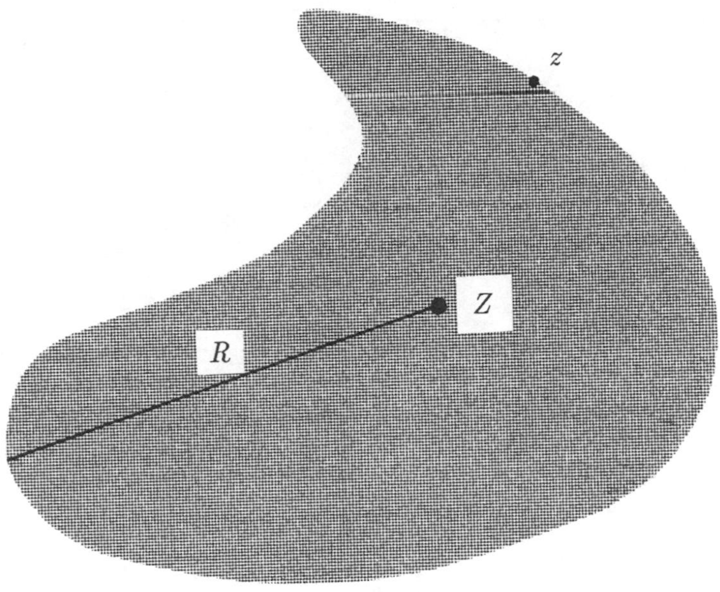

Fig. 2. Graphical definition of coherence: the vortex (shaded grey) must occupy at least a fraction c of the containing disk $|z - Z| \leq R$ centered on the vortex centroid $Z = X + iY$.

size is related to the fact that smaller vortices are, on average, closer to bigger vortices than bigger vortices are to themselves, and hence feel greater average strain (Dritschel 1991b).

A further difference between contour surgery and fixed-grid calculations of turbulence is shown up in the decay of enstrophy in time. Figure 6 shows the total enstrophy (the solid curve marked $c = 0$) as well as the enstrophy associated only with the coherent vortices ($c = 0.2$ and 0.1). All measures decay as $t^{-1/4}$ over the last half of the calculations. This is much slower than the $t^{-3/8}$ observed in the pseudo-spectral and corresponding simple point-vortex model calculations of Carnevale *et al.* (1991), suggesting that numerical dissipation in fixed-grid calculations eliminates much of the coherent vorticity by eroding vortex edges and eliminating small vortices.

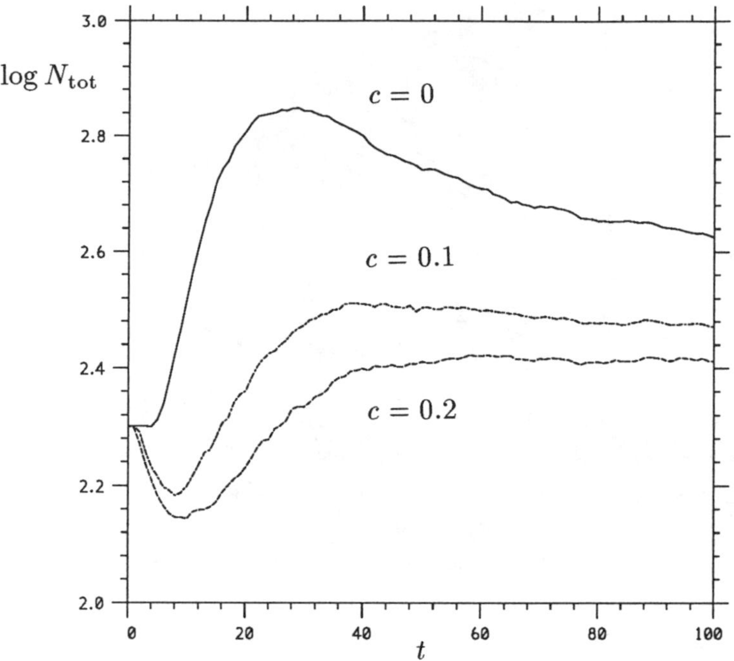

Fig. 3. Number of contours averaged over the 10 datasets, $N_{\rm tot}$, versus time. A logarithmic scale (base 10) is used along the abscissa.

3. Conclusions

The striking differences observed between contour surgery and fixed-grid simulations of turbulence include:

- *No decay of the total number of coherent vortices on account of the emergence of a population of small-scale coherent vortices*;
- *No power-law dependence of vortex number on vortex size*;
- *Significantly non-circular vortex shapes*; and
- *A slower decay in time of the enstrophy.*

These differences appear to be the result of a much greater total effect of numerical dissipation in long-time simulations at *fixed* spatial resolution than was previously realized. Fixed resolution causes vortices to erode at a rate on the order of one grid length per eddy turnaround time (Dritschel & Legras 1991). This is a consequence of numerical dissipation that continually replenishes low-level vorticity being stripped dynamically from the periphery of vortices by nearby vortices. Dissipation accounts for the absence of small-scale vortices, which by this argument erode quicker, and for the greater

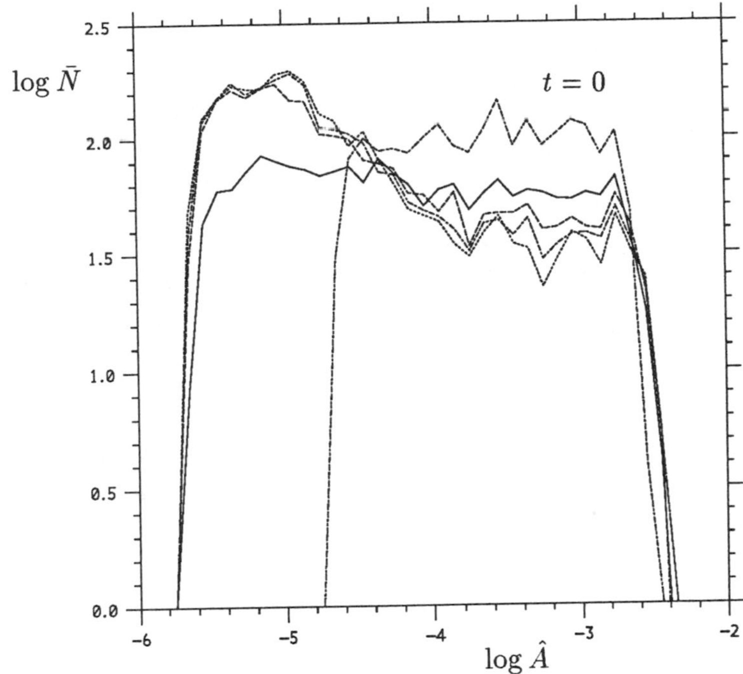

Fig. 4. Number of vortices versus fractional area ($\hat{A}$, area divided by 4π) on a log-log plot, obtained by data averages in 40 evenly distributed bins (for coherence level $c = 0.2$). Times shown are: $t = 0$ (dash-dash-dot-dot), $t = 24$ (solid), $t = 48$ (dashes), $t = 72$ (dash-dot), and $t = 96$ (dotted). The averages were taken over the 10 data sets as well as over intervals of time. The data were saved at unit intervals of time, and the data for times $t - 4$ to $t + 4$ and for the 10 data sets were averaged to produce the results shown for time t. No time average was performed for the $t = 0$ results.

decay rate of enstrophy. Also, the smearing of vortex edges damps out the eccentricity of vortices by aiding the axisymmetrization process (Melander *et al.* 1987, Dritschel 1989a).

Much work remains to quantify nearly-inviscid two-dimensional turbulence. A theory depends heavily on understanding the fundamental inelastic interactions (e.g. merger and straining-out) responsible not only for vortex growth but more generally for the distribution of vortices both in size and in space. This is not a problem that can be answered *a priori* by a simple point vortex model with merger rules (Benzi *et al.* 1991, Carnevale *et al.* 1991), even if these rules respect observed results for *isolated* inelastic interactions (Dritschel & Waugh 1991), since such a model cannot be internally verified. It is also now evident that any such rules need to account for

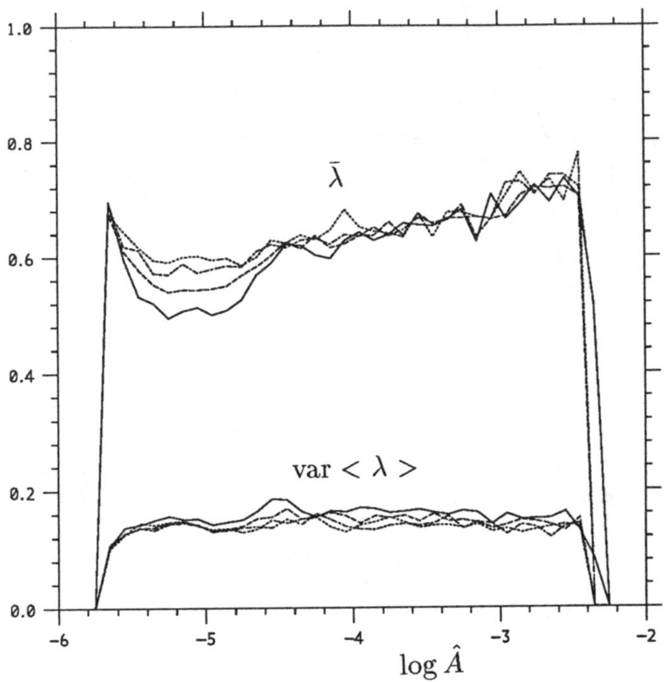

Fig. 5. Inverse aspect ratio (upper curves) and variance (lower curves) of coherent vortices versus log fractional area ($c = 0.2$). The line styles are the same as used in Fig. 4. At $t = 0$, all the vortices have unit aspect ratio.

external strain forcing (longer-range interactions with other vortices), and preliminary results paint a complex picture (Waugh 1991).

Attention in this paper has focused on what is believed to be the simplest problem of turbulence — dilute vorticity of just one positive and one negative, opposite value. Such a configuration is extremely convenient for contour surgery simulations, but it is also a plausible one for representing the state of nearly-inviscid turbulence a long time beyond the initial conditions. This is because the effect of vortex stripping, and of inelastic vortex interactions generally, is to produce greater and greater vorticity gradients in time — weaker vorticity is more susceptible to being strained-out than stronger vorticity. After a great many close range inelastic interactions, the peak positive and negative levels of vorticity stand the greatest chance to remain within the coherent vortices.

This leads on to an interesting, potentially useful model for the late stages of turbulence, when vortices are very well separated except for rare close-range interactions. Well-separated vortices interact, to a good approxima-

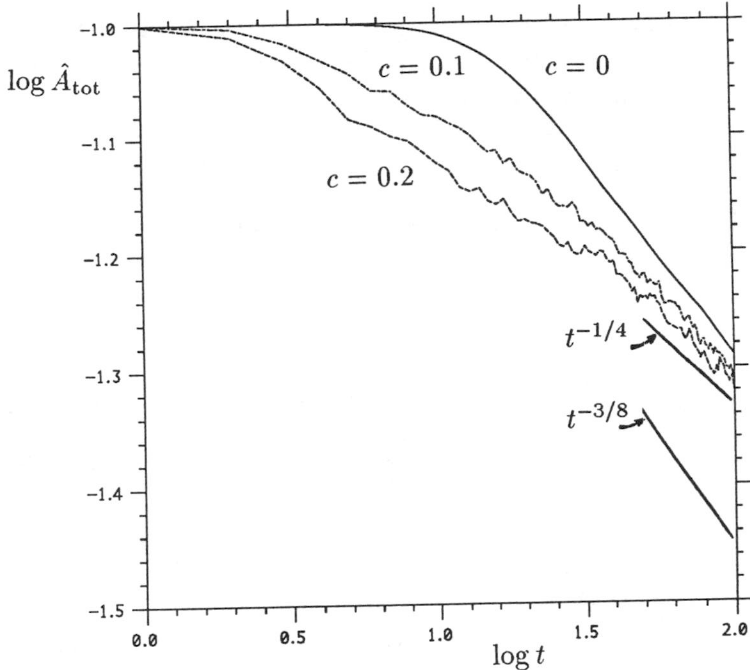

Fig. 6. Total and coherent enstrophy decay with time (log-log plot). Shown is the proportional quantity, the fractional area occupied by vorticity versus time (all vortices have the same vorticity magnitude).

tion, like point vortices, but of course such a description breaks down for close separations. While it is possible to compute the flow efficiently using the accelerated contour surgery code in this limit, one can immediately get an idea of what to expect just by considering what is required to bring vortices close enough together for an inelastic interaction. With odds that increase to unity as the fraction of coherent vorticity decreases to zero, close-range interactions will require *three* vortices, since this is the only way to maintain sufficient strain to push together vortices of the same sign of vorticity. Also, all three vortices cannot have the same sign of vorticity; moreover, the odd vortex must have an area close to the product divided by the sum of the areas of the two like-signed vortices and lie along a portion of a circular arc of radius proportional to the separation of the two like-signed vortices. Under these conditions, three point vortices collapse (self-similarly) to a point in finite time (Aref 1979). Thus, three finite-sized vortices, initially well separated, can only come close enough together for an inelastic interaction if their strengths and positions approximately satisfy the conditions for the collapse of three point vortices. Quantifying the result of all possible collapse and near-collapse sequences for vortex patches is in progress and may lead to a simple model of dilute turbulence.

Acknowledgements

Many ideas in this paper were fertilized in discussions with B. Legras, D.W. Waugh, and in particular N.J. Zabusky.

References

AREF, H., 1979 Motion of three vortices. *Phys. Fluids*, **22**, pp. 393-400.

BASDEVANT, C., LEGRAS, B., SADOURNY, R. & BÉLAND, M., 1981 A study of barotropic model flows: intermittency, waves and predictability. *J. Atmos. Sci.*, **38**, pp. 2305-2326.

BENZI, R., BRISCOLINI, M., COLELLA, M & SANTANGELO, P., 1991 A simple point vortex model for two-dimensional decaying turbulence. (preprint).

BENZI, R., PATARNELLO, S. & SANTANGELO, P., 1989 Self-similar coherent structures in two-dimensional decaying turbulence. *J. Phys. A: Math. Gen.*, **21**, pp. 1221-1237.

CARNEVALE, G.F., MCWILLIAMS, J.C., POMEAU, Y., WEISS, J.B. & YOUNG, W.R., 1991 Evolution of vortex statistics in two-dimensional turbulence. *Phys. Rev. Lett.*, **66**, pp. 2735-2737.

DRITSCHEL, D.G., 1989a Contour dynamics and contour surgery: numerical algorithms for extended, high-resolution modelling of vortex dynamics in two-dimensional, inviscid, incompressible flows. *Comp. Phys. Rep.*, **10**, pp. 77-146.

DRITSCHEL, D.G., 1989b On the stabilization of a two-dimensional vortex strip by adverse shear. *J. Fluid Mech.*, **206**, pp. 193-221.

DRITSCHEL, D.G., 1991a A moment-accelerated contour surgery algorithm for many-vortex, two-dimensional flow simulations. (in preparation).

DRITSCHEL, D.G., 1991b Vortex properties of two-dimensional turbulence. *Phys. Fluids A*, (submitted).

DRITSCHEL, D.G., HAYNES, P.H., JUCKES, M.N. & SHEPHERD, T.G., 1991 The stability of a two-dimensional vorticity filament under uniform strain. *J. Fluid Mech.*, **230**, pp. 647-665.

DRITSCHEL, D.G. & LEGRAS, B., 1991 Vortex stripping. *Phys. Fluids A*, (submitted). See also: DRITSCHEL, D.G., 1989 Strain-induced vortex stripping. In: "Mathematical aspects of vortex dynamics" (ed. R.E. Caflisch), SIAM.

DRITSCHEL, D.G. & WAUGH, D.W., 1991 Quantification of the inelastic interaction of unequal vortices in two-dimensional vortex dynamics. *Phys. Fluids A*, (submitted).

LEGRAS, B. & DRITSCHEL, D.G., 1991 A comparison of the contour surgery and pseudo-spectral methods. *J. Comput. Phys.*, (submitted).

MCWILLIAMS, J.C., 1990 The vortices of two-dimensional turbulence. *J. Fluid Mech.*, **219**, pp. 361-385.

MELANDER, M.V., MCWILLIAMS, J.C. & ZABUSKY, N.J., 1987 Axisymmetrization and vorticity-gradient intensification of an isolated two-dimensional vortex through filamentation. *J. Fluid Mech.*, **178**, pp. 137-159.

ROBERT, R. & SOMMERIA, J., 1991 Statistical equilibrium states for two-dimensional flows. *J. Fluid Mech.*, **229**, pp. 291-310.

WAUGH, D.W. & DRITSCHEL, D.G., 1991 The stability of filamentary vorticity in two-dimensional geophysical vortex-dynamics models. *J. Fluid Mech.*, **231**, pp. 575-598.

ON VORTEX RECONNECTION AND TURBULENCE

Oluş N. BORATAV *
Institute for Theoretical Physics
University of California
Santa Barbara, California 94106-4030, U.S.A
Richard B. PELZ **
Institute for Theoretical Physics
University of California
Santa Barbara, California 94106-4030, U.S.A
and
Norman J. ZABUSKY
Rutgers University
Department of Mechanical and Aerospace Engineering and CAIP
Piscataway, New Jersey 08855-0909, U.S.A

ABSTRACT. We present a summary of the recent simulations on vortex reconnection in orthogonally interacting vortex tubes. We investigate the relation between the shape change of the vortices during reconnection and the space-averaged mid-eigenvalue of the strain rate tensor $<\beta>$ and find that $<\beta>$ becomes positive as the vortices are flattened. We also find that the velocity derivative skewness values S_{uu} and S_{vv} approach the typical turbulent flow values as small-scales are produced during reconnection. The eigenanalysis of the velocity gradient tensor shows that as the vortices get more and more flattened, the number of complex eigenvalues increases. Also, among these eigenvalues, the ones with the positive real part are the group whose number increases the most. These results show similarities with turbulent flow simulations. We also find that rate of change of helicity changes enormously during reconnection. We propose that the main contribution to the helicity change during reconnection is due the change of the twistedness of the vortex lines which can be captured by the quantity 'the vortical helicity' ('volicity') defined as $\omega \cdot \nabla \times \omega$.

1. Introduction

Vortex reconnection is a generic process in three-dimensional vortex interaction problems. Recent quantifications (Boratav *et al.* 1991a,b) have shown

* Permanent address: Rutgers University, Department of Mechanical and Aerospace Engineering, Piscataway, NJ 08855-0909
** Permanent address: Rutgers University , Department of Mechanical and Aerospace Engineering and CAIP, Piscataway, NJ 08855-0909

H. K. Moffatt et al. (eds.), Topological Aspects of the Dynamics of Fluids and Plasmas, 363–376.

that the process results in the rapid increase (bursting) of local quantities such as mean square strain rate and vorticity. Extremely high gradient regions are formed while the vortices are being flattened. Tube-like, vortical regions change their shape as a result of strong deformation and become pancake-like. At the end of the process, the shape of the vortices changes back to tube-like while small-scale structures which we call *'remnants'* (Zabusky *et al.*, 1991, Boratav *et al.*, 1991b) are produced.

More detailed analysis is essential due to various reasons: **First**, it has *not* been established whether the reconnection processes take place in turbulent flow fields. There have been recent models which attempt to relate the 'intermittency' and deviation from Gaussianity in turbulent flows to high-amplitude, transient structures which undergo strong local interactions (She, 1991, She *et al.*, 1991). Surprising results have been found such as the significant contribution of low wavenumber scales to velocity-derivative skewness values (She *et al.*, 1988). A more detailed quantification of the reconnection process will lead to understanding whether such processes are associated with the intermittency events in turbulent flow evolutions. **Second**, there is the unsettled question of whether the interaction of tubular vortices could result in the formation of a finite-time singularity in three dimensional Euler flows or not. The simulations of Pumir and Siggia (1990) and Kerr (1991) present conflicting results. While viscous simulations can *not* predict the existence of a finite time singularity in three-dimensional Euler equations, they might still give us some hints of the singular (or non-singular) behaviour as Reynolds number is increased and the dependence of the evolution of some local quantities on Reynolds number is thoroughly examined. **Third**, the reconnection process is an ideal problem in which one can examine the evolution of the strain rate, velocity gradient tensor invariants, vorticity-strain correlations as the vortices rapidly change their shape. If one can develop reliable criteria to understand how the topological changes in interacting vortices are related to the changes in deformation and rotation tensor invariants, one can possibly extend these results to turbulent flow simulations. So far in turbulent flow literature, there is no consistent method to examine whether the turbulent flow structures are tube-like or sheet-like.

2. Recent Results

The database used is generated by solving the three-dimensional Navier-Stokes equations (with hyperviscosity regularization) using a Fourier pseudospectral method (Zabusky *et al.* 1991, Boratav *et al.*, 1991a,b). Initially, the vortices are orthogonally-offset having a compact (or Gaussian) vorticity distributions in the core. The maximum Reynolds number (Γ/ν) in our runs is 2100. We performed runs having 64^3, 96^3, 256^3 gridpoints. We used the 256^3 run results mainly to compare the evolution of local and total quantities with the lower resolution runs. The comparisons showed that the 96^3 runs are adequately resolved. In our previous work (Zabusky *et al.*, 1991, Boratav *et al.*, 1991a,b), we presented the sequence of physical processes as: i) The early, sudden formation of vortex filaments from each vortex (fingering). ii) The formation of a local dipole. iii) Formation of new bridges perpendicular to the axis of the dipole. iv) Extreme flattening of the dipole cores. v) Moving away of the newly formed bridges. vi) Dissipation of the flattened dipole (named as remnants at this stage). We found that the early fingering and the flattening of the dipole are the two processes which result in the formation of smaller scales. These can be regarded as cascading mechanisms in three-dimensional vortex interactions. We have shown that both processes result in the increase of enstrophy (Boratav *et al.*, 1991b, Figs. 24, 25).

In our previous work, in order to define the time scale of reconnection, we examined the evolution of local quantities and defined the time sequence of different events during the interaction. The sequence and the notation we used are as follows: t_1 shows the time when there is a sharp drop in the distance between the maximum $\pm\omega$ values (formation of the dipole), t_2 is the time when the local velocity peak is attained (the location corresponds to the dipole jet velocity), t_3 is the time when the maximum mean square strain rate is reached, and t_4 is the time of the vorticity peak. We defined the reconnection time scale (T_R) as: $T_R = t_4 - t_1$. We present these results once again in Table 1 which gives the time sequence of 4 different runs having 96^3 resolution:

Table No. I
Time Sequence of Physical Processes During Reconnection

Run	Γ/ν	t_1	t_2	t_3	t_4	T_R
a	696	3.25	3.6	4.60	5.10	1.85
b	1044	3.06	3.6	4.15	4.60	1.54
c	1392	3.05	3.6	4.00	4.30	1.25
d	2088	3.00	3.5	3.80	4.06	1.06

Based on the results given in Table I, we make the following comments on the time scale of reconnection: i) The time scale of reconnection is faster than a viscous scale. ii) There are some estimates in literature which give the reconnection time to scale as $\ln Re/\alpha$ where α is the extensional (positive) straining value. This scaling can only be consistent with our results if the strong Reynolds number dependence of α is taken into account (Kambe, private communication).

We present in Fig.1, the evolution of vortices at four different instances. Note that t=3 is just before t_1, t=4.00 is when the peak strain rate is

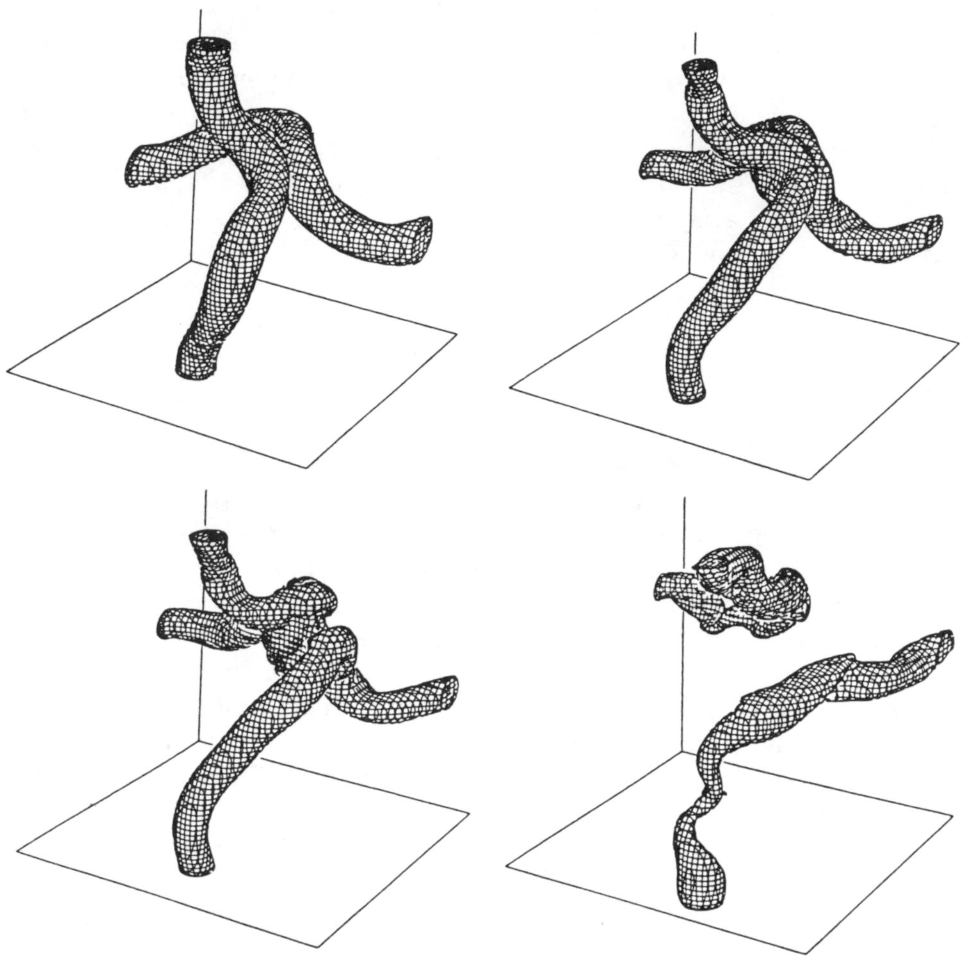

Fig. 1. Vorticity magnitude isosurfaces ($75\%\omega_0$) at t=3.0, 4.0, 4.5, 6.0. Resolution is 256^3. All the parameters are the same as in Run c.

attained (t_4), t=4.5 is after t_3 and t=6 is after t_4. The run in this figure has a resolution of 256^3 all the other parameters are the same as of Run c.

In our previous work, we also examined the time evolution of the eigenvalues α, β, γ of the rate of strain tensor S_{ij}. We summarize our results as follows: i) The growth rate of local maximum vorticity is faster than a linear rate. ii) The growth rate of the mid-eigenvalue β is Reynolds number dependent and faster than a linear rate. iii) α and γ strains are time dependent. iv) The peaks in local maximum ω and ϵ are reached earlier as Reynolds number is increased. vi) The reconnection time scale decreases as Reynolds number is increased. vii) All of the above contradicts with the assumptions and the results of Saffman's model (1990). We recall that Saffman's recent model of reconnection makes use of the 2D Kirchoff solution for a rotating elliptical vortex. It was shown by Moore and Saffman (1971) that a steady state solution of the ellipse is only possible when the ratio of maximum extension to core vorticity, $\alpha/\omega_0 < 0.15$. In our runs, this ratio is around 0.35-0.5. No steady Kirchoff ellipse solutions exist in this range.

We close this section by discussing the relevance of our results to a possible singularity formation in Euler equations. We find in our simulations that; i) The peak maximum vorticity increases as Reynolds number is increased. ii) The rate of increase of maximum vorticity is at least exponential in lower Reynolds number runs and larger than exponential in the highest Reynolds number run. iii) The time scale of reconnection decreases as the Reynolds number is increased. iv) All the above results are consistent with the trends that would be observed if there is a local singularity in vorticity as $Re \rightarrow \infty$. We believe that higher Reynolds number runs are needed to investigate the problem further.

3. Vortex shape change during reconnection

In this section, we will investigate in detail how the rapid shape change described as; *tube-like* $\longrightarrow$ *pancake-like* $\longrightarrow$ *tube-like*; during the reconnection process can be quantified. Motivated by Betchov's arguments (1956) on homogeneous isotropic turbulence, we investigate the evolution of β eigenvalue during reconnection. We present in Table II, the percentage of points having positive/negative β and the space averaged beta, $<\beta>$, at different instances for Run c.

Table No. II

Sign Distribution percentage of β and the values of space averaged mid eigenvalue $<\beta>$:

Time	$\%\beta > 0$	$\%\beta < 0$	$<\beta>$
3.00	38.6	61.4	-0.1260
3.25	42.6	57.4	-0.0001
3.60	54.2	45.8	0.1602
4.00	58.3	41.7	0.2611
4.30	63.2	36.8	0.3883
4.60	56.7	43.3	0.2695
5.20	42.1	57.9	-0.0769

In this table, we see that as the vortices get flattened, the average β value changes sign and becomes positive. Note that β sign change takes place around t=3.25, and the peak β is attained at t=4.3 which also corresponds to t_4 when the maximum vorticity is attained. At this instant, the dipole structure is at its most flattened stage. We also find that at the end of the reconnection process, $< \beta >$ becomes negative again.

Next we consider the time evolution of the velocity derivative skewness S_{uu} and S_{vv} during reconnection. These are defined as:

$$S_{uu} = -\frac{< (\partial u/\partial x)^3 >}{< (\partial u/\partial x)^2 >^{3/2}} \tag{3.1}$$

$$S_{vv} = -\frac{< (\partial v/\partial y)^3 >}{< (\partial v/\partial y)^2 >^{3/2}} \tag{3.2}$$

where $< \;\; >$ denotes space averaging. Figs. 2 and 3 give the time evolution of S_{uu} and S_{vv} for all runs. (Note that S_{vv}=S_{ww} from symmetry). Here for Runs b-d, we see sudden increases in skewness values during the evolution. (We will *not* discuss the results of Run a which seems to be strongly viscous). For these runs, the peak values are obtained at a time after the peak vorticity is attained. (See Table III for t_{uu} and t_{vv} the instances when the peak S_{uu} and S_{vv} are attained). For homogeneous isotropic turbulence, the skewness can be expressed in terms of the rate of enstrophy production and is an indication of the cascading process and also the stretching of vortex filaments. We note in Figs. 2, 3 and Table III that the peak skewness values are in the range of typical turbulent flow values (See Kerr, 1981 for details). Our previous results also showed enstrophy to increase in these intervals during which small-scales are produced in the form of extremely flattened dipole which gets dissipated at later stages.

Table No. III
Time of maximum S_{uu} and S_{vv} and their values:

Run	t_{uu}	S_{uu}	t_{vv}	S_{vv}
b	5.70	0.163	4.90	0.386
c	5.40	0.293	4.60	0.426
d	5.00	0.407	4.30	0.490

4. Evolution of strain-rate, velocity-gradient tensor invariants during reconnection

The analysis of strain-vorticity correlations in our previous work revealed a very interesting result. Namely, when the peak in mean square strain rate is attained, the vorticity at high straining regions gets aligned with the eigenvector s_β of the strain-rate tensor corresponding to the middle eigenvalue β. The other two eigenvectors, s_α and s_γ are perpendicular to the vorticity at high straining regions. We present in Fig. 4, the scatter plots of the angle between the three eigenvalues and the vorticity ($\cos\theta_\alpha, \cos\theta_\beta, \cos\theta_\gamma$) versus

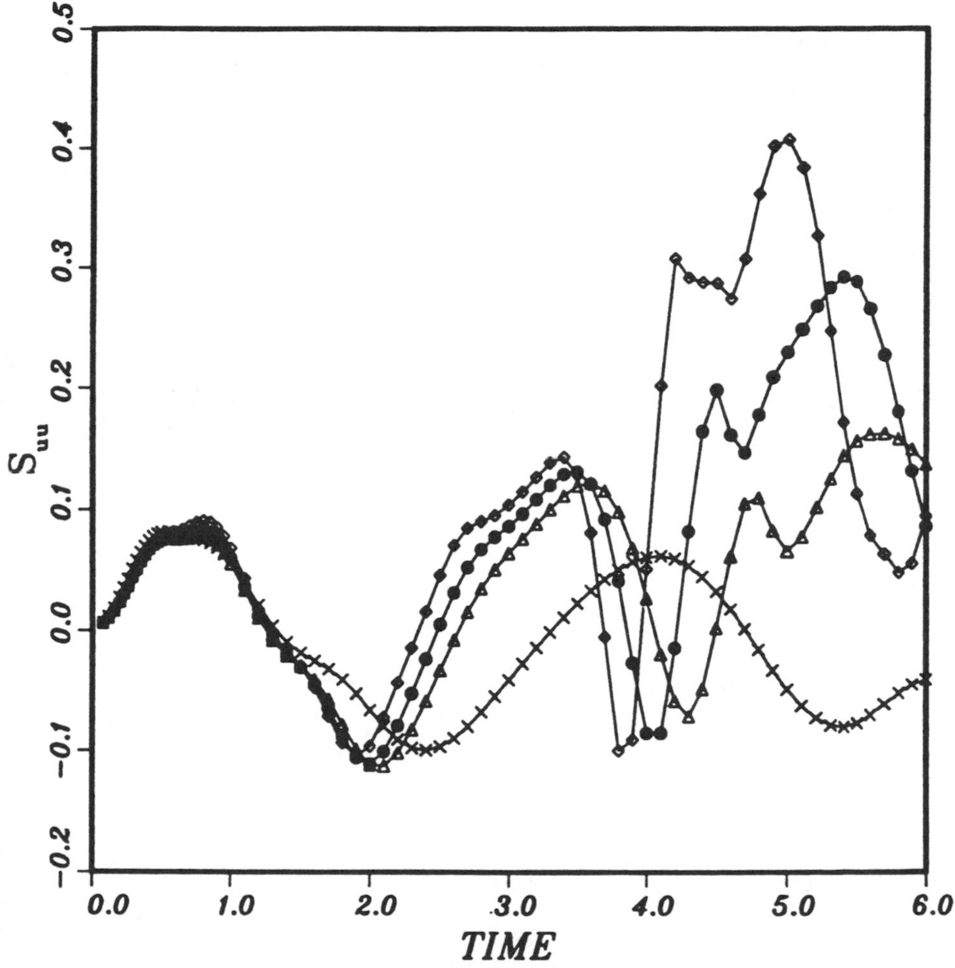

Fig. 2. Velocity derivative skewness S_{uu} evolution for all runs. Crosses: Run a, triangles: Run b, solid circles: Run c, losenges: Run d.

the strain rate (ϵ) for Run c, at t=4. This result is important due to the following: **First**, similar alignment properties have been found in turbulent flow simulations (Kerr 1987, Ashurst *et al.* 1987, She *et al.* 1991) and the process is examined in detail by Dresselhaus and Tabor (1992). However, it is not completely evident whether such a trend is due to the local kinematics of vortices or strongly dependent on the dynamical equations. **Second**, at the instant of perfect alignment between s_β and ω, the vortices are extremely flattened and appear to be sheet-like. One might argue that the flow in this sheet-like structure is almost two-dimensional and the alignment properties are a natural consequence of such a topology (Pumir, private com-

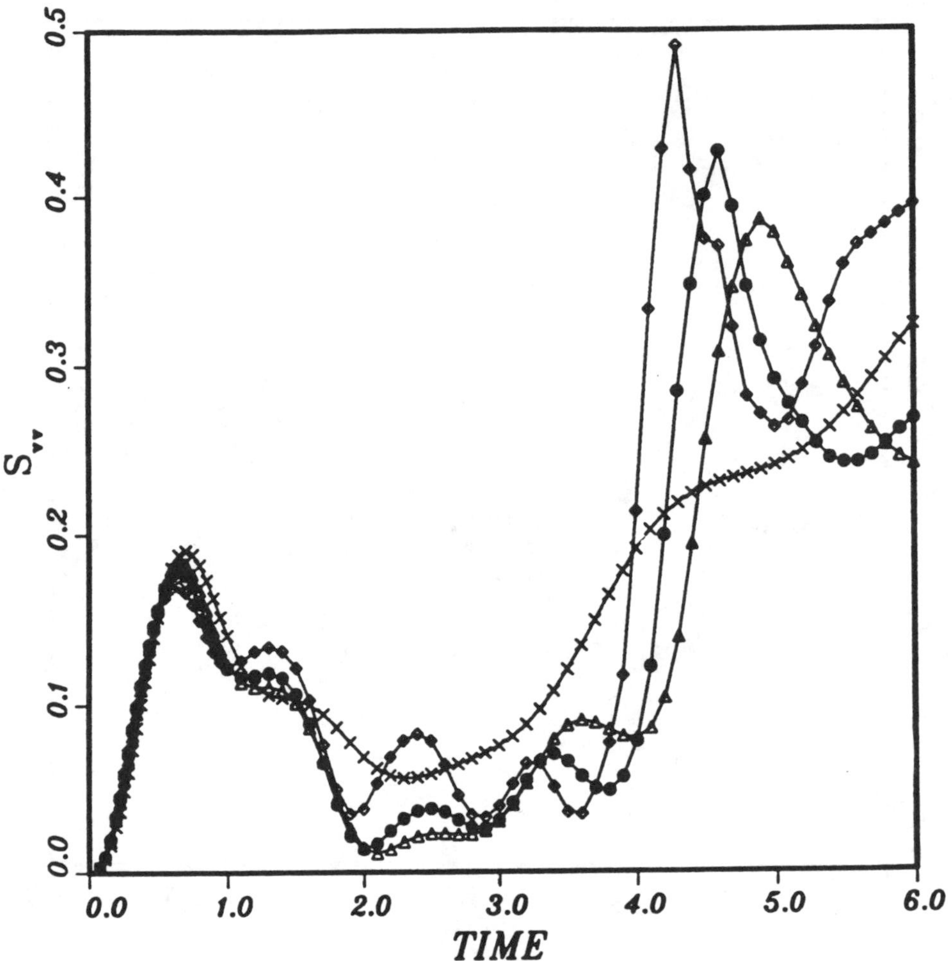

Fig. 3. Velocity derivative skewness S_{vv} evolution for all runs. Crosses: Run a, triangles: Run b, solid circles: Run c, losenges: Run d.

munication). In fact, there is also some recent work by Jimenez (1992) who relates the alignment characteristics to the kinematical constraints of a two-dimensional Burgers vortices. Similar work is also done by Majda (1992). So the main question is whether there is a quasi-two dimensional flow field at the instant when we observe the alignment properties and a sheet-like (pancake-like) topology. For this, we investigated whether the middle eigenvalue approaches zero in these high straining regions of interest. The results for Run c, t=4 are presented in Table IV which gives the average α, β and γ values at different straining bands ($\epsilon > 0.8\epsilon_{max}$, $\epsilon > 0.6\epsilon_{max}$, $\epsilon > 0.4\epsilon_{max}$, and $\epsilon > 0.2\epsilon_{max}$) and also the whole flow field:

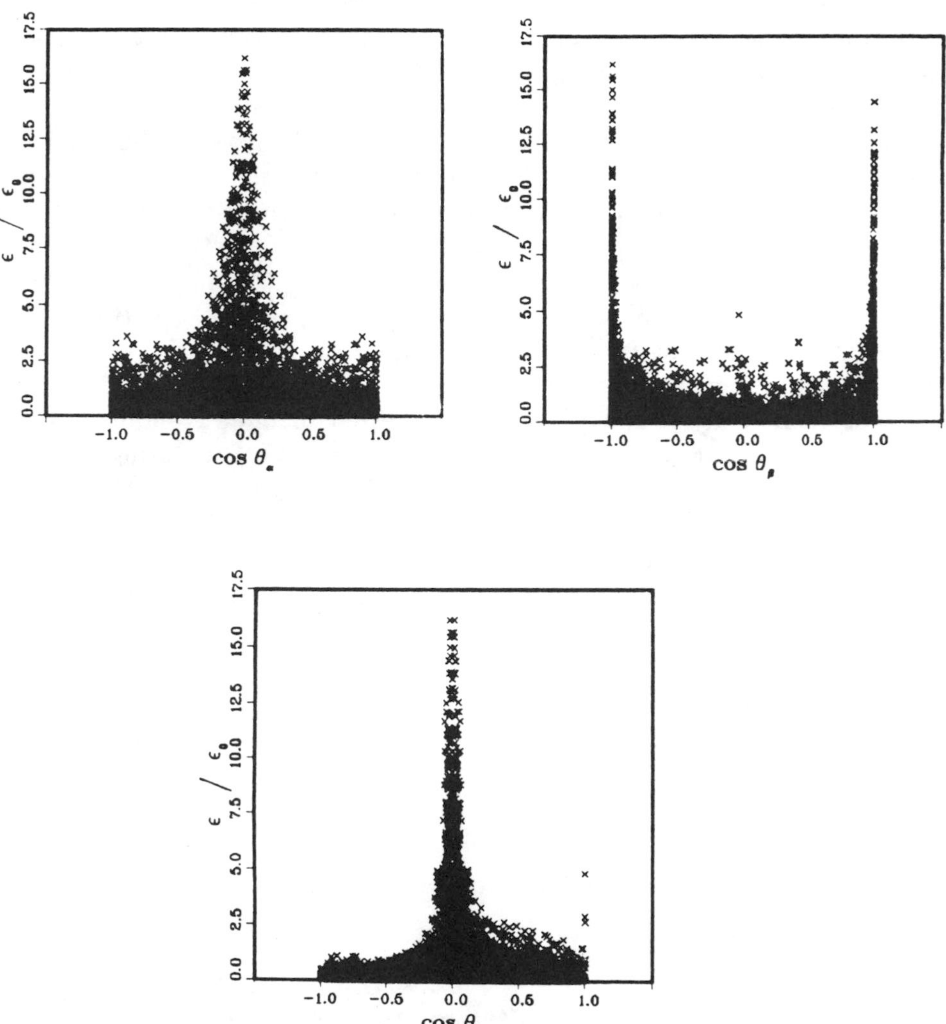

Fig. 4. Scatter plots of $\cos\theta_\alpha, \cos\theta_\beta, \cos\theta_\gamma$ vs ϵ for Run c at t=4.

Table No. IV
Average α, β, γ values at different straining bands:

Band	$<\alpha>, <\beta>, <\gamma>$	Ratio
$> 0.8\epsilon_{max}$	17.8, 5.1, -22.9	3.5, 1.0, -4.5
$> 0.6\epsilon_{max}$	16.2, 4.5, -20.7	3.6, 1.0, -4.6
$> 0.4\epsilon_{max}$	14.2, 4.1, -18.3	3.5, 1.0, -4.5
$> 0.2\epsilon_{max}$	11.5, 3.1, -14.6	3.7, 1.0, -4.7
Whole flow field	3.2, 0.3, -3.5	-10.7, 1.0, 11.7

We note two important results in Table IV: i) The mid-eigenvalue β is strictly non-zero at high straining bands. The trend becomes stronger as one goes to the higher straining regions. Hence, we do *not* find a local two-dimensional structure. ii) The turbulent flow simulations give a distribution in the range (3, 1, -4) and (4, 1, -5) at moderate to high straining regions (Jimenez 1992, Sondergaard *et al.* 1991) which is close to our results.

Next, we recall an important result discussed in the recent work by Dresselhaus and Tabor (1992): The main reason why the vorticity vector is *not* aligned with the largest straining direction (s_α) is the influence of the rotation terms. The rotation terms can be subgrouped as the vorticity term, principal axes rotation term, the non-local effects due to the pressure term and the viscous effects. Here, our search will be limited to the effect of the vorticity term only. Namely, we are interested in understanding how the straining and enstrophy terms compete with one another during reconnection. The problem, to a certain extent, is equivalent to investigating how the eigenvalues of the velocity gradient tensor A_{ij} ($= \partial u_i/\partial x_j$) change during reconnection. To see this clearly, the velocity gradient tensor can be decomposed into its symmetric (strain rate) and skew symmetric (rotation) parts, A=S+R. Also note that,

$$tr(A^2) = tr(S^2) + tr(R^2) \tag{4.1}$$

The first term on the right hand side is simply the mean square strain rate given by

$$tr(S^2) = \alpha^2 + \beta^2 + \gamma^2 \tag{4.2}$$

and the second term is equivalent to the enstrophy:

$$tr(R^2) = -\frac{1}{4}\omega^2 \tag{4.3}$$

So we note that the influence of the rotation (enstrophy) term is to make the $tr(A^2)$ negative (to decrease it). $tr(A^2)$ term can only be negative if the eigenvalues of the A tensor are complex. (In fact, the opposite statement is not necessarily true, that is, the eigenvectors may be complex but $tr(A^2)$ can still be positive). So, one mechanism causing the $s_\beta - \omega$ alignment might be due to the gradual dominance of the vorticity term over straining during reconnection. If this is the case, we would expect the number of points in the whole flow field which have complex eigenvalues of the tensor A to increase as the vortices get flattened. We checked this expectation for Run c between t=3 and 4. We found that between t=3 and 4 as the vortices get flattened, the number of complex eigenvalues of the velocity gradient tensor A increases by 10 percent. This 10 percent of the points are in the region of reconnection. After t=4 until t=5.4, the number decreases by almost 10 percent.

We also examined the type of the eigenvalue distribution and its evolution during the reconnection process. The method is inspired by Chong *et al.*'s (1990) general classification of three-dimensional flow fields. In this method, one has to compute the Q-R invariants of the velocity gradient tensor and depending on the location of each point in the Q-R plane, one can classify the type of the critical points. Our analysis showed that the main group of complex eigenvalues whose number is increasing during the flattening of the cores belong to the group 10b (Chong *et al.*, 1990). This is the group with the real part of the eigenvalue being positive. In critical point terminology, it corresponds to the 'unstable focus' type (Chong *et al.*, 1990). On the other hand, the analysis on the real eigenvalues of A showed the largest decrease in real eigenvalues to be the group having two negative and one positive eigenvalue (type 6a, stable node, saddle, saddle) whereas the group with two positive and one negative eigenvalues (type 6b, unstable node, saddle, saddle) increased in number. These results are quite interesting because the eigenvalue distributions around maximum straining time show remarkable similarities with the turbulent flow simulations (for example, see Chen *et al.*, Soria *et al.*, Sondergaard *et al.*, 1990). The detailed results will be published elsewhere.

5. Helicity during reconnection

Helicity integral defined as $\int u \cdot \omega dV$ is an invariant in an inviscid flow for vortex filaments (Moffatt, 1969). In viscous vortex interactions having distributed vorticity, helicity is neither an invariant nor does it give direct information about the knottedness of the vortex lines. The rate of change of helicity, for the viscous flows, can be expressed as (Moffatt, 1969):

$$\frac{dI}{dt} = -2\nu \int \omega \cdot (\nabla \times \omega) dV \tag{5.1}$$

The integrand of the right hand side of Eqn. 6, $\omega \cdot \nabla \times \omega$, is a measure of how a vortex line is twisted (spiraling). We shall call this quantity *'vortical helicity'* or *'volicity'*. A decrease in helicity with time can be topologically interpreted as an increase in magnitude and/or the number of the right-hand-twisted vortex lines in the overall volume. Note that the relation between helicity density and streamlines is analogous to that between volicity and vortex lines. Volicity, however, has the benefit of being Galilean invariant.

We present in Fig. 5 the total helicity evolution in our two higher Reynolds number runs (Run c and Run d). We see in Fig. 5 (for two runs) three regions with different helicity change rates. The first interval (before t=4 for both runs) shows a rapid decay (faster than a linear rate) in the total helicity. Recall that the vortices are pushed towards one another during this interval and fast vorticity cancellation (faster than a viscous scale) is taking place.

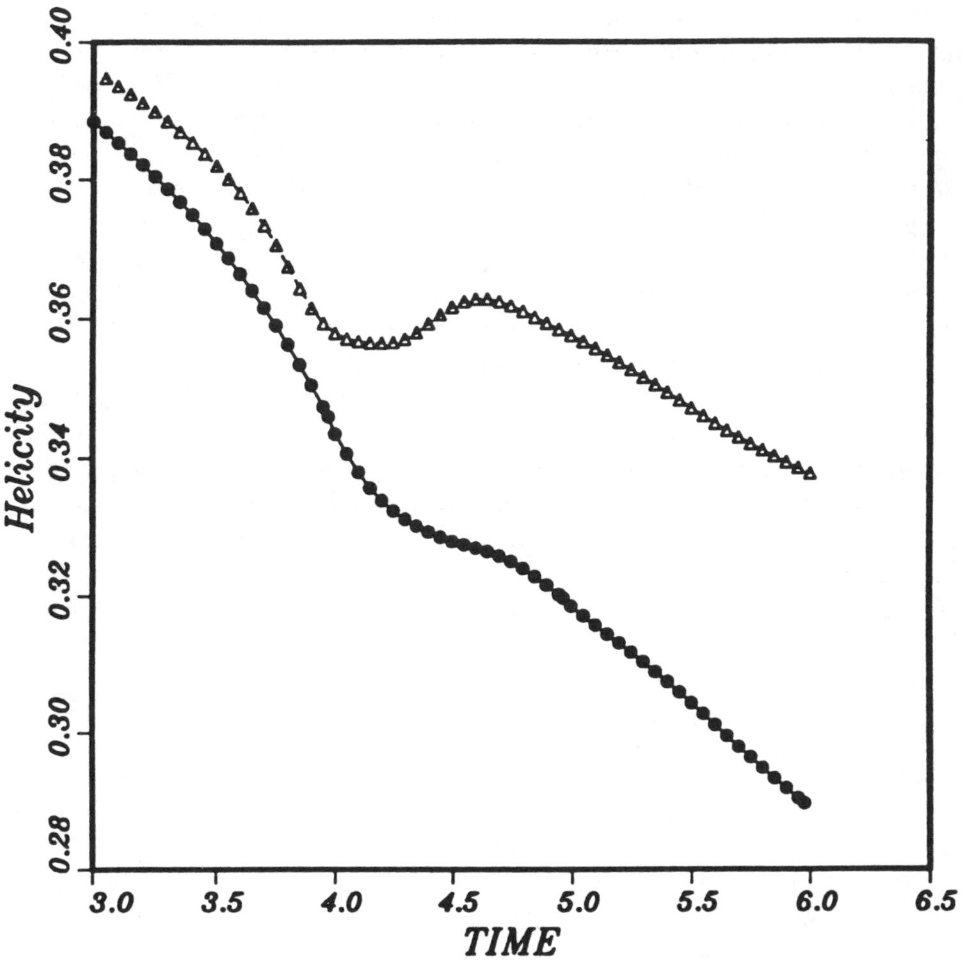

Fig. 5. Total helicity evolution in Run c (solid circles) and Run d (triangles).

After t=4, decay rate in helicity slows down and for the higher Reynolds number run, we observe an helicity increase. This interval corresponds to the later stage of reconnection when the newly-formed bridges increase in strength.

The increase in helicity might be interpreted as an increase in the right-hand twist component of the vortex lines. The subsequent decrease in helicity (after t=4.75) is at a slower rate compared to that before t=4. It can be attributed to the formation of a left-hand twist in vortex lines and possibly to the influence of viscosity. We may infer from isovorticity plots (Fig. 1) the behaviour of volicity in the context of helicity evolution. In the collapse, the tubes seem to form a right-handed spiral. The twist seems to decrease as the reconnection process proceeds. As the vortical bridge increases in strength, it shows a left-handed twist. A detailed analysis of the vorticity vectors and vortex lines during reconnection is in progress and the results will be presented elsewhere.

6. Conclusions

The main conclusion in this work is that the reconnection event can be considered as a rapid deformation process during which the vortices change their shapes from tubular to pancake-like and back to tubular. When the vortices are at the intermediate pancake-like state (extremely flattened), many characteristics show similarities with the turbulent flow simulation results. One interesting feature at this state is that the average β is positive. So far in literature, there was *not* any clear evidence that the evolution of $< \beta >$ can truly show the shape change of the vortices. As the vortices change back to their tubular state, small-scale cascading process takes place which is also reflected to the sudden increase of velocity derivative skewness values (around 0.4). We also find that rate of change of helicity changes sharply during reconnection. We suggest that the volicity ($\omega \cdot \nabla \times \omega$) is a measure of the twistedness of the vortex lines.

Our future direction will be to investigate whether such processes take place in turbulent flow simulations and whether their vorticity dynamics is reflected into the intermittency statistics.

Acknowledgements

The authors would like to thank the director of the I.T.P. James Langer and the directors of the NATO workshop on topological fluid dynamics, Keith Moffatt and Michael Tabor for inviting them to the workshop and to the Institute. Discussions with Keith Moffatt, Tsutomu Kambe, Javier Jimenez, Amitava Bhattacharjee, Alain Pumir, Robert Kerr, David Dritschel, John Greene, Zhen-Su She and Michael Tabor have been exciting and inspiring. The computations were done on the Pittsburgh Cray Y-MP, and NRL Connection Machine. OB and RP acknowledge the support of NSF grant (88-08780) and the Air Force Grant (AFOSR-91-0248). NZ acknowledges the partial support of the NSF Grant (DMS 89-01900) and the Office of Naval Research Contract (N00014-89-J1320).

References

ASHURST, WM. T., KERSTEIN, R., KERR, R. M., GIBSON, C. H., 1987 Alignment of vorticity and scalar gradient with strain rate in simulated Navier-Stokes turbulence. *Phys. Fluids*, **30(8)**, pp. 2343-2354.

BETCHOV, R., 1956 An inequality concerning the production of vorticity in isotropic turbulence. *J. Fluid Mech.*, **1**, pp. 497-504.

BORATAV, O. N., PELZ, R. B. & ZABUSKY, N. J., 1991a Winding and reconnection mechanisms of closely interacting vortex tubes in three dimensions. In "Vortex dynamics and vortex methods", Greengard and Anderson (eds.), *Lectures in Applied Mathematics, AMS*, **Vol. 28**, in press.

BORATAV, O. N., PELZ, R. B. & ZABUSKY, N. J., 1991b Reconnection in orthogonally interacting vortex tubes: Direct numerical simulations and quantifications. *Phys. Fluids*, in press.

BORATAV, O. N., PELZ, R. B., & ZABUSKY, N. J., 1991c Evolution during reconnection of the strain rate, velocity gradient tensor invariants. *Bull. Am. Phys. Soc.*, **36(10)**, pp. 2644.

CHEN, J., SORIA, J., SONDERGAARD, R., CHONG, M., ROGERS, M., MOSER, R., PERRY, A. & CANTWELL, B., 1990 A study of the topology of dissipating motions in direct numerical simulations of a time-developing compressible mixing layer. *Bull. Am. Phys. Soc.*, **35(10)**, pp. 2254.

CHONG, M. S., PERRY, A. E. & CANTWELL, B. J., 1990 A general classification of three-dimensional flow fields. *Phys. Fluids*, **2(5)**, pp. 765-777.

DRESSELHAUS, E. & TABOR, M., 1992 The kinematics of stretching and alignment of material elements in general flow fields. *J. Fluid Mech.*, in press.

JIMENEZ, J., 1992 Kinematic alignment effects in turbulent flows. *Phys. Fluids*, in press.

KERR, R. M., 1981 Theoretical investigation of a passive scalar such as temperature in isotropic turbulence. Ph. D. Thesis. *Cornell University*, pp. 55.

KERR, R. M., 1987 Histograms of helicity and strain in numerical turbulence. *Phys. Rev. Let.*, **59(7)**, pp. 783-786.

KERR, R. M. & HERRING, J. R., 1990 Inviscid simulations of turbulence. *Bull. Am. Phys. Soc.*, **35(10)**, pp. 2306.

KERR, R. M. & HERRING, J. R., 1991 Inviscid and high Re simulations of turbulence. *Bull. Am. Phys. Soc.*, **36(10)**, pp. 2703.

MAJDA, A., 1992 Vorticity, turbulence and acoustics in fluid flow. *1990 John von Neumann Lecture of the S.I.A.M.*, in press.

MOFFATT, H. K., 1969 The degree of knottedness of tangled vortex lines. *J. Fluid Mech.*, **35(1)**, pp. 117-129.

MOORE, D. W. & SAFFMAN, P. G., 1971 Structure of a line vortex in an imposed strain. In "Aircraft wake turbulence and its detection.", Olsen, Goldburg and Rogers (eds.), Plenum Press.

PUMIR, A. & SIGGIA, E., 1990 Collapsing solutions to the 3-D Euler equations. *Phys. Fluids*, **2(2)**, pp. 220-241.

SAFFMAN, P. G., 1990 A model of vortex reconnection. *J. Fluid Mech.*, **212**, pp. 395-402.

SHE, Z-S., JACKSON, E. & ORSZAG, S.A., 1988 Scale-dependent intermittency and coherence in turbulence. *J. Sci. Comp.*, **3(4)**, pp. 407-434.

SHE, Z-S., 1991 Physical model of intermittency in turbulence: Near-dissipation-range non-Gaussian statistics. *Phys. Rev. Let.*, **66(5)**, pp. 600-603.

SHE, Z-S., JACKSON, E., & ORSZAG, S. A., 1991 Structure and dynamics of homogeneous turbulence: models and simulations. *Proc. R. Soc. Lond. A*, **434**, pp. 101-124.

SORIA, J., CHEN, J., SONDERGAARD, R., CHONG, M., ROGERS, M., MOSER, R., PERRY, A. & CANTWELL, B., 1990 A study of the topology of fine scale motions in direct numerical simulations of a time-developing incompressible mixing layer. *Bull. Am. Phys. Soc.*, **35(10)**, pp. 2254.

SONDERGAARD, R., CHEN, J., & CANTWELL, B., 1991 Local vorticity-strain and pressure-strain distributions in direct numerical simulations of compressible and incompressible flow, *Bull. Am. Phys. Soc.*, **36(10)**, pp. 2674.

SONDERGAARD, R., SORIA, J., CHEN, J., CHONG, M., ROGERS, M., MOSER, R., PERRY, A. & CANTWELL, B., 1990 A study of the topology of fine scale motions in direct numerical simulations of a time-developing iwake. *Bull. Am. Phys. Soc.*, **35(10)**, pp. 2254.

ZABUSKY, N. J., BORATAV, O. N., PELZ, R. B., GAO, M., SILVER, D. & COOPER, S. P., 1991 Emergence of coherent patterns of vortex stretching during reconnection: A scattering paradigm. *Phys. Rev. Lett.*, **67(18)**, pp. 2469-2472.

NEW ASPECTS OF VORTEX DYNAMICS: HELICAL WAVES, CORE DYNAMICS, VISCOUS HELICITY GENERATION, AND INTERACTION WITH TURBULENCE

FAZLE HUSSAIN and MOGENS V. MELANDER*
Department of Mechanical Engineering
University of Houston
Houston, TX 77204-4792, U.S.A.

ABSTRACT. By direct numerical simulation of the Navier-Stokes equations we explore new aspects of viscous vortex dynamics. Core dynamics is shown to play a crucial role in vortex dynamics, vortex reconnection and enstrophy production. We propose and demonstrate that complex helical wave decomposition is a powerful new tool for explaining core dynamics, coherent structure evolution and interaction, and coupling of coherent structures with turbulence. We also discuss viscous generation of total helicity and how nontrivial topology of vortex lines develops from initially trivial topology.

1. Introduction

There appears to be an evolving consensus on the proposal that coherent structures in fluid turbulence should be characterized by coherent vorticity. Here, we build on the suggestion that vortex dynamics is a tractable avenue for understanding evolutionary dynamics of coherent structures, their role in turbulent transport phenomena, and their interaction with fine-scale turbulence. We show how this can be achieved by exploring in direct numerical simulations the evolution of idealized vortex modules and their interactions, and delve into new aspects and details of vortex dynamics, currently inaccessible via state-of-the-art measurement technology. We emphasize that core dynamics, almost completely ignored in previous research, is very important in vortex dynamics and vortex reconnection, and can be understood much better by the method of *complex helical wave decomposition* than by the traditional approaches (such as coupling between swirl and meridional flow). This decomposition provides considerable new analytical capability and insight into vortex evolution, cascade, and interaction with other vortices and turbulence.

* Permanent address: Dept. of Mathematics, SMU, Dallas, TX.

H. K. Moffatt et al. (eds.), Topological Aspects of the Dynamics of Fluids and Plasmas, 377–399.

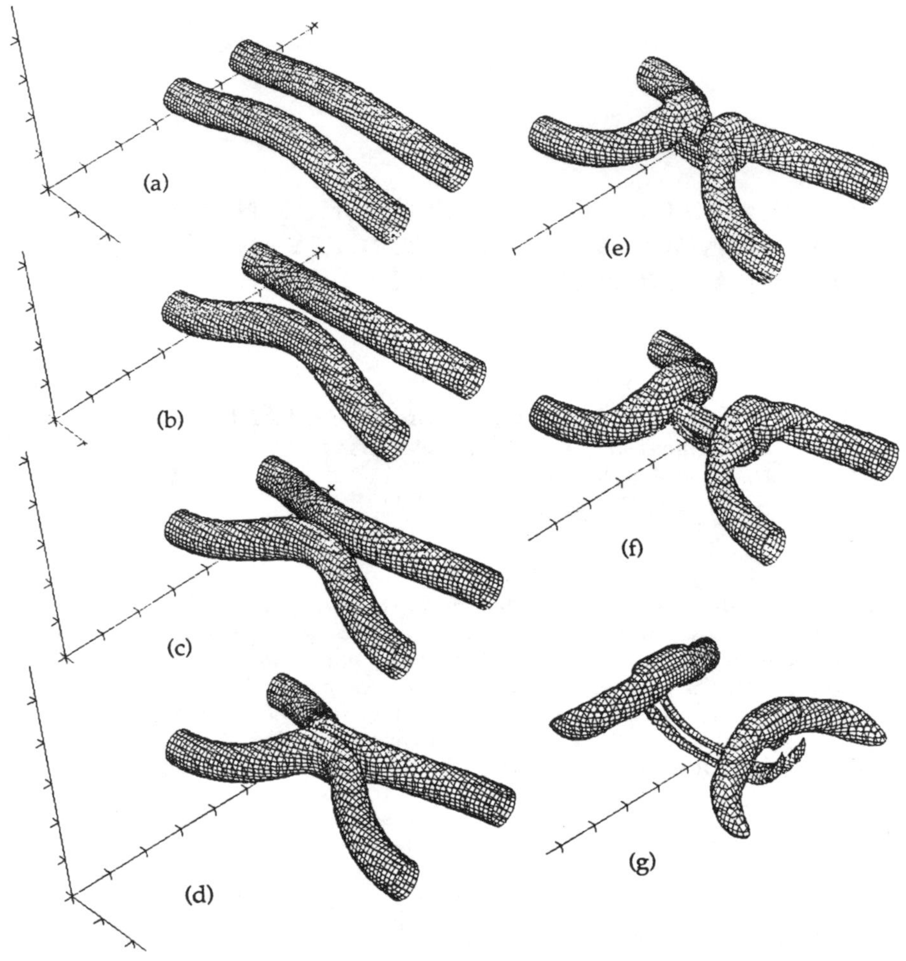

Fig. 1. Time evolution of $|\boldsymbol{\omega}|$ surface during reconnection of two antiparallel vortex tubes at $Re = 1000$.

2. Reconnection

Our interest in vortex core dynamics was motivated by our study of reconnection of two antiparallel vortex tubes of compact support and essentially Gaussian core, simulated via a spectral method. A sinusoidal perturbation of the vortex axes assured continued mutual approach of the two vortices by self-induction and hence reconnection. A sequence of vorticity level surfaces (at 30% of the initial peak value) for a Reynolds number $\Gamma/\nu = 1000$ is shown in figure 1. The reconnection evolves through *three* phases: (i) collision and core deformation, (ii) 'bridging' marked by a dramatic change in topology due to cross-linking of vortex lines, and (iii) 'threading' during which a remnant (unreconnected part) of the original vortex pair is sustained by stretching by the newly formed bridges. The crucial mechanism is *bridging* – formation of a dipole orthogonal to the initial vortex pair – as a result of rapid transfer of circulation due to reconnection of vortex lines by cross-diffusion. Bridging is a result of a complex interaction of three competing effects: self-induction that sustains collision, vorticity augmentation by stretching by the vortex parts away from the interaction region, and viscous cross diffusion (see Melander & Hussain 1988,1990, referenced hereinafter as MH, for detailed explanation). The accumulated orthogonal vorticity bundles, or bridges, induce a 'downwash' which arrests mutual approach of the remnants of the two initial vortices by reversing their curvature, and thus creating and stretching the long-lasting threads. By their self-induction the bridges move far apart, weakening the downwash and permitting the thread dipole to reverse curvature again, halfway between the bridges, and may then collide again by self-induction and start the next stage of the reconnection. Hence we speculate that reconnection occurs in bursts with a frequency that increases with increasing Re, but reaches a constant value as $Re \to \infty$. The circulation transferred into bridges within each burst depends on Re and becomes infinitesimal in the high Re-limit. Hence the circulation transfer tends to zero as $Re \to \infty$, but this convergence is nonuniform in t .

There are some interesting features of the bridging mechanism, as revealed by the geometry of the vortex lines. As a result of stretching experienced by the bridges (which accumulate increasing circulation as reconnection progresses), they have significant axial variation in their vorticity distribution (see figure 2). The peak $|\omega|$ is much higher near the interaction region than elsewhere. Also, the cross section of the vortex core varies from crescent-shaped in the reconnection region (section A) to circular (section B) away from it. These variations between A and B result in helical vortex lines within the bridges, causing an axial flow directed toward A. Hence results an inviscid vorticity "smoothing mechanism" that tends to reduce the axial variation of $|\omega|$. Incidentally, this is the reason why enstrophy production $P_\omega = \omega_i S_{ij} \omega_j$ is negative over some part of the bridge cross-section, but positive over another part of the cross section (because of vortex stretching induced by the threads (see MH)); *i.e.* in the same cross-section we have both positive and negative P_ω.

3. Core Dynamics

In order to focus on the axial flow in the bridges, we have idealized this problem to that of an axisymmetric vortex with a sinusoidal variation of core size. This way we eliminate the complications of curvature and the flow induced by the threads. Because of the axial variation of the core diameter, there is also an axial variation of swirl and hence twisting of vortex

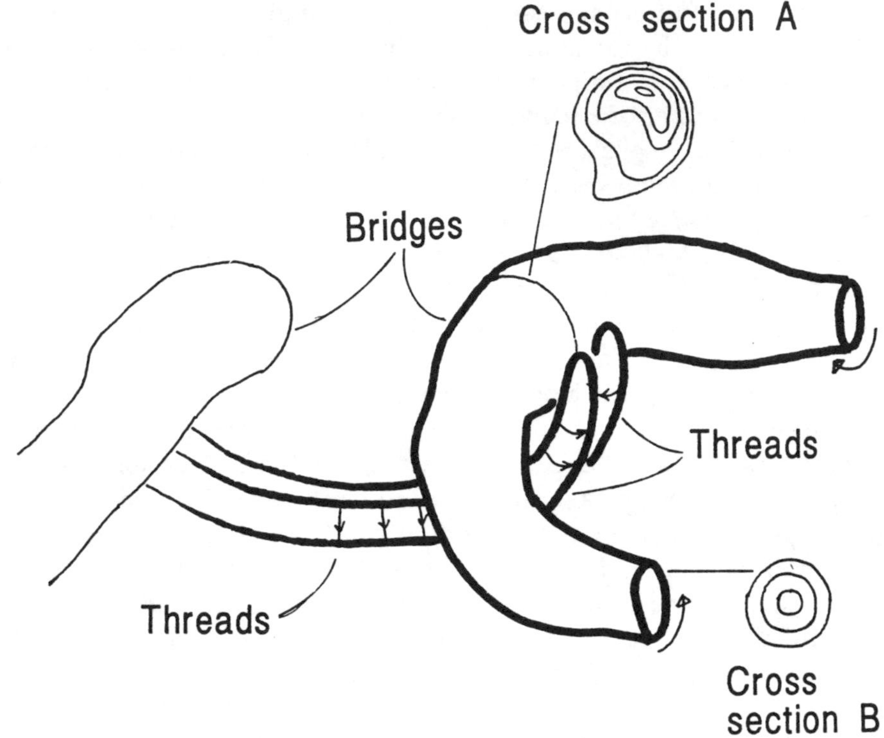

Fig. 2. Axial variation of vorticity in bridges.

lines soon after $t = 0$. Figure 3 shows a schematic of the vortex at a time $t > 0$ with vortex lines lying on the shaded surface denoting an axisymmetric vortex surface (not a vorticity level surface, but rv = constant); a 90° cutaway is introduced to enhance comprehension of meridional streamlines, and swirl and axial velocity profiles. The vortex evolves as a result of the interaction between meridional flow and swirl. The velocities u, v, w in the (r, θ, z) coordinates are governed by the following equations

$$D_t\zeta \;= \text{viscous term}; \qquad \zeta \equiv rv; \tag{3.1}$$

$$D_t\eta \;= r^{-4}\partial\zeta^2/\partial z + \text{viscous term}; \qquad \eta \equiv \omega_\theta/r; \tag{3.2}$$

where $D_t \equiv \partial_t + u\partial_r + w\partial_z$ denotes material derivative in the meridional flow, whose streamfunction Ψ is related to u, w and ω_θ as follows: $u = \Psi_z/r, w = -\Psi_r/r$ and $r\omega_\theta = \Psi_{rr} - \Psi_r/r + \Psi_{zz}$.

The vortex evolution in time is depicted by $|\omega|$ contours in figure 4 and meridional streamlines in figure 5. The axial variation of the swirl, ζ , produces coiling of vortex lines and hence axial pumping of fluid in a direction

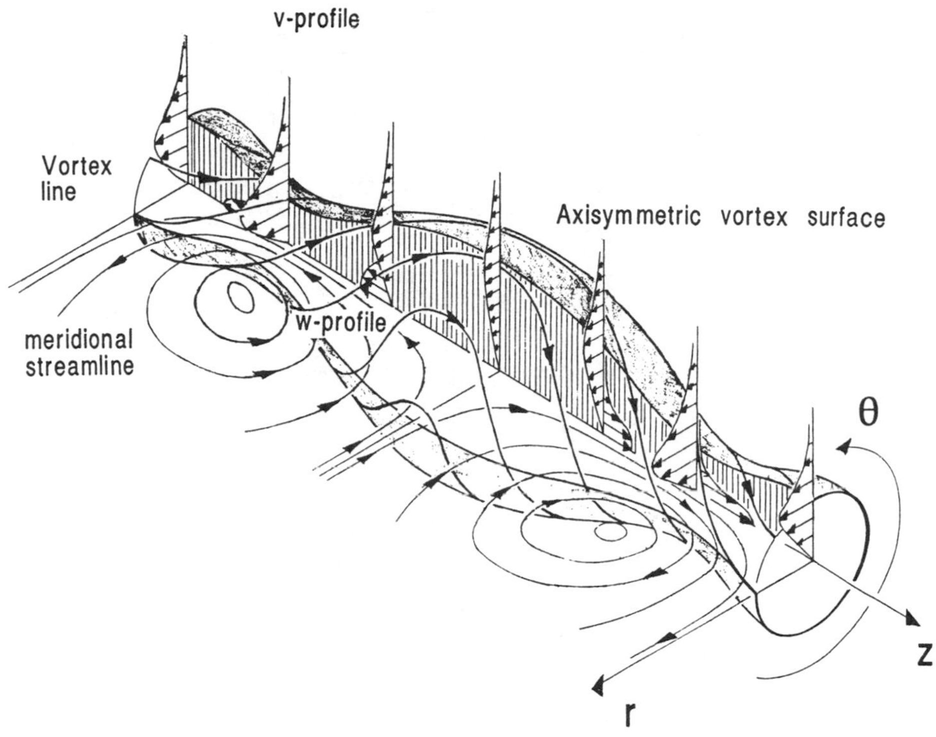

Fig. 3. Axisymmetric vortex surface with a 90° cutaway on which vortex lines lie. Also shown are streamlines in a meridional plane, and $v(r)$ and $w(r)$ profiles.

that will tend to smooth out core size variation and hence opposite to the axial variation of $|\omega|$ (*i.e.*, w is opposite to the vorticity flux). Axial variation of swirl is large where core size is small. Also, the twisting is stronger near the axis than at the outer edges. Note that both swirl and meridional flow would have undergone uneventful decay were it not for the coupling term $r^{-4}\partial\zeta^2/\partial z$ between the two fields. Azimuthal vorticity, ω_θ, initially zero, is thus immediately created by axial expansion and contraction of the vortex tube via the coupling term. This is the basic mechanism for motion of waves along the vortex axis. Maximum twisting of the vortex lines occurs at the inflexion points of $\zeta^2(z)$. Creation of ω_θ is necessary to start the meridional flow. Since the meridional flow first forms counter circulating cells (we call them primary cells), the vorticity peak (at the smallest initial core) bifurcates into two which travel in opposite directions. We may also think of this phenomenon as wavepackets which bounce back and forth while undergoing viscous decay. The first rebound induces a meridional flow reversal within each primary cell, thereby generating a secondary cell, which soon dominates by pushing out the primary cell. The wavepacket behavior of the vorticity

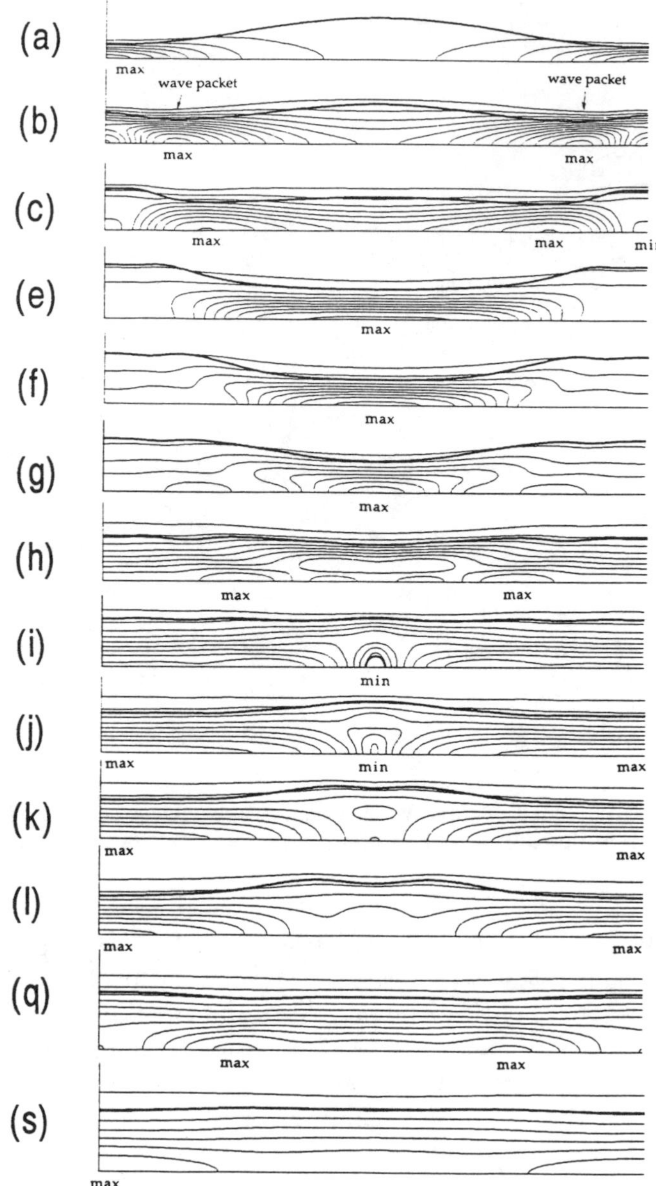

Fig. 4. Time-evolution of $|\boldsymbol{\omega}|$ contours. The thick line denotes the contour $N_k = 1$, where $N_k = [(\boldsymbol{\omega} \cdot \boldsymbol{\omega})/2S_{ij}S_{ij}]^{1/2}$ is the kinematic vorticity number.

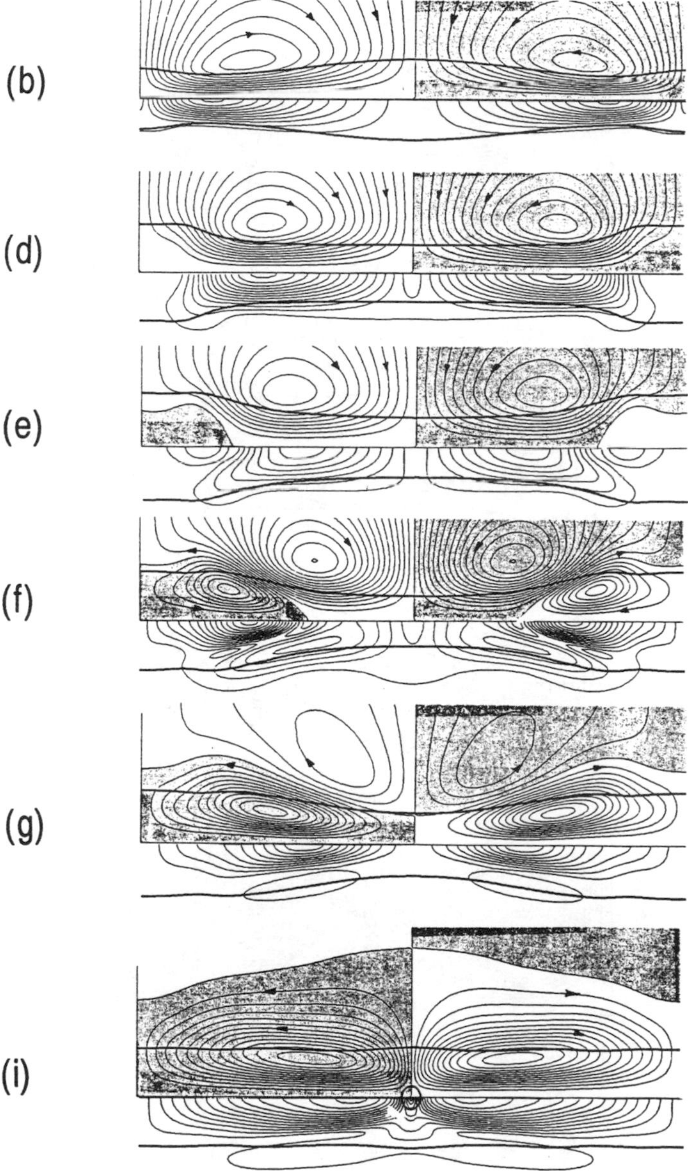

Fig. 5. See caption next page.

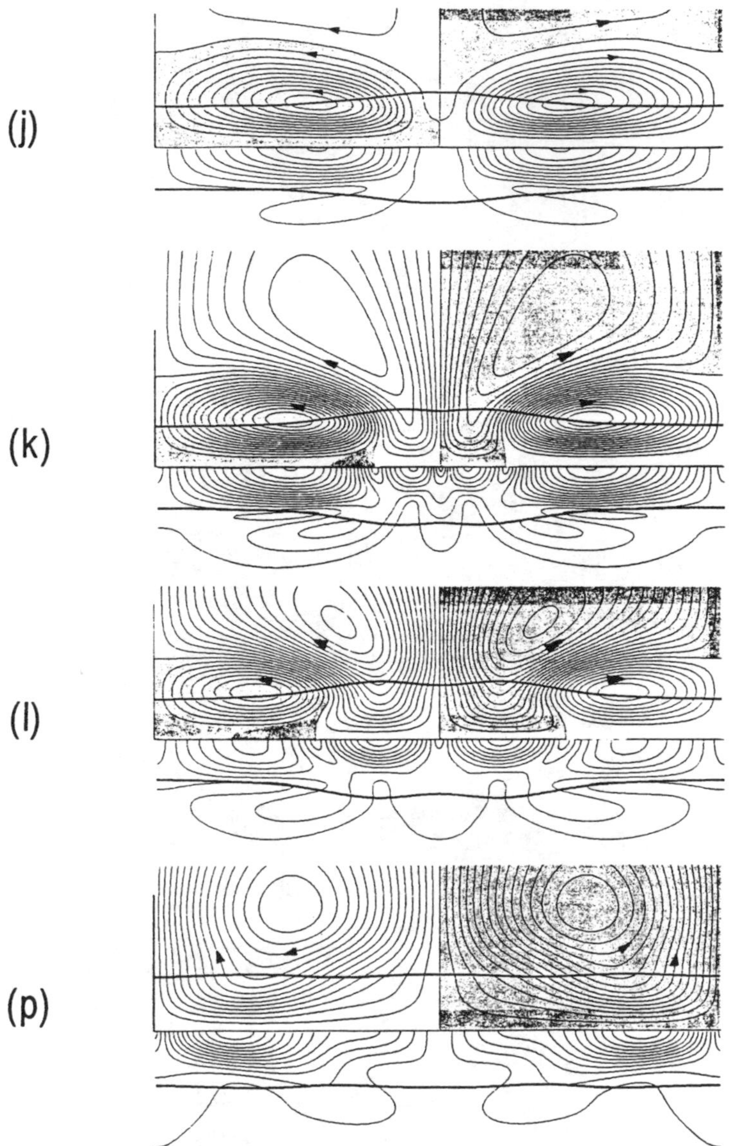

Fig. 5. Time evolution of the meridional streamfunction Ψ (above axis). Meridional plane vorticity magnitude $\sqrt{u^2 + w^2}$ is shown below axis. The shaded regions denotes negative Ψ. Also shown is the contour $N_k = 1$.

is associated with enstrophy production P_ω which can be related to Ψ for inviscid flows as

$$P_\omega \approx -\frac{4\omega_z^2}{r^2}\frac{\partial \Psi}{\partial z} \tag{3.3}$$

Thus P_ω can be inferred at any time from the meridional flow streamline pattern (figure 5).

It is important to focus on the generation of meridional flow, hence $\eta = \omega_\theta / r$. We find that for inviscid flow, $D_t\eta$ = axial vorticity $\times$ differential rotation along an axisymmetric vortex surface; thus the instantaneous value of η does not affect instantaneous material generation of η, *i.e.* $D_t\eta$. Because of axial transport of vorticity as wavepackets, η and $D_t\eta$ also oscillate in time, η leading $D_t\eta$ by a phase shift of about 90°; this phase shift controls η-transport. The two η-transport mechanisms (convection of η by meridional flow and η-generation by differential rotation along axis) are in opposite directions (hence slower η-transport) in the primary cell, but in the same direction (hence faster η-transport) in the secondary cell.

By investigating this flow for increasingly higher Re we have found that the core size variations do not decay in the high Re-limit. Moreover, we find that the wavemotion is more rapid at higher Re, albeit the wavespeed remains finite as $Re \to \infty$. This result has significant implications for computational vortex filament models (*e.g.* Moore & Saffman 1972) where it is assumed that core variations are smoothed out by the above wavemotion. This assumption is suspect in the light of our findings.

4. Helicity and Helical Wave Decomposition

The helical nature of flows has become the focus of considerable interest in recent years (Moffatt 1969). Of the three quantities: helicity density $h = \mathbf{u}\cdot\boldsymbol{\omega}$, relative helicity $h_r = h/(|\mathbf{u}||\boldsymbol{\omega}|)$ and helicity integral $H = \int \mathbf{u}\cdot\boldsymbol{\omega} dV$, only the last is Galilean invariant and is a conserved quantity for inviscid flow. From $|\boldsymbol{\omega}\wedge\mathbf{u}|^2 = u^2\omega^2(1-h_r^2)$ and the Navier-Stokes equations

$$\mathbf{u}_t + \boldsymbol{\omega}\wedge\mathbf{u} = -\nabla(P/\rho + u^2/2) + \nu\Delta\mathbf{u}, \tag{4.1}$$

it is common to claim that cascade is small if $h_r \simeq 1$. But such thinking is fraught with many pitfalls. First, although $h_r \simeq \pm 1$ does imply small $|\boldsymbol{\omega}\wedge\mathbf{u}|$, it does not imply small $\nabla\wedge(\boldsymbol{\omega}\wedge\mathbf{u})$ as high frequency oscillations of $\boldsymbol{\omega}\wedge\mathbf{u}$ are possible. In some frames, $\boldsymbol{\omega}\wedge\mathbf{u}$ can be large but purely potential so that vorticity obeys purely diffusion equation and hence there is no cascade. A rectilinear vortex is a glaring example. Thus, for cascade suppression, Beltramization (*i.e.*, $h_r = 1$) is sufficient, but not necessary. An example of cascade suppression for which $\nabla\wedge(\boldsymbol{\omega}\wedge\mathbf{u}) = 0$ is not satisfied is a laminar vortex ring which travels at a constant speed.

Thus, while H is a topological property (discussed later), h and h_r are of questionable use. But recognizing that local helical structure in a flow contains some essential physics, we propose that complex helical wave decomposition (Moses 1971; Lesieur 1990) is the relevant tool for understanding vortex dynamics and interactions. This decomposition in essence expands the field variables (*e.g.*, $\mathbf{u}, \boldsymbol{\omega}$) in terms of the eigenfunctions of the curl operator, which has only real eigenvalues in a periodic box or in infinite space. Moreover, the eigenfunctions can be chosen to form a complete set of orthonormal basis functions. In Fourier space, for each wavevector $\mathbf{k}$ there are two eigenmodes corresponding to the positive and negative eigenvalues k and -k. All functions which are linear combinations of eigenfunctions of positive eigenvalues can be called *right-handed* as trajectories or vector lines locally form right-handed helices. Similarly, we call linear combinations of eigenfunctions corresponding to negative eigenvalues *left-handed.* Thus for an incompressible velocity field $\mathbf{u} = \mathbf{u}_R + \mathbf{u}_L + \nabla\phi$. Since $\nabla\phi$ is the projection of $\mathbf{u}$ onto the eigenspace corresponding to the eigenvalue $k = 0$ of the curl operator and since $\nabla \cdot \mathbf{u} = 0$, we have $\Delta\phi = 0$. If the potential part of the flow is constant at infinity, or for a periodic box, $\nabla\phi$ is a constant. Thus $\mathbf{u}_R$ and $\mathbf{u}_L$ are unique, and both translationally and rotationally invariant. Similarly, $\boldsymbol{\omega} = \boldsymbol{\omega}_R + \boldsymbol{\omega}_L$ (assuming no rotation at infinity).

The decomposition of a flow field into polarized (*i.e.*, right- and left-handed) components is rooted in the intrinsic physical nature of the Navier-Stokes equations: the eigenmodes of the curl operator are exact solutions of the Euler equation and constitute a special class of solutions called Beltrami flows with constant abnormality (*i.e.* $\boldsymbol{\omega} = constant\ \mathbf{u}$). Because of orthogonality of the eigenfunctions of the curl operator, helicity takes the simple form

$$H = H_R + H_L \tag{4.2}$$

where

$$H_R = \int \boldsymbol{\omega}_R \cdot \mathbf{u}_R dV \geq 0; \qquad H_L = \int \boldsymbol{\omega}_L \cdot \mathbf{u}_L dV \leq 0. \tag{4.3}$$

The decomposition also provides a clearer insight into the flow physics: cascade is inhibited wherever $(\boldsymbol{\omega}_R + \boldsymbol{\omega}_L) \wedge (\mathbf{u}_R + \mathbf{u}_L)$ and its first spatial derivative are small. One can also easily show that cascade from interactions of modes is small when the modes are of the same parity (*i.e.* same handedness) but large when they are of opposite parity. One can thus infer that polarized vortices are persistent, hence "coherent", features of the flow.

We apply the decomposition to obtain some understanding of our axisymmetric vortex in the physical space. Remember that vector lines of $\boldsymbol{\omega}_R$ ($\boldsymbol{\omega}_L$) are right-handed (left-handed) helices. In general the physical space distributions of $\boldsymbol{\omega}_R$ and $\boldsymbol{\omega}_L$ overlap. The $\boldsymbol{\omega}_R$ and $\boldsymbol{\omega}_L$ distributions of our vortex are shown as functions of time in figure 6 (note identical distributions of $|\omega_R|$ and $|\omega_L|$ at $t = 0$ in figure 6a). This evolution is much more revealing than the evolution of h_r field (not shown). The polarized components move in opposite directions. Aside from the nonlinear interaction between ω_R and ω_L, there is an obvious resemblance to d'Alembert formula for the 1D wave equation.

Note that $|\omega_R|$ is antisymmetric to $|\omega_L|$ in z with respect to $z = l$ at all times, coinciding in frame (a). The helical decomposition thus gives unique separation of ω into ω_R and ω_L components, a feature we think is essential for deeper understanding of coherent structures in turbulent flows. Note that $|\omega_R|$ peak moves to the left instead to the right as one would expect; this is a consequence of nonlinear interaction (discussed below). In fact, in the case of an isolated vortex with only right-handed polarization, $|\omega_R|$ peak does indeed move to the right (figure 7). In figure 6 we see that as they move, the wavepackets deform: they broaden by diffusion, but also elongate and form a bubble by nonlinear effects. The nonlinearity is also responsible for the breakup of the initial front-back symmetry, *i.e.* steepening at the front with a tail at the back.

To understand the nonlinear interactions between polarized components, we extract each component by the projection operators (P^+, P^-). Here we discuss only ω_R (by symmetry, ω_L equation is obvious)

$$\frac{\partial \omega_R}{\partial t} = \underbrace{-\nabla \wedge (\omega_R \wedge \mathbf{u}_R)}_{I} + \underbrace{P^-[\nabla \wedge (\omega_R \wedge \mathbf{u}_R)]}_{II} - \underbrace{P^+[\mathbf{u}_L \cdot \nabla \omega_R]}_{III} \tag{4.4}$$

$$+ \underbrace{P^+[\omega_R \cdot \nabla \mathbf{u}_L]}_{IV} - \underbrace{P^+[\nabla \wedge (\omega_L \wedge \mathbf{u})]}_{V} + \underbrace{\frac{1}{Re}\Delta \omega_R}_{VI} . \tag{4.5}$$

Term I is the inviscid self-evolution; term II is the generation of ω_L by evolution of ω_R; term III is the contribution to ω_R by its advection by $\mathbf{u}_L$; term IV is stretching of ω_R by $\mathbf{u}_L$; term V is the ω_R generation by ω_L, consisting of self-interaction of ω_L and the right-handed part of stretching and advection of ω_L by $\mathbf{u}_R$; term VI is the viscous diffusion term. Terms II–V represent coupling between the left- and right-handed components. We have evaluated these terms for the vortex and find that in general their dominance in decreasing order are as follows: III, I, V, IV, VI and II.

The evolution of ω_R can be largely understood in terms of two effects: self-evolution and advection by the $\mathbf{u}_L$ field. The self-evolution (term I) attempts to move the ω_R-packet to the right by adding vorticity at the front and annihilating it at the back. But because $\mathbf{u}_L$ moves the ω_R-packet to the left, the leftward motion dominates, although it is slower than $\mathbf{u}_L$. Also, because the velocity field has the largest axial component on the axis,

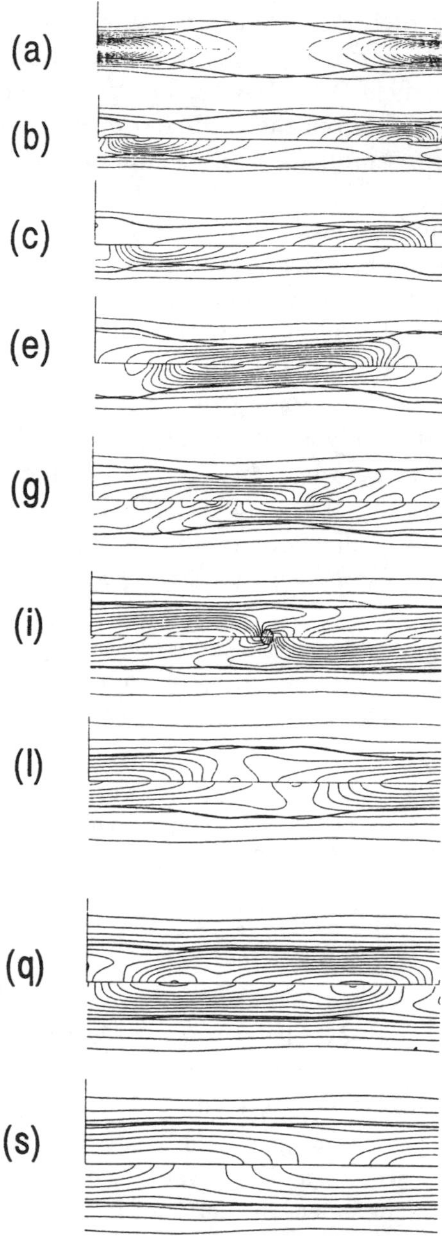

Fig. 6. Evolution of $|\omega_R|$ (above) and $|\omega_L|$ (below) the axis, with the $N_k = 1$ contour overlaid.

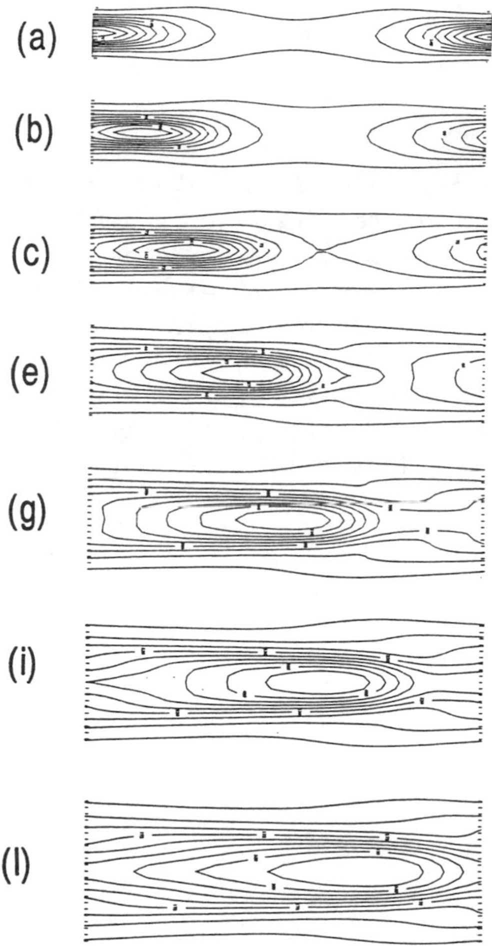

Fig. 7. Evolution of $|\boldsymbol{\omega}_R|$ for an initially right-handed vortex; *i.e.* the initial condition for this simulation is $\boldsymbol{\omega} = \boldsymbol{\omega}_R$.

it pushes the front of the packet backward at a faster rate near the axis (figure 6g), hence the formation of a low enstrophy bubble (figure 6i). The enhanced elongation is a combined effect of terms III and V.

The above gives a glimpse of how helical wave decomposition provides a new tool for analysis of core dynamics in terms of tractable interaction of polarized velocity and vorticity wave packets (see Melander & Hussain 1991a for details). Such physical insight into the core dynamics is impossible via the usual tools of vortex dynamics such as Biot-Savart law and local induction approximation. We hope many other important vortex phenomena (two discussed briefly below) can be understood better by this approach.

Vortex breakdown is another obvious candidate (subject of a separate study by us).

5. Viscous Generation of Helicity

A net helicity value (H) results when H_R and H_L do not balance each other exactly. A fascinating consequence of the orthogonality of the polarized components is that the value of H – and for that matter energy $E = 1/2 \int \mathbf{u}\cdot\mathbf{u} dV$ and enstrophy $Z = \int \boldsymbol{\omega}\cdot\boldsymbol{\omega} dV$ as well – is independent of the component's relative position in physical space. As an example, consider two given vortex rings, one completely right-handed and the other completely left-handed. These two rings may be placed anywhere in the physical space, far from each other, in close proximity, overlapping, on top of each other, or linking each other, but the values of H, E and Z are the same.

The evolution of H can be expressed in terms of the polarized components, namely,

$$\frac{dH_R}{dt} = \dot{H}_{R_{|inviscid}} + \dot{H}_{R_{|viscous}} \;; \quad \frac{dH_L}{dt} = \dot{H}_{L_{|inviscid}} + \dot{H}_{L_{|viscous}} \,, \tag{5.1}$$

where

$$\dot{H}_{R_{|inviscid}} = -\dot{H}_{L_{|inviscid}} \,, \tag{5.2}$$

$$\dot{H}_{R_{|viscous}} = -2\nu \int \omega_R \cdot (\nabla \wedge \omega_R) dV \leq 0, \tag{5.3}$$

and

$$\dot{H}_{L_{|viscous}} = -2\nu \int \omega_L \cdot (\nabla \wedge \omega_L) dV \geq 0. \tag{5.4}$$

Equations (5.3), (5.4) show that viscosity causes decay of H_R and H_L. However, H_R and H_L can increase in magnitude by inviscid effects, but their increase must equal. Note that although H is conserved when $\nu = 0$, there are no equivalent conservation laws for H_R and H_L. An interesting consequence of (5.1)–(5.4) is that H can increase or decrease by viscous effects in a flow which is not dynamically evolving. As a trivial example consider an inviscidly steady vortex ring with helical vortex lines; here we have $\dot{H}_{R_{|inviscid}} = \dot{H}_{L_{|inviscid}} = 0$ while (5.3) and (5.4) will be non-zero and different in magnitude; consequently, the value of H changes by pure diffusion. A more profound example involves two well separated vortical structures, one right-handed and the other left-handed, both being steady in the inviscid limit. Now suppose that they initially have equal but opposite helicity integrals and that they are of different sizes or shapes such that their decay rates are different; then H grows even though there is no dynamic interaction between the two rings.

The above analysis shows that non-trivial topology – for which $H \neq 0$ is a sufficient condition – results from the difference in the evolutions of the left- and right-handed components of the vorticity field. It is clear that they may do so even if $H = 0$ initially. H measures the overall balance between the left- and right-handedness of the flow field, but a much stronger condition than $H = 0$ initially is required to guarantee that H remains zero. Our following computational example clearly demonstrates that initially trivial topology is not sufficient to guarantee that H remains zero. What is required is of course that the decays given by (5.1)–(5.4) match at all times. The presence of reflectional spatial symmetry is sufficient to ensure that. Based on our numerical simulations we now believe that such symmetries may also be necessary for H to remain zero.

We now focus on how non-trivial vortex line topology can result from simple vortex interactions. For this purpose we consider non-pathological vortex interactions that may easily be realized in laboratory experiments. Moreover, we assume that such interactions start from initially trivial vortex line topology; collisions of vortex rings under asymmetric conditions clearly satisfy these requirements. Here the asymmetry is not only realistic, but also crucial in breaking geometric constraints that would otherwise make the left- and right-handed fields evolve symmetrically and thus result in $H_R \equiv -H_L$ (and consequently $H = H_R + H_L \equiv 0$).

We performed several simulations of asymmetric ring collision, and all of them exhibited generation of helicity. Out of these simulations we will discuss only one, namely the one that produced the largest value of $H_N \equiv H/\sqrt{2EZ}$; see figure 8. The initial condition for this simulation is shown in figures 9(a,f). The Reynolds number ($Re = \Gamma/\nu$) equals 672, which is sufficiently low as our results are independent of resolution (see figure 8).

The physical space evolution is illustrated by iso-vorticity surfaces at five instants in figure 9 (a–e); the contour level changes between frames so as to best illustrate the overall vortical structure. Comparison with figure 8 shows that H_N remains zero until the two rings collide (figure 9b). During the succeeding interaction where the two rings reconnect and combine into one looped vortical structure (figures 9c,d) there is a substantial increase in H_N. The increase in helicity is however small after this compound structure has formed and before it starts to split in two. The second major rise in H_N occurs as a second reconnection splitting the structure into two commences (figure 9d,e). Note, however, that the second reconnection is not complete as the two pieces remain connected by two circulation rich "threads". The so-called "bridges" for the second reconnection are not visible in figures 9 (e,j); this, however, is so only because of the chosen contour level; they are,

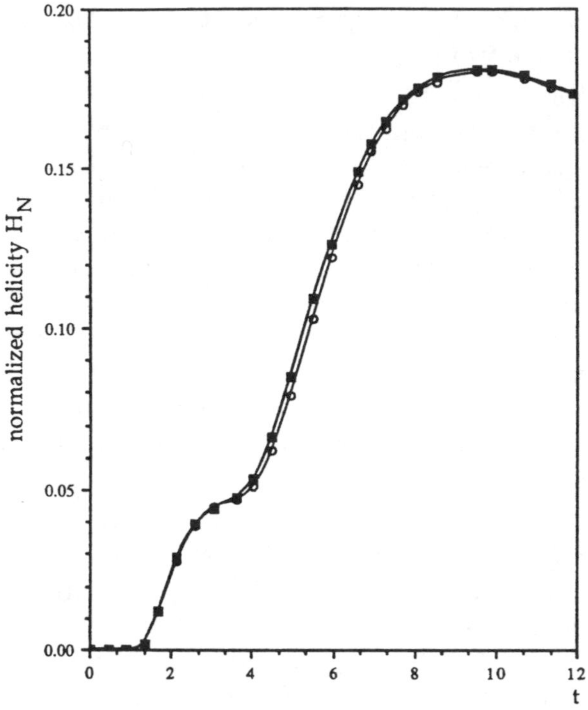

Fig. 8. H_N for the evolution shown in figure 9 (a–e, f–j). Two different simulations $(64^3, 128^3)$ are shown.

in fact, seen at lower levels. Also note that the two reconnections observed in figure 9 are more messy (especially at lower ω levels not shown here) than those occurring in the highly symmetric simulation of Kida *et al.* (1989), for which H remains identically zero, and moreover the vortex line topology stays trivial due to symmetry. While generation of helicity during the reconnections in figure 9 is evident from figure 8, the vortical structure as a whole is not knotted or linked in filament-like fashion. Instead, the helicity results from the instantaneous vortex lines *winding chaotically* within the vortical structure itself. More precisely, the helicity generation results from differences in the evolution of the polarized vorticity components (figures 9k-n). The difference in the spatial distributions of $|\omega_R|$ and $|\omega_L|$ is clearest at the larger ω; we have chosen the contour levels in figures 9 (k-n) accordingly. The polarized components gradually separate between frames (k) and (n), in spite of the initial pointwise match. The last frame (n) shows that the second reconnection produces two predominantly right-handed structures with predominantly left-handed "threads" in between. Quantitative dominance of right-handed part is discussed in Melander & Hussain (1991b).

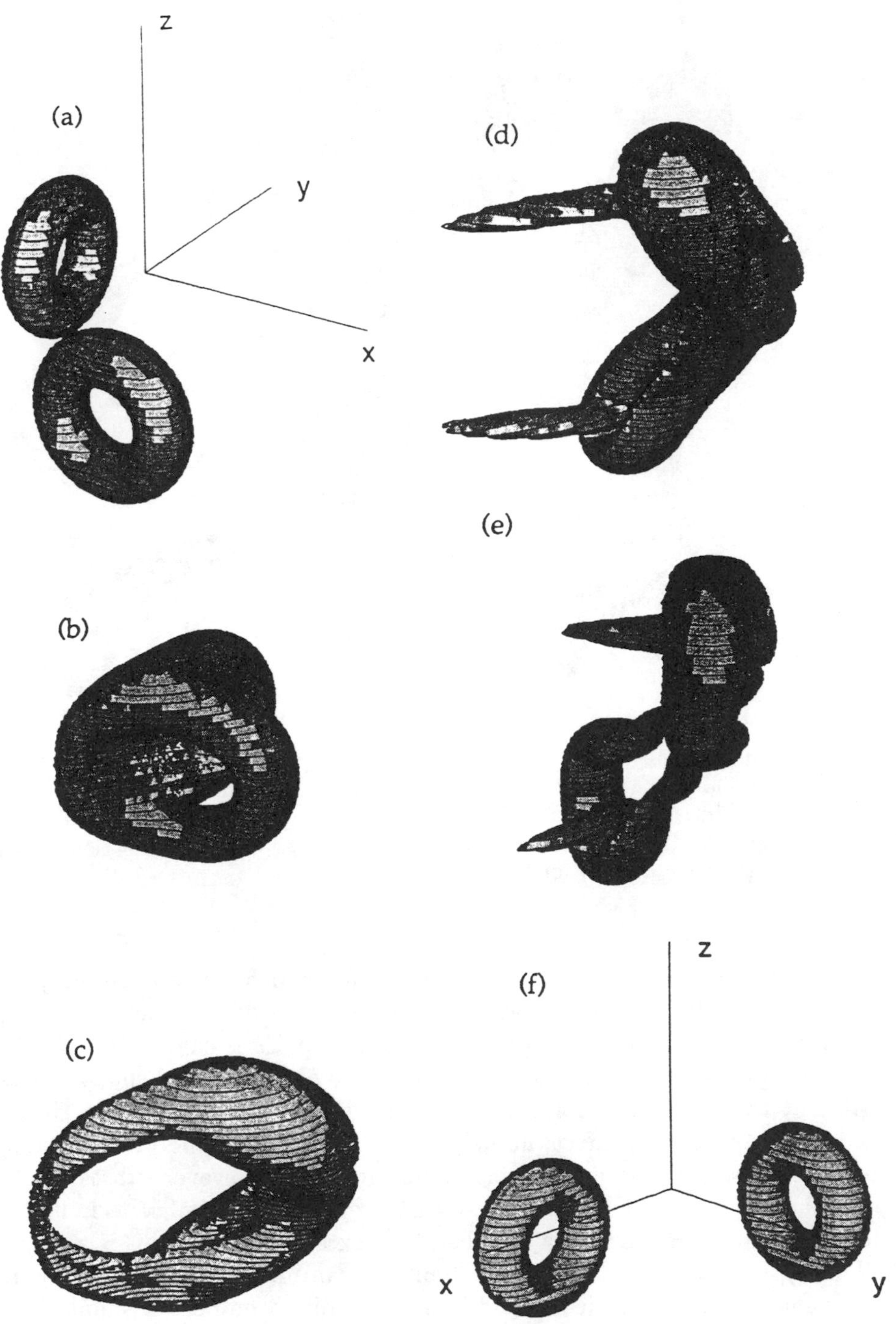

Fig. 9. See caption next page.

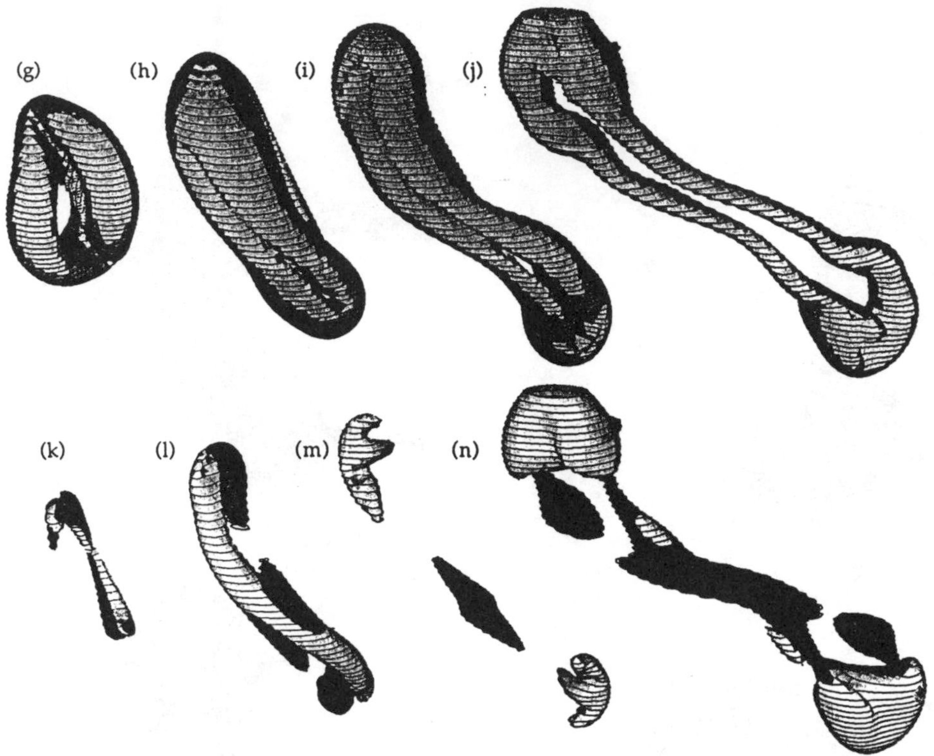

Fig. 9. Asymmetric collision of two vortex rings at $Re = 672$ with 128^3 simulation. (a)–(e) correspond to one view and (f)–(j) correspond to another. $|\omega|$ levels are shown for $t = 0$ in (a), (f); for $t = 2$ in (b), (g); $t = 3.5$ in (c), (h); $t = 6$ in (d), (i); $t = 10$ in (e), (j). The polarized vorticity components $|\omega_L|$ (darkened) and $|\omega_R|$ shown in (k), (l), (m) and (n) for $t = 2, 3.5, 6$ and 10.

Generation of helicity is a viscous phenomenon which depends not only on ν, but also on vorticity and its spatial derivatives. This dependence is, in fact, a sensitive one for when in the previous simulation Re is changed by $\pm 50\%$ the helicity production changes significantly. We observe that helicity starts to grow at the same time for three different Re (*viz.*, 336, 672 and 1008). This is thus an inviscid effect; namely, the vortex rings collide at the same time. The subsequent evolutions of the helicity differ however substantially among the three cases. For the highest Re the evolution is similar as in figure 8, but the helicity H_N is always smaller. The extent to which this is a result of H being conserved in the inviscid limit we can not answer, because it is not possible to approach the inviscid limit using direct numerical simulation. Only through large eddy simulations is such possible, but they would not treat the vortex line topology correctly. The results described above are typical for collision of vortex rings under asymmetric conditions; only the amount of helicity produced varies.

When this study started we considered it highly unlikely that a filament-like knotting or linking of entire vortices could be achieved in a direct numerical simulation at practically possible spatial resolution, as the vorticity field produced during finite core vortex interactions tends to get rather messy. However, in an independent recent study, Aref & Zawadzki (1991) employed a vortex-in-cell simulation (not a direct simulation), to produce a fairly clean linking. With direct 128^3 simulation and equivalent initial conditions, we were able to obtain a similar linking. In spite of the overall similarity between the vortical structures evolving in their and our simulations, we find considerable discrepancy in the production of the total helicity H during the evolution (see Melander & Hussain (1991b) for details).

We did not find any substantial production of H. Contrary to Aref & Zawadzki (1991) who obtained 70% of the filament linkage value $2\Gamma^2$, we obtained only 10%. Moreover, this peak value is reached while vortices are not linked. The fact that H remains nearly zero, while the linking of the vortices nevertheless occurs, has the implication that the "external" helicity due to linking of the large scale vortices is matched by opposite "internal" helicity arising from winding of vortex lines within the vortices.

The significance of our finding is as follows. While the actual generation of helicity H is a viscous phenomenon, it requires that the physical space distributions of ω_R and ω_L be different. For flows which are not subject to constraining reflectional geometric symmetries, there is a natural inviscid tendency for the polarized vorticity components to evolve differently and to separate in physical space, hence resulting in different distribution of ω_R and ω_L. Since coherent structure in turbulent flows are not subject to any such geometrical constraints, the implication is that the coherent structures become at least partially polarized and hence helical (as in figure 9). In fact, a separate numerical study by us of the breakdown of circular jet into turbulence shows separation of the polarized vorticity components to occur simultaneously with transition to turbulence.

6. Interaction with Fine-scale Turbulence

In an attempt to understand the coupling between large-scale coherent structures and fine-scale turbulence, we have repeated the simulation reported earlier with the axisymmetric sinusoidal coherent vortex (called CV) immersed in a background of isotropic fine-scale incoherent turbulence. Care was taken to assure that the initial rms level was sufficiently high for the fine-scale turbulence to survive long enough to dynamically interact with CV. The fine-scale size was chosen to be the smallest permitted by the resolution, but still with a spectral gap between it and that of CV. We find that the scales in the background incoherent vorticity grow progressively, so that

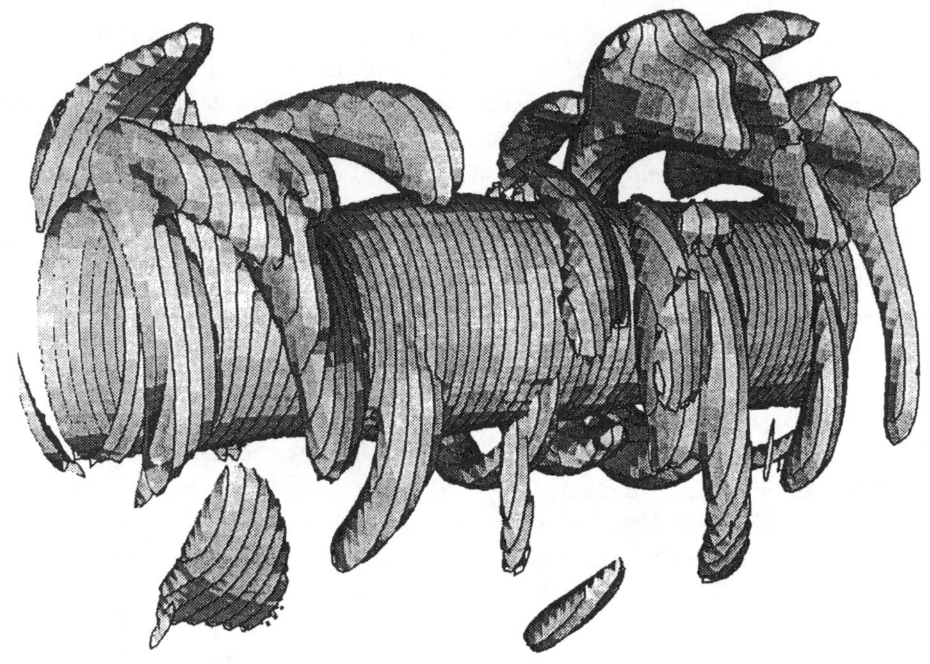

Fig. 10. $|\omega|$ surface at 10% of the peak value.

at the end of the simulation the largest incoherent scales are comparable to CV. The spectral gap is gradually filled up not by diffusive mechanisms, but by interaction between incoherent turbulence and CV. While the size of the incoherent turbulence remains uniform in space, the strength is higher at the boundary of CV where it is wrapped around CV and energized by stretching. This is not dissimilar from the growth of axisymmetric vortices on an impulsively rotated rod except that in our case the curved vortices around CV are intrinsically asymmetric, induced by the shear in the boundary of CV.

The spiral structures formed from organization of the incoherent turbulence are indeed quite different from the spiral sheet-like structures proposed by Lundgren (1982). The structures we find are rod-like spiral patterns wrapped around CV with senses aligned or opposed to the swirl of CV (figure 10). The helical wave decomposition is also helpful in understanding the coupling between CV and incoherent turbulence. The spiral structures are found to have a tendency to be highly polarized: they are either predominantly right-handed or predominantly left-handed. We indeed find a strong correspondence between peaks of incoherent turbulence in the boundary of CV and high values of the degree of polarization. The incoherent turbulence spirals undergo growth through merger by (truly inviscid) pairing mecha-

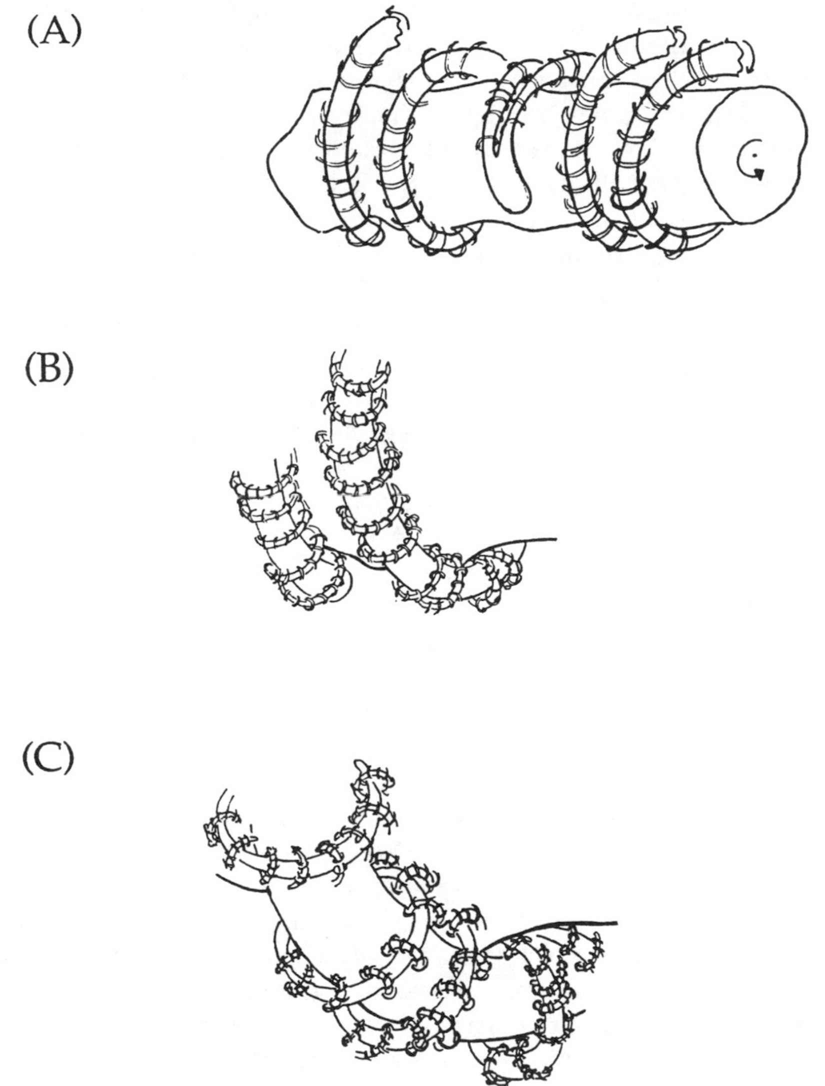

Fig. 11. Schematic of the conjectured fractal vortex with successive magnification.

nism and not by fusion through viscous decay and diffusion. In order for the spiral vortices to pair, their transport (axial to CV) is essential. There are two distinct mechanisms for such transport: i) transport of nearby opposite-signed vortices as dipoles, and ii) self-induced transport due to curvature of the azimuthally aligned vortices (not present in the 2D case). Because of the shear of CV, the small-scale vortex spiral undergoes differential stretching which provides the spiral an intrinsic tendency to become polarized.

In virtually all turbulence theory, local isotropy, which also forms the cornerstone of Kolmogorov's hypothesis, is universally assumed and implies decoupling of large and fine scales. By extrapolating our results (figure 10), we conjecture, for the case of very high Re, a hierarchy of structures of successively smaller scales, *i.e.* a fractal cascade as shown in figure 11. This supports our long-held doubt regarding the validity of the concept of local isotropy. Of course local isotropy does not require individual fine scales to be isotropic, but merely the lack of any statistical preference for the fine-scale structure's orientation. We feel that the fine scales retain some preferred orientation with respect to the coherent structures.

The CV-in-turbulence of course decays, and these unforced flows will eventually laminarize, in a time scale depending on both Re and the initial rms turbulence level. If the level is too low, CV recovers axisymmetry; if too high, CV is completely disrupted. For intermediate values interactions discussed above occur. However, in this case there is the interesting possibility of excitation of bending waves on CV. Without bending waves, as the spiral structures wrap around CV, they tend to be axisymmetric, thus diminishing their stretching by CV. When bending waves are excited, CV ceases to be axisymmetric and is then in a position to continue the stretching of the spiral incoherent turbulence. This is an example of feedback (or backscatter) from the smaller scales to CV. Thus we have here a mechanism of coherent structure/fine-scale interaction: the former organizes the latter by vortex stretching and the latter, if Re is high enough, induces bending waves in the former to generate the mechanism of its own survival.

This is a progress report of our continuing research activities on vortex dynamics and their connection to coherent structures in shear flow turbulence. This research is funded by the Office of Naval Research and the Air Force Office of Scientific Research.

References

AREF, H. & ZAWADZKI, I., 1991 Linking of vortex rings (manuscript).

KIDA, S., TAKAOKA, M. & HUSSAIN, F., 1989 Reconnection of the two vortex rings. *Phys. Fluid A* 1, 630-632.

LESIEUR, M., 1990 *Turbulence in Fluids* (Kluwer Acad. Publ.).

LUNDGREN, T. S., 1982 Strained spiral vortex model for turbulent fine structure. *Phys. Fluids.* **25**, 2193.

MELANDER, M. V. & HUSSAIN, F., 1988 Cut-and-connect of two antiparallel vortex tubes. CTR Report S88, Stanford U., 254-286.

MELANDER, M. V. & HUSSAIN, F., 1990 Topological Aspects of Vortex Reconnection. In *Topological Fluid Mechanics*, eds. H.K. Moffatt & A. Tsinober, 485-499 (Cambridge U. Press).

MELANDER, M. V. & HUSSAIN, F., 1991a Vortex core dynamics, helical waves and organization of fine-scale turbulence (submitted).

MELANDER, M. V. & HUSSAIN, F., 1991b Viscous generation of helicity, and non-trivial topology of vortex lines in coherent structures (submitted).

MOFFATT, H. K., 1969 The degree of knottedness of tangled vortex lines. *J. Fluid Mech.* **35**, 117.

MOORE, D. W. & SAFFMAN, P. G., 1972 The motion of a vortex filament with axial flow. *Phil. Trans. Roy. Soc. A* **272**, 403-429.

MOSES, H. E., 1971 Eigenfunction of the curl operator, rotationally invariant Helmholtz theorem and application to electromagnetic theory and fluid mechanics. *SIAM J. Appl. Math.* **21**, 114-144.

INTERMITTENCY GROWTH IN 3D TURBULENCE

Y. KIMURA
Center for Nonlinear Studies
Los Alamos National Laboratory
Los Alamos, New Mexico 87545, U.S.A.
and
National Center for Atmospheric Research
P.O. Box 3000
Boulder, Colorado 80307, U.S.A.

1. Introduction

Recently increasing attention has been paid to the non-Gaussian properties of small scales in turbulent flows as a manifestation of intermittency.[1] Past numerical studies were, however, almost always restricted to simulations with external forces or to results after several eddy-turn-over times. This is because the equilibrium shapes of the Probability Density Functions (PDFs) of velocity and velocity gradients were the main concern (Kerr, 1985; Yamamoto and Hosokawa, 1988; She, Jackson and Orszag, 1988; Kida and Murakami, 1989; Métais and Herring, 1989; Vincent and Meneguzzi, 1991).

As a theoretical tool to analyze non-Gaussianity, Kraichnan and his coworkers developed a systematic technique called mapping closure. The working hypothesis of the technique is that the shape of PDF is determined by a balance between advection which produces active small eddies and dissipation which wipes them out. As these processes have different time scales (i.e., dissipation becomes effective later than advection for fields initially at large scales), different shapes of PDF are possible as a result of combinations of the processes. Accurate statistical calculations for decaying turbulence with the initial Gaussian distribution is vital, in the examination of the hypothesis.

[1] The reader is referred to a recent query about the prevalence of exponentials as non-Gaussian addressed by Narasimha and discussed by Herring in *"Whither Turbulence? Turbulence at the Crossroads"* ed. Lumley, J.L. (Springer, 1989).

H. K. Moffatt et al. (eds.), Topological Aspects of the Dynamics of Fluids and Plasmas, 401–413.

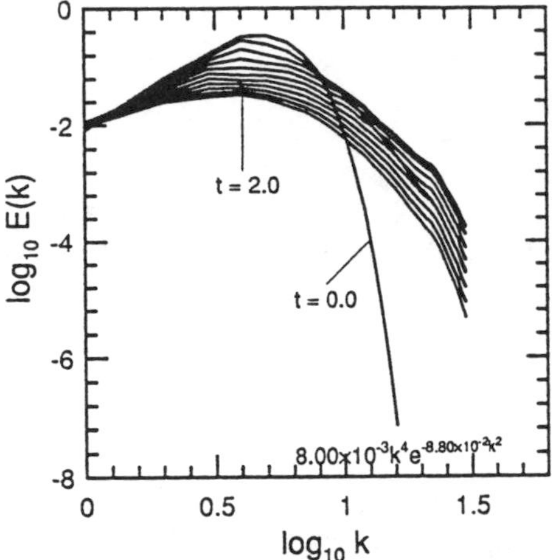

Fig. 1. Development of the kinetic energy spectrum. The initial spectrum is $E(k,0) = 0.008k^4exp(-0.088k^2)$, and others are plotted with a time interval of 0.2 up to $t = 2.0$.

In this article, we shall present numerical results on the development of the shape of PDFs of the velocity components and transverse velocity gradients of Navier-Stokes turbulence (section 2). The possibility of controlling intermittency will be discussed in section 3. We used three-dimensional pseudospectral simulations with 64^3 periodic grid points. To get clean statistical information, averages over a large number of ensembles of different initial conditions satisfying the same energy spectrum were taken.

2. Numerical Results

First we shall see the time scale difference in various quantities such as developments of the energy spectrum, dissipation rate of energy, skewness factor of longitudinal velocity gradients, and kurtosis factor of transverse velocity gradients.

The development of the energy spectrum is shown in Fig.1. The initial spectrum is

$$E(k,0) = 0.008k^4exp(-0.088k^2) \ , \tag{1}$$

and subsequent $E(k,t)$ are plotted with a time interval of 0.2 up to $t = 2.0$. (The time is not normalized by the eddy-turnover time of about 0.2.) The initial Taylor's microscale Reynolds number is about 40. The spectrum first changes to a self-similar stage and then decays uniformly, with the transition to self-similarity finished around $t \sim 0.4$.

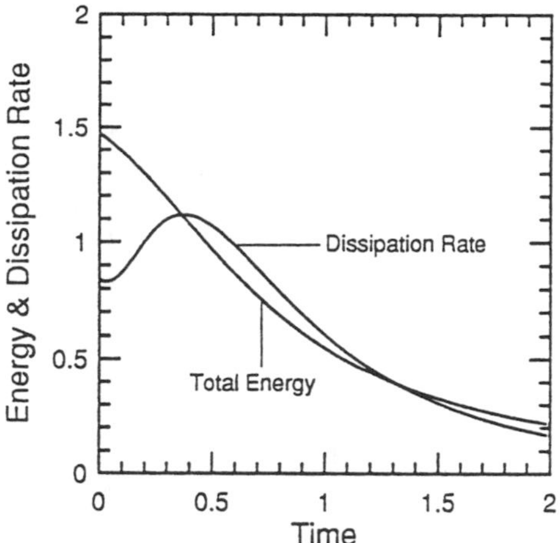

Fig. 2. The total kinetic energy $E_{tot} = \frac{1}{2}\sum_{\mathbf{k}} |\mathbf{v}(\mathbf{k})|^2$ and the energy dissipation rate, $\mathcal{E} = -\frac{d}{dt}E_{tot}$ as functions of time.

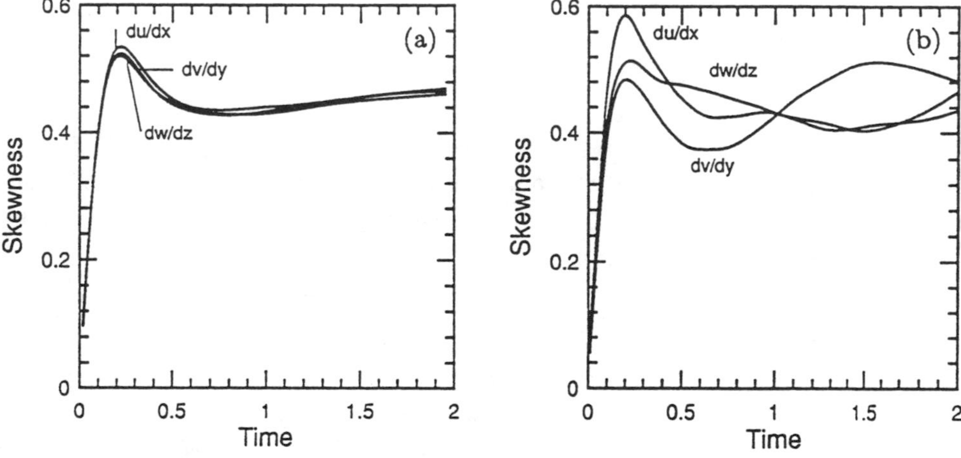

Fig. 3. The skewness factor of longitudinal velocity gradients. (a) ensemble average of 100 realizations; (b) single realization.

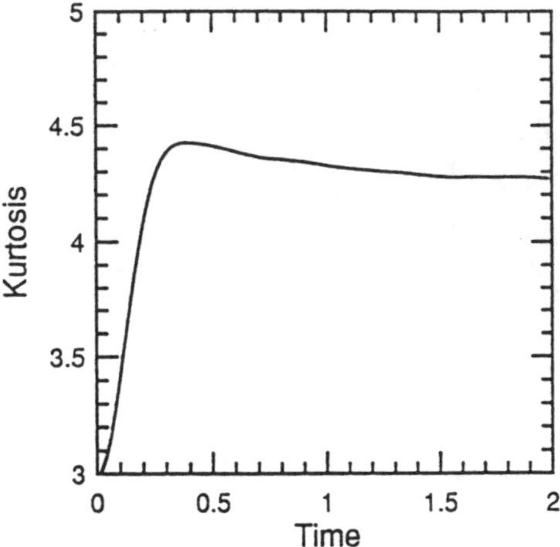

Fig. 4. The kurtosis factor of transverse velocity gradients.

In Fig. 2 we plot the total kinetic energy,

$$E_{tot} = \frac{1}{2} \sum_{\mathbf{k}} |\mathbf{v}(\mathbf{k})|^2 \ , \tag{2}$$

and the energy dissipation rate,

$$\mathcal{E}(t) = -\frac{d}{dt} E_{tot} \ , \tag{3}$$

as functions of time. The energy dissipation rate (which is proportional to the enstrophy for isotropic turbulence) has a maximum value at $t \sim 0.4$, and this time coincides with that of the appearance of the self-similar stage in the energy spectrum development.

The following relationship is known for the enstrophy D (Lesieur 1987),

$$\frac{dD}{dt} = \sqrt{\frac{98}{135}} S(t) D^{3/2} - 2\nu P(t) \ , \tag{4}$$

where $D(t)$ is the enstrophy,

$$D(t) \equiv \int_0^\infty k^2 E(k,t) dk \ , \tag{5}$$

$S(t)$ is the skewness factor of longitudinal velocity gradients,

$$S(t) \equiv -\frac{\langle (\partial u_i / \partial x_i)^3 \rangle}{\langle (\partial u_i / \partial x_i)^2 \rangle^{3/2}} \ , \ \ (i: 1, 2, 3) \ , \tag{6}$$

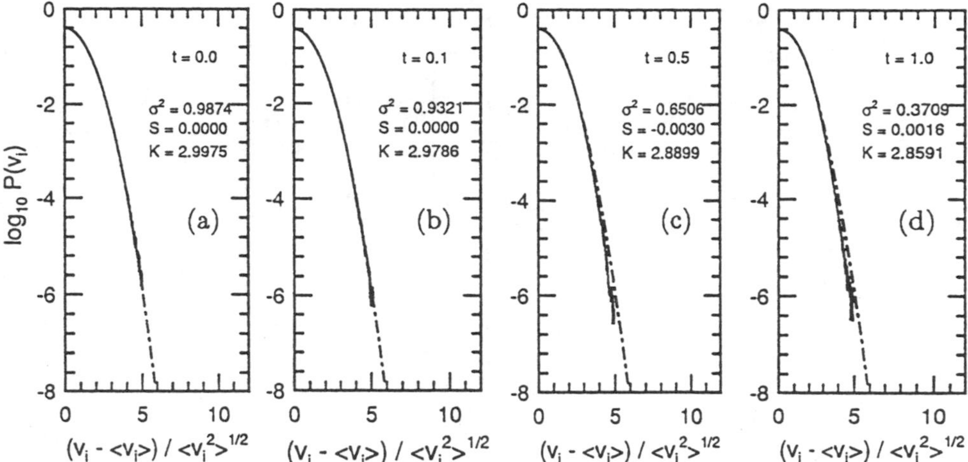

Fig. 5. PDFs of velocity components (a) t=0.0, (b) t=0.1, (c) t=0.5, (d) t=1.0.

(no summation rule is adopted), and $P(t)$ is the palinstrophy,

$$P(t) \equiv \int_0^\infty k^4 E(k,t) dk \ . \tag{7}$$

The initial slight decrease in $\mathcal{E}(t)$ is due to the viscous term in (4) because $S(t)$ is zero for the Gaussian distribution. The nonlinear term then becomes gradually effective, and $\mathcal{E}(t)$ (or $D(t)$) increases until the viscous term begins to be dominant again because of the energy transfer to smaller scales.

If ν is 0 and $S(t)$ is positive (-definite), the enstrophy will blow up at a certain finite time, giving a real-time singularity for this case. When ν is not zero, on the other hand, the enstrophy will be de-singularized, and we might have a conjugate pair of complex-time singularities.

The skewness factor of the three longitudinal velocity gradients are plotted from the average of 100 realizations (a), and a single realization (b), in Fig. 3. The difference between (a) and (b) clearly shows that the ensemble average is needed to extract an isotropic feature of the skewness factor from the data to enable us to discuss a possible enstrophy blow-up by virtue of (4).

Figure 3 shows a sizable overshoot around $t \sim 0.22$, and a stable stage appears after that which agrees with the theoretical predictions (for example, the E.D.Q.N.M. theory by André and Lesieur, 1977), and also with previous simulations (for example, Brachet *et al.*, 1983).

There is a time difference, albeit small, between the peak of the dissipation rate of energy and that of the skewness.[2] Intuitively speaking, $S(t)$ is

[2] This was pointed out by R.M. Kerr.

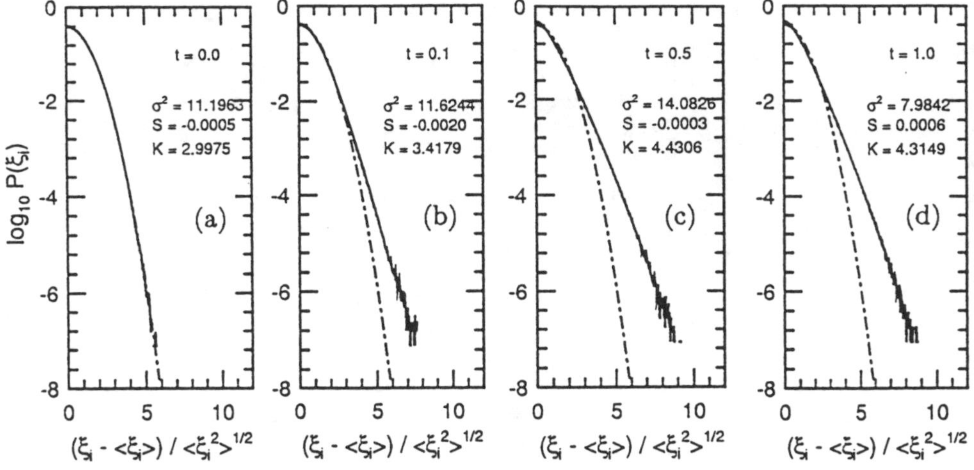

Fig. 6. PDFs of the transverse velocity gradients (a) t=0.0, (b) t=0.1, (c) t=0.5, (d) t=1.0.

an indicator of the energy transfer, and the enstrophy is produced by the energy transfer. Apparently, it seems natural that $S(t)$ (*cause*) has an earlier peak than $\mathcal{E}(t)$ (*effect*), but the true mechanism is still an unsolved problem.

Figure 4 shows the kurtosis factor of transverse velocity gradients $K(t)$,

$$K(t) \equiv \frac{\langle(\partial u_i/\partial x_j)^4\rangle}{\langle(\partial u_i/\partial x_j)^2\rangle^2} \ , \quad (i \neq j : 1, 2, 3) \ , \tag{8}$$

which is another crucial quantity indicating a departure from Gaussianity, i.e., intermittency. The plot is an average kurtosis of 6 different kinds of transverse gradient, and each is obtained as an ensemble average of 10 realizations of initial conditions.

Starting from 3.0, characteristic for the Gaussian distribution, $K(t)$ increases rapidly and maximizes at $t \sim 0.4$, then decreases slowly to a stationary value. The time for the maximum kurtosis coincides with that for the dissipation rate, therefore the biggest (spatial) intermittency appears when the enstrophy becomes maximum. This suggests that if the maximum enstrophy is associated with a complex-time singularity so is the maximum kurtosis factor. Thus the coincidence of the peak times implies a crucial roll of complex-time singularities in spatial intermittency as well as in temporal intermittency, which was proposed a decade ago by Frisch and Morf (1981).

PDFs of velocity components and their transverse gradients are shown in Figs. 5 and 6. The data of the PDFs were sampled at t=0.0, 0.1, 0.5, 1.0, and accumulated from 100 realizations. To get higher accuracy, we combined the three components of velocity and the six components of transverse gradients

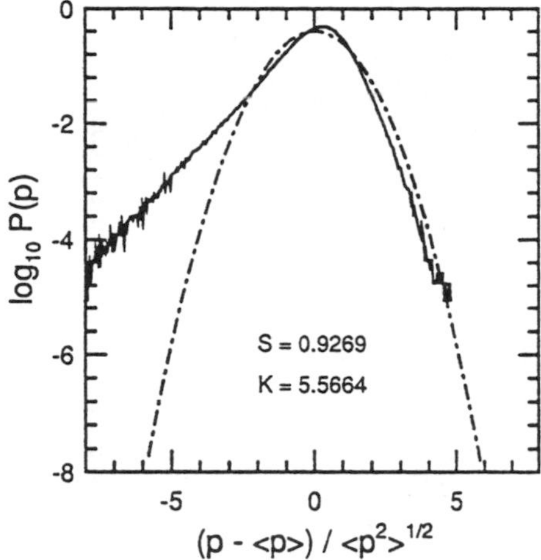

Fig. 7. PDFs of the pressure in Navier-Stokes turbulence.

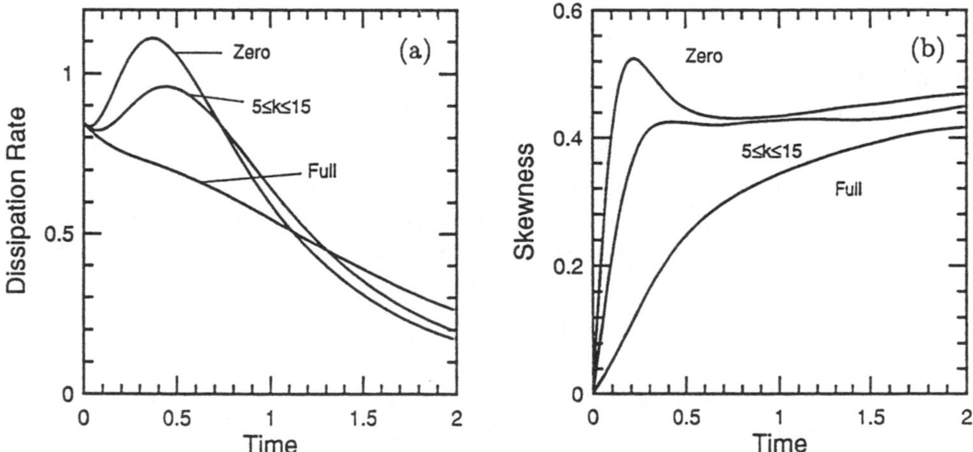

Fig. 8. Effect of helicity input on (a) the dissipation rate, and (b) the skewness factor of longitudinal velocity gradients. "Zero", "$5 \leq k \leq 15$", and "Full" mean the input of helicity in "no wave band (exactly zero helicity)", "a wave band of $5 \leq k \leq 15$", and "full wave band", respectively.

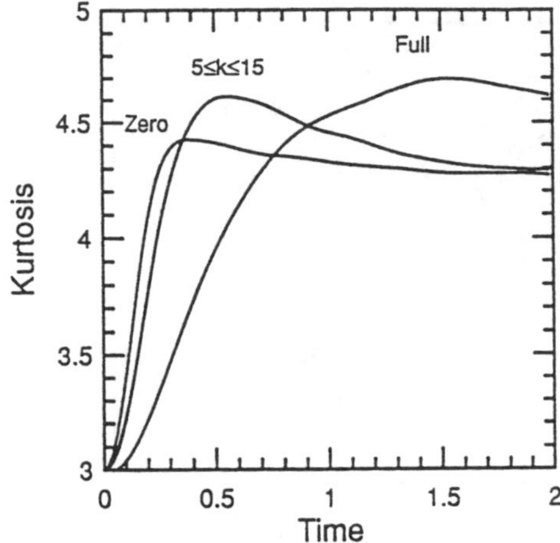

Fig. 9. Effect of helicity input on the kurtosis factor of transverse velocity gradients.

together, respectively, and folded at the vertical axis because both kinds of PDFs should be symmetric with this axis.

The PDFs of velocity components show thinner shapes than the Gaussian distribution (dashed line) in Fig. 5. This small inward defect from Gaussian was already reported in Fig. 4 of Vincent and Meneguzzi (1991). But they ignored this defect to conclude that the PDF of velocity components is close enough to Gaussian, which is well-known from many experiments (for example, Van Atta and Chen, 1968). The result in Fig. 5 suggests a reconsideration of the Gaussianity of velocity components in turbulence.[3]

The PDFs of transverse gradients show exponential-like tails, as already reported from both experiments (Van Atta and Chen, 1970; Castaing, Gagne and Hopfinger, 1990) and simulations (Yamamoto and Hosokawa, 1988; She, Jackson and Orszag, 1988; Kida and Murakami, 1989; Métais and Herring, 1989; Vincent and Meneguzzi, 1991) (Fig. 6). The departure from Gaussian is already saturated at $t = 0.5$. From Fig. 4, the saturation time seems to be $t \sim 0.4$ when K(t) obtains the maximum value. Though some work has been done to elucidate the exponential-like tails (Kraichnan, 1991; She, 1991), treatment of non-local quantities like the pressure or its gradient remains open to conjecture. Figure 7 shows a typical PDF of the pressure of Navier-Stokes turbulence. The asymmetric feature of the pressure PDF has been

[3] In his text book (*The theory of homogeneous turbulence*, Cambridge University Press, 1953, p. 170), Batchelor cited an experimental result by R.W. Stewart on the kurtosis factor of velocity difference. When a spatial distance is large, two velocities are statistically independent and the kurtosis factor is reduced to $\frac{3}{2} + \frac{\langle u_i^4 \rangle}{2\langle u_i^2 \rangle^2}$. Stewart's plot shows a lower value than 3 asymptotically, which indicates a lower kurtosis for velocity components.

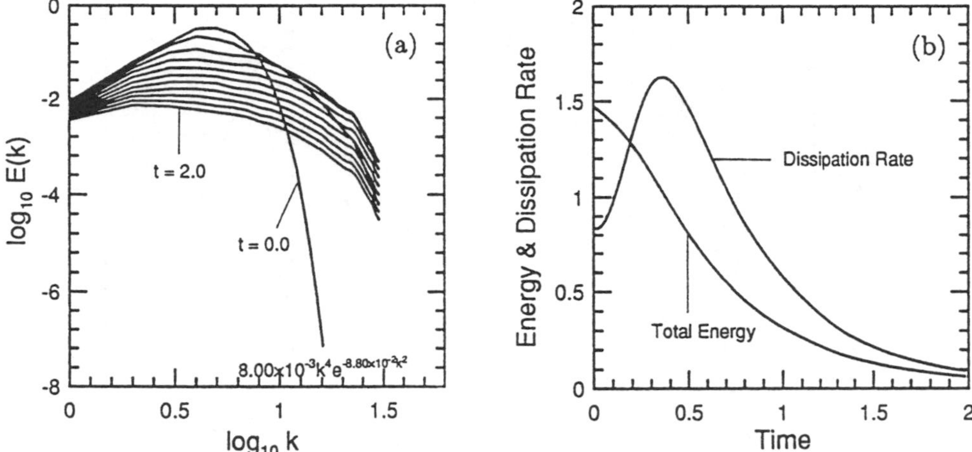

Fig. 10. (a) The energy spectra for the passive-vector equation under a frozen random advecting velocity; (b) the dissipation rate of energy.

already observed (for isotropic turbulence, Métais and Lesieur, 1991; for a mixing layer, Comte, Lesieur and Lamballais, 1991), and this may play an important role in clarifying the mechanism of the intermittency growth.

3. Control of intermittency

We have seen that the developments of PDFs are intimately related to the balance between advection and dissipation in turbulence. Thus intermittency might be controllable if we can somehow adjust the balance. In order to adjust the balance we propose two heuristic numerical methods. One would input the helicity initially and the other would modify the Navier-Stokes equation.

3.1. INITIAL HELICITY INPUT

Helicity is a conservative quantity for an inviscid flow which measures the degree of knottedness of the vortex lines (Moffatt, 1969). As was suggested by Kraichnan (1973), the helicity (or the partial alignment of velocity $\mathbf{u}$ with vorticity $\boldsymbol{\omega}$) reduces the effect of the nonlinearity $\mathbf{u} \times \boldsymbol{\omega}$ and depresses the overall turbulent energy transfer. When a flow is viscous, the helicity itself decays, and the decay rate of the ratio of the total helicity and the total energy plays an important role. If the helicity can be sustained until the peak time of enstrophy, there seems to be a significant effect on vortex stretching.

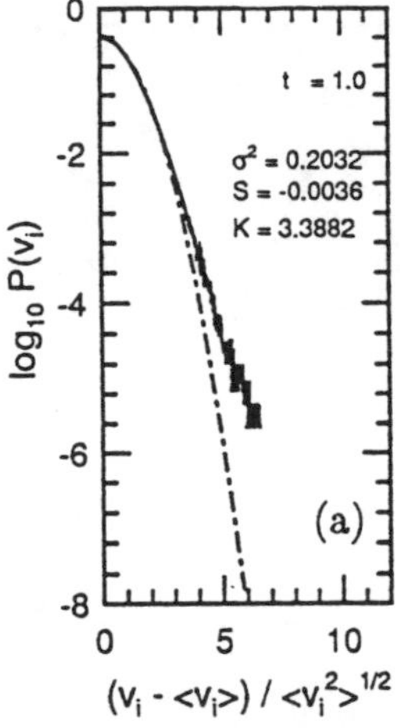

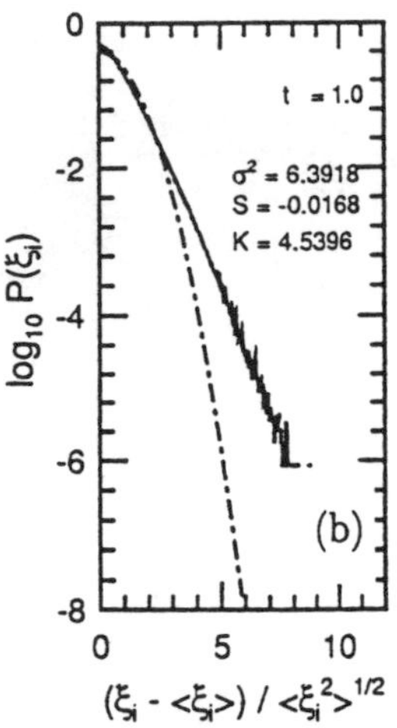

Fig. 11. PDFs of (a) the components and (b) the transverse gradients, of **u** (passive vector equation). They both show exponential-like non-Gaussian tails.

Helicity H in the k-space expression is

$$H = \sum_{\mathbf{k}} \mathbf{u}(\mathbf{k}).\omega(-\mathbf{k}) = 2\sum_{\mathbf{k}} \mathbf{k}.[\mathbf{u}_r(\mathbf{k}) \times \mathbf{u}_i(\mathbf{k})] \ , \tag{9}$$

where $\mathbf{u}(\mathbf{k}) = \mathbf{u}_r(\mathbf{k}) + i\mathbf{u}_i(\mathbf{k})$ (Polifke, 1991). Thus, by adjusting the angles between the real and imaginary vectors of complex velocities in the wave space, we can input the helicity to the wave modes. To maintain the incompressibility, first a set of real vectors is generated such that each vector is perpendicular to a wave vector and is properly scaled for the energy spectrum. Then an imaginary vector, also perpendicular to the wave vector, is formed making a given angle with the real vector.

Figures 8a and 8b show the enstrophy and the skewness factor under three different initial values of helicity. We observed that the peaks of the enstrophy and the skewness factor became lower and shifted downward with the helicity input. The vortex stretching was eventually delayed and depressed. In Fig. 9, the corresponding kurtosis factors are plotted. Though the kurtosis factor was depressed at an early stage of development, it achieved a higher value as a greater helicity was input initially. The result seems paradoxical because the helicity which was supposed to suppress generation of small scales on the contrary enhanced intermittency. Further careful consideration of the generation of (helical) structures is necessary by checking other types of initial conditions (for example, anisotropic, or structured), or by changing the Reynolds number.

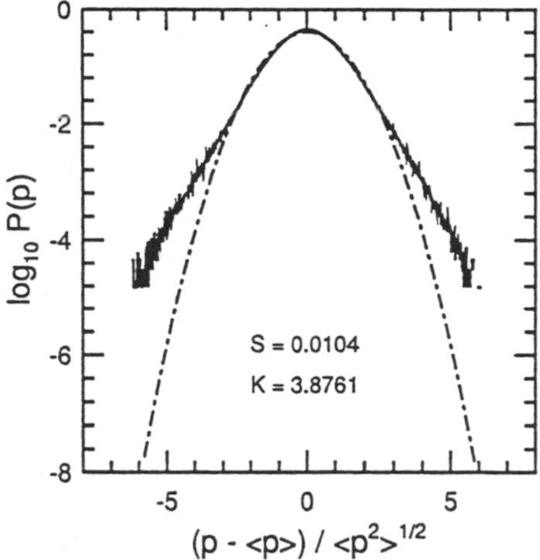

Fig. 12. PDFs of the pressure in the incompressible passive vector equation.

3.2. MODIFIED NAVIER-STOKES EQUATION

As a tool to analyze the intermittency growth in Navier-Stokes turbulence, we propose to study the following modified Navier-Stokes equation,

$$\frac{\partial \mathbf{u}}{\partial t} + (\mathbf{v}.\nabla)\,.\mathbf{u} = -\alpha \nabla p + \nu \nabla^2 \mathbf{u} \ , \tag{10}$$

$$\nabla.\mathbf{v} = 0 \ . \tag{11}$$

The points of modification are summarized as follows:

- 1. The advecting velocity ($\mathbf{v}$) and the advected velocity ($\mathbf{u}$) are separated. We have a variety of choices for $\mathbf{v}$, such as,
  (i) frozen velocity field in time,
    (a) independent of $\mathbf{u}$ (random mixing),
    (b) initial field of $\mathbf{u}(0)$,
  (ii) refreshed after a period Δt (white noise approximation when $\Delta t \to 0$),
  (iii) velocity field governed by other equations.
- 2. A prefactor α is put before the pressure term,
  (i) $\alpha = 0$ (passive-vector equation),
  (ii) $\alpha = 1$ (incompressible passive-vector equation).

Here we concentrated only on cases in which $\mathbf{v}$ is frozen and independent of $\mathbf{u}$. (A detailed report is to be published soon, Kimura and Kraichnan.) Figures 10a and 10b plot the energy spectra and the dissipation rate of energy, respectively, when $\alpha = 0$ (passive vector equation). An obvious difference from those in the Navier-Stokes equation is that the slope of the middle wave-range is less steep. In accordance with the observation about the energy spectrum, which suggests a bigger energy transfer from large to small scales, the dissipation rate of energy shows a higher hump than the previous case. In contrast with the Navier-Stokes turbulence, however, the PDF of the components of $\mathbf{u}$ shows a much wider tail than in the Gaussian distribution (Fig. 11a), and the PDF of the transverse gradients attains even greater non-Gaussianity (Fig. 11b). The non-Gaussianity of the passive-vector equation as well as the passive-scalar equation are from a higher order effect because the Gaussian distribution should be expected when there is either only advection or diffusion. Research on the non-Gaussian distribution is still underway. As an example in which $\alpha = 1$ (incompressible passive vector equation), we raise the PDF of the pressure in Fig. 12. The most significant discrepancy from Fig. 7 is that the shape is symmetric, which may manifest a different mechanism in producing different intermittency.

Acknowledgements

Most of the present work is the result of collaboration with Bob Kraichnan. The author expresses his cordial thanks for his encouragement and discussions. The pseudo-spectral code was originally developed by R. Panda (IBM), and all calculations were done on a IBM 3090-300E with vector processors at Los Alamos National Laboratory. Special thanks go to C. Whittaker (IBM) for the computation. The author also thanks J. R. Herring, R. M. Kerr, D. Montgomery, and R. Pelz for fruitful suggestions, and the participants in the workshop on Topological Fluid Dynamics at the Institute for Theoretical Physics at the University of California, Santa Barbara, which is supported by NSF Grant PHY89-04035.

NCAR is sponsored by the National Science Foundation.

References

André, J.C. & Lesieur, M., 1977 Influence of helicity on high Reynolds number isotropic turbulence. *J. Fluid Mech.*, **81**, pp. 187-207

Brachet, M., Meiron, D.I., Orszag, S.A., Nickel, B.G., Morf, R.H. & Frisch, U., 1983 Small-scale structure of the Taylor-Green vortex. *J. Fluid Mech.*, **130**, pp. 411-452.

Castaing, B., Gagne. Y. & Hopfinger, E.J., 1990 Velocity probability density functions of high Reynolds number turbulence. *Physica*, **D 46**, pp. 177-200.

Chen, H., Chen, S. & Kraichnan, R.H., 1989 Probability distribution of a stochastically advected scalar field. *Phys. Rev. Lett.*, **63**, pp. 2657-2660.

Comte, P., Lesieur, M. & Lamballais, E., 1991 Large and small-scale stirring of vorticity and a passive scaler in a 3D temporal mixing layer. *Phys. Fluids*, **A**, submitted.

Frisch, U. & Morf, R., 1981 Intermittency in nonlinear dynamics and singularities at complex times. *Phys. Rev.*, **A23**, pp. 2673-2705.

KERR, R., 1985 Higher-order derivative correlations and the alignment of small-scale structures in isotropic numerical turbulence. *J. Fluid Mech.*, **153**, pp. 31-58.

KIDA, S. & MURAKAMI, Y., 1989 Statistics of velocity gradients in turbulence at moderate Reynolds numbers. *Fluid Dyn. Res.*, **4**, pp. 347-370.

KIMURA, Y. & KRAICHNAN, R.H., in preparation.

KRAICHNAN, R.H., 1973 Helical turbulence and absolute equilibrium *J. Fluid Mech.*, **59**, pp. 745-752.

KRAICHNAN, R.H., 1991 Turbulent cascade and intermittency growth. *Proc. R. Soc. London*, Turbulence and stochastic processes: Kolmogorov's ideas 50 years on. Eds. J.C.R. Hunt, O.M.Phillips and D. Williams, **A 434**, pp. 65-78.

LESIEUR, M., 1987 Turbulence in Fluids *Kluwer Academic Publishers*, p. 93.

MÉTAIS, O. & HERRING, J.R., 1989 Numerical studies of freely decaying homogeneous stratified turbulence. *J. Fluid Mech.*, **202**, pp. 117-148.

MÉTAIS, O. & LESIEUR, M., 1991 Spectral large-eddy simulation of isotropic and stably-stratified turbulence. *J. Fluid Mech.*, submitted.

MOFFATT, H.K., 1969 The degree of knottedness of tangled vortex lines. *J. Fluid Mech.*, **35**, pp. 117-129.

POLIFKE, W., 1991 Statistics of helicity fluctuations in homogeneous turbulence. *Phys. Fluids*, **A3**, pp. 115-129.

SHE, Z.-S., 1991 Physical model of intermittency in turbulence: Near-dissipation-range non-Gaussian statistics *Phys. Rev. Lett.*, **66**, pp. 600-603.

SHE, Z.-S., JACKSON, E. & ORSZAG, S.A., 1988 Scale-dependent intermittency and coherence in turbulence. *J. Sci. Comp.*, **3**, pp. 407-434.

VAN ATTA, C.W. & CHEN, W.Y., 1968 Correlation measurements in grid turbulence using digital harmonic analysis. *J. Fluid Mech.*, **34**, pp. 497-515.

VAN ATTA, C.W. & CHEN, W.Y., 1970 Structure functions of turbulence in the atmospheric boundary layer over the ocean. *J. Fluid Mech.*, **44**, pp. 145-159.

VINCENT, A. & MENEGUZZI, M., 1991 The spatial structure and statistical properties of homogenous turbulence, *J. Fluid Mech.*, **225**, pp. 1-20.

YAMAMOTO, K. & HOSOKAWA, I., 1988 A decaying isotropic turbulence pursued by the spectral method, *J. Phys. Soc. Japan*, **57**, pp. 1532-1535.

DYNAMICAL MECHANISMS FOR INTERMITTENCY EFFECTS IN FULLY DEVELOPED TURBULENCE

Zhen-Su SHE
Program in Applied and Computational Mathematics
Princeton University
Princeton, NJ 08544, U.S.A.

ABSTRACT. Intermittency effects in fully developed turbulence are manifested in strongly non-gaussian statistics associated with small-scale fluctuations and in anomalous scaling exponents characterizing the decay of high-order velocity correlation functions over inertial-range distances. Phenomenological models developed for the understanding of such intermittent behavior are discussed, with particular emphasis on multifractal models and a two-fluid model. Multifractal models give a thermodynamic description of anomalous exponents, but do not offer any dynamical picture underlying the physical phenomenon. The two-fluid model performs a heuristic modeling of the Navier-Stokes dynamics in terms of stochastic dynamics of random eddies and self-stretching of coherent vortex structures. Quantitative measures of intermittency obtained by the two-fluid model are in good agreement with laboratory and computer simulation results for both non-gaussian distribution functions and anomalous scaling exponents.

1. Introduction

Among the many lines of research in turbulence, two primary categories of problems may be distinguished: the complex physics of large-scale dynamics and the universal statistical mechanics of small-scale fluctuations. The first category includes traditional fluid mechanics and has been particularly active in recent years because of the development of mathematical theories for nonlinear systems. In most fluid flows, even in presence of turbulence, large-scale coherent features, usually called coherent sructures (Hussain 1986), are consistently present. There is no doubt that these structures play an important role in the dynamics of the flow system; however, they seem to exhibit a very wide variety of characteristics, depending on geometrical properties of the system. Such non-universality leads to difficulties in reaching a consensus on such basic issues as the very definition of coherent structures. Various approaches have been suggested, from a dynamical systems point of view. The fundamental question is: what are the *effective* dynamical equa-

H. K. Moffatt et al. (eds.), Topological Aspects of the Dynamics of Fluids and Plasmas, 415–426.

tions governing the dynamics of large-scale features? Efforts in this direction may have direct bearing on engineering applications, as well as on the theoretical understanding of the statistical mechanics of complex systems (see She & Jackson 1991). In this note, however, I will not address this category of problems; instead, I will concentrate on the second category: universal small-scale dynamics.

A half-century ago, Kolmogorov (1941) proposed a particularly simple and elegant phenomenological theory of turbulence. Although there is doubt as to whether small-scale dynamics is completely universal and independent of large-scale conditions, there are strong indications that certain statistical measures, such as velocity correlation functions at two nearby space points, do exhibit scaling behavior independent of flow systems. We are thus led to believe that, at least where these statistical measures are concerned, there exist simpler statistical models (than the full Navier-Stokes equations) which allow for theoretical calculation of basic statistical quantities. This spirit has been carried through in closure modeling of fully developed turbulence, pioneered by Kraichnan some 30 years ago (Kraichnan 1958, 1959). Renormalized perturbation theories starting from the Navier-Stokes equations, although usually combined with some *ad hoc* assumptions, have led to a number of successful predictions consistent with Kolmogorov's assertion. Generally speaking, both Kolmogorov's 1941 phenomenology and statistical closure models of turbulence give a good description of low-order velocity correlations.

However, it was later discovered experimentally (Kuo & Corrsin 1972, Anselmet et al. 1983) that small-scale turbulent fluctuations exhibit strongly non-gaussian statistics and high-order velocity correlation functions decay with scaling exponents significantly different from those predicted by the Kolmogorov 1941 theory. This behavior,[1] commonly referred to as intermittency effects, casts doubt on the fundamental assumption in earlier theories that turbulence is a homogeneously random medium across spatial scales. In particular, the existence of intermittency implies that a greater spatial concentration of turbulent activity appears at smaller scales than at large scales.

The question then arises of how to modify Kolmogorov's phenomenology in order to account for intermittent fluctuations. There have been attempts to search for alternative explanations of the seemingly universal Kolmogorov energy spectrum $E(k) \sim k^{-5/3}$ in terms of particular geometrical structures (Lundgren, 1982, Hunt & Vassilicos 1991). If these geometrical structures are the *primary* cause of inertial range scalings, however, Kolmogorov's phenomenology would be completely irrelevant. So far, no experimental or nu-

[1] Throughout this note, we assume that these experimental measurements are, at least, qualitatively correct. This assumption has not been fully confirmed.

merical evidence supports this point of view, and it is difficult to conceive that geometrical structures such as spiral vortices could dominate even the low-order correlations in the presence of strong turbulence. Somewhat more popular, but rather abstract are the multifractal models (Mandelbrot 1974, Parisi & Frisch 1983, Meneveau & Sreenivasan 1987a, see also Sreenivason 1991). Instead of being a stochastic self-similar field with a unique scaling exponent as Kolmorogov suggested, turbulence actually contains a continuous range of scaling exponents, according to the multifractal picture (see Frisch 1991). A Legendre transform relates the spectrum of fractal dimensions of various exponents to scaling exponents of the nth-order velocity correlation functions, so that anomalous scaling exponents of high-order correlation functions can be explained by invoking an appropriate set of fractal dimensions.

2. Multifractal Models

The multifractal model of Parisi and Frisch is a static picture, with no connection to turbulence dynamics. The one-to-one correspondence between fractal dimensions and high-order velocity correlation exponents implies a mere shift of the object to be explained. Moreover, there is a basic question whether it is ever possible to define local scaling exponents of the velocity field in physical space. There have been attempts to measure such local exponents from experimentally recorded velocity signals using wavelet transforms (Bacry et al. 1990), but Benzi et al. (1991) showed that the histogram of exponents depends on the form of the wavelet used. Specifically, a more nonlocal wavelet packet detects a more restricted set of exponents. This can be understood by the uncertainty principle: a self-similar process having a well-defined scaling exponent (a locally defined property in wavenumber space) is a global property in physical space, and any local measurement in physical space unavoidably suffers from statistical fluctuations. Thus, it seems impossible to separate statistical fluctuations of scaling exponents (false exponents) from real ones belonging to the multifractal.

The above remark suggests giving up the idea of a strict geometrical interpretation of multifractals as a collection (set) of space points having a certain scaling exponent. In other words, multifractal is no longer fractal. Instead, one should emphasize such global measures as the nth order moments. In this case, fractal dimensions are just a different set of variables obtained through a Legendre transform. We may call this the thermodynamic interpretation of multifractals. In this case, the usefulness of the concept of multifractality depends on whether it is easier to construct realizable multifractal models containing few parameters and having a clear physical meaning. Models satisfying the first criterion have been proposed; two examples are the random β-model (Benzi et al. 1984) and the p-model (Meneveau

& Sreenivasan 1987b). In these models, a multifractal of the local dissipation rate, $\epsilon = \sum(\frac{\partial u_i}{\partial x_j} + \frac{\partial u_j}{\partial x_i})^2$, is generated via a random or deterministic multiplicative process mimicing the energy cascade.

It should be noted that the physical meaning of the proposed multiplicative cascade is not clear. First, the cascade is characterized by a transformation of a locally averaged dissipation rate $\epsilon_r = \int_0^r \epsilon(x)dx$ to $\epsilon_{r'}$ with r' and r both being inertial range distances $r' < r$. However, ϵ_r is not a pure inertial range quantity (Kraichnan 1974), so that theoretically such a transformation cannot be regarded as a pure inertial range process. This implies, in addition, that the connection of the scaling exponents of ϵ_r to exponents of velocity correlation functions is questionable, because the latter is a purely inertial range quantity. These arguments do not, of course, rule out the possibility that the inertial range energy transfer process is actually a multiplicative process with amplifying fluctuations as the scale decreases. Observe, however, that different multifractal cascade models propose different scenarios of cascade, which adds a high degree of arbitrariness. Let us mention also the fact that anomalous scaling exponents of velocity correlation functions are observed already in moderately high Reynolds number flows, where a multistage cascade picture seems to be inadquate. These considerations suggest that current multifractal models do not contain a clear physical picture of how intermittency effects are developed in turbulent flows.

Finally, let us mention a different set of intermittency models (Andrews et al. 1989, Castaing 1989, Kida 1991), which are extensions of the log-normal model by Kolmogorov (1962). Most of these models also work with the locally averaged dissipation rate ϵ_r, and assume either an *ansatz* distribution function $P(\epsilon_r)$ or $P(\sigma_{rr'})$ where $\sigma_{rr'} = \epsilon_r/\epsilon_{r'}$ is usually called the breakdown coefficient. These distributions can be parameterized in such a way that n-th order moments of ϵ_r vary with r in power-law fashion: $\langle \epsilon_r^n \rangle \sim r^{\tau_n}$. This is generally possible if $P(\sigma_{rr'})$ belongs to the class of infinitely divisible distributions. In the case of $P(\sigma_{rr'})$, one can again roughly speak about a multiplicative process; however its connection to inertial range cascade is also not clear, for the reason stated in the last paragraph. From a statistical point of view, the two multifractal models mentioned above are just particular realizations of $P(\sigma_{rr'})$ which are non-zero only for discrete values of $\sigma_{rr'}$. Physically, a continuous distribution function $P(\sigma_{rr'})$ seems more realistic; on the other hand, the picture of cascade (if ϵ_r can be interpreted as inertial range energy flux) would not be intuitive. Since many of the above models generate reasonable scaling exponents (close to experimental measurements) by adjusting free parameters, there is no basis to prefer one over another. After all, none of the proposals seems to be related to the Navier-Stokes dynamics, which, I believe, is their most severe shortcoming; in other words, justification of the validity of those models for Navier-Stokes turbulence seems virtually impossible inside of their own framework.

3. Two-Fluid Model

In this section, we discuss a very different picture of universal small-scale dynamics consistent with intermittency features. Our cartoon view is that turbulent fields contain a mixture of highly random eddies and isolated coherent structures. To be more specific, consider the vorticity field $\omega(x)$ in a fully developed turbulent flow. The typical fluctuating amplitude of the vorticity is the rms value: $\omega_{rms} = \sqrt{\langle\omega^2\rangle}$. Random eddies correspond to those vorticity elements having low to moderate amplitude: $\omega \sim \omega_{rms}$. Structures correspond to high amplitude fluctuation events: $\omega \gg \omega_{rms}$. It has been observed both numerically (Siggia 1981, Hosokawa & Yamamoto 1989, She, Jackson & Orszag 1990, Vincent & Meneguzzi 1991) and experimentally (Douady et al. 1991) that these high amplitude fluctuation events form filaments, and low amplitude fluctuations seem to display a high degree of randomness (She, Jackson & Orszag 1991). This justifies our picture. Similar classification of random eddies versus structures can be made for inertial range fluctuations by considering the velocity increments across an inertial range distances as the statistical field (instead of the vorticity).

While the idea that turbulence consists of inter-mixed random and coherent parts is not surprising, the particular classification described above allows for a simple description of key dynamical features. The specification of these dynamical features is the essence of the **two-fluid model**, which we will describe now.

The nonlinear interaction in turbulence can be equivalently described by interactions among vortex elements distributed throughout the entire space. Vortex elements interact with each other through their strain fields generated according to the Biot-Savart law. The total strain amplitude acting on a particular element is then the sum of actions from surrounding elements plus its self-action or self-generated strain field. We now observe that the relative importance of the self-interaction compared to the external interaction part should be an increasing function of the amplitude of the considered vortex element, because of different rates of convergence of the Biot-Savart integral as the domain of integration is enlarged. In other words, low-amplitude vortex elements (random eddies) have negligible self-interaction; only high-amplitude vortex elements (structures) experience a significant self-stretching process. This self-stretching process is characterized by the fact that the strain amplitude depends on the vorticity amplitude of the element. In the generic case, both amplitudes are linearly proportional[2]. Next, the external interaction involving only surrounding elements will generate a

[2] The only exception is when the vorticity is perfectly aligned with one of the principal axes of the strain tensor. This is not the generic case.

strain amplitude independent of the vorticity amplitude of the element upon which it acts. Since vortex elements in fully developed turbulence are typically randomly distributed, there are generally cancelations of contributions from different elements. In the case of space homogeneity, the net contribution may be considered constant in space. We call this last assumption a mean-field approximation. In summary, the two-fluid model postulates that random eddies in turbulence interact with an effective mean strain field, while structures undergo mainly a nonlinear self-stretching process.

The concept of mean versus self-interaction, associated respectively with random eddies and structures, is the essence of the two-fluid model. As we will see below, it is precisely the self-interaction of structures that is responsible for the generation of non-gaussian statistics for small-scale fluctuations and anomalous scaling exponents of velocity correlation functions. Self-interacting structures are the ingredient missing in Kolmogorov's phenomenology as well as in classical closure theories. Thus, we suggest that the dynamic mechanism for the development of intermittency effects in fully developed turbulence is the self-interaction of highly excited vortex structures, which introduces more order and coherence into the otherwise immense random sea.

4. Mapping Technique and Quantitative Description

Quantitative calculation of statistical measures characterizing intermittency effects based on the two-fluid phenomenology uses a noval technique first introduced by Kraichnan (1990, Chen, Chen & Kraichnan 1989), called mapping closure. The mapping technique was proposed to model non-gaussian probability density function (PDF) resulting from a dynamical process. A mapping function is introduced to link the dynamically evolving field to a reference gaussian field, and the development of a non-gaussian PDF is modeled by deriving a closure equation describing the evolution of the mapping function. Kraichnan (1990) successfully applied the technique to the 1-D Burgers equation, and suggested that it may be promising for the Navier-Stokes equations. A full implementation of the mapping closure for the Navier-Stokes equation, taking into account the solenoidal property of the three-dimensional vector field and non-local effects due to pressure, has not yet been done.

The two-fluid model (She 1991a, She & Orszag 1991) employs a phenomenological setup of the mapping closure idea, and generalizes the notion of the mapping function. This leads to the calculation of the PDF of velocity derivatives and velocity increments, in terms of a parameter, say h, which is an exponent characterizing the stretching dynamics. Comparison with direct numerical simulation results suggests that h is a function of the Reynolds number where the PDF of the velocity gradients is concerned, and is a function of the separation distance when we deal with velocity increments (assuming very large Reynolds number). A self-similar argument is then developed to constrain the function $h = h(R)$ (or $h = h(\ell)$), which gives predictions of ratios of scaling exponents for high-order flatness. These ratios can be directly compared to both experimental and numerical measurements. For a detailed description of this setup, readers are referred to She (1991b). Here, we give a brief presentation of the strategy, but concentrate more on discussion of the model's implications and predictions.

We assume that the dynamics of random eddies follows closely the description by Kolmogorov (1941) which can be modeled by a Langevin dynamical equation:

$$\partial_t v_0 = f - \frac{\nu}{\eta^2} v_0, \tag{4.1}$$

where v_0 is the velocity of eddies of size η, the Kolmogorov dissipation length. Analytical closure theories (see for example Kraichnan 1977) provide a theoretical basis for this modeling, in which the force f is statistically independent of v_0 and has gaussian statistics. Thus, the PDF of v_0 is gaussian:

$$P_G(v_0) = \frac{1}{\sqrt{2\pi\sigma}} \exp\left(-\frac{v_0^2}{2\sigma}\right), \qquad \sigma = \langle v_0^2 \rangle. \tag{4.2}$$

We now consider the modification of such dynamics due to the presence of structures (large amplitude eddies). We introduce a mapping function J describing the ratio of the change in eddy size under a stretching process: $J = \eta/\ell$, where ℓ is the size of the considered eddy (the smallest dimension of a fluid element) after the stretching. The velocity v_ℓ also changes during the stretching of the fluid element; this is modelled by a power law, $v_\ell = v_0(\ell/\eta)^h$, with a dynamical exponent h. The exponent h may be connected to the geometry of local structures (see Frisch & She 1991). Thus, we can relate the vorticity $\omega = v_\ell/\ell$ to the gaussian field $\omega_0 = v_0/\eta$ by

$$\omega = \omega_0 J^{1-h}. \tag{4.3}$$

Since the length scale is reduced by a factor J, the viscous action should be enhanced by a factor J^2, which provides a simple way to achieve a closure description[3].

The two-fluid model described in the last section simplifies the Navier-Stokes equation for the vorticity in the following way[4]:

$$\begin{aligned} D_t\omega &= S\omega + \nu\Delta\omega \\ &= S_0\omega + S_1(\omega)\omega + \nu J^2\omega \end{aligned} \tag{4.4}$$

[3] A formal mapping closure theory should justify these simple considerations.

[4] We only consider the scalar version of the equation

where S_0 is the mean stretching rate (external action) and $S_1(\omega)$ is the self-stretching rate (self-interaction), and D_t is the substantial derivative (following fluid elements). Here, a more concrete modeling of $S_1(\omega)$ is needed; we assume that

$$\frac{S_1(\omega)}{S_0} = \frac{|\omega|}{\omega_c} H(\frac{|\omega|}{\omega_c}) = \frac{|\omega|}{\omega_c}\left(1 - e^{-\frac{|\omega|}{\omega_c}}\right), \tag{4.5}$$

meaning that the self-stretching rate is asymptotically proportional to the vorticity amplitude for $|\omega| \gg \omega_c$, where ω_c characterizes the typical amplitude of structures. The weighting function $H(\frac{|\omega|}{\omega_c})$ is a crude modeling of the behavior of the self-action compared to the external action hidden in the Biot-Savart integral. One simple argument to support this formula is to observe that as $\omega \to 0^{\pm}$, $S_0 + S_1$ should be continuous and differentiable.

Substituting (4.3) and (4.5) into (4.4) and properly renormalizing coefficients (see She 1991b), we obtain an equation for the mapping function J:

$$\partial_t J = \frac{\omega_0}{\omega_c} H(\frac{|\omega|}{\omega_c}) J^{2-h} + J - J^3. \tag{4.6}$$

The steady solution of (4.6) has a geometrical intepretation: it gives the equilibrium eddy size η/J in the presence of the self-stretching as a function of the initial (gaussian) vorticity amplitude . As $\omega_0 \to 0$, $J = 1$ so that eddies maintain the same size as the Kolmogorov dissipation length. However, for $\omega_0 \gg \omega_c$, the self-stretching dominates and the eddy sizes are given by

$$\ell(\omega_0) = \eta\left(\frac{\omega_0}{\omega_c}\right)^{-1/1+h}, \qquad \omega_0 \gg \omega_c \tag{4.7}$$

Correspondingly, we can relate the equilibrium vorticity amplitude to the initial amplitude through (4.3):

$$\omega = \omega_0\left(\frac{\omega_0}{\omega_c}\right)^{\frac{1-h}{1+h}} \sim \omega_0^{\frac{2}{1+h}}, \qquad \omega_0 \gg \omega_c \tag{4.8}$$

This last result is the mapping relation we desire to obtain. In general, we solve the nonlinear algebraic equation on the r.h.s. of (4.6) to obtain a function $\omega = F(\omega_0; h, \omega_c)$, from which the PDF of the vorticity can be readily calculated:

$$P(\omega) = P_G(\omega_0)\frac{\partial \omega_0}{\partial \omega} \tag{4.9}$$

In this modeling, we have two phenomenological parameters: ω_c and h. The tail (asymptotic) behavior of $P(\omega)$ depends critically on h: $P(\omega) \sim \exp(-\omega^{1+h})$. In the modeling published so far (She 1991a, She 1991b), the parameter ω_c has been fixed, and it is found that by only changing h, the

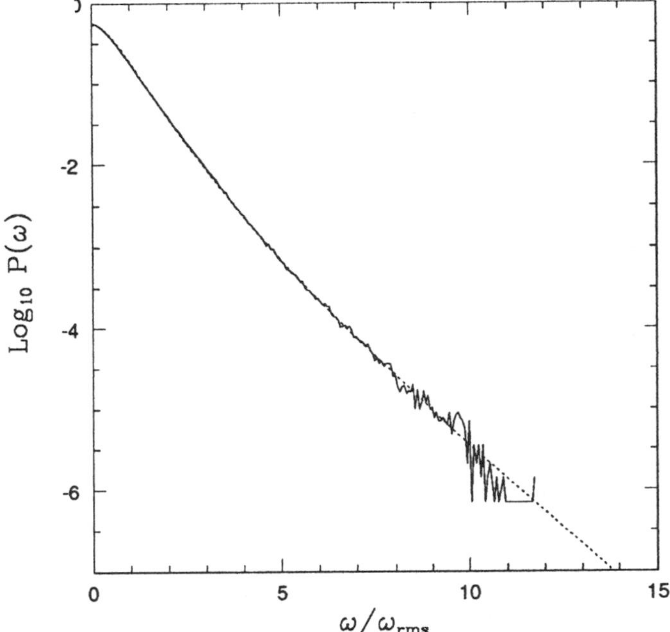

Fig. 1. A comparison of the model PDF (dashed line) obtained by setting $h = -0.14$ to the measured PDF (solid line) in direct numerical simulation of isotropic turbulence at $R_\lambda = 77$.

PDF (4.9) with the solution of (4.6) mimics very closely those measured from direct numerical simulations of isotropic Navier-Stokes turbulence at different Reynolds numbers. Fig. 1 is an illustration of the fit.

It is worth mentioning that the model, with parameters h and ω_c fixed, also predicts the PDF of high-order gradients such as Δv, $\Delta\omega$ etc. at the same Reynolds number. Instead of (4.3), we now have

$$\Delta v = \Delta v_0 J^{2-h}, \qquad \Delta\omega = \Delta\omega_0 J^{3-h}, \qquad \text{etc.} \tag{4.10}$$

The PDFs of these high-order gradients have recently been measured in turbulence fields from high resolution simulations and agree very well with the model's prediction; this will be reported elsewhere.

The same analysis can be carried out for the PDFs of velocity increments (see She & Orszag 1991). A striking property of the model is that there is

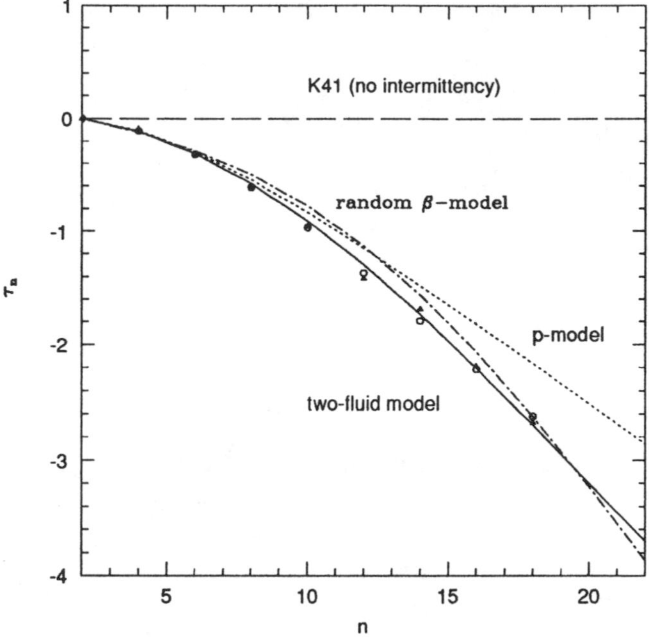

Fig. 2. Comparison of numerically (circle) and experimentally (triangle) measured scaling exponents τ_n for flatness factors of order n, to multifractal models (p-model: dotted line; random β-model: dotted dash line) and to the two-fluid model (solid line).

a function $h = h(\ell)$ for fixed v_ℓ^c (the analogue of ω_c) such that n-th order flatness factors F_n vary with ℓ in power law fashion *for all* n:

$$F_n(\ell) = \frac{\langle v_\ell^n \rangle}{\langle v_\ell^2 \rangle^{n/2}} \sim \ell^{\tau_n}. \tag{4.11}$$

These exponents are determined in the present model up to a multiplicative coefficient. Comparison to measured scaling exponents is presented in Fig. 2, together with two other multifractal models.

5. Conclusion

We attempt here to address the question, what dynamical features are responsible for observed statistical behavior in fully developed turbulence? It is clear that neither the Kolmogorov 1941 theory nor the multifractal phenomenology make any connection to Navier-Stokes dynamics. Thus, the dynamics of energy transfer and intermittency effects is virtually absent in these models. It is physically conceivable that in turbulence becomes fully developed in incompressible flows, the energy transfer is governed by a general principle such as the maximum stochasticity postulated by Kraichnan (1958). As a consequence, low-order correlations exhibit highly universal behavior. However, high-order correlation functions are very sensitive to specific dynamics, and their understanding requires a deep analysis of the underlying mathematical equation. The two-fluid model is a first (crude) step to do this. But the underlying picture can be used to develop a detailed statistical mechanical treatment similar to the work by Chorin (1991).

So far, we have restricted ourselves to the incompressible Navier-Stokes equation, but the spirit can be carried over to the analysis of other turbulent flows, such as thermal convection. Finally, we like to point out that, despite its success in quantitative descriptions, the model is of a phenomenological nature; further development involving a more systematic derivation of the heuristic ideas presented here is vital to furnish a theory of the universal small-scale dynamics of turbulence.

Acknowledgements

I thank E. Jackson, R.H. Kraichnan, S.A. Orszag and V. Yakhot for continuous discussions of ideas presented here. I wish also to thank the Institute for Theoretical Physics at University of Saint Barbare for the hospitality during the NATO advanced workshop "Topological Fluid Mechanics".

References

ANDREWS, L. C., PHILLIPS, R. L., SHIVAMOGGI, B. K., BECK, J. K. & JOSHI, M. L. 1989 *Phys. Fluids A* **1**, 999.

ANSELMET, F., GAGNE, Y., HOPFINGER, E. J. & ANTONIA, R. A. 1984 *J. Fluid Mech.* **140**, 63.

BACRY, E., ARNEODO, A., FRISCH, U., GAGNE, Y., & HOPFINGER, E. 1990 In *Turbulence and Coherent Structures* (ed. M. Lesieur & O. Métais), Kluwer.

BENZI R., PARADIN, G., PARISI, G. & VULPIANI, A. 1984 *J. Phys. A* **17**, 3521.

BENZI, R. & VERGASSOLA, M. 1991 *Fluid Dynamics Res.* **8**.

CASTAING, B. 1989 Conséquences d'un principe d'extremum en turbulence. *J. Phys. (Paris)* **50**, 147.

CHEN, H., CHEN S. & KRAICHNAN, R. H. 1989 Probability distribution of a stochastically advected scalar field. *Phys. Rev. Lett.* **63**, 2657.

CHORIN, A. J. 1991 Equilibrium statistics of a vortex filament with applications. *Commun. Math. Phys.* pp. 686.

DOUADY, S., COUDER, Y. & BRACHET, M. E. 1991 Direct Observation of the Intermittency of Intense Vorticity Filaments in Turbulence. *Phys. Rev. Lett.* **67**, 983.

FRISCH, U., 1991 From global scaling, *à la* Kolmogorov, to local, multifractal scaling in fully developed turbulence. *Proc. Roy. Soc. Lond.* **434**, 89.

FRISCH, U. & SHE, Z.-S. 1991 On the Probability Density Function of Velocity Gradients in Fully Developed Turbulence. *Fluid Dynamics Research* **8**, 139.

HOSOKAWA, I. & YAMAMOTO, K. 1989 *J. Phys. Soc. Japan* **58**, 20.

HUNT, J. C. R. & VASSILICOS, J. C. 1991 Kolmogorov's contributions to the physical and geometrical understanding of small-scale turbulence and recent developments. *Proc. Roy. Soc. Lond.* **434**, 183.

HUSSAIN, A.K.M.F., 1986 *J. Fluid Mech.* **173**, 303.

KIDA, S. 1991 Log-stable distribution and intermittency of turbulence. *J. Phys. Soc. Japan*, to appear.

KOLMOGOROV, A. N. 1941 The Local Structure of Turbulence in Incompressible Viscous Fluid for Very Large Reynolds Numbers. *C. R. Acad. Sci. URSS* **30** (4), 301.

KOLMOGOROV, A. N. 1962 A refinement of previous hypotheses concerning the local structure of turbulence in a viscous incompressible fluid at high Reynolds number. *J. Fluid Mech.* **13**, 82.

KRAICHNAN, R. H. 1958 *Phys. Rev.* **109**, 1407.

KRAICHNAN, R. H. 1959 The structure of isotropic turbulence at very high Reynolds numbers. *J. Fluid Mech.* **5**, 497.

KRAICHNAN, R. H. 1974 *J. Fluid Mech.* **62**, 305.

KRAICHNAN, R. H. 1977 Eulerian and Lagrangian renormalization in turbulence theory. *J. Fluid Mech.* **83**, 349.

KRAICHNAN, R. H. 1990 *Phys. Rev. Lett.* **65** 575.

KUO, A. Y. S. & CORRSIN S. 1972 *J. Fluid Mech.* **56**, 447.

LUNDGREN, T. S. 1982 *Phys. Fluids* **25**, 2193.

MANDELBROT, B. 1974 Intermittent Turbulence in Self-Similar Cascades: Divergence of High Moments and Dimension of the Carrier. *J. Fluid Mech.* **62**, 331.

MENEVEAU, C. & SREENIVASAN, K. R. 1987a The multifractal spectrum of the dissipation field in turbulent flows. In *Physics of Chaos and Systems far from Equilibrium* (ed. Minh-Duong Van & B. Nichols), Nucl. Phys. B (Proc. Suppl.). North-Holland, Amsterdam, pp. 2–49.

MENEVEAU, C. & SREENIVASAN, K. R. 1987b Simple multifractal cascade model for fully developed turbulence. *Phys. Rev. Lett.* **59**, 1424.

PARISI, G. & FRISCH, U. 1985 In *Turbulence and Predictability in Geophysical Fluid Dynamics and Climate Dynamics* (ed. M. Ghil, R. Benzi & G. Parisi), North-Holland, Amsterdam, p. 71.

SHE, Z.-S. 1991a Physical model of intermittency in turbulence: near-dissipation-range non-gaussian statistics. *Phys. Rev. Lett.* **66**, 600.

SHE, Z.-S. 1991b Intermittency and non-gaussian statistics in turbulence. *Fluid Dynamics Res.* **8**, 142.

SHE, Z.-S. & JACKSON, E. 1991 A constrained Euler system for Navier-Stokes turbulence. *Phys. Rev. Lett.*, submitted.

SHE, Z.-S., JACKSON, E. & ORSZAG, S. A. 1990 Intermittent vortex structures in homogeneous isotropic turbulence. *Nature* **344**, 226.

SHE, Z.-S., JACKSON, E. & ORSZAG, S. A. 1991 Structure and dynamics of homogeneous turbulence: models and simulations. *Proc. Roy. Soc. Lond.* **434**, 101.

SHE, Z.-S. & ORSZAG, S. A. 1991 Physical model of intermittency in turbulence: inertial-range non-gaussian statistics. *Phys. Rev. Lett.* **66**, 1701.

SIGGIA, E. D. 1981 *J. Fluid Mech.* **107**, 375.

SREENIVASAN, K. R. 1991 Fractals and multifractals in fluid turbulence. *Ann. Rev. Fluid Mech.* **23**, 539.

VINCENT, A. & MENEGUZZI, M. 1991 *J. Fluid Mech.* **225**, 1.

THE MULTISPIRAL MODEL OF TURBULENCE AND INTERMITTENCY.

J.C. VASSILICOS
Department of Applied Mathematics and Theoretical Physics
University of Cambridge
Silver Street
Cambridge CB3 9EW
U.K.

ABSTRACT. A spiral vortex sheet modelled as in Moffatt (1992) yields similarity exponents ξ_p for the velocity field's statistics – $\langle(\delta u_{\|}(\mathbf{r}))^p\rangle \sim r^{\xi_p}$ – which have the same properties as the ξ_p derived in the β - model (linear in p plus a constant). A superposition of such spiral vortex sheets with different Kolmogorov capacities D_K yields results for ξ_p that resemble closely the results of the multifractal model of turbulence (*e.g.* ξ_{2p} is the Legendre transform of a function $D_K(\sigma)$ characteristic of the multispiral structure of the velocity field). Unlike the multifractal and the β - models, the multispiral and the spiral β - models of turbulence carry an important difference between even and odd statistics; a breaking of symmetry (left/right) is needed for the odd p statistics not to vanish and for the properties of intermittency and non-linear energy dissipation to be incorporated in the model.

The method used here for the solution of these two models of homogeneous and isotropic turbulence (the spiral β - model and the multispiral model) is a generalisation of the method used in Vassilicos & Hunt (1991) to derive the relation between the power spectrum ($p = 2$ statistics) and the fractal dimension (Kolmogorov capacity D_K) of interfaces. It is more general than is needed here (*e.g.* it does not depend on the specific *spiral* structure of the velocity field); it is of use, in particular, when a field's structure is not fractal in the Hausdorff sense, but has nevertheless non-trivial Kolmogorov capacities.

1. Introduction

In 1941 Kolmogorov published two papers dealing, respectively, with the second and the third order statistics of the small scale relative velocities $\delta\mathbf{u}(r) = \mathbf{u}(\mathbf{x}+\mathbf{r}) - \mathbf{u}(\mathbf{x})$ between two points of space $\mathbf{x}$ and $\mathbf{x}+\mathbf{r}$ in a homogeneous and isotropic turbulent flow ($r = |\mathbf{r}|$). (These papers can be found most easily in a special issue of the Proceedings of the Royal Society of London, series A, vol. 434 (1991), celebrating Kolmogorov's ideas 50 years on). Under the assumptions that the Reynolds number is large enough, that the small scale velocity field is homogeneous, isotropic and

H. K. Moffatt et al. (eds.), Topological Aspects of the Dynamics of Fluids and Plasmas, 427–442.

universal–independence from large scale forcing, and from boundary and initial conditions–and that the mean rate of energy dissipation ϵ of the turbulent fluctuations is the *same everywhere* in space (sufficiently far from the boundaries), Kolmogorov recognised that the relative velocity statistics within the inertial range of length scales can only depend on ϵ and r (the second hypothesis of similarity), and used dimensional analysis to show that

$$\langle(\delta\mathbf{u}(\mathbf{r}))^2\rangle \approx C_2\epsilon^{2/3}r^{2/3} \tag{1a}$$

when $\eta \ll r \ll L$ (L is an integral length scale and η is the Kolmogorov viscous length scale). The value of the constant C_2 in (1a) is not specified by Kolmogorov's theory, but it is expected to be universal.

From the Navier-Stokes equations he was able to deduce that, for homogeneous and isotropic turbulence,

$$\langle(\hat{\mathbf{r}}\cdot\delta\mathbf{u}(\mathbf{r}))^3\rangle \approx -\frac{4}{5}\epsilon r, \tag{1b}$$

($\hat{\mathbf{r}}$ is a unit vector along $\mathbf{r}$), in that same inertial range of length scales where his dimensional arguments lead, in fact, to the conclusion that

$$\langle(\delta u_{||}(\mathbf{r}))^p\rangle \approx C_p(\epsilon r)^{p/3} \tag{1c}$$

for $p = 2, 3, 4, \ldots$ ($\delta u_{||} = \hat{\mathbf{r}}\cdot\delta\mathbf{u}$). According to the theory, the dimensionless constants C_p are all universal (note: C_2 in (1a) is not the same as in (1c)).

A consequence of this theory–specifically of (1c)–is that the small scale turbulent fluctuations are statistically self-similar in space; this is because the statistics of $\lambda^{1/3}\delta u_{||}(\mathbf{r}')$ are the same as these of $\delta u_{||}(\mathbf{r})$ when $\mathbf{r}' = \lambda^{-1}\mathbf{r}$ for any positive real number λ which represents, therefore, a dilation of space.

The assumption of statistical independence of the small scales from the large energy containing scales has recently been questioned (*e.g.* see Hunt et al (1988), Brasseur & Yeung (1991)); this raises the related question of whether the statistics of the small scale turbulence are indeed universal or not. Frisch (1991) argues that all C_p–except for $p = 3$–cannot be universal. Nevertheless, he recovers formulae (1c) by turning Kolmogorov's theory on its head and assuming that the small scale turbulence is statistically self-similar in space; it is now an assumption (not a consequence as in Kolmogorov) that there is a single exponent h such that the statistics of $\lambda^h\delta u_{||}(\mathbf{r}')$ are the same as those of $\delta u_{||}(\mathbf{r})$ when $\mathbf{r}' = \lambda^{-1}\mathbf{r}$ ($\lambda \in \mathrm{R}_+$). The value $h = 1/3$ is then deduced from (1b).

The question of the universality of C_p will not be discussed in the present paper. Here we will concentrate on obtaining scaling laws of the type

$$\langle(\delta u_{||}(\mathbf{r}))^p\rangle \sim r^{\xi_p}, \tag{2}$$

for inertial range values of r. The experimental measurements of Anselmet et al (1984) have shown that such scaling laws do indeed exist in ranges of length scales within the inertial range of small scale turbulence, but that the powers ξ_p may not equal $p/3$ as predicted by Kolmogorov's theory. There have been a few attempts to derive ξ_p theoretically by 'correcting' several aspects of Kolmogorov's theory (log-normal model, β-model, multifractal model; none of these approaches provide a universal means of estimating C_p, and neither will the one presented here).

The present paper investigates how, if the properties of self-similarity and space-fillingness (the property that ϵ is the same everywhere in space) are redefined,

(i) the exponents ξ_p may match the experimental findings and

(ii) the structure of small scale turbulence is consistent with the hypotesis that it is dominated by vortex tubes with a spiral internal structure.

She *et al.* (1991) find that, in DNS of homogeneous isotropic turbulence, the fine scales are dominated by vortex tubes; the internal structure of these vortex filaments seems to be spiral with a few turns (She, private communication). In similar numerical calculations, Ruetsch & Maxey (1992) have identified vortex sheets in a turbulent velocity field which undergo Kelvin-Helmholtz instability and lead to isolated, single spiral structures. Spiral structures of vorticity have also been recently seen in similar numerical simulations by Brasseur & Lin (private communication).

The statistical self-similarity in space of the small scale turbulent fluctuations aroused Mandelbrot's interest who suggested that turbulent velocity fields may have a fractal structure with a non-integer Hausdorff dimension (see Mandelbrot (1982) and references therein). A pattern of spirals with smaller spirals on them–and so on to increasingly smaller scales–is a good example of a fractal. Experimental measurements of the Kolmogorov capacity (or box dimension) using the box-counting algorithm were made for various interfaces in various turbulent flows (see Sreenivasan (1991)); the non-integer value which was obtained does not imply that the structure of these interfaces (and by inference, the structure of the velocity field distorting them) is fractal. Some single spirals can have a non-integer Kolmogorov capacity even though they are smooth objects and their Hausdorff dimension equals their topological dimension (see Vassilicos & Hunt (1991)). In fact, the spiral that arises from numerical integration of a vortex sheet's instability driven evolution by Krasny (1986) does not have a cascade of smaller spirals on it–it is not fractal–and it is such that its intersections with the x-axis (corresponding to the sheet's initial configuration) are at a distance $x_n \sim n^{-2}$ from the centre of the spiral (Moffatt (1992)) (where n numbers the successive coils of the spiral and its intersections with the x-axis from large x inwards–towards $x \to 0$). The Kolmogorov capacity of these intersections is $D_K = 1/3$ (if $x_n \sim n^{-a}$, $D_K = \frac{1}{1+a}$, see Vassilicos &

Hunt (1991)), which is non-integer. If, as Ruetsch & Maxey (1992) seem to observe, Kelvin-Helmholtz instability of vortex sheets is an important mechanism in the generation of small scale turbulence, then it may be safer to assume that the small scale turbulence has a spiral structure characterised by non-integer Kolmogorov capacities D_K, than to assume it to have a fractal structure characterised by non-integer Hausdorff dimensions D_H.

The assumption that the mean rate of dissipation is the same everywhere in space (the space-filling property of Kolmogorov turbulence) was criticised very early on by Landau (*e.g.* see Frisch (1991)). The intermittent character of small scale turbulence, *i.e.* the fact that the turbulent activity, and therefore the energy dissipation, do not occur equally everywhere in space, contradicts Kolmogorov's assumption of space-fillingness.

The first attempt to remedy this shortcoming of Kolmogorov's 1941 theory came in 1961 from Obukhov and Kolmogorov themselves (see Frisch (1991) for a discussion and references) who proposed to replace the original theory with the so-called log-normal model. A central role in this model is played by ϵ_r, the spatial average of the mean rate of dissipation over a ball of radius r; the logarithm of ϵ_r is assumed to have a gaussian (normal) distribution with variance $\sigma_r^2 = A + \mu \ln(r_0/r)$ where μ is a positive parameter. This assumption replaces the assumption of space-fillingness under which ϵ_r would equal ϵ for all r, and both A and μ would vanish. Clearly, what is done in the log-normal model, is to broaden the distribution of $\ln \epsilon_r$ from a delta function (implicit in Kolmogorov's assumption of space-fillingness) to a gaussian with a finite variance that is a function of $\ln r$.

It is precisely the dependence of σ_r^2 on $\ln r$ which leads to divergent results from Kolmogorov's original theory; in the log-normal model,

$$\xi_p = p/3 - \frac{\mu}{18} p(p-3). \tag{3}$$

The intermittency is incorporated in the theory because μ is positive, and therefore extreme events carry more weight than otherwise.

A particular consequence of (3) is that ξ_p decreases with p for large enough values of p; this is shown by Frisch (1991) to be in contradiction with the basic physics of incompressible flow. Frisch (1991) lists and discusses a few more problems of the lognormal model.

A second attempt to replace the assumption of space-fillingness by something that would allow the property of intermittency to tie in with inertial range scaling laws has been made with the β-model (see, again, Frisch (1991) for references). The β-model addreses the assumption of space-fillingness more directly than the log-normal model does; rather than a change in the probability distribution of ϵ_r, it involves a change in the spatial distribution of turbulent activity. Specifically, one starts (as Frisch (1991) does in order to recover Kolmogorov's 1941 results without having to assume universality) from an assumption of self-similarity in space, but modified as follows: there exists a single exponent h and a *fractal* sub-set S_h of $\mathbf{x}$-space, such that if $\mathbf{x} \in S_h$, then $\delta u_{||}(\mathbf{x}, \mathbf{r}) \sim r^h$ as $r \to 0$; the Hausdorff dimension of S_h is $2 + D_H$ $(0 < D_H \leq 1)$. It appears easy to estimate the statistics (averages over space) of such $\delta u_{||}$:

$$\langle (\delta u_{||}(\mathbf{r}))^p \rangle \sim r^{ph} r^{3-(2+D_H)} = r^{ph+1-D_H} \tag{4}$$

as $r \to 0$ ($\mathbf{x}$ is omitted because of homogeneity). Kolmogorov's assumption of space-fillingness corresponds to the extreme situation where the Hausdorff dimension of S_h is 3 (*i.e.* $D_H = 1$). In this case one recovers $\xi_p = p/3$ ($h = 1/3$ is, like previously, derived from (1b)). Otherwise, the β-model gives

$$\xi_p = ph + 1 - D_H. \tag{5}$$

The experimental values of ξ_p obtained by Anselmet *et al.* (1984) may not agree with either the log-normal or the β-model. They find that ξ_p increases with p over the entire accessible range of p, but that $\frac{d\xi_p}{dp}$ seems to decrease with p. This led to the introduction of the multifractal model as a way of keeping with the idea that the turbulent activity is concentrated on fractal sets within the flow, and yet fit the experimental data. Similarly to the β-model, the multifractal model of inertial range turbulence starts from an assumption of self-similarity in space (see Frisch (1991)): there exists a range of exponents $h \in (h_{min}, h_{max})$, and different fractal sets S_h of Hausdorff dimension $2 + D_H(h)$ for each of these exponents h, such that if $\mathbf{x} \in S_h$, then $\delta u_{||}(\mathbf{x}, \mathbf{r}) \sim r^h$ as $r \to 0$ $(0 < D_H(h) \leq 1$ for all $h)$. Equation (4) is now replaced by

$$\langle (\delta u_{||}(\mathbf{r}))^p \rangle \sim \int d\mu(h)\, r^{ph+1-D_H(h)} \tag{6}$$

as $r \to 0$. The measure $\mu(h)$ corresponds to the weight of the different scalings. Using the method of steepest descents it then follows that ξ_p is the Legendre transform of $D_H(h)$, *i.e.*

$$\xi_p = min_h(ph + 1 - D_H(h)). \tag{7}$$

One can always find a function $D_H(h)$ for which (7) fits the experimental values of ξ_p. Unfortunately, Hausdorff dimensions D_H cannot be measured directly in practice, so that it is difficult to check experimentally whether the multifractal model is just another irrelevant way to fit the data, or whether it really captures some fundamental property of small scale turbulent structure.

On the other hand, Kolmogorov capacities are easily accessible in practice via the box-counting algorithm. But, as we have already pointed out previously, a non-integer value of D_K is by no means a proof that the examined geometry is fractal, and if it is not, $D_K \geq D_H = 0$, and the above multifractal and β-models do neither apply nor work. The exponent h, directly sensitive to the singular (non-differentiable) structure of the turbulence, cannot be used to derive (4), (5) and (6), (7) when S_h are nearly everywhere *smooth*, non-fractal sets characterised by non-integer values of D_K. Here we introduce an exponent σ which characterises the sparse–isolated–singular behaviour of the flow by holding information about the spatial extent over which portions of the velocity field are smooth and non-singular.

The object of the present paper is to derive the structural exponents ξ_p from the assumption that the small scale turbulence is predominantly smooth and spiral structured rather than fractal (or multifractal). The analysis of this paper applies to both 2-d and 3-d turbulence: the picture to hold in mind in the 2-d case is one of weak patches of vorticity wrapping around stronger patches of vorticity (homogeneously and isotropically distributed about the plane), thus producing spiral patterns (see Gilbert (1988)); in the 3-d case, the picture we refer to is one of a homogeneous and isotropic distribution of vortex filaments with a spiral internal vortex sheet structure. The analysis in the sequel is essentially 1-dimensional, but the results are valid in more than 1 dimensions because of homogeneity and isotropy. The results in section 2 are written for 3-d turbulence. To obtain the 2-d turbulence results, replace the velocity u by the vorticity ω in the sequel's formulae.

2. The Spiral β-Model and Multispirals

2.1. THE SPIRAL β-MODEL

The turbulence is assumed to be homogeneous and isotropic as $Re \to \infty$, so that a line in any direction through the flow should cut across a sufficiently large number of spiral vortex sheet structures, some of which very near the centre of the spiral accumulation. The points of intersection of these spiraling sheets with the line are assumed to have a non-trivial Kolmogorov capacity D_K ($D_K < 1$). It is also a consequence of the homogeneity and the isotropy of the small scale turbulence that the value of D_K is independent of the 1-d cut chosen to probe the flow.

The velocity field $u_{||}$ sampled along a 1-d cut is assumed to be *smooth* between the points where the cut intersects the spiral vortex sheet structure of the flow. Across these points of discontinuity, $u_{||}$ undergoes sudden jumps. A simplifying assumption in this paper is made with the replacement of the word 'smooth' by the word 'constant'. We take $u_{||}$ to be effectively constant between the points of discontinuity; $\frac{du_{||}}{dx}$ is indeed very small in the regions between jumps (where $u_{||}$ is smooth) compared to the sudden increase in $\frac{du_{||}}{dx}$ on the points where vortex sheets intersect the cut (x is the coordinate along the 1-d cut).

Between two consecutive such points we assume $u_{||}$ to be proportional to l^σ when the consecutive points of discontinuity are at a distance l from each other; specifically, as $l \to 0$,

$$|u_{||}(l)| \sim l^\sigma. \tag{8}$$

This is the assumption which replaces the $\beta - model$. It is not totally ad hoc; it is a reformulation of Moffatt's (1992) modeling of the velocity field inside a spiraling vortex sheet (see Appendix A). The singularity exponent σ which replaces h, is defined with reference to the regions where the flow is smooth–constant. The exponent h used in the β-model is defined instead *on* the set S_h, *i.e.* at these points where the flow is singular. In order to obtain (4) and (5) it is therefore essential to assume S_h to be fractal. This assumption is not made here.

It is shown in Vassilicos & Hunt (1991) (see also Appendix B) that the probability density function $n_0(l)$ for two consecutive points of discontinuity to be at a distance l from each other is

$$n_0(l) \sim l^{-D_0} \tag{9a}$$

as $l \to 0$, when these points have a non-trivial Kolmogorov capacity D_K, and

$$D_0 = D_K. \tag{9b}$$

We want to calculate $\langle(\hat{\mathbf{r}} \cdot \delta\mathbf{u}(\mathbf{x},\mathbf{r}))^p\rangle$, where the average is either taken over the entire 3-d $\mathbf{x}$-space, or over many realisations of the 3-d turbulent flow, or both. Because of homogeneity and isotropy,

$$\langle(\hat{\mathbf{r}} \cdot \delta\mathbf{u}(\mathbf{x},\mathbf{r}))^p\rangle = \langle(u_{||}(x+r) - u_{||}(x))^p\rangle, \tag{10}$$

where x and $x+r$ are points on an arbitrary 1-d cut through the flow at a distance $r = |\mathbf{r}|$ from each other, and where the average on the right hand side of (10) is either taken over the entire 1-d x-space of the cut, or over many realisations of $u_{||}$ on that cut, or both.

As is well known,

$$\langle(u_{||}(x+r) - u_{||}(x))^p\rangle = \langle u_{||}^p\rangle(1+(-1)^p) + \sum_{j=1}^{p-1}(-1)^{p-j}C_p^j\langle u_{||}^j(x+r)u_{||}^{p-j}(x)\rangle. \tag{11}$$

For $1 \le j \le p-1$, the contributions to these statistics can be decomposed as follows:

$$\langle u_{||}^j(x+r)u_{||}^{p-j}(x)\rangle = \sum_{q=0}^{+\infty} T_q(r;p,j), \tag{12}$$

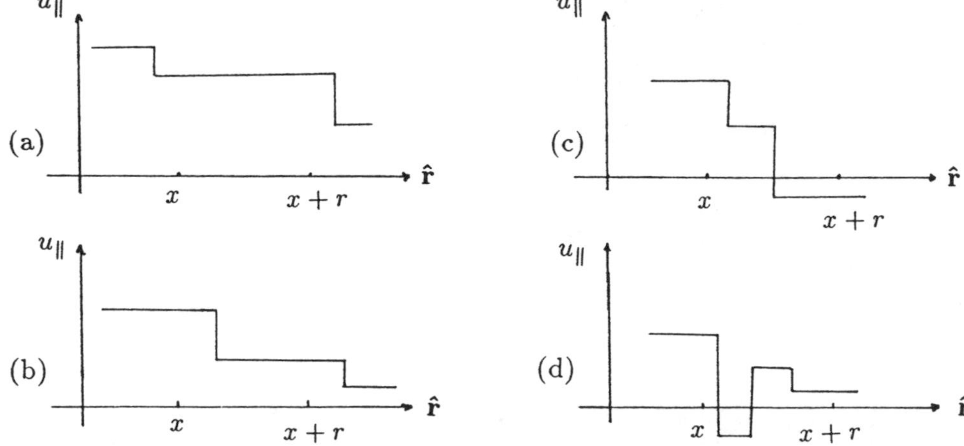

Fig. 1. $T_0(r;p,j)$ carries the contribution from situations like (a). $T_1(r;p,j)$, $T_2(r;p,j)$ and $T_3(r;p,j)$ correspond respectively to (b), (c) and (d).

where $T_q(r;p,j)$ is the average of $u_{\|}^j(x+r)u_{\|}^{p-j}(x)$ over all configurations where there are exactly q points of discontinuity between x and $x+r$ (see figure 1).

The probability density $n_0(l)$ does not hold enough information to enable a calculation of T_q when $q \geq 1$. For $q=0$ though, one can easily see that

$$T_0(r;p,j) = \int_r^{+\infty} n_0(l)\overline{u_{\|}^j(l)u_{\|}^{p-j}(l)}\,dl; \tag{13}$$

the overline indicates an average over the distribution of signs (+ or −) of $u_{\|}$ which has not been specified, and which holds essential information, as will clearly appear in the sequel, as to the difference between odd-p statistics and even-p statistics.

In order to evaluate the integral (13) one needs to introduce an integral length scale. This can be done on the grounds that beyond a certain length scale L (*i.e.* for $l \geq L$), different physics take over, which may be dominated by boundary conditions and which would imply a much faster fall off of $n_0(l)$ than a power law. For simplicity we set $n_0(l)=0$ when $l \geq L$, and therefore obtain, from (8) and (9a), that

$$T_0(r;p,j) \sim \frac{L^{p\sigma+1-D_0}}{p\sigma+1-D_0}\left(1-(r/L)^{p\sigma+1-D_0}\right) \tag{14}$$

as $r/L \to 0$. Also, if $p\sigma+1-D_0 \geq 0$,

$$T_0(r;p,j) = T_0(0;p,j)\left(1-(r/L)^{p\sigma+1-D_0}\right). \tag{15}$$

Note that $D_0 \leq 1$, and therefore it is sufficient that $\sigma \geq 0$ for (15) to be valid for all positive integers p.

In order to calculate $T_q(r;p,j)$ when $q \geq 1$, we need to introduce the probability density functions $n_q(l)$ which determine the chance for exactly q points of discontinuity to be between two other such points that are at a distance l from each other. One can show (see Appendix B) that for spiral accumulation patterns of the form $x_n \sim n^{-a}$,

$$n_q(l) \sim l^{-D_q} \tag{16}$$

as $l \to 0$, and $D_q = \frac{1}{1+a} < 1$. In fact, it appears that in general (see Appendix B), when $q \geq 2$, $D_{q-1} = D_{Kq}$; D_{Kq} are the generalised dimensions (generalised Kolmogorov capacities in fact) introduced by Hentschel & Procaccia in 1983 ($D_{K0} = D_K = D_0$).

If the generic structure of the turbulence is not only characterised by D_0, but also by D_1, D_2, D_3, *etc...* as defined by (16) (which are not greater than 1, and may not be in general equal to each other–see Appendix B), then $T_1(r;p,j)$ can be estimated as follows:

$$T_1(r;p,j) = \int_r^L n_1(l)\,dl \int_0^l \overline{u_{\|}^{j}(l-l_1)u_{\|}^{p-j}(l_1)n_0(l_1)}\,dl_1, \tag{17}$$

and a trivial calculation leads to:

$$T_1(0;p,j) - T_1(r;p,j) \sim (r/L)^{p\sigma+2-D_0-D_1} \tag{18}$$

for small r/L.

When $q \geq 2$, $T_q(r;p,j)$ can be shown to be of $O[(r/L)^{p\sigma+3-D_0-D_q-D_{q-2}}]$ because

$$T_q(r;p,j) \leq \int_r^L n_q(l)\,dl \int_0^r n_{q-2}(l_1)\,dl_1 \int_0^{l-l_1} \overline{u_{\|}^{j}(l_2)u_{\|}^{p-j}(l-l_1-l_2)n_0(l_2)}\,dl_2. \tag{19}$$

It follows, therefore (see (B9)), that to leading order in r/L,

$$\langle u_{\|}^{j}(x+r)u_{\|}^{p-j}(x)\rangle \approx T_0(r;p,j) + T_1(r;p,j), \tag{20}$$

which implies that if $p\sigma + 1 - D_0 \geq 0$,

$$\langle u_{\|}^{p}\rangle - \langle u_{\|}^{j}(x+r)u_{\|}^{p-j}(x)\rangle \sim (r/L)^{p\sigma+1-D_0} \tag{21}$$

as $r/L \to 0$.

This result applies whether p is even or odd. It reflects the fact that at small distances, a space varying quantity like $u_{\|}$ is well autocorrelated with itself if it remains, on average, approximately constant over these small distances. This is the meaning of the approximation in (20).

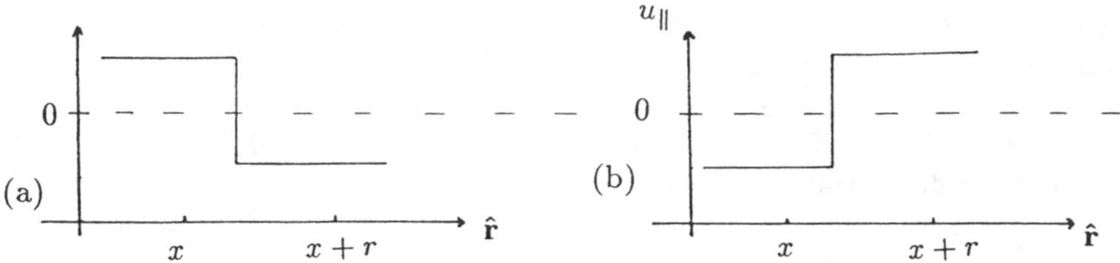

Fig. 2. If (a) is more likely than (b) then the odd $-p$ statistics of the relative velocities are strictly negative.

In order to recover turbulence statistics in terms of relative velocities (which are traditionally linked to the notion of 'turbulent eddies') we need to combine (21) with (11). The result is different for odd and even p; when p is even,

$$\langle (u_{\|}(x+r) - u_{\|}(x))^p \rangle \sim (r/L)^{\xi_p} \tag{22a}$$

as $r/L \to 0$, and

$$\xi_p = p\sigma + 1 - D_0. \tag{22b}$$

When p is odd though, the small-scale relative velocity statistics depend vitally on the distribution of the signs of $u_{\|}$. From (11) it is easily seen, for example, that if this distribution is such that $\langle u_{\|}^j(x+r)u_{\|}^{p-j}(x)\rangle = \langle u_{\|}^{p-j}(x+r)u_{\|}^j(x)\rangle$ for all positive integers $j \le p-1$, then $\langle (u_{\|}(x+r) - u_{\|}(x))^p \rangle = 0$ when p is odd.

One can check that $T_0(r;p,j) = T_0(r;p,p-j)$ irrespective of the distribution of signs of $u_{\|}$ (see (13)). It follows that

$$\langle u_{\|}^j(x+r)u_{\|}^{p-j}(x)\rangle - \langle u_{\|}^{p-j}(x+r)u_{\|}^j(x)\rangle \approx T_1(r;p,j) - T_1(r;p,p-j) \tag{23}$$

at leading order in r/L. For $\langle (u_{\|}(x+r) - u_{\|}(x))^p \rangle$ to be strictly negative when p is odd, it is enough to assume (see (11) and (23)) that the velocity $\hat{\mathbf{r}}\cdot\mathbf{u} = u_{\|}$ is more often negative on the right (towards $\hat{\mathbf{r}}$) of a point of discontinuity and positive on the left (towards $-\hat{\mathbf{r}}$) than positive on the right and negative on the left of such a point (see figure 2). This is an assumption about the skewness of the velocity field.

This additional assumption concerning the spatial distribution of signs of $u_{\|}$ is essential in order to incorporate intermittency in the model. The spiral assumption is not enough in itself, even though it represents a very natural departure from space-fillingness. There can obviously be no intermittency if the odd-p moments of the relative velocity vanish. Furthermore, it is a consequence of the Navier-Stokes equations that $\langle (u(x+r) - u(x))^3 \rangle < 0$

(see (1b)). Here all odd statistics are negative. Specifically, when p is odd,

$$\langle (u_{\|}(x+r) - u_{\|}(x))^p \rangle \approx C_p (r/L)^{\xi_p} \qquad (24a)$$

as $r/L \to 0$ (from (11), (18) and (23)); $C_p < 0$ (whilst it is obvious that $C_p > 0$ for even integers p), and

$$\xi_p = p\sigma + 2 - D_0 - D_1. \qquad (24b)$$

The spiral structure is said to be space-filling when $D_0 = D_1 = 1$. This corresponds to a spiral that is as slow to wind in as possible, and fills in this way the space around its centre of accumulation. In the limit where D_0 and D_1 tend to 1, ξ_p tends to $p\sigma$; in the context of the spiral β-model, the space-fillingness of the energy dissipation, which is assumed in Kolmogorov 1941, is interpreted as being the limit when the spiral sheets of dissipation (where the velocity derivatives are high) has such a slow inwards winding (or accumulating) pattern that $D_0 = D_1 = 1$. One indeed recovers the Kolmogorov expression for ξ_p in that limit, and the spiral β-model allows, in general, for the spirals not to be space-filling, *i.e.* for D_0, $D_1 < 1$. Thus, the turbulent velocity field is intermittent.

It is not clear from the literature why the conventional β-model makes no distinction between odd and even p statistics. In Appendix A we show how Moffatt's (1992) result for $p = 2$ can be recovered within the framework of the spiral β-model. Note, before closing this subsection, that (22) is valid when $p\sigma + 1 - D_0 \geq 0$, and (24) is valid when $p\sigma + 2 - D_0 - D_1 \geq 0$. All the p-statistics of the relative velocities may or may not converge as $r/L \to 0$; there may be an upper bound p_{max}, such that ξ_p is positive when $p \leq p_{max}$, and such that when $p > p_{max}$, these p-statistics diverge.

2.2. MULTISPIRALS

The small scale turbulence is now assumed to be made of different spiral vortex sheet structures of different Kolmogorov capacities D_K ($D_K < 1$). For a spiral of a given D_K, the magnitude of the velocity $u_{\|}$ between two consecutive points of discontinuity is still given by (8), but now σ is a function of D_K. In other words, we assume the existence of a spectrum of exponents σ (as in the multifractal model where a spectrum of exponents h is assumed) which correspond to spiral singularities of Kolmogorov capacity $D_K(\sigma) = D_0(\sigma)$. We also have to assume the existence of a function $D_1(\sigma)$ for the calculation of the odd-p statistics to be feasible. Equations (21), (22a,b) and (24a,b) are now respectively replaced by:

$$\langle u_{\|}^p \rangle - \langle u_{\|}^j(x+r) u_{\|}^{p-j}(x) \rangle \sim \int d\mu(\sigma) (r/L)^{p\sigma + 1 - D_0(\sigma)}, \qquad (25)$$

$$\langle (u_{\|}(x+r) - u_{\|}(x))^p \rangle \approx C_p \int d\mu(\sigma)(r/L)^{p\sigma+1-D_0(\sigma)} \tag{26}$$

where $C_p > 0$ when p is even, and

$$\langle (u_{\|}(x+r) - u_{\|}(x))^p \rangle \approx C_p \int d\mu(\sigma)(r/L)^{p\sigma+2-D_0(\sigma)-D_1(\sigma)} \tag{27}$$

where $C_p < 0$ when p is odd. The measure $\mu(\sigma)$ corresponds to the weight of the different local spiral velocity fields. As usual, the three formulas above are valid when $r/L \to 0$.

Using the method of steepest descents we obtain the following results:

$$\xi_p = min_\sigma(p\sigma + 1 - D_0(\sigma)) \tag{28a}$$

when p is even.

$$\xi_p = min_\sigma(p\sigma + 2 - D_0(\sigma) - D_1(\sigma)) \tag{28b}$$

when p is odd. And for all positive integers p, and $1 \leq j \leq p-1$, we have that

$$\langle u_{\|}^p \rangle - \langle u_{\|}^j(x+r) u_{\|}^{p-j}(x) \rangle \sim (r/L)^{min_\sigma(p\sigma+1-D_0(\sigma))}. \tag{29}$$

2.3. SCALAR INTERFACES: THE CASE $\sigma = 0$

When a scalar F is released in the turbulence, and before molecular diffusion has had time to act (we assume the molecular diffusivity of the scalar to be much smaller than the viscosity of the fluid), we may effectively represent F by an extremely contorted interface such that $F = +1$ on one side and $F = -1$ on the other side of the interface. Turbulent interfaces are known by experiment (see Sreenivasan (1991)) to have non-trivial Kolmogorov capacities D_K^i; experimental measurements appear to imply that $D_K^i = 1/3$ above, and $D_K^i = 1$ below the Kolmogorov viscous length scale.

The analysis of subsection 2.1 applies here with $\sigma = 0$ (it would not apply if F did not change signs across the interface). One can indeed write that

$$\langle (F(x+r) - F(x))^p \rangle \sim (r/L)^{\xi_p} \tag{30}$$

as $r/L \to 0$, with $\xi_p = 1 - D_k^i$ when p is even. (The relative p-statistics of F vanish for odd p). As noted by Vassilicos & Hunt (1991), the value $D_K^i = 1/3$ implies $\xi_2 = 2/3$, and $D_K^i = 1$ implies $\xi_2 = 0$, in agreement with theory and the measured statistics of passive scalars. ξ_p does not depend on p for passive interfaces, but only on whether p is even or odd; one may expect $\xi_p = 2/3$ in the inertial range, for all even positive integers p (in the limit of infinite Prandtl number and subject to the right initial conditions where a passive scalar can indeed be regarded as a step function F equal to either +1 or -1). It is not known how D_K^i relates to the generalised Kolmogorov capacities $D_{K0}(\sigma)$ and $D_{K2}(\sigma)$ of the turbulent velocity field.

3. Conclusion

The multispiral model gathers in a single picture the fractal properties of the turbulence, the inertial range scaling laws of the turbulent statistics and the observed smoothness of the velocity field with localised singularities which appear in the form of thin and elongated vortex tubes that have an internal vortex sheet spiral structure.

The property of intermittency can also be incorporated provided that the following assumption is made: as one samples the velocity $\hat{\mathbf{r}} \cdot \mathbf{u} = u_{||}$ along a linear cut through the flow in the direction $\hat{\mathbf{r}}$, $u_{||}$ jumps more often from a positive to a negative value than from a negative to a positive one at those points where the cut crosses a vortex sheet. This implies that the odd-p statistics of the turbulence do not vanish, which is a necessary condition for the average turbulent energy dissipation not to vanish either, and for the turbulence to be skewed. The turbulent velocity field is then more or less intermittent as the accumulation patterns of the turbulent structures are more or less space-filling.

The main results of this paper are equations (28) and (29); the singularity exponents σ characterise the spatial extent over which the velocity field is smooth, and $D_0(\sigma)$ and $D_1(\sigma)$ are, respectively, the generalised Kolmogorov capacities $D_{K0}(\sigma)$ and $D_{K2}(\sigma)$ (see Appendix B) characterising the accumulation pattern of the spiral structures corresponding to the exponents σ. Formally, the results of this paper do not depend on the existence of spiral structures, but on the existence of isolated accumulation patterns of some kind. Spiral structures seem to be the most plausible accumulation patterns, and a class of them are known to have the properties assumed here (non-trivial generalised Kolmogorov capacities–see Appendix B).

Formulae (28) can be tested by experiment; ξ_p, $D_{K0}(\sigma)$ and $D_{K2}(\sigma)$ are measurable unlike the Hausdorff dimensions of the multifractal model which are not. It is indeed surprising that multifractals make no distinction between even and odd p statistics. Such a distinction is central if one aims at an understanding of the dynamics (*e.g.* the energy dissipation) of the turbulence.

Acknowledgements

I am grateful to Keith Moffatt for suggesting that the next thing to do with spirals was to calculate higher moments. My constant interaction with Julian Hunt has had a decisive imprint in my research, and is also gratefully acknowledged. Their comments and also those of Uriel Frisch and Alan Kerstein have brought improvements to the text. During the time of this work, I have been supported by CEC grant 4120-90-10ED.

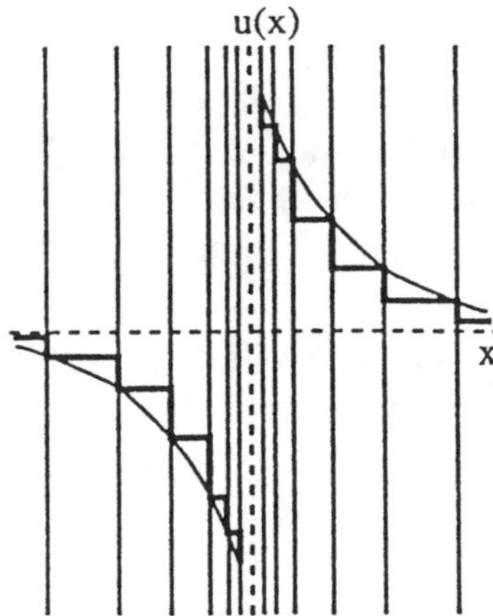

Fig. 3. From Moffatt (1992); the velocity profile on a transversal through the centre of the spiral vortex sheet.

Appendix A: Moffatt's (1992) Spiral Model

Moffatt (1992) assumes the following velocity profile on a transversal through the centre of a spiral vortex sheet (see figure 3): between two intersections of the sheet with the linear cut, the velocity is $u_n \sim n^b$, and these intersections are located at $x_n \sim n^{-a}$ $(n = 1, 2, 3, ...)$. Let $\delta x_n = x_n - x_{n+1} \sim n^{-a-1}$; $u_n \sim \delta x_n^{-\frac{b}{1+a}}$, so that $\sigma = -\frac{b}{1+a}$ (see (8)). It is shown in Vassilicos & Hunt (1991) that $D_0 = \frac{1}{1+a}$, and therefore a straightforward application of the results in subsection 2.1 gives

$$\xi_2 = 2\sigma + 1 - D_0 = 1 - \frac{2b+1}{1+a} \qquad (A1)$$

provided that $2\sigma + 1 - D_0 = \frac{a}{2} - b \geq 0$. (A1) has indeed been obtained by Moffatt (1992) using Fourier methods under the condition $b \leq \frac{a}{2}$.

Appendix B: The Generalised Kolmogorov capacities

One can find in Vassilicos & Hunt (1991) the following relation, valid as $\epsilon \to 0$;

$$N(0, \epsilon)\epsilon + L \int_{\epsilon}^{L} n_0(l)\, dl \approx L \qquad (B1)$$

where $N(0, \epsilon)$ is the minimum number of boxes of size ϵ needed to cover the points of a set (here the points of discontinuity of $u_{||}$), and $n_0(l)$ is the probability density function for two *consecutive* points of discontinuity to be at a distance l from each other.

(B1) can be generalised as follows ($q \geq 2$): for $\epsilon \to 0$,

$$N(q,\epsilon)\epsilon + L\int_{\epsilon}^{L} n_{q-1}(l)\,dl \approx L; \tag{B2}$$

$N(q,\epsilon)$ is the minimum number of boxes of size ϵ needed to cover part of the set so that each box covers at least q points of that set; $n_{q-1}(l)$ is the probability density which determines the chance for exactly $q-1$ points of the set to be between two other points of that set at a distance l from each other.

It is shown in Vassilicos & Hunt (1991) that for (spiral) 1-d accumulation patterns of the form $x_n \sim n^{-a}$ ($a > 0$),

$$N(0,\epsilon) \sim \epsilon^{-D_{K0}} \tag{B3a}$$

and

$$D_{K0} = \frac{1}{1+a}. \tag{B3b}$$

The same proof can be used to show that for $q \geq 2$,

$$N(q,\epsilon) \sim \epsilon^{-D_{Kq}} \tag{B4}$$

and

$$D_{Kq} = \frac{1}{1+a}. \tag{B5}$$

If we now assume a set of points to have non-trivial generalised Kolmogorov capacities D_{K0} and D_{K2}, D_{K3}, *etc.* as defined by (B4) (which were first introduced under the denomination of generalised dimensions by Hentschel & Procaccia (1983) in the more restricted context of fractals), then by differentiating (B1) and (B2) with respect to ϵ we obtain ($q \geq 1$)

$$n_{q-1}(l) \sim l^{-D_{q-1}}, \tag{B6}$$

$$D_0 = D_{K0} \tag{B7a}$$

and for $q \geq 2$

$$D_{q-1} = D_{Kq}. \tag{B7b}$$

It is clear that $N(q,\epsilon) \geq N(q',\epsilon)$ if $q \leq q'$, and therefore

$$D_{K0} \geq D_{K2} \geq D_{K3} \geq D_{K4} \geq \ldots \tag{B8}$$

which implies, in particular, that all generalised Kolmogorov capacities are not greater than 1 as $D_{K0} \leq 1$. From (B7) and (B8),

$$1 \geq D_0 \geq D_1 \geq D_2 \geq \ldots \tag{B9}$$

References

ANSELMET, F., GAGNE, Y., HOPFINGER, E.J. & ANTONIA, R.A. 1984 High-order velocity functions in turbulent shear flows. *J. Fluid Mech.* **140**, 63.

BRASSEUR, J.G. & YEUNG, P.K. 1991 Large and small-scale coupling in homogeneous turbulence: analysis of the Navier-Stokes equations in the asymptotic limit. In the proceedings of the eighth symposium on turbulent shear flows. September 1991, Munich, Germany.

FRISCH, U. 1991 From global scaling, a la Kolmogorov, to local multifractal scaling in fully developed turbulence. *Proc. R. Soc. Lond.* A**434**, 89.

GILBERT, A.D. 1988 Spiral singularities and spectra in two-dimensional turbulence. *J. Fluid Mech.* **193**, 474.

HENTSCHEL, H.G.E. & PROCACCIA, I. 1983 The infinite number of dimensions of probabilistic fractals and strange attractors. *Physica* **8D**, 435.

HUNT, J.C.R., KAIMAL, J.C. & GAYNOR, J.E. 1988 Eddy structure in the convective boundary layer: new measurements and new concepts. *Q. J. R. Meteorol. Soc.* **114**, 827.

KRASNY, R. 1986 Desingularisation of periodic vortex sheet roll-up. *J. Comp. Phys.* **65**, 292

MANDELBROT, B.B. 1982 *The Fractal Geometry of Nature.* W.H. Freeman, New York.

MOFFATT, H.K. 1992 Spiral structures in turbulent flow. In the proceedings of the I.M.A. conference: 'Fractals, wavelets and Frourier transforms: new developments and new applications'. December 1990, Cambridge, England, (ed. M. Farge, J.C.R. Hunt & J.C. Vassilicos). Clarendon Press, Oxford.

RUETSCH, G.R. & MAXEY, M.R. 1992 The evolution of small-scale turbulence in homogeneous isotropic turbulence. *Phys. Fluids* A (submitted).

SHE, Z.-S., JACKSON, E. & ORSZAG, S.A. 1991 Structure and dynamics of homogeneous turbulence: models and simulations. *Proc. R. Soc. Lond.* A**434**, 101.

SREENIVASAN, K.R. 1991 Fractals and multifractals in fluid turbulence. *Annu. Rev. Fluid Mech.* **23**, 539.

VASSILICOS, J.C. & HUNT, J.C.R. 1991 Fractal dimensions and spectra of interfaces with application to turbulence. *Proc. R. Soc. Lond.* A**435**.

MEASUREMENTS OF LOCAL SCALING OF TURBULENT VELOCITY FIELDS AT HIGH REYNOLDS NUMBERS

G.M.Zaslavsky*, and A.A.Chernikov**
Institute for Theoretical Physics
University of California
Santa Barbara
California 93106-4030
USA
A.A.Praskovsky, and M.Yu.Karyakin
Central Aerohydrodynamic Institute
Zhukovsky-3
Moscow region 140160
USSR
and
D.A.Usikov**
Laboratory for Plasma Research
University of Maryland
College Park
Maryland 20742
USA

ABSTRACT. We present the results of direct measurement of the local scaling exponent for two-point velocity differences in high Reynolds number turbulent flow ($R_\lambda = 3.2 \cdot 10^3$). The measurements of all velocity components are obtained in the return channel of a big wind tunnel at the Central Aerohydrodynamic Institute. The wavelet transform is applied to the evaluation of local scaling exponents. The degree of locality is restricted by averaging over the length corresponding to maximal scales in the inertial range. From analysis of the measurements we obtain a scaling exponent distribution histogram corresponding to scaling inhomogeneity. The mean and most probable value of the scaling exponent is close to 1/3 which follows from Kolmogorov-Obukhov theory. The range of local scaling exponents is $0 < \alpha < 0.6$. We present detailed analysis of the limitations of using the wavelet transform to determine local scaling exponents.

* Permanent address: Courant Institute of Mathematical Sciences, New York University, 251 Mercer Str., New York, NY 10012
** Permanent address: Space Research Institute, Profsoyuznaya 84/32 Moscow 117810, USSR

H. K. Moffatt et al. (eds.), Topological Aspects of the Dynamics of Fluids and Plasmas, 443–461.

1. Introduction

The modern understanding of the structure of fully developed turbulent flows is based on the qualitative idea of Richardson (1922) concerning the cascade character of energy transfer, *i.e.*, on the fact that small scale vortices acquire energy as a result of successive breakdown of large-scale vortices. It is assumed that energy transfers from large-scale vortices to small-scale vortices and that dissipation takes place only in the smallest scales. Formally Navier-Stokes equations with zero viscosity ($\nu = 0$) are invariant under the simultaneous transformations of scale $\mathbf{x} \to \lambda\mathbf{x}$, velocity $\mathbf{v} \to \lambda^{\alpha}\mathbf{v}$ and time $t \to \lambda^{1-\alpha}t$ where α is an arbitrary scale exponent.The Kolmogorov (1941) and Obukhov (1941) model which is based on several physically reasonable hypotheses allows us to eliminate this uncertainty and obtain results about local properties of fully developed turbulence using dimensional analysis. It is assumed that the quantity characterizing the process is the mean rate of energy transfer, which in equilibrium conditions is equal to the mean rate of energy dissipation $\langle\epsilon\rangle$ in small scales. Particularly in the inertial range where $L >> r >> \eta$, a well known "two-thirds" law is obtained

$$\langle(\delta\mathbf{v})^2\rangle \sim (\langle\epsilon\rangle r)^{2/3}, \quad \delta\mathbf{v} = \mathbf{v}(\mathbf{x}+\mathbf{r}) - \mathbf{v}(\mathbf{x}) \tag{1.1}$$

where $r = |\mathbf{r}|$ and $v = |\mathbf{v}|$ are the characteristic scale and velocity of turbulent fluctuations, L is the integral scale, $\eta = \nu^{3/4}\langle\epsilon\rangle^{-1/4}$ is the Kolmogorov dissipative scale and angular brackets $\langle\rangle$ denote space averaging.

Landau (1954) (see Landau and Lifschitz, 1987) pointed out that this model should be improved by taking into consideration the random character of energy dissipation in turbulent flows and the growth of its fluctuations with decreasing scale. It particularly leads to the phenomenon of intermittency. In 1962 Kolmogorov and Obukhov modified their theory to include this effect. A simple dynamical model in which the energy dissipation is considered as a very uneven function in space was proposed by Novikov and Stewart (1964) and later developed by Frisch, Sulem and Nelkin (1978). Mandelbrot (1974) suggested a geometrical interpretation of these models which is based on the fractal representation of 3D subsets in which the dissipation takes place.

A more general multifractal model of fully developed turbulence was developed by Parisi and Frisch (1985). In this model the global scale invariance (1.1) is replaced by the local scale invariance

$$\delta v(r) \sim r^{\alpha} \tag{1.2}$$

where the scaling exponent α is a function of position. The set of points having the same singularity exponent α is a fractal set having dimension $d(\alpha)$. It is shown that the structure function

$$\langle(\delta v)^p\rangle \sim r^{\zeta(p)} \tag{1.3}$$

grows with r according to a power law and the function ζ has a nonlinear dependence on p

$$\zeta(p) = min_{\alpha}(\alpha p + 3 - d(\alpha)) \tag{1.4}$$

Equation (1.4) connects the local property of the turbulent flow, *i.e.* its fractal dimension $d(\alpha)$ with the function $\zeta(p)$ which describes the deviations from linear scaling for large p. Note that particularly for $p = 2$ there could be deviations from the "two-thirds" law *i.e.* $\zeta(2) \neq 2/3$. Numerous experiments showed that the deviation is negligibly small. As mentioned by Yaglom (1981) there is now no reason to doubt the universal applicability of this law to all turbulent flows with high enough Reynolds numbers.

Modern experiments allow us to define local scaling exponents α in different regions of the turbulent flow. The wavelet transform which was first applied by Gouppilaud, Grossman and Morle (1984) to seismological data analysis is used for this purpose. The application of the wavelet transform to turbulent data analysis was started by Farge and Rabreau (1988) and Argoul *et al.* (1989) and developed by Emerson and Sirovich (1989), Bacry *et al.* (1989) and Emerson *et al.* (1990) for a wide class of 2D and 3D flows.

The wavelet analysis is based on a 2D integral transformation $T_g(a, x)$ of a function $f(y)$ and it depends on the coordinate x and scale a:

$$T_g(a, x) = a^{-1/2} \int f(y) \cdot g(\frac{x - y}{a}) dy, \qquad a > 0 \tag{1.5}$$

The kernel function g (analyzing wavelet) has a vanishing integral

$$\int g(y) dy = 0.$$

For a given position x_0, the amplitude $T_g(a, x)$ is maximized when the scale a is of the same order as the characteristic scale of the signal $f(y)$ near x_0. This property of the wavelet transform is very important for the visualization of self-similar properties of fractal and multifractal objects (Arneodo *et al.* (1988)). Recently it has been pointed out by Holschneider (1988) and Argoul *et al.* (1989) that if a function $f(x)$ has the self-similar property

$$|f(x_0 + \lambda x) - f(x_0)| \sim \lambda^{\alpha(x_0)} |f(x_0 + x) - f(x_0)| \tag{1.6}$$

its wavelet transform (1.5) will also scale like

$$T_g(\lambda a, x + \lambda b) \sim \lambda^{\alpha(x_0)} T_g(a, x + b) \tag{1.7}$$

This property allows us to use the wavelet transform as a tool for the investigation of local scaling of a turbulent velocity field in the inertial range. The first results of such an investigation were presented by Argoul et al. (1989)

and Bacry *et al.* (1989) for the 3D turbulent velocity signals obtained in the large wind tunnel at Modane. The Reynolds number based on the Taylor microscale is $R_\lambda = 2720$ ($R_\lambda = u'\lambda/\nu$ where u' is the *rms* turbulent velocity, λ is the Taylor microscale) and the turbulence has an inertial range of approximately two orders of magnitude. The results indicate that the most frequent exponents are close to the Kolmogorov value of 1/3 but violent rare events are however found for negative exponents at -0.1 or less.

Now we would like to mention two principal questions about this straightforward definition of the scaling exponents. The first is connected to the determination of the scaling exponent distribution histogram $N(\alpha)$ over the whole length of turbulent signal, and particularly to the width of the histogram which defines the degree of multifractality. The second question is about wavelet analysis itself and its ability to answer the first question. Equation (1.7) assumes the existence of a sufficiently large range of scaling for a reliable definition of the scaling exponents $\alpha(x)$. The inertial range is defined by the minimal L_{min} and maximal L_{max} scales of length. The quantity L_{max} is restricted from above by the size of a wind tunnel and L_{min} is restricted from below by the Kolmogorov dissipative scale l_ν. The following inequality usually holds $L_{max}/L_{min} << L/l_\nu \simeq R_L^{3/4}$, where R_L is the Reynolds number. The fact that R_L is finite restricts the possibilities of the wavelet analysis and in what follows we discuss these restrictions.

The minimal scale a of the transformation $T_g(a, x)$ in (1.5) should not be smaller than the Kolmogorov dissipative scale l_ν when we study the scaling property in the inertial range. For the same reason instead of the continous process $f(x)$ we usually consider a sequence of values $f(x_n)$ with minimal interval $|x_{n+1} - x_n| \geq l_\nu$. The maximal scale of a set $x_0, x_1, \ldots$ is restricted from above by the low boundary of the spectrum of inertial interval, *i.e.* by the value L_{max}. Thus the number of points in a set $x_0, x_1, \ldots$ should not exceed L_{max}/l_ν or $R_L^{3/4}$. In the following we will use the scale transformation as a power of 2 so $a_n = 2^n a_0$ where $a_0 \geq l_\nu$. Hence the following inequality is obeyed

$$max(n) < (3/4) log_2 R_L \tag{1.8}$$

For the experiments described below $max(n) = 13$. This number $max(n)$ defines the precision of the scale exponent approximation using the wavelet transform and restricts the range of the scaling exponent distribution histogram. A further increase of the accuracy in α and subsequently more precise analysis of the deviations from the average scale $\langle\alpha\rangle = 1/3$ (Kolmogorov-Obukhov law) could be reached by increasing the Reynolds number R_L according to (1.8).

The measurements of all the velocity components were obtained in the return channel of a big wind tunnel at high Reynolds numbers $R_\lambda \simeq 3.2 \cdot 10^3$ at the Central Aerohydrodynamic Institute. These experiments are analogous to the experiments described by Argoul *et al.* (1989) and Bacry *et al.* (1989). We continue by presenting the results of the measurements and by applying the wavelet transform to evaluate the local scaling exponents. The data obtained confirm the existence of scaling inhomogeneity and we describe the detailed analysis of the limitations of using the wavelet transform to determine local scaling exponents.

2. Fractal properties of streamlines

Among different properties of the wavelet transform, the definition of scaling exponents α in (1.2) is, to some extent, insensitive to some rough variations of the shape of analyzing wavelet. In this section we discuss a simple demonstration of this fact by analyzing the fractal dimension of a curve. Under certain conditions this curve may represent a streamline. The procedure is, in some sense, a calibration of the measuring process. A further comparison of the wavelet transform of the curve with the wavelet transform of a turbulent signal allows us to reveal the set of common features which could be used in what follows.

It was shown by Beloshapkin *et al.* (1989) that the stream function

$$\Psi(x,y) = \Psi_0 \sum_{j=1}^{q} cos(\mathbf{r}\mathbf{e_j}) \tag{2.1}$$

is an exact solution of 2D stationary Navier-Stokes equations under the appropriate forcing. Here $\mathbf{r} = (\mathbf{x}, \mathbf{y})$ and $\mathbf{e_j} = [cos(2\pi j/q), sin(2\pi j/q)]$ is a set of unit vectors which form a regular q-star. If $q \neq 2, 3, 4, 6$ these structures correspond to patterns with quasicrystal symmetries. In the general case for a special value of the streamfunction $\Psi = h_c(q)$ there exists a set of long streamlines with selfsimilar properties as $h \to h_c$. Among these streamlines there are some which are localized near separatrices, *i.e.* near streamlines passing through saddle points of the relief $\Psi = h = const.$

For the "calibration" of the wavelet transform a set of streamlines of the flow (1.2) with $q = 8$ is considered surrounding the central point. These streamlines cross the y axis in the vicinity of the saddle points $y = \pm\pi Q_n$, where Q_n are integers satisfying the recurrence relation

$$Q_{n+1} = 2Q_n + Q_{n-1}, \qquad Q_0 = 0, \quad Q_1 = 2 \tag{2.2}$$

Equation (2.2) defines a ratio Q_{n+1}/Q_n, which in turn, is the best rational approximation to $1+\sqrt{2}$. In Fig.1 the subsequent streamlines are shown for $Q_n = 10, 24, 58$ ($n = 3, 4, 5$). Let us neglect the small curvature of the small elements in Fig.1 and replace the curves by polygons. It is not difficult to see that the ratio of two subsequent streamline perimeters is equal to 3 (it is proportional to the number of sides of the closed polygons). The successive

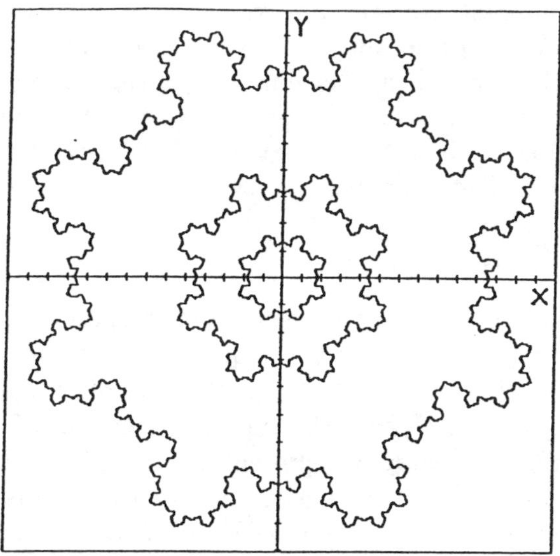

Fig. 1. Streamlines of stationary flow with 8-fold symmetry.

ratios of the y coordinates of these curves at $x = 0$ is equal to $1 + \sqrt{2}$. Thus as $Q_n \to \infty$ the fractal dimension D of the long streamlines is equal to

$$D = \log 3/\log(1 + \sqrt{2}) \simeq 1.246... \tag{2.3}$$

The corresponding critical value $h_c = 0$.

The time-dependent trajectory $y(t)$ is used as a test signal for the wavelet transform. It is obtained as a result of the numerical integration of the equations for liquid particles

$$\dot{\mathbf{r}} = \mathbf{e_z} \times grad\Psi$$

Here the streamfunction $\Psi \equiv \Psi_8(x, y), \Psi_0 = 1$ and the initial conditions are: $x(t = 0) = 0, y(t = 0) = -187.17 \simeq -58\pi$, *i.e.* the trajectory of a liquid particle is analysed as it moves along the streamline. The streamline corresponds to the curve of maximal size in Fig.1. The correspondent function $y(t)$ (the signal oscillogram) is shown in Fig.2.

The wavelet transform of the function $y(t)$ is shown in Fig.3a and Fig. 3b with two different kinds of analyzing wavelet g. The analyzing wavelet for the first case (Fig.3a) corresponds to the so-called "french hat"

$$g(x) = \begin{cases} 1 & \text{if} \quad |x| < 1 \\ -1/2 & \text{if} \; 1 < |x| < 3 \\ 0 & \text{if} \quad |x| > 3 \end{cases} \tag{2.4}$$

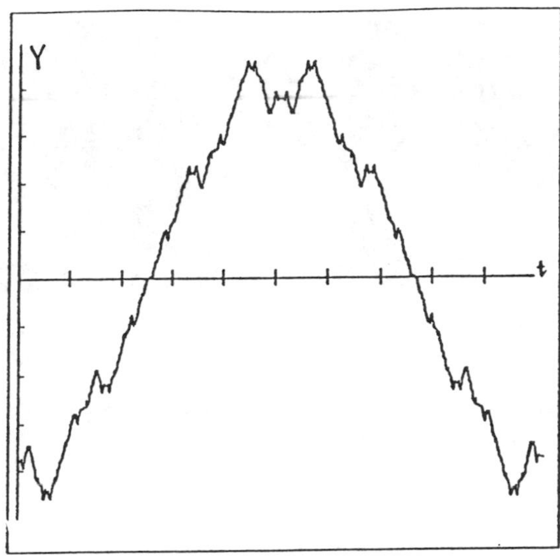

Fig. 2. Lagrangian particle trajectory for the streamline of maximal size in Fig.1.

For the second case the analyzing wavelet is defined as follows

$$g(x) = \begin{cases} 1 & \text{if} \quad 0 < x < 1 \\ -1 & \text{if} \ -1 < x < 0 \\ 0 & \text{if} \qquad |x| > 1 \end{cases} \tag{2.5}$$

The amplitude of the wavelet transform $T_g(a, x)$ defines the "relief" of a surface of two variables: a (the scale of the transform) and x (the coordinate which is proportional to the time shift).

For the relief image the following standard procedure is used. An equidistant sequence of amplitude values $T_g(a, x)$ is fixed with interval ΔT and all regions of the relief are colored successively in black and white. The corresponding pictures of the wavelet relief visualization are shown in Fig.3a,b. A relief section $T_g(a, x)$ at $x = const$ allows us to determine the local scaling exponent α of the function $y(t)$ using equation (1.7). This result should be averaged over many sections (this procedure is discussed in detail in section 4). We average over 5120 sections. Using the correspondence $D = 2 - \alpha$ between the fractal dimension of the curve D and the scaling exponent α (Mandelbrot (1982)), numerical analysis gives us in both cases the same value of the fractal dimension $D = 1.24 \pm 0.02$. Thus these examples demonstrate a weak dependence of the scaling on the chosen shapes of the analyzing wavelets (2.4) and (2.5) and establish the reliability of the used forms of signal analysis.

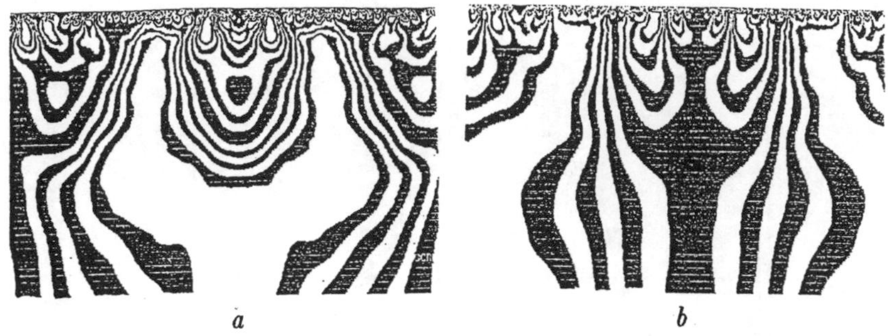

Fig. 3. Wavelet transform of the Lagrangian particle trajectory: a) with kernel function (2.4), b) with kernel function (2.5).

3. Turbulent signal from the wind tunnel experiments

The turbulent flow studied was that in the wind tunnel at the Central Aero-hydrodynamic Institute. The main purpose of the experiment is to analyse the turbulent velocity fluctuations which are measured in the return channel at high Reynolds numbers (the turbulence parameters are presented in what follows). The return channel has length $175m$ and width $22m$, and its height rises linearly from $20m$ at the entrance to $32m$ at the exit. The probe was mounted on a support of height $5m$ in the central vertical plane of the channel at a distance of $120m$ from its entrance. The direction of x-axis is chosen along the main flow, y and z-axes are correspondingly along horizontal and vertical directions. Three probes are used. In the first probe the sensitive element is a wire of platinized tungsten of diameter $2.5\mu m$ and length $0.25mm$, while in the second probe the length is $0.5mm$. The third probe is X-shaped. The length of wires for it are $0.5mm$, they are perpendicular to each other and the distance between them is $0.5mm$. Two components of the velocity are measured simultaneously (u and v or u and w). Here and subsequently (U, V, W) and (u, v, w) are the components of the averaged velocity and its fluctuations.

The signals from the probes are passed through a low-pass filter with an upper frequency of $8kHz$ and characteristic sharpness $36dB/octave$ and recorded on the magnetic tape recorder. The duration of every record is $5min$. The signals recorded on the magnetograph are passed through a low-pass filter with an upper frequency of $f_c = 3.7kHz$ and characteristic sharpness $36dB/octave$, and are fed into a computer with interrogation frequency $f_s = 16kHz$. The ratio of the useful signal dispersion to the the noise dispersion at the studied points is not less than 400. The data comprise 4,320,000 numbers, which correspond to a duration of $4.5min$.

For the wavelet analysis the data are fed into a computer with characteristic frequences $f_c = 1.7kHz$ and $f_s = 7kHz$. The duration of the signal is $37.45s$ which corresponds to 262,144 numbers. The decreasing of f_s allows us to cut the volume of data twice conserving its duration. The purpose of our study is to apply the wavelet analysis to turbulent fluctuations in the inertial range of scales. Thus the exclusion of dissipative scales from the consideration does not influence the final results.

The transformation from temporal characteristics of the signals to spatial characteristics are made by means of the Taylor hypothesis of "frozen" turbulence

$$\begin{gathered} r = x^{(2)} - x^{(1)} = U(t^{(2)} - t^{(1)}), \\ u(x,t) = u(x - U\tau, t + \tau), \\ \partial u/\partial x = -(\partial u/\partial t)/U \end{gathered} \tag{3.1}$$

where t is time, τ is a time interval and r is the distance between two points in the x direction.

The dissipation of energy is calculated under the assumption of local isotropy of the turbulence; *i.e.*, it is assumed that

$$\langle\epsilon\rangle = 15\nu\langle(\partial u/\partial x)^2\rangle \tag{3.2}$$

where angular brackets <> denote time averaging or, in accordance with (3.1), space averaging. The integral scale L and the Taylor microscale λ are determined by the well-known expressions

$$L = \frac{U}{< u^2 >} \int_0^\infty \langle u(t)u(t+\tau)\rangle d\tau, \tag{3.3}$$

and

$$\lambda = (\langle u^2\rangle / \langle(\partial u/\partial x)^2\rangle)^{1/2} \tag{3.4}$$

The measurements of the mean longitudinal velocity gave $U = 11.1m/s$ for the first and second probes, and $U = 10.8m/s$ and $U = 11.0m/s$ for the third X-shaped probe located in a position corresponding to (u,v) or (u,w) measurements. Thus the uncertainties of the wind tunnel experiments do not exceed 2 percent. The perpendicular components of the velocity are $V \simeq 0$ and $W \simeq 0.96m/s$. The rms levels of velocity fluctuations are $u' = 1.0m/s, v' = 0.96m/s$ and $w' = 0.83m/s$. The turbulent energy dissipation estimated according to (3.2) is equal to $\langle\epsilon\rangle = 0.12m^2/s^3$. The corresponding Kolmogorov dissipative scale and Taylor microscale are $\eta = 0.41mm$ and $\lambda = 46mm$. The latter corresponds to a Reynolds number $R_\lambda = 3.2 \cdot 10^3$.

The ratio l/η is 1.2 for the probes with $l = 0.5mm$ and $l/\eta = 0.6$ for the probe with $l = 0.25mm$. Hence the spatial resolution of the measurements is

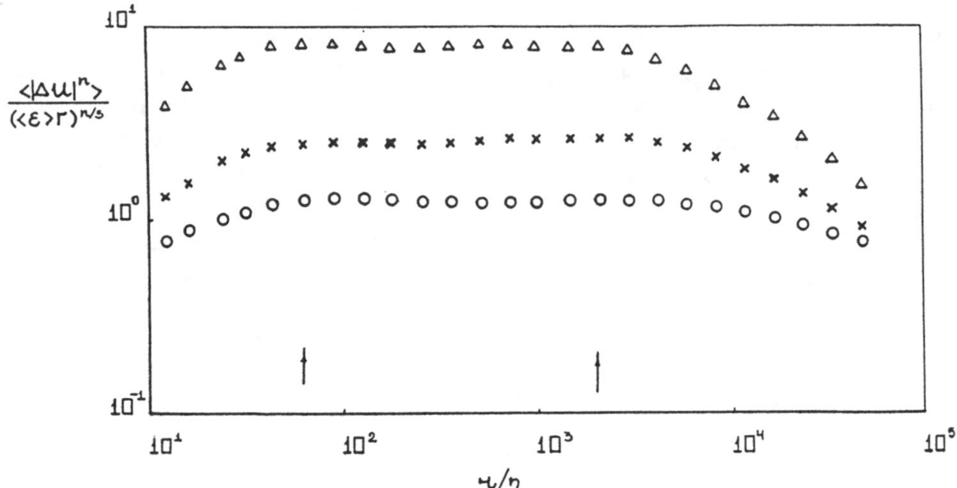

Fig. 4. Structure functions of longitudinal velocity: $o - n = 1, x - n = 2, \Delta - n = 3$.

good enough though the cut frequency is below the Kolmogorov frequency $f_k = U/2\pi\eta = 4.3kHz$. The cut frequency is chosen from the condition that the resolution in longitudinal direction is approximately equal to the resolution in transverse direction which, in turn, is restricted by the length of the wire: $U/2\pi l \simeq 3.5kHz$ for $l = 0.5mm$. However this restiction of the frequency range does not influence the characteristics of the inertial interval since the minimal size of vortices in this interval exceeds the Kolmogorov scale by an order of magnitude.

The integral scale of the turbulence measured by the probe with $l = 0.25mm$ is equal to $L = 4.82m$, *i.e.* $L/\eta = 1.18 \cdot 10^4$ and $R_L = u'L/\nu = 3.3 \cdot 10^5$. Such high Reynolds numbers $R_\lambda \simeq 3.2 \cdot 10^3$ and $R_L = 3.3 \cdot 10^5$ achieved in the experiment allow us to assume the existence of a wide enough inertial range in the spectrum of turbulence. The structure functions of all velocity components of the first, second and third orders are determined for the evaluation of the inertial range boundaries

$$\langle |\Delta u_i|^n \rangle = \langle |u_i(x) - u_i(x+r)|^n \rangle, \qquad u_i = u, v, w; \qquad n = 1, 2, 3.$$

The typical results recorded by the probe with $l = 0.25mm$ are shown in Fig.4.

The behaviour of all normalized structure functions as functions of r appears to be approximately the same. It is not difficult to see that there exists a range of scales $\eta << r << L$ where the structure functions are approximated well enough by the power law (1.3). However the choice of the boundaries contains some uncertainty. In this paper the range of scales

$60 << r/\eta << 2000$ is accepted as the inertial interval (these boundaries are shown in Fig.4 by the arrows). The estimation of the value $\zeta(1)$ and the errors in (1.3) are obtained inside this interval for the first order structure function. The estimation results are listed in Table 1 (the meaning of $\langle\alpha\rangle$ and σ is explained in section 4). The value $\zeta(1)$ for all measurements corresponds to the Kolmogorov-Obukhov law with high precision.

4. Wavelet analysis of the velocity fluctuations

The wavelet transform (1.5) with the kernel (2.4) is applied to the results of the measurements of the velocity components u, v and w. The lower parts of Fig. 5a,b,c represent the images of such signals, while the upper parts correspond to the visualization of the wavelet reliefs obtained the same way as in section 2. The relief in Fig.5a covers the part of a signal with the length $\Delta x = 32.5m$ (the horizontal axis) and the scale range from $a_{min} = 0.1m$ to $a_{max} = 35.5m$(the vertical axis).

Fig.5b is a four-fold magnification of the region in the vicinity of the arrows in Fig.5a, $\Delta x = 8.12m$, $a_{min} = 0.025m, a_{max} = 8.88m$. Fig 5c is obtained the same way relatively to Fig.5b with the same magnification coefficient. The wavelet transform images for all oscillograms (1-4) (see section 3) are analogous to the images shown in Fig.5 which are in qualitative agreement with the multifractal model. A typical image of the relief represents a fractal tree with bifurcated branches of different widths and large numbers of cone-like structures at small scales. The appearance of such structures gives evidence concerning the existence and selfsimilarity of singularities in the oscillogram of velocity fluctuations. Note some similarity between Fig.3 for the fractal curve and Fig.5 for the turbulent signal.

The analysis of the relief sections in different parts of the oscillogram shows us that there exist two kinds of the typical sections (Fig.6) The sections with the clear scaling in the inertial range $60\eta < a < 2000\eta$ belong to the first type (curve a). The deviations from the power law (1.7) are small for this type of sections and were discussed by Bacry *et al.* (1989). The second type (curve b), also mentioned by Bacry *et al.* (1989), does not allow us to reliably define a scaling exponent.

The authors assume the deviations from the pure power law to be of physical nature. The scaling relation (1.2) takes place, generally speaking, in the inertial range of scales. Thus the measurements of the scaling exponent α should naturally be carried out by averaging over a length x_m close to the length L_{max} corresponding to the maximal scale in the inertial interval. In practice, this means that the logarithm in (1.5) is replaced by its averaged

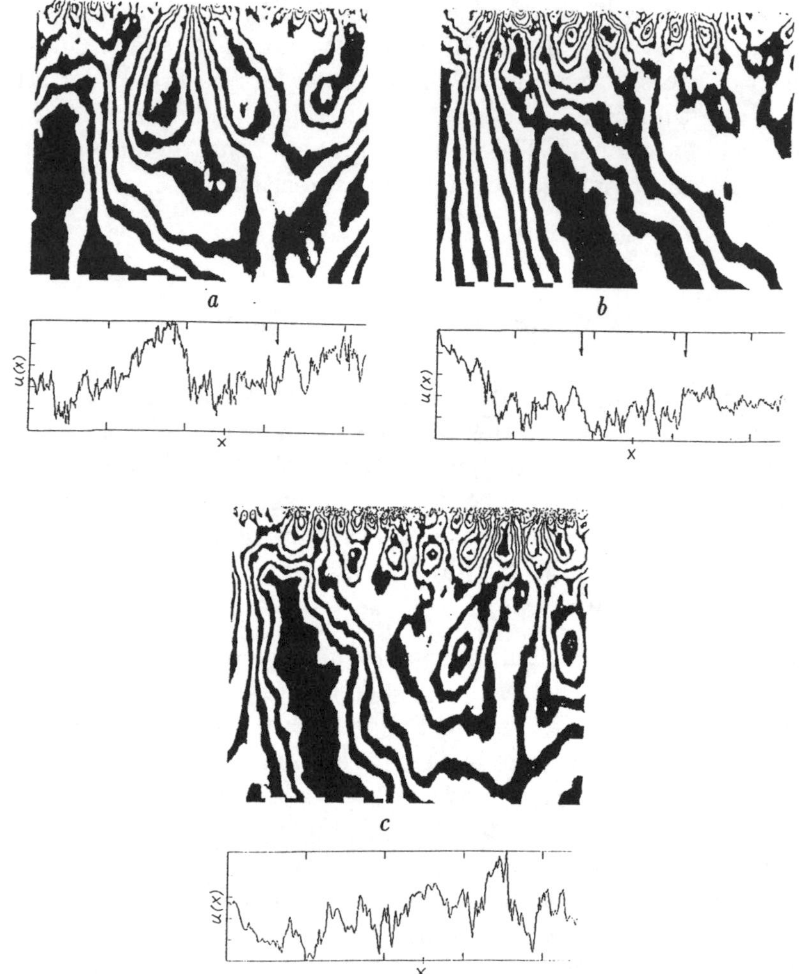

Fig. 5. Wavelet transforms and signal oscillograms. a) large scales, b) intemediate scales (inside arrows in fig.5a), c) small scales (inside arrows in Fig.5b).

value:

$$< \log T_g(a,x) >= \frac{1}{x} \int_{x-\frac{x_m}{2}}^{x+\frac{x_m}{2}} \log T_g(a,x)dx$$

This replacement means the whole space is partitioned in cells of characteristic size $x_m \leq L_{max}$ and α is now defined in every cell. The sections of the relief amplitudes, averaged over different intervals x_m, are shown in Fig.7. It is clear from Fig.7 that the larger x_m is, the more distinctly the power law can be seen. All the results in this section are obtained by averaging over the interval $x_m = L_{max}$. Every oscillogram is divided into nonintersecting pieces

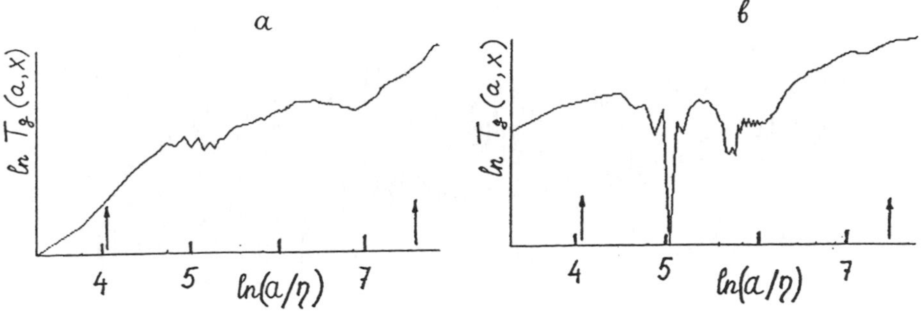

Fig. 6. Typical relief sections (the arrows show the boundaries of the inertial range).

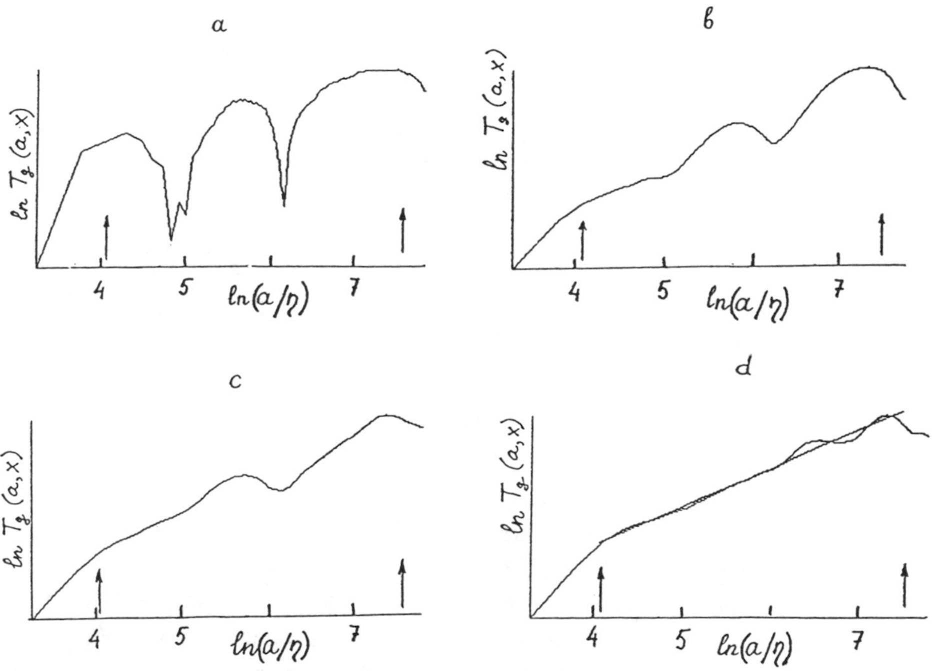

Fig. 7. Relief section with different averaging (the arrows show the boundaries of the inertial range): a) without averaging, b) the interval of averaging is 500η, c) the interval of averaging 1000η, d) the interval of averaging is 2000η.

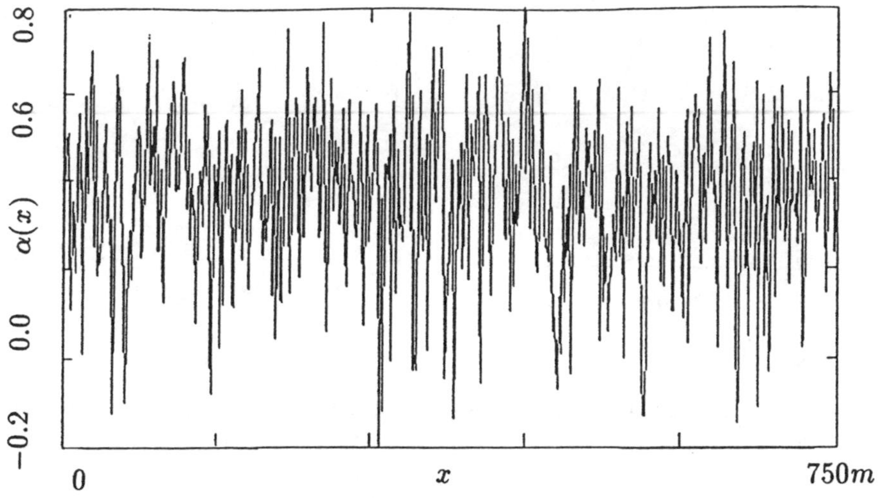

Fig. 8. Graph of the dependence of the local scaling exponent α on x.

Probe	I-shaped		X-shaped	
	$l = 0.25mm$	$l = 0.5mm$	position (u, v)	position (u, w)
Oscillogram	1	2	3	4
Velocity component	u	u	v	w
$\zeta(1)$	0.33 ± 0.02	0.33 ± 0.01	0.33 ± 0.02	0.33 ± 0.03
$< \alpha >$	0.32 ± 0.02	0.30 ± 0.02	0.30 ± 0.02	0.29 ± 0.02
σ	0.20	0.21	0.20	0.20

TABLE I

of length $L_{max} = 2000\eta$, which cover the oscillogram densely. As a result of such a procedure 915 values of the scaling exponents $\alpha(x_j)$ are obtained for every oscillogram. The distance $x_{j+1} - x_j = 2000\eta$. The dependences $\alpha(x)$ are obtained for all studied oscillograms. As an illustration, the graph $\alpha(x)$ for the oscillogram 2 is shown in Fig.8. One can see from this graph that the values of $\alpha(x_j)$ belong to the sufficiently wide interval from 0 to approximately 1, but rare events are found for negative exponents. These results are in qualitative agreement with the data presented by Bacry *et al.* (1989). The averaged values of $\langle\alpha\rangle$ are listed in Table 1. For the estimation of the errors in $\langle\alpha\rangle$ the fact that all the values $\alpha(x_j)$ are noncorrelated is taken into consideration. This statement is checked straightforwardly by correla-

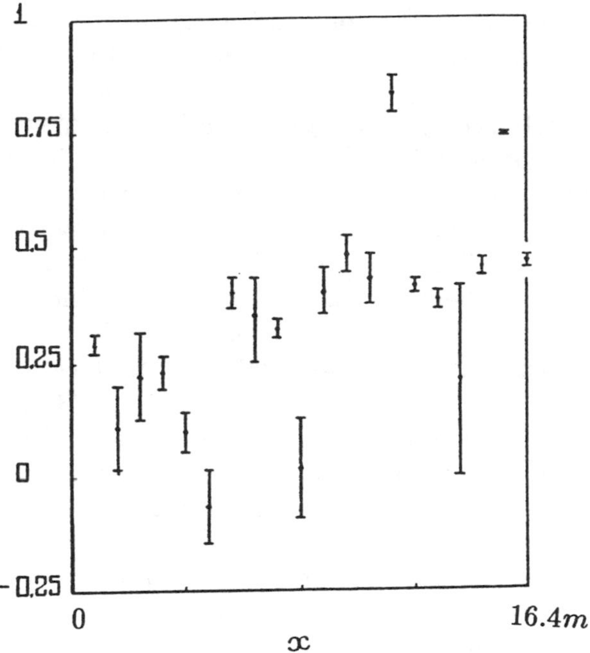

Fig. 9. Local scaling exponents and errors in the approximation (1.7).

tion analysis. The absence of correlations between different values of α from different intervals follows from the fact that the positions x_j are divided by the distance which is equal to the upper boundary of the inertial interval. It follows from Table 1 that values of $\langle\alpha\rangle$ obtained by using wavelet analysis are in satisfactory agreement with the values of $\zeta(1)$ and are close to the averaged value $\zeta(1) = 1/3$ which the Kolmogorov-Obukhov theory predicts. The *rms* deviations

$$\sigma = \sqrt{\frac{1}{J}\sum_{j=1}^{J}(\alpha(x_j) - \langle\alpha\rangle)^2}$$

which are defined from the sequence $\alpha(x_j)$ and are shown in Table 1 characterize the degree of variability of the scaling exponents α.

The deviations from power law in many cases are considerable in spite of the averaging procedure. That is why the *rms* errors $\Delta\alpha(x_j)$ in the power law approximation (1.7) are estimated at every point x_j in addition to the determination of $\alpha(x_j)$ in the inertial interval (see Fig.7d). For the illustration of these results a small piece of the graph $\alpha(x_j)$ with the errors $\Delta\alpha(x_j)$ is shown in Fig.9 for the oscillogram 4. It is not difficult to see that excluding the cells where $\Delta\alpha(x_j)$ is small enough there exist cells where the scaling relation is not fulfilled. The cells from the first group belong to the volume where the scaling relation (1.2) is realised in a definite moment of time, while the cells from the second group belong to the volume where such scaling is absent.

The results obtained may be interpreted by the following picture of the velocity fluctuations behaviour in fully developed turbulence. At an arbitrary time, independent cascade processes of energy transfer are excited only in parts of the liquid volume. The values of the scaling exponents α are different in different cascades. The most probable value of α is close to 1/3, however there exist cascades where the values of α differ significantly from 1/3, and there exist cells where the existence of the scaling is doubtful.

The scaling exponent histograms, which are obtained for all oscillograms prove the existence of α values which differ from 1/3. The functional dependence of all obtained oscillograms appears to be similar. The histograms $N(\alpha)$ for the longitudinal and transversal components of the turbulent velocity fluctuations are shown in Fig.10a and Fig.11a. These oscillograms are in qualitative agreement with the results presented by Bacry *et al.* (1989). The *rms* values of the errors in the α definition are also determined (they are absent in the paper of Bacry *et al.* (1989)). These errors are obtained by the following algorithm. For every i-th subinterval the values α_k are chosen in the range $\alpha_i \leq \alpha_k \leq \alpha_{i+1}$, $k = 1, 2, \ldots N_i$. Then using the corresponding value $\Delta\alpha_k$ the *rms* error of the i-th subinterval is defined by

$$\Delta\alpha_i = \sqrt{\frac{1}{N_i}\sum_{k=1}^{N_i}(\Delta\alpha_k)^2} \tag{4.1}$$

The results shown in Fig.10b and Fig.11b indicate that the errors increase strongly with deviation of α from its averaged value $\langle\alpha\rangle = 1/3$.

It is evident from Fig.10 and Fig.11 that the distribution $N(\alpha)$ is measured accurately in the range of values $0 \leq \alpha \leq 0.6$. But if $\alpha < 0$ or $\alpha > 0.6$ the errors are so large that the function $N(\alpha)$ in these ranges appears to be unreliable. Such high values of errors $\Delta\alpha$ give evidence concerning either the insufficiently large Reynolds number, which as mentioned in section 1, restricts the number of points, or the absence of scaling in the corresponding volumes.

5. Conclusion

This study of the local structure of the velocity field is carried out in fully developed turbulence at high Reynolds number $R_\lambda \simeq 3.2 \cdot 10^3$. Wavelet analysis is used to determine local scaling exponents. In contrast to previous papers, *e.g.* Bacry *et al.* (1989), the meaning of "locality" is analyzed for the definition of scaling exponents α. Because of the impossibility of determining α at a point, the minimal interval is introduced in which a local scaling

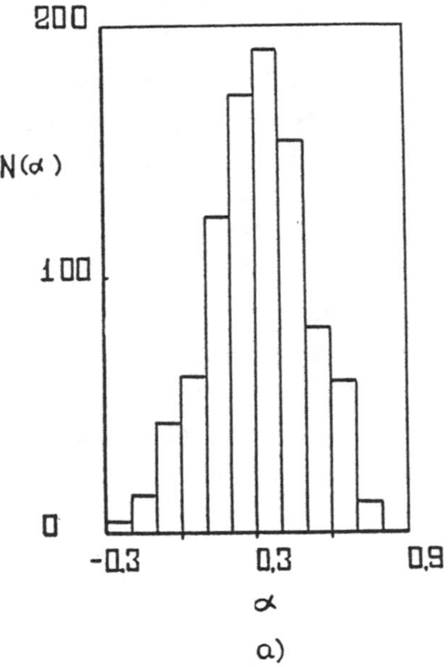

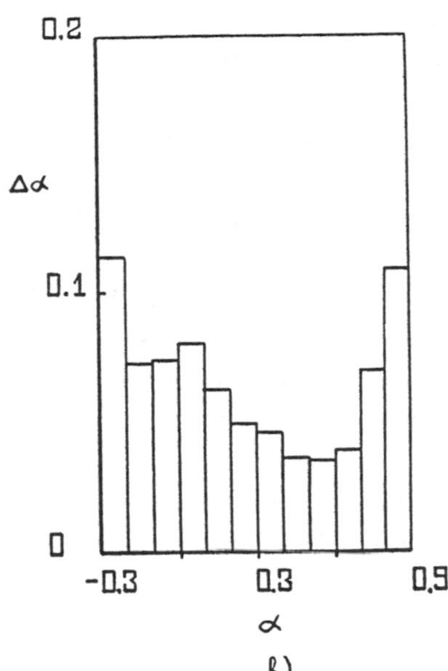

Fig. 10. a) Scaling distribution histogram for the longitudinal velocity fluctuations (oscillogram 1), b) errors in $N(\alpha)$.

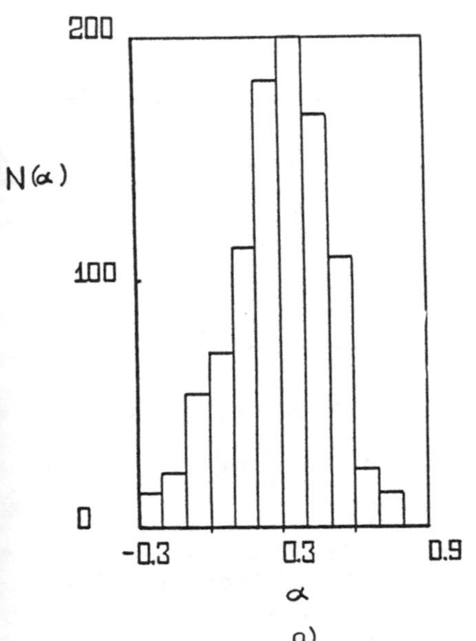

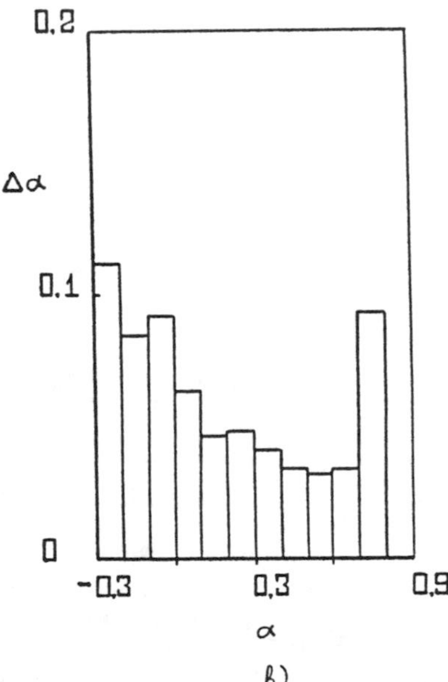

Fig. 11. a) Scaling distribution histogram for the transversal velocity fluctuations (oscillogram 4), b) errors in $N(\alpha)$.

exponent α may be considered as "local". Such an interval is defined by the maximal size of the vortices in the inertial range, *i.e.* $L_{max} = 2000\eta$

The large number of points (equal to 915) ensure sufficient precision in the definition of the local scaling exponent α using the wavelet transform. A further increase in the degree of locality could be achieved by increasing L_{max} or decreasing η, that means, according to (1.8), increasing the Reynolds number. Thus the histograms $N(\alpha)$ in Figs.10a and 11a represent a compromise between the degree of "nonlocality" in the set of 512 points in the α definition and the distribution of these α' values over different sets. The results of measurements show, that the mean and most probable value of the scaling exponent is close to 1/3 which follows from the Kolmogorov-Obukhov theory. The domain of local scaling exponents is $0 < \alpha < 0.6$. These results are in agreement with the data obtained by Bacry *et al.* (1989). However in contrast to Bacry *et al.* (1989), it is shown that the natural restrictions on the wavelet analysis lead to large errors in the $N(\alpha)$ determination out of the range $0 < \alpha < 0.6$ because of the existence of the dissipative scale and finite size of the probes. The origin of these errors might be both connected to insufficient accuracy of the measurements and the absence of scaling in some regions of the turbulent flow.

Acknowledgements

The authors thank D. Linardatos for his help in preparing the English version of the article.

This article was prepared at the Space Research Institute of the USSR Academy of Science and at the Institute for Theoretical Physics at UCSB and partly supported by NSF Grant PHY89-04035 and by the US Department of Energy.

References

ARGOUL, F., ARNEODO, A., GRASSEAU, G., GAGNE, Y., HOPFINGER, E.J. & FRISCH, U., 1989 *Nature*, **338**, pp. 51-53.

ARGOUL, F., ARNEODO, A., ELIZGARAY, J., GRASSEAU, G. & MURENZI, R., 1989 *Phys. Lett.*, **135A**, pp. 327-335.

ARNEODO, A., GRASSEAU, G., & HOLSCHNEIDER, M., 1988 *Phys. Rev. Lett.*, **61**, pp. 2281-2284.

BACRY, E., ARNEODO, A., FRISCH, U., GAGNE, Y. & HOPFINGER, E., 1989 In *Turbulence and Coherent Structures*, eds. Lesieur, M. & Metais, O., Kluwer Publishers.

BELOSHAPKIN, V.V., CHERNIKOV, A.A., NATENZON, M.YA., PETROVICHEV, B.A., SAGDEEV, R.Z. & ZASLAVSKY, G.M., 1989 *Nature*, **337**, pp. 133-137.

EMERSON, R. & SIROVICH, L., 1989 A survey of wavelet analysis applied to turbulent data. *Report 89-182*, Center for Fluid Mechanics, Brown University.

EMERSON, R., SIROVICH, L. & SREENIVASAN K.R., 1990 *Phys. Lett.*, **145A**, pp. 314-322.

FARGE, M. & RABREAU, G.C., 1988 *C. R. Acad. Sci.*, **307 II**, pp. 1479-1486.

FRISCH, U., SULEM, P.L. & NELKIN, M., 1978 *J. Fluid. Mech.*, **87**, pp. 719-736.

GOUPPILAUD, P., GROSSMANN, A. & MORLET, J., 1984 *Geoexploitation*, **23**, pp. 85-102.

HOLSCHNEIDER, M., 1988 *J. Stat. Phys.*, **50**, pp. 963-993.

KOLMOGOROV, A.N., 1941 *Dokl. Akad. Nauk SSSR*, **30**, pp.299-303.

KOLMOGOROV, A.N., 1962 *J. Fluid Mech.*, **13**, pp. 82-85.

LANDAU, L.D. & LIFSHITZ, E.M., 1987 *Fluid Mechanics*, 2nd edition, Pergamon Press.

MANDELBROT, B.B., 1974 *J. Fluid Mech.*, **62**, pp. 331-358.

MANDELBROT, B.B., 1982 *The fractal geometry of nature*, San Francisco: Freeman.

NOVIKOV, E.A. $ STEWART, R.W., 1964 *Izv. Akad. Nauk SSSR, ser. Geophys.*, **3**, pp. 408-411.

OBUKHOV, A.M., 1941 *Dokl. Akad. Nauk SSSR*, **32**, pp.22-24.

OBUKHOV, A.M., 1962 *J. Fluid Mech.*, **13**, pp. 77-81.

PARISI, G. & FRISCH, U., 1985 In *Turbulence and predictability in geophysical fluid dynamics and climate dynamics*, pp. 71-78, Eds. Ghil, M., Benzi, R. & Parisi, G., Amsterdam: North Holland.

RICHARDSON, R., 1922 *Weather prediction by numerical process*, Cambridge Univ. Press.

YAGLOM, A.M., 1981 *Izv. Akad. Nauk SSSR, ser. ser. Atmosph. and Ocean*, **17**, pp. 1235-1257.

ON THE DETERMINATION OF UNIVERSAL MULTIFRACTAL PARAMETERS IN TURBULENCE

D. LAVALLÉE
Earth-Space Research Group, University of California
5276 Hollister ave., suite 260 , Santa Barbara, Ca. 93111, USA
S. LOVEJOY
Physics Department, McGill University
3600 University St., Montréal, Qué., H3A 2T8, CANADA
and
D. SCHERTZER, and F. SCHMITT
EERM/CRMD, Météorologie Nationale
2 Av. Rapp, 75007, Paris, FRANCE

ABSTRACT. The scaling behavior observed in turbulent flows has lately been the object of growing interest. Scaling exponents are fundamental since they describe the statistical properties over wide ranges of length scales. Multifractals are characterized by their scaling exponents and when generated by canonical cascade processes they generally will belong to specific universal classes. In this case the scaling exponents are specified by two parameters, the Lévy index α and the codimension of the mean singularity C_1. In particular, these parameters are believed to characterize the probability distribution of the singularities of the Navier-Stokes equations. The first data analysis technique specifically designed to directly estimate these parameters, the "Double Trace Moment" is described. The methods are then used to analyse the scaling behaviour of turbulent velocity data, providing estimates of their universal multifractal parameters: $\alpha \approx 1.3 \pm 0.1$ and $C_1 \approx 0.25 \pm 0.05$.

1. Introduction

In fully developed turbulence in three dimensions, scale invariance is a symmetry well known since at least Kolmogorov (1941). High variability or intermittency of velocity fields is characteristic of the signal measured in turbulent flows (Fig. 1). This behavior is observed over a wide range of length scales.[1] The question of how to characterize and relate these two fundamental properties has mainly been considered using the cascade phenomenology. The latter provides a mechanism for the transfer of energy

[1] In the atmosphere, the range of scales goes from $\approx 10,000$ km to ≈ 1 mm.

H. K. Moffatt et al. (eds.), Topological Aspects of the Dynamics of Fluids and Plasmas, 463–478.

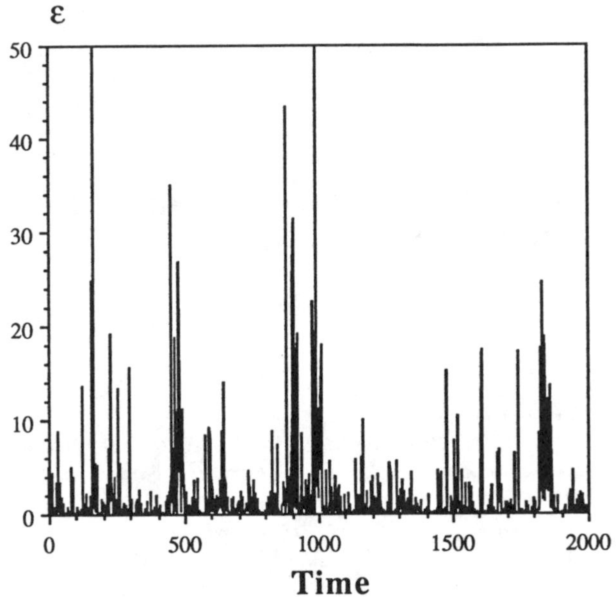

Fig. 1. Illustration of the measured signal of the velocity in turbulent flow.

flux from large to small structures. It also appears to be the generic process of scale invariant fields, characterized by a scaling exponent function: its (multi) fractal dimensions (Grassberger (1983), Hentschel & Proccacia (1983), Schertzer & Lovejoy (1983, 1985a, b), Benzi *et al.* (1984) and Parisi & Frisch (1985)). Thus, initial attempts to characterize turbulence by one parameter—a unique fractal dimension (Mandelbrot (1974) and Frisch *et al.* (1978)) have been followed by several empirical studies[2] of the infinite hierarchy of dimensions (Schertzer & Lovejoy[3] (1985b, 1987a), Meneveau & Sreenivasan (1987a, b, 1989) and Chhabra & Jensen (1989)).

However (canonical) multiplicative cascade processes for which stable attractive multifractal generators exist have the functional behavior of their scaling exponents completely governed by two universal parameters, α and C_1 (Schertzer & Lovejoy (1987a, b, 1989, 1991), Schertzer *et al.* (1988, 1991), Brax & Peshanski (1991) and Kida (1991)). The Lévy index α characterizes the degree of multifractality ($\alpha = 0$ corresponds to monofractal and $\alpha = 2$—the maximum—to "lognormal" multifractals[4]) and C_1 is the codimension which characterizes the sparseness/inhomogeneity of the mean of the process

[2] Using the strange attractor formalism of multifractal.

[3] The multiple scaling and multiple fractal dimensions discussed in these papers was for the highly turbulent (and well measured) radar reflectivity field of rain in 1, 2, 3, 4 and 1.5 dimensions (a fractal network).

[4] As pointed out in Schertzer & Lovejoy (1987a, b), a more correct expression is multifractal with a normal generator.

($C_1 = 0$) is space filling). These two parameters along with the Kolmogorov scaling exponents H give a complete description of the scaling properties of turbulence. Two independent analyses of turbulent flows, Kida (1991) and Schertzer *et al.* (1991), Schmitt *et al.* (1991a, b) have provided the first estimates for the universal velocity parameters[5] (with respectively $\alpha = 1.65$ and 1.3). After summarizing the basic properties of multiplicative cascade processes (section 2), we describe the Double Trace Moment (DTM) technique for estimating the fundamental parameters α and C_1 in section 3 as well as its application to turbulence data (section 4).

2. Cascade Processes and the Turbulent Multifractal Formalism

Since the work of Richardson and Kolmogorov, the cascade phenomenology[6] of turbulence has become a basis for the investigation and simulation of intermittency and scale invariance. Cascade processes were invoked to explain how the energy is transferred from large to small structures in turbulence. An illustration of a (continuous) multiplicative cascade processes is given in Fig. 2. In such processes a multiplicative increment modulates the transfer of energy flux from parent eddy to each subeddy. After several iterations the energy flux is concentrated over smaller and smaller volumes, a typical manifestation of intermittency. The multiple scaling properties of the result (best described as mathematical measures[7]) is given by the following relations;

$$\varepsilon_\lambda \approx \lambda^\gamma, \tag{2.1}$$

$$Pr(\varepsilon'_\lambda \geq \lambda^\gamma) \approx \lambda^{-c(\gamma)}, \tag{2.2}$$

where ε_λ is a field intensity threshold at resolution λ, ε'_λ a random field intensity and λ the ratio of the largest scale of interest L (*e.g.*, the integration scale) to the smallest scale of homogeneity ℓ (*e.g.*, the dissipation scale): $\lambda = L/\ell$. When $\lambda \to \infty$ (or $\ell \to 0$), $\gamma > 0$ is the order of the singularity and $\gamma < 0$ is the "order of the regularity". The probability distribution $Pr(\varepsilon'_\lambda \geq \lambda^\gamma)$ of the field intensities is also multiple scaling. Futhermore, examining low dimensional cut of the probability space (dimension D), $c(\gamma)$ has the geometric interpretation of a codimension as long as $c(\gamma) \leq D$. The corresponding fractal dimensions are obtained by subtracting: $D - c(\gamma)$.

[5] Gabriel *et al.* (1988) test the universality of the turbulent cloud radiance fields and (Lovejoy & Schertzer (1990)) estimate the parameters at infrared and visible wavelengths.

[6] This idea has initiated various cascade models of turbulence: the "pulse in pulse" in Novikov & Stewart (1964); the log-normal model in Yaglom (1966); the "β-model" in Mandelbrot (1974) and Frisch *et al.* (1978); the "α-model" in Schertzer & Lovejoy (1983); the "p-model" in Meneveau & Sreenivasan (1987a,b); the continuous cascade process in Schertzer & Lovejoy (1978a,b) (see also Fig. 2).

[7] A classification of multifractals according to their extreme singularities is discussed in Schertzer & Lovejoy (1991) and Schertzer *et al.* (1991).

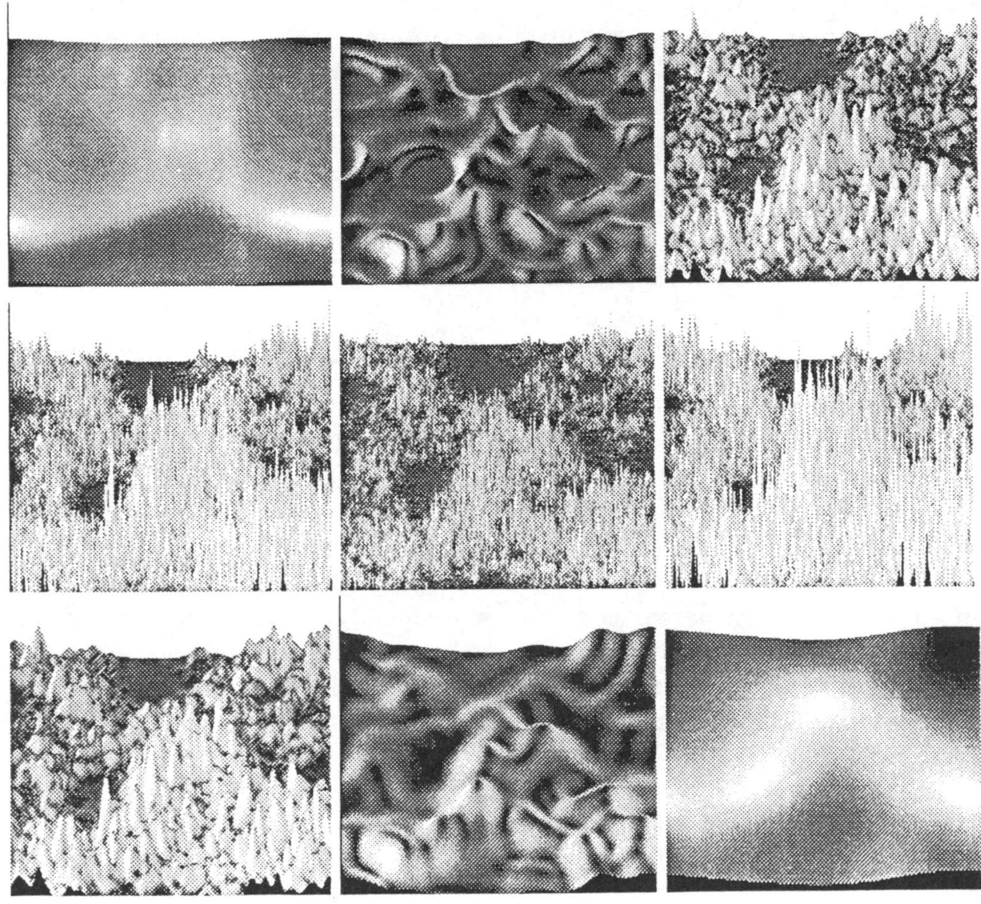

Fig. 2. The three top images and the two images on the second row correspond to five stages in the construction of a continuous multiplicative process. The resolutions are, respectively, one half of the image size, $1/8^{th}$, $1/32^{th}$, $1/128^{th}$ and, finally, $1/256^{th}$ for the center image. These correspond to the "bare" quantities. (Notice that the vertical axis for the central image has been compressed for visualization purposes.) The third image on the second row and the three images on the last row, correspond to the "dressed" quantities, obtained by averaging. Their resolutions are, respectively, $1/128^{th}$ of the image size, $1/32^{th}$, $1/8^{th}$ and, finally, one half for the last image (bottom right-hand corner). The parameters used for these images, $H = 1/2$, $C_1 = 0.25$ and $\alpha = 1.3$, are those estimated for the turbulent velocity field (Schertzer *et al.* (1991), Schmitt *et al.* (1991a, b), see also below).

The function $c(\gamma)$ is a codimension since the probability measures the fraction of the (infinite dimensional) probability space occupied by the singularities exceeding the order γ. Unlike the singularity spectrum function, $f(\alpha) = D - c(\gamma)$ with $\alpha = D - \gamma$, (introduced in Halsey *et al.* (1986) for studying the multifractal probability measures associated with strange attractors), the codimension function is intrinsic to the process: it is independent of the observing space dimension D. The strange attractor notation can not be used in turbulence since the relevant cascades are stochastic processes, *i.e.*, $D \to \infty$, the corresponding multifractal measures (ε) are not probability measures. Equalities in (2.1) and (2.2) are valid up to slowly varying functions of λ (*e.g.*, log's) and normalization constants.

Using the definition of statistical moments, the corresponding scaling behavior of the qth order statistical moments is characterized by a scaling exponent $K(q)$:

$$\langle \varepsilon_\lambda^q \rangle = \lambda^{K(q)} \propto \int \lambda^{q\gamma} \lambda^{-c(\gamma)} d\gamma. \tag{2.3}$$

The brackets $\langle \cdots \rangle$ in (2.3) indicates the statistical ensemble average, and the integral is equivalent to the Laplace transform of the probability distribution. In the limit $\lambda \to \infty$, the scaling exponents[8] $K(q)$ and $c(\gamma)$ are related by a Legendre transformation (Parisi & Frisch (1985) and Halsey *et al.* (1986)). The monotonically increasing function $C(q) \equiv K(q)/(q-1)$ is the (dual) codimension function associated with the statistical moments of the field intensities. The codimensions $c(\gamma)$ and $C(q)$, are not only scale invariant but also independent of the dimension D of the space in which the field is embedded. They are the fundamental functions characterizing the scaling properties of the processes under study (Schertzer *et al.* (1991). When the mechanism of redistribution of the energy flux in the cascade process is not homogeneous, the process leads to multifractal measures described by the equation given above. Without any *a priori*, the functions $c(\gamma)$ or $C(q)$ could be any (respectively convex or increasing) functions. However, when the conservation principle that governs the redistribution of the energy flux is canonical[9] and either the turbulence is "mixed" or the cascade is "densified" by the excitation of a continuum of intermediate scales, the process

[8] The scaling exponent $\tau(q) = (q-1)D - K(q)$ is also used in the literature (Halsey *et al.* (1986)).

[9] As in statistical physics, the canonical conservation indicates that the statistical ensemble averages of the processes $\langle \varepsilon_\lambda \rangle$ is constant (taken for convinience as 1). On the other hand, microcanonical cascade processes correspond to an exact conservation of the energy flux from the eddy to its sub-eddies (see Mandelbrot (1974) and Meneveau & Sreenivasan (1987)). Despite their apparent simplicity the microcanonical cascade constraint is quite complex and restrictive since it applies at all scales and prohibits large fluctuations. It is still not known how to generally make continuous microcanonical cascades nor are there believed to be universality classes.

leads to universal multifractals (Schertzer & Lovejoy (1987a, b), Schertzer *et al.* (1991)). In this case the scaling exponents have their functional behavior specified by two parameters α and C_1:

$$c(\gamma) = \begin{cases} C_1 \left(\frac{\gamma}{C_1 \alpha'} + \frac{1}{\alpha}\right)^{\alpha'}, \text{ with } \frac{1}{\alpha} + \frac{1}{\alpha'} = 1, (\alpha \neq 1, 0 \leq \alpha \leq 2) \\ C_1 \exp\left(\frac{\gamma}{C_1} - 1\right), \text{ if } \alpha = 1; \end{cases} \tag{2.4}$$

$$K(q) = \begin{cases} \frac{C_1}{\alpha - 1}(q^\alpha - q), \text{ if } \alpha \neq 1; \\ C_1 q \ln(q), \quad \text{ for } \alpha = 1. \end{cases} \quad \text{with } q \geq 0 \text{ for } \alpha < 2 \tag{2.5}$$

The derivation of these equations relies on the application of the (generalized) central limit theorem for the addition of random variables to the generator of the cascade (Schertzer &Lovejoy (1987a, b, 1989)). The most basic parameter, α, takes values between 0 and 2 and indicates the class to which the probability distribution belongs. When $\alpha = 2$, the process has a Gaussian generator whereas $0 < \alpha < 2$ corresponds to the less well known infinite variance cases of the central limit theorem: the (stable) "Lévy" distributions with index α. Decreasing the value of α allows us to identify three qualitatively different cases; first a regime with unbounded singularities ($1 < \alpha < 2$), followed by the (asymmetric) Cauchy generator multifractals ($\alpha = 1$) and finally the third situation with $0 < \alpha < 1$ corresponding to Lévy generators[10] with bounded singularities.[11] The case $\alpha = 0$ coincides with the popular monofractal "β-model" (Mandelbrot (1974) and Frisch *et al.* (1978)). The second parameter, C_1, is the fractal codimension of the singularities contributing to the average values of the field. It is related to the sparsity of the average level of intensity.

3. The Signature of Universality

Scaling exponents are usually determined by examining the statistical properties of the fields of interest at different length scales. This can be done either by directly estimating the probability distribution of the field intensities or by looking at the behavior of their statistical moments. However, two important considerations must be taken into account to perform the analysis on simulated or empirical data.

First, the singular behavior of ε_λ as the cascade proceeds to its small scale limit ($\lambda \to \infty$) requires the introduction of the distinction between "bare"

[10] The expressions "log Lévy" (Brax & Peshanski (1991)) or "log stable" (Kida [1991]) are inexact due to the problem of "dressing" described below (Schertzer & Lovejoy (1987a)).

[11] The upper bound is given by $C_1/(1 - \alpha)$ when $\alpha < 1$.

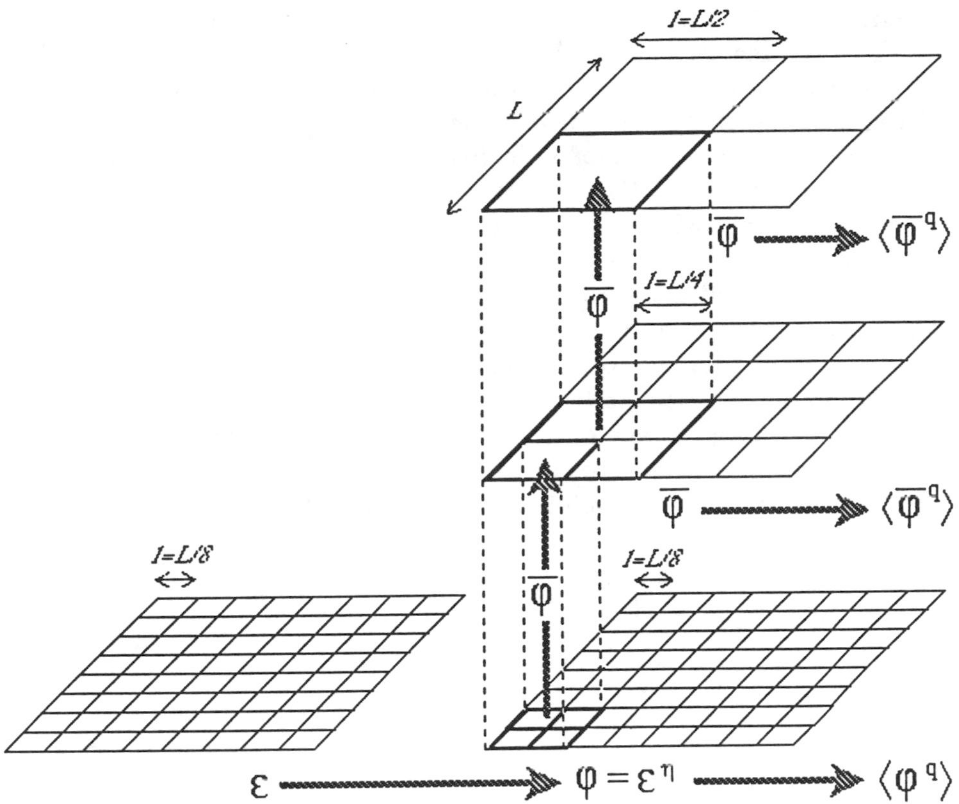

Fig. 3. A schematic diagram illustrating the different averaging scales in the DTM.

and "dressed" quantities (Schertzer & Lovejoy (1987a, b). The "bare" quantities are essentially theoretical and are obtained after a cascade process has proceeded over a finite range of scales λ (see Fig. 2). The results (2.4) and (2.5) for $c(\gamma)$, $K(q)$ hold for the "bare" quantities and in the limit $\lambda \to \infty$. The "dressed" quantities are those obtained after integrating a completed cascade[12] ($\lambda \to \infty$) over a finite scale λ'. Since they are strongly dependent on the singular behavior of the smallest length scale from which the cascade was developed, they are more variable than their "bare" counterparts at the same scale ratio. This dependence on the small scale limit is also responsible for the divergence of the high order statistical moments of the "dressed" quantities (leading to the "pseudo scaling" discussed in Schertzer & Lovejoy (1983), Lavallée (1991) and Schertzer *et al.* (1991b) now understood as a multifractal phase transition). The critical values of moments and

[12] More precisely, quantities are said to be finitely "dressed" when λ is kept finite and infinitely "dressed" when $\lambda \to \infty$ (Lavallée (1991)).

singularities contributing to the divergence are q_D, γ_D where $C(q_D) = D$, $\gamma_D = K'(q_D)$. The experimental quantities are best approximated by the "dressed quantities", since they are usually obtained with measuring devices by integrating inhomogeneous fields over scales much larger than the inner scale of the processes observed. Finally, a "dressing" operation is needed to obtain (or transform) the field at various lower resolutions in order to analyze and characterize their scaling properties.

The exponents $c(\gamma)$—or $K(q)$—are then given by the scaling behavior of the probability distribution—or of the qth statistical moments—of "dressed" quantities as functions of the scale ratio λ' on log–log graphs, for various values of γ—or q. The scaling exponents determined by such analysis techniques correctly characterize the multifractal processes as long as γ—or q—did not exceed either γ_D, q_D or the critical values γ_s or q_s. Introducing the sampling dimension D_s, related to the number of samples N by $D_s \equiv \log_\lambda(N)$, then the critical value γ_s—solution of $c(\gamma_s) \approx D + D_s$—is the maximum order of singularities that can be observed in the sample size N. When $q > q_c = \min(q_s, q_D)$ the asymptotic behavior of $K(q)$ is then linear and given by (Lavallée *et al.* (1991a)):

$$K(q) \approx q\gamma_c - c(\gamma_c), \quad q > q_c. \tag{3.1}$$

The qth moment q_s corresponding, by the Legendre transform, to γ_s is given by the following relation:

$$q_s = \left.\frac{dc(\gamma)}{d\gamma}\right|_{\gamma=\gamma_s} = \left(\frac{D + D_s}{C_1}\right)^{1/\alpha}. \tag{3.2}$$

For small N, $q_s < q_D$, hence increasing the values of D or D_s provides a larger interval of values $c(\gamma)$ or $K(q)$ to estimate their universal parameters α and C_1 until the second critical values γ_D or q_D is reached (this cannot be avoided when the Lévy index α is greater than 1). In any case the determination of the universal parameters from $c(\gamma)$ or $K(q)$ will depend on nonlinear regression techniques and on the ability to apply them to the appropriate range of values of γ or q.

The DTM method allows a direct estimate of the values of α. The main idea behind it is to study how the scaling properties of fields are modified when a transformation or operation is performed on them, and in particular, to determine the functional behavior of the scaling exponents and their dependence on α and C_1. The operation considered here consists in taking the ηth power[13] of $\varepsilon_{\lambda'}$ at the largest scale ratio λ' (smallest size). Then a dressing operation is performed on $\varepsilon^\eta_{\lambda'}$ and various q^{th} order statistical moments are estimated at decreasing values of the scale ratio $\lambda \leq \lambda'$ (see

[13] This corresponds to the change $\gamma \rightarrow \eta\gamma$.

also Fig. 3). The q, η DTM at resolution λ' and λ has the following multiple scaling behavior :

$$Tr_\lambda \left[\varepsilon_{\lambda'}^{\eta}\right]^q = \left\langle \sum_i \left(\int_{B_{\lambda i}} d^D x \, \varepsilon_{\lambda'}^{\eta} \right)^q \right\rangle \approx \lambda^{K(q,\eta)-(q-1)D}, \tag{3.3}$$

where the sum is over all the events $\varepsilon_{\lambda'}^{\eta}$ in the disjoint boxes $B_{\lambda i}$ at scale λ (of volume λ^{-D}) and corresponds to the "dressing" operation. This is equivalent to averaging the fields over the boxes B_λ. The sum is over all the disjoint boxes B_λ needed to completely cover the field. For $\eta = 1$ the right hand side of (3.3) reduces to the usual trace moment (the ensemble average of the usual partition function). The scaling exponent $K(q,\eta)$ is related to the usual scaling exponent $K(q,1) \equiv K(q)$ by the following relation (Lavallée (1991) and Lavallée *et al.* (1991b, c)):

$$K(q,\eta) = K(q\eta) - qK(\eta). \tag{3.4}$$

The first term on the right side of (3.4) is the usual scaling exponent of $\langle \varepsilon_{\lambda}^{\eta q} \rangle$ (see (2.4)), corresponding to the scaling properties of the "bare" quantities at the same scale length λ. The second term assures the preservation of the two scaling symmetries characterizing the operation performed to "dress" the fields at different scale ratios λ'. First, the "dressing" operator conserves the first moment[14] , $q = 1$, of the $\varepsilon_{\lambda}^{\eta}$ which is constant (scale independent), this implies that $K(1,\eta) = 0$. Second, for $q = 0$, the statistical moments must be independent of the multifractal processes[15] at any scale ratio λ, hence $K(0,\eta) = 0$. Use of the universality classes in (2.5) gives the expression for $K(q,\eta)$:

$$K(q,\eta) = \eta^\alpha K(q,1) = \begin{cases} \frac{C_1}{\alpha-1}\eta^\alpha(q^\alpha - q), & \alpha \neq 1; \\ \\ C_1 \, \eta q \log(q), & \alpha = 1, \end{cases} \tag{3.5}$$

with $0 \leq \alpha \leq 2$ and $q > 0$ for $\alpha \neq 2$.

The main feature of this relation is that $K(q,\eta)$ factorizes into the product of two functions, one for each of the independent variables η and q. The η-dependent part has a simple dependence on α. Thus, by keeping q fixed (but different from the special values 0 or 1) the slope of the curve of $|K(q,\eta)|$ as a function of η on a log-log graph gives the value of the parameter α. Universality assures us that for different values of q these curves

[14] The sum of averaged $\varepsilon_{\lambda}^{\eta}$ (over boxes of different sizes) is constant. For $q = 1$ the sum and the integral commutes in (3.3).

[15] In this case the dependence in λ is only through the (multiple) scaling properties of the support (dimension D) of the processes.

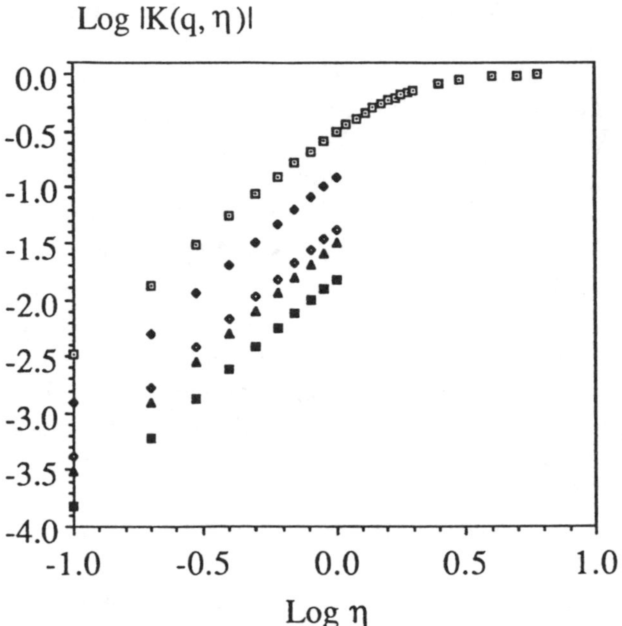

Fig. 4. From top to bottom, the curves of the log $|K(q,\eta)|$ as functions of the log η, with $\alpha = 2$ (lognormal) and $C_1 = 0.18$, are given for the following values of q: 2., 1.5, 0.5, 0.25 and 0.9. All the curves are parallel as predicted by (3.5). When their slopes are used to estimate α, with η taking values between 0.1 and 1, we obtained respectively: 1.97, 1.99, 2.00, 1.99 and 2.00. The values of C_1, obtained by solving the expression for the log $\eta = 0$ intercept given by the log $|K(q,1)|$ and using (3.5), are respectively: 0.16, 0.17, 0.17, 0.17 and 0.17. For $q = 2$ and η large enough the curve $K(2,\eta)$ becomes constant. Other numerical results are discussed in Lavallée (1991) and Lavallée *et al.* (1991b).

will be parallel and will have the same slopes. Thus, this typical behavior of the $K(q,\eta)$ is the signature of universality as illustrated in Fig. 4. On the log-log plot the values of C_1 can be estimated by solving the expression for the intercept, which is a function of C_1 and α, or alternatively by solving the expression of the slope of $K(q,\eta)$ as a function of η^α on a log–log graph. It is important to notice that in both cases the accuracy of the estimates of C_1 will depend on the accuracy of the estimated α. Whenever $\max(q, q\eta) > \min(q_s, q_D)$ expression (3.4) breaks down and the scaling exponent $K(q,1)$ becomes a linear function of q and (3.4) indicates then that $K(q,\eta)$ becomes independent of η (see also Fig. 4).

The particular scaling behavior of the DTM provides two advantages when compared with the existing multifractal analysis methods. First, the estimated scaling exponent $K(q,\eta)$ is independent of the normalization[16] of

[16] This is also true of trace moments on single realizations which is then equivalent to a

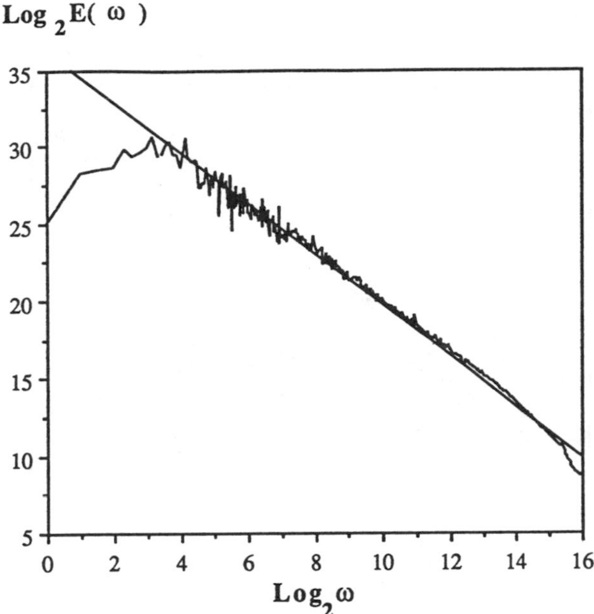

Fig. 5. The velocity spectrum averaged over 2^9 samples each of length 2^{13} times the finest resolution (which corresponds to 5kHz). The spectrum $E(\omega) \approx \omega^{-\beta}$ (ω is the frequency) with $\beta \approx 1.65$ is shown for comparison.

the data (when $\gamma \to \gamma + b$). The second is that when a multiplicative change of γ is made ($\gamma \to a\gamma$) the scaling exponent is transformed as $K(q,\eta) \to K(q,a\eta) = a^\alpha K(q,\eta)$ (where a corresponds to a contraction in the γ space, but is also equivalent to the raising of the fields $\varepsilon_{\lambda'}$ to an unknown power a by the experimental apparatus). This implies that the determination of α will also be independent of the power a to which the process is raised.[17] In other words the universality has been exploited to give a method to determine α which is invariant under the affine transformation $\gamma \to a\gamma + b$. Since (3.5) is invariant under the transformation $\eta \to a\,\eta$ and $C_1 \to a^{-\alpha} C_1$, if a is unknown, the parameter C_1 can only be determined up to a multiplicative constant a^α.

4. Universality in Turbulent Flows

The velocity measurements were performed by Gagne at the ONERA wind tunnel, and by Modane with a high resolution hot wire anemometer, sampling at 5kHz. The number of samples analyzed is $N = 2^9$, each of scale

partition function.

[17] The notation a is reserved to the factors of the γ generated in the cascade process or resulting from the integration over the ath power of the fields

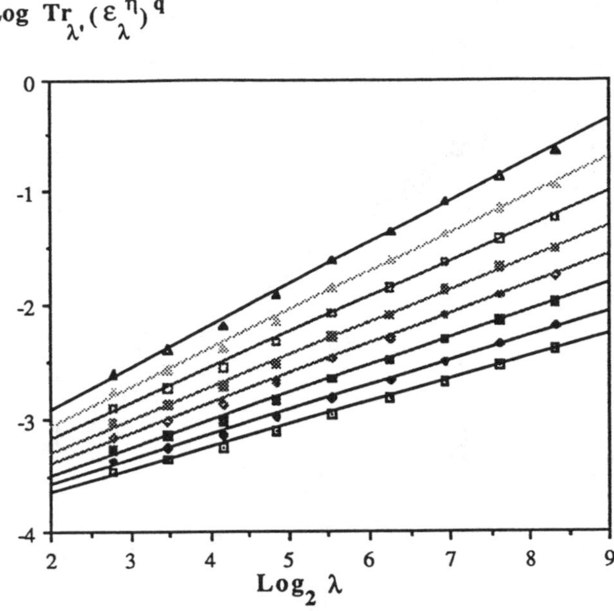

Fig. 6. Log of the DTM as a function of λ for various values of q, η, showing that the scaling is well respected. The extreme four octaves which are part of the dissipation range were not used.

ratio $\approx 2^9$ in the inertial (scaling) regime and $\approx 2^4$ in the dissipation region (hence $D_s = 1$). The time series had empirical energy spectra of the form $E(\omega) \approx \omega^{-\beta}$, with $\beta \approx 1.65$, ω is the frequency as displayed in Fig. 5. It was then power law filtered by $\omega^{1/3}$. This removes the $\lambda^{-1/3}$ Kolmogorov scaling yielding the roughly stationary quantity $\varepsilon_\lambda^{1/3}$ related to the velocity by:

$$\Delta v_\lambda \approx \varepsilon_\lambda^{1/3} \lambda^{-1/3}. \tag{4.1}$$

The amount of filtering required to yield an exactly stationary process is not exactly $\omega^{1/3}$. It also depends on the exponent involved in the statistical space/time transformation required to transform from temporal to spatial statistics (in the atmosphere the usual Taylor's hypothesis needs an anisotropic generalization), but also on the (initially unknown) values of C_1, α, as well as on possible deviations from the theoretical behavior. Fortunately, the DTM technique is not adversely affected by the precision of the filtering (Lavallée (1991)).

The DTM technique is applied to the series—transformed as indicated above—for various values of the parameters q, η. The results are shown in Fig. 6. As long as η and $q\eta$ are below $q_s \approx 5$, the plots of $\log|K(q,\eta)|$ vs. $\log \eta$ are very straight, as expected for universal multifractals, with slopes

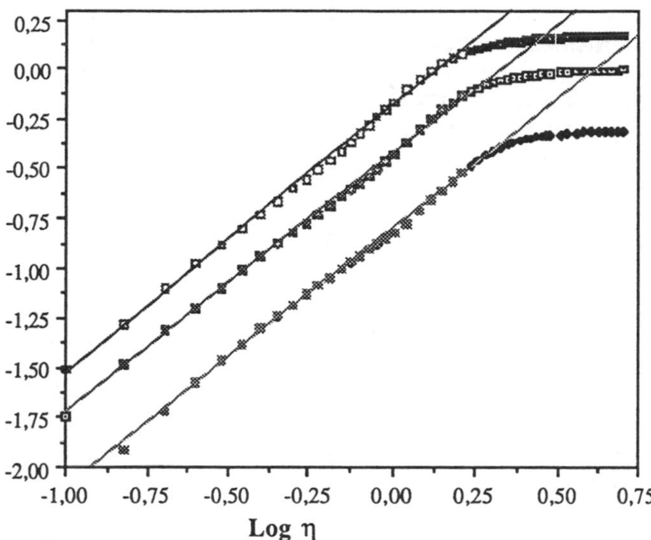

Fig. 7. Curves of $\log|K(q,\eta)|$ vs $\log\eta$, for $q = 2.5, 2.0, 1.5$ (top to bottom). The curves are nearly parallel and for $\eta q < q_s \approx 5$ have slopes $\alpha \approx 1.3\pm0.1$. Furthermore, the values of $\log|K(q,1)|$ give $C_1 \approx 0.25 \pm 0.05$.

$\alpha \approx 1.3 \pm 0.1$, and intercepts yielding $C_1 \approx 0.25 \pm 0.05$. For comparison with other empirical results, the standard intermittency parameter[18] μ is calculated. For the lognormal model ($\alpha = 2$), $\mu = 2C_1$, whereas for the β-model ($\alpha = 0$), $\mu = C_1$. Here, with $\alpha = 1.3$, we obtain $\mu \approx 1.55C_1 = 0.35 \pm 0.1$ which is exactly in the middle of the conventionally accepted range 0.2 (Meneveau & Sreenivasan (1987b)) to 0.5 (Anselmet *et al.* (1984)). Using (3.2) we obtained the critical values $q_s \approx 5.0$ which is in agreement with the breakdown of (3.5) for large q found in Fig. 7.

A summary of the estimated α and C_1, when using the DTM to study and characterize the scaling properties of a wide variety of experimental data, are given in table I.

5. Conclusion

The results discussed above confirm the importance of scale invariance in characterizing the extremely variable nature of turbulence. Because $\alpha \geq 1$ turbulence is an unconditionally "hard" multifractal process: *i.e.*, high enough order statistical moments will diverge when energy fluxes are averaged over

[18] This is the autocorrelation exponent for ε_λ, $\mu = K(2,1)$.

The empirical data		α	C_1	References
Temperature in turbulent flows		1.2	0.35	Schertzer *et al.* (1991)
Wind velocity in turbulent flows		1.3	0.25	Schmitt*et al.* (1991a)
Rain§	vertical direction	1.35	0.15	Tessier *et al.* (1991)
	time	0.5	0.6	Tessier *et al.* (1991)
Rain§§	horizontal direction	1.35	0.15	Tessier *et al.* (1991)
	time	0.5	0.6	Tessier *et al.* (1991)
Hadron jets		0.5–0.7	0.5–0.25	Ratti *et al.* (1991)
Sea surface infrared reflectivities		1.75	0.15	Lavallée *et al.* (1991d)
Landscape topography		1.8	0.05	Lavallée *et al.* (1991c)
Cloud radiances (visible wavelength)		1.35	0.10	Tessier *et al.*(1991)
Cloud radiances (IR wavelength)		1.35	0.15	Tessier *et al.* (1991)

§Radar reflectivities over Montréal
§§Daily rainfall accumulation on the entire earth for 1983

TABLE I
This table summarizes the values of α and C_1 obtained from various experimental data using the DTM.

spaces with arbitrary dimensions (as long as the average is over scales much larger than the dissipation scale). This is in qualitative accord with empirical evidence from atmospheric data[19] (Schertzer & Lovejoy (1985b)). The "hard" nature of the turbulence is also confirmed by an independent estimate $\alpha \approx 1.65$ by Kida (1991).[20] The result given in section 4 holds for length scales much smaller than the resolution of the data set (*i.e.*, as long as the scaling symmetry is preserved) and hence provide a complete statistical description of the hierarchy of singularities of the Navier-Stoke equations.

[19] The atmospheric data suggested $q_D \approx 5$, not q_s. It is possible that the discrepancy is due to different C_1 values in the atmosphere and wind tunnels. Experiments are currently underway to check this possibility.

[20] Neither the problems of the sampling dimension nor the problems of the "dressed" divergence of moments were considered in this study. Since the estimate was based on moments up to order 18, the exact numerical value is questionable.

Acknowledgements

We thank A. Davis, C. Gautier, C. Hooge, D. Jordan, P. Ladoy, P. Peterson, K. Pflug, G. Sarma, Y. Tessier, R. Viswanathan, B. Watson and J. Wilson for helpful comments and discussions. F. Bégin is thanked for the simulations presented in Fig. 2. We are grateful to Y. Gagne and E. Hopfinger, U. Frisch and the DRET for providing the velocity wind data, and the Atmospheric Radiation Measurement program contract #DE-FG03-90ER61062 for partial financial support.

References

ANSELMET, F., GAGNE, Y., HOPFINGER, E.J.& ANTONIA, R. A., 1984 High order velocity structure functions in turbulent shear flows. *J. Fluid Mech.*, **140**, pp. 6375.

BENZI, R., PALADIN, G. PARISI, G. & VULPIANI, A., 1984 On the multifractal nature of fully developed turbulence and chaotic systems. *J. Phys. A*, **17**, pp. 3521–3531.

BRAX, P. & PESHANSKI, R., 1991 Lévy stable law description of intermittent behaviour and quark gluon plasma phase transitions. *Phys. Lett. B*, **253**, pp. 225–230.

CHHABRA, A. & JENSEN, V., 1989 Direct determination of the $f(\alpha)$ singularity spectrum. *Phys. Rev. Lett.*, **62**, pp. 1327–1330.

FRISCH, U., SULEM, P. L. & NELKIN, M., 1978 A simple dynamical model of intermittency in fully developed turbulence. *J. Fluid Mech.*, **87**, pp. 719–724.

GABRIEL, P., LOVEJOY, S., SCHERTZER, D., AUSTIN, G.L., 1988 Multifractal analysis of satellite resolution dependence.

GRASSBERGER, P., 1983 Generalized dimensions of strange attractors. *Phys. Lett. A*, **97**, pp. 227–230.

HALSEY, T.C., JENSEN, M.H., KADANOFF, L.P., PROCACCIA, I. & SHRAIMAN, B., 1986 Fractal measures and their singularities: the characterization of strange sets. *Phys. Rev. A.*, **33**, pp. 1141–1151.

HENTSCHEL, H.G.E. & PROCCACIA, I., 1983 The infinite number of generalized dimensions of fractals and strange attractors. *Physica*, **8D**, pp. 435–444.

KIDA, S., 1991 Logstable distribution and intermittency of turbulence. *J. Phys. Soc. Jpn.*, **60**, pp. 58.

KOLMOGOROV, A.N., 1941 Local structure of turbulence in an incompressible liquid for very large Reynolds numbers. *Dokl. Acad. Sci. USSR*, **30**, pp. 299–303

LAVALLÉE, D., 1991 Multifractal techniques: Analysis and simulations of turbulents fields. Ph. D. thesis, University McGill, Montréal, Canada.

LAVALLÉE, D., SCHERTZER, D. & LOVEJOY, S., 1991a On the determination of the codimension function. *Scaling, Fractals and Non-Linear Variability in Geophysics*, edited by D. Schertzer and S. Lovejoy, Kluwer, Holland.

LAVALLÉE, D., SCHERTZER, D. & LOVEJOY, S., 1991b Turbulence and Universal Multifractals II: A new method for determining multifractal indices: Double trace moments. Submitted to *C. R. Acad. des Sci. Paris*, 11/90.

LAVALLÉE, D., LOVEJOY, S., SCHERTZER, D. & LADOY, P., 1991c Nonlinear variability, multifractal analysis and simulation of landscape topography. *Fractals in Geography*, L. De Cola, N. Lam eds., Kluwer, Dordrecht-Boston (in press).

LAVALLÉE, D., LOVEJOY, S. & SCHERTZER, D., 1991d Nonlinear variability, multifractal analysis and simulation of landscape topography. *S.P.I.E. Proceedings*, **1558**, pp. 68–75.

LOVEJOY, S., SCHERTZER, D., 1990 Multifractals, universilaty classes and radar measurements of clouds and rain. *J. Geophys. Res.*, **95**, 201–2034.

MANDELBROT, B., 1974 Intermittent turbulence in self-similar cascades: Divergence of high moments and dimension of the carrier. *J. Fluid Mech.*, **62**, pp. 331–350.

MENEVEAU, C. & SREENIVASAN, K. R., 1987a Simple multifractal cascade model for fully developed turbulence. *Phys. Rev. Lett.*, **59**, pp. 1424–1427.

MENEVEAU, C. & SREENIVASAN, K. R., 1987b The multifractal spectrum of the dissipation field in turbulent flows. *Nucl. Phys. B*, Proc. Suppl., **2**, pp. 49–76.

MENEVEAU, C. & SREENIVASAN, K. R., 1989 Measurement of $f(\alpha)$ from scaling of histograms, and applications to dynamical systems and fully developed turbulence. *Phys. Lett. A*, **137**, pp. 103–112.

PARISI, G. & FRISCH, U., 1985 A multifractal model of intermittency. *Turbulence and predictability in geophysical fluid dynamics and climate dynamics.* Eds. M. Ghil, R, Benzi, G. Parisi, North-Holland, pp. 84–88.

RATTI *et al.* 1991 Universal multifractal analysis of multiparticle production in hadron-hadron colisions at $\sqrt{s} = 16.7$ GeV. *Proceedings of Hadron Physics Conf.*, Wuhan, China, Sept. 1991.

SCHERTZER, D. & LOVEJOY, S., 1983 On the dimension of atmospheric motions. *Preprint Vol., IUTAM Symp. on Turbulence and Chaotic Phenomena in Fluids*, pp. 141–144.

SCHERTZER, D. & LOVEJOY, S., 1985a Generalized Scale Invariance in turbulent phenomena. *P. Chem. Hydrodynamics J.*, **6**, 623–635.

SCHERTZER, D. & LOVEJOY, S., 1985b The dimension and intermittency of atmospheric dynamics, *Turbulent Shear Flow 4*, B. Launder ed., Springer, pp. 7–33.

SCHERTZER, D. & LOVEJOY, S., 1987a Singularités anisotropes, et divergence de moments en cascades multiplicatives. *Annales Math. du Qué.*, **11**, pp. 139–181.

SCHERTZER, D. & LOVEJOY, S., 1987b Physically based rain and cloud modeling by anisotropic, multiplicative turbulent cascades. *J. Geophys. Res.*, **92**, pp. 9693–9714.

SCHERTZER, D. & LOVEJOY, S., 1989 Nonlinear variability in geophysics: multifractal analysis and simulations. *Fractals: Physical Origin and Consequences*, Ed. L. Pietronero, Plenum, New York, pp. 49–79.

SCHERTZER, D., LOVEJOY, S., VISVANATHAN, R., LAVALLÉE, D. & WILSON, J., 1988 Universal Multifractals in Turbulence. *Fractal Aspects of Materials: Disordered Systems*, Ed. D.A. Weitz, L.M. Sander, B.B. Mandelbrot, Materials Research Society, Pittsburg, pp. 267–269.

SCHERTZER, D., LOVEJOY, S., LAVALLE, D. & SCHMITT, F., 1991 Universal hard multifractal turbulence: theory and observations. *Nonlinear Dynamics of Structures*, R.Z. Sagdeev, U. Frisch, A.S. Moiseev, A. Erokhin Eds., North Holland., pp. 213–235

SCHERTZER, D., LAVALLÉE, D. & LOVEJOY, S., 1991b Hard multifractal phase transitions (to be published).

SCHMITT, F., LOVEJOY, S., SCHERTZER, D., LAVALLÉE, D. & HOOGE, C., 1991b Turbulence and Universal Multifractals III: First estimates of multifractal indices for velocity and temperature fields. Submitted to *C. R. Acad. des Sci. Paris*, 11/90.

SCHMITT, F., LAVALLÉE, D., SCHERTZER, D. & LOVEJOY, S., 1991b Empirical determination of universal multifractal exponents in turbulent velocity fields. In press, *Phys. Rev. Lett.*, 6/91.

TESSIER, Y., LOVEJOY, S., SCHERTZER, D. 1991 Universal multifractal in rain anf clouds: theory and observations. Submitted to *J. Appl. Meteor.*

PART VI

CHAOS, INSTABILITY, AND DYNAMO THEORY

ANOMALOUS TRANSPORT AND FRACTAL KINETICS

G.M. ZASLAVSKY*
Institute for Theoretical Physics
University of California
Santa Barbara, California 94106-4030, U.S.A.

ABSTRACT. An orbit of the passive particle transport is associated with a streamline of a flow, if the flow is inviscid, stationary and incompressible. In a general situation 3-D flows have a nontrivial topology of streamlines due to their chaotic behaviour. As a result of the chaos the real space has a complicated structure of islands and stochastic sea of fractal or multifractal nature. Transport of a passive particle can be described as a random walk in a multifractal medium. It is important that the random walk is also fractal in time . Altogether, this produces nontrivial exponents of the asymptotic law of the average particle displacement. A kinetic equation of particle transport is described in many details to draw the attention to new nontrivial aspects of the statistical irreversibility. The multifractal transport creates the possibility of anomalous diffusion due to the Lévy process of a particle random walk or due to some analog of the Lévy process. Coherent structures in the random walk of a passive particle and its intermittency are features of the fractal or multifractal space-time nature.

1. Introduction

Before the insight into phenomenon of chaos led to a new understanding of the particle dynamics, the problem of kinetic equations for a system was mainly technical. Various kinds of tools to find an appropriate non-stationary statistical description of motion are known, such as the Boltzman equation, master equation, Fokker-Planck-Kolmogorov (FPK) equation, *etc.* The main feature of these tools was introducing some randomness in addition to the equations of motion. Sometimes the randomness was looking like a formal equivalent suggestion (for example, some special limit properties in the Kolmogorov theory (Kolmogorov, 1938) of the Fokker-Planck equation). The fairly simple form of the FPK equation justified its great applicability to a wide class of physical problems (Chandrasekhar, 1943). The general form of the FPK equation, has been derived by Kolmogorov (1938):

* Permanent address: Courant Institute of Mathematical Sciences, NYU, 251 Mercer St., New York, NY 10012

H. K. Moffatt et al. (eds.), Topological Aspects of the Dynamics of Fluids and Plasmas, 481–491.

$$\frac{\partial P(x,t)}{\partial t} = -\frac{\partial}{\partial x}(\mathcal{A}P(x,t)) + \frac{1}{2}\frac{\partial^2}{\partial \chi^2}(\mathcal{B}P(x,t)) \tag{1.1}$$

where $P(x,t)$ is the probability density to find a particle in the point x at time t and $\mathcal{A}$ and $\mathcal{B}$ are some coefficients. Simplification of (1.1) can be found in Lifshitz and Pitaevsky (1981) where the divergent form of the FPK equation:

$$\frac{\partial P(x,t)}{\partial t} = \frac{1}{2}\frac{\partial}{\partial x}\mathcal{D}\frac{\partial P(x,t)}{\partial x} \tag{1.2}$$

was derived for the Hamiltonian systems under appropriate conditions. The diffusion coefficient $\mathcal{D}$ can be expressed in the form:

$$\mathcal{D} = \frac{\overline{(\Delta x)^2}}{\Delta t}, \tag{1.3}$$

where the bar means averaging over some "local" distributions of different possibilities (different "local" parameters) and the limit

$$\Delta x \to 0,\ \ \Delta t \to 0,\ \ \overline{(\Delta x)^2}/\Delta t = \text{const} \tag{1.4}$$

is taken. The value Δx is a small change of x during a small time interval Δt in accordance with the equations of motion.

If $\mathcal{D} = \text{const}$, for simplicity, then the asymptotic equation

$$\langle x^2 \rangle = \mathcal{D}t\ \ (t \to \infty) \tag{1.5}$$

comes from (1.2), where the angular brackets $\langle \ldots \rangle$ mean averaging over the distribution function $P(x,t)$. There is well update understanding of all expressions (1.1)–(1.5), including condition of their origin, from different points of view. Nevertheless the next several comments are worth making because of a new spirit of the random walk process that forms the main point of this article. If the motion of the variable x is governed by a smooth Hamiltonian, then its infinitesimal displacement Δt during the time interval Δt will be proportional to Δt. Nevertheless the random walk process is "organized" in such way that only the special limit (1.4) is finite. This limit results in a typical Gaussian form of asymptotic (1.5), which implies the existence of at least two time scales: τ_c — a correlation time, over which the averaging in (1.3) is done, and τ_0 such that the asymptotics (1.5) is valid if

$$t \gg \tau_0 \gg \tau_c \tag{1.6}$$

Actually the inequalities (1.6) are not as trivial as it can be imagined at first glance, this will be discussed later. Let us mention here only the fact that (1.6) means the equation (1.5) can be considered as a symbolic mapping of the microscopic properties of the motion (*i.e.*, (1.3), (1.4)) onto its macroscopic behaviour (*i.e.*, the diffusion law (1.5)).

Nontrivial sense of the last statement is evident when one considers a system with the chaotic dynamics. There is no possibility of using Gaussian random process *a priori.* On the contrary local properties of the random walk should be obtained from the first principles (instead of (1.4)) and a macroscopic law should be derived from local properties. We will write a macroscopic behaviour in the form

$$\langle |x| \rangle \sim \text{const.}\, t^{\mu} (t \to \infty) \tag{1.7}$$

where $\mu = 1/2$ corresponds to the normal diffusion (1.5), and $\mu \neq 1/2$ to anomalous diffusion. For the free motion without acceleration $x \sim$ const. t and $\mu = 1$, whereas the local acceleration process yields $\mu > 1$. It seems that the Richardson law of turbulent diffusion $\mu = 3/2$ just means the existence of local acceleration. The phenomenon of dynamical chaos reveals a fractal, or even multifractal local properties of phase space and time.

There is a good understanding that the anomalous properties of particle transport are connected with fractal behaviour of a leading random walk or turbulent process (Mandelbrot 1982, Montroll and Shlesinger 1984). The crucial part of this understanding is based on the so-called Lévy's flights process (Lévy, 1937) and its connection to the fractal properties of the space-time behaviour of a particle (Montroll and Shlesinger 1984, Shlesinger 1989). The case in which chaos is not too strong and "channels" of chaos form a connected web in the phase space of the system (Zaslavsky *et al.* 1986) may be a good example of the anomalous transport (Afanas'ev *et al.* 1991). Lévy-like processes were observed in a numerical simulation and a theoretical study of possible values of the exponent μ in (1.7) were considered. It was also shown in the number of publications (Chernikov *et al.* 1990, Petrovichev *et al.* 1990 and references therein) that the same phenomenon of anomalous ($\mu \neq 1/2$) diffusion along the stochastic web is possible in the 3-D stationary fluid flows. A more general discussion of the problem was given by Zaslavsky (1991). The Montroll-Weiss kinetic equation (Montroll and Weiss, 1965) was taken as a starting point. Nevertheless there was some dissatisfaction because of impossibility to use such a well known way to derive the kinetic equation, like the FPK-equation derivation is.

The aim of this paper is to adjust the Kolmogorov scheme for the kinetic equation to the new circumstances of chaotic motion in a fractal space-time situation. It will be shown how a new form of the micro (local) macroscopic mapping will appear to give the main asymptotic property of the random walk which yields the value of μ in (1.8) as a function of the fractal dimension of a space-time random walk.

2. Fractional Integration And Differentiation

This chapter is a brief mathematical introduction to the following derivation of the kinetic equation for a fractal random walk. The main point concerns the fractional integration and differentiation. In the sense of Gelfand and Shilov (1958) the main idea in introducing an arbitrary order integration is to use the Cauchy formula:

$$g_n(x) = \int_{-\infty}^{x} \int_{-\infty}^{x_{n-1}} \dots \int_{-\infty}^{x_1} g(\xi) d\xi dx_1 \dots dx_{n-1} =$$

$$= \frac{1}{(n-1)!} \int_{-\infty}^{x} g(\xi)(x-\xi)^{n-1} d\xi \tag{2.1}$$

which can be rewritten in the form of convolution

$$g_n(x) = g(x) * \frac{x^{n-1}}{\Gamma(n)}. \tag{2.2}$$

The definition of convolution is clear from (2.1). Functions should be correspondently integrable. The idea of the fractional integration is in generalizing (2.2) to the arbitrary order α of integration

$$g_\alpha = g(x) * \frac{x^\alpha}{\Gamma(\alpha)}, x \geq 0 \tag{2.3}$$

where x^α should be considered as a generalized function on the positive semiaxis. Fractal differentiation can be introduced as a reverse operation to (2.3), *i.e.*,

$$g^{(\alpha)} = \frac{d^\alpha g}{dx^\alpha} \equiv g_{-\alpha} = g(x) * \frac{x^{-\alpha}}{\Gamma(-\alpha)} \tag{2.4}$$

Let us mention two important formulae which are proved by Gelfand and Shilov (1958). The first one is

$$\frac{d^\alpha}{dx^\alpha}\left(\frac{d^\beta g}{dx^\beta}\right) = \frac{d^{\alpha+\beta}}{dx^{\alpha+\beta}} g \tag{2.5}$$

and the second one concerns the differentiation of δ–functions

$$\frac{d^\alpha}{dx^\alpha}\left(\delta^{(k)}(x)\right) = \frac{x^{-k-\alpha-1}}{\Gamma(-k-\alpha)}, x \geq 0 \tag{2.6}$$

where k is an integer.

We shall need one more property of the fractional differentiation. Consider an integral

$$\left(g^{(\alpha)}(x)\cdot f(x)\right)\equiv\int_{-\infty}^{\infty}dx g^{(\alpha)}(x)f(x).$$

and support that this integral exists. Definition (2.3) yields

$$\left(g^{(\alpha)}(x)f(x)\right)=\int_{-\infty}^{\infty}dx\int_{-\infty}^{x}d\xi g(\xi)\frac{(x-\xi)^{\alpha-1}}{\Gamma(\alpha)}f(x)$$

Changing the order of integrations gives

$$\left(g^{(\alpha)}(x)\cdot f(x)\right)=\int_{-\infty}^{\infty}d\xi\int_{\xi}^{\infty}dx g(\xi)\frac{(x-\xi)^{\alpha-1}}{\Gamma(\alpha)}f(x)=$$

$$=\int_{-\infty}^{\infty}d\xi g(\xi)\int_{-\infty}^{-\xi}dx'\frac{(-x'-\xi)^{\alpha-1}}{\Gamma(\alpha)}f(-x')\ =\int_{-\infty}^{\infty}d\xi g(\xi)f^{(\alpha)}(-\xi).$$

This means that

$$\left(g^{(\alpha)}(x)\cdot f(x)\right)=\left(g(x)\cdot\frac{d^{\alpha}}{d(-x)^{\alpha}}f(x)\right)\tag{2.7}$$

Let us note that the formula (2.7) is similar to the standard one in case when α is an integer. More information on fractal calculus can be found in Ross (1965). See also the application to the polymer interactions (Douglas, 1989).

3. Kinetic Equation

There are several ways of obtaining the kinetic description of a fractal random walk (Montrol and Weiss, 1965; Montrol and Shlesinger, 1984). Here the Chapman-Kolmogorov equation is the starting point

$$W(x_3,t_3|x_1,t_1)=\int dx_2W(x_3,t_3|x_2,t_2)W(x_2,t_2|x_1,t_1)\tag{3.1}$$

where $W(x\ ,t\ |x_1,t_1)$ is the probability density of having a particle at the position x , at the time t if at the time t_1 the particle was at the point x_1. An analog of the Kolmogorov method of deriving the kinetic equation will be used.

Let us consider the case when

$$W(x',t'|x,t)=W(x',x;t'-t)\tag{3.2}$$

In an infinitesimal time interval Δt the regular change of the probability density W is

$$W(y,x;t+\Delta t)=W(y,x;t)+\Delta t\frac{dW(y,x;t)}{dt}+0(\Delta t),\tag{3.3}$$

which, provided that there exists a limit, is equal to:

$$\lim_{\Delta t \to 0} \frac{1}{\Delta t} \{W(y,x;t+\Delta t) - W(y,x;t)\} = \frac{\partial W(y,x;t)}{\partial t}, \tag{3.4}$$

Meanwhile a fractal property of the random walk could imply a $\Delta t \to 0$ limit different from (3.3) or (3.4)

$$\lim_{\Delta t \to 0} \frac{1}{(\Delta t)^\beta} \{W(y,x;t+\Delta t) - W(y,x;t)\} \equiv$$

$$\equiv \frac{\partial^\beta}{\partial t^\beta} W(y,x;t) \tag{3.5}$$

where β is fractional and the derivative should be considered in the sense of (2.4). The limit (3.5) for $\beta \neq 1$ means the existence of a singularity at a point t in contrast to an analytical expansion for $\beta = 1$.

The difference in brackets in (3.5) should be infinitesimal too and the next step will be to estimate it. Let us use the equation (3.1). Then (3.5) can be rewritten in the form

$$\frac{\partial^\beta}{\partial t^\beta} P(y,t) = \lim_{\Delta t \to 0} \frac{1}{(\Delta t)^\beta} \left\{ \int_{-\infty}^{\infty} dz \cdot W(y,z;\Delta t) P(z,t) - P(y,t) \right\} \tag{3.6}$$

where a simplified notation is used for the density probability of finding a particle at the position ξ at the moment t

$$P(\xi,t) \equiv W(\xi,x;t) \tag{3.7}$$

if the initial position was x. Transitional probability $W(y,z;\Delta t)$ during the infinitesimal time interval Δt should be concentrated at z close to y. Suppose, as usual, that

$$W(y,z;\Delta t) = W(y-z,\Delta t). \tag{3.8}$$

Then the expansion can be taken for infinitesimal $y - z$

$$W(y-z,\Delta t) = \delta(y-z) + A|y-z|^\alpha \partial^{(\alpha)}(y-z) + \ldots \tag{3.9}$$

where the exponent α reflects the order of singularity and A is some function of z. The usual expansion is performed over integer exponents. Inserting of (3.9) into (3.6) gives

$$\frac{\partial^\beta P(y,t)}{\partial t} = \lim_{\Delta t \to 0} \frac{1}{(\Delta t)^\beta} \int_{-\infty}^{\infty} dz A(z) |y-z|^\alpha \partial^{(\alpha)}(y-z) \cdot P(z,t)$$

or using the properties (2.3), (2.6) and (2.7) we have

$$\frac{\partial^\beta P(x,t)}{\partial t^\beta} = \frac{\partial^\alpha}{\partial(-x)^\alpha}\left(\mathcal{D}(x)P(x,t)\right) \tag{3.10}$$

where

$$\mathcal{D}(x) \equiv \lim_{\Delta t\to 0}\frac{1}{(\Delta t)^\beta}\int_{-\infty}^{\infty} dz A(z)|x-z|^\alpha \equiv$$

$$\equiv \lim_{\Delta t\to 0}\frac{\overline{|\Delta x|^\alpha}}{(\Delta t)^\beta}. \tag{3.11}$$

and bar means averaging over $A(z)$ which is some characteristic of $W(\Delta x, \Delta t)$. Equations (3.10) and (3.11) can be considered as a generalization of the FPK equation for the case of a fractal space-time random walk.

The crucial role belongs to the equation (3.11) which corresponds to the main non-vanishing finite term in the limit $\Delta t \to 0$. In a standard case we have $\beta = 1$ and $\alpha = 2$. Here the values of $\beta \neq 1$ means that there is a singularity in the sequence of values $x(t_1), x(t_2), \ldots$ which depends on how the set $t_1, t_2, \ldots$ is chosen, which intervals $\Delta t_j = t_{j+1} - t_j$ are taken and how the limit $\Delta t \to 0$ is performed. All these properties are important in the fractal (singular) situation. Some difference exists for fractal properties of the displacements Δx. Formula (3.11) shows that exponent α reflects fractal properties of the average random walk rather than properties of the real coordinates space. Nevertheless, it should be mentioned that the average is taken over some scale l_c which corresponds to the local time scale t_c (see 1.7). In analogy with (1.7) the large scale behaviour l_D can be introduced

$$l_D \gg l_c \tag{3.12}$$

which comes from the solution of equation (3.10), and which corresponds to the distance of a random walk, after which the asymptotics of large t from (1.7) can appear. The scale l_c indicates the characteristic width of the distribution $W(\Delta x, \Delta t)$ as well as the level of smoothing of the real singular subspace in which the random walk takes place.

4. Asymptotic Properties

If $\mathcal{D}$ is a constant the kinetic equation (3.10) gives:

$$\langle |x|^\alpha \rangle = \text{const.}\, t^\beta (t \to \infty), \tag{4.1}$$

or due to self-similarity of the definition (2.3)

$$\langle |x| \rangle = \text{const.}\, t^{\beta/\alpha} (t \to \infty). \tag{4.2}$$

For $\beta = 1$ and $\alpha = 2$ the usual laws follow from (4.1) and (4.2). But for fractional α and β there is a new property. It is clear that asymptotics, in general case, has an exponent β/α, which includes the characteristics of both space and time. In this case spatial and temporal properties are inseparable. The result (4.2) was obtained also by Afanas'ev *et al.* (1991) from the Montroll-Weiss equation.

The process of diffusion for real the dynamical systems with chaos can be pictured in such a way (Afanas'ev *et al.* 1991). The full phase space is filled by clusters with different dimensions. In the process of wandering a particle switches from one cluster to another. The kinetic equation (3.10) can be rewritten in a more general form:

$$\frac{\partial^\beta P}{\partial x^\beta} = \sum_\alpha \frac{\partial}{\partial(-x)^\alpha}(\mathcal{D}_\alpha P), \tag{4.3}$$

where coefficients $\mathcal{D}_\alpha$ include a probability of cluster with singularity α and transition probability from one cluster to another is neglected. Existence of different exponents α will change the asymptotic law (4.2). It seems that the value max $\mu \equiv$ max β/α will determine the main asymptotic behaviour of the displacement $\langle 1|x|\rangle$. But coefficients for different μ_j can be very small for all $\mu_j > \mu_0$ where μ_0 is a "leading" exponent. Then for different time scales the crossover from one asymptotic behaviour to another one will appear and the particle transport should display different intermediate asymptotics.

5. Examples

We can demonstrate the existence of the anomalous exponents μ in the asymptotic law

$$\langle |x| \rangle \sim t^\mu, (t \to \infty) \tag{5.1}$$

on different examples. One such example is described by Afanas'ev *et al.* (1991) where a dynamical system of 4-dimensional mapping was studied. The system describes the particle motion in a magnetic field and in a wave packet. The multidimensional phase space creates an appropriate situation for the anomalous diffusion. Another example, which will be presented below in more detail, is taken from the articles by Chernikov *et al.* (1990) and Petrovichev *et al.* (1990).

Consider a passive particle motion in the velocity field with hexagonal symmetry

$$\mathbf{r} = \mathbf{u} = (-\frac{\partial \psi}{\partial y} + \varepsilon \sin z, \frac{\partial \psi}{\partial x} + \varepsilon \cos z, \psi) \tag{5.2}$$

where

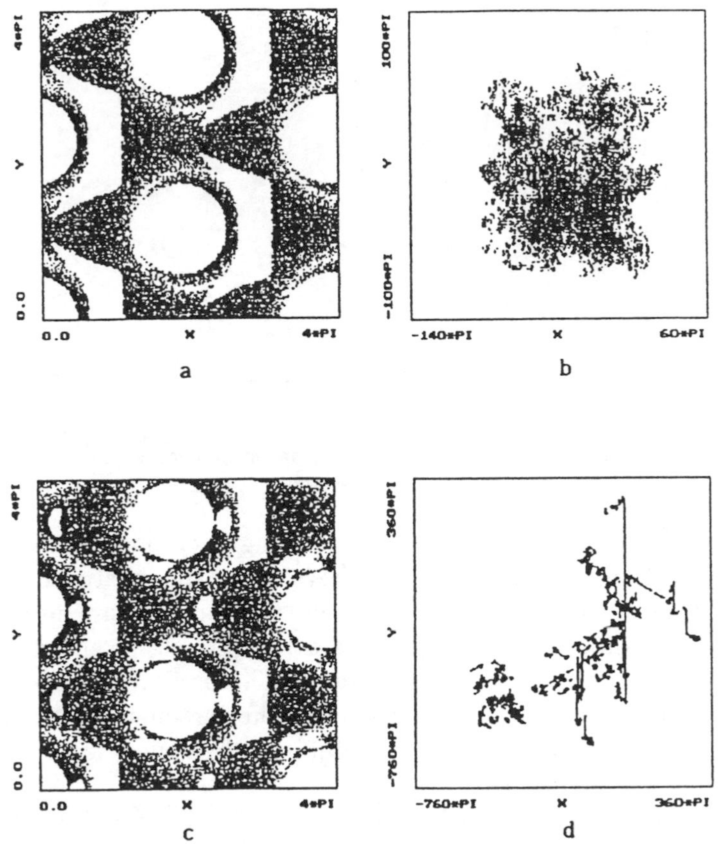

Fig. 1. Poincare sections of a streamline by the plane $z = 0 \pmod{2\pi}$ with $q = 3$ and: $\varepsilon = 0.7(a, b)$ and $\varepsilon = 1.1(c, d)$ for small (a, c) and large (b, d) scales.

$$\psi = \cos x + \cos \frac{x + \sqrt{3}y}{2} + \cos \frac{x - \sqrt{3}y}{2} \tag{5.3}$$

and ε is a parameter. Then there exists a stochastic web of finite width of order $\varepsilon(\varepsilon \lesssim 1)$ which has hexagonal symmetry in the (x, y)–plane and is periodic in z (Zaslavsky *et al.* 1988). The stochastic web is a connected net of channels in space (x, y, z) along which a passive particle will wander if the particle's initial position is taken inside the web. An example of the stochastic web and of the particle random walk pattern is shown in fig. 1. One can observe long flights in fig. 1(d). The appearance of flights is due to the particle orbit 'stickyness' in the vicinity of the boundaries of small islands in fig. 1. The phase space close to the islands' boundary has a fractal structure because of cantori. A cantorus is a cantor-like manifold the points of which belong to an unstable periodic orbit. To take a random walk

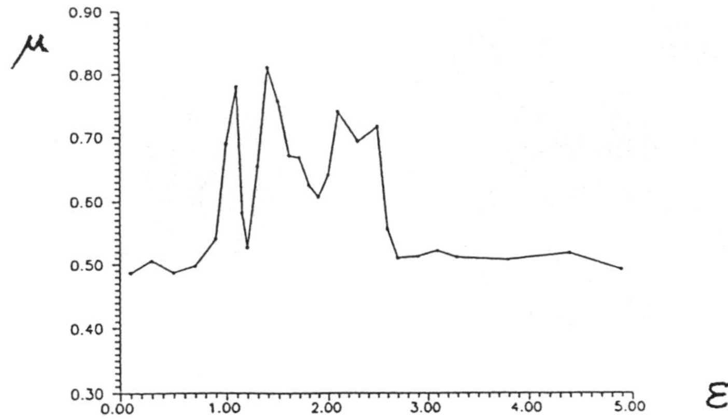

Fig. 2. Plot of the diffusion exponent μ versus ε.

a particle should pass through the cantori using their numerous gaps. We may assert (see the papers mentioned above) that the existence of cantori and their special structure is responsible for the local fractal properties of space and of the anomalous transport as a result. Fig. 2 shows a plot of the anomalous transport exponents versus ε. The oscillations of μ coincide with the bifurcations of the phase portrait of the system and that confirms a strong correlation between large scale diffusion behaviour and small scale phase portrait local structure.

Another example to consider is the Poincare Recurrence Time Statistics (PRTS) for the same passive particle motion (5.2) with hexagonal symmetry. Let us take a small sphere S inside the stochastic web and initial condition for a particle orbit inside the S. The particle will come back repeatedly. Let $T_1, T_2, \ldots$ will be a set of consequent periods of recurrences. One can consider a density probability to have recurrence time T. There is some evidence that this distribution will be of Poissonian form. It was shown (Zaslavsky and Tippett, 1991) that PRTS depends on how the diffusion process is realized. In order to demonstrate this two values of ε were taken. One value was $\varepsilon > 2.6$, when diffusion is normal (see fig. 2), and the second was $\varepsilon < 2.6$ when diffusion is anomalous. The first case gives Poissonian distribution for periods of recurrences, as it was expected. The second case gives a power tail distribution which correlates with suggestion of the influence of local fractal properties of the phase space (or real space) in the vicinity of islands' boundaries.

6. Conclusion

It seems that a simplification has been done if using the FPK-like equation (3.10) with fractional derivatives in time and space. Nevertheless at least two important comments should be made. The first one concerns the origin of the exponents α and β. So far we cannot find them directly from the equations of motion. More detailed analysis of the phase space will be desirable to get such an understanding. The second comment concerns the problem of irreversibility. This problem can be considered in a very natural way for the FPK equation. The structure of this equation guarantees the irreversibility and of the entropy growth with time. The situation for the equation (3.10) is unclear. Introducing the local fractal structures in (3.5), (3.9) does not imply any symmetry properties due to the time reflection $t \to -t$. This problem seems to be technical and it can be solved if better understanding of the dynamics on fractals is achieved.

Acknowledgements

This research was supported in part by the National Science Foundation under Grant No. PHY89-04035.

References

AFANAS'EV, V.V., SAGDEEV, R.Z., & ZASLAVSKY, G.M., 1991 *Chaos*, **1**, 143.
CHANDRASEKHAR, S., 1943 *Rev. Mod. Phys.*, **15**, 1.
CHERNIKOV, A.A., PETROVICHEV, B.A., ROGALSKY, A.V., SAGDEEV, R.Z., & ZASLAVSKY, G.M., 1990 *Phys. Lett. A*, **144**, 127.
DOUGLAS, J.F., 1989 *Macromolucules* **22**, 1786.
GELFAND, I.M. & SHILOV, G.E., 1958 Generalized Functions, vol. 1 (Moscow, Fizmat); translation 1964 (Academic Press, New York.)
KOLMOGOROV, A.N., 1938 *Uspelchi Matem. Nauk.*, **5**, 5.
LÉVY, P., 1937 Théorie de l'Addition des Variables Aléatoires (Gauthier-Villars, Paris 1937).
LIFSHITZ, E.M. & PITAEVSKY, L.P., 1981 Physical Kinetics, (Pergamon Press, Oxford.)
MANDELBROT, B.B., 1982 The fractal nature of geometry, (Freeman, San Francisco).
MONTROLL, E.W., & SHLESINGER, M., 1984 In: Studies in statistical mechanics, v. **11**, p. 1, Eds. J. Lebowitz and E.W. Montroll (North-Holland, Amsterdam).
MONTROLL, E.W., & WEISS, G.M., 1965 *Journ. Math. Phys.*, **6**, 167.
PETROVICHEV, B.A., ROGALSKY, A.V., SAGDEEV, R.Z., & ZASLAVSKY, G.M., 1990 *Phys. Lett. A*, **150**, 391.
ROSS, B., ed., 1975 *Fractional Calculus and Its Applications, Lecture Notes* **457** (Springer-Verlag, New York).
SHLESINGER, M.F., 1989 *Physica D*, **38**, 304.
ZASLAVSKY, G.M., 1991 Monte Verita Colloquium on Turbulence (to be published).
ZASLAVSKY, G.M. & TIPPETT, M., 1991 *Phys. Rev. Lett.* **67**, 3251.
ZASLAVSKY, G.M., SAGDEEV, R.Z. & CHERNIKOV, A.A., 1988 *Sov. Phys.* -JETP, **67**, 270.
ZASLAVSKY, G.M., ZAKHAROV, M. YU., SAGDEEV, R.Z., USIKOV, D.A., & CHERNIKOV, A.A., 1986 *Sov. Phys.* -JETP **64**, 294.

KINEMATICAL INSTABILITY AND LINE-STRETCHING IN RELATION TO THE GEODESICS OF FLUID MOTION

T. KAMBE, F. NAKAMURA, and Y. HATTORI
Department of Physics
University of Tokyo
Hongo, Bunkyo-ku, Tokyo 113, Japan

ABSTRACT. Motion of an incompressible ideal fluid is represented as geodesics on the group of all volume preserving diffeomorphisms. For the fluid motions in a cubic space with periodic boundary conditions, the geodesic equation, covariant derivative and curvature tensors are explicitly presented. This formulation can be applied immediately to a couple of simple flows with the Beltrami property, including the ABC flows. The curvature in the section of any two different ABC flows is found to be negative. The L^2-distance between two states with slightly different initial velocity fields is related to the curvature in a section of the two fields. Negative curvature, which is likely to occur in many flows, means an enhanced deviation of the two states (*i.e.* the kinematical instability). Geodesic variations are governed by the Jacobi equation, which leads to the time evolution of magnitude of a line element in terms of the section curvature. Nearly exponential growth of line elements is shown by some numerical examples.

1. Introduction

Hydrodynamics of an ideal (incompressible and inviscid) fluid is connected with a problem of finding *geodesics* on the group of all volume preserving diffeomorphisms (Arnold 1966, 1978). For two dimensional flows on the torus $\mathsf{T}^2 = \mathsf{R}^2/(2\pi\mathsf{Z})^2$ (R: space of real numbers; Z: set of integers), Arnold gave explicit formulas for the commutator, inner product, riemannian connection and geodesic equation. These formulas allow us to calculate the riemannian curvature tensors on the geodesics of any two-dimensional cross-section. By the Jacobi equation in the differential geometry, the stability of the geodesics is determined by the curvatures on them. By the stability it is meant here kinematically how the difference of particle positions between two flows of different initial conditions develops with time. Negative curvature leads to an enhanced growth of the difference from that in initial conditions with time.

Consider a flow of an ideal fluid of uniform density in a three dimensional

H. K. Moffatt et al. (eds.), Topological Aspects of the Dynamics of Fluids and Plasmas, 493–504.

flow domain (manifold) M. Given the velocity $v_t(y)$ of fluid motion for $y \in$M and $t \in$R, the particle motion $y(t)$ is described by the ordinary differential equation, $dy/dt = v_t(y)$ with $y(0) = x \in$ M, where x is the initial position of the particle. The solution is written in the form: $y = g_t(x)$, which describes a smooth curve in M starting from x, *i.e.* a Lagrangian particle path. The manifold M is provided with the metric given by the inner product $(U \cdot V)$ and the covariant derivative $\nabla_U V$ (or riemannian connection for this metric, *e.g.* see Kobayashi & Nomizu 1969) at any point $x \in$M for any two vector fields U and V on M. The covariant derivative is a differentiation of the vector field V in the direction of U.

For each t, the mapping g_t : M $\to$ M is an auto-diffeomorphism carrying every particle of the fluid in M from the place it was at time 0 to the place it is at time t. In other words, the curve g_t for the parameter t describes time development of the configuration of particles, *i.e. a flow.* In case that $v_t(y)$ is divergence-free, the diffeomorphism is volume preserving. The volume preserving diffeomorphisms from M to itself form an infinite dimensional group, which is denoted as $\mathcal{D}_v$(M). The flow g_t is a *curve* in the group $\mathcal{D}_v$, and it is generated by v_t in the sense $\dot{g}_t(x) \equiv dg_t/dt = v_t(g_t(x)) = v_t \circ g_t(x)$, where $v_t \circ g_t(x)$ is a right translation of the field v_t by the element g_t for $x \in$ M. It is noted that $v_t(y)$ is the Eulerian description of the velocity field for the space coordinate y, whereas $v_t \circ g_t(x)$ is its Lagrangian counterpart for the coordinate x.

Consider two divergence-free vector (velocity) fields on $\mathcal{D}_v$ given by $\tilde{U} = U \circ g_t$ and $\tilde{V} = V \circ g_t$ for $U,\ V \in T_e\mathcal{D}_v$ (the tangent space at the origin e, e being an identity map). The vectors $\tilde{U}$ and $\tilde{V}$ are *right invariant* vector fields, that is, they satisfy the invariance of the metric,

$$< \tilde{U},\, \tilde{V} >|_{g_t} = \int_{\mathsf{M}} (U(g_t(x)) \cdot V(g_t(x)))\, dx \;\; = < U,\, V >|_e \quad . \tag{1}$$

In addition, the group $\mathcal{D}_v$ is endowed with a connection for the two divergence-free vector fields $\tilde{U}$ and $\tilde{V}$ defined by

$$\tilde{\nabla}_{\tilde{U}}\tilde{V}\;|_{g_t} = \mathrm{P}[\nabla_U V] \circ g_t$$

(Ebin & Marsden 1970), where the operation P[ ] denotes the projection to divergence-free part since the covariant derivative $\nabla_X Y$ is not necessarily divergence-free for two divergence-free vector fields $X,\ Y$ on M.

The principle of least action asserts that the motion of an ideal fluid is a geodesic. The action I is defined as

$$I = \int \; < \dot{g}_t,\, \dot{g}_t >|_{g_t}\, dt = \int dt \int (v_t(y) \cdot v_t(y))\, dx \quad ,$$

where $y = g_t(x)$, and the variational problem leads to

$$\tilde{D}\dot{g}_t/dt = 0 \ , \tag{2}$$

where

$$\tilde{D}\dot{g}_t/dt \equiv (\partial v_t/\partial t + \mathrm{P}[\nabla_{v_t} v_t]) \circ g_t \ . \tag{3}$$

The curve g_t is a *geodesic, i.e.* $\tilde{D}\dot{g}_t/dt = 0$, because this means that the tangent vector $\dot{g}_t$ is parallel along the curve g_t (Milnor 1963).

For the covariant derivative $\nabla_{v_t} v_t$, we have the orthogonal decomposition,

$$\nabla_{v_t} v_t \mid_e = \{\mathrm{P}[\nabla_{v_t} v_t] - \mathrm{grad}\, p\} \mid_e \ , \tag{4}$$

where the function p is a smooth scalar function on M. Thus, the right translation of equation (2) with (3) leads to

$$\partial v_t/\partial t + \mathrm{P}[\nabla_{v_t} v_t] = 0 \ , \tag{5}$$

or in view of (4), $\partial v_t/\partial t + \nabla_{v_t} v_t = -\mathrm{grad}\, p$. This is the Euler equation for an ideal fluid on riemannian manifold M. For a flat cartesian space, the covariant derivative reduces to the form

$$\nabla_v v = (\mathbf{v} \cdot \mathrm{grad})\mathbf{v} \ , \tag{6}$$

for the velocity field $\mathbf{v}$ with cartesian components (v_i) at $\mathbf{x} = (x_i)$ and grad$= (\partial/\partial x_i), (i = 1, 2, 3)$. Thus in the hydrodynamic notations, we have recovered the Euler equation,

$$\partial \mathbf{v}/\partial t + (\mathbf{v} \cdot \mathrm{grad})\mathbf{v} = -\mathrm{grad}\, p \ , \quad \mathrm{div}\ \mathbf{v} = 0 \tag{7}$$

where p is the pressure divided by the uniform fluid density.

In the Lie algebra of the group $\mathcal{D}_v$, the commutator $[\ ,\ ]_*$ is defined for two divergence-free vector fields X, Yas

$$[X, Y]_* = \tilde{\nabla}_X Y - \tilde{\nabla}_Y X \tag{8}$$

where $\tilde{\nabla}$ stands for $\mathrm{P}[\nabla \quad]$. The right invariant property of the metric and connection allows us to make corresponding calculation at $e \in \mathcal{D}_v$.

Given divergence-free vector fields X, Y, Z and W, a new divergence-free vector field $\tilde{R}(X,Y)Z$ called the curvature tensor is defined by

$$\tilde{R}(X, Y)\, Z = -\tilde{\nabla}_X \tilde{\nabla}_Y Z + \tilde{\nabla}_Y \tilde{\nabla}_X Z + \tilde{\nabla}_{[X,Y]_*} Z \ , \tag{9}$$

and then the curvature (tensor) $\tilde{R}_{XYZW}$ is given by

$$\tilde{R}_{XYZW} =< \tilde{R}(X, Y)\, Z, W > \ . \tag{10}$$

The sectional curvature for the section $\sigma \subset \mathrm{T}_e\mathcal{D}_v$ spanned by X and Y is

$$\tilde{K}(\sigma) =< \tilde{R}(X, Y)\, X, Y > /(< X, X >< Y, Y > - < X, Y >^2) \tag{11}$$

where the denominator is the square of the area of parallelogram spanned by X and Y.

Stability of a geodesic g_t is determined by the Jacobi field A_t along g_t. The norm $|A_t| \equiv \sqrt{< A_t, A_t >}$ denotes the evolution of distance between two geodesics g_t^0 and g_t^s per unit variation of s which start from the same origin e with different initial tangent vectors X and $X + sY$ respectively (Hicks 1965):

$$|A_t| = |Y|t - \tfrac{1}{6}\tilde{K}(\sigma)t^3 + O(t^4) \quad . \tag{12}$$

Thus negative $\tilde{K}(\sigma)$ means enhanced deviation of two geodesics in the section σ spanned by X and Y.

Recently we (Nakamura, Hattori & Kambe 1992) have given an explicit form of the geodesic equation for the motion of an ideal fluid on a three-torus T^3 and the *riemannian curvature* of the group of volume preserving diffeomorphisms. The curvature turns out to be negative for the two-dimensional section consisting of a particular vector (velocity) field (*i.e.* a simple flow with Beltrami property) and a general vector field. These are shown in the sections 2 and 3. In the final section 4, we present a study of particle motion, based on the Jacobi field.

2. Geodesics in Fourier space

We will now investigate a three-dimensional fluid motion on the flat three-torus $\mathsf{T}^3 = \mathsf{R}^3/(2\pi\mathsf{Z})^3$, *i.e.* $\mathbf{x} = \{(x_1, x_2, x_3);\ \mathrm{mod}\ 2\pi\}$ for $\mathbf{x} \in \mathsf{T}^3$, and the curvature of the group $\mathcal{D}_v(\mathsf{T}^3)$. The elements of the Lie algebra of the group $\mathcal{D}_v(\mathsf{T}^3)$ can be thought of as real periodic vector fields on T^3 with the divergence-free property. Such a periodic fields are represented by the real part of corresponding complex Fourier forms. The Fourier base $e^{i\mathbf{k}\cdot\mathbf{x}}$ is denoted by $e_{\mathbf{k}}$, where $\mathbf{k} = (k_i)$ for $i = 1, 2, 3$. Complexified forms of the Lie algebra, inner product, commutator and the riemannian connection and curvature tensor are given such that all these functions become linear (or multi-linear) in the complex vector space of the complexified Lie algebra. The functions $e_{\mathbf{k}}$ ($\mathbf{k} \in \mathsf{Z}^3, \mathbf{k} \neq 0$) form a basis of this vector space.

The velocity field is represented as

$$v_t(x) \equiv \mathbf{u}(t, \mathbf{x}) = \sum_{\mathbf{k}} \mathbf{u}_{\mathbf{k}}(t)\, e_{\mathbf{k}} \quad , \tag{13}$$

where $\mathbf{u}(t, \mathbf{x})$ and $\mathbf{u}_{\mathbf{k}}(t)$ have three components, the latter being written as $u^i(\mathbf{k})$ too occasionally ($i = 1, 2, 3$). The Fourier components must satisfy the two properties,

$$(\mathbf{k} \cdot \mathbf{u}_{\mathbf{k}}) = 0, \qquad \mathbf{u}_{-\mathbf{k}} = \mathbf{u}_{\mathbf{k}}^* \tag{14}$$

to express the solenoidal and real conditions, respectively. Where the asterisk * denotes the complex conjugate. It should be noted that $\mathbf{u_k}$ has two independent polarizations consistent with the first condition.

Let us take four vector fields satisfying the conditions (14) : $\mathbf{u_k}\, e_\mathbf{k}$, $\mathbf{v_l}\, e_\mathbf{l}$, $\mathbf{w_m}\, e_\mathbf{m}$, $\mathbf{z_n}\, e_\mathbf{n}$. Then we have the following expressions. From (1), the inner product is

$$< \mathbf{u_k}\, e_\mathbf{k}\,,\ \mathbf{v_l}\, e_\mathbf{l} >= (2\pi)^3\, (\mathbf{u_k}\cdot\mathbf{v_l})\delta_{\mathbf{0},\mathbf{k+l}}\ , \tag{15}$$

which is non-zero only for $\mathbf{k}+\mathbf{l}=\mathbf{0}$. Using (6) and operating the projection P, the covariant derivative is

$$\tilde{\nabla}_{\mathbf{u_k}e_\mathbf{k}}\mathbf{v_l}e_\mathbf{l} \;=\; i\,(\mathbf{u_k}\cdot\mathbf{l})\frac{\mathbf{k}+\mathbf{l}}{|\mathbf{k}+\mathbf{l}|}\times\left(\mathbf{v_l}\times\frac{\mathbf{k}+\mathbf{l}}{|\mathbf{k}+\mathbf{l}|}\right)e_\mathbf{k+l}\ . \tag{16}$$

The equation (8) gives the commutator:

$$[\mathbf{u_k}\, e_\mathbf{k}\,,\ \mathbf{v_l}\, e_\mathbf{l}]_* = i\left(\ (\mathbf{u_k}\cdot\mathbf{l})\mathbf{v_l} - (\mathbf{v_l}\cdot\mathbf{k})\mathbf{u_k}\ \right)e_\mathbf{k+l}\ . \tag{17}$$

From the definitions (9) and (10), the curvature tensor is

$$\begin{aligned}\tilde{R}_\mathbf{klmn} \;&=\; < \tilde{R}(\mathbf{u_k}\, e_\mathbf{k}\,,\ \mathbf{v_l}\, e_\mathbf{l})\,\mathbf{w_m}\, e_\mathbf{m}\,,\ \mathbf{z_n}\, e_\mathbf{n} > \\ &=\; (2\pi)^3\left(\frac{(\mathbf{u_k}\cdot\mathbf{m})(\mathbf{w_m}\cdot\mathbf{k})}{|\mathbf{k}+\mathbf{m}|}\frac{(\mathbf{v_l}\cdot\mathbf{n})(\mathbf{z_n}\cdot\mathbf{l})}{|\mathbf{l}+\mathbf{n}|}\right.\\ &\qquad\left. -\ \frac{(\mathbf{v_l}\cdot\mathbf{m})(\mathbf{w_m}\cdot\mathbf{l})}{|\mathbf{l}+\mathbf{m}|}\frac{(\mathbf{u_k}\cdot\mathbf{n})(\mathbf{z_n}\cdot\mathbf{k})}{|\mathbf{n}+\mathbf{k}|}\right)\ .\end{aligned} \tag{18}$$

The curvature $\tilde{R}_\mathbf{klmn}$ takes nonzero value only for $\mathbf{k}+\mathbf{l}+\mathbf{m}+\mathbf{n}=\mathbf{0}$ and nonzero $\mathbf{k},\mathbf{l},\mathbf{m},\mathbf{n}$. (The terms with zero denominator should be deleted.) This formula is valid for $\mathcal{D}_v(\mathsf{T}^n)$ too if the factor $(2\pi)^3$ is replaced by $(2\pi)^n$.

The above expression of the curvature tensor is of the same form as that of Arnold for two-dimensional case. In the latter case the direction of the vector $\mathbf{u_k}$ is uniquely determined due to the condition (14), that is, for $\mathbf{k}=(k_1,\,k_2,\,0)$ we have $\mathbf{u_k}=i(k_2,\,-k_1,\,0)$ and similar expressions for the other vectors. Then it can be readily shown that our formulas just reduce to those of Arnold where our $\mathbf{u_k}e_\mathbf{k}$ corresponds to his $e_\mathbf{k}$.

An analysis for the distance of each particle convected by two different velocity fields of T^2, which is related to the Jacobi field (12), has been carried out by Hattori (1990). It is found by the Taylor series expansion that an L^2-distance between two mappings g_t^0 and g_t^s (with small values of s) defined by

$$d(g_t^0, g_t^s) = (\int_{T^2}|g_t^0(\mathbf{x}) - g_t^s(\mathbf{x})|^2 dx dy)^{1/2}\ , \qquad \mathbf{x}=(x,y) \tag{19}$$

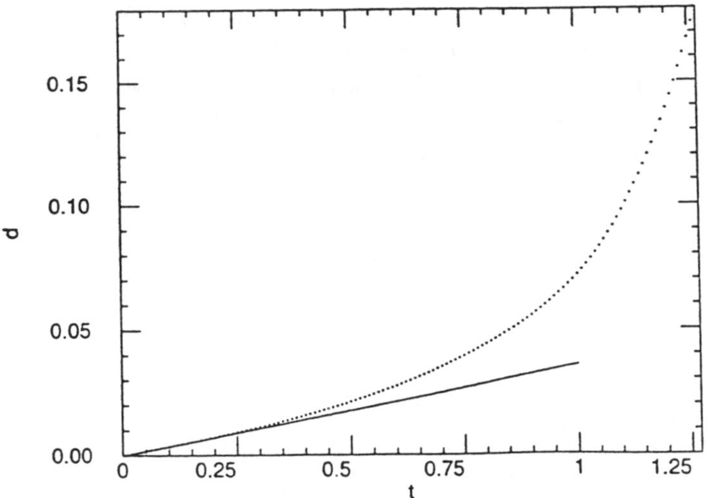

Fig. 1. Time evolution of the distance d in the case of $a = 1$.

evolves in the same way as (12). This analysis can be extended to T^3 without difficulty.

We show an example of the effect of the curvature in the two-dimensional case. Instead of g_t^0 and g_t^s, we consider the two velocity fields given as

$$\dot{g}_t^A|_e = v_o + s\, v_p \ , \qquad \dot{g}_t^B|_e = v_o - s\, v_p \ .$$

where v_o and v_p are given below. The series expansion of $d(g_t^A, g_t^B)$ with respect to a small t is represented by the terms of odd powers. The first linear term is given as $2s|v_p|\,t$, and the coefficient of the cubic term is found to be proportional to the curvature $\tilde{K}(v_o, v_p)$ with a negative sign. To illustrate the influence due to the different signs of $\tilde{K}$, we take a particular case in which the velocity fields v_o and v_p are represented by the streamfunctions,

$$\psi_o = 2\,\mathrm{Re}[e_{\mathbf{k}} + e_{\mathbf{l}}] \ , \qquad \psi_p = 2\,\mathrm{Re}[i(\mathbf{k}\cdot\mathbf{p})e_{\mathbf{k}} + i(\mathbf{l}\cdot\mathbf{p})e_{\mathbf{l}}]$$

respectively. Then assuming that $\mathbf{k} = (a, 1), \mathbf{l} = (a, -1)$ and $\mathbf{p} = (1, 0)$ for an integer a, we find that $\tilde{K}(v_o, v_p)$ is proportional to $(a^2-2)/(a^2+1)$ with a positive constant coefficient (Hattori 1991). Hence $\tilde{K}$ is negative or positive according as a^2 is less or greater than 2. Figures 1 and 2 show the time evolutions of $d(g_t^A, g_t^B)$ for a =1 and 3 respectively, where the solid straight lines illustrate the linear behavior due to the first term.

The geodesic equation (5) reduces to

$$\frac{\partial u^m(\mathbf{k})}{\partial t} + i \sum_{\mathbf{p}+\mathbf{q}=\mathbf{k}} \sum_{j,l} k_j(\delta_{ml} - \frac{k_m k_l}{k^2}) u^j(\mathbf{p}) u^l(\mathbf{q}) = 0 \qquad (20)$$

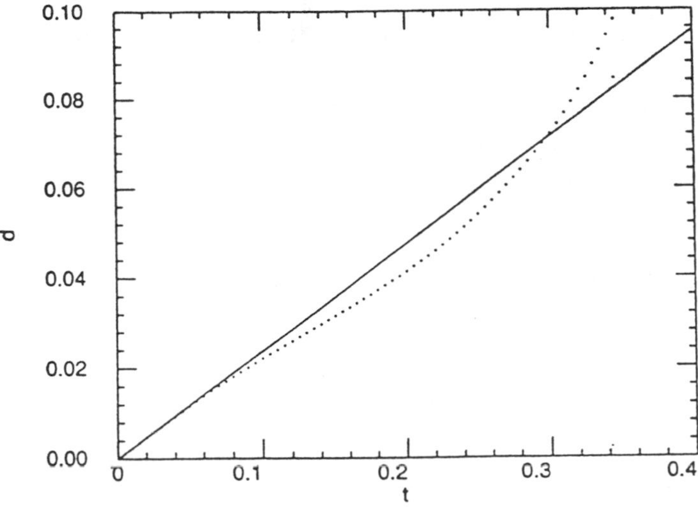

Fig. 2. Time evolution of d in the case of $a = 3$.

by using (16) where δ_{ij} is the Kronecker's delta. It is quite natural that this is exactly identical to the Fourier representations of the Euler equation (Kraichnan 1959), or the Navier-Stokes equation when the viscosity coefficient ν vanishes.

3. Beltrami flows and ABC flows

As an application, we consider a flow with Beltrami property, that is, we assume that the velocity field $\mathbf{U_p} = \mathbf{u_p}\, e_\mathbf{p} + \mathbf{u_{-p}}\, e_{-\mathbf{p}}$, *i.e.* $2\mathrm{Re}[\mathbf{u_p}\, e^{i\mathbf{p}\cdot\mathbf{x}}]$, satisfies the condition, curl $\mathbf{U_p} = \lambda \mathbf{U_p}$, for a parameter $\lambda \in \mathsf{R}$. This eigenvalue problem can be solved with $\lambda^2 = |\mathbf{p}|^2$. It is readily shown that $\mathbf{U_p}$ is a steady-state solution. Let $\mathbf{X} = \sum \mathbf{v_l}\, e_\mathbf{l}$ be any velocity field satisfying (14). Using the above formulas, we obtain

$$
\begin{aligned}
&< \tilde{R}(\mathbf{U_p}, \mathbf{X})\mathbf{U_p}, \mathbf{X} > \\
&= -(2\pi)^3 \sum_{\mathbf{l}} \frac{|(\mathbf{u_p}\cdot\mathbf{l})|^2}{|\mathbf{p}+\mathbf{l}|^2} \left| \frac{(\mathbf{u_p}\cdot\mathbf{l})^2}{|(\mathbf{u_p}\cdot\mathbf{l})|^2}(\mathbf{v_l}\cdot\mathbf{p}) - (\mathbf{v_{l+2p}}\cdot\mathbf{p}) \right|^2 \le 0. \qquad (21)
\end{aligned}
$$

This property of non-positivity is considered to be a three dimensional version of the Arnold's result for the curvature of the group $\mathcal{D}_v(\mathsf{T}^2)$ in any two-dimensional section containing the direction ξ represented by the stream function $\frac{1}{2}(e_\mathbf{p} + e_{-\mathbf{p}})$.

The above result can be extended to a two-mode Beltrami flow $\mathbf{U_{p,q}}$ which is defined by the linear combination $\mathbf{U_p} + \mathbf{U_q}$ of two Beltrami flows $\mathbf{U_p}$ and $\mathbf{U_q}$. Although it is not difficult to derive the curvature formula in the section of $\mathbf{U}_{\mathbf{p},\mathbf{q}}$ and the general direction $\mathbf{X}$, we show here only the case in which $\mathbf{X}$ is another Beltrami flow represented by $\mathbf{V}_{\mathbf{p},\mathbf{q}} = \mathbf{V}_{\mathbf{p}} + \mathbf{V}_{\mathbf{q}}$ where $\mathbf{V}_{\mathbf{p}} = \mathbf{v}_{\mathbf{p}}\, e_{\mathbf{p}} + \mathbf{v}_{-\mathbf{p}}\, e_{-\mathbf{p}}$. The two polarization amplitudes $\mathbf{v}_{\mathbf{p}}$ and $\mathbf{u}_{\mathbf{p}}$ are linearly independent and the Beltrami condition is satisfied with $\lambda = |\mathbf{p}|$. Then the sectional curvature is

$$
\begin{aligned}
&< \tilde{R}(\mathbf{U}_{\mathbf{p},\mathbf{q}}, \mathbf{V}_{\mathbf{p},\mathbf{q}})\mathbf{U}_{\mathbf{p},\mathbf{q}}, \mathbf{V}_{\mathbf{p},\mathbf{q}} > /(2\pi)^3 \\
&= \; -N_+ \, |(\mathbf{u}_{\mathbf{p}} \cdot \mathbf{q})(\mathbf{v}^*_{\mathbf{q}} \cdot \mathbf{p}) - (\mathbf{u}^*_{\mathbf{q}} \cdot \mathbf{p})(\mathbf{v}_{\mathbf{p}} \cdot \mathbf{q})|^2 \\
&\quad - 4N_+ \, \mathrm{Im}[(\mathbf{u}_{\mathbf{p}} \cdot \mathbf{q})(\mathbf{v}^*_{\mathbf{p}} \cdot \mathbf{q})] \cdot \mathrm{Im}[(\mathbf{u}_{\mathbf{q}} \cdot \mathbf{p})(\mathbf{v}^*_{\mathbf{q}} \cdot \mathbf{p})] \\
&\quad - N_- \, |(\mathbf{u}_{\mathbf{p}} \cdot \mathbf{q})(\mathbf{v}_{\mathbf{q}} \cdot \mathbf{p}) - (\mathbf{u}_{\mathbf{q}} \cdot \mathbf{p})(\mathbf{v}_{\mathbf{p}} \cdot \mathbf{q})|^2 \\
&\quad - 4N_- \, \mathrm{Im}[(\mathbf{u}_{\mathbf{p}} \cdot \mathbf{q})(\mathbf{v}^*_{\mathbf{p}} \cdot \mathbf{q})] \cdot \mathrm{Im}[(\mathbf{u}^*_{\mathbf{q}} \cdot \mathbf{p})(\mathbf{v}_{\mathbf{q}} \cdot \mathbf{p})] \; ,
\end{aligned}
$$

where $N_{\pm} = 2/|\mathbf{p} \pm \mathbf{q}|^2$.

Next we show an interesting example, that is an application to the ABC flow (Dombre *et al.* 1986) represented by

$$
\begin{aligned}
\mathbf{U}_{\mathrm{ABC}} = \quad & \mathrm{A}\Big[(i,1,0)\, e^{ix_3} + (-i,1,0)\, e^{-ix_3}\Big] \\
+ \; & \mathrm{B}\Big[(0,i,1)\, e^{ix_1} + (0,-i,1)\, e^{-ix_1}\Big] \\
+ \; & \mathrm{C}\Big[(1,0,i)\, e^{ix_2} + (1,0,-i)\, e^{-ix_2}\Big] \; ,
\end{aligned}
$$

where A,B,C $\in$ R. This is a three-mode Beltrami flow, *i.e.* each term on the right hand side satisfies the Beltrami property with $\lambda = -1$. We have another ABC flow $\mathbf{U}_{\mathrm{A'B'C'}}$ for $(\mathrm{A'}, \mathrm{B'}, \mathrm{C'}) \neq (\mathrm{A}, \mathrm{B}, \mathrm{C})$. It is straightforward to show that

$$
\begin{aligned}
&< \tilde{R}(\mathbf{U}_{\mathrm{ABC}}, \mathbf{U}_{\mathrm{A'B'C'}})\mathbf{U}_{\mathrm{ABC}}, \mathbf{U}_{\mathrm{A'B'C'}} > \\
&= -2\,(2\pi)^3\{(\mathrm{AB'} - \mathrm{BA'})^2 + (\mathrm{BC'} - \mathrm{CB'})^2 + (\mathrm{CA'} - \mathrm{AC'})^2\} \; ,
\end{aligned}
$$

and that the denominator of the equation (11), $< X, X >< Y, Y > - < X, Y >^2$, is also given by

$$
4^2(2\pi)^6\{(\mathrm{AB'} - \mathrm{BA'})^2 + (\mathrm{BC'} - \mathrm{CB'})^2 + (\mathrm{CA'} - \mathrm{AC'})^2\} \; .
$$

Thus the curvature in the section σ spanned by any pair in the family of ABC flows takes the uniform nagative value,

$$
\tilde{K}(\sigma) = -\frac{1}{64\pi^3} \; . \tag{22}
$$

In other words, any two-dimensional section of the three-mode Beltrami flows is characterized by a negative constant curvature.

It is interesting to see that even in the case where both $(\mathrm{A}, \mathrm{B}, \mathrm{C})$ and $(\mathrm{A'}, \mathrm{B'}, \mathrm{C'})$ are close and the streamlines are not chaotic, the particle motion by $\mathbf{U}_{\mathrm{A'B'C'}}$ will deviate from that of $\mathbf{U}_{\mathrm{ABC}}$ and not be predicted from the particle motion by $\mathbf{U}_{\mathrm{ABC}}$ in the course of time. It is expected that the present method may be applied to the analysis of rate of growth of line-element or surface-element in various flow fields.

4. Particle motions in relation to the Jacobi field

Let us consider a one-parameter family of geodescis h_t^s for $s \in (-\epsilon, \epsilon)$ and some $\epsilon > 0$ where $h_t^0 = g_t$. Slightly different initial conditions $(h_0^s, \dot{h}_0^s)$ determine a new geodesic h_t^s which is different slightly from g_t initially. The Jacobi field along g_t is defined by the derivative,

$$\tilde{W}_t = \frac{\partial h_t^s}{\partial s}|_{s=0} \tag{23}$$

which is a vector field of geodesic variation. The field $\tilde{W}_t$ satisfies the second order differential equation called the Jacobi equation,

$$\tilde{D}^2 \tilde{W}_t / dt^2 + \tilde{R}(\dot{g}_t, \tilde{W}_t)\dot{g}_t = 0 \quad . \tag{24}$$

The field $\tilde{W}_t$ is represented as $W_t \circ g_t$ in terms of the Eulerian Jacobi field W_t.

Let c_s be any curve in $\mathcal{D}_v$ for $s \in (-\epsilon, \epsilon)$ and $c_0 = e$. Then the curve $h_t^s = g_t \circ c_s$ is a geodesic variation of g_t. In fact, the tangent vector to h_t^s is

$$\frac{\partial h_t^s}{\partial t} = \dot{g}_t \circ c_s = v_t \circ g_t \circ c_s = v_t \circ h_t^s \ . \tag{25}$$

Accordingly the covariant derivative of $\partial h_t^s / \partial t$ is

$$\frac{\tilde{D}}{dt}\left(\frac{\partial h_t^s}{\partial t}\right) = (\partial v_t / \partial t + \mathrm{P}[\nabla_{v_t} v_t]) \circ h_t^s = \frac{\tilde{D}\dot{g}_t}{dt} \circ c_s \tag{26}$$

by (3). Evidently this vanishes due to (2).

Thus the Jacobi field (23) is represented as

$$\tilde{W}_t = \frac{\partial}{\partial s} g_t \circ c_s|_{s=0} = T_{g_t} \cdot A_s \tag{27}$$

where

$$A_s = \frac{\partial c_s}{\partial s}|_{s=0} \quad \in \ T_e \mathcal{D}_v$$

and T_{g_t} denotes the Jacobian matrix of the map g_t from $g_0 = e$. The geodesic variation h_t^s and the Jacobi field $\tilde{W}_t$ thus defined are said left translated.

To obtain a differential equation for W_t, we use the well-known formula of the riemannian geometry. For any map $f : \mathrm{R}^2 \to \mathcal{D}_v$, *i.e.* $f : (t,\ s) \mapsto f(t, s)$, it holds that

$$\frac{\tilde{D}}{dt}\frac{\partial f}{\partial s} = \frac{\tilde{D}}{ds}\frac{\partial f}{\partial t} \tag{28}$$

where $\tilde{D}/dt$ and $\tilde{D}/ds$ are the covariant derivatives along the t-curve ($t \mapsto f(t,s)$) and s-curve ($s \mapsto f(t,s)$) respectively.

The above formula can be applied to $f = h_t^s = g_t \circ c_s$. To this end, we remind that the vector field $\tilde{W}_t^s = \partial h_t^s/\partial s$ along the t-curve is expressed as $W_t^s \circ h_t^s$ in terms of the Eulerian field W_t^s, and the vector field $\partial h_t^s/\partial t$ along s-curve by $v_t \circ h_t^s$. Then the left hand side of the formula (28) is given by

$$(\frac{\partial}{\partial t}W_t^s + \mathrm{P}[\nabla_{v_t} W_t^s]) \circ h_t^s \quad ,$$

and the right hand side is given by

$$\mathrm{P}[\nabla_{W_t^s} v_t] \circ h_t^s \quad .$$

Thus, right translating the both expressions and setting $s = 0$ with $W_t^s|_{s=0} = W_t$, one obtains the once-integrated Jacobi equation:

$$\frac{\partial}{\partial t}W_t + \mathrm{P}[\nabla_{v_t} W_t] = \mathrm{P}[\nabla_{W_t} v_t] \quad , \tag{29}$$

or equivalently adding equal gradient components to both sides, one has

$$\frac{\partial}{\partial t}W_t + \nabla_{v_t} W_t = \nabla_{W_t} v_t \quad . \tag{30}$$

This type of equaiton is familiar in the fluid mechanics, *e.g.* the vorticity equation and the induction equation in MHD without diffusivities.

Now some remarks are to be given. Consider a vector field $A(x)$ at any point $x \in \mathsf{M}$ at an initial instant and take a curve $c_s(x)$ in M such that

$$c_0(x) = x \quad \text{and} \quad \frac{dc_s}{ds}|_{s=0} = A(x) \quad .$$

Then we may assign a line element,

$$dl = A(x)\, ds = (dl_x,\ dl_y,\ dl_z)$$

to the particle at x. The curve c_s is given by the equation $dc_s(x)/ds = A(c_s(x))$ for any $x \in \mathsf{M}$. Thus we have a family of integral curves. Time evolution of the line element is given by

$$\frac{\partial}{\partial s}\, g_t \circ c_s(x) \quad .$$

When $A(x)$ is divergence-free, c_s is a curve in $\mathcal{D}_v$ such that $c_0 = e$. Therefore the evolution of line elements is a left translated Jacobi field along g_t in $\mathcal{D}_v$. The vorticity field $\omega_t(x)$, *i.e.* $\omega = \mathrm{curl}\mathbf{v}$ in the hydrodynamic notation, defined at each point $x \in \mathsf{M}$ is a typical example of such Jacobi field. Its evolution equation is given by (30), its cartesian form is

$$\frac{\partial}{\partial t}\omega + (\mathbf{v}\cdot \mathrm{grad})\omega = (\omega \cdot \mathrm{grad})\mathbf{v} \quad . \tag{31}$$

The integral curves corresponding to c_s are the vortex lines.

The squared norm of the Jacobi field multiplied by $\frac{1}{2}$ is defined as

$$\Omega_t = \frac{1}{2} < \tilde{W}_t, \tilde{W}_t >|_{g_t} = \frac{1}{2}\int_M (W_t \cdot W_t)\, dx \; . \tag{32}$$

Its first time derivative is given by

$$\frac{d}{dt}\Omega_t = < W_t, \nabla_{W_t} v_t > \tag{33}$$

and the second time derivative is

$$\frac{d^2}{dt^2}\Omega_t = -\tilde{K}(v_t, W_t) + < \tilde{\nabla}_{W_t} v_t, \tilde{\nabla}_{W_t} v_t > \tag{34}$$

(Anosov 1967). The second term on the right hand side is the squared norm of the covariant derivative $\tilde{\nabla}_{W_t} v_t$ which is obviously non-negative, whereas the first is the curvature in the section spanned by v_t (velocity) and W_t (Jacobi) fields. If the curvature $\tilde{K}(v_t, W_t)$ is negative, the second time derivative becomes positive. When W_t is the vorticity field, then the Ω_t is the enstrophy $\frac{1}{2} < \omega_t, \omega_t >$ and its second time derivative will be positive if $\tilde{K}(v_t, \omega_t) < 0$. In the theory of turbulence the average of the right hand side of (33) is proportional to $-S$, S being the skewness factor of the distribution of the longitudinal derivative of velocity in isotropic flows. The value of S is known to be negative in the experiments and numerical simulations. Thus $d\Omega_t/dt$ is expected to be positive in turbulent flows.

Another example is the two-dimensional vorticity field or passive scalar field. Suppose that $\omega(x, y)$ is such a field in the cartesian (x, y) plane. A Jacobi field A_t (or $\mathbf{A}$) is defined by

$$A_t \; : \quad \mathbf{A} = \nabla \times (0, 0, \omega(x, y)) = (\omega_y, -\omega_x, 0) \tag{35}$$

Obviously this field is divergence-free. Time development of $\mathbf{A}$ in the divergence-free velocity field, $\mathbf{v} = (u, v, 0) = (\psi_y, -\psi_x, 0)$ where ψ is a stream-function, is governed by the equation (31) with $\mathbf{A}$ in place of ω where $(\omega = v_x - u_y = -\nabla^2\psi)$, which is rewritten as

$$\frac{\partial}{\partial t}\mathbf{A} + \nabla \times (\mathbf{A} \times \mathbf{v}) = 0$$

by using $\mathrm{div}\mathbf{A}=0$ and $\mathrm{div}\mathbf{v}=0$. Introducing the expression (35) leads to

$$\nabla \times [\, 0,\ 0,\ (\frac{\partial}{\partial t}\omega + (\mathbf{v}\cdot \mathrm{grad})\omega)\,] = 0 \; .$$

Integrating once, one obtains

$$\frac{\partial}{\partial t}\omega + (\mathbf{v} \cdot \mathrm{grad})\omega = 0 \quad . \tag{36}$$

This is just the two-dimensional vorticity equation without diffusion. In this case the squared norm (32) is the palinstrophy.

Acknowledgements

We are grateful to the Institute for Theoretical Physics, University of California, Santa Barbara where the final form of this article was prepared when one of the authors, T. Kambe, was staying.

References

ANOSOV, D.V., 1967 *Proc. Steklov Inst. Math.* (Am. Math. Soc., 1969) No. 90 .

ARNOLD, V.I., 1966 *Ann. Inst. Fourier*, Grenoble **16**, 1, 319; 1978 *Mathematical Methods of Classical Mechanics,* Appendix 2, Springer-Verlag.

DOMBRE, T., FRISCH, U., GREENE, J.M., Hénon, H., MEHR, A., & SOWARD, A.M., 1986 *J. Fluid Mech.* **167**, 353.

EBIN, D. G. & MARSDEN, J., 1970 *J. Ann. Math.* **90**, 102 .

HATTORI, Y., 1990 Master thesis, Dept. of Phys., Univ. of Tokyo.

HICKS, N. J., 1965 *Notes on Differential Geometry* Van Nostrand, Princeton.

KOBAYASHI, S. & NOMIZU, K., 1969 *Foundations of Differential Geometry* Vol. 1, John Wiley & Sons.

KRAICHNAN, R., 1959 *J. Fluid Mech.* **5**, 497.

MILNOR, J., 1963 *Morse Theory* Princeton University.

NAKAMURA, F., HATTORI, Y. & KAMBE, T., 1992 *J. Physics A: Mathematical and General*, to appear.

THE BEHAVIOR OF ACTIVE AND PASSIVE PARTICLES IN A CHAOTIC FLOW

M. D. DAHLEH
Department of Mathematics
University of California
Los Angeles, California 90024-1555, U.S.A.

ABSTRACT. In this paper we examine the behavior of particles in the exterior flow of a Kida vortex which is known to have a relatively large chaotic zone surrounding the entire ellipse. The variability of the finite time spreading rate of particles located in the chaotic region is presented. It is shown that the particles which have a high spreading rate are particles which pass near a stagnation point. We also show that the tendency of these particles to separate can be significantly reduced or even eliminated by increasing the circulation of each particle.

1. Introduction

Coherent structures are of current interest because they are known to emerge from randomly initialized two dimensional flows (McWilliams, 1984), and because these features are observed in fluids. Examples of coherent features include Gulf Stream Rings and the Great Red Spot of Jupiter. A simple model for a coherent vortex in a complicated flow is given by the Kida vortex, which consists of an elliptical patch of constant vorticity embedded in a uniform shear flow (Kida, 1981). It is the only known analytic nonlinear solution of the two dimensional Euler equations. One can use the behavior of particles in the exterior flow of the Kida ellipse as a model of more complicated coherent structures in two dimensional turbulent background flows.

In (Polvani and Wisdom, 1990) it is shown that a Kida vortex can advect passive tracers chaotically. Passive particles are particles whose vorticity is given by the vorticity of the background flow at that point. These particles do not interact with the flow. They are merely advected by it. Active particles, on the other hand, are particles which have concentrated vorticity in addition to the vorticity of the background flow. These particles can interact with each other and with the background flow (*i.e.* they are point vortices). In

H. K. Moffatt et al. (eds.), Topological Aspects of the Dynamics of Fluids and Plasmas, 505–515.

this paper, we study both active and passive particles.

The paper is organized as follows. First, the ordinary differential equations which govern the behavior of the Kida ellipse are presented. These equations describe the angle of inclination, ϕ, and the aspect ratio, λ, of the ellipse. Second, the equations for the Lagrangian motion of an active particle in the exterior flow are presented. Third, in order to gain a better understanding of the variability of the chaotic zones, finite time spreading rates for particles in a chaotic region of a Kida vortex are calculated. Finally, the effect of the addition of circulation on the behavior of two particles and four particles in the exterior flow is discussed. The two and four particle configurations represent a basic model of isolated eddies in a background flow field.

2. Governing Equations

The behavior of the Kida vortex is determined by the aspect ratio of the ellipse which is not constant in time and the angle of inclination which is the angle between the x-axis and the major axis of the ellipse (Kida, 1981). In the presence of the following background shear flow:

$$\Psi_B = \frac{1}{4}(w+s)x^2 + \frac{1}{4}(w-s)y^2, \tag{2.1}$$

the equations for the time evolution of the aspect ratio and the angle of inclination are given by

$$\frac{d\phi}{dt} = \frac{\lambda\kappa}{(1+\lambda)^2} + \frac{1}{2}(w + s(\frac{1+\lambda^2}{1-\lambda^2})\cos(2\phi)), \tag{2.2}$$

$$\frac{d\lambda}{dt} = -s\lambda\sin(2\phi), \tag{2.3}$$

where κ is the (uniform) vorticity within the ellipse, w is the rotation and s is the strain. The specific Kida vortex which we will study has $w = -.2$, $s = .1$, and $\lambda = .9$ when $\phi = 0$. In (Polvani and Wisdom, 1990), it is shown that this ellipse has a relatively large and uniformly dense chaotic region surrounding the elliptical patch of constant vorticity. Hence it is well suited for our purposes. In addition, the calculations presented have κ equal to 1 and the area of the ellipse equal to π. Hence, the Kida vortex has a circulation of π.

The above equations do not directly give the behavior of particles in the exterior of the flow. However one can use these equations, following the method presented in (Polvani and Wisdom, 1990), to derive the equations for the Lagrangian motion of an active tracer. If we had an expression for the total exterior stream function, denoted by Ψ_T, then the position of a particle would be given by

$$\frac{dx}{dt} = \frac{\partial \Psi_T}{\partial y} \quad \text{and} \quad \frac{dy}{dt} = -\frac{\partial \Psi_T}{\partial x}. \tag{2.4}$$

For the Kida vortex the total exterior stream function consists of the stream function for the background flow denoted Ψ_B, the stream function for particles in the exterior flow of an elliptical patch of constant vorticity denoted by Ψ_v (Lamb, 1945), and the stream function for particles with vorticity denoted by Ψ^a. For passive particles, $\Psi^a(x,y) = 0$. Since it is easier to convert Cartesian coordinates into elliptic coordinates than the other way around, the equations for the position of a particle are derived in elliptic coordinates. The appropriate coordinates for this problem are given by

$$x = c\cosh\rho\cos\theta\cos\phi - c\sinh\rho\sin\theta\sin\phi, \tag{2.5}$$

$$y = c\cosh\rho\cos\theta\sin\phi + c\sinh\rho\sin\theta\cos\phi, \tag{2.6}$$

where $c = \frac{(1-\lambda^2)}{\lambda}$. After straightforward algebraic calculations with equation (2.4), one obtains the following equations for the time evolution of ρ and θ:

$$\frac{d\rho}{dt} = \frac{h^2}{2}\Big(-(\Omega_k \sin 2\theta + 2c^{-2}\frac{\partial(\Psi_v + \Psi^a)}{\partial\theta}) + \frac{1}{2}sF(\rho,\theta,\phi)\Big), \tag{2.7}$$

$$\frac{d\theta}{dt} = \frac{h^2}{2}\Big(-(\Omega_k \sinh 2\rho - 2c^{-2}\frac{\partial(\Psi_v + \Psi^a)}{\partial\rho}) + \frac{1}{2}sG(\rho,\theta,\phi)\Big), \tag{2.8}$$

where $h^2 = (\cosh^2\rho - \cos^2\theta)^{-1}$,

$$\begin{aligned} F &= (\cosh 2\rho\sin 2\theta\cos 2\phi + \sinh 2\rho\cos 2\theta\sin 2\phi) \\ &- \Lambda(\cos 2\phi\sin 2\theta + \sinh 2\rho\sin 2\phi), \end{aligned} \tag{2.9}$$

$$\begin{aligned} G &= (\sinh 2\rho\cos 2\theta\cos 2\phi - \cosh 2\rho\sin 2\theta\sin 2\phi) \\ &+ \Lambda(\sin 2\phi\sin 2\theta - \sinh 2\rho\cos 2\phi), \end{aligned} \tag{2.10}$$

and $\Lambda = \frac{(1+\lambda^2)}{(1-\lambda^2)}$.

The motion of a particle in the exterior flow of the Kida ellipse is determined by integrating equations (2), (3), (7) and (8) with a fourth order predictor-corrector. The Cartesian positions can be recovered from the solutions of these equations by equations (5) and (6).

3. Finite time spreading rate

In this section we look at the separation of passive particles in the chaotic exterior flow over a finite time. One characteristic of chaotic regions is that the trajectories of particles which are initially very close to each other diverge exponentially. The Liapunov exponent is a measure of the exponential rate of divergence. However, it only exists in the long time limit. Since we are interested in the spreading of particles over a finite time, we need a method which measures the spreading for any time.

In order to determine the spreading rate for a given point in the flow, three particles are needed. One is initially placed at the point of interest, say the origin, and the other two are initially placed along orthogonal vectors passing through the original point. The matrix of initial positions is given by

$$S = \begin{pmatrix} x_1^i & x_2^i \\ y_1^i & y_2^i \end{pmatrix}, \tag{3.1}$$

where x_1^i and y_1^i represent the difference between the first point and the origin and similarly for x_2^i and y_2^i. As time progresses the initial vectors are acted on by the flow to form new vectors. The column vectors for the particle positions at a given time form the following matrix:

$$M = \begin{pmatrix} x_1^f & x_2^f \\ y_1^f & y_2^f \end{pmatrix}. \tag{3.2}$$

where x_1^f and y_1^f represent the difference between the current location of the first point and the current location of the origin and similarly for x_2^f and y_2^f. The multiplication of matrix M and S^{-1} gives the following matrix:

$$L = \frac{1}{DS}\begin{pmatrix} x_1^f y_2^i - x_2^f y_1^i & x_1^i x_2^f - x_1^f x_2^i \\ y_1^f y_2^i - y_2^f y_1^i & x_1^i y_2^f - x_2^i y_1^f \end{pmatrix} \tag{3.3}$$

where $DS = x_1^i y_2^i - x_2^i y_1^i$. In analogue to the Liapunov exponent, the natural log of the magnitude of the eigenvalues of the matrix L is defined to be the finite time spreading rate. Initially the two vectors are normalized so that the product of the eigenvalues of L is one. Since the flow is incompressible, the product of these eigenvalues equals one throughout the calculation. This fact is used to test the accuracy of the calculation.

The finite time spreading rates are calculated for a small region of the flow for the Kida vortex specified by $w = -.2$, $s = .1$ and $\lambda = .9$ when $\phi = 0$. Since we are calculating the particle positions using the ordinary differential equations of section 2, the calculation is carried out in (ρ, θ) coordinates. The particle separation needs to be reinitialized whenever it becomes too large. When the particle separation exceeds a given distance, the vector connecting the particles is deflated towards the center. The deflation in ρ and θ corresponds to a deflation in x and y. The Cartesian coordinate deflation is given by

$$\begin{pmatrix} \Delta x \\ \Delta y \end{pmatrix} = \begin{pmatrix} \frac{\partial x}{\partial \rho} & \frac{\partial x}{\partial \theta} \\ \frac{\partial y}{\partial \rho} & \frac{\partial y}{\partial \theta} \end{pmatrix}_{\rho_m, \theta_m} \begin{pmatrix} \Delta \rho \\ \Delta \theta \end{pmatrix} \tag{3.4}$$

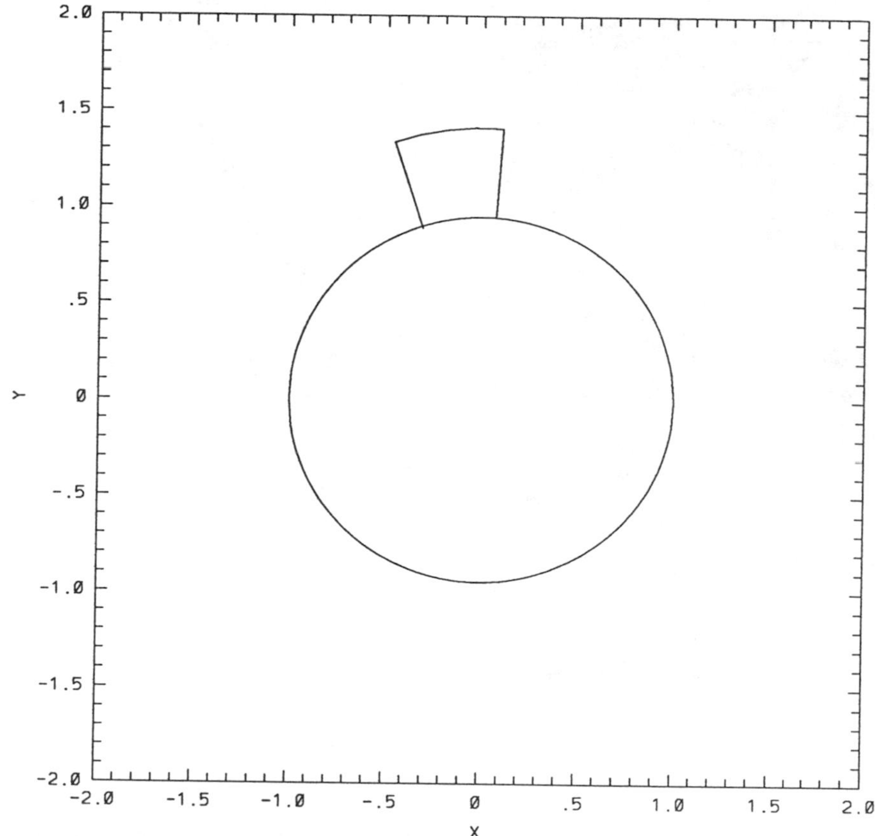

Fig. 1a. A Kida ellipse with $\lambda = .9$ when $\phi = 0$. The small box is the region for which we calculate the finite time spreading rate. The region is bounded by $\rho = 2.2$, $\rho = 2.49$, $\theta = 1.5$ and $\theta = 1.99$.

where

$$\Delta\rho = \rho(2) - \rho(1), \tag{3.5}$$

$$\Delta\theta = \theta(2) - \theta(1), \tag{3.6}$$

and ρ_m and θ_m are the coordinates of the midpoint. The deflation in the x and y coordinates is computed so that the vector may be restored to the correct length at the end of the calculation. This procedure gives the separation distance (Lichtenberg and Liebermann, 1983).

The region for which we calculate the finite time spreading rate is bounded by $\rho = 2.2$, $\rho = 2.49$, $\theta = 1.5$, and $\theta = 1.99$ (see figure 1a). Figures (1b-1c) show the variability of the spreading rate after two revolutions of the inner ellipse.

These calculations have a spatial resolution of 50x30. One can clearly see a banded structure of high spreading in the figures. Figure 1b has been

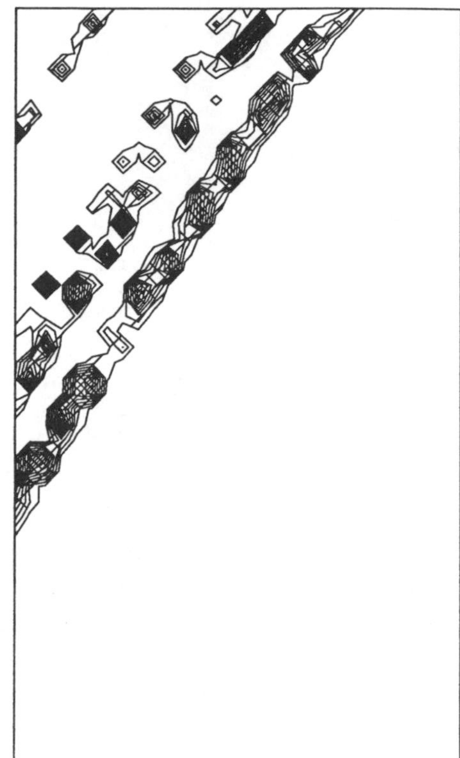

Fig. 1b, 1c. Spatial distribution of the finite time spreading rate after two revolutions of the Kida ellipse with vorticity 1, area π, and $\lambda = .9$ when $\phi = 0$. The background shear flow is given by eqn. (2.1) with $w = -.2$ and $s = .1$. The spatial resolution is 50x30. (b) The contour levels have been chosen to show regions of similar spreading rates. The higher spreading rates are indicated by the darker shading. (c) In order to see the details of the banded region it is contoured separately. The darker shading represents the higher spreading rates.

scaled so that regions which have similar spreading rates can be identified. In large regions of the rest of the flow, the spreading rates are close to zero which indicates that the particles do not separate for the duration of the calculation.

In order to understand the behavior of particles in this banded region, individual particle trajectories are examined. The particle trajectories for three particles which are located near each other in the banded region and whose Liapunov exponents differ greatly are shown in figure 2–figure 4. One can see from figure 2, that this particle came close to a stagnation point of the flow. This particle has a spreading rate of 6.0282. The particle next to it does not come as close to a stagnation point (see figure 3) and its Liapunov exponent is 2.93 which is smaller. In figure 4, the spreading rate is close to

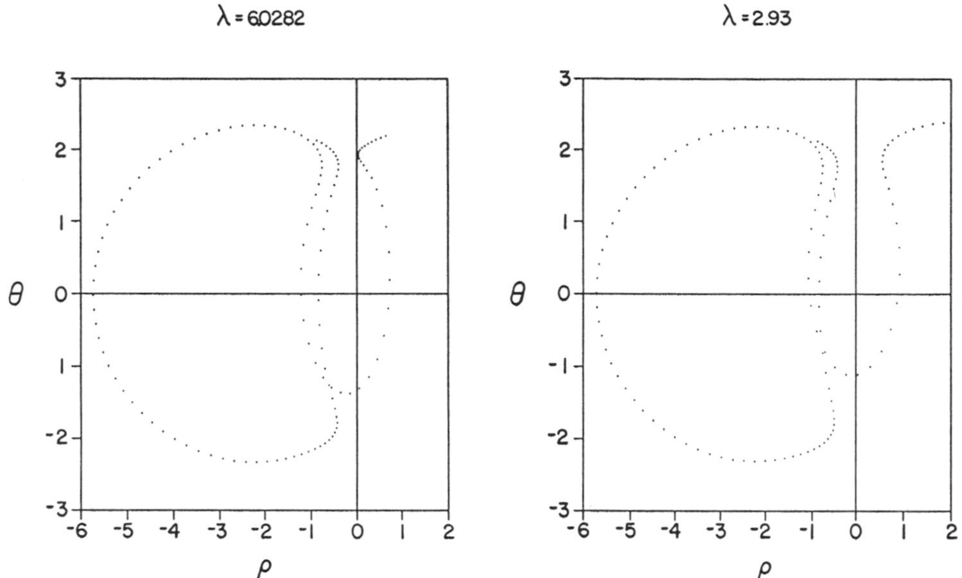

Fig. 2, 3. Particle trajectory for two revolutions of the Kida ellipse. After two revolutions the particle has a finite time Liapunov exponent of (2) 6.0282 and (3) 2.93.

1. This particle is three gridpoints away from the particle in the previous figure and its particle trajectory traces out a path which does not come as close to a stagnation point. Figure 5 shows a periodic particle trajectory. This particle has a spreading rate of 1.8782×10^{-3}. By examining particle trajectories, we see that a large spreading rate is associated with passing near a stagnation point.

4. Active particles

In this section we examine the effect of circulation on the behavior of the particles in the same region of the exterior flow as above. We consider two configurations of particles. The first consist of two vortices which are located at the same θ coordinate, but which are .01 apart in ρ. The second configuration consists of four vortices. Two are the same as in the previous case, and two are located at the ρ midpoint and separated from the original two points by .005 in θ. As we will see, the amount of circulation needed to prohibit the particles from separating is surprisingly small in comparison to the circulation of the ellipse. The circulation prohibits separation because the mutually induced rotation of the particles averages out the strain field that is trying to pull them apart.

From the previous finite time Liapunov exponent calculation for the Kida ellipse, we know that in this region not all of the passive particles will sep-

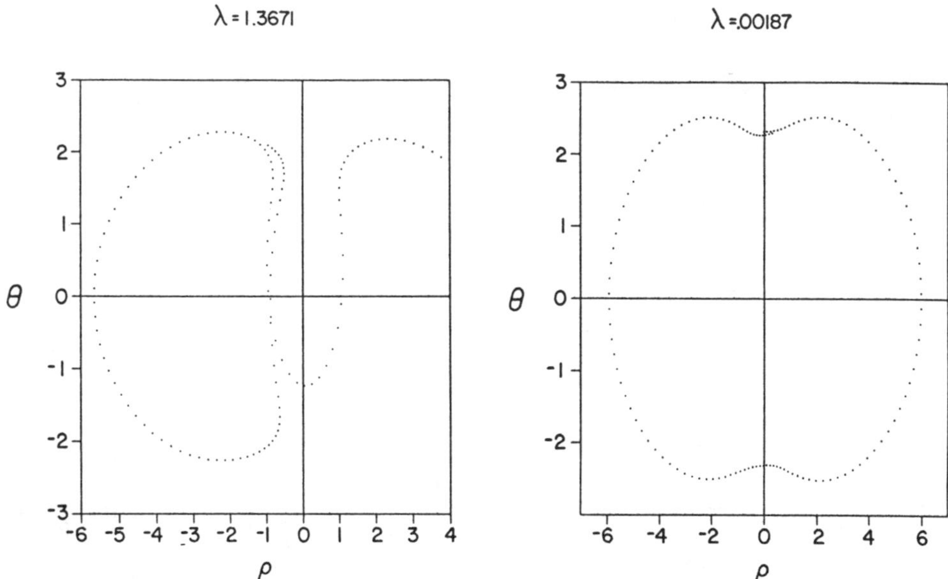

Fig. 4, 5. Particle trajectory for two revolutions of the Kida ellipse. (4) The particle has a finite time Liapunov exponent of 1.3671. (5) Periodic trajectory of a particle with the finite time Liapunov exponent of .00187.

arate after two revolutions of the inner ellipse. In order to calculate the distribution of separation times, the particle trajectories of 375 particles, which are uniformly distributed in the region, are compared. In either the two or four particle configuration, particles are considered to have separated once the distance between the particles is greater than .2. The particles which do not separate are recorded in column 21.

First we will examine the behavior of the two vortex configuration. In figure 6, the distribution of separation times for passive particles for this configuration is given. One can see that all but 100 of the particles separate. With the addition of even a small amount of circulation the number of particles which separate decreases significantly, and many of the particles separate at a later time. Figure 7 and figure 8 show the separation time for particles which have the circulation equal to .0008 and .001 respectively. It should not be surprising that the active particles which separate when the circulation is equal to .001 lie in regions where the passive particles have high finite time spreading rates (see figure 9). When the circulation of each particle is increased to .006, all of the pairs of particles studied stay together for two revolutions of the Kida ellipse.

The behavior of the four vortex configuration is very similar to the two vortex case. In the four particle configuration, all of the passive particles separate before the inner ellipse has completed two revolutions. Since the

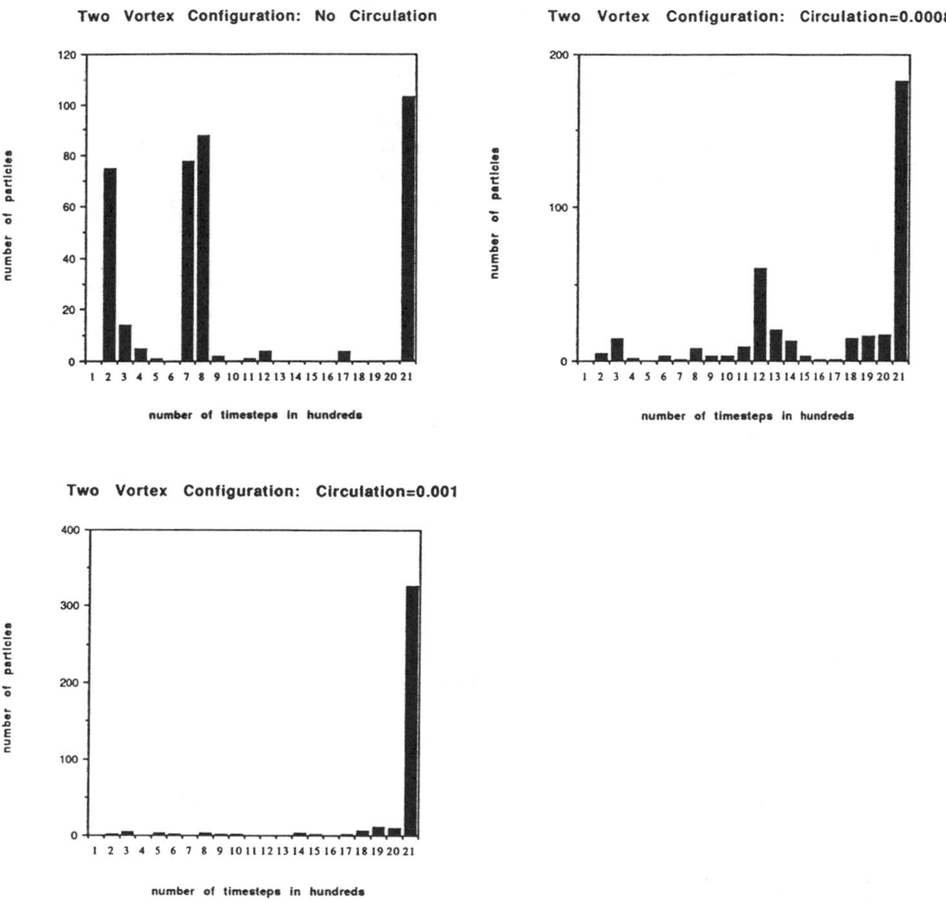

Fig. 6–8. Histogram for the two particle configuration indicating the time at which the two particles become a distance greater than .2. The number of particles in box 21 represents the particles which do not separate for two revolutions of the Kida ellipse. (6) Particles with no circulation. (7) Particles with the circulation equal to .0008. (8) Particles with the circulation equal to .001.

four particle configuration has two directions in which the particles can separate, it is more likely that the passive particles will diverge away from each other in this configuration than in the previous one. However, this configuration requires a smaller amount of circulation to hold the particles together (see figure 6b). When the circulation of each particle is .0004, providing a total circulation of .0016, many fewer particles separate. A circulation of .001 added to each of the particles, will prevent any of the particles from separating. This is less circulation than is needed for the two vortex case. The actual amount of circulation needed to prevent a configuration from breaking up depends on the number of vortices and the distance between the vortices. In both the two and four vortex cases the amount of circulation

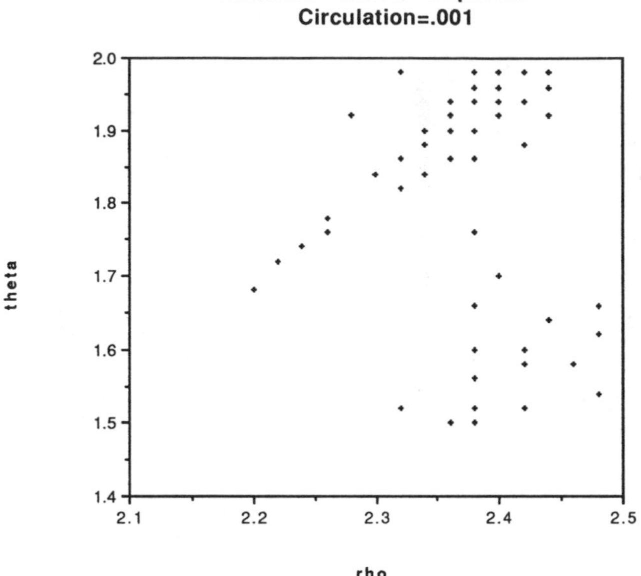

Fig. 9. The location in ρ-θ coordinates of particles which have circulation .001 and which do not separate during the calculation.

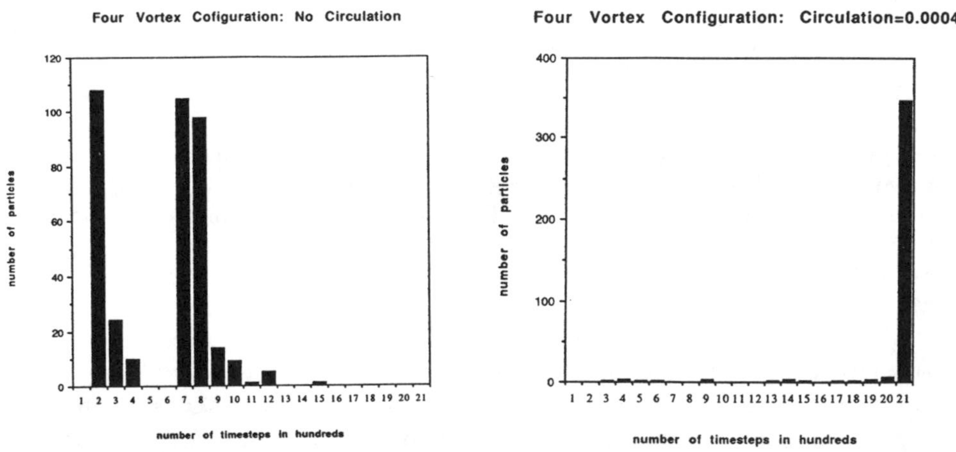

Fig. 10, 11. Histogram for the four particle configuration indicating the time at which any two particles become a distance greater than .2. The number of particles in box 21 represents the particles which do not separate for the duration of the calculation. (10) Particles with no circulation. (11) Particles with the circulation equal to 0.0004.

needed is surprisingly small in comparison to π which is the circulation of the inner ellipse.

These results suggest that sufficiently intense vortices can survive the tendency for large scale velocity fields to disperse them. However, the vortex breakup problem shows very sensitive dependence on initial conditions, so the length of time a vortex survives depends very much on its initial location.

5. Conclusions

In an effort to understand the behavior of passive particles in the chaotic exterior region of a Kida ellipse, the spreading rate of particles after two revolutions of the inner ellipse is presented. One sees a large variability of these rates in the region studied. By examining individual particle trajectories, we conclude that particles which experience a high spreading rate are particles which pass near a stagnation point. The ability of particles to separate can be eliminated by the addition of small amounts of circulation to the particles. The amount of circulation needed is on the order of 10^3 less than the circulation of the ellipse.

Acknowledgements

Part of this work was conducted while the author was at the National Center for Atmospheric Research which is funded by the National Science Foundation.

References

KIDA, S., 1981 Motion of an elliptic vortex in a uniform shear flow. *Journal of the Physical Society of Japan*, **50** (10), 3517-3520.

LAMB, H., 1945 *Hydrodynamics* (Dover, New York).

LICHTENBERG, A. AND LIEBERMANN, M., 1983 *Regular and Stochastic Motion* (Springer-Verlag, New York).

MCWILLIAMS, J., 1984 The emergence of isolated coherent vortices in turbulent flow, *Journal of Fluid Mechanics* **146**, 21-43.

POLVANI, L. AND WISDOM, J. 1990 Chaotic Lagrangian trajectories around an elliptical patch embedded in a constant and uniform background shear flow. *Phys. Fluids A* **2** (2), 123-126.

CHAOS ASSOCIATED WITH FLUID INERTIA.

K. BAJER* & H.K. MOFFATT **
Institute for Theoretical Physics
University of California
Santa-Barbara, California 93106-4030, U.S.A.

ABSTRACT. The low Reynolds number flow between two concentric steadily rotating spheres is considered. The pattern of streamlines is explained in terms of an adiabatic invariant. It is shown that, when the two rotation vectors are not parallel, the streamlines become chaotic when the Reynolds number Re is increased and the onset of global chaos occurs near $Re = 20$.

1. Introduction

The objective of this work is to determine whether, and to what extent, fluid inertia is responsible for the presence of chaos, and associated particle dispersion, in the streamlines of steady flow of an incompressible fluid. In a parallel study (Bajer & Moffatt 1990, Bajer, Moffatt & Nex, 1990) it has been shown that a wide family of inertialess (Stokes) flows in a spherical domain are characterised by chaotic streamlines; however these flows can be generated only by appropriate, and rather artificial, conditions imposed on the tangential velocity at the boundary, and would be difficult to realise in practise. Stone, Nadim & Strogatz (1991) have studied similar Stokes flows inside a spherical droplet immersed in a general linear flow. Such flows have chaotic streamlines and are possible to realise in practice. The problem of the flow inside a droplet is similar to the problem of the electromagnetic stirring of molten metals (Moffatt, 1991). High frequency external magnetic fields together with the shape of the domain occupied by liquid metal determine the velocity components tangent to the surface of the domain. There exists a unique Stokes flow compatible with this imposed surface velocity, but in metallurgical applications the Reynolds number is, usually, high and

* Permanent address: University of Warsaw, Institute of Geophysics, ul. Pasteura 7, 02-093 Warszawa, Poland
** Permanent address: DAMTP, University of Cambridge, Silver Street, Cambridge CB3 9EW, UK.

H. K. Moffatt et al. (eds.), Topological Aspects of the Dynamics of Fluids and Plasmas, 517–534.

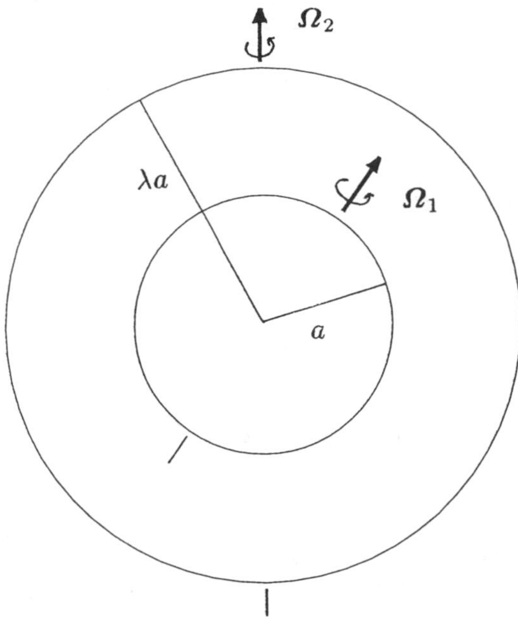

Fig. 1. The domain of the flow - spherical annulus between two concentric solid spheres rotating steadily.

fluid inertia is important. The flow that we consider in this paper is easily realisable. In the Stokes limit, in which inertia is neglected, it has closed streamlines. The question that we adress is: what happens to the streamline pattern when account is taken of fluid inertia, even when this is weak in relation to viscous effects.

Specifically, we consider the flow in the spherical annulus between two concentric spheres of radii a, λa ($\lambda > 1$), rotating with angular velocities $\boldsymbol{\Omega}_1$, $\boldsymbol{\Omega}_2$ respectively (figure 1). In the Stokes regime, the solution to this problem is well-known (Landau & Lifshitz, 1987, p.65): the steady velocity field $\mathbf{u}^{(0)}(\mathbf{x})$ driven by viscous stresses alone is given by

$$\mathbf{u}^{(0)}(\mathbf{x}) = f_1(r)\boldsymbol{\Omega}_1\wedge\mathbf{x} + f_2(r)\boldsymbol{\Omega}_2\wedge\mathbf{x}, \tag{1.1}$$

where r is distance from the centre measured in units of a (i.e. $1 < r < \lambda$ in the annulus), and

$$f_1(r) = \frac{(\lambda/r)^3 - 1}{\lambda^3 - 1} \quad , \qquad f_2(r) = \frac{\lambda^3 - (\lambda/r)^3}{\lambda^3 - 1} = 1 - f_1(r) \tag{1.2}$$

The flow (1.1) satisfies the condition $\mathbf{x}\cdot\mathbf{u}^{(0)}$, i.e. it is a purely *toroidal* flow, and may be expressed in the form

$$\mathbf{u}^{(0)} = \nabla\wedge\left(\mathbf{x}T^{(0)}(\mathbf{x})\right) \tag{1.3}$$

where

$$T^{(0)}(\mathbf{x}) = (f_1(r)\boldsymbol{\Omega}_1 + f_2(r)\boldsymbol{\Omega}_2)\cdot\mathbf{x} \quad , \tag{1.4}$$

the streamlines of the flow being given by the intersections of the surfaces $T^{(0)} = cst.$ with spheres $r = cst.$ These streamlines are circles, the circles on a sphere of radius r all lying in planes with normal parallel to the vector $f_1(r)\boldsymbol{\Omega}_1 + f_2(r)\boldsymbol{\Omega}_2$. This vector varies continuously as a function of r from $\boldsymbol{\Omega}_1$ on $r = 1$ to $\boldsymbol{\Omega}_2$ on $r = \lambda$. Obviously the Stokes flow (1.3) has no simple symmetry, and it may be expected that this simple streamline pattern may be severly disrupted when account is taken of inertia.

In section 2 we calculate the secondary flow induced by the rotation of only one sphere, while a general formula for arbitrary $\boldsymbol{\Omega}_1$ and $\boldsymbol{\Omega}_2$ is given in section 3. In order to understand the pattern of streamlines in the limit $Re = \Omega_1 a^2/\nu \to 0$ we derive an equation for the adiabatic drift (section 4) and consider the influence of the departures from axisymmetry on the drift surfaces (section 5).

2. Secondary flow associated with rotation of one sphere

It is a straightforward matter to calculate the secondary flow $\mathbf{u}^{(1)}(\mathbf{x})$ driven by (weak) inertia forces when only one of the spheres rotates. Specifically, suppose that $\boldsymbol{\Omega}_1 \neq 0$, $\boldsymbol{\Omega}_2 = 0$. Then the resulting flow will be axisymmetric about the direction of $\boldsymbol{\Omega}_1$. Let (r,θ,φ) be spherical polar coordinates based on this axis of symmetry, and let $\psi^{(1)}(r,\theta)$ be the Stokes streamfunction of the secondary flow in the meridian plane; then the streamlines of the composite flow $\mathbf{u}^{(0)} + \mathbf{u}^{(1)}$ will be helices wound on the family of nested tori $\psi^{(1)}(r,\theta) = cst.$ (figure 2), the helicity being antisymmetric about the plane $\theta = \pi/2$.

With $\mathbf{u}^{(0)}(\mathbf{x}) = f_1(r)\boldsymbol{\Omega}_1\wedge\mathbf{x}$ the convective (centrifugal) acceleration associated with $\mathbf{u}^{(0)}$ is

$$\begin{aligned}\mathbf{u}^{(0)}\cdot\nabla\mathbf{u}^{(0)} &= f_1(r)(\boldsymbol{\Omega}_1\wedge\mathbf{x})\cdot\nabla\{f_1(r)\boldsymbol{\Omega}_1\wedge\mathbf{x}\}\\ &= f_1^2(r)[(\mathbf{x}\cdot\boldsymbol{\Omega}_1)\boldsymbol{\Omega}_1 - \boldsymbol{\Omega}_1^2\mathbf{x}],\end{aligned} \tag{2.1}$$

and the curl of this quantity, which is responsible for the generation of vorticity in the secondary flow, is given by

$$\begin{aligned}\nabla\wedge\{\mathbf{u}^{(0)}\cdot\nabla\mathbf{u}^{(0)}\} &= (\mathbf{x}\cdot\boldsymbol{\Omega}_1)G_{11}(r)(\mathbf{x}\wedge\boldsymbol{\Omega}_1)\\ &= -\tfrac{1}{2}\nabla\wedge\{(\boldsymbol{\Omega}_1\cdot\mathbf{x})^2 G_{11}(r)\mathbf{x}\}\end{aligned} \tag{2.2}$$

where

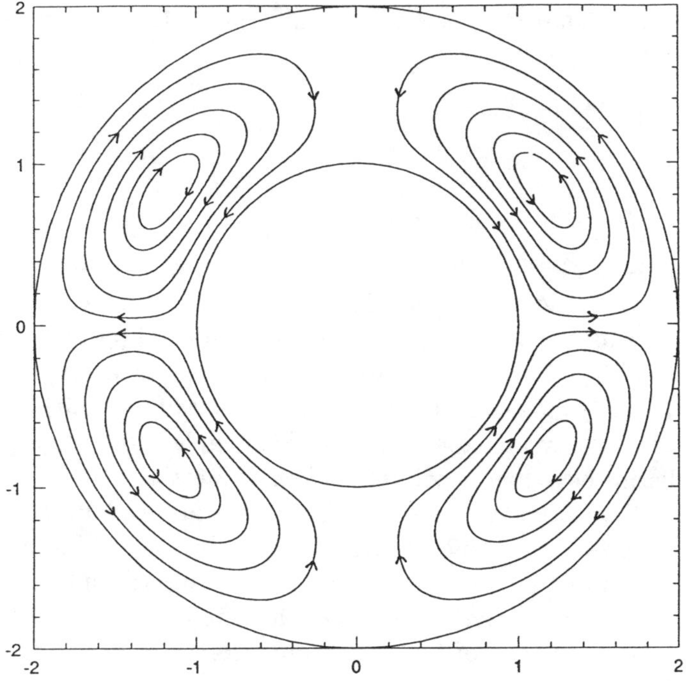

Fig. 2. Surfaces of constant $\Psi^{(1)}$ for $\boldsymbol{\Omega}_1 = 1$, $\boldsymbol{\Omega}_2 = 0$, $\lambda = 2$.

$$G_{11}(r) = 2r^{-1} f_1(r) f_1'(r) = \frac{6\lambda^3(1 - (\lambda/r)^3)}{(\lambda^3 - 1)^2 r^5} \quad . \tag{2.3}$$

The vorticity field $\boldsymbol{\omega}^{(1)} = \boldsymbol{\nabla} \wedge \mathbf{u}^{(1)}$ driven by (2.2) is toroidal in character; this means that, as expected, $\mathbf{u}^{(1)}$ is poloidal (i.e. meridional in this axisymmetric situation). Writing $\mathbf{u}^{(1)}$ in the form

$$\mathbf{u}^{(1)} = \boldsymbol{\nabla} \wedge \boldsymbol{\nabla} \wedge (\mathbf{x} P^{(1)}(\mathbf{x})) \tag{2.4}$$

we have

$$\boldsymbol{\omega}^{(1)} = \boldsymbol{\nabla} \wedge \mathbf{u}^{(1)} = -\boldsymbol{\nabla} \wedge (\mathbf{x} \boldsymbol{\nabla}^2 P^{(1)}). \tag{2.5}$$

Note that $P^{(1)}$ and $\psi^{(1)}$ are related by

$$\psi^{(1)}(r, \theta) = r \sin\theta \frac{\partial P^{(1)}}{\partial \theta}. \tag{2.6}$$

Morover

$$\nabla^2 \boldsymbol{\omega}^{(1)} = -\boldsymbol{\nabla} \wedge (\boldsymbol{\nabla} \wedge \boldsymbol{\omega}^{(1)}) = -\boldsymbol{\nabla} \wedge (\mathbf{x} \nabla^4 P^{(1)}), \tag{2.7}$$

Now, to first order in the Reynolds number Re, $\boldsymbol{\omega}^{(1)}$ satisfies the equation

$$\nabla^2\boldsymbol{\omega}^{(1)} = \nu^{-1}\nabla\wedge(\mathbf{u}^{(0)}\cdot\nabla\mathbf{u}^{(0)}) \quad , \tag{2.8}$$

where ν is the kinematic viscosity of the fluid. Comparison of (2.3) and (2.7) then provides the equation for $P^{(1)}$:

$$\nabla^4 P^{(1)} = -\tfrac{1}{2}\nu^{-1}(\boldsymbol{\Omega}_1\cdot\mathbf{x})^2 G_{11}(r) + F(r) \quad . \tag{2.9}$$

where $F(\cdot)$ is an arbitrary function. The no-slip boundary conditions are satisfied provided

$$P^{(1)} = 0 \quad , \qquad \partial P^{(1)}/\partial r = 0 \qquad \text{on} \qquad r = 1, \lambda \quad . \tag{2.10}$$

The required solution has the form

$$P^{(1)}(\mathbf{x}) = \alpha(r)(\boldsymbol{\Omega}_1\cdot\mathbf{x})^2 + \beta(r)\boldsymbol{\Omega}_1^2 \tag{2.11}$$

and then from (2.6),

$$\psi^{(1)}(r,\theta) = 2\boldsymbol{\Omega}_1^2 r\alpha(r)\cos\theta\sin^2\theta \quad . \tag{2.12}$$

Substitution of (2.11) in (2.9) leads to a fourth-order equation for $\alpha(r)$, namely

$$\left\{\frac{1}{r^6}\frac{d}{dr}r^6\frac{d}{dr}\right\}^2\alpha(r) = -\frac{1}{2\nu}G_{11}(r) \quad . \tag{2.13}$$

It is straightforward, but tedious, to find the solution of this equation ($G_{11}(r)$ being given by 2.3), with boundary conditions (from 2.10)

$$\alpha = 0 \quad , \qquad \partial\alpha/\partial r = 0 \qquad \text{on} \qquad r = 1, \lambda \quad . \tag{2.14}$$

Figure 2 shows the surfaces of constant $\Psi^{(1)}$ for $\boldsymbol{\Omega}_1 = \hat{\mathbf{i}}_z$, $\boldsymbol{\Omega}_2 = 0$, $\lambda = 2$.

3. Spheres spinning around different axes

Let us now consider $\boldsymbol{\Omega}_1 \neq \boldsymbol{\Omega}_2$. The Stokes flow (1.1) is no longer axisymmetric and the curl of the convective acceleration (which is not centrifugal) takes form:

$$\nabla\wedge\left(\mathbf{u}^{(0)}\cdot\nabla\mathbf{u}^{(0)}\right) = \nabla\wedge\left(\mathbf{u}^{(0)}\wedge\boldsymbol{\omega}^{(0)}\right) \tag{3.1}$$

$$= \nabla\wedge\left\{\left[\frac{(f_1^2)'}{2r}(\boldsymbol{\Omega}_1\cdot\mathbf{x})^2 + \frac{(f_2^2)'}{2r}(\boldsymbol{\Omega}_2\cdot\mathbf{x})^2 + \frac{(f_1f_2)'}{r}(\boldsymbol{\Omega}_1\cdot\mathbf{x})(\boldsymbol{\Omega}_2\cdot\mathbf{x})\right]\mathbf{x}\right\}$$

The equation for $P^{(1)}$, analogous to (2.9, 2.10) becomes:

$$\nabla^4 P^{(1)} = \frac{(f_1^2)'}{2r}(\boldsymbol{\Omega}_1\cdot\mathbf{x})^2 + \frac{(f_2^2)'}{2r}(\boldsymbol{\Omega}_2\cdot\mathbf{x})^2 + \frac{(f_1f_2)'}{r}(\boldsymbol{\Omega}_1\cdot\mathbf{x})(\boldsymbol{\Omega}_2\cdot\mathbf{x}) + F(r)$$

$$P^{(1)} = 0, \qquad \frac{\partial P^{(1)}}{\partial r} = 0 \qquad \text{on} \quad r = 1, \lambda \tag{3.2}$$

In spherical geometry this double-harmonic equation can be solved by a general procedure of expanding $P^{(1)}$ and the right-hand side in a series of spherical harmonics. With an appropriate choice of $F(\cdot)$ the solution is:

$$P^{(1)} = \frac{3}{(\lambda^{-3}-1)^2}\left[\alpha_1(r)(\boldsymbol{\Omega}_1\cdot\mathbf{x})^2 + \alpha_2(r)(\boldsymbol{\Omega}_2\cdot\mathbf{x})^2 + \alpha_3(r)(\boldsymbol{\Omega}_1\cdot\mathbf{x})(\boldsymbol{\Omega}_2\cdot\mathbf{x})\right],$$

where α_i satisfy the following ODEs:

$$\left(\hat{O}_6\right)^2 \alpha_1(r) = -\frac{1}{r^8} + \frac{\lambda^{-3}}{r^5} \tag{3.3a}$$

$$\left(\hat{O}_6\right)^2 \alpha_2(r) = -\frac{1}{r^8} + \frac{1}{r^5} \tag{3.3b}$$

$$\left(\hat{O}_6\right)^2 \alpha_3(r) = +\frac{2}{r^8} - \frac{1+\lambda^{-3}}{r^5} \tag{3.3c}$$

$$\alpha_i(1) = \alpha_i(\lambda) = \alpha_i'(1) = \alpha_i'(\lambda) = 0 \tag{3.3d}$$

$$\hat{O}_6 = \frac{1}{r^6}\frac{d}{dr}r^6\frac{d}{dr} \tag{3.4}$$

A general solution of (3.3) takes form:

$$\alpha_i = \frac{(r-1)^2(r-\lambda)^2}{r^5}\left[C_i^0 + C_i^1 r + C_i^2 r^2 + C_i^3 r^3\right] \tag{3.5}$$

and the coefficients C_i^j are determined by the boundary conditions. Finally we obtain the following expression for $P^{(1)}$:

$$P^{(1)} = L(\lambda)R(r,\lambda)\sum_{i,j=1}^{2}\sum_{k=0}^{3} W_{ij}^k(\lambda)r^k(\boldsymbol{\Omega}_i\cdot\mathbf{x})(\boldsymbol{\Omega}_j\cdot\mathbf{x}) \tag{3.6}$$

where:

$$L(\lambda) = \frac{-\frac{1}{8}\lambda^3(\lambda^3-1)^{-2}}{4\lambda^6 + 16\lambda^5 + 40\lambda^4 + 55\lambda^3 + 40\lambda^2 + 16\lambda + 4} \tag{3.7}$$

$$R(r,\lambda) = \frac{(r-1)^2(r-\lambda)^2}{r^5} \tag{3.8}$$

$$W_{11}^0(\lambda) = 2\lambda^7 + 8\lambda^6 + 20\lambda^5 + 16\lambda^4 + \lambda^3 - 2\lambda^2 \tag{3.9}$$

$$W_{11}^1(\lambda) = 4\lambda^6 + 16\lambda^5 + 17\lambda^4 - 6\lambda^3 - 18\lambda^2 - 8\lambda$$

$$W_{11}^2(\lambda) = 2(1+\lambda)(3\lambda^4 + 2\lambda^3 - 5\lambda^2 - 8\lambda - 2)$$

$$W_{11}^3(\lambda) = 3\lambda^4 + 2\lambda^3 - 5\lambda^2 - 8\lambda - 2$$

$$W_{22}^0(\lambda) = -2\lambda^7 + \lambda^6 + 16\lambda^5 + 20\lambda^4 + 8\lambda^3 + 2\lambda^2$$

$$W_{22}^1(\lambda) = -8\lambda^7 - 18\lambda^6 - 6\lambda^5 + 17\lambda^4 + 16\lambda^3 + 4\lambda^2$$

$$W_{22}^2(\lambda) = 2(1+\lambda)\lambda^2(-2\lambda^4 - 8\lambda^3 - 5\lambda^2 + 2\lambda + 3)$$

$$W_{22}^3(\lambda) = \lambda^2(-2\lambda^4 - 8\lambda^3 - 5\lambda^2 + 2\lambda + 3)$$

$$W_{12}^0(\lambda) = W_{21}^0(\lambda) = -\frac{1}{2}\lambda^3(9\lambda^3 + 36\lambda^2 + 36\lambda + 9)$$

$$W_{12}^1(\lambda) = W_{21}^1(\lambda) = -\lambda(-4\lambda^6 - 7\lambda^5 + 5\lambda^4 + 17\lambda^3 + 5\lambda^2 - 7\lambda - 4)$$

$$W_{12}^2(\lambda) = W_{21}^2(\lambda) = 2(1+\lambda)(\lambda^6 + 4\lambda^5 + \lambda^4 - 2\lambda^3 + \lambda^2 + 4\lambda + 1)$$

$$W_{12}^3(\lambda) = W_{21}^3(\lambda) = \lambda^6 + 4\lambda^5 + \lambda^4 - 2\lambda^3 + \lambda^2 + 4\lambda + 1$$

When $\boldsymbol{\Omega}_1 = \boldsymbol{\Omega}_2$ the flow in a spherical annulus is a solid body rotation. It is then an exact solution of the Navier-Stokes equation, and hence the secondary flow (and the higher order corrections) vanish identically. This implies the following identity:

$$\sum_{i,j=1}^{2} W_{ij}^k(\lambda) = 0 \qquad \text{for all} \quad \lambda \quad ; \tag{3.10}$$

which can be verified directly from (3.9).

Substituting (3.6) into (2.4) we finally obtain an expression for the secondary flow:

$$\begin{aligned} \mathbf{u}_1 &= A\boldsymbol{\Omega}_1 + B\boldsymbol{\Omega}_2 - C\mathbf{x} \quad , \\ A &= \left(3C_{11} + r^2 D_{11}\right)(\boldsymbol{\Omega}_1 \cdot \mathbf{x}) + \left(3C_{12} + r^2 D_{12}\right)(\boldsymbol{\Omega}_2 \cdot \mathbf{x}) \quad , \\ B &= \left(3C_{12} + r^2 D_{12}\right)(\boldsymbol{\Omega}_1 \cdot \mathbf{x}) + \left(3C_{22} + r^2 D_{22}\right)(\boldsymbol{\Omega}_2 \cdot \mathbf{x}) \quad , \\ C &= D_{11}(\boldsymbol{\Omega}_1 \cdot \mathbf{x})^2 + 2D_{12}(\boldsymbol{\Omega}_1 \cdot \mathbf{x})(\boldsymbol{\Omega}_2 \cdot \mathbf{x}) + D_{22}(\boldsymbol{\Omega}_2 \cdot \mathbf{x})^2 \\ &\quad + C_{11}\boldsymbol{\Omega}_1^2 + 2C_{12}\boldsymbol{\Omega}_1 \cdot \boldsymbol{\Omega}_2 + C_{22}\boldsymbol{\Omega}_2^2 \quad ; \end{aligned} \tag{3.11}$$

where C_{ij}, D_{ij} are functions of r and λ:

$$C_{ij} = 2LR \sum_{k=0}^{3} W_{ij}^k r^k \quad , \tag{3.12a}$$

$$D_{ij} = 2L \left[r^{-1} \frac{dR}{dr} \sum_{k=0}^{3} W_{ij}^k r^k + \frac{R}{r^{-2}} \sum_{k=0}^{3} k W_{ij}^k r^k \right] \quad ; \tag{3.12b}$$

which satisfy an identity analogous to (3.10):

$$\sum_{i,j=1}^{2} C_{ij} = \sum_{i,j=1}^{2} D_{ij} = 0 \quad . \tag{3.13}$$

When $\boldsymbol{\Omega}_1$, $\boldsymbol{\Omega}_2$ are parallel, for example when

$$\boldsymbol{\Omega}_1 = \omega_1 \hat{\mathbf{i}}_z \quad , \qquad \boldsymbol{\Omega}_2 = \omega_2 \hat{\mathbf{i}}_z \quad , \tag{3.14}$$

we obtain:

$$P^{(1)} = L(\lambda) R(r, \lambda) r^2 \cos^2 \theta \sum_{i,j=1}^{2} \sum_{k=0}^{3} W_{ij}^k(\lambda) r^k \omega_i \omega_j \tag{3.15}$$

The streamlines of $\mathbf{u}^{(0)} + Re\mathbf{u}^{(1)}$ lie on the surfaces of constant $\Psi^{(1)}$, which can be easily derived from (2.6). The double sum in (3.15) is a cubic polynomial in r. For $\omega_2/\omega_1 \in I = [-0.641, -0.251]$ one of its roots satisfies $1 \leq r_* \leq 2$. Then the secondary flow given by $\Psi^{(1)}$ is a 'two-cell' flow (figure 3), while for $\omega_2/\omega_1 \notin I$ it is a 'single-cell' flow similar to that in figure 2. In particular $\omega_1 = -\omega_2$ is the latter case.

The situation when $(\omega 2 - \omega 1)/\omega 1 \ll 1$ and $Re \gg 1$ was considered by Proudman (1956); in this limit the flow is dominated by Coriolis forces and the structured is determined by boundary layers and internal shear layers. This limit is quite different from the flow considered here.

4. Adiabatic invariants associated with a weak secondary flow

When the streamlines of the basic flow are closed loops, then a small perturbation $\mathbf{u}^{(1)}$ typically causes a drift across closed orbits (Bajer & Moffatt, 1990). The drift is slow compared with the periods of the unperturbed orbits. Every fluid particle, drifting slowly, 'selects' a 1-parameter family of unperturbed orbits, i.e. it outlines a two-dimensional surface [1]. In our case a basic (Stokes) flow is given by (1.1) and its (closed) streamlines are determined by two invariants:

$$I_1 = \boldsymbol{\Omega}(r) \cdot \mathbf{x} = const. \quad , \qquad I_2 = r^2 = \mathbf{x}^2 = const. \quad ; \tag{4.1}$$

[1] In the vicinity of stagnation points the time-scale of an unperturbed flow is infinite. Hence, for any perturbation $\mathbf{u}^{(1)}$ of finite magnitude there exists a small neighbourhood of the stagnation points where the surfaces do not exist

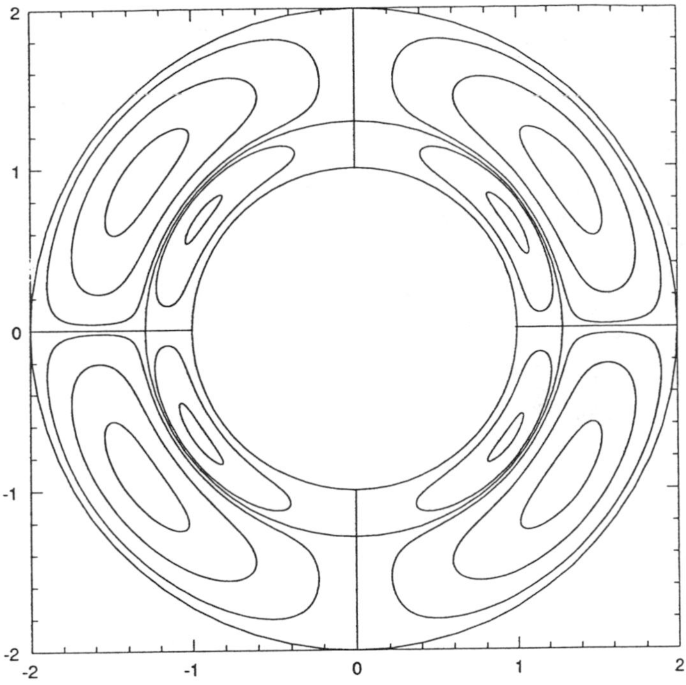

Fig. 3. The axisymmetric 'two-cell' flow.

$$\boldsymbol{\Omega} = f_1(r)\boldsymbol{\Omega}_1 + f_2(r)\boldsymbol{\Omega}_2 \quad .$$

For small Reynolds number the secondary flow (3.11) acts as a small perturbation and the invariants I_1, I_2 change slowly as a result of an adiabatic drift:

$$\left\langle \frac{dI_1}{dt} \right\rangle = Re\left\{ r^{-1}\frac{df_1}{dr}(\boldsymbol{\Omega}_1 - \boldsymbol{\Omega}_2)\cdot\left\langle \mathbf{x}(\mathbf{x}\cdot\mathbf{u}^{(1)})\right\rangle + \left\langle \boldsymbol{\Omega}\cdot\mathbf{u}^{(1)}\right\rangle \right\}, \qquad (4.2a)$$

$$\left\langle \frac{dI_2}{dt} \right\rangle = 2Re\left\{ \langle A\boldsymbol{\Omega}_1\cdot\mathbf{x}\rangle + \langle B\boldsymbol{\Omega}_2\cdot\mathbf{x}\rangle - \langle C\rangle r^2 \right\}. \qquad (4.2b)$$

The angular brackets denote the average over an unperturbed streamline:

$$\mathbf{x}_{\mathbf{u}}(t) = I_1\Omega^{-1}\hat{\mathbf{k}} + \sigma\cos(\Omega t)\hat{\mathbf{i}} + \sigma\sin(\Omega t)(\hat{\mathbf{k}}\wedge\hat{\mathbf{i}}), \qquad (4.3)$$

$$\sigma = \sqrt{I_2 - I_1^2/\Omega^2} \quad , \qquad \hat{\mathbf{k}} = \boldsymbol{\Omega}/\Omega \quad ;$$

where $\hat{\mathbf{i}}$ is any unit vector perpendicular to $\boldsymbol{\Omega}$.

Evaluating these averages we obtain the equations of the slow drift:

$$\frac{1}{Re}\frac{dI_1}{dt} = F_1(I_2)I_1^3 + F_2(I_2)I_1 \quad , \tag{4.4a}$$

$$\frac{1}{Re}\frac{dI_2}{dt} = G_1(I_2)I_1^2 + G_2(I_2) \quad ; \tag{4.4b}$$

with the following functions F_1, F_2, G_1, G_2:

$$\begin{aligned} F_1(I_2) &= \frac{3df_1/dr}{2\sqrt{I_2}\Omega^3}\sum_{i,j=1}^{2}\left\{C_{ij}[5p_ip_j - \boldsymbol{\Omega}_i\cdot\boldsymbol{\Omega}_j]\hat{\mathbf{k}} - p_i\boldsymbol{\Omega}_j - p_j\boldsymbol{\Omega}_i\right\}\cdot(\boldsymbol{\Omega}_1 - \boldsymbol{\Omega}_2 \\ &\quad -\frac{1}{2\Omega^2}\sum_{i,j=1}^{2}\{D_{ij}[3p_ip_j - \boldsymbol{\Omega}_i\cdot\boldsymbol{\Omega}_j]\} \quad , \\ F_2(I_2) &= \frac{-\sqrt{I_2}}{2\Omega}\sum_{i,j=1}^{2}\left\{C_{ij}[9p_ip_j - \boldsymbol{\Omega}_i\cdot\boldsymbol{\Omega}_j]\hat{\mathbf{k}} - 3p_i\boldsymbol{\Omega}_j - 3p_j\boldsymbol{\Omega}_i\right\}\cdot(\boldsymbol{\Omega}_1 - \boldsymbol{\Omega}_2 \\ &\quad +\sum_{i,j=1}^{2}\{[C_{ij} + \tfrac{1}{2}I_2D_{ij}][3p_ip_j - \boldsymbol{\Omega}_i\cdot\boldsymbol{\Omega}_j]\} \quad , \\ G_1(I_2) &= \frac{3}{\Omega^2}\sum_{i,j=1}^{2}C_{ij}[3p_ip_j - \boldsymbol{\Omega}_i\cdot\boldsymbol{\Omega}_j] \quad , \\ G_2(I_2) &= -\tfrac{1}{3}I_2\Omega^2G_1(I_2) \quad , \end{aligned}$$

$$p_i = \hat{\mathbf{k}}\cdot\boldsymbol{\Omega}_i \quad .$$

The two-dimensional, autonomous system (4.4) is integrable and its first integral is an adiabatic invariant of the flow $\mathbf{u}^{(0)} + Re\mathbf{u}^{(1)}$.

5. Departures from axisymmetry

When $\boldsymbol{\Omega}_1 \parallel \boldsymbol{\Omega}_2$ the flow is axisymmetric and has one or two 'cells' in each quadrant, depending on ω_2/ω_1. Now suppose the angle between the directions of $\boldsymbol{\Omega}_1$ and $\boldsymbol{\Omega}_2$ is small but finite. We shall demonstrate that the topology of the stream-surfaces undergoes a discontinuous change in the limit of $\epsilon \to 0$. Let us first consider two different small perturbations of a simple case $\boldsymbol{\Omega}_1 = \boldsymbol{\Omega}_2$ ($\mathbf{u}^{(1)} = 0$). We take:
a) an axisymmetric perturbation,

$$\boldsymbol{\Omega}_1 = (1-\epsilon)\hat{\mathbf{i}}_z \quad , \qquad \boldsymbol{\Omega}_2 = (1+\epsilon)\hat{\mathbf{i}}_z \quad ; \tag{5.1a}$$

b) a non-axisymmetric perturbation,

$$\boldsymbol{\Omega}_1 = \epsilon\hat{\mathbf{i}}_z + (1 - \tfrac{1}{2}\epsilon^2)\hat{\mathbf{i}}_z \quad , \qquad \boldsymbol{\Omega}_2 = \hat{\mathbf{i}}_z \quad ; \tag{5.1b}$$

and examine the radial component of the slow drift (see 4.4b):

$$
\begin{aligned}
r^{-1}\left\langle \mathbf{u}^{(1)}\cdot\mathbf{x}\right\rangle &= \frac{Re}{2r}(I_1^2-\tfrac{1}{3}I_2\Omega^2)G_1(I_2)\\
&= \frac{3Re}{r\Omega^4}L(\lambda)R(r,\lambda)S(\mathbf{x}) \quad , \qquad (5.2)\\
S(\mathbf{x}) &= \sum_{i,j=1}^{2}\sum_{k=0}^{2} W_{ij}^k[3(\boldsymbol{\Omega}_i\cdot\boldsymbol{\Omega})(\boldsymbol{\Omega}_j\cdot\boldsymbol{\Omega})-\boldsymbol{\Omega}_i\cdot\boldsymbol{\Omega}_j\boldsymbol{\Omega}^2]r^k \quad .
\end{aligned}
$$

Using (3.10) we obtain the following expressions for $S(\mathbf{x})$:

$$
\text{a)}\qquad S(\mathbf{x}) = 8\epsilon\sum_{k=0}^{3}[(f_1-1)W_{11}^k+f_1W_{22}^k]r^k+O(\epsilon^2) \quad , \qquad (5.3\text{a})
$$

$$
\text{b)}\qquad S(\mathbf{x}) = \epsilon^2\sum_{k=0}^{3}[(3f_1-2)W_{11}^k+(1-3f_1)W_{22}^k]r^k+O(\epsilon^4) \qquad (5.3\text{b})
$$

Taking, for example, $\lambda = 2$ one can easily verify that in case a) the leading order term in $S(\mathbf{x})$ has no zeros in the interval $1 \leq r \leq \lambda$. In the case b) there is one root $r_0(\lambda)$ for every λ, and $r_0(2) = 1.25259689....$

This means that a small axisymmetric departure from $\boldsymbol{\Omega}_1 = \boldsymbol{\Omega}_2$ results in a 'one-cell flow', while a non-axisymmetric deflection of type b) yields two cells with the helicities of opposite sign. In the latter case the boundary between the two cells is at $r = r_0$ when $\epsilon \to 0$.

The surfaces of constant adiabatic invariant (or drift surfaces) can be obtained by numerical integration of (4.4), see figure 4. The periods of the orbits of $\mathbf{u}^{(0)}$ are bounded. Hence, for small Re the streamlines of $\mathbf{u}^{(0)} + Re\mathbf{u}^{(1)}$ lie on the drift surfaces.

The above analysis shows a discontinuous change of the drift *pattern*, but $\mathbf{u}^{(1)} \equiv 0$ when $\boldsymbol{\Omega}_1 = \boldsymbol{\Omega}_2$, so the *drift* changes continuously in the limit $\epsilon \to 0$. Considering a small non-axisymmetric departure from $\boldsymbol{\Omega}_1 = (1-\delta)\boldsymbol{\Omega}_2$, $\delta \ll 1$, we find a continuous change of the drift pattern. As the angle between $\boldsymbol{\Omega}_1$ and $\boldsymbol{\Omega}_2$ tends to zero, one of the two cells in each quadrant disappear into the inner (outer) boundary when $\delta > 0$ ($\delta < 0$).

So far we discussed only the case of $\boldsymbol{\Omega}_1$ and $\boldsymbol{\Omega}_2$ being (nearly) parallel. When $|\boldsymbol{\Omega}_1| = |\boldsymbol{\Omega}_2|$ and the angle α between $\boldsymbol{\Omega}_1$ and $\boldsymbol{\Omega}_2$ is increased the inner and the outer cells change their relative position. Figure 5 shows the drift surfaces for $\alpha = 45°$. If $\boldsymbol{\Omega}_1$ and $\boldsymbol{\Omega}_2$ are anti-parallel, then a small departure from axial symmetry leads to an 'almost discontinuous' change of the drift flow. When $\boldsymbol{\Omega}_1 = -\boldsymbol{\Omega}_2$ the flow $\mathbf{u}^{(0)} + Re\mathbf{u}^{(1)}$ has one cell in each quadrant (see sec. 3). Yet, as can be easily verified, $< \mathbf{u}^{(1)}\cdot\mathbf{x} >= 0$ on a spherical surface $r = r_1$ where r_1 is such that $\boldsymbol{\Omega}(r_1) = f_1\boldsymbol{\Omega}_1 + f_2\boldsymbol{\Omega}_2 = (1-2f_1)\boldsymbol{\Omega}_2 = 0$. This is a special surface where $\mathbf{u}^{(0)} = 0$, i.e. where the

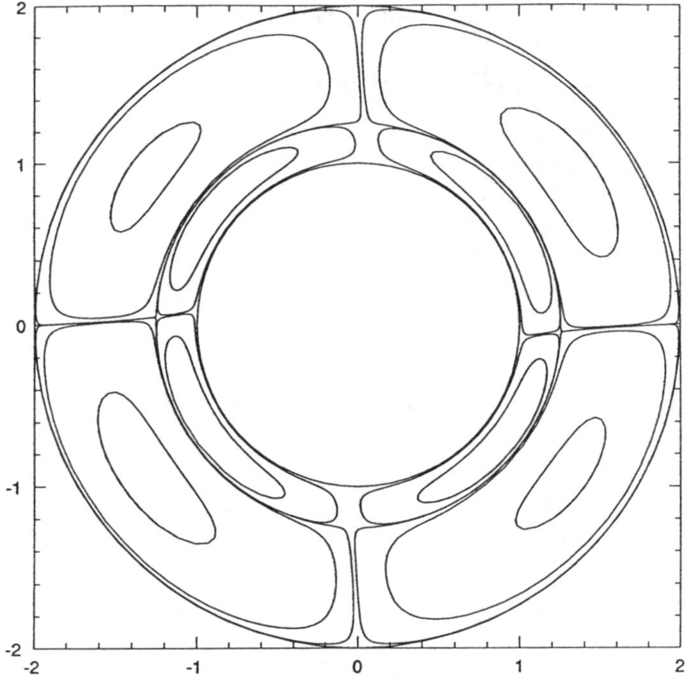

Fig. 4. The surfaces of constant adiabatic invariant for $\lambda = 2$, $|\boldsymbol{\Omega}_1| = |\boldsymbol{\Omega}_2|$ and the angle between $\boldsymbol{\Omega}_1$ and $\boldsymbol{\Omega}_2$ equal 5°.

adiabatic approximation is not valid because the periods of the unperturbed orbits are unbounded.

In figure 6a) we show the drift surfaces computed from (4.4) with $\boldsymbol{\Omega}_1 = -\boldsymbol{\Omega}_2$. The two cells in each quadrant have helicities of the *same* sign, and, away from $r = r_1$, the surfaces match the stream-surfaces of $\mathbf{u}^{(0)} + Re\mathbf{u}(1)$ as given by (2.5), (3.15). In figure 6b) we show, qualitatively, the structure of the drift. The X-type stagnation points and the separatrices joining them make it possible to have a two-cell drift with the same sense of rotation in both cells.

However, the special surface $\boldsymbol{\Omega}(r) = 0$ exists *only* when α is exactly zero. For any finite α we have $\boldsymbol{\Omega} \neq 0$ everywhere and the periods of the unperturbed orbits are bounded. Hence, for any small α and sufficiently small Re the streamlines of $\mathbf{u}^{(0)} + Re\mathbf{u}^{(1)}$ are on the drift surfaces similar to those shown in figure 6. [2] Then the fluid in the two cells is separated and does not mix.

[2] 'Sufficiently small' means such that the time-scale of $Re\mathbf{u}^{(1)}$ is everywhere small compared with $2\pi/|\Omega(r)|$.

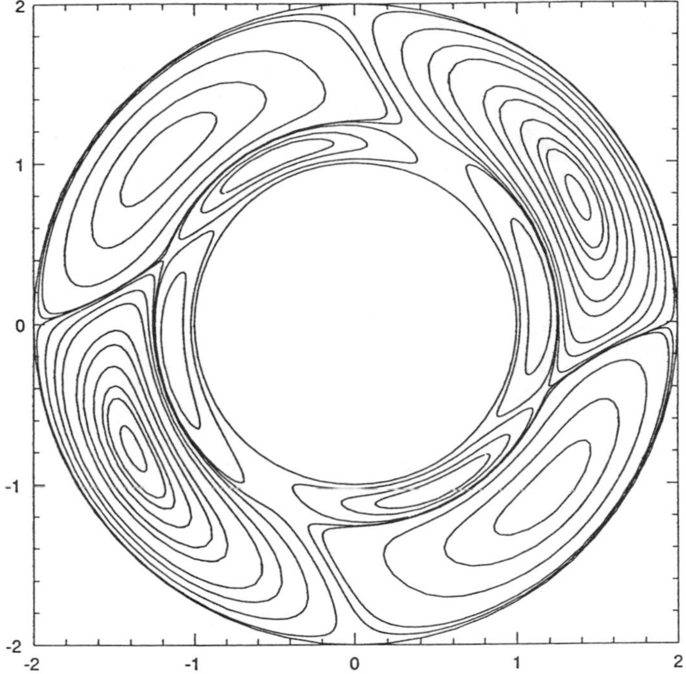

Fig. 5. The drift surfaces for $\boldsymbol{\Omega}_1 = (1/\sqrt{2}, 0, 1/\sqrt{2})$, $\boldsymbol{\Omega}_2 = (0, 0, 1)$.

If the (small) Reynolds number is fixed and we decrease α, then, in the limit $\alpha \to 0$ the drift surfaces tend to the limit shown in figure 6, but when α is too small the streamlines of $\mathbf{u}^{(0)} + Re\mathbf{u}^{(1)}$ begin to 'jump' across the surface $\boldsymbol{\Omega}(r) = 0$ and they switch to a 'single-cell'pattern. This transition occurs at a critical value of $\alpha = \alpha_c(Re)$ which tends to zero when $Re \to 0$. Therefore, for small Re the change from a single-cell drift to a double-cell drift is 'almost discontinuous': the boundary between the two cells does not emerge from either wall but stays on the surface $\boldsymbol{\Omega}(r) = 0$.

When ω_2/ω_1 is such that the axisymmetric flow has two cells then a small deflection creates a three-cell drift. In figure 7 we show the drift surfaces for $\boldsymbol{\Omega}_2 = \hat{\mathbf{i}}_z$, $|\boldsymbol{\Omega}_1| = 3.3$ and the angle between $\boldsymbol{\Omega}_1$ and $\boldsymbol{\Omega}_2$ equal 175°.

6. Transition to chaos

If the averaging procedure is valid in the entire domain, then, in the limit of $Re \to 0$, the surfaces of constant adiabatic invariant are identical with the stream-surfaces of $\mathbf{u}^{(0)} + Re\mathbf{u}^{(1)}$. For finite Re we obtain the Poincare sections of the streamlines by numerically solving the equations:

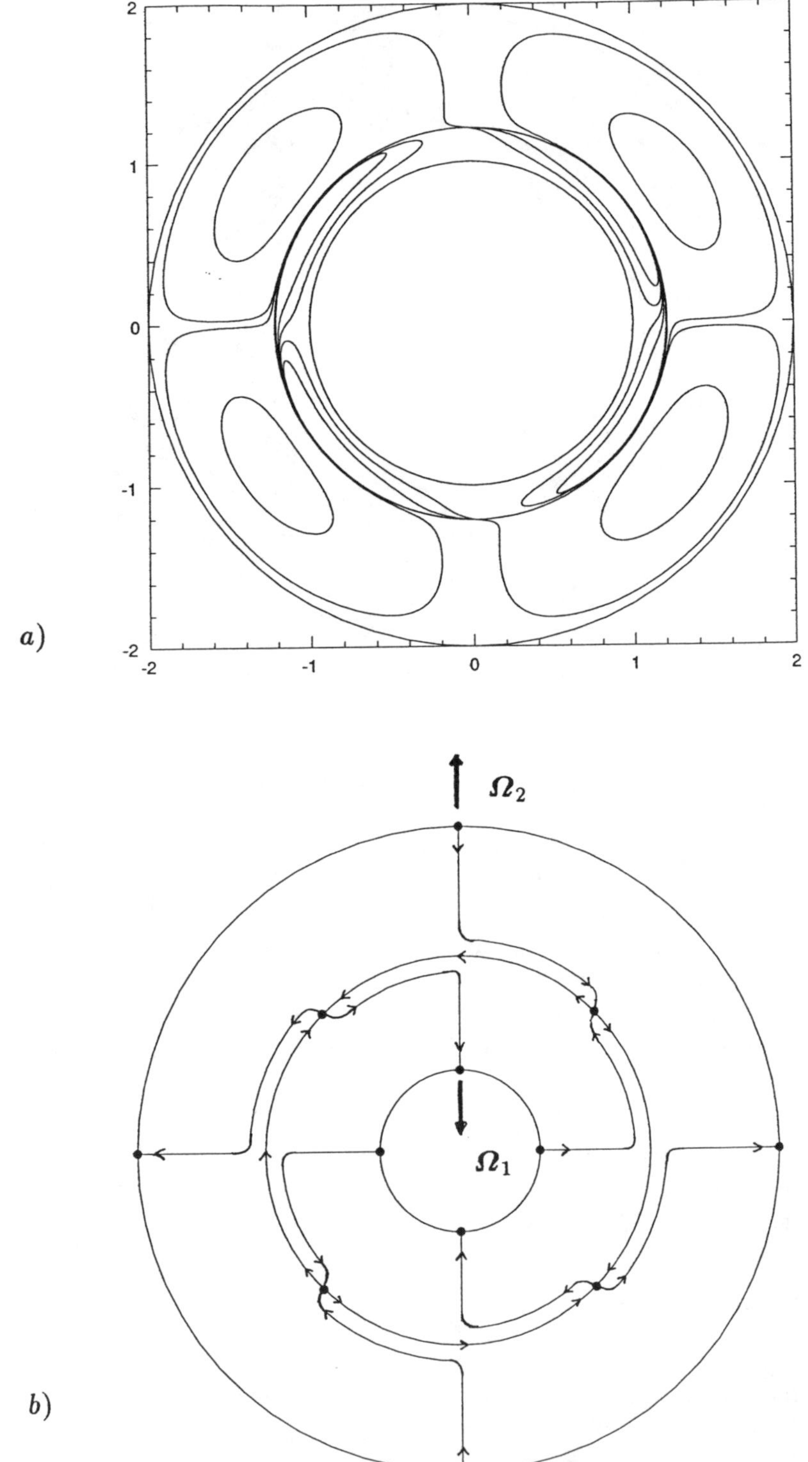

Fig. 6. (a) The drift surfaces (solutions of 4.4) for $\lambda = 2$, $\Omega_1 = -\Omega_2$.
(b) The qualitative structure of a two-cell drift with the same sign of helicity in both cells.

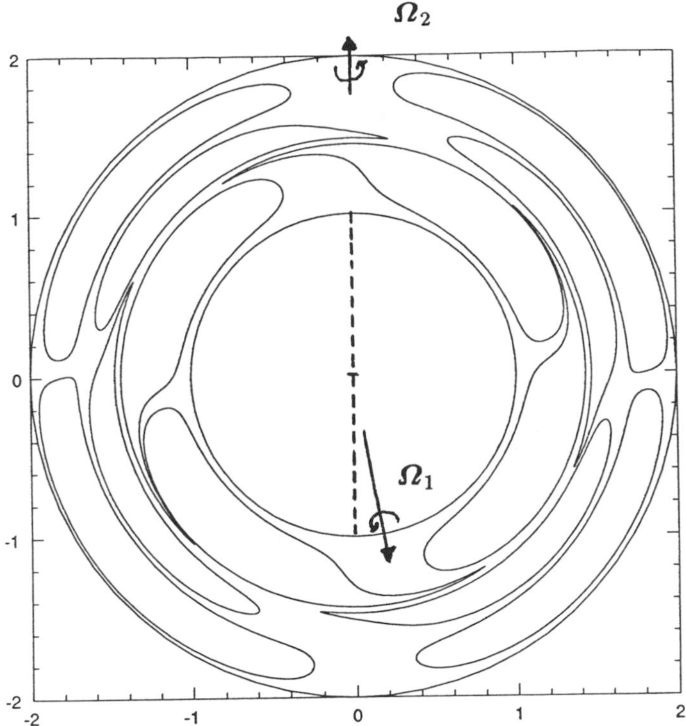

Fig. 7. The three-cell drift for $\lambda = 2$, $\boldsymbol{\Omega}_2 = \hat{\mathbf{i}}_z$, $|\boldsymbol{\Omega}_1| = 3.3$ and the angle between $\boldsymbol{\Omega}_1$ and $\boldsymbol{\Omega}_2$ equals 175°

$$\frac{d\mathbf{x}}{dt} = \mathbf{u}^{(0)} + Re\mathbf{u}^{(1)} \quad . \tag{6.1}$$

We find that the streamlines stay near the surfaces of constant adiabatic invariant for the Reynolds number as high as 10. In figure 8 we show the drift surfaces for $\boldsymbol{\Omega}_1 = (\frac{2}{5}, 0, \frac{1}{5})$, $\boldsymbol{\Omega}_2 = (0, 0, 1)$, and in figure 9 two streamlines of $\mathbf{u}^{(0)} + Re\mathbf{u}^{(1)}$ for $Re = 1, 10$. For $Re = 1$ the streamlines are practically indistinguishable from the adiabatic surfaces. When $Re = 10$ the inner streamline is still very close to a drift surface. The outer one, which is close to a separatrix, is 'out of focus', i.e. it generally follows an adiabatic surface but also has a significant chaotic transverse scatter. When the Reynolds number is further increased we observe, as might be expected, a transition to global chaos. In figure 10 we show the Poincaré section of a single streamline for $Re = 20$. The chaotic streamline penetrates most of the spherical annulus. Four islands can be seen near the elliptic stagnation points of the drift. The transverse chaotic scatter is smallest there and this region is the last to participate in global chaos.

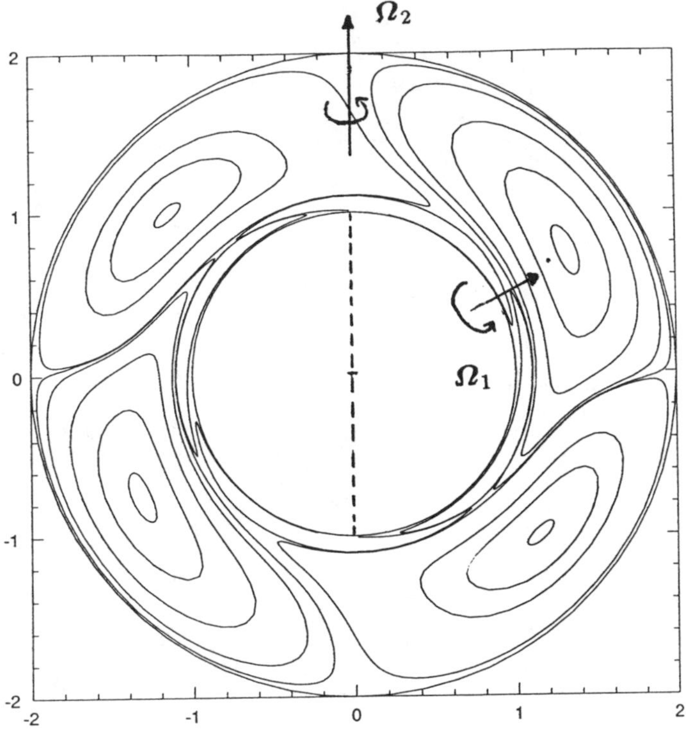

Fig. 8. The drift surfaces for $\lambda = 2$, $\boldsymbol{\Omega}_1 = (\frac{2}{5}, 0, \frac{1}{5})$, $\boldsymbol{\Omega}_2 = (0, 0, 1)$.

7. Conclusions

We have shown that the streamline pattern in the low Reynolds number flow between two rotating solid spheres can be explained in terms of an adiabatic invariant. The level surfaces of this invariant have to be computed numerically, but some important information about the topology of the flow, e.g. the number of drift cells and the positions of their boundaries, can be derived analytically. The Reynolds number is a parameter which can be controlled in a laboratory experiment and hence, the above analysis could, in principle, be verified experimentally.

We have also shown that the low Reynolds number flow obtained by adding the first order correction to the Stokes flow (i.e. $\mathbf{u}^{(0)} + Re\mathbf{u}^{(1)}$) has strongly chaotic streamlines for $Re \sim 20$. For such Re the first order approximation may still be valid, as it is, for example, in a two-dimensional flow around a solid cylinder. It is possible that in a spherical annulus, at $Re = 20$, higher order corrections must be added, but this is unlikely to change the chaotic nature of the flow.

The analysis of the drift surfaces in a weakly non-axisymmetric case en-

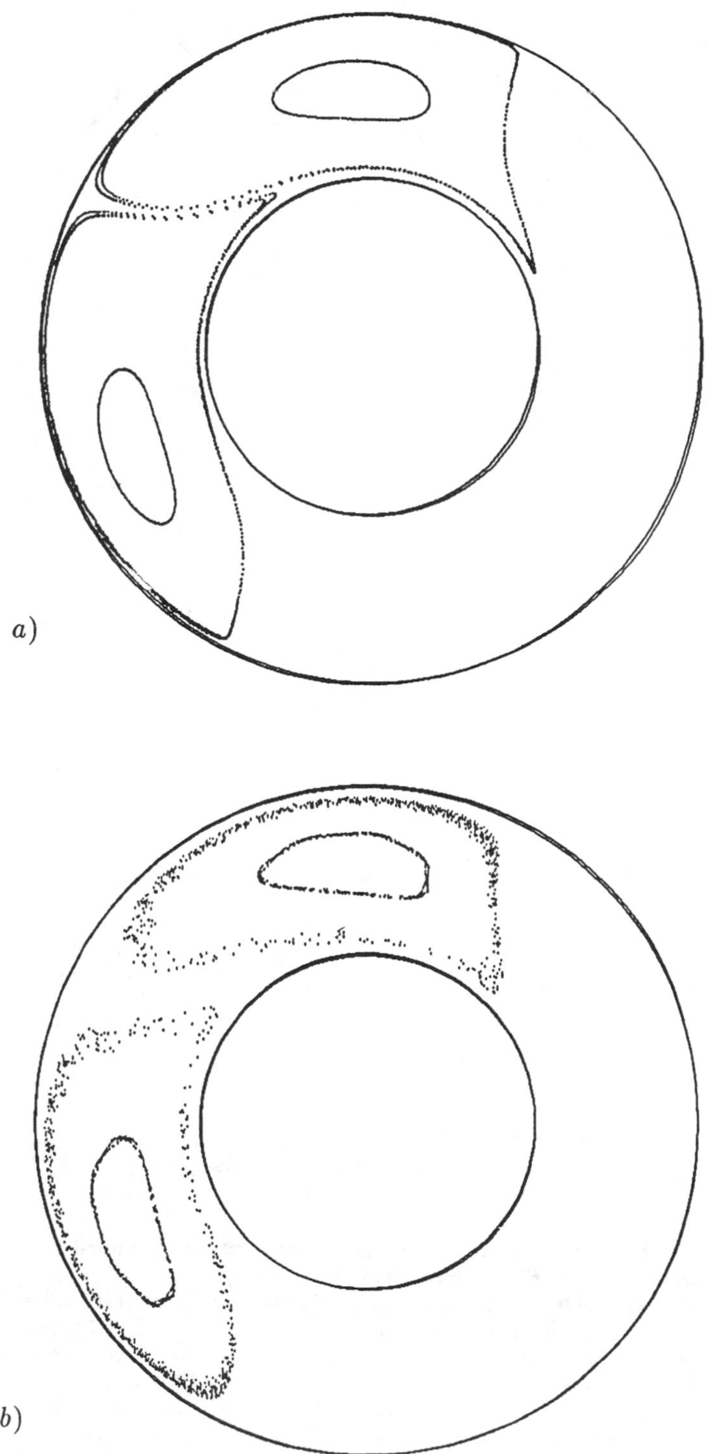

Fig. 9. The Poincaré sections of the two streamlines of $\mathbf{u}^{(0)} + Re\mathbf{u}^{(1)}$ with α, $\boldsymbol{\Omega}_1$, $\boldsymbol{\Omega}_2$ as in fig. 8 and (a) $Re = 1$, (b) $Re = 10$.

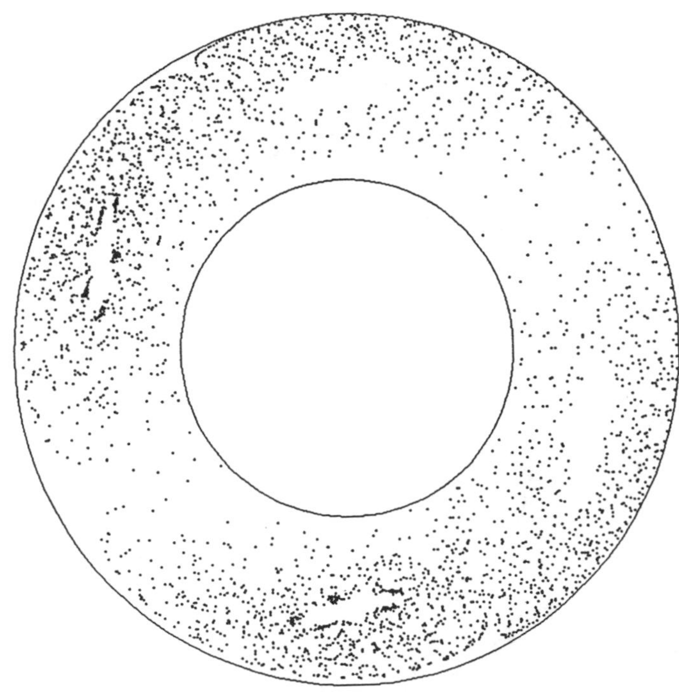

Fig. 10. The Poincaré section of a single streamline of $\mathbf{u}^{(0)} + Re\mathbf{u}^{(1)}$ with α, $\boldsymbol{\Omega}_1$, $\boldsymbol{\Omega}_2$ as in fig.8 and $Re = 20$.

abled us to predict a qualitative change in the streamline topology in the case when $\boldsymbol{\Omega}_1 \cdot \boldsymbol{\Omega}_2 < 0$ and $\boldsymbol{\Omega}_1 \wedge \boldsymbol{\Omega}_2$ is small. The drift flow changes from the two-cell to one-cell pattern. This change is realised by the process of 'diffusion' of streamlines across the surface $f_1(r)\boldsymbol{\Omega}_1 + (1 - f_1(r))\boldsymbol{\Omega}_2 = 0$. A study of this (anomalous) 'diffusion' presents a challenging theoretical problem.

References

BAJER, K., MOFFATT H.K. & NEX, F.H., 1990 Steady confined Stokes flows with chaotic streamlines. *Topological Fluid Mechanics*, Proceedings IUTAM Symposium, Cambridge, August 13-18, 1989, Eds: H.K. Moffatt & A. Tsinober, Cambridge University Press, 1990.

BAJER, K., & MOFFATT H.K., 1990 On a class of steady confined Stokes flows with chaotic streamlines. *J. Fluid Mech.*, **212**, pp. 337-363.

LANDAU, L.D., & LIFSHITZ, E.M., 1987 *Fluid Mechanics* (2nd Edn.), Pergamon Press.

MOFFATT, H.K., 1991 Electromagnetic stirring. *Phys. Fluids A*, **3**(5), pp. 1336-1343.

PROUDMAN, I., 1956 The almost rigid rotation of viscous fluid between concentric spheres. *J. Fluid Mech.*, **1**, pp. 505-516.

STONE, H.A., NADIM, A., STROGATZ, S.H., 1991 Chaotic streamlines inside drops immersed in steady linear Stokes flows. *J. Fluid Mech.* (to appear)

INSTABILITY CRITERIA IN FLUID DYNAMICS

SUSAN FRIEDLANDER
Department of Mathematics (M/C 249)
University of Illinois at Chicago
Chicago, Illinois 60680 U.S.A.

and

MISHA M. VISHIK
Department of Mathematics
University of Chicago
Chicago, Illinois 60637

ABSTRACT. The following two results are presented concerning linear instability criteria in fluid dynamics. The first result states that an instability in the underlying Euler equation is necessary for the existence of an instability in the Navier-Stokes equations in the limit of high Reynolds number. The second result gives a sufficient condition for instability of the Euler equations in terms of a geometric quantity that can be considered as an analogue of a Lyapunov exponent for fluid motion. The effectiveness of this criterion is illustrated in several examples including the instability of a flow with a hyperbolic stagnation point and a generalized baroclinic instability.

1. Introduction

There is considerable structural similarity between the dynamo equations for the generation of a magnetic field by the motion of an electrically conducting fluid and the equations governing the motion of an incompressible fluid. Inspired by techniques and results from dynamo theory, we describe how the Lagrangian trajectories and quantities calculated via the Lagrangian trajectories play an important role in hydrodynamical stability. Two main results, whose proofs are given in a companion paper (Vishik and Friedlander, 1992), are presented in this paper. Related instability results are also to be found in another paper appearing in this volume (Lifschitz and Hameiri, 1992).

In Section 2, we consider the fluid dynamical analogy of the so-called "fast" dynamo problem. We discuss the result that some instability in the underlying Euler equation is necessary for the existence of an instability in the Navier-Stokes equation in the high Reynolds number limit. This necessary condition is analogous to the condition of exponential stretching in the fast dynamo problem. Such a condition is not completely obvious since viscosity is known to provide instability under certain circumstances. We note that our result is consistent with the well known fact that the growth rate of the instability that occurs in Poiseuille flow vanishes in the limit of high Reynolds number. However this example is not formally covered by our theorem because we consider only periodic or free space boundary conditions.

In Section 3 we discuss a second result that is also proved by asymptotic methods. We present a sufficient condition for (linear) instability of the Euler equations in terms of a universal geometric estimate from below on the growth rate of a small perturbation of a 3-dimensional steady inviscid flow. This estimate, which can be calculated by solving an ODE along each Lagrangian trajectory, appears to govern instabilities in an inviscid fluid in the same way that Lyapunov exponents govern dynamo instabilities in an infinitely conducting fluid. Thus we present a "dynamical systems type" quantity that can be considered as a Lyapunov exponent for fluid motion. We also discuss another instability criteria which takes into account the smoothness of the initial perturbation. In accordance with intuition, stretching of higher gradients enhances instability in a metric with higher derivatives.

In Section 4 we illustrate the effectiveness of the instability criteria for the Euler equations presented in Section 3. We describe how any flow with a hyperbolic stagnation point is unstable. We discuss the extension of the instability criteria to the case of an inhomogeneous Boussinesq fluid. We can show that the stabilizing effects of buoyancy cannot overcome the instability due to stretching at a hyperbolic stagnation point. We also apply the instability criteria to a configuration relevant to geophysical fluid dynamics, namely a stratified fluid rotating with a spatially varying angular velocity, and we obtain a sufficient condition for a generalized baroclinic instability.

2. Navier-Stokes Equations

The kinematic dynamo concerns the generation of a magnetic field $\mathbf{H}(\mathbf{x},t)$ by the motion in 3D of an electrically conducting fluid with prescribed steady velocity $\mathbf{u}(\mathbf{x})$. The magnetic field satisfies the induction equations:

$$\dot{\mathbf{H}} = \nabla \times (\mathbf{u} \times \mathbf{H}) + \epsilon \nabla^2 \mathbf{H} \equiv M_\epsilon \tag{1}$$

$$\nabla \cdot \mathbf{H} = 0 \tag{2}$$

with initial conditions $\mathbf{H}(\mathbf{x},0) = \mathbf{H}_0(\mathbf{x})$ and appropriate boundary conditions (e.g. free space or periodic boundary conditions).[1] Here ϵ denotes the

[1] Spec M_ϵ denotes the spectrum of the operator M_ϵ which is discrete. For the simplicity of exposition, in this paper we mainly restrict ourselves to periodic boundary conditions, however the results are also valid for the free space problem.

inverse magnetic Reynolds number. Dynamo action is said to occur for flows $\mathbf{u}(\mathbf{x})$ such that

$$\sup_{\sigma_\epsilon \,\in\, \text{Spec } M_\epsilon} \text{Re } \sigma_\epsilon > 0. \tag{3}$$

The dynamo is called "fast" if

$$\overline{\lim_{\epsilon \to 0}} \sup_{\sigma_\epsilon \,\in\, \text{Spec } M_\epsilon} \text{Re } \sigma_\epsilon > 0 \tag{4}$$

The possible growth of the magnetic field on the convective time scale (i.e. fast dynamo) is of importance in astrophysics as well as being of intrinsic mathematical interest because of the delicate nature of the singular limit $\epsilon \to 0$ (c.f. Zeldovich and Ruzmaikin, 1980; Arnold, Zeldovich, Ruzmaikin and Sokoloff, 1982; Soward and Childress, 1986; Bayly, 1986; Soward, 1987; Gilbert and Childress, 1990). It was proved by Vishik (1989) that for a smooth flow $\mathbf{u}(\mathbf{x})$, exponential stretching at least somewhere in the flow is a necessary condition for fast dynamo action; i.e.

$$\overline{\lim_{\epsilon \to 0}} \sup_{\sigma_\epsilon \,\in\, \text{Spec } M_\epsilon} \text{Re } \sigma_\epsilon \leq \sup_{\mathbf{x}_0, \eta_0} \chi(\mathbf{x}_0, \eta_0). \tag{5}$$

The quantity $\chi(\mathbf{x}_0, \eta_0)$ is the Lyapunov exponent for the flow

$$\dot{\mathbf{x}} = \mathbf{u}(\mathbf{x}), \quad \mathbf{x}(0) = \mathbf{x}_0 \tag{6}$$

and it is defined by the expression

$$\chi(\mathbf{x}_0, \eta_0) = \overline{\lim_{t \to \infty}} \frac{1}{t} \log \left| \frac{\partial \mathbf{x}}{\partial \mathbf{x}_0} \eta_0 \right| \tag{7}$$

for any vector η_0 that is advected by the flow. The concept of exponential stretching is quantified by the positivity of the Lyapunov exponent. For example, in the case of a flow where the growth is only polynomial, all the Lyapunov exponents vanish and hence from (5) such a flow cannot sustain a fast dynamo.

The Navier-Stokes equations governing the flow of a viscous, incompressible fluid are

$$\dot{\mathbf{v}} = -(\mathbf{v} \cdot \nabla)\mathbf{v} - \nabla P + \epsilon \nabla^2 \mathbf{v} + \epsilon \mathbf{F}(\mathbf{x}) \tag{8}$$

$$\nabla \cdot \mathbf{v} = 0 \tag{9}$$

where $\mathbf{v}(\mathbf{x}, t)$ is the Eulerian velocity, $P(\mathbf{x}, t)$ the pressure, $\mathbf{F}(\mathbf{x})$ is an external force and ϵ is the inverse Reynolds number. Taking the curl of (8) gives the equation for the evolution of vorticity $\nabla \times \mathbf{v}$:

$$\nabla \times \dot{\mathbf{v}} = \nabla \times [\mathbf{v} \times (\nabla \times \mathbf{v})] + \epsilon \nabla^2 (\nabla \times \mathbf{v}) + \epsilon \nabla \times \mathbf{F}. \tag{10}$$

Equation (10) is clearly very similar to the dynamo equation (1) with the vorticity replacing the magnetic field. However, unlike the kinematic dynamo problem, we cannot treat $\mathbf{v}$ and $\nabla \times \mathbf{v}$ as independent. To exploit the analogy between the vorticity equation (10) and the magnetic induction equation (1), we consider the linearized Navier-Stokes equations for a small perturbation $\mathbf{w}(\mathbf{x}, t)$ about a steady flow $\mathbf{u}(\mathbf{x})$:

$$\dot{\mathbf{w}} = \mathbf{w} \times (\nabla \times \mathbf{u}) + \mathbf{u} \times (\nabla \times \mathbf{w}) - \nabla P + \epsilon \nabla^2 \mathbf{w} \tag{11}$$

$$\nabla \cdot \mathbf{w} = 0 \tag{12}$$

with initial condition

$$\mathbf{w}(\mathbf{x}, 0) = \mathbf{w}_0(\mathbf{x}). \tag{13}$$

The antidynamo theorem, stated in (5), was proved by Vishik using WKB type asymptotics to construct an approximate Green's function for M_ϵ. Vishik and Friedlander (1992) use similar techniques to analyze the stability of the system (11)–(13) with appropriate boundary conditions. The fluid problem is considerably more complex because of the presence of the pressure term which makes the operator nonlocal. For purely technical reasons it is more convenient to consider the stability of the adjoint problem

$$\dot{\mathbf{w}} = (\mathbf{u} \cdot \nabla)\mathbf{w} - \left(\frac{\partial \mathbf{u}}{\partial \mathbf{x}}\right)^T \mathbf{w} - \nabla P + \epsilon \nabla^2 \mathbf{w} \equiv L_\epsilon \mathbf{w} \tag{14}$$

$$\nabla \cdot \mathbf{w} = 0 \tag{15}$$

$$\mathbf{w}(\mathbf{x}, 0) = \mathbf{w}_0(\mathbf{x}). \tag{16}$$

Let $G(t)$ be the Greens function (evolution operator) for (14)–(16), i.e. $(G(t)\mathbf{w}_0)(\mathbf{x}) = \mathbf{w}(\mathbf{x}, t)$. In the cases of free space or periodic boundary conditions, Friedlander and Vishik use asymptotic expansions to estimate the evolution operator (i.e. Greens function) for the operator L_ϵ. It is useful to partition the Greens function into high frequency and low frequency parts with the symbol of the high frequency part being non-smooth at the origin. The analysis centers on the high frequency terms and it is shown that the growth rate of the high frequency part of $G(t)$ in the limit of large Reynolds number is bounded by the growth of the semi-group defined by the linearized Euler equation.

In analogy with the term "fast" dynamo, we refer to instability for the problem (14)–(16) in the limit of large Reynolds number, as "fast" vorticity generation. The following theorem, which we state for the case of periodic boundary conditions, is the Navier-Stokes analogue of the necessary condition for a fast dynamo given by Vishik (1989). The theorem states that some instability in the underlying Euler equation is a necessary condition for "fast" vorticity generation.

THEOREM 1.

$$\overline{\lim_{\epsilon\to 0}} \max_{\sigma_\epsilon \in \text{ Spec } L_\epsilon} \text{Re } \sigma_\epsilon \leq \lim_{t\to\infty} \frac{1}{t} \log \|\mathcal{G}(t)\|_2 \tag{17}$$

where $\mathcal{G}(t)$ is the evolution operator (Greens function) for the corresponding linearized Euler equation

$$\dot{\mathbf{w}} = (\mathbf{u}\cdot\nabla)\mathbf{w} - \left(\frac{\partial \mathbf{u}}{\partial \mathbf{x}}\right)^T \mathbf{w} - \nabla P, \quad \nabla\cdot\mathbf{w} = 0. \tag{18}$$

The group of operators $\mathcal{G}(t)$ is well defined in the Hilbert space

$\mathcal{H}^0 = (L^2(T^3))^3 \cap \{(\nabla\cdot\ ) = 0\}$.

We briefly sketch the main ideas in the proof of Theorem 1. We construct the full asymptotic expansion as $\epsilon \to 0$ of the Greens function $G(t)$ of the initial value problem (14)–(16). This construction consists of two steps. First we use a cut off function ψ to separate low frequencies from high frequencies in the Fourier expansion of the initial perturbation $\mathbf{w}_0$. We write

$$\begin{aligned}\mathbf{w}_0(\mathbf{x}) &= \frac{1}{(2\pi\sqrt{\epsilon})^3}\int e^{\xi\cdot(\mathbf{x}-\mathbf{y})/\sqrt{\epsilon}}[\psi(\xi/\epsilon^{1/4}) \\ &+ (1-\psi(\xi/\epsilon^{1/4}))]\mathbf{w}_0(\mathbf{y})d^3\xi d^3y \\ &= \left(\psi\left(\frac{\epsilon^{1/4}}{i}\frac{\partial}{\partial\mathbf{x}}\right)\mathbf{w}_0\right)(\mathbf{x}) + \left(1-\psi\left(\frac{\epsilon^{1/4}}{i}\frac{\partial}{\partial\mathbf{x}}\right)\mathbf{w}_0\right)(\mathbf{x}).\end{aligned} \tag{19}$$

where ψ is a smooth function that satisfies

$$\psi(\xi) = \begin{cases} 1, & |\xi| < 1 \\ 0, & |\xi| > 2\ . \end{cases}$$

The evolutions of the high and low frequency parts are constructed in separate ways. We first consider the high frequency part of the Greens function. Following the approach used in the dynamo problem we (formally) write

$$\begin{aligned}(G^{hf}(t)w_0)(\mathbf{x},t) &= \frac{1}{(2\pi\sqrt{\epsilon})^3}\int\Big\{e^{(iS(\mathbf{x},\xi,t)-i\mathbf{y}\cdot\xi)/\sqrt{\epsilon}}\Big(\sum_{j=0}^{\infty} A_j(\mathbf{x},\xi,t)\sqrt{\epsilon}^{\,j}\Big) \\ &\qquad (1-\psi(\xi/\epsilon^{1/4})\mathbf{w}_0(\mathbf{y})\Big\}d^3\xi d^3y.\end{aligned} \tag{20}$$

We obtain the eikonal equation for S and the system of transport equations for the 3×3 matrix coefficients $A_j(\mathbf{x},\xi,t)$.

$$\dot{S} = (\mathbf{u}\cdot\nabla)S, \qquad S(\mathbf{x},\xi,0) = \mathbf{x}\cdot\xi. \tag{21}$$

For simplicity we write down only the first transport equation

$$\begin{aligned}\dot{A}_0(\mathbf{x},\xi,t)\eta &= (\mathbf{u}\cdot\nabla)A_0\eta - \left(\frac{\partial\mathbf{u}}{\partial\mathbf{x}}\right)^T A_0\eta \\ &+ \frac{\nabla S}{|\nabla S|^2}[\nabla S\cdot(\nabla\times\mathbf{u})\times A_0\eta] - |\nabla S|^2 A_0\eta, \end{aligned} \tag{22}$$

where $\eta \in \mathbf{R}^3$ is an arbitrary vector, with

$$A_0(\mathbf{x},\xi,0) = 1 - \frac{\xi\otimes\xi}{|\xi|^2}.$$

The vector ∇S satisfies the "cotangent" ODE along the Lagrangian trajectory of the reversed flow:

$$(\nabla S)^{\cdot} = \left(\frac{\partial\mathbf{u}}{\partial\mathbf{x}}\right)^T \nabla S. \tag{23}$$

Hence (22) could be solved as an ODE along each Lagrangian trajectory separately. We note that (21)–(23) imply that

$$\nabla S(\mathbf{x},\xi,t)\cdot A_0(\mathbf{x},\xi,t)\eta = 0$$

which is the main order part of the solenoidality condition.

We now turn to the low frequency terms. We note that the diffusion part of (14) is not important in the main order of approximation on the spatial scale defined by the cut off function in the limit $\epsilon\to 0$. We therefore apply the evolution operator $\mathcal{G}(t)$ acting in $\mathcal{H}^0$ to the first term of the expansion (19).

Finally, the key approximation theorem (which we formulate only for the main order of the approximation) states that

$$\|G(t) - \mathcal{G}(t)\circ\chi\left(\frac{\epsilon^{1/4}}{i}\frac{\partial}{\partial\mathbf{x}}\right) - G^{hf}(t)\|_2 = 0(t\epsilon^{1/4}) \tag{24}$$

uniformly with respect to $t\in[0,T]$. Both terms approximating $G(t)$ in (24) could eventually be estimated via the growth of the semigroup $\mathcal{G}(t)$. This leads to Theorem 1.

3. Euler Equations

We now turn to the situation where dissipation is neglected. The magnetic induction equation for an infinitely conducting fluid is equation (1) with $\epsilon = 0$. This problem,

$$\dot{\mathbf{H}} = \nabla\times(\mathbf{u}\times\mathbf{H}), \qquad \nabla\cdot\mathbf{H} = 0 \tag{25}$$

$$\mathbf{H}(\mathbf{x},0) = \mathbf{H}_0(\mathbf{x}) \tag{26}$$

admits the so called Cauchy (or Alfvén) solution corresponding to the advection of the initial field $\mathbf{H}_0$ which is "frozen" into the fluid, i.e.

$$\mathbf{H}(\mathbf{x},t) = \left(\frac{\partial \mathbf{x}}{\partial \mathbf{x}_0}\right) \mathbf{H}_0. \tag{27}$$

From the definition of the Lyapunov exponent given by (7), it follows that a sufficient condition for dynamo instability in(25)–(27) is the positivity of the Lyapunov exponent $\chi(\mathbf{x}_0, \mathbf{H}_0)$ for some $\mathbf{x}_0$ and $\mathbf{H}_0$. In other words, some exponential stretching in the flow $\mathbf{u}$ ensures dynamo action for an infinitely conducting fluid.

The motion of a non-dissipative, i.e. inviscid, incompressible fluid is governed by the Euler equations,

$$\dot{\mathbf{v}} = -(\mathbf{v}\cdot\nabla)\mathbf{v} - \nabla P, \quad \nabla\cdot\mathbf{v} = 0 \tag{28}$$

Motivated by the magnetic dynamo problem (25)–(26), Vishik and Friedlander (1992), Friedlander and Vishik (1991a,b) use WKB techniques to obtain sufficient conditions for instability of a small perturbation $\mathbf{w}(\mathbf{x},t)$ of an inviscid steady flow $\mathbf{u}(\mathbf{x})$. For technical reasons it again proves useful to study the adjoint problem

$$\dot{\mathbf{w}} = (\mathbf{u}\cdot\nabla)\mathbf{w} - \left(\frac{\partial \mathbf{u}}{\partial \mathbf{x}}\right)^T \mathbf{w} - \nabla P \equiv L_0\mathbf{w} \tag{29}$$

$$\nabla\cdot\mathbf{w} = 0 \tag{30}$$

$$\mathbf{w}(\mathbf{x},0) = \mathbf{w}_0(\mathbf{x}). \tag{31}$$

We note that in contrast with the Navier-Stokes operator L_ϵ, the Euler operator L_0 has a continuous spectrum. We study the stability of (20)–(31) not by an analysis of the spectrum but rather by examining the growth rate of the Greens function $\mathcal{G}(t)$ for L_0 as $t\to\infty$. Asymptotic expansions yield an approximate Green's function for (29)–(31) via which the following theorem is proved.

THEOREM 2.

$$\overline{\lim_{t\to\infty}}(1/t)\log \sup_{\substack{\mathbf{x}_0,\xi_0,\mathbf{b}_0 \\ |\xi_0|=1,\ \mathbf{b}_0\cdot\xi_0=0}} |\mathbf{b}(\mathbf{x}_0,\xi_0,t)| \le \sigma \tag{32}$$

where σ is the exact growth rate of a small perturbation, i.e.

$$\sigma = \lim_{t\to\infty} \frac{1}{t}\log\|\mathcal{G}(t)\|_2. \tag{33}$$

The vector $\mathbf{b}(\mathbf{x}_0, \xi_0, t)$ is the first term in an expansion of the amplitude of a high frequency wavelet localized at $\mathbf{x}_0$. The vector ξ is the spatial wave number vector for the wavelet. The vectors $\mathbf{b}$ and ξ satisfy the following system of ODE:

$$\dot{\mathbf{x}} = -\mathbf{u}(\mathbf{x}) \tag{34}$$

$$\dot{\xi} = \left(\frac{\partial \mathbf{u}}{\partial \mathbf{x}}\right)^T \xi \tag{35}$$

$$\dot{\mathbf{b}} = -\left(\frac{\partial \mathbf{u}}{\partial \mathbf{x}}\right)^T \mathbf{b} - [(\nabla \times \mathbf{u}) \times \mathbf{b} \cdot \xi]\xi/|\xi|^2 \tag{36}$$

with initial conditions at $t = 0$

$$\mathbf{x} = \mathbf{x}_0, \ \xi = \xi_0, \mathbf{b} = \mathbf{b}_0 \ \text{ with } \ \xi_0 \cdot \mathbf{b}_0 = 0.$$

The sufficient condition for instability given by (32) is a precise mathematical formulation of the concept of local instability for inviscid fluid flow widely discussed in the physical literature.

The use of ODE to obtain instability criteria for PDE appears to be a powerful technique where it can be appropriately applied. We briefly mention some of the work whose techniques are related to those that we employ. Eckhoff and Storesletten (1980) and Eckhoff (1981) showed that local instability problems for hyperbolic systems can be essentially reduced to a local analysis involving ODE. Bayly (1988) studied the stability of quasi-two dimensional steady flows via an analysis of a Floquet system of ODE. Lifschitz and Hameiri (1991) use WKB methods, similar in spirit to the work earlier described in Friedlander and Vishik (1991a,b), Vishik and Friedlander (1992), to obtain instability criteria for the Euler equations.

We briefly sketch the derivation of the system of ODE (34)–(36) from an asymptotic expansion in powers of a small parameter δ. We expand the initial perturbation velocity $\mathbf{w}_0(\mathbf{x})$ in the form

$$\mathbf{w}_0(\mathbf{x}) = \mathbf{b}_0(\mathbf{x})e^{i(\mathbf{x}\cdot\xi_0)/\delta} + \ldots \tag{37}$$

and write $\mathbf{w}(\mathbf{x}, t)$ and $P(\mathbf{x}, t)$ in the compatible forms,

$$\mathbf{w}(\mathbf{x}, t) = e^{iS(\mathbf{x},\xi_0,t)/\delta}\mathbf{b}(\mathbf{x}, \xi_0, t) + \ldots \tag{38}$$

$$P(\mathbf{x}, t) = \delta e^{iS(\mathbf{x},\xi_0,t)/\delta}Q(\mathbf{x}, \xi_0, t) + \ldots \tag{39}$$

We substitute (38) and (39) into (29) to obtain the eikonal and transport equations:

$$\dot{S} = (\mathbf{u} \cdot \nabla)S, \quad S|_{t=0} = \mathbf{x} \cdot \xi_0 \tag{40}$$

$$\dot{\mathbf{b}} = (\mathbf{u} \cdot \nabla)\mathbf{b} - \left(\frac{\partial \mathbf{u}}{\partial \mathbf{x}}\right)^T \mathbf{b} - iQ\nabla S. \tag{41}$$

From the solenoidality equation (30)

$$\mathbf{b} \cdot \nabla S = 0, \tag{42}$$

hence from (41)

$$Q = [(\nabla \times \mathbf{u}) \times \mathbf{b} \cdot \nabla S]\nabla S / i|\nabla S|^2 \tag{43}$$

We write $\nabla S = \xi$ and substitute for (43) into (41) to obtain the system of ODE for $\mathbf{b}$ and ξ given by (34)–(36). We note that $\mathbf{b}$ differs from $A_0\mathbf{b}_0$ by a decaying scalar multiplier along the Lagrangian trajectory:

$$e^{-\int_0^t |\xi(\tau)|^2 d\tau}\mathbf{b}(\mathbf{x}_0, \xi_0, t) = A_0(\mathbf{x}, \xi, t)\mathbf{b}_0.$$

It is of interest to note that (32) gives a fluid dynamic analogue of the Lyapunov exponent criteria for a sufficient condition for dynamo instability in an infinitely conducting fluid. We can consider the LHS of (32) to be a Lyapunov type exponent for the Euler equations. We note that the evolution operator for (36) defines a Lyapunov-Oseledets cocycle over the natural extension of the initial dynamical system to the cotangent bundle to the configuration space, i.e. the evolution of the "frozen" covector over $\dot{\mathbf{x}} = -\mathbf{u}(\mathbf{x})$. It is only the direction of ξ (without orientation) that enters into equation (30). Hence (30) represents a cocycle over the projectivisation of the cotangent bundle. In particular, there is a solution ξ to (35) in the direction of the maximal Lyapunov exponent at each point $\mathbf{x}$ (where, and if, such a direction exists). We could therefore apply Oseledets ergodic theorem (Ruelle, 1979) to this cocycle and hence prove the existence of the limit in (32) for such initial directions. We emphasize that the corresponding "Lyapunov-Oseledets" exponents for the linearized Euler equation are analogous to the usual Lyapunov exponents corresponding to the inviscid dynamo operator (25). One would expect that the mean value of these exponents is an important "dynamical systems type" quantity which is related to the linearized Euler equation in the same way that topological entropy is related to (25).

It was suggested to us by P. Constantine that there are different "degrees" of instability for different degrees of smoothness of the initial data. The following theorem shows that this is in fact the case. We consider the stability problem for the linearized Euler equations with an initial perturbation $\mathbf{w}_0$ such that

$$\mathbf{w}_0(\mathbf{x}) \in (\mathcal{H}^s)^3 \cap \{(\nabla \cdot\ ) = 0\}.$$

For the case of periodic boundary conditions we write

$$\mathbf{w}_0(\mathbf{x}) = \sum_{\mathbf{k} \in \mathbf{Z}^3} \tilde{\mathbf{w}}_0(\mathbf{k}) e^{i\mathbf{k}\cdot\mathbf{x}}, \tag{44}$$

with

$$\|\mathbf{w}_0\|_s^2 = \sum_{\mathbf{k}} (1 + |\mathbf{k}|^2)^s |\mathbf{w}_0(\mathbf{k})|^2 < \infty.$$

Let $\mathcal{G}_s(t)$ be the evolution operator for

$$\dot{\mathbf{w}} = -(\mathbf{u} \cdot \nabla)\mathbf{w} - (\mathbf{w} \cdot \nabla)\mathbf{u} - \nabla P, \ \nabla \cdot \mathbf{w} = 0, \ \mathbf{w}|_{t=0} = \mathbf{w}_0(\mathbf{x}). \tag{45}$$

The growth rate σ_s is bounded from below by the following quantity.

THEOREM 3.

$$\overline{\lim_{t\to\infty}} (1/t) \log \sup_{\substack{\mathbf{x}_0, \xi_0, \mathbf{c}_0 \\ |\xi_0| = 1, \mathbf{c}_0 \cdot \xi_0 = 0}} \|(1 + |\xi|^2)^{s/2} \mathbf{c}(\mathbf{x}_0, \xi_0, t)\| \le \sigma_s \tag{46}$$

where

$$\dot{\mathbf{x}} = \mathbf{u}(\mathbf{x}) \tag{47}$$

$$\dot{\xi} = -\left(\frac{\partial \mathbf{u}}{\partial \mathbf{x}}\right)^T \xi \tag{48}$$

$$\dot{\mathbf{c}} = -\left(\frac{\partial \mathbf{u}}{\partial \mathbf{x}}\right)\mathbf{c} + 2\left(\frac{\partial \mathbf{u}}{\partial \mathbf{x}}\,\mathbf{c} \cdot \xi\right)\xi/|\xi|^2, \tag{49}$$

with initial conditions at $t = 0$

$$\mathbf{x} = \mathbf{x}_0, \ \xi = \xi_0, \ \mathbf{c} = \mathbf{c}_0 \ \text{ with } \ \mathbf{c}_0 \cdot \xi_0 = 0.$$

We note that for $s = 0$ equations (47)–(49) give the "adjoint" formulation of Theorem 2. Loosely speaking, the implication of Theorem 3 is that the adjoint Lyapunov exponent is added to the growth rate of an instability with increasing smoothness of $\mathbf{w}_0$ (i.e. s being increased by 1). This means that, in accordance with physical intuition, stretching of higher gradients (which takes place in any nontrivial flow) provides a mechanism for instability in a metric with higher derivatives. The presence of the term in (46) involving the growth of ξ enhances the instability with increasing values of s.

4. Examples

The fact that Theorem 2 gives effective sufficient conditions for instability for the Euler equations is illustrated in Friedlander and Vishik (1991 a, b). We emphasize that the LHS of (32) involves the supremum, hence <u>any</u> Lagrangian trajectory of the flow could provide a positive value on the bound from below for σ and thus imply instability. Since it is sufficient to find one trajectory for which the LHS of (26) is positive, a natural place to seek such a trajectory is at a stagnation point for the velocity $\mathbf{u}(\mathbf{x})$. Equations (35) and (36) are then constant coefficient ODEs and the exponential growth with time of $\mathbf{b}$ follows from the existence of an eigenvalue of an appropriate matrix with positive real part. For example, it is easy to check instability by this criteria for the so called ABC flow

$$u_1 = \cos x_2 - \sin x_3, \; u_2 = \cos x_3 - \sin x_1, \; u_3 = \cos x_1 - \sin x_2. \tag{50}$$

In fact, the existence of <u>any</u> hyperbolic stagnation point in the flow $\mathbf{u}(\mathbf{x})$ gives rise to a positive lower bound for σ from criteria (32) and hence such a flow is unstable. The following arguments justify this statement. We denote by A the matrix $(\frac{\partial \mathbf{u}}{\partial \mathbf{x}})^T$ at a hyperbolic stagnation point $\mathbf{x_S}$. A has at least one real positive eigenvalue λ with an eigenvector that we denote by $\mathbf{a}$. We choose $\xi = \mathbf{a}e^{\lambda t}$. It is easy to see that the matrix Λ satisfying

$$\eta = -A\eta + (A\eta \cdot \mathbf{a})\mathbf{a}/|\mathbf{a}|^2 \tag{51}$$

has its spectrum in the invariant subspace of vectors perpendicular to $\mathbf{a}$ given by the negative of the pair of eigenvalues of A distinct from λ. Hence equations (34)–(36) admit a solution $\mathbf{b}$ that grows exponentially with time and thus from (32) the flow is unstable.

The stability criteria given in Section 2 for a homogeneous fluid can easily be extended to the case of an unbounded inhomogeneous fluid in a gravitational field. For a Boussinesq fluid the statement of Theorem 2 is valid with $\mathbf{b}$ and ξ satisfying the extended system:

$$\dot{\mathbf{x}} = -\mathbf{u}(\mathbf{x}) \tag{52}$$

$$\dot{\xi} = \left(\frac{\partial \mathbf{u}}{\partial \mathbf{x}}\right)^T \xi \tag{53}$$

$$\dot{\mathbf{b}} = -\left(\frac{\partial \mathbf{u}}{\partial \mathbf{x}}\right)^T \mathbf{b} - [(\nabla \times \mathbf{u}) \times \mathbf{b} \cdot \xi]\xi/|\xi|^2 + r[\nabla\rho - (\xi \cdot \nabla\rho)\xi/|\xi|^2] \tag{54}$$

$$\dot{r} = -\mathbf{b} \cdot \nabla\Phi \tag{55}$$

with initial conditions at $t = 0$

$$\mathbf{x} = \mathbf{x}_0, \; \xi = \xi_0, \; \mathbf{b} = \mathbf{b}_0, \; r = r_0; \;\; \xi_0 \cdot \mathbf{b}_0 = 0.$$

The symbols $\rho(\mathbf{x})$, $\Phi(\mathbf{x})$ denote the density distribution of the steady flow $\mathbf{u}(\mathbf{x})$ and the gravitational potential respectively. We assume that $\mathbf{u}$, $\nabla\rho$ and $\nabla\Phi$ and their derivatives are bounded as $|\mathbf{x}| \to \infty$. The quantity r is the main term in a WKB asymptotic expansion for the perturbation density. In the instability criteria analogous to (32), $\mathbf{b}$ is replaced by the four-component vector $(\mathbf{b}, r)$ and the supremum is taken over $\mathbf{x}_0, \xi_0, \mathbf{b}_0, r_0$; $\xi_0 \cdot \mathbf{b}_0 = 0$.

We now consider the case of a stratified flow with a hyperbolic stagnation point $\mathbf{x}_s$. We denote by A the matrix $(\frac{\partial \mathbf{u}}{\partial \mathbf{x}})^T$ at $\mathbf{x}_s$. For simplicity we consider the case where A has a positive real eigenvalue which we denote by λ. The case where A has one negative and two complex conjugate eigenvalues can be treated by similar arguments. We choose $\xi = \xi_0 e^{\lambda t}$. We denote by B the matrix corresponding to the first two terms on the RHS of (54); i.e. we write (54) in the form

$$\dot{\mathbf{b}} = B\mathbf{b} + r[\nabla\rho - (\xi_0 \cdot \nabla\rho)\xi_0/|\xi_0|^2]. \tag{56}$$

We seek solutions to (55)–(56) in the form $\mathbf{b} = \mathbf{b}_0 e^{\gamma t}$, $r = r_0 e^{\gamma t}$. Both the vector $\mathbf{b}$ and the vector $[\nabla\rho - (\xi_0 \cdot \nabla\rho)\xi_0/|\xi_0|^2]$ lie in the subspace perpendicular to ξ_0. We consider equations (55)–(56) restricted to this subspace, and write the characteristic equation for γ in the form

$$\left|(B_\perp - \gamma I) - \frac{\nabla_\perp\rho \otimes \nabla_\perp\Phi}{\gamma}\right| = 0, \tag{57}$$

where $\perp$ denotes an operator on the subspace perpendicular to ξ. Since

$$\mathrm{Tr}\ B_\perp + (-\lambda) = \mathrm{Tr}\ B = 0, \tag{58}$$

it follows that Tr $B_\perp$ is positive. Hence at least one root γ given by (57) has positive real part. In the two-dimensional problem γ can be easily written in terms of λ as

$$\gamma = \left[\lambda \pm \sqrt{\lambda^2 - 4\nabla_\perp\rho \cdot \nabla_\perp\Phi}\,\right]\Big/2. \tag{59}$$

Thus the presence of buoyancy cannot remove the instability due to stretching associated with the existence of a stagnation point even if the density gradient itself is stabilizing. In the case where the density gradient is destabilizing, the growth rate of the stagnation point instability will be enhanced by the effects of gravitational overturning.

Our final example illustrating the effectiveness of the instability criteria in Theorem 2 is relevant to problems arising in geophysical fluid dynamics, namely the question of stability of a stratified fluid rotating about a fixed axis with a sheared angular velocity. We write the velocity $\mathbf{u}$ in a coordinate system rotating with constant angular velocity Ω_0 as

$$\mathbf{u} = s\Omega(s,z)\widehat{\phi}. \tag{60}$$

The steady state equation, known as the "thermal wind" equation in meteorology, requires that Ω and ρ satisfy:

$$s\frac{\partial}{\partial z}(\Omega + \Omega_0)^2 + \rho_z\Phi_s - \rho_s\Phi_z = 0 \tag{61}$$

[(s, ϕ, z) denote cylindrical polar coordinates]. In the rotating frame, equation (54) is modified by the addition of the term

$$2\Omega(\widehat{\mathbf{k}} \times \mathbf{b} - (\xi \cdot \widehat{\mathbf{k}} \times \mathbf{b})\xi/|\xi|^2). \tag{62}$$

We consider a stagnation point $\mathbf{x}_s$ where Ω_0 is chosen to be equal to the local angular velocity of the rotating fluid. (i.e. the velocity $\mathbf{u} = 0$ in the rotating frame). In this example

$$\left(\frac{\partial \mathbf{u}}{\partial \mathbf{x}}\right)^T = \begin{pmatrix} 0 & s\Omega_s & 0 \\ 0 & 0 & 0 \\ 0 & s\Omega_z & 0 \end{pmatrix} \quad \text{at } \mathbf{x} = \mathbf{x}_s \tag{63}$$

and equation (53) admits the solution

$$\xi = \begin{pmatrix} c_1 \\ 0 \\ c_3 \end{pmatrix} \tag{64}$$

where c_1 and c_3 are arbitrary constants. We seek a solution for $\mathbf{b}$ of the form $\mathbf{b} = \mathbf{b}_0 e^{\gamma t}$. We substitute for ξ from (64) into (54) and combine (54) and (55) to obtain the following equation for the growth rate γ of $\mathbf{b}(t)$:

$$\begin{vmatrix} -\gamma(c_1^2 + c_3^2) & -(2\Omega_0 + s\Omega_s)c_3^2 + s\Omega_z c_1 c_3 & 0 & \rho_3 c_3^2 - \rho_z c_1 c_3 \\ 2\Omega_0(c_1^2 + c_3^2) & -\gamma(c_1^2 + c_3^2) & 0 & 0 \\ 0 & -c_1^2 s\Omega_z + c_1 c_3(2\Omega_0 + s\Omega_s) & -\gamma(c_1^2 + c_3^2) & c_1^2\rho_z - \rho_s c_1 c_3 \\ -\Phi_3(c_1^2 + c_3^2) & 0 & -\Phi_z(c_1^2 + c_3^2) & -\gamma(c_1^2 + c_3^2) \end{vmatrix}$$

$$= 0 \tag{65}$$

We expand the determinant in (65) and use the thermal wind equation (61) to obtain the equation

$$\begin{aligned} \gamma^2[\gamma^2(c_1^2 + c_3^2) \; + \; & \{\Phi_s\rho_s c_3^2 + \Phi_s\rho_z c_1^2 - c_1 c_3(\rho_s\Phi_z + \rho_z\Phi_s) \\ & + \; 2\Omega_0(c_3^2(2\Omega_0 + s\Omega_s) - c_1 c_3 s\Omega_z)\}] = 0. \end{aligned} \tag{66}$$

Hence there exists a positive root γ, implying instability by Theorem 2, when

$$\begin{aligned} \{c_3^2[\Phi_s\rho_s \; + \; & 2\Omega_0(2\Omega_0 + s\Omega_s)] - c_1 c_3[\rho_s\Phi_z + \rho_z\Phi_s + 2\Omega_0 s v_z] \\ & + \; c_1^2\Phi_z\rho_z\} < 0. \end{aligned} \tag{67}$$

The ratio of c_1 to c_3, which determines the direction of the wave number vector ξ, is arbitrary. Hence (67) gives a sufficient condition for instability with respect to perturbations associated with the direction of ξ. For example when $c_1 = 0$, and hence from the solenoidality condition $b_3 = 0$, (67) implies that the flow is unstable when

$$\Phi_s\rho_s + 2\Omega_0(2\Omega_0 + s\Omega_s) < 0 \tag{68}$$

which is a combination of the Rayleigh criteria for instability of a rotating fluid with angular velocity $\Omega(s)$ and gravitational instability. When $c_3 = 0$, and hence $b_1 = 0$, (67) gives the gravitational instability criteria

$$\rho_z \Phi_z < 0. \tag{69}$$

The condition (67) implies that the flow will be unstable even if the inequalities (68) or (69) are not satisfied. There exists a direction (c_1/c_s) such that (67) is satisfied provided

$$(\rho_s \Phi_z - \rho_z \Phi_s + 2\Omega_0 s \Omega_z)^2 - 4\Phi_z \rho_z (\Phi_s \rho_s + 2\Omega_0(2\Omega_0 + s\Omega_s)) > 0. \tag{70}$$

Using the thermal wind equation (61), converts (70) into the criteria

$$\Omega_0 \Phi_z (s\Omega_z \rho_s - (2\Omega_0 + s\Omega_s)\rho_z) > 0. \tag{71}$$

The inequality (71) is a generalized criteria for baroclinic instability that occurs when the energy due to the horizontal density gradient is large enough to overcome the stabilizing effects of rotation and buoyancy.

Acknowledgements

S. Friedlander acknowledges NSF Grant No. DMS 9000137. M. Vishik acknowledges NSF Grant No. DMS 9105688.

References

ARNOLD, V. I., ZELDOVICH, YA. B., RUZMAIKIN, A. A. & SOKOLOFF, D. D., 1982. A magnetic field in a stationary flow with stretching in Riemannian space. *J. Exp. Theor. Phys.*, **81** no. 6, pp. 2052-2058.

BAYLY, B., 1986 Fast magnetic dynamos in chaotic flows. *Phys. Rev. Lett.*, **57** no. 22, pp. 2800.

BAYLY, B., 1988 Three dimensional instabilities in quasi two dimensional inviscid flows. *Phys. Fluids*, **31**, pp. 56.

ECKHOFF, K. S., 1981 On stability for symmetric hyperbolic systems. *J. Differential Equations*, **40**, pp. 94-115.

ECKHOFF, K. S. AND STORESLETTEN, L., 1980 On the stability of rotating compressible inviscid fluids. *J. Fluid Mech.*, **99**, pp. 433-448.

FRIEDLANDER, S. & VISHIK, M. M., 1991a Instability criteria for the flow of an inviscid incompressible fluid. *Phys. Rev. Lett.*, **66** no. 17, pp. 2204-2206.

FRIEDLANDER, S. & VISHIK, M. M., 1991b Dynamo theory, vorticity generation and exponential stretching. *Chaos*, **1** no. 2, pp. 198-205.

GILBERT, A. D. & CHILDRESS, S., 1990 Evidence for fast dynamo action in a chaotic web. *Phys. Rev. Lett.*, **65** no. 17, pp. 2133-2136.

LIFSCHITZ, A. & HAMEIRI, E., 1991 Local stability conditions in fluid dynamics. *Phys. Fluids A*, **3**, pp. 2644-2651.

LIFSCHITZ, A. & HAMEIRI, E., 1992 Localized instabilities in fluids. Submitted for publication to the Proc. NATO Workshop on Topological Fluid Dynamics.

RUELLE, D., 1979 Ergodic theory of differentiable dynamical systems. *Pub. Math. I.H.E.S.*, **50**, pp. 275-306.

SOWARD, A. M., 1987 Fast dynamo action in a steady flow. *J. Fluid Mech.*, **180**. pp. 267-295.

SOWARD, A. M. & CHILDRESS, S., 1986 Analytic theory of dynamos. *Adv. Space Res.*, **6** no. 8, pp. 7-18.

VISHIK, M. M., 1989 Magnetic field generation by the motion of a highly conducting fluid. *Geophys. Astrophys. Fluid Dyn.*, **48**, pp. 151-167.

VISHIK, M. M. & FRIEDLANDER, S., 1992 Dynamo theory methods for hydrodynamic stability. To appear in *J. Math. Pure et Appliques.*

ZELDOVICH, YA. B. & RUZMAIKIN, A. A., 1980 The magnetic field in a conducting fluid in two dimensional motion. *J. Exp. Theor. Phys.*, **78**, pp. 980-986.

LOCALIZED INSTABILITIES IN FLUIDS

ALEXANDER LIFSCHITZ
Department of Mathematics
University of Illinois at Chicago
Chicago, Illinois 60680, U.S.A.

and

ELIEZER HAMEIRI
Courant Institute of Mathematical Sciences
New York University
New York, New York 10012, U.S.A.

ABSTRACT. Three-dimensional flows of an inviscid incompressible fluid are considered and it is demonstrated how the geometrical optics method can be used for investigating their stability. The evolution of localized short wavelength envelopes is analyzed and it is shown that their growth rate can be evaluated in terms of the growth rate of solutions of characteristic equations along stream lines of the basic flow. Analyzing the corresponding characteristic equations, a local stability condition is obtained. By virtue of this condition it is shown that steady flows are unstable if they have points of stagnation; it is also shown that vortex rings without swirl with convex stream lines are unstable when the circulation decreases outward.

1. Introduction

Investigating the stability of a flow of an inviscid incompressible fluid is important in order to understand the most fundamental features of the fluid motion. In the present paper we describe short wavelength instabilities of spatially periodic (or asymptotically uniform), three-dimensional flows. We call these instabilities localized because they develop in thin stream tubes along certain stream lines. Short wavelength instabilities play an important role in many situations; in particular, it is widely believed that they are responsible for the transition from two-dimensional coherent structures (such as vortex rings with swirl) to three-dimensional chaotic ones, see, e.g., Bayly *et al.* (1988). During the last decade or so short wavelength instabilities have been studied by a number of researchers (cf. Bayly 1988; Eckhoff and Storesletten 1978, 1980; Friedlander and Vishik 1991a,b; Hameiri 1985; Hameiri and Lifschitz 1990; Lifschitz 1984, 1991a,b; Lifschitz and Hameiri

H. K. Moffatt et al. (eds.), Topological Aspects of the Dynamics of Fluids and Plasmas, 551–561.

1991a,b). In the present paper we discuss our recent results on these instabilities; related stability results are also given in another paper appearing in this volume (Friedlander and Vishik 1992).

We concentrate on the linear problem. Some concluding remarks are devoted to the nonlinear problem. We call a spatially periodic (or asymptotically uniform) flow stable in the velocity norm if the L_2 norm of the velocity in the perturbed flow is uniformly bounded in time, otherwise we call it unstable. We base our approach to the analysis of short wavelength instabilities on the geometrical optics method and consider the behavior of solutions of the initial value problem for the perturbed Euler's equations with the initial data chosen in the form of localized short wavelength envelopes. As a measure of the wavelength we choose a dimensionless small parameter ε. Emphasize that for the inviscid fluid the corresponding initial value problem does not have a typical internal length scale, so that this small parameter can be made arbitrarily small. Due to this fact the analysis of the short wavelength perturbations provides reliable results concerning the stability of basic flows. The geometrical optics approach to the stability problem is very flexible; it allows us to consider the stability of time dependent basic flows and investigate not only exponentially growing instabilities but also instabilities having algebraic growth. Our main result, a local stability condition, can be stated as follows. A basic flow of an incompressible fluid is unstable in the velocity norm if for certain stream line the magnitude of the amplitude which is governed by the transport equation associated with the wave advected by the flow is growing in time without bound. This instability condition is very effective from a practical point of view because it allows one to estimate from below the growth rate of solutions of the initial value problem for a partial differential equation in terms of the growth rate of solutions of the initial value problem for an ordinary differential equation. Our result is a natural generalization of the well-known results of Eckhoff and Storesletten (1978, 1980) concerning the stability of helical and circular flows of a compressible gas. It is worth mentioning that unstable ballooning modes which are localized near magnetic lines of force are well known in magnetohydrodynamics (cf. Connor *et al.* 1979). These modes are responsible for the most dangerous instabilities of toroidal equilibrium configurations. In our present study we broadly use the magnetohydrodynamic technique and extend it to hydrodynamics.

In order to find geometrical optics solutions of the Euler's equations we need to construct formal asymptotic solutions and to prove that these solutions are close to actual ones. It can easily be shown that for the Euler's equations rays coincide with stream lines of the basic flow; for this reason they do not intersect and we are able to find global formal asymptotic solutions. Moreover, using appropriately modified energy inequalities, we can prove that these solutions are close to actual ones. As a result, we are able to estimate the growth rate of actual solutions in terms of the behavior of the leading-order terms of the corresponding asymptotic solutions. These terms are governed by the so-called transport equations, which are ordinary differential equations along rays (stream lines of the flow), so that we obtain a local stability condition for general time-dependent, three-dimensional flows.

In order to demonstrate how to apply this local stability condition in practice, we consider two important classes of steady three-dimensional flows: (a) flows having points of stagnation, (b) axisymmetric vortex rings with swirl. Using our stability condition we prove that steady basic flows having points of stagnation are linearly and nonlinearly unstable with respect to short wavelength instabilities. (One can expect that near such points parameters of nearby flows depend on time in a complicated fashion). For axisymmetric vortex rings we obtain explicit stability conditions for the core. For such a ring we can distinguish instabilities of two types: (a) instabilities having Floquet behavior and exponential growth rate, (b) instabilities growing algebraically in time. Conditions guaranteeing the stability with respect to exponentially growing perturbations can be formulated in terms of the properties of the monodromy matrices associated with the corresponding transport equations, while stability conditions with respect to algebraically growing perturbations can be reduced to certain inequalities for stream surface averages of equilibrium quantities.

2. Instabilities of Three-Dimensional Flows

Consider a spatially periodic (or asymptotically uniform), three-dimensional flow of an inviscid incompressible fluid. The velocity $\mathbf{V}(t,\mathbf{x})$ and the reaction pressure $P(t,\mathbf{x})$ for such a flow are governed by the Euler's equations supplied with appropriate initial data. In order to analyze the stability of the basic flow $(\mathbf{V},P)$ we consider a perturbed flow $(\mathbf{V}^*,P^*)$ and study the behavior of the difference $(\mathbf{V}^*-\mathbf{V},P^*-P)=(\mathbf{v},p)$. This difference satisfies the linearized Euler's equations

$$\frac{D\mathbf{v}}{Dt}+L\mathbf{v}+\nabla p=0, \quad \nabla\cdot\mathbf{v}=0, \tag{2.1}$$

with appropriate boundary conditions and initial data. Here and below we use the notation L for the tensor $\nabla\mathbf{V}$, $L^i_j=\partial V^i/\partial x^j$.

In order to obtain the local stability conditions sought we choose the initial data $\mathbf{v}(0,\mathbf{x})=\mathbf{v}_0(\mathbf{x})$ in the form of a short wavelength localized envelope

$$\mathbf{v}_0(\mathbf{x})=\exp(i\Phi_0(\mathbf{x})/\varepsilon)\hat{\mathbf{v}}_0(\mathbf{x})=\exp(i\Phi_0(\mathbf{x})/\varepsilon)[\nabla\alpha_0(\mathbf{x})\times\nabla\Phi_0(\mathbf{x})], \tag{2.2}$$

where ε is a small dimensionless parameter, and Φ_0, α_0 are smooth localized functions.

It can be shown that the corresponding initial value problem has a global geometrical optics solution of the form

$$(\mathbf{v}, p) = \exp(i\Phi/\varepsilon)[(\mathbf{a}, 0) + \frac{i\varepsilon}{|\nabla\Phi|^2}((\nabla\cdot\mathbf{a})\nabla\Phi, 2L\mathbf{a}\cdot\nabla\Phi)] + \varepsilon(\mathbf{v}^{(r)}, p^{(r)}), \quad (2.3)$$

where Φ is the phase which is advected by the fluid, $\mathbf{a}$ is the amplitude which is governed by the transport equation, and $(\mathbf{v}^{(r)}, p^{(r)})$ is the corresponding remainder. The eikonal equation and the initial conditions for the phase have the form

$$\frac{D\Phi}{Dt} = \frac{\partial\Phi}{\partial t} + \mathbf{V}\cdot\nabla\Phi = 0, \quad \Phi(0, \mathbf{x}) = \Phi_0(\mathbf{x}). \quad (2.4a, b)$$

The transport equation and the initial conditions for the amplitude can be written as

$$\frac{D\mathbf{a}}{Dt} + L\mathbf{a} - 2L\mathbf{a}\cdot\nabla\Phi\frac{\nabla\Phi}{|\nabla\Phi|^2} = 0, \quad \mathbf{a}(0, \mathbf{x}) = \hat{\mathbf{v}}_0(\mathbf{x}). \quad (2.5a, b)$$

The velocity $\mathbf{v}^{(r)}$ can be estimated uniformly in ε on every finite time interval. Emphasize that by virtue of the incompressibility condition $\mathbf{a}$ and $\mathbf{k}$ should be orthogonal to each other. The transport equation preserves the corresponding scalar product, so that the orthogonality condition is satisfied at any time because it is satisfied initially.

The eikonal equation and the transport equation can be written in the characteristic form

$$\frac{d\mathbf{x}}{d\tau} = \mathbf{V}, \quad \frac{d\mathbf{k}}{d\tau} = -L^T\mathbf{k}, \quad (2.6a, b)$$

$$\mathbf{x}(0) = \mathbf{x}_0, \quad \mathbf{k}(0) = \mathbf{k}_0 = \nabla\Phi_0(\mathbf{x}_0), \quad (2.6c, d)$$

$$\frac{d\mathbf{a}}{d\tau} = -L\mathbf{a} + 2L\mathbf{a}\cdot\mathbf{k}\frac{\mathbf{k}}{|\mathbf{k}|^2}, \quad \mathbf{a}(0) = \mathbf{a}_0 = \hat{\mathbf{v}}_0(\mathbf{x}_0). \quad (2.7a, b)$$

Equations (2.6a,b) are Hamiltonian with the corresponding Hamiltonian $\Omega = \mathbf{V}\cdot\mathbf{k}$. We denote solutions of the problems (2.6), (2.7) by $\mathbf{x}(\tau; \mathbf{x}_0)$, $\mathbf{k}(\tau; \mathbf{x}_0, \mathbf{k}_0)$, $\mathbf{a}(\tau; \mathbf{x}_0, \mathbf{k}_0, \mathbf{a}_0)$ in order to emphasize their dependence on the initial data. Solutions of the problems (2.4), (2.5) can be represented as

$$\Phi[t, \mathbf{x}(t; \mathbf{x}_0)] = \Phi_0(\mathbf{x}_0), \quad \mathbf{a}[t, \mathbf{x}(t; \mathbf{x}_0)] = \mathbf{a}[t; \mathbf{x}_0, \nabla\Phi_0(\mathbf{x}_0), \mathbf{a}_0(\mathbf{x}_0)]. \quad (2.8)$$

It should be emphasized that characteristic equations are written in a geometrically invariant form and can be used in general curvilinear coordinates.

Assuming that at the initial moment our wave envelope is localized in a small vicinity of a point $\mathbf{x}_0$, its wave front is perpendicular to a vector $\mathbf{k}_0$, and its amplitude is parallel to a vector $\mathbf{a}_0$, $\mathbf{a}_0 \cdot \mathbf{k}_0$, we can prove that the basic flow is unstable if the corresponding solution $\mathbf{a}(\tau; \mathbf{x}_0, \mathbf{k}_0, \mathbf{a}_0)$ is unbounded in time. Considering all possible choices of $\mathbf{x}_0, \mathbf{k}_0, \mathbf{a}_0$ we obtain the local stability condition sought and prove that the basic flow of an incompressible inviscid fluid is unstable if

$$\sup_{\mathbf{x}_0, \mathbf{k}_0, \mathbf{a}_0, \mathbf{a}_0 \cdot \mathbf{k}_0 = 0} \overline{\lim}_{\tau \to \infty} |\mathbf{a}(\tau; \mathbf{x}_0, \mathbf{k}_0, \mathbf{a}_0)| = \infty. \tag{2.9}$$

In the following sections we demonstrate how this condition can be used in order to analyze particular classes of flows.

3. Instabilities of Steady Flows Having Points of Stagnation

In this section we use the stability condition (2.9) in order to prove that a steady three-dimensional flow of an inviscid incompressible fluid having a nondegerate point of stagnation is unstable. Consider such a flow and assume that $\mathbf{X}$ is a nondegenerate stagnation point such that $\mathbf{V}(\mathbf{X}) = 0$ and $L(\mathbf{X}) = \nabla \mathbf{V}(\mathbf{X}) \neq 0$. Following Friedlander and Vishik (1991a) who studied hyperbolic points we choose $\mathbf{X}$ as the initial data for equation (2.6a). The corresponding particle trajectory consists of the point $\mathbf{X}$ itself. Equation (2.6b) for $\mathbf{k}$ is a linear equation with constant coefficients and can be solved explicitly. Coefficients of the transport equation (2.7a) for $\mathbf{a}$ can be written in terms of exponential (or trigonometric) functions. In general, this equation cannot be solved in the closed form but we can always choose initial conditions in such a way that the corresponding solution is unbounded and prove that the flow in question is unstable. Emphasize that due to the local character of the geometrical optics approach we can restrict ourselves to the consideration of an arbitrarily small vicinity of the point $\mathbf{X}$ (cf. Friedlander and Vishik 1991a). This observation allows us to use the results on the stability of flows with linear velocity profiles (cf. Bayly 1986; Craik and Criminale 1986; Kelvin 1887; Lagnado *et al.* 1984).

The equilibrium conditions restrict the form of the matrix L. We can distinguish the following cases: (a) $\mathbf{X}$ is a three-dimensional hyperbolic point, (b) $\mathbf{X}$ is a two-dimensional hyperbolic point, (c) $\mathbf{X}$ is a two-dimensional elliptical point, (d) $\mathbf{X}$ is a two-dimensional shear point.

In case (a) stream lines locally are three-dimensional hyperbolas, the matrix L is symmetric and

$$L\mathbf{x} = \lambda_1 \mathbf{x} \cdot \mathbf{e}_1 \mathbf{e}_1 + \lambda_2 \mathbf{x} \cdot \mathbf{e}_2 \mathbf{e}_2 + \lambda_3 \mathbf{x} \cdot \mathbf{e}_3 \mathbf{e}_3, \tag{3.1}$$

where λ_i are nonzero real eigenvalues of the matrix L and $\mathbf{e}_i$ are the corresponding orthonormal eigenvectors. The incompressibility condition yields $\lambda_1 + \lambda_2 + \lambda_3 = 0$, so that we can assume that $\lambda_1 > 0, \lambda_3 < 0$. Choosing $\mathbf{k}_0 = \mathbf{e}_1, \mathbf{a}_0 = \mathbf{e}_3$ we obtain

$$\mathbf{k}(\tau; \mathbf{X}, \mathbf{e}_1) = \exp(-\lambda_1 \tau)\mathbf{e}_1, \quad \mathbf{a}(\tau; \mathbf{X}, \mathbf{e}_1, \mathbf{e}_3) = \exp(-\lambda_3 \tau)\mathbf{e}_3, \tag{3.2}$$

so that we have an exponential instability.

In case (b) zero is a simple eigenvalue of the matrix L with the corresponding eigenvector $\mathbf{e}$. In the plane orthogonal to $\mathbf{e}$ stream lines are two-dimensional hyperbolas and we can choose a basis $\mathbf{f}_1, \mathbf{f}_2$ and a biorthogonal basis $\mathbf{g}_1, \mathbf{g}_2$ in such a way that

$$L\mathbf{x} = \lambda[\mathbf{x} \cdot \mathbf{g}_1 \mathbf{f}_1 - \mathbf{x} \cdot \mathbf{g}_2 \mathbf{f}_2], \tag{3.3}$$

where λ is a characteristic frequency. Taking $\mathbf{k}_0 = \mathbf{e}, \mathbf{a}_0 = \mathbf{f}_2$ we obtain

$$\mathbf{k}(\tau; \mathbf{X}, \mathbf{e}) = \mathbf{e}, \quad \mathbf{a}(\tau; \mathbf{X}, \mathbf{e}, \mathbf{f}_2) = \exp(\lambda \tau)\mathbf{f}_2, \tag{3.4}$$

so that our two-dimensional flow is exponentially unstable with respect to three-dimensional perturbations. It is interesting to note that it is unstable with respect to two-dimensional perturbations as well. Indeed, for $\mathbf{k}_0 = \mathbf{g}_1, \mathbf{a}_0 = \mathbf{f}_2$ the corresponding solutions have tbe form

$$\mathbf{k}(\tau; \mathbf{X}, \mathbf{g}_1) = \exp(-\lambda \tau)\mathbf{g}_1, \quad \mathbf{a}(\tau; \mathbf{X}, \mathbf{g}_1, \mathbf{f}_2) = \exp(\lambda \tau)\mathbf{f}_2, \tag{3.5}$$

and we have an exponential instability. Nevertheless, the vorticity of this perturbation does not grow in time in agreement with the fact that the two-dimensional enstrophy is a conserved quantity.

In case (c) zero is also a simple eigenvalue of L. Denote by $\mathbf{e}$ the corresponding eigenvector and consider the plane orthogonal to $\mathbf{e}$. In this plane stream lines are ellipsis and we can choose an orthonormal basis $\mathbf{f}_1, \mathbf{f}_2$ in such a way that

$$L\mathbf{x} = \lambda[\frac{1}{E}\mathbf{x} \cdot \mathbf{f}_2 \mathbf{f}_1 - E\mathbf{x} \cdot \mathbf{f}_1 \mathbf{f}_2], \tag{3.6}$$

where λ is a characteristic frequency, and $0 < E \leq 1$ is a parameter characterizing ellipticity of the flow. Case $E = 1$ corresponds to rigid body rotation. It is convenient to decompose vectors $\mathbf{k}$ and $\mathbf{a}$ in the orthonormal basis $\mathbf{e}$, $\mathbf{f}_1$, $\mathbf{f}_2$, and represent them as $(k_{||}, \mathbf{k}_\perp) = (k_{||}, k_1, k_2)$, $(a_{||}, \mathbf{a}_\perp) = (a_{||}, a_1, a_2)$. A general solution of equation (2.6b) can be represented as $(\cos\Theta, E\sin\Theta\sin\tau, \sin\Theta\cos\tau)$, where Θ is the angle between $\mathbf{e}$ and $\mathbf{k}$ at the initial moment. The corresponding transport equation for $\mathbf{a}$ cannot be solved explicitly, nevertheless, after considerable amount of algebra it can be reduced to the equation for $\mathbf{a}_\perp$,

$$\frac{d\mathbf{a}_\perp}{d\tau} + Q\mathbf{a}_\perp = 0, \tag{3.7}$$

where Q is a 2 by 2 periodic matrix function. Denote by $M(E, \Theta)$ the corresponding monodromy matrix, and by $\zeta_1(E, \Theta), \zeta_2(E, \Theta)$ its characteristic exponents. It can be shown (cf. Bayly 1986; Waleffe 1990) that for each E (except of $E = 1$) there is an interval $(\Theta_-(E), \Theta_+(E))$ such that $\Re\zeta_1(E, \Theta) > 0$ when $\Theta_-(E) < \Theta < \Theta_+(E)$. According to the Floquet theorem, it means that for appropriate inclination angles the corresponding amplitudes grow in time exponentially and we have an instability. Note that in contrast to the previous case this instability is necessarily three-dimensional.

In case (d) zero is an eigenvalue of L of multiplicity three and we can choose an orthonormal basis $\mathbf{e}, \mathbf{f}, \mathbf{g}$ in such a way that

$$L\mathbf{x} = \lambda \mathbf{x} \cdot \mathbf{f}\mathbf{g}, \tag{3.8}$$

where λ is a characteristic frequency, so that stream lines are straight. We can take $\mathbf{k}_0 = \mathbf{e}, \mathbf{a}_0 = \mathbf{g}$ and obtain a growing solution of the transport equation

$$\mathbf{k}(\tau; \mathbf{X}, \mathbf{e}) = \mathbf{e}, \quad \mathbf{a}(\tau; \mathbf{X}, \mathbf{k}, \mathbf{g}) = -\lambda\tau\mathbf{g} + \mathbf{f}. \tag{3.9}$$

Thus, the shear flow under consideration is algebraically unstable. This instability is less dangerous than exponential instabilities obtained in the previous cases.

Finally, we can conclude that any steady flow of an inviscid incompressible fluid having a point of stagnation is unstable. In certain cases we can expand the area of applicability of this result by means of a Galilean transformation. In particular, we can prove that any plane-parallel flow is (algebraically) unstable with respect to short wavelength perturbations. It should be emphasized that our result is valid only for steady flows. Dependence of the matrix $L(\mathbf{X})$ on time can have a strong stabilizing impact. In particular, it can be shown that in a rotating coordinate system flows with elliptical stagnation points become stable when angular velocity is chosen in a proper way. (Note in parenthesis that the Kirchhoff vortex cannot be stabilized by the Coriolis force).

4. Instabilities of Axisymmetric Vortex Rings with Swirl

Our analysis of the stability of steady flows having points of stagnation described in the previous section is based on the fact that the corresponding characteristic equations can be solved explicitly. There exist other classes of steady flows having similar properties. In particular, this is true for vortex rings. In this section we discuss the stability of the core of a general axisymmetric vortex ring with swirl.

Consider an axisymmetric vortex ring with swirl. In appropriate cylindrical coordinates (r, ϕ, z) the velocity $\mathbf{V}$ and the pressure P are independent of ϕ. We can introduce a stream function $\Psi(r, z)$ and represent the velocity field $\mathbf{V}(r, z)$ in the form

$$\mathbf{V} = \mathbf{V}_P + \mathbf{V}_T = \nabla\phi \times \nabla\Psi + f\nabla\phi, \tag{4.1}$$

where $\mathbf{V}_P, \mathbf{V}_T$ are the poloidal and toroidal velocities respectively, and $f = f(\Psi)$. Stream surfaces coincide with level surfaces of Ψ. We assume that within the vortex core Ψ has a unique minimum, so that within the core stream surfaces are toroidal, outside the core they are cylindrical.

It is more convenient to describe the core of the vortex ring in question in natural coordinates (Ψ, θ, ϕ), where θ is an appropriate poloidal angle chosen in such a way that the velocity $\mathbf{V}$ has the form

$$\mathbf{V} = (V^\Psi, V^\theta, V^\phi) = \frac{1}{r^2}(0, g(\Psi), f(\Psi)), \tag{4.2}$$

so that stream lines are straight. Natural coordinates have been used in magnetohydrodynamics for a long time, recently they became popular in fluid dynamics. For each stream surface we can define the poloidal turnaround time $T = T(\Psi)$, the velocity shear $q = q(\Psi)$, and the circulation $C = C(\Psi)$,

$$T = \oint \frac{dl}{|\mathbf{V}_P|}, \quad q = \frac{f}{g} = \frac{f}{2\pi}\oint \frac{1}{r^2}\frac{dl}{|\mathbf{V}_P|}, \quad C = \oint |\mathbf{V}_P| dl. \tag{4.3}$$

If $q = q(\Psi)$ is rational, then stream lines on the corresponding stream surface are closed, otherwise they cover this surface ergodically.

In natural coordinates the eikonal equations for $\mathbf{x}, \mathbf{k}$ can be solved explicitly, while the transport equation for $\mathbf{a}$ can be analyzed in detail. Using θ as a stream line parameter and assuming that $\theta_0 = 0, \phi_0 = 0$ we can represent the solution of the problem (2.6) as

$$\Psi = \Psi_0, \quad \phi = q(\Psi_0)\theta, \quad \mathbf{k} = \Pi\hat{\mathbf{k}}(\theta; \Psi_0, \hat{\Lambda}, \hat{\Omega}), \tag{4.4}$$

where $\hat{\mathbf{k}}$ is a known function of its arguments. Here $\Pi = k_{\phi 0}$, $\hat{\Lambda} = k_{\Psi 0}/\Pi$, $\hat{\Omega} = \Omega/\Pi = \mathbf{V}\cdot\mathbf{k}/\Pi$. It should be emphasized that Ω and Π are conserved quantities.

Introducing the conserved quantity $\kappa = \Omega\dot{T}/2\pi - \Pi\dot{q}$, where the dot denotes the derivative with respect to Ψ, we can distinguish the following possibilities: (a) $\kappa = 0$, $\mathbf{k}$ is a periodic function of θ, (b) $\kappa \neq 0$, $\mathbf{k}$ is an asymptotically linear function of theta. Broadly speaking, cases (a) and (b) are analogous to the elliptical and shear cases considered in the previous section. We concentrate on the first possibility, the second one can be treated following similar lines. In order to analyze the stability we need to study solutions of the transport equation (2.7). In natural coordinates this equation has a complicated form because its coefficients explicitly depend on elements of the metric tensor and Christoffel symbols. In order to obviate this difficulty and reveal the geometric nature of the transport equation we use the following construction. At each point we consider the (nonorthogonal) basis consisting of vectors $\mathbf{V}_P$, $\nabla\Psi$, $\mathbf{k}$. Using the fact that the amplitude $\mathbf{a}$ is orthogonal to $\mathbf{k}$ we present this amplitude in the form

$$\mathbf{a} = c_1\mathbf{e}_1 + c_2\mathbf{e}_2 \quad \text{with} \quad \mathbf{e}_1 = \mathbf{V}_P - \frac{\mathbf{V}_P \cdot \mathbf{k}}{|\mathbf{k}|^2}\mathbf{k}, \quad \mathbf{e}_2 = \nabla\Psi - \frac{\nabla\Psi \cdot \mathbf{k}}{|\mathbf{k}|^2}\mathbf{k}. \quad (4.5)$$

Plugging this expansion into equation (2.7) we obtain the following system of ordinary differential equations for the unknown vector $\mathbf{c} = (c_1, c_2)$

$$\Sigma\frac{d\mathbf{c}}{d\tau} + \Gamma\mathbf{c} = 0, \quad (4.6)$$

where Σ and Γ are 2 by 2 periodic matrix functions with elements of the form

$$\Sigma_{ij} = \mathbf{e}_j \cdot \mathbf{e}_i, \quad \Gamma_{ij} = (\mathbf{V} \cdot \nabla)\mathbf{e}_j \cdot \mathbf{e}_i + L\mathbf{e}_j \cdot \mathbf{e}_i. \quad (4.7)$$

Equation (4.6) is simpler than equation (2.7) because its coefficients do not depend explicitly on Christoffel symbols. In particular, the operator of differentiation acts on different components of the vector $\mathbf{c}$ independently. Elements of the matrices Σ and Γ are periodic functions of θ. They can be written explicitly in terms of geometric quantities characterizing the vortex ring in question, such as the normal curvature of stream lines, etc., but the corresponding expressions are too cumbersome to be presented here. Denote by $M(\Psi_0, \hat{\Lambda})$ the monodromy matrix corresponding to equation (4.6), and by $\zeta_1(\Psi_0, \hat{\Lambda})$, $\zeta_2(\Psi_0, \hat{\Lambda})$ its characteristic exponents. It can be shown that for the stability of a vortex ring it is necessary that

$$\sup_{\Psi_0, \hat{\Lambda}} \Re\zeta_i(\Psi_0, \hat{\Lambda}) \leq 0, \quad i = 1, 2. \quad (4.8)$$

In general, the characteristic exponents can be found only via numerical methods. Nevertheless, for vortex rings without swirl certain conclusions can be drawn analytically. For such a ring equation (4.6) assumes the form

$$\frac{d}{d\tau}\begin{pmatrix} |\mathbf{V}_P|^2 c_1 \\ c_2 \end{pmatrix} + \begin{pmatrix} 0 & \dot{h}|\nabla\Psi|^2 \\ 2\frac{\sigma_N}{|\nabla\Psi|} & 0 \end{pmatrix}\begin{pmatrix} |\mathbf{V}_P|^2 c_1 \\ c_2 \end{pmatrix} = 0, \quad (4.9)$$

where $h = h(\Psi)$ is the Bernoulli function, $h = \frac{1}{2}|\mathbf{V}|^2 + P$, and σ_N is the normal curvature of stream lines. Assuming for simplicity that stream lines are convex, so that $\sigma_N < 0$, we can write the stability condition sought as

$$\dot{h} > 0. \tag{4.10}$$

In terms of the circulation C this condition can be written as

$$\dot{C} = \dot{h}T > 0. \tag{4.11}$$

Thus, it is necessary for the stability of a vortex ring without swirl with convex stream lines that the magnitude of the circulation be a function increasing outward. This condition can be considered as a natural generalization of the classical Rayleigh condition for cylindrical flows, and the Bayly condition for two-dimensional flows.

5. Concluding Remarks

In the present paper we demonstrated how the geometrical optics method can be used for studying short wavelength instabilities of three-dimensional flows of an incompressible, inviscid fluid. Our analysis can be extended in several directions. First, instabilities of barotropic and compressible fluids can be analyzed. Second, viscosity effects can be studied. Third, the L_2 velocity norm used as a measure of the perturbation magnitude can be replaced by the L_2 vorticity norm. Although estimates for the remainder in the vorticity norm are not currently available, formal analysis of short wavelength perturbations provides reliable results concerning the stability of basic flows due to the fact that the dimensionless parameter ε characterizing the wavelength can be made arbitrarily small. Moreover, this analysis can be extended to the nonlinear case. It can be shown that stability conditions derived via the weakly nonlinear geometrical optics method of Hunter and Keller (1983) coincide with their linear counterparts due to the remarkable fact that the nonlinear acceleration term automatically vanishes in the weakly nonlinear short wavelength limit.

The relationship between our approach to the stability problem and the conventional spectral one is not straightforward. For circular flows Lebovitz and Lifschitz (1991) recently described the short wavelength part of the spectrum and we refer the interested reader to their paper.

Numerical verification of the present theory is the important problem left by this work. We hope to be able to present results of our numerical investigation in the near future.

Acknowledgements

AL thanks Keith Moffatt, Michael Tabor, and George Zaslavsky for their kind invitation to participate in the NATO Advanced Research Workshop on Topological Fluid Dynamics. This work was supported by the National Science Foundation under grant DMS-9100327, and by the Department of Energy under grant DE-FG02-86ER-53223.

References

BAYLY, B.J., 1986 Three-dimensional instability of elliptical flow. *Phys. Rev. Lett.*, **57**, pp. 2160-2163.

BAYLY, B.J., 1988 Three-dimensional centrifugal-type instabilities in inviscid two-dimensional flows. *Phys. Fluids*, **31**, pp. 56-64.

BAYLY, B.J., ORSZAG, S.A. & HERBERT, T., 1988 Instability mechanisms in shear-flow transition. *Ann. Rev. Fluid Mech.*, **20**, pp. 359-391.

CONNOR, J.W., HASTIE, R.J. & TAYLOR, J.B., 1979 High mode number stability of an axisymmetric toroidal plasma. *Proc. R. Soc. London Ser. A*, **365**, pp. 1-17.

CRAIK, A.D.D. & CRIMINALE, W.O., 1986 Evolution of wavelike disturbances in shear flow: a class of exact solutions of the Navier-Stokes equations. *Proc. R. Soc. London Ser. A*, **406**, pp. 13-26.

ECKHOFF, K.S. & STORESLETTEN, L., 1978 A note on the stability of steady inviscid helical gas flows. *J. Fluid Mech.*, **89**, pp. 401-411.

ECKHOFF, K.S. & STORESLETTEN, L., 1980 On the stability of rotating compressible and inviscid fluids. *J. Fluid Mech.*, **99**, pp. 433-448.

FRIEDLANDER, S. & VISHIK, M.M., 1991a Instability criteria for the flow of an inviscid incompressible fluid. *Phys. Rev. Lett.*, **66**, pp. 2204-2206.

FRIEDLANDER, S. & VISHIK, M.M., 1991b Dynamo theory, vorticity generation, and exponential stretching. *Chaos*, **1**, pp. 198-205.

FRIEDLANDER, S. & VISHIK, M.M., 1992 Instability criteria in fluid dynamics. This volume.

HAMEIRI, E., 1985 Ballooning modes in fluid dynamics. Proceedings of the Sherwood Theory Conference, Madison, Wisconsin, USA.

HAMEIRI, E. & LIFSCHITZ, A., 1990 Ballooning modes in plasmas and in classical fluids. Proceedings of the Sherwood Theory Conference, Williamsburg, Virginia, USA.

HANTER, J.K. & KELLER, J.B., 1983 Weakly nonlinear high frequency waves. *Commun. Pure Appl. Math.*, **36**, pp. 547-569.

KELVIN, LORD, 1887 Stability of fluid motion: rectilinear flows of viscous fluid between two parallel plates. *Phil. Mag.* **24** (5), pp. 188-196.

LAGNADO, R.R., PHAN-THIEN, N. & LEAL, L.G., 1984 The stability of two-dimensional linear flows. *Phys. Fluids*, **27**, pp. 1094-1101.

LEBOVITZ, N. & LIFSCHITZ, A., 1991 Short wavelength instabilities of rotating, compressible fluid masses. Submitted to *Proc. R. Soc. London Ser. A*.

LIFSCHITZ, A., 1984 On the continuous spectrum in some problems of mathematical physics. *Soviet Phys. - Doklady*, **29**, pp. 625-627.

LIFSCHITZ, A., 1991a Essential spectrum and local stability conditions in hydrodynamics. *Phys. Lett. A*, **152**, pp. 199-204.

LIFSCHITZ, A., 1991b Short wavelength instabilities of incompressible three-dimensional flows and generation of vorticity. *Phys. Lett. A*, **157**, pp. 481-487.

LIFSCHITZ, A. & HAMEIRI, E., 1991a Local stability conditions in fluid dynamics. *Phys. Fluids A*, **3**, pp. 2644-2651.

LIFSCHITZ, A. & HAMEIRI, E., 1991b Localized instabilities of vortex rings with swirl. Submitted for publication.

WALEFFE, F., 1990 On the three-dimensional instability of strained vortices. *Phys. Fluids A*, **2**, pp. 76-80.

KINEMATIC FAST DYNAMO ACTION IN A TIME-PERIODIC CHAOTIC FLOW

I. Klapper*
Courant Institute of Mathematical Sciences
New York University
New York, NY 10012

ABSTRACT.

Magnetic dynamo action is the generation and maintenance of magnetic field through the motions of conducting fluid. In the kinematic approximation (where magnetic field feedback on velocity is ignored) we consider velocity fields **u** which result in exponential growth of magnetic field **B**. A fast dynamo is a dynamo which maintains strictly positive exponential growth in the limit of magnetic diffusivity going to zero. Here a three dimensional chaotic flow consisting of perturbed helical cells is considered for kinematic fast dynamo action. A dynamo mechanism common to many chaotic flows is identified, and magnetic field growth rate is measured by numerically solving the magnetic induction equation using Brownian motion techniques. To better understand the dynamo mechanism, the scale of the initial magnetic field relative to shear magnitude is varied and resulting growth rates are compared with a model dynamo.

1. Introduction

By combining the pre-Maxwell equations with Ohm's law we arrive at the magnetic induction equation

$$\frac{\partial \mathbf{B}}{\partial t} = \nabla \times (\mathbf{u} \times \mathbf{B}) + \eta \nabla^2 \mathbf{B}. \tag{1.1}$$

Here **B** is the magnetic field, **u** is the velocity field, and η is the magnetic diffusivity. Non-dimensionalized, (1.1) takes the form

$$\frac{\partial \mathbf{B}}{\partial t} = \nabla \times (\mathbf{u} \times \mathbf{B}) + \frac{1}{R_m} \nabla^2 \mathbf{B} \tag{1.2}$$

where $R_m \equiv UL/\eta$ is the magnetic Reynolds number. The velocity field **u** satisfies the equation

* Current address: Department of Mathematics, University of Arizona, Tucson, Arizona 85721

H. K. Moffatt et al. (eds.), Topological Aspects of the Dynamics of Fluids and Plasmas, 563–571.

$$\frac{\partial \mathbf{u}}{\partial t} + \mathbf{u}.\nabla \mathbf{u} = -\nabla p + \mathbf{J} \times \mathbf{B} + \mathbf{F}, \ \nabla.\mathbf{u} = 0 \tag{1.3}$$

where $\mathbf{J} = \nabla \times \mathbf{B}$. Here we will make the kinematic approximation of dropping the quadratic (in $\mathbf{B}$) term $\mathbf{J} \times \mathbf{B}$. Then $\mathbf{u}$ is independent of $\mathbf{B}$, and given a prescribed velocity field $\mathbf{u}$, (1.2) becomes a linear equation for $\mathbf{B}$. The kinematic dynamo problem is then to find velocity fields $\mathbf{u}$ that allow exponentially growing solutions of (1.2).

Observing that the sun's magnetic field exhibits activity on characteristic time scales many orders of magnitude faster than the diffusive time scale and thus apparently dynamo action independent of diffusive mechanisms is occuring, Vainshtein & Zeldovich (1972) introduced the concept of the kinematic fast dynamo. In order to quantify the notion of a dynamo that operates without a direct role for diffusion, we consider a fast dynamo to be one for which

$$\lim_{R_m \to \infty} p(R_m) > 0 \tag{1.4}$$

where $p(R_m)$ is the dynamo growth rate.

Setting $R_m = \infty$, (1.2) reduces to

$$\frac{\mathbf{D}\mathbf{B}}{\mathbf{D}t} = \mathbf{B}.\nabla \mathbf{u} \tag{1.5}$$

which has the well-known Cauchy solution

$$\mathbf{B}(\mathbf{x}(\mathbf{a}, t), t) = J(\mathbf{x}, \mathbf{a}, t)\mathbf{B}(\mathbf{a}, 0). \tag{1.6}$$

Here $\mathbf{x}(\mathbf{a}, t)$ is the Lagrangian particle coordinate defined by

$$\mathbf{dx} = \mathbf{u}dt, \ \mathbf{x}(\mathbf{a}, 0) = \mathbf{a} \tag{1.7}$$

and $J(\mathbf{x}, \mathbf{a}, t)$ is the Jacobian matrix of the particle trajectory from $\mathbf{a}$ to $\mathbf{x}$. Thus the Cauchy solution can be found simply by advecting magnetic field vectors as material vectors under the action of $\mathbf{u}$. This suggests that flows which generate Jacobian matrices with exponentially growing eigenvalues will result in exponentially growing magnetic field. Hence chaotic flows seem like promising candidates for fast dynamo action. However, exponential growth in bounded regions will in general be accompanied by folding and the production of exponentially small scales. Thus for large but finite R_m, dissipation may become important.

2. Small Diffusivity in the Fast Dynamo Process

In order to have fast dynamo action, we don't want diffusion to have any direct role in magnetic field amplification. One way this could happen is by limiting diffusion to the role of smoothing of fine scales. This leads to the "flux growth" hypothesis (Bayly & Childress (1988), Finn & Ott (1989)): in the limit of $R_m \to \infty$, the magnetic field growth rate is equal to the growth rate of flux of the $R_m = \infty$ (Cauchy) solution in a small region. In other words, as the magnetic flux in a small region is a local average, then for large magnetic Reynolds number the effect of diffusion is limited to smoothing out small scaled structure developed in the Cauchy solution. If true, this is a useful concept as the Cauchy solution (1.6) can be determined by solving a set of ODE's. However, the flux growth hypothesis has been proven to be true in only a very few special cases.

To study the flux growth hypothesis, we need for the purposes of comparison to calculate the exact solution of (1.2). Solution (1.6) can be "generalized" to finite magnetic Reynolds number by introducing Brownian motion to particle trajectories. That is, for $R_m < \infty$ (1.2) has the solution

$$\mathbf{B}(\mathbf{x}(\mathbf{a},t),t) = < J(\mathbf{x},\mathbf{a},t)\mathbf{B}(\mathbf{a},0) > \tag{2.1}$$

where $< \cdot >$ represents averaging over noisy trajectories defined by

$$\mathbf{dx} = \mathbf{u}dt + \sqrt{2R_m^{-1}}\mathbf{dw}, \ \ \mathbf{x}(\mathbf{a},0) = \mathbf{a} \tag{2.2}$$

beginning from a random point $\mathbf{a}$ and ending at the given non-random point $\mathbf{x}$ (see e.g., McKean (1969)). Here $\mathbf{w}(t)$ is Gaussian noise. The Jacobian matrix for each of these noisy trajectories is calculated using the equation $dJ = \nabla \mathbf{u}dt$ without noise along the noisy trajectory. This method of incorporating diffusion into numerical calculations has a long history (e.g., Chorin (1973), Drummond et al. (1984)).

We now introduce a flow $\mathbf{u}$ called the chaotic helical cell flow as an example to be studied for fast dynamo action (Klapper 1992). The velocity field for this flow is defined by

$$\begin{aligned} u(x,y) &= (-1)^{n+m+1}(x-n) \\ v(x,y) &= (-1)^{n+m}(y-m) + \epsilon \sin t \\ w(x,y) &= (-1)^{n+m}\alpha(x-n)(y-m) \end{aligned}$$

for $n-0.5 \le x < n+0.5$ and $m-0.5 \le y < m+0.5$. Particles flow along helical trajectories in the $\pm z$ direction while undergoing a time-periodic perturbation in the y direction caused by the term $\epsilon \sin t$. This perturbation causes the separatrices $y = m$ to break forming a chaotic web in which particles are free to wander in the y direction (figure 1). The chaotic helical cell flow is a piecewise smooth version of the time-periodically perturbed Roberts' cell flow (Roberts (1971)). Although not globally smooth, the chaotic helical cell flow has the computational advantage that it can be effectively piecewise integrated into the form of a map. We note that the velocity field is discontinuous along the lines $x = n + 0.5$, $y = m + 0.5$ resulting in the

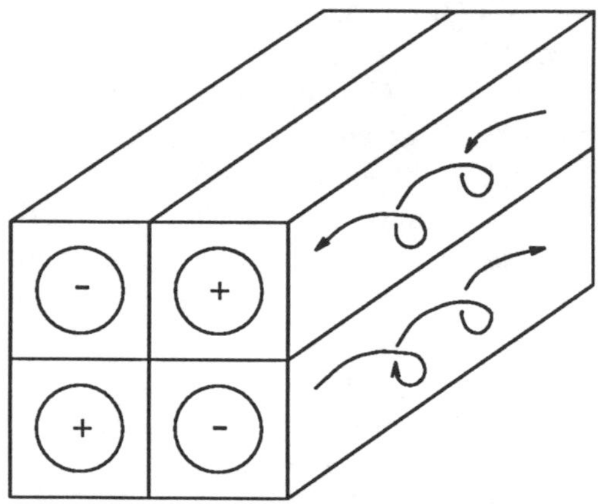

Fig. 1. One periodicity section of the helical cell flow consisting of four helical cells. The perturbation breaks horizontal separatrices resulting in a chaotic web forming around cell edges.

formation of vorticity sheets there, but it can be easily shown that the Jacobian matrices remain well behaved in the sense that they are bounded despite the singularity formed in $\nabla \mathbf{u}$. Thus the solutions (1.6) and (2.1) still make sense. In fact, the Jacobians have a jump discontinuity across the lines $x = n + 0.5$ and $y = m + 0.5$ but it easy to imagine thickening these lines into small regions and defining the flow there so that the Jacobian matrices remain continuous.

In figure 2 we show the growth rate of magnetic field versus time at the point $\mathbf{p} = (-.5, -.005, 0.)$ for several different magnetic Reynolds numbers with initial magnetic field $\mathbf{B}_0 = (exp(3.5\pi iz), 0, 0)$. $O(10^7)$ noisy trajectories were used for each curve. There is fairly clear evidence of dynamo action, and the individual curves seem to be converging on each other indicating fast dynamo action (i.e., the growth rate appears to be independent of R_m). In figure 3 we see the growth rate calculated using the flux growth method compared to the exact calculation at (a) $R_m = 10^5$ and (b) $R_m = 10^6$. Up until the time when the flux growth method loses accuracy, the two methods are consistent with each other, and the consistency seems to improve with the large magnetic Reynolds number. This is a sign that for large R_m it is indeed true that the role of diffusion is becoming limited to passive smoothing.

3. A Fast Dynamo Mechanism

In order to see how fast dynamo action might occur it is instructive to look at the region where the separatrices $y = m$ are broken. If R_m is very

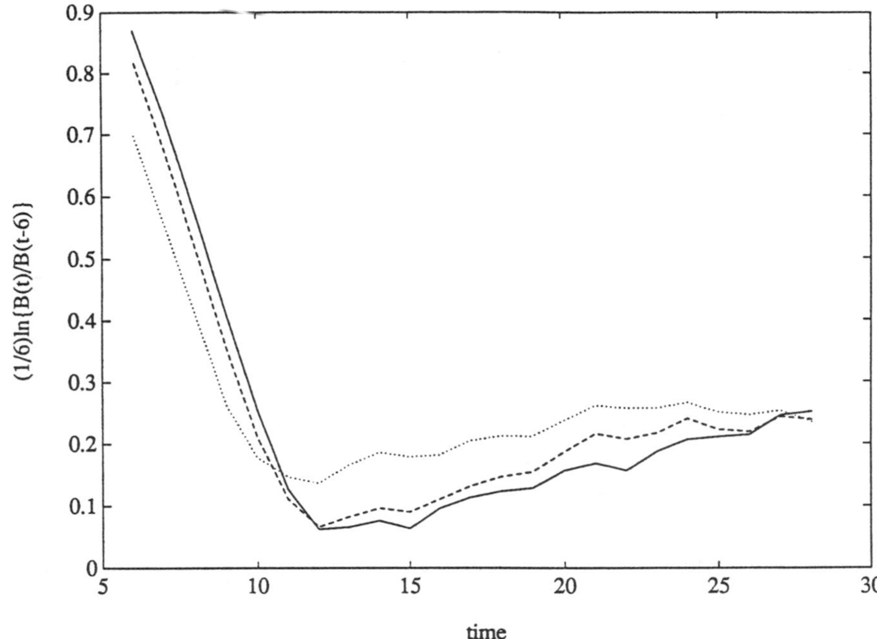

Fig. 2. Growth rates of the magnetic field $\mathbf{B}_0 = (e^{3.5\pi iz}, 0, 0)$ at the point $\mathbf{p} = (-.5, .-005, 0.)$ for $R_m = 2 * 10^4$ (dot), $R_m = 10^5$ (dash), and $R_m = 2 * 10^5$ (solid). The shear parameter is set at $\alpha = 2$.

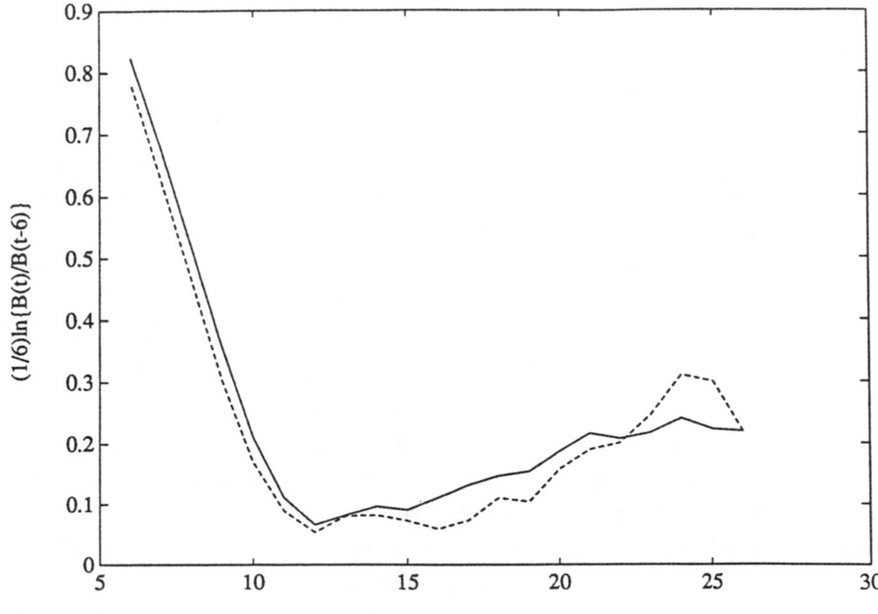

Fig. 3. Comparison of the magnetic field growth rate at the point $\mathbf{p} = (-.5, -.005, 0.)$ calculated at $R_m = 10^5$ using the the noisy trajectory method (exact) (solid) and the flux growth method (approximate) (dash).

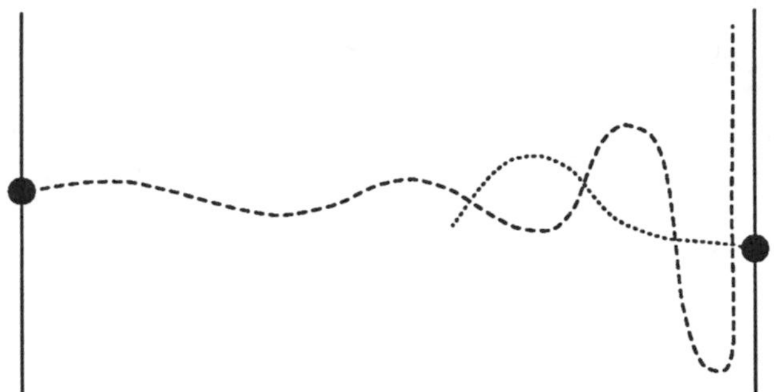

Fig. 4. Schematic projection of the unstable manifolds of two hyperbolic points of the chaotic helical cell flow in the x-y plane. Solid lines are unbroken vertical separatrices, the dashed line is an unstable manifold, and the dotted line is a stable manifold.

large, we can think of the magnetic field as almost being frozen to the fluid. Thus magnetic field lines will be stretched approximately along the unstable manifolds of the chaotic flow. The unstable manifold that comes out of one of the periodic points that forms a corner of a helical cell meets the stable manifold of the next corner transversely resulting in a stretching and folding process in the x-y plane (figure 4). This process combines with shearing in the z direction to result in what is called the stretch-fold-shear (SFS) fast dynamo mechanism (Bayly & Childress (1988)). We now present a model of the SFS mechanism called the random stretch-fold-slide (SFSl) map. (this map was considered in a different context by Finn et al. (1991)). The random SFSl map operates on a unit cube of conducting fluid with magnetic field through the following three steps (figure 5). First, the cube is stretched to double its length in the x direction and contracted to half its height in the y direction. Second, the resulting structure is cut along the plane $x = 1$ and one half is folded back onto the other. Third, the top half of the resulting cube is slid a distance α in the z direction over the bottom half. The number α is chosen randomly with uniform distribution between 0 and 1.

We consider initial magnetic field of the form $\mathbf{B}_0 = (e^{2\pi ikz}, 0, 0)$. When $R_m = \infty$ this form is preserved by the random SFSl map. As it turns out, the expected growth rate of magnetic flux (at $R_m = \infty$) is easy to calculate, and in figure 6 we show the flux growth rate p versus initial field wavenumber k. The maximum growth at any iteration is to double the field; the number to which the wavy curve in figure 6 converges is approximately $(1/2)\ln 2$. To understand this, consider the case when the wavenumber k is large. Then for a given iteration, a typical slide will be very long compared to the scale of the magnetic field variation so that the phase of magnetic field in the

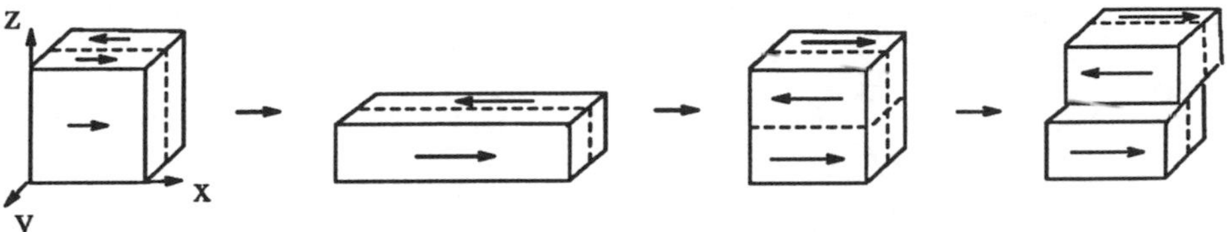

Fig. 5. The random stretch-fold-slide map consisting of three steps. First the conducting cube is stretched double in the x-direction and then, second, is folded back into a cube. Finally, third, the upper half of the cube is slid a random distance α in the z-direction.

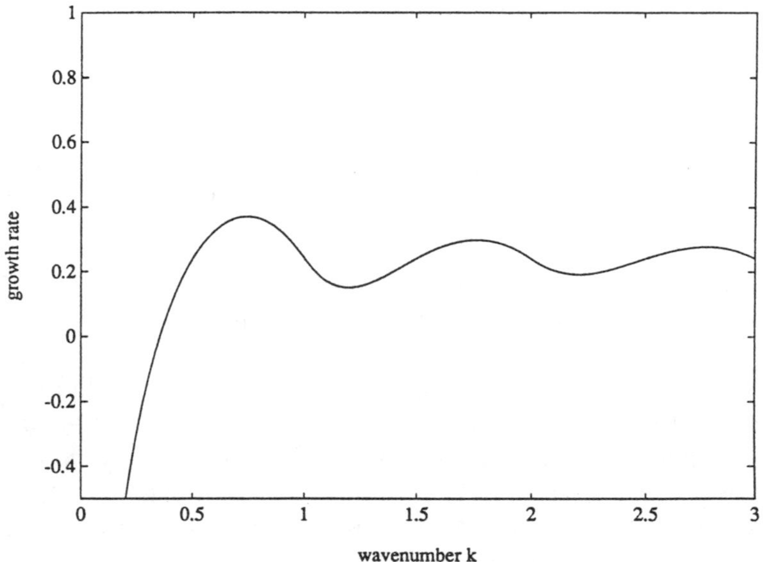

Fig. 6. The flux growth rate p versus initial magnetic field wavenumber k for the SFSl map.

top half of the cube will be essentially random with respect to the phase of field in the bottom half. Then when calculating the new flux, we will be adding two contributions that have doubled in strength (because of the stretch step) but have random phases with respect to each other. This is just like taking a random walk in the complex plane with two vectors of length 1 but random phase. In this case the mean squared length of the sum is $\sqrt{2}$. Thus for large wavenumber compared to the shear scale and hence essentially random phases, we can expect the growth rate to be about $1/2$ the maximum (remember, a logarithm has been taken).

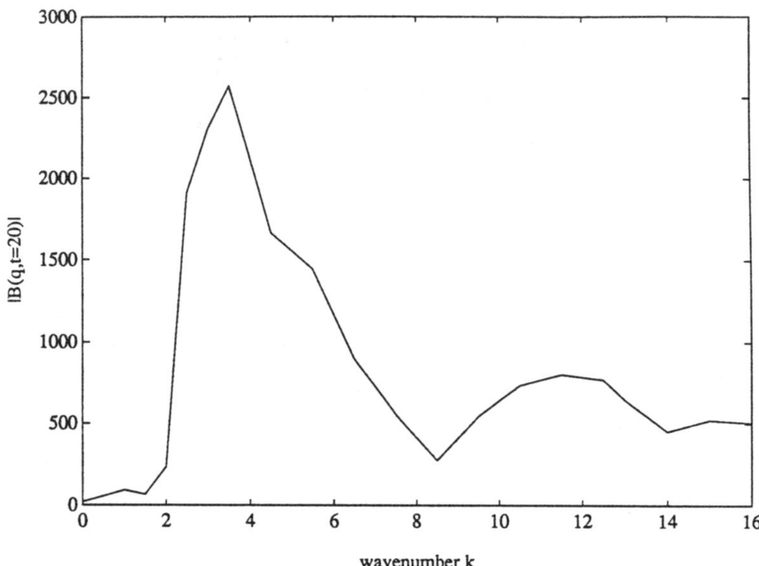

Fig. 7. The magnitude of the magnetic field at the point $q = (-.5, -.005, 0)$ at time 20 versus initial magnetic field wavenumber k for the chaotic helical cell flow.

As a comparison in figure 7 we plot the magnitude of magnetic field in the chaotic helical cell at a certain point after 20 time units ($R_m = 10^5$) against wavenumber of the initial field (no logarithm is taken) again choosing the initial field to be of the form $\mathbf{B}_0 = (e^{2\pi iz}, 0, 0)$. We see a characteristic waviness similar to the random SFSl map and, in fact, the magnitude of field for the larger wave numbers is not far from the square root of the maximum possible (i.e., the growth rate of finite line segments which have no cancellation). This suggests that for large wavenumbers, dynamo action by random comstructive interference is occuring.

Acknowledgements

The author would like to thank S.Childress and A.D. Gilbert for their considerable assistance, the organisers for a fine conference, and the staff of the ITP for their hospitality. This work was supported by a Fulbright grant and by NSF grants ATM-9011772 and DMS-8922676.

References

Bayly, B. & Childress, S., "Construction of fast dynamos using unsteady flows and maps in three dimensions, " *Geophys. Astrophys. Fluid Dyn.* **44**, 207-40 (1988).

Chorin, J.C., "Numerical study of slightly viscous flow," *J. Fluid Mech.* **57**, 785-796 (1973).

DRUMMOND, I.T., DUANE, S., & HORGAN, R.R., "Scalar diffusion in simulated helical turbulence with molecular diffusivity," *J. Fluid Mech.* **138**, 75-91 (1984).

FINN, J. M. & OTT, E., "Chaotic flows and fast magnetic dynamos," *Phys. Fluids* **31**, 2992-3011 (1988).

FINN, J.M., HANSON, J.D., KAN, I., & OTT, E., "Steady fast dynamo flows," *Phys. Fluids B* **3**, 1250-69 (1991).

KLAPPER, I., "A study of fast dynamo action in chaotic helical cells," to appear *J. Fluid Mech.*, (1992).

MCKEAN, H.P., *Stochastic Integrals*, Academic Press, New York (1969).

ROBERTS, P.H. & SOWARD, A.M., "Dynamo theory," *Annu. Rev. Fluid Mech.*, to appear (1992).

VAINSHTEIN, S.I. & ZELDOVICH, YA. B., "Origin of magnetic fields in astrophysics," *Sov. Phys. Usp.* **15**, 159-72 (1972).

AN EXACT TURBULENT CLOSURE FOR THE HYDROMAGNETIC DYNAMO

HUBERT H. SHEN
Center for Turbulence Research
Stanford/NASA-Ames
MS202A-1
Moffett Field, CA 94035, USA

ABSTRACT. Exact statistical solutions are exhibited for the steady-state MHD equations with viscosity, buoyancy, differential rotation, scalar diffusion and source and arbitrarily-large (but finite) electrical conductivity. The incorporation of compressibility effects is also outlined. These solutions are not necessarily unique or complete; however, they constitute an exact analytic turbulent closure and may be viewed as generalizations of the usual ideal static or force-free deterministic equilibria to turbulent, driven, dissipative ensembles. Statistics for velocity, magnetic field, scalar and scalar source are derived rather than postulated; no perturbative, phenomenological or variational arguments are invoked. Detailed balance, realizability and time dependence are also discussed. This provides a more rigorous foundation for previous (*e.g.*, Mean Field Electrodynamics) approaches.

1. Introduction

The problem of turbulence in an electrically-conducting fluid, although central to controlled fusion and many geo- and astrophysical processes, is in its infancy compared to (nonconducting) hydrodynamics. "The understanding and manipulation of magnetohydrodynamic (MHD) turbulence is more in need of theory than is the case for Navier-Stokes fluids..." (Montgomery 1989). As pointed out by Montgomery, there is a surprising, almost embarassing gap in our understanding of what the elementary or equilibrium states of a driven, dissipative MHD fluid are. Much work has been done in the past couple decades (as reviewed by *e.g.*, Moffatt 1978, Soward & Childress 1986, Roberts & Soward 1992) on the closely- related "dynamo" problem, namely, how fluid motion can overcome Ohmic dissipation to induce and maintain magnetic fields in geophysical and astrophysical contexts. Dynamo models, however fruitful and illuminating, are often constrained to rely upon phenomenological, statistical or perturbative assumptions or to limit themselves to the so-called "kinematic" dynamo problem, in which

H. K. Moffatt et al. (eds.), Topological Aspects of the Dynamics of Fluids and Plasmas, 573–597.

one ignores the back- reaction of the generated magnetic field upon the velocity. Work on the self-consistent and fully-developed "hydromagnetic dynamo, though difficult and fraught with uncertainties, needs to be extended. Dynamo theory is far from reaching its final form." (Cowling 1981)

In what follows we consider the hydromagnetic dynamo from the Hopf functional point of view. This has recently been developed in the context of Navier-Stokes turbulence (Shen & Wray 1991); here we incorporate buoyancy, rotation, electrical conductivity, scalar diffusion and source. No perturbative, phenomenological or statistical assumptions or variational arguments are invoked; we seek an exact steady-state turbulent closure of the MHD equations. This leads to closed-form analytic expressions for correlation functions (such as the mean electromotive force (emf)) and moment-generating functionals for the velocity, magnetic field and scalar which generalize the usual ideal, static, nonstatistical solutions. Equilibrium, stationary nonequilibrium solutions and time-dependent solutions are proposed. The incorporation of compressibility and realizability are outlined. This method of testing the validity of so-called alpha models (Moffatt 1978) for the generation of large-scale fields should also be applicable to testing the validity of analogous non-MHD models (Frisch *et al.* 1987 and references therein) for large- scale turbulent structure generation in (non-magnetic) anisotropic, helical, or compressible flows. No claim is made for uniqueness or completeness; however, the fact that (1) exact statistical solutions can be obtained at all, and that (2) in fact more than one class of solutions appears to emerge from this approach, seems sufficiently promising to warrant further investigation.

2. Functional MHD Equations

The equations which we consider here are the MHD equations consisting of the Navier-Stokes equations with Lorentz force and buoyancy

$$\left(\frac{\partial}{\partial t} - \nu\nabla^2\right)\mathbf{u}(\mathbf{x},t) = -\mathbf{u}\cdot\nabla\mathbf{u}(\mathbf{x},t) - \frac{\nabla p}{\rho} + \hat{g}c(\mathbf{x},t) + \frac{\mathbf{J}(\mathbf{x},t)\times\mathbf{B}}{\rho} \tag{2.1}$$

the induction equation in the usual nonrelativistic low-frequency regime

$$\left(\frac{\partial}{\partial t} - \frac{1}{\mu\sigma}\nabla^2\right)\mathbf{B}(\mathbf{x},t) = \nabla\times(\mathbf{u}\times\mathbf{B}) \tag{2.2}$$

and the scalar (temperature) equation with diffusion and source term

$$\frac{\partial c}{\partial t} + \nabla\cdot(\mathbf{u}c) - \nabla\cdot(D\nabla c) = Q \tag{2.3}$$

supplemented by incompressibility, Ohm's law, the definition of the Lorentz force and the "pre-Maxwell" equations:

$$\nabla \cdot \mathbf{u}(\mathbf{x},t) = 0 \qquad \mathbf{J} = \sigma(\mathbf{E} + \mathbf{u} \times \mathbf{B}) \qquad \mathbf{F} = (q\mathbf{E} + \mathbf{J} \times \mathbf{B}) \tag{2.4}$$

$$\nabla \times \mathbf{B} = \mu \mathbf{J} \qquad \nabla \cdot \mathbf{B} = 0 \qquad \nabla \times \mathbf{E} = -\frac{\partial \mathbf{B}}{\partial t} \qquad \nabla \cdot \mathbf{E} = \frac{q}{\varepsilon} \tag{2.5}$$

Coriolis and baroclinic effects will also be briefly discussed. The pure thermal convection problem (in the Boussinesq approximation) is of course recovered by setting the magnetic field and electrical conductivity to zero and going to a potential temperature formulation (Busse 1981).

We define the moment-generating functional

$$\phi \equiv \left\langle e^{\int_{-\infty}^{+\infty} d\mathbf{x}\, [\mathbf{f}(\mathbf{x})\cdot\mathbf{u}(\mathbf{x}) + \mathbf{g}(\mathbf{x})\cdot\omega(\mathbf{x}) + \mathbf{h}(\mathbf{x})\cdot\mathbf{B}(\mathbf{x}) + \mathbf{l}(\mathbf{x})\cdot\mu\mathbf{J}(\mathbf{x}) + z(\mathbf{x})c(\mathbf{x}) + q(\mathbf{x})Q(\mathbf{x})]} \right\rangle \tag{2.6}$$

where the brackets indicate ensemble average over all realizations of $\mathbf{u}(\mathbf{x})$, $\mathbf{B}(\mathbf{x})$, $c(\mathbf{x})$ and $Q(\mathbf{x})$. $\phi[\mathbf{f}(\mathbf{x}), \mathbf{g}(\mathbf{x}), \mathbf{h}(\mathbf{x}), \mathbf{l}(\mathbf{x}), z(\mathbf{x}), q(\mathbf{x})]$ is the functional Fourier transform of the joint probability density. (The customary factor of "i" in the exponent has been absorbed into the dummy functions $\mathbf{f}$, $\mathbf{g}$, etc. for notational simplicity.) ϕ evolves under the above dynamics to a stationary state given by the Hopf equations:

$$\nabla \times \left[\left(\frac{\delta}{\delta \mathbf{f}(\mathbf{x})} + \nu \nabla \right) \times \frac{\delta \phi}{\delta \mathbf{g}(\mathbf{x})} + \frac{\delta}{\delta \mathbf{l}(\mathbf{x})} \times \frac{1}{\mu\rho} \frac{\delta \phi}{\delta \mathbf{h}(\mathbf{x})} + \hat{g} \frac{\delta \phi}{\delta z(\mathbf{x})} \right] = 0 \tag{2.7}$$

$$\nabla \times \left[\frac{\delta}{\delta \mathbf{f}(\mathbf{x})} + \frac{1}{\mu\sigma} \nabla + \frac{\lambda \nabla}{\nabla^2} \right] \times \frac{\delta \phi}{\delta \mathbf{h}(\mathbf{x})} = 0 \tag{2.8}$$

$$\nabla \cdot \left(\frac{\delta}{\delta \mathbf{f}(\mathbf{x})} - D\nabla \right) \frac{\delta \phi}{\delta z(\mathbf{x})} = \frac{\delta \phi}{\delta q(\mathbf{x})} \tag{2.9}$$

where $\lambda \equiv$ steady-state frequency or growth rate of $\mathbf{B}$. Solutions have been obtained for the special case of isotropic flow neglecting buoyancy, the scalar equation and either all nonlinear or all dissipative terms (Stanisic 1985). We impose no such restrictions; we first solve the vorticity and magnetic functional equations without buoyancy and then incorporate buoyancy and the scalar equation.

3. Equilibrium Solutions

In order to find a solution, we add and subtract "ghost" torques

$$\nabla \times (\mathbf{u} \times \alpha_4 \mathbf{B}) + \nabla \times \nabla \times \alpha_5 \mathbf{B}/\mu\sigma \tag{3.1}$$

whose purpose is to "interpolate" between terms in the Hopf equations, so that the resulting adjacent terms in the equations differ from each other by changing only one ordinary or functional derivative. We recognize this type of equation as essentially wavelike or convective in nature, thereby enabling us to write down a general functional ansatz for its solution. Portions of each term are balanced pairwise by portions of other terms to achieve an overall statistical steady-state. (Those who prefer to visualize in topological terms may view this method as analogous to the familiar decomposition of a knotted vortex tube into 2 or more linked tubes by the insertion of equal and opposite flux tube elements between 2 points. The condition of detailed balance corresponds to solenoidality of vorticity, *i.e.*, the condition that the knot form a closed loop.)

Balancing a fraction α_1 of the dissipative term against a fraction $(1-\alpha_2)$ of the transfer term yields

$$\nabla \times \nabla \times \nu\alpha_1 \frac{\delta\phi}{\delta \mathbf{g}(\mathbf{x})} = \nabla \times \left(\frac{\delta}{\delta \mathbf{f}(\mathbf{x})} \times (1-\alpha_2) \frac{\delta\phi}{\delta \mathbf{g}(\mathbf{x})} \right) \tag{3.2}$$

The remainder of the transfer term is in turn balanced by a fraction $(1-\alpha_3)$ of the magnetic term, mediated by one of the ghost terms:

$$\nabla \times \frac{\delta}{\delta \mathbf{l}(\mathbf{x})} \times \frac{(1-\alpha_3)}{\mu\rho} \frac{\delta\phi}{\delta \mathbf{h}(\mathbf{x})} = -\nabla \times \left(\frac{\delta}{\delta \mathbf{f}(\mathbf{x})} \times \alpha_4 \frac{\delta\phi}{\delta \mathbf{h}(\mathbf{x})} \right) \tag{3.3}$$

$$\nabla \times \left(\frac{\delta}{\delta \mathbf{f}(\mathbf{x})} \times \alpha_2 \frac{\delta\phi}{\delta \mathbf{g}(\mathbf{x})} \right) = \nabla \times \left(\frac{\delta}{\delta \mathbf{f}(\mathbf{x})} \times \alpha_4 \frac{\delta\phi}{\delta \mathbf{h}(\mathbf{x})} \right) \tag{3.4}$$

Finally, the remainder of the magnetic term is balanced by the remainder of the dissipative term, mediated by the other ghost term:

$$\nabla \times \nabla \times \nu(1-\alpha_1) \frac{\delta\phi}{\delta \mathbf{g}(\mathbf{x})} = \nabla \times \nabla \times \frac{\alpha_5}{\mu\sigma} \frac{\delta\phi}{\delta \mathbf{h}(\mathbf{x})} \tag{3.5}$$

$$\nabla \times \frac{\delta}{\delta \mathbf{l}(\mathbf{x})} \times \frac{\alpha_3}{\mu\rho} \frac{\delta\phi}{\delta \mathbf{h}(\mathbf{x})} = -\nabla \times \nabla \times \frac{\alpha_5}{\mu\sigma} \frac{\delta\phi}{\delta \mathbf{h}(\mathbf{x})} \tag{3.6}$$

The fractions α_j are to be determined by imposing self-consistency upon the dynamics, as we will see.

This pairwise balancing procedure enables us to organize previous deterministic or linearized solutions (Taylor 1986, Shercliff 1965) and generalize them to resistive, nonperturbative and statistical (nonfactoring correlation functions) solutions. For example, the usual magnetostatic equilibria in which velocity vanishes and the Lorentz force balances pressure would correspond to setting $\alpha_5 = 0$. The kinematic dynamo, for which one neglects magnetic backreaction but has in general nonvanishing transfer and dissipation, would correspond to setting $\alpha_4 = \alpha_5 = \alpha_2 = 0, \alpha_1 = 1$. Alfven waves, for which $\mathbf{u}$ is parallel to $\mathbf{B}$, would correspond to setting $\alpha_3 = 1, \alpha_2 = 0$. Hartmann flows or weak-field dynamos, in which magnetic forces balance viscosity, and Eulerized flows (exhibiting depressed nonlinearity) would be generalized by setting $\alpha_1 = \alpha_4 = 0, \alpha_3 = 1$, while Stokes flows would correspond to the case $\alpha_2 = 1, \alpha_5 = \alpha_3 = 0$. Magnetostrophic flows or strong field dynamos (in the absence of buoyancy), in which Coriolis torques balance Lorentz torques, would follow from decomposing $\mathbf{g} \cdot \omega$ in equation (2.6) into mean and fluctuating vorticity (with a separate dummy function for each) and setting $\alpha_1 = 1 - \alpha_2$. Ideal flows (Gilbert & Sulem 1990) would correspond to setting $\alpha_2 = \alpha_3 = 1$; in order to satisfy $\nabla \times (\mathbf{u} \times \mathbf{B}) = 0$, the ghost in equations (3.3) and (3.4) would have to be replaced by a nonvanishing ghost (no pun intended) such as $\pm \nabla \times \left(\frac{\delta}{\delta \mathbf{l}(\mathbf{x})} \times \alpha_4 \frac{\delta \phi}{\delta \mathbf{g}(\mathbf{x})} \right)$.

Recognizing equations (2.8) and (3.2)-(3.6) as equations for characteristic curves (albeit in function space) leads us to immediately write down forms for their solution. For equation (3.2),

$$\frac{\delta \phi}{\delta \mathbf{g}(\mathbf{x})} = \frac{1}{\nu \alpha_1} \mathbf{G} \left(H(\mathbf{x}) + \int_{-\infty}^{+\infty} d\mathbf{x}\, \mathbf{f}(\mathbf{x}) \cdot \frac{\nu \alpha_1}{1 - \alpha_2} \left[\nabla H + \mathbf{Z}^{(1)} \right] \right) \tag{3.7}$$

where the "phase shift" $\mathbf{Z}^{(1)}$ of the traveling wave solution ("propagating" in $\mathbf{f}$ and $\mathbf{x}$ rather than in $\mathbf{x}$ and t) is governed by

$$\nabla \times (\mathbf{Z}^{(1)} \times \mathbf{G}') = 0 \tag{3.8}$$

Equations (3.3) and (3.6) imply

$$\begin{aligned} \frac{\delta \phi}{\delta \mathbf{h}(\mathbf{x})} = \frac{\mu \sigma}{\alpha_5} \mathbf{W} \Big(M(\mathbf{x}) + \int_{-\infty}^{+\infty} d\mathbf{x}\, \mathbf{l}(\mathbf{x}) \cdot \frac{\alpha_5 \rho}{-\sigma \alpha_3} \left[\nabla M + \mathbf{Z}^{(3)} \right] \\ + \int_{-\infty}^{+\infty} d\mathbf{x}\, \mathbf{f}(\mathbf{x}) \cdot \frac{(1 - \alpha_3) \alpha_5}{\mu \sigma \alpha_3 \alpha_4} \left[\nabla M_1 + \mathbf{Z}^{(3)} \right] \Big) \end{aligned} \tag{3.9}$$

where

$$\nabla \times (\mathbf{Z}^{(3)} \times \mathbf{W}') = 0 \tag{3.10}$$

Equation (2.8) yields

$$\frac{\delta \phi}{\delta \mathbf{h}(\mathbf{x})} = \mathbf{L} \left(N(\mathbf{x}) + \int_{-\infty}^{+\infty} d\mathbf{x}\, \mathbf{f}(\mathbf{x}) \cdot \left[\nabla \left(\frac{N}{\mu \sigma} + \frac{\lambda N}{\nabla^2} \right) + \mathbf{Z}^{(5)} \right] \right) \tag{3.11}$$

where

$$\nabla \times (\mathbf{Z}^{(5)} \times \mathbf{L}') = 0 \tag{3.12}$$

(The arguments of $\mathbf{G}$ and $\mathbf{L}$ in the above equations have additional contributions from $\mathbf{l}(\mathbf{x})$ as in (3.9).) "Prime" denotes derivative with respect to argument. Equation (3.4) implies

$$\alpha_2 \frac{\delta\phi}{\delta\mathbf{g}(\mathbf{x})} = \alpha_4 \frac{\delta\phi}{\delta\mathbf{h}(\mathbf{x})} + \mathbf{Z}^{(4)} \tag{3.13}$$

where

$$\nabla \times \frac{\delta}{\delta\mathbf{f}(\mathbf{x})} \times \mathbf{Z}^{(4)} = 0 \tag{3.14}$$

while equation (3.5) implies

$$(1-\alpha_1)\frac{\delta\phi}{\delta\mathbf{g}(\mathbf{x})} = \frac{\alpha_5}{\nu\mu\sigma}\frac{\delta\phi}{\delta\mathbf{h}(\mathbf{x})} + \mathbf{Z}^{(2)} \tag{3.15}$$

where

$$\nabla \times \nabla \times \mathbf{Z}^{(2)} = 0 \tag{3.16}$$

Solenoidality of velocity and magnetic field will be guaranteed if

$$\nabla \cdot \frac{\mathbf{G}^{(\mathbf{n})}}{\alpha_1} = \nabla \cdot \frac{\mathbf{W}^{(\mathbf{n})}}{\alpha_5} = 0, \quad n = 0, 1, \cdots \tag{3.17}$$

implying

$$\nabla \cdot \frac{\alpha_1}{1-\alpha_2}\left[\nabla H + \mathbf{Z}^{(1)}\right] = \nabla \cdot \frac{\alpha_5}{\alpha_3}\left[\nabla M + \mathbf{Z}^{(3)}\right] = 0 \tag{3.18}$$

These equations have a solution:

$$\frac{1-\alpha_1}{\alpha_1}\mathbf{G} = \mathbf{W} \qquad \mathbf{L} = \frac{\mu\sigma}{\alpha_5}\mathbf{W} \tag{3.19}$$

$$\nabla\left(\frac{M_1}{\mu\sigma} + \frac{\lambda M_1}{\nabla^2}\right) = \frac{\alpha_5(1-\alpha_3)}{\mu\sigma\alpha_4\alpha_3}\nabla M_1 \tag{3.20}$$

$$H = M_1 = N, \qquad \mathbf{Z}^{(2)} = \mathbf{Z}^{(4)} = 0 \tag{3.21}$$

$$\mathbf{Z}^{(1)} = \mathbf{Z}^{(3)} = \mathbf{Z}^{(5)}\frac{-\sigma\alpha_3}{\rho\alpha_5} \tag{3.22}$$

Self-consistency of (3.7) with (3.9) and (3.13) with (3.15) respectively constrains the fractions α_j to satisfy

$$\alpha_1\alpha_2\alpha_3 = (1-\alpha_1)(1-\alpha_2)(1-\alpha_3) \tag{3.23}$$

$$\nu\mu\sigma\frac{\alpha_4}{\alpha_5} = \frac{\alpha_2}{1-\alpha_1} \tag{3.24}$$

If $\mathbf{B}$ is stationary ($\lambda = 0$, DC dynamo) we have the further condition

$$\frac{\alpha_4}{\alpha_5} = \frac{1-\alpha_3}{\alpha_3} \tag{3.25}$$

4. Effect of Scalar Equation

4.1. NO SCALAR DIFFUSION OR EXPLICIT SOURCE

In the absence of scalar diffusion and source, the stationary scalar Hopf equation (2.9) becomes

$$\frac{\delta}{\delta \mathbf{f}(\mathbf{x})} \cdot \nabla \frac{\delta \phi}{\delta z(\mathbf{x})} = 0 \tag{4.1}$$

Hence the buoyancy term in the vorticity Hopf equation becomes

$$\nabla \times \hat{g} \frac{\delta \phi}{\delta z(\mathbf{x})} = -\hat{g} \times \left(\frac{\delta}{\delta \mathbf{f}(\mathbf{x})} \times \mathbf{n} \right) \tag{4.2}$$

for some functional $\mathbf{n}$ where

$$\nabla \times \frac{\delta}{\delta \mathbf{f}(\mathbf{x})} \times \mathbf{n} = 0 \tag{4.3}$$

in order for $\frac{\delta}{\delta \mathbf{f}(\mathbf{x})} \times \mathbf{n}$ to represent the gradient $\nabla \frac{\delta \phi}{\delta z(\mathbf{x})}$. Add and subtract a "ghost" torque

$$\alpha_6 \hat{g} \times \left(\frac{\delta}{\delta \mathbf{f}(\mathbf{x})} \times \frac{\delta \phi}{\delta \mathbf{g}(\mathbf{x})} \right) \tag{4.4}$$

where α_6 depends upon $\mathbf{x}$ and replace $(1 - \alpha_2)$ by $(2 - \alpha_2)$ above. Then we may obtain a solution by imposing the condition that buoyancy is balanced by part of the transfer term (mediated by the ghost term), yielding

$$\nabla \times \left(\frac{\delta}{\delta \mathbf{f}(\mathbf{x})} \times \frac{\delta \phi}{\delta \mathbf{g}(\mathbf{x})} \right) = -\alpha_6 \hat{g} \times \left(\frac{\delta}{\delta \mathbf{f}(\mathbf{x})} \times \frac{\delta \phi}{\delta \mathbf{g}(\mathbf{x})} \right) \tag{4.5}$$

$$\hat{g} \times \left(\frac{\delta}{\delta \mathbf{f}(\mathbf{x})} \times \mathbf{n} \right) = \alpha_6 \hat{g} \times \left(\frac{\delta}{\delta \mathbf{f}(\mathbf{x})} \times \frac{\delta \phi}{\delta \mathbf{g}(\mathbf{x})} \right) \tag{4.6}$$

This will be satisfied if

$$(\nabla H + \mathbf{Z}^{(1)}) \times \mathbf{G}' = \nabla \psi_1 \exp(-\int^{\mathbf{x}} \alpha_6 \hat{g} \cdot d\mathbf{l}) \tag{4.7}$$

$$\hat{g} \times \left(\frac{\delta}{\delta \mathbf{f}(\mathbf{x})} \times \mathbf{n} \right) = \hat{g} \times \nabla \psi_1 \exp(-\int^{\mathbf{x}} \alpha_6 \hat{g} \cdot d\mathbf{l}) \tag{4.8a}$$

Equation (4.8) is readily satisfied by $\mathbf{n}$ of the form

$$\mathbf{n} \equiv -\frac{1}{3} \int_{-\infty}^{+\infty} d\mathbf{x}\, \mathbf{f}(\mathbf{x}) \times \left[\nabla \psi_1 \exp(-\int^{\mathbf{x}} \alpha_6 \hat{g} \cdot d\mathbf{l}) + \hat{g} \psi_2(\mathbf{x}) \right] \tag{4.8b}$$

The 3 scalar functions in the above expression are determined by the 3 equations (4.3).

If Coriolis forces are present, we obtain instead that

$$\nabla \times \left(\frac{\delta}{\delta \mathbf{f}(\mathbf{x})} \times \Omega\phi\right) = -\alpha_6 \hat{g} \times \left(\frac{\delta}{\delta \mathbf{f}(\mathbf{x})} \times \Omega\phi\right) \tag{4.9}$$

$$\hat{g} \times \left(\frac{\delta}{\delta \mathbf{f}(\mathbf{x})} \times \mathbf{n}\right) = \alpha_6 \hat{g} \times \left(\frac{\delta}{\delta \mathbf{f}(\mathbf{x})} \times \Omega\phi\right) \tag{4.10}$$

This will be satisfied if

$$-\Omega \times \mathbf{G}^{(-1)} = \nabla\psi_1 \exp(-\int^{\mathbf{x}} \alpha_6 \hat{g} \cdot d\mathbf{l}) \tag{4.11}$$

$$\hat{g} \times \left(\frac{\delta}{\delta \mathbf{f}(\mathbf{x})} \times \mathbf{n}\right) = \hat{g} \times \nabla\psi_1 \exp(-\int^{\mathbf{x}} \alpha_6 \hat{g} \cdot d\mathbf{l}) \tag{4.12}$$

where $\mathbf{G} = \nabla \times \mathbf{G}^{(-1)}$ defines $\mathbf{G}^{(-1)}$.

4.2. SCALAR DIFFUSION AND SOURCE

In the presence of scalar diffusion, we let part of the diffusion be balanced by convection and part by the scalar source. The former condition may be expressed as

$$D\alpha_6 \nabla \frac{\delta\phi}{\delta z(\mathbf{x})} = \frac{\delta}{\delta z(\mathbf{x})} \frac{\delta\phi}{\delta \mathbf{f}(\mathbf{x})} + \nabla \times \psi \tag{4.13}$$

Letting $\psi = 0$ for simplicity yields a buoyancy term

$$\nabla \times \hat{g} \frac{\delta\phi}{\delta z(\mathbf{x})} = \frac{\hat{g}}{D\alpha_6} \frac{\delta}{\delta z(\mathbf{x})} \times \frac{\nabla}{\nabla^2} \times \frac{\delta\phi}{\delta \mathbf{g}(\mathbf{x})} \tag{4.14}$$

Replacing $(1-\alpha_1)$ by $(2-\alpha_1)$ and letting the buoyancy be balanced by part of viscosity yields

$$\frac{\delta}{\delta z(\mathbf{x})} \sim \frac{D\alpha_6 \nu}{|\hat{g}|^2} \hat{g} \cdot \nabla\nabla^2 \tag{4.15a}$$

The first moment is given by

$$\hat{g} \frac{\delta\phi}{\delta z(\mathbf{x})} = \nu \nabla^2 \frac{\delta\phi}{\delta \mathbf{f}(\mathbf{x})} \tag{4.15b}$$

In conjunction with (4.15) and (3.2), this allows us to determine α_6. Alternatively, if Coriolis forces are present to balance the buoyancy (geostrophic balance), we have

$$\frac{\delta}{\delta z(\mathbf{x})} = \frac{\hat{g}}{|\hat{g}|^2} \cdot \Omega \times \frac{\delta}{\delta \mathbf{f}(\mathbf{x})} \tag{4.16a}$$

This gives us a prescription for the scalar moments. Compare with the prescription implied by (4.2) and (4.6):

$$\nabla \frac{\delta \phi}{\delta z(\mathbf{x})} = \frac{\delta}{\delta \mathbf{f}(\mathbf{x})} \times \mathbf{n} \tag{4.16b}$$

$$\nabla \frac{\delta}{\delta z(\mathbf{x})} = \alpha_6 \frac{\delta}{\delta \mathbf{f}(\mathbf{x})} \times \frac{\delta}{\delta \mathbf{g}(\mathbf{x})} \tag{4.16c}$$

Similarly taking into account the source term yields

$$\frac{\delta}{\delta q(\mathbf{x})} \sim \nabla \cdot \left(D(1-\alpha_6) \nabla \frac{\delta}{\delta z(\mathbf{x})} \right) \tag{4.17}$$

which provides us with a prescription for the source moments. If there is no explicit source, we may let $\alpha_6 = 1$. The scenario in which buoyancy is balanced by Lorentz forces to yield a guiding-center drift current which is perpendicular to gravity and the magnetic field is already implicit in equations (3.3, 3.4, 4.5, 4.6).

4.3. COMPRESSIBILITY

If there is a baroclinic term $\nabla P \times \nabla \rho$ in the vorticity equation, the statistical description must be augmented to include the pressure and density explicitly. Let us add the term $[n_1(\mathbf{x})P(\mathbf{x}) + n_2(\mathbf{x})\rho(\mathbf{x})]$ to the integrand in the exponent in equation (2.6). Suppose that the equation of state prescribes P as a position-dependent functional $P_1[\mathbf{u}, c, \rho, \mathbf{x}]$ of velocity, scalar and density. Then the argument of $\mathbf{G}$ in equation (3.7) has the additional integrals

$$\int_{-\infty}^{+\infty} d\mathbf{x} \Big\{ z(\mathbf{x})q(\mathbf{x}) + n_2(\mathbf{x})P_2(\mathbf{x}) + n_1(\mathbf{x})P_1 \left[\frac{\nu\alpha_1}{1-\alpha_2} [\nabla H + \mathbf{Z}^{(1)}], \right.$$
$$\left. q, P_2, \mathbf{x} \right] \Big\} \tag{4.18}$$

P_2 is related to the first (vector) argument of P_1 in the same way that density is related to velocity in the steady-state continuity equation $\nabla \cdot (\rho \mathbf{u}) = 0$ (which holds here in a statistical sense):

$$\nabla \cdot (P_2[\nabla H + \mathbf{Z}^{(1)}]) = 0 \tag{4.19}$$

(see equations (6.10, 6.17-6.19) for one explicit formal solution.) Balancing the baroclinic term solely against buoyancy $\hat{g} \times \nabla\rho$ would yield

$$\nabla P_1 = -\hat{g} + n_3(\mathbf{x})\nabla P_2 \tag{4.20}$$

This gives us 3 additional equations for 3 additional unknowns n_3, q (appearing in equation (4.20) through P_1) and H (appearing in P_1 and implicitly in P_2, through equation (4.19)). The incompressibility condition (3.18) used previously to determine H and the Boussinesq equation of state $\rho \sim T$ used previously to determine q through equations (4.15) or (4.16) no longer apply, of course. More generally, buoyancy is balanced against a combination of baroclinic, transfer, Coriolis and viscous forces, leading to more unknowns and equations via the procedure outlined above. The adiabatic and isothermal subcases, in which the steady-state equations of state take the form $\mathbf{u}\cdot\nabla(P/\rho^\gamma) = 0$ or $\mathbf{u}\cdot\nabla(P/\rho) = 0$ respectively, may be treated by a procedure analogous to §4.1.

4.4. ARBITRARY EXPLICIT FORCE

If the dynamo is driven by an arbitrary explicit force (Braginsky 1964) instead of the Boussinesq-type buoyancy forces described above, other approaches may be useful. For example, if the force is solenoidal, the external torque may be written as $\nabla\times\nabla\times\psi$ for some vector function ψ. If this force is balanced by viscosity, we obtain that the ψ moments are proportional to the corresponding moments of $\nu\omega$ (within a function whose curl curl vanishes) where ω is the vorticity. If the force is nonsolenoidal, we may let the dummy field which is conjugate to the force (in the definition of the generating functional) be a pseudovector "angle" $\boldsymbol{\theta}(x)$. Then using the angular momentum identity

$$\frac{\delta}{\delta\boldsymbol{\theta}(\mathbf{x})} = \frac{\delta}{\delta\boldsymbol{\theta}(\mathbf{x})} \times \frac{\delta}{\delta\boldsymbol{\theta}(\mathbf{x})} \tag{4.21}$$

and balancing the force against part of transfer by adding and subtracting a ghost torque

$$\nabla \times \frac{\delta}{\delta\boldsymbol{\theta}(\mathbf{x})} \times \frac{\delta}{\delta\mathbf{g}(\mathbf{x})} \tag{4.22}$$

yields

$$\frac{\nu\alpha_1}{1-\alpha_2}\nabla M_1 + \mathbf{Z}^{(1)} = \text{Beltrami} \tag{4.23}$$

This particular choice of dummy field is awkward to implement in practice because of the noncommutativity of the functional derivative $\frac{\delta}{\delta\boldsymbol{\theta}(\mathbf{x})}$. However, it may be viewed as justification for the use of Beltrami flows for the velocity in kinematic dynamo models.

5. Closure

From the above solution, one can write down exact expressions for correlation functions. For example, the mean emf is given by

$$\begin{aligned}\langle \mathbf{u}(\mathbf{x}) \times \mathbf{B}(\mathbf{x}) \rangle &= -\frac{1-\alpha_3}{\alpha_3\alpha_4}\left[\nabla M_1 + \mathbf{Z}^{(1)}\right] \times \mathbf{W}' \\ &= [\mathbf{u_0} - \frac{(1-\alpha_3)\alpha_5}{\mu\sigma\alpha_4\alpha_3}\nabla] \times \langle \mathbf{B}(\mathbf{x}) \rangle \end{aligned} \tag{5.1}$$

since $\mathbf{W}'$ is parallel to $\mathbf{W}$ by solenoidality. (Equation (5.1) may be taken as the definition of $\mathbf{u_0}$.) This may be viewed as an anisotropic inhomogeneous α effect (*e.g.*, Krause & Rädler 1980). One may compare this exact result with the conventional MFE prediction:

$$\begin{aligned}&-\frac{\tau}{3}\left[\langle \mathbf{u}\cdot\omega\rangle\langle \mathbf{B}\rangle + \langle u^2\rangle\langle \nabla \times \mathbf{B}\rangle\right] = \\ &\qquad -\frac{\tau\mu\sigma}{3\alpha_5(1-\alpha_2)}\left[\nabla H + \mathbf{Z}^{(1)}\right]\cdot\left[\mathbf{G}'\mathbf{W} + \mathbf{G}^{(-1)'}\nabla \times \mathbf{W}\right]\end{aligned} \tag{5.2}$$

where $\tau \equiv$ phenomenological correlation time of turbulent velocity. If we superpose solutions with different (nonparallel) $\mathbf{W}$, then $\mathbf{u_0} \times \langle \mathbf{B}(\mathbf{x})\rangle$ will be replaced by a term which is in general not perpendicular to $\langle \mathbf{B}(\mathbf{x})\rangle$, permitting generation of toroidal current (and hence poloidal field) from toroidal field, as desired for dynamo action.

The size of the magnetic fluctuations

$$\frac{\langle B^2\rangle}{\langle B\rangle^2} = \frac{\alpha_2\alpha_5^2(1-\alpha_3)}{\alpha_3(\alpha_4\mu\sigma)^2}\nabla \times \mathbf{Z}^{(1)} \cdot \frac{\mathbf{W}'}{W^2} \tag{5.3}$$

Whether this ratio is $\gg 1$ or $\ll 1$ determines the regimes of validity of Ohmic diffusion and first-order smoothing, respectively. Other correlation functions (*e.g.*, $\langle \mathbf{B}\cdot\nabla\omega\rangle$ and $\langle \mathbf{B}(\mathbf{x})\mathbf{B}(\mathbf{x}')\rangle$) may also be computed to shed light upon questions of quenching (*e.g.*, Malkus & Proctor 1975, Kraichnan 1979), inverse cascade (*e.g.*, Frisch *et al.* 1975) or the formation of current sheets (*e.g.*, Parker 1989).

6. Steady State Without Detailed Balance

Consider again the vorticity Hopf equation without buoyancy. We rewrite it schematically as:

$$\left(\frac{\partial}{\partial\xi} + \frac{\partial}{\partial\eta} + \frac{\partial}{\partial\zeta}\right)\phi = 0 \tag{6.1}$$

where the 3 terms correspond to the viscous, transfer and Lorentz force terms respectively (equations (7.6) - (7.8)). This has the general solution

$$\phi(\xi, \eta, \zeta) = \int db_1 db_2 \, \rho(b_1, b_2) \, \phi(b_1\xi + b_2\eta - (b_1 + b_2)\zeta) \tag{6.2}$$

However, one may gain much more insight by decomposing equation (6.1) into 3 simultaneous equations:

$$\left(\alpha_1 \frac{\partial}{\partial \xi} + (1 - \alpha_2)\frac{\partial}{\partial \eta}\right) \phi = F_1[\phi] \tag{6.3}$$

$$\left(\alpha_2 \frac{\partial}{\partial \eta} + (1 - \alpha_3)\frac{\partial}{\partial \zeta}\right) \phi = F_2[\phi] \tag{6.4}$$

$$\left(\alpha_3 \frac{\partial}{\partial \zeta} + (1 - \alpha_1)\frac{\partial}{\partial \xi}\right) \phi = F_3[\phi] \tag{6.5}$$

where the $F_j[\phi]$ are in general operators on ϕ satisfying the stationarity condition

$$\sum_{j=1}^{3} F_j[\phi] = 0 \tag{6.6}$$

For

$$F_j[\phi] = 0 \tag{6.7}$$

this reduces to the earlier, detailed balance case. If in addition we restrict ourselves to

$$\frac{\partial \phi}{\partial \xi} = \frac{\partial \phi}{\partial \eta} = \frac{\partial \phi}{\partial \zeta} = 0 \tag{6.8}$$

we recover the ideal magnetostatic deterministic solution, which is most often used as a starting point for stability studies but whose fundamental inadequacy has been discussed at length (Montgomery & Phillips 1989). Hence we see that the sought-after generalization of this static solution to resistive, turbulent, nonequilibrium solutions can be achieved within the framework of the decomposition (6.3-6.5) with nonzero α_j and F_j.

The physical interpretation of this decomposition may be further clarified by going to a probability (rather than moment-generating functional) description. We may write the stationary Hopf equation in "Vlasov-equation" format (dropping the magnetic and scalar variables for notational simplicity) as

$$\frac{\partial \phi}{\partial t} = \int \prod_{\mathbf{x}} d\mathbf{u}(\mathbf{x})\, d\mathbf{x}\, i\mathbf{f} \cdot \dot{\mathbf{u}}\, P[\mathbf{u}] e^{i\int \mathbf{f}\cdot\mathbf{u}}$$
$$= \int \prod_{\mathbf{x}} d\mathbf{u}(\mathbf{x})\, d\mathbf{x}\, \dot{\mathbf{u}}\, P[\mathbf{u}] \cdot \frac{\delta}{\delta \mathbf{u}} e^{i\int \mathbf{f}\cdot\mathbf{u}}$$
$$= -\int \prod_{\mathbf{x}} d\mathbf{u}(\mathbf{x})\, d\mathbf{x}\, e^{i\int \mathbf{f}\cdot\mathbf{u}} \frac{\delta}{\delta \mathbf{u}} \cdot (\dot{\mathbf{u}}\, P[\mathbf{u}]) = 0 \tag{6.9}$$

Inverse functional Fourier transforming with respect to $\mathbf{f}(\mathbf{x})$ yields

$$\frac{\partial P}{\partial t} = -\int d\mathbf{x}\, \frac{\delta}{\delta \mathbf{u}} \cdot \left(\dot{\mathbf{u}} P[\mathbf{u}]\right) = 0 \tag{6.10}$$

Separating the dissipative (D), transfer (T) and magnetic (M) contributions yields

$$0 = \sum_{j=D,T,M} \left(\frac{dP}{dt}\right)_j \quad , \quad \dot{\mathbf{u}} = \sum_{j=D,T,M} (\dot{\mathbf{u}})_j \tag{6.11}$$

The $(\dot{\mathbf{u}})_j$ are just the functional Fourier transforms of $\frac{\partial}{\partial \xi_j}$. We now rewrite this as

$$0 = \frac{\partial}{\partial t} \sum_{j=D,T,M} P_j \tag{6.12}$$

where the "partial probabilities"

$$P_j(t) \equiv \int_{-\infty}^{t} dt' \left(\frac{dP}{dt'}\right)_j + P_j(-\infty)$$
$$= -\int_{-\infty}^{t} dt' \int d\mathbf{x}\, \frac{\delta}{\delta \mathbf{u}} \cdot (\dot{\mathbf{u}})_j\, P[\mathbf{u}] + P_j(-\infty) \tag{6.13}$$

The constants of integration $P_j(-\infty)$ are arbitrary; if we choose them to satisfy

$$\sum_{j=D,T,M} P_j(-\infty) = 1 \tag{6.14}$$

then

$$\sum_{j=D,T,M} P_j(t) = 1 \tag{6.15}$$

for all t. Similarly, $P_j(t)$ will remain bounded in the interval [0,1] for all t if initially so bounded, just as

$$P[\mathbf{u}(\mathbf{x},t)]\prod_{\mathbf{x}} d\mathbf{u}(\mathbf{x},t) = P[\mathbf{u}(\mathbf{x},0)]\prod_{\mathbf{x}} d\mathbf{u}(\mathbf{x},0) \tag{6.16}$$

guarantees the boundedness of P[**u**].

Hence we see that the *time rate of change of the probability* contributed by dissipation, transfer and magnetic processes (with total probability change = 0 in steady state) is formally equivalent to the time rate of change of the *probability contributed by dissipative, transfer and magnetic "states"* (with total probability change = 0 in a closed system.) In other words, we have shifted our perspective slightly, from solving for stationary probabilities with independent variable **u** and parameter j, to solving for nonstationary probabilities with independent variable j and parameter **u**. The F_j's represent the net probability flux or transition rate between pairs of states. Note that when $\frac{\partial \phi}{\partial \xi}$ vanishes, we obtain that $\nu \nabla^2 \omega$ weakly vanishes (its ensemble average with any n-point function of **u**, **B**, c and their derivatives vanishes) implying that P_D vanishes. In other words, P_D may be interpreted as a measure of the intensity of vortex reconnection, P_T as a measure of vortex stretching and P_M as a measure of magnetic stretching.

One could try to solve the Vlasov-type equation (6.10) directly. For example, for the one-point velocity pdf, one could transform to the local principal axes frame where

$$\frac{\partial \dot{u}_i}{\partial u_j} \sim \delta_{ij} \tag{6.17}$$

Then one may verify that

$$P(\mathbf{u}(\mathbf{x})) = \prod_{i=1}^{3} \frac{1}{\dot{u}_i} + \int^{\mathbf{u}} d\mathbf{v} \cdot \mathbf{n} \times \dot{\mathbf{u}} \tag{6.18}$$

where

$$n_j(\mathbf{u}) \equiv \prod_{i \neq j} \frac{1}{\dot{u}_i} + \text{jth component of null eigenvector of} \tag{6.19}$$

$$\begin{pmatrix} \dot{u}_3 k_1 & \dot{u}_3 k_2 & -\dot{u}_1 k_1 - \dot{u}_2 k_2 \\ \dot{u}_2 k_1 & -\dot{u}_1 k_1 - \dot{u}_3 k_3 & \dot{u}_2 k_3 \\ -\dot{u}_2 k_2 - \dot{u}_3 k_3 & \dot{u}_1 k_2 & \dot{u}_1 k_3 \end{pmatrix} e^{\mathbf{k}\cdot\mathbf{u}} \tag{6.20}$$

However, this becomes intractable (and not particularly illuminating) if one is interested in more than just the velocity at one point. Moreover, this solution satisfies equation (6.10) pointwise; one may have to consider, not the local condition (integrand of equation (6.10)) but the global one (equation (6.10)) in order to obtain stationary solutions of interest.

7. Driven Steady State for Particular Generating Functional

Consider the case in which the $F_j[\phi]$ are arbitrary (subject to equation (6.6)) but $\phi = \phi(B_1(\xi)B_2(\eta)B_3(\zeta))$. Then equations (6.3)-(6.5) become

$$\begin{pmatrix} \alpha_1 & 1-\alpha_2 & 0 \\ 0 & \alpha_2 & 1-\alpha_3 \\ 1-\alpha_1 & 0 & \alpha_3 \end{pmatrix} \begin{pmatrix} k_1 \\ k_2 \\ k_3 \end{pmatrix} = \begin{pmatrix} F_1 \\ F_2 \\ F_3 \end{pmatrix} \frac{1}{\phi' B} \tag{7.1}$$

$$k_j \equiv \frac{B_j'(\xi_j)}{B_j(\xi_j)} = \text{constant by eqn.(6.1)}$$

$$B \equiv B_1(\xi)B_2(\eta)B_3(\zeta)$$

If the determinant of the α matrix vanishes, detailed balance can occur. More generally, the stationary ensemble exhibits a net probability flux between pairs of states or a nonzero cyclic flow of probability through the 3 states. Equivalently, one may write the matrix equation as

$$\begin{pmatrix} k_1 & -k_2 & 0 \\ 0 & k_2 & -k_3 \\ -k_1 & 0 & k_3 \end{pmatrix} \begin{pmatrix} \alpha_1 \\ \alpha_2 \\ \alpha_3 \end{pmatrix} = \begin{pmatrix} F_1/\phi' B - k_2 \\ F_2/\phi' B - k_3 \\ F_3/\phi' B - k_1 \end{pmatrix} \tag{7.2}$$

Because the determinant of the k matrix vanishes, one may solve for the net transition rates

$$\begin{pmatrix} F_1 \\ F_2 \\ F_3 \end{pmatrix} = \begin{pmatrix} k_2 \\ k_3 \\ k_1 \end{pmatrix} \phi' B \tag{7.3}$$

$$\text{when} \quad \begin{pmatrix} \alpha_1 \\ \alpha_2 \\ \alpha_3 \end{pmatrix} = \text{null eigenvector of k matrix} \tag{7.4}$$

In general, one cannot set the net transition rates equal to zero because $\mathbf{k} \neq$ null eigenvector of the α matrix (*i.e.*, regardless of the choice of k_j, one cannot solve for α_j given vanishing F_j because the k matrix is noninvertible.) Hence, these solutions differ fundamentally (Graham 1973) from the previous ones which exhibited detailed balance because the entropy production

$$\dot{S} = \sum_{j=D,T,M} \dot{P}_j \log P_j \neq 0 \quad \text{for } F_j \neq 0 \tag{7.5}$$

Translating back into hydromagnetic language, we have

$$\frac{\partial}{\partial \xi} \rightarrow \nabla \times \nabla \times \nu \frac{\delta}{\delta \mathbf{g}(\mathbf{x})} \tag{7.6}$$

$$\frac{\partial}{\partial \eta} \to \nabla \times \frac{\delta}{\delta \mathbf{f}(\mathbf{x})} \times \frac{\delta}{\delta \mathbf{g}(\mathbf{x})} \tag{7.7}$$

$$\frac{\partial}{\partial \zeta} \to \nabla \times \frac{\delta}{\delta \mathbf{l}(\mathbf{x})} \times \frac{1}{\mu\rho} \frac{\delta}{\delta \mathbf{h}(\mathbf{x})} \tag{7.8}$$

By equation (7.1), $B_j = \exp(k_j \xi_j)$ where

$$k_1 \xi \to \int_{-\infty}^{+\infty} d\mathbf{x}\, \mathbf{g}(\mathbf{x}) \cdot \nabla \times A_D \tag{7.9}$$

$$k_2 \eta \to \int_{-\infty}^{+\infty} d\mathbf{x}\, [\mathbf{g}(\mathbf{x}) \cdot \nabla \times (A_T - A_D) + \mathbf{f}(\mathbf{x}) \cdot A_T] \tag{7.10}$$

$$k_3 \zeta \to \int_{-\infty}^{+\infty} d\mathbf{x}\, [\mathbf{h}(\mathbf{x}) \cdot A_M + \mathbf{l}(\mathbf{x}) \cdot \nabla \times A_M] \tag{7.11}$$

These expressions are the minimum required to exhibit nonvanishing $\partial B_j(\xi_j)/\partial \xi_j$; they incorporate Ampere's Law and the definition of vorticity.

By imposing that the derivatives satisfy

$$\frac{\partial (k_m \xi_m)}{\partial \xi_n} = k_m \delta_{mn} \tag{7.12}$$

we obtain

$$\mathbf{k}_1 \equiv \nabla \times \nabla \times \nu \nabla \times A_D \tag{7.13}$$

$$\mathbf{k}_2 \equiv \nabla \times (A_T \times \nabla \times (A_T - A_D)) \tag{7.14}$$

$$\mathbf{k}_3 \equiv \nabla \times (\nabla \times \frac{A_M}{\mu\rho}) \times A_M \tag{7.15}$$

$$\nabla \times \nabla \times \nabla \times (A_T - A_D) = 0 \tag{7.16}$$

$$\nabla \times (A_T \times \nabla \times A_D) = 0 \tag{7.17}$$

"Uncurling" equations (7.13)-(7.17) yields "constants" of integration:

$$-\nabla^2 A_D = \nabla \phi_D + \mathbf{k}_1 \times \mathbf{x}/3\nu \tag{7.18}$$

$$-\nabla^2 A_T = \nabla \phi_T + \mathbf{k}_1 \times \mathbf{x}/3\nu \tag{7.19}$$

$$A_T = n_1(\mathbf{x}) \nabla \times A_D + P_1, \quad P_1 \times \nabla \times A_D = \nabla \psi_1 \tag{7.20}$$

$$A_T = n_2(\mathbf{x}) \nabla \times A_T + P_2, \quad P_2 \times \nabla \times A_T = \nabla \psi_2 + \mathbf{k}_2 \times \mathbf{x}/3 \tag{7.21}$$

$$A_M = n_3(\mathbf{x})\nabla \times A_M + P_3, \quad P_3 \times \nabla \times A_M = \nabla\psi_3 + \mu\rho\mathbf{k}_3 \times \mathbf{x}/3 \tag{7.22}$$

while the magnetic equation (2.8) with $\lambda = 0$ becomes

$$A_T \times A_M = \nabla \times A_M + \nabla\psi_4 \tag{7.23a}$$

(Compare this with the MFE prediction:)

$$A_T \times A_M = \alpha A_M + \eta\nabla \times A_M \tag{7.23b}$$

This gives us enough unknown fields to satisfy the equations (7.18)-(7.23).

Solenoidality will be satisfied if

$$\nabla \cdot A_T = \nabla \cdot A_M = 0 \tag{7.24}$$

which suggests a vector potential or stream function representation for A_M and A_T. For the case in which ϕ is linear in B, the stationarity (vorticity) condition is simply

$$k_1 + k_2 + k_3 = 0 \tag{7.25}$$

For ϕ analytic in B or containing negative powers of B, cancellation of each power of B requires that additional conditions be imposed, which we will not discuss further here.

In the context of our "partial probability" picture, a solution with nonzero transition rates corresponds to a probability packet cycling consecutively between ω stretching, B stretching (and reconnection, by the induction equation) and ω reconnection. If instead one writes the Hopf equations in terms of $\mathbf{u} \cdot \nabla\mathbf{u}$ and $\mathbf{B} \cdot \nabla\mathbf{B}$, depletion of state "M" implies zero magnetic tension, which may have implications for coronal mass ejection (Low 1990).

8. Driven Steady State for Particular Detailed Imbalance

Consider the case in which there is no detailed balance and the (hence nonvanishing) $F_j = F_j(\phi)$. Then equations (6.3)-(6.5) become

$$\left(\alpha_1\frac{\partial}{\partial\xi} + (1-\alpha_2)\frac{\partial}{\partial\eta}\right)\phi = F_1(\phi) \tag{8.1}$$

$$\left(\alpha_2\frac{\partial}{\partial\eta} + (1-\alpha_3)\frac{\partial}{\partial\zeta}\right)\phi = F_2(\phi) \tag{8.2}$$

$$\left(\alpha_3\frac{\partial}{\partial\zeta} + (1-\alpha_1)\frac{\partial}{\partial\xi}\right)\phi = F_3(\phi) \tag{8.3}$$

again with the stationarity condition

$$\sum_{j=1}^{3} F_j(\phi) = 0 \tag{8.4}$$

Rewrite this as

$$\alpha_1 \frac{\partial S_1}{\partial \xi} + (1-\alpha_2)\frac{\partial S_1}{\partial \eta} = 1 \tag{8.5}$$

$$\alpha_2 \frac{\partial S_2}{\partial \eta} + (1-\alpha_3)\frac{\partial S_2}{\partial \zeta} = 1 \tag{8.6}$$

$$\alpha_3 \frac{\partial S_3}{\partial \zeta} + (1-\alpha_1)\frac{\partial S_3}{\partial \xi} = 1 \tag{8.7}$$

where

$$S_j(\phi) \equiv \int \frac{d\phi}{F_j(\phi)} \tag{8.8}$$

Solving, we obtain

$$S_1 = S_1\left(\int^{\xi} d\xi' \,(1-\alpha_2) - \int_0^{\eta} d\eta' \,\alpha_1\right) + \int^{\xi} \frac{d\xi'}{\alpha_1} \tag{8.9}$$

$$S_2 = S_2\left(\int^{\eta} d\eta' \,(1-\alpha_3) - \int_0^{\zeta} d\zeta' \,\alpha_2\right) + \int^{\eta} \frac{d\eta'}{\alpha_2} \tag{8.10}$$

$$S_3 = S_3\left(\int^{\zeta} d\zeta' \,(1-\alpha_1) - \int_0^{\xi} d\xi' \,\alpha_3\right) + \int^{\zeta} \frac{d\zeta'}{\alpha_3} \tag{8.11}$$

$$\text{where} \quad \frac{\partial \alpha_1}{\partial \xi} = \frac{\partial \alpha_2}{\partial \eta} = \frac{\partial \alpha_3}{\partial \zeta}, \qquad \alpha_j(0) = 1 \tag{8.12}$$

and where α_1, α_2 and α_3 are independent of η, ζ and ξ respectively. We translate back into hydromagnetic language as in the previous section except that we choose

$$k_1\xi \to \int_{-\infty}^{+\infty} d\mathbf{x}\, \mathbf{g}(\mathbf{x}) \cdot \nabla \times A_D \tag{8.13}$$

$$k_2\eta \to \frac{1}{2}\left(\int_{-\infty}^{+\infty} d\mathbf{x}\ [\mathbf{g}(\mathbf{x}) \cdot \nabla \times (A_T - A_D) + \mathbf{f}(\mathbf{x}) \cdot A_T]\right)^2 \tag{8.14}$$

$$k_3\zeta \to \frac{1}{2}\left(\int_{-\infty}^{+\infty} d\mathbf{x}\, [\mathbf{h}(\mathbf{x}) \cdot A_M + \mathbf{l}(\mathbf{x}) \cdot \nabla \times A_M]\right)^2 \tag{8.15}$$

The quadratic powers are the minimum required to give nonvanishing $\partial(k_2\eta)/\partial\eta$ and $\partial(k_3\zeta)/\partial\zeta$. Stationarity implies

$$\sum_{j=1}^{3}\frac{\partial\phi}{\partial S_j}=0 \tag{8.16}$$

which is satisfied if

$$\phi(S_1,S_2,S_3)=\phi(b_1S_1+b_2S_2-(b_1+b_2)S_3) \tag{8.17}$$

or more generally

$$\phi(S_1,S_2,S_3)=\int db_1db_2\;\rho(b_1,b_2)\;\phi(b_1S_1+b_2S_2-(b_1+b_2)S_3) \tag{8.18}$$

Normalization of probability imposes the constraint

$$\phi(0,0,0)=1 \tag{8.19}$$

For the case of constant α_j, condition (7.12) implies that S_j is linear in the arguments displayed in (8.9)-(8.12) (not to be confused with (8.8)). The resulting solutions are analogous to the secular solutions of the wave equation, just as the solutions of §3,4 are analogous to the bounded solutions of the wave equation.

9. Realizability

For the solution described in §7, in the case that ϕ is linear in B, the probability density for the velocity and magnetic fields is a sum of (functional) delta functions, each of the form $\prod_{\mathbf{x}}\delta(\mathbf{u}(\mathbf{x})-\mathbf{A}_T(\mathbf{x}))\delta(\mathbf{B}(\mathbf{x})-\mathbf{A}_M(\mathbf{x}))$ multiplied by similar factors for the vorticity, current, scalar and scalar source (and pressure and density, if the flow is compressible). Realizability imposes the constraint that the coefficients multiplying the delta functions be positive. For the solution described in §8, for the case of constant α_j, realizability constrains the functional dependence of ϕ upon the argument displayed in equation 8.18. For example, if ϕ is chosen to be exponential, the probability density for the velocity, vorticity, magnetic field and current would be joint Gaussian, hence positive everywhere as desired. Realizability remains to be verified for the solutions exhibiting detailed balance.

More generally, it has been suggested (Kraichnan 1991) that realizability could be imposed upon the probability $P[\mathbf{u}]$ by introducing a complex probability amplitude $\psi[\mathbf{u}]$ and its complex conjugate such that

$$P[\mathbf{u}]=\psi^*[\mathbf{u}]\psi[\mathbf{u}] \tag{9.1}$$

which is positive semidefinite as desired. Although the Hopf equation could be written in terms of $\psi[\mathbf{u}]$ or its functional Fourier transform $\tilde{\psi}[\mathbf{f}]$ where

$$\phi[\mathbf{f}] = \sum_{\{\mathbf{g}(\mathbf{x})\}} \tilde{\psi}^*[\mathbf{g}-\mathbf{f}]\tilde{\psi}[\mathbf{g}] \tag{9.2}$$

it is not immmediately obvious how to solve this infinite system of coupled equations or, equivalently, under what circumstances a given solution of the Hopf equation can be represented in the form (9.2).

One may, however, consider the equation of motion for $\tilde{\psi}[\mathbf{f}]$ rather than for $\phi[\mathbf{f}]$. If one imposes the single- valuedness of evolution under equations (2.1)-(2.3) upon the probability *amplitude* (a condition analogous to but slightly-more stringent than equation (6.16)), then, letting T denote the time evolution operator, we have (symbolically)

$$\begin{aligned} T\tilde{\psi}[\mathbf{f}] &= \int T\{d\mathbf{u}\ \psi[\mathbf{u}]\}e^{-i\int \mathbf{f}\cdot\mathbf{u}} = \int d\{T^{-1}\mathbf{u}\}\ \psi[T^{-1}\mathbf{u}]e^{-i\int \mathbf{f}\cdot\mathbf{u}} \\ &= \int d\mathbf{v}\ \psi[\mathbf{v}]e^{-i\int \mathbf{f}\cdot T\mathbf{v}} \end{aligned} \tag{9.3}$$

Hence, $\tilde{\psi}[\mathbf{f}]$ obeys the same Hopf equation as $\phi[\mathbf{f}]$. In other words, any solution of the Hopf equation (*e.g.*, the solutions exhibiting detailed balance) can be inserted into equation (9.2) to obtain a realizable ϕ. Guaranteeing realizability in this manner does have a drawback, however; the functional integration in equation (9.2) must be performed (at least in the vicinity of $\mathbf{f} = 0$) in order to evaluate moments (unless the moments are of at least first order in all vector components of all physical fields in the exponent of equation (2.6)).

A certain class of moments can be evaluated without performing functional integration. Consider moments obtained by taking functional derivatives of

$$\tilde{P}[\mathbf{f}] \equiv \tilde{\psi}^*[\mathbf{f}]\tilde{\psi}[\mathbf{f}] \tag{9.4}$$

rather than of $\phi[\mathbf{f}]$. $\tilde{P}[\mathbf{f}]$ may be interpreted as the probability density for the conjugate field $\mathbf{f}$; its density in velocity space is analogous to the Wigner distribution function. One may verify that

$$i\frac{\delta\tilde{P}[\mathbf{f}]}{\delta\mathbf{f}(\mathbf{x})}|_{\mathbf{f}=0} = \sum_{\{\mathbf{u}(\mathbf{x})\}}\sum_{\{\mathbf{v}(\mathbf{x})\}} \mathbf{u}(\mathbf{x})\psi^*[\mathbf{u}+\mathbf{v}]\psi[\mathbf{v}] \tag{9.5}$$

$$= \mathrm{Tr}[\tilde{\rho}\Delta\mathbf{u}(\mathbf{x})] = \langle\Delta\mathbf{u}(\mathbf{x})\rangle \tag{9.6}$$

$$(\tilde{\rho})_{u,v} \equiv \psi[\mathbf{u}]\psi^*[\mathbf{v}], \qquad (\Delta\mathbf{u})_{u,v} \equiv \mathbf{u}-\mathbf{v} \tag{9.7}$$

If the (probability) density matrix $\tilde{\rho}$ is diagonal in the velocity-realization basis, then there is no phase correlation between states with different $\mathbf{u}(\mathbf{x})$ and

$$\langle \Delta \mathbf{u}(\mathbf{x}) \rangle = 0 \tag{9.8}$$

In other words, $\langle \Delta \mathbf{u}(\mathbf{x}) \rangle$ is a measure of the *coherent* velocity spread at a given point or of the phase coherence or interference between different realizations. The probability amplitudes satisfy the Hopf equation and hence can exhibit "propagative" behavior in the variables $\mathbf{f}$ and $\mathbf{x}$, analogous to the propagation of conventional wavefunctions in $\mathbf{x}$ and t. However, because the conjugate fields are dummy variables, it is not clear that interference between different realizations (analogous to the Aharonov-Bohm effect) would be physically observable.

In order to compute moments without having to perform a functional integration, we require that $\phi[\mathbf{f}]$ be local in $\mathbf{f}$, yet that $P[\mathbf{u}]$ be positive semidefinite. Consider (in one dimension for notational simplicity) the following piecewise prescription:

$$\begin{aligned} P[u] &\equiv \int_0^{\infty} df\, \phi[f]\, e^{-fu} \text{ for } u > 0 \\ &\equiv \int_{-\infty}^{0} df\, \phi[f]\, e^{-fu} \text{ for } u < 0 \end{aligned} \tag{9.9}$$

where $\phi[f]$ takes the form given in equation (9.4). This expression for $P[u]$ is positive semidefinite for all real u. At $u = 0$, $P[u]$ is undefined; however, this does not affect moments since $P[0]$ is weighted by $u = 0$ in the integral over u. One may verify (using the Cauchy theorem) that taking functional derivatives of $\phi[f]$ and then setting $f = 0$ yields moments of the ***absolute value*** of u. Given the evolution equation (6.10) for $P[u]$, ϕ will satisfy the Hopf equation if surface terms $\phi[f = 0]$ and $\delta\phi[f = 0]/\delta f$ vanish (these arise from the functional integration by parts which is implicit in $\dot{u}P$.) The vanishing of the surface terms can be achieved by subtracting the constant $\psi[f = 0]$ from $\psi[f]$, which does not alter the validity of the Hopf equation for ψ. ϕ will satisfy the Hopf equation if ψ does and if cross terms involving functional derivatives of ψ and ψ^* vanish, which in turn can be achieved if equation (3.19) is satisfied and if one requires that the real part

$$Re\{\nabla \times (\mathbf{G}^{(-1)} \times \mathbf{G})\} = 0 \tag{9.10}$$

Normalization of probability (8.19) is replaced by the condition

$$\int_{-\infty}^{+\infty} df\, \frac{\phi[f]}{|f|} = 1 \tag{9.11}$$

which, by linearity of the Hopf equation, can be satisfied by multiplying ψ by the appropriate constant factor. Whether this approach can be modified to easily generate moments of non-absolute- valued quantities remains to be seen.

10. Time dependence

Time dependence may be incorporated into the framework of §6-8 by adding a term proportional to t to the arguments of ϕ in equations 6.2 and 8.18 and by relaxing the stationarity constraints (6.6, 7.25, 8.16). The coefficient of t in the argument of ϕ is chosen to be minus the sum of the coefficients of ξ, η and ζ.

Alternatively, one may take a simpler approach for the class of flow ensembles exhibiting "limited statistical linearity" (defined below). Consider the incompressible MHD equations without the buoyancy term or scalar equation. If we add a multiple β of the induction equation to the velocity equation, we obtain

$$-\frac{\partial}{\partial t}[\mathbf{u}(\mathbf{x},t)+\beta\mathbf{B}(\mathbf{x},t)] = \mathbf{u}\cdot\nabla\mathbf{u}-\mathbf{B}\cdot\nabla\mathbf{B}+\beta\mathbf{u}\cdot\nabla\mathbf{B}-\beta\mathbf{B}\cdot\nabla\mathbf{u} + \frac{\nabla p}{\rho}-\frac{1}{R}\nabla^2\mathbf{u}-\frac{\beta}{R_M}\nabla^2\mathbf{B} \tag{10.1}$$

Setting $\beta \equiv \pm 1$ and defining

$$\mathbf{v}\equiv\mathbf{u}+\mathbf{B}, \quad \mathbf{w}\equiv\mathbf{u}-\mathbf{B} \tag{10.2}$$

we obtain

$$-\frac{\partial}{\partial t}\mathbf{v}(\mathbf{x},t) = \mathbf{w}\cdot\nabla\mathbf{v}+\frac{\nabla p}{\rho}-a\nabla^2\mathbf{v}-b\nabla^2\mathbf{w} \tag{10.3}$$

$$-\frac{\partial}{\partial t}\mathbf{w}(\mathbf{x},t) = \mathbf{v}\cdot\nabla\mathbf{w}+\frac{\nabla p}{\rho}-a\nabla^2\mathbf{w}-b\nabla^2\mathbf{v} \tag{10.4}$$

where

$$a\equiv\frac{1}{2}\left(\frac{1}{R}+\frac{1}{R_M}\right), \quad b\equiv\frac{1}{2}\left(\frac{1}{R}-\frac{1}{R_M}\right) \tag{10.5}$$

Pressure may be eliminated by using incompressibility and the solenoidality of the magnetic field and inserting the projection operator $\{\mathbf{1}-(\nabla\nabla/\nabla^2)\}$ in front of the nonlinear terms.

A solution of the Hopf equations corresponding to equations (10.3-4) may be obtained by choosing a moment-generating functional

$$\phi[\mathbf{f},\mathbf{g},t]\equiv\left\langle e^{i\int_{-\infty}^{+\infty}d\mathbf{x}\,[\mathbf{f}(\mathbf{x})\cdot\mathbf{v}(\mathbf{x})+\mathbf{g}(\mathbf{x})\cdot\mathbf{w}(\mathbf{x})]}\right\rangle \tag{10.6}$$

which manifestly closes the equations, *i.e.*, a ϕ whose off-diagonal second functional derivatives are linear combinations of its first functional derivatives. A moment-generating functional for $\mathbf{v}, \mathbf{w}$ which satisfies this requirement is

$$\phi[\mathbf{f}, \mathbf{g}, t] \equiv \phi_1[\mathbf{f}, t]\, e^{\int_{-\infty}^{+\infty} d\mathbf{x}\, \mathbf{g}\cdot\mathbf{c_1}} + \phi_2[\mathbf{g}, t]\, e^{\int_{-\infty}^{+\infty} d\mathbf{x}\, \mathbf{f}\cdot\mathbf{c_2}} \tag{10.7}$$

where ϕ_1 and ϕ_2 are arbitrary. One may readily verify for this ϕ that

$$\frac{\delta^2\phi}{\delta f_i(\mathbf{x})\delta g_j(\mathbf{x})} = c_{1j}\frac{\delta\phi}{\delta f_i(\mathbf{x})} + c_{2i}\frac{\delta\phi}{\delta g_j(\mathbf{x})} - c_{1j}c_{2i}\phi \tag{10.8}$$

This implies that

$$\begin{aligned}&\langle \mathbf{u}(\mathbf{x}, t) \times \mathbf{B}(\mathbf{x}, t)\rangle \\ &\quad = (\mathbf{c_2} - \mathbf{c_1}) \times \langle \mathbf{u}(\mathbf{x}, t)\rangle - (\mathbf{c_2} + \mathbf{c_1}) \times \langle \mathbf{B}(\mathbf{x}, t)\rangle + \mathbf{c_1} \times \mathbf{c_2}\end{aligned} \tag{10.9}$$

which offers another rigorous alternative to the conventional MFE model. Note that diagonal second functional derivatives such as $\delta^2\phi/\delta f_i \delta f_j$ do not reduce to linear functional derivatives; correlation functions such as $\langle u_i(\mathbf{x})u_j(\mathbf{x})\rangle$ or $\langle B_i(\mathbf{x})B_j(\mathbf{x})\rangle$ cannot in general be written in terms of $\langle u_i(\mathbf{x})\rangle$ and $\langle B_i(\mathbf{x})\rangle$.

The resulting closed system of 2 linear equations for the first functional derivatives of ϕ is readily diagonalized. For constant $\mathbf{c_1}$ and $\mathbf{c_2}$, the term in the Hopf equation corresponding to $\mathbf{w}\cdot\nabla\mathbf{v}+\frac{\nabla p}{\rho}$ in equation (10.3) simplifies:

$$\{1 - (\nabla\nabla/\nabla^2)\}\nabla \cdot \frac{\delta}{\delta\mathbf{g}}\frac{\delta\phi}{\delta\mathbf{f}} \to \mathbf{c_1} \cdot \nabla\frac{\delta\phi}{\delta\mathbf{f}} \tag{10.10}$$

(similarly for equation (10.4)) Hence, one obtains propagating diffusive modes (an admixture of velocity and magnetic field) governed by a dispersion relation $\omega(\mathbf{k})$ which satisfies

$$(\omega - \mathbf{c_1} \cdot \mathbf{k} - iak^2)(\omega - \mathbf{c_2} \cdot \mathbf{k} - iak^2) + b^2k^4 = 0 \tag{10.11}$$

For $\mathbf{c_1} = \mathbf{c_2}$, the eigenmodes reduce to excitations which are purely kinetic or magnetic, diffusing on purely viscous or resistive time scales, respectively. For $R = R_M$, $\mathbf{v}$ and $\mathbf{w}$ decouple, corresponding to modes whose velocity and magnetic field oscillate in and out of phase, respectively. The case of nonconstant $\mathbf{c_1}$ and $\mathbf{c_2}$ and higher-order correlation functions may be computed by solving analogous but larger, inhomogeneous closed systems of equations, involving a diffusive kernel. Whether these modes are purely statistical (appearing only in the ensemble-averaged flow) or play a role in individual realizations remains to be clarified.

11. Conclusion

The Hopf functional approach offers a new exact method for obtaining stationary MHD states, both with and without detailed balance, which generalize the usual ideal static or force-free, equilibrium, nonturbulent states. The treatment of time dependence beyond the usual, initial linear stability regime also appears possible. Closed-form analytic expressions for the mean emf and other correlation functions emerge without making perturbative, phenomenological or statistical assumptions. Recognizing the wavelike character of the functional differential equations enables one to reduce them to a system of ordinary differential equations, a reduction of the number of degrees of freedom in the problem from N^{3L^3} to L^3 where N and L are large numbers on the order of the number of permitted values for velocity at a given point and the spatial extent of the system, respectively. The solutions obtained are not necessarily unique or complete but merely illustrative. The possibility of superposition indicates that matching to the mean flow may be necessary to determine relative strengths of different solutions, unless it is possible to close the equations by substituting, for example, the mean emf and Lamb vector back into the stationary vorticity equation. Initial conditions or a variational criterion may play a role in selecting the correct ensemble or discarding spurious ones. Computation of moments and of the probability of arbitrary realizations of the velocity, magnetic field, scalar and scalar source is currently under investigation.

Acknowledgements

The author is grateful to Alan Wray for stimulating discussions and Stephen Childress for informative comments. He also wishes to thank Paul Roberts and David Montgomery for preprints of their work.

References

BRAGINSKY, S.I., 1964 Self-excitation of a magnetic field during motion of a highly-conducting fluid. *Soviet Phys. JETP* **20**, 726.

BUSSE, F.H., 1981 *Hydrodynamic Instabilities and Transition to Turbulence*, §5, Springer Verlag.

COWLING, T.G., 1981 Present status of dynamo theory. *Ann. Rev. Astron. Astrop.* **19**, 115.

FRISCH, U., POUQUET, A., LEORAT, J., AND MAZURE, A., 1975 Possibility of an inverse cascade of magnetic helicity in MHD turbulence. *J. Fluid Mech.* **68**, 769.

FRISCH, U., SHE, Z.S., AND SULEM, P.L., 1987 Large-scale flow driven by anisotropic kinetic alpha effect. *Physica* **28D**, 382.

GILBERT, A.D., AND SULEM, P.L., 1990 On inverse cascade in alpha- effect dynamos. *Geophys. Astrophys. Fluid Dyn.* **51**, 243.

GRAHAM, R., 1973 *Quantum Statistics in Optics and Solid State Physics*, 32, Springer Verlag.

KRAICHNAN, R.H., 1979 Consistency of the alpha effect turbulent dynamo. *Phys. Rev. Lett.* **42**, 1677; private communication, 1991.

KRAUSE, F., AND RÄDLER, K.H., 1980 *Mean Field Magnetohydrodynamics and Dynamo Theory*, Pergamon Press.

LOW, B.C., 1990 Equilibrium and dynamics of coronal magnetic fields. *Ann. Rev. Astron. Astrop.* **28**, 491.

MALKUS, W.V.R., AND PROCTOR, M.R.E., 1975 Macrodynamics of alpha effect dynamos in rotating fluids. *J. Fluid Mech.* **67**, 417.

MOFFATT, H.K., 1978 *Magnetic Field Generation in Electrically Conducting Fluids*, Cambridge University Press.

MONTGOMERY, D., AND PHILLIPS, L., 1989 MHD turbulence: Relaxation processes and variational principles. *Physica* **D37**, 215.

MONTGOMERY, D., 1989 *Trends in Theoretical Physics*, Vol. I, 239-262.

PARKER, E.N., 1989 Spontaneous discontinuities and the optical analogy for stationary magnetic fields. *Geophy. Astrop. Fluid Dyn.* **46**, 105.

ROBERTS, P.H., AND SOWARD, A.M. 1992 *Ann. Rev. Fluid Mech.* (in press).

SHEN, H.H., AND WRAY, A.A., 1991 *J. Statistical Phys.* (in press).

SHERCLIFF, J.A.,1965 *A Textbook of Magnetohydrodynamics*, Pergamon Press.

SOWARD, A.M., AND CHILDRESS, S., 1986 Analytic theory of dynamos. *Adv. Space Res.* **6**, 7.

STANISIC, M.M., 1985 *Mathematical Theory of Turbulence*, Springer Verlag.

TAYLOR, J.B.,1986 Relaxation and magnetic reconnection in plasmas. *Rev. Mod. Phys.* **58**, 741.

AUTHOR INDEX

SUBJECT INDEX